ASPECTS OF MODERN RADAR

ASPECTS OF MODERN RADAR

Eli Brookner, Editor

Artech House
Boston • London

Library of Congress Cataloging-in-Publication Data

Advances in modern radar/Eli Brookner, editor
p. cm.
Includes bibliography: p.
Includes index.
ISBN 0-89006-263-3
1. Radar. I. Brookner, Eli, 1931-
TK6575.A376 1988
621.3848 dc19 88-11624

ARTECH HOUSE, INC.
685 Canton Street
Norwood, MA 02062

International Standard Book Number: 0-89006-263-3
Library of Congress Catalog Card Number: 88-11624

10 9 8 7 6 5 4 3 2 1

Contents

The Authors

Eli Brookner
Consulting Scientist
Raytheon Company
Wayland, Massachusetts

Richard J. Doviak
Senior Scientist
National Severe Storms Laboratory
Norman, Oklahoma

Robert T. Hill
Board of Governors and
Radar Systems Panel
AESS/IEEE

Eugene F. Knott
Senior Principal Engineer
Boeing Aerospace
Seattle, Washington

David G. Laighton
Advanced Microwave Integrated Circuit Development Organization
Raytheon Company
Wayland, Massachusetts

Edward K. Reedy
Georgia Tech Research Institute
Georgia Institute of Technology
Atlanta, Georgia

Helmut E. Schrank
Advisory Engineer
Westinghouse Electric Corporation
Baltimore, Maryland

Lieutenant Joseph Sensi, Jr.
United States Navy

Samuel M. Sherman
Electronic System Consultant
Cherry Hill, New Jersey

James C. Wiltse
Georgia Tech Research Institute
Georgia Institute of Technology
Atlanta, Georgia

D. S. Zrnić
Managers, Doppler Radar Project
National Severe Storms Laboratory
Norman, Oklahoma

Preface

This book is aimed at the working engineer and the student, written in simple terms so that engineers who are not specialists in the topics covered can quickly become familiar with them. This is true for the beginning engineer, the specialist from another area, or the manager who needs to talk to the specialist in a knowledgeable way. Some of the specialties covered are radar cross section (RCS) prediction and reduction, solid-state bipolar and monolithic microwave integrated circuits (MMIC), monopulse principles and techniques, ultra-low-sidelobe techniques, the basics of weather radar, and millimeter-wave (MMW) systems and technology. Also covered are radar basics for the engineer entirely unfamiliar with radar. A very extensive survey of radar systems, technology, and their trends is given, also geared to a wide audience—the specialists, the engineering managers, and the novice. The book provides as well an extremely useful and extensive table of the parameters of over 200 surface and airborne radars.

The literature relative to the specialties covered here is difficult to read, and it becomes very time consuming to obtain a basic understanding. The writing in this book is made simple, avoiding complicated mathematics and undefined jargon. Also, the coverages of these specialties are kept to a reasonable length. As a result, a basic understanding can be obtained in a reasonable time by both the nonspecialist and specialist from another area.

This book is a complement to the author's very successful book, *Radar Technology* (Artech House, 1979). The present book covers topics not included there and updates such topics as solid-state power devices, tubes, digital technology, and the recent developments in radar systems. Also, the catalogue of radar parameters given in *Radar Technology* is updated to include recently developed radars and radars not covered there—including secondary radars, airborne attack, air-surveillance, and bomber radars. The parameters for more than 200 radars are listed in this updated catalogue (Appendix 2A). A summary of the highlights of the various chapters is given next.

Chapter 1, Radar—An Overview by R.T. Hill, is written for the engineer trying to get off to a good start in radar or just trying to get a quick understanding of the various concepts of radar. The reader is introduced to the radar equation, different types of radars, and their various applications (e.g., continuous wave (CW), pulsed doppler (PD), two-dimensional, three-dimensional, tracking, and phased array radars). Also introduced are the subjects of noncoherent and coherent processing, the different target fluctuating models, pulse compression, the airborne radar clutter problem, constant false-alarm rate (CFAR) detection, phased array scanning and beam formation, and automatic detection and tracking. Finally, the challenges of the future are presented.

Chapter 2, authored by this writer, gives a very extensive survey of recent developments and future trends in radar systems, technology, and techniques. Radar systems covered are three-dimensional radars (phase-phase scanned, one-dimensional electronically scanned in elevation and mechanically scanned in azimuth, and stacked-beam), two-dimensional radars, bistatic radars, over-the-horizon (OTH), and hemispherical coverage radar systems. Technologies covered are solid-state hybrid and monolithic microwave integrated circuits (MMIC) components, microstrip antennas, digital signal processing (including very high speed integrated circuits (VHSIC), systolic arrays, residue arithmetic, gigahertz GaAs technology, silicon compilers, and digital signal processing (DSP) chips), ultra-low-sidelobe antennas (ULSA), high-power tubes, analog signal processing, flat displays, and fiber optics.

Chapter 3, by J. Sensi, Jr., covers the AEGIS system from the overall system point of view. Then, Appendix 3A gives detailed descriptions of the AEGIS radar.

Chapter 4, by D. Laighton, gives an excellent tutorial on MMIC technology, including the power output limits from UHF to X-band, yields, and cost trends. Also covered are air-bridge technology, the effect of VSWR mismatch at the output of the power amplifier, a rule of thumb for determining the output of a device based on its gate periphery width, passive components—distributed and lumped—and where and why each type is used, the different classes of power amplifier (PA) operations—class A, AB, B, C, and D—and the PAs used for MMIC and why those are used.

Chapter 5, by S.M. Sherman, gives a clear and concise tutorial on monopulse—a technique of wide practical importance as well as theoretical interest. The chapter covers the advantages and disadvantages of monopulse relative to sequential lobing; amplitude and phase monopulse; how it is obtained with reflectors, lens, and phased array antennas; its realization with different types of feeds (four- and five-horn feeds and with multiple propagation modes) and how the performances obtained with these feeds compare with the optimum achievable pattern; microwave circuits for obtaining the sum and difference signals; different processing techniques (quadrature, dot-product, and the use of amplitude and phase); the difference between azimuth and traverse angle and the secant correction to compensate for this difference; the problem of obtaining monopulse measurements for unresolved targets; target angle glint; the effects of multipath on low angle monopulse measurements and techniques for coping with it (e.g., the use of low-E, offset nulling, a symmetrical (even) difference-over-sum pattern, double nulling, and the complex indicated angle technique); the "nose-diving" phenomenon; finally, the various sources of angle errors.

Chapter 6, Low-Sidelobe Phased Array and Reflector Antennas by H. E. Schrank, covers the increasingly important subject of how to obtain very low antenna sidelobes, or ultra-low antenna sidelobes, with array and reflector antennas. Ultra-low sidelobes are needed for achieving invulnerability to jamming through the sidelobes and for coping with sidelobe clutter, especially for airborne radars. The historic breakthrough in achieving ultra-low antenna sidelobes for a slotted waveguide array is described. This work led to the successful AWACS array antenna. The principles for achieving low antenna sidelobes are reviewed, the various sources of errors that led to sidelobe degradation are detailed, and extensive examples of successful array antennas are presented. The achievement of very low antenna sidelobes for reflector antennas through the use of absorber-lined rims, castellated rims, absorber-lined shrouds (tunnelling) are discussed, and the use of a polarization-twist Cassegrain antenna to avoid the sidelobe degradation resulting from feed blockage is presented. The author also shows that very low sidelobes can be achieved through the use of an offset-fed reflector. Finally, he gives the illumination that provides the best sidelobes in the presence of blockage.

Chapter 7, Fundamentals of MMW Radar Systems by E.K. Reedy and J.C. Wiltse, gives an excellent overview of the systems issues for millimeter-wave radars and the current technology of millimeter-wave components. Also given are a brief history of millimeter-wave radars together and a listing of early millimeter-wave radars (prior to 1977). Finally, the authors present examples of recent and future applications for millimeter-wave systems, and cover the environmental factors—attenuation *versus* frequency; backscatter from ground, rain, snow, dust, and smoke; turbulence effects, multipath effects; typical target cross sections; the advantages accrued by the use of millimeter-wave systems. Finally, potential applications are considered for millimeter-wave systems—for surveillance, instrumentation, and measurement; missile seekers and terminal guidance systems; fire control and tracking systems.

Chapter 8, Radar Cross Section by E.F. Knott, for the first time, simplifies a very difficult subject. Covered in great detail are the subjects of how to predict the cross section of simple and complex targets, and the extremely important subject of how to reduce target cross section, which has received increasing interest since the announcement of the "stealth" bomber by the US Department of Defense.

Knott's chapter starts with a coverage of the definition of target cross section. This is followed by extensive useful examples of the cross section of many simple as well as some complex targets—corner reflectors, cylinders, spheres, prolate spheroids, ogives, doubly and singly curved surfaces, curved edges, straight edges, and apexes. Also given are example cross section values for ships, ground vehicles, aircraft, birds and insects. This is followed by a discussion of the different methods for predicting the target cross section for variously shaped targets, starting with the exact methods, such as the well known method of moments (MOM) and followed by various high-frequency approximation methods of increasing accuracy. The high-frequency approximate methods covered are the methods of geometric optics (GO), physical optics (PO), geometrical theory of diffraction (GTD), physical theory of diffraction (PTD), method of equivalent currents (MEC), and incremental length diffraction coefficient (ILDC).

Next, Chapter 8 covers the four methods of reducing target cross section. These are through shaping, radar absorbing materials (RAM), passive cancelation, and active cancelation. The radar absorbing techniques covered are the Salisbury screen, Dällenbach layer, Jaumann multiple resistive sheets, circuit analog absorber, and the use of magnetic materials. The latter three techniques can provide target cross section reductions over a wide bandwidth. Knott also covers the use of pyramidal absorbers, graded absorbers, and dipped honeycomb for achieving low cross sections over a wide bandwidth. Finally, the application of stealth techniques to the reduction of the cross section of an automobile is discussed for illustration.

In Chapter 9, The Doppler Weather Radar by R.J. Doviak and D.S. Zrnić, the use of microwave radar for remote sensing of weather conditions is discussed. Derived are the reflectivity from rain as a function of the raindrop size distribution, the radar wavelength, and the refractive index of the water drops. The issue of measuring rainfall rate using a microwave radar is discussed in detail, with curves and relationships for the attenuation *versus* rainfall rate provided for various radar carrier frequencies. The radar doppler spectrum and its principle moments are related to the spatial distribution of weather, wind, and water. Doppler radar measurements allow the remote determination of weather conditions. Examples of doppler measurements for different weather conditions are given and analyzed. Among the examples presented are those for thunderstorms, downbursts, and microbursts (which were the cause of the disastrous Delta Airlines L-1011 crash at Dallas-Fort Worth Airport in 1985), cyclones, tornadoes, gust fronts, and turbulence. The astounding success achieved with weather radars in thunderstorm cyclone detection well in advance of tornado formation is leading to the incorporation of doppler capabilities in new radars throughout North America (see Section 2.3.2 of Chapter 2).

The chapters can be read in any order desired by the reader, as his or her interests may guide. This book is an outgrowth of a Boston Section IEEE and Boston IEEE/AESS jointly sponsored radar lecture series titled "Radar—Past,

BROOKNER
PREFACE

Present and Future'' which was given in 1986. This course was very well received, with some 400 engineers attending.

Acknowledgments

I am indebted to a large number of people who helped me during the preparation of this book. I would like to thank T. Mahoney (of Raytheon), who edited my Chapter 2; W. Comisky, who provided funding for obtaining the radar photographs and photostats of the figures for Chapter 2; Ernest E. Witschi of the Boston IEEE office, for administering the business end of the evening radar lecture series, which led to this book; L. Lavallee and R. Punchard (both of Raytheon), who gave me ready access to their extensive collection of radar data, which was of help in compiling the radar catalogue of Chapter 2; Dr. L. Schwartzman (Unisys, formerly Sperry), who provided me with data on Unisys radars as well as photos; R. Rosien (ITT-Giffillan), who provided photos and data on ITT-Giffillan radars; F. Nelson (Hughes), who proofread the signal processing material in my chapter; D.A. Ethington and P.J. Kahrilas (both of Hughes), who provided extensive photos and data on their radars; Dr. M. Barkan (Zoran) who provided data on Zoran's chips and also proofread the material relating to digital signal processor chips in my chapter; J.J. Janda and E.J. Rzepala (both of Raytheon) for proofreading several of the chapters; S. McAuliffe (Artech House) for helping to get the chapter authors to meet their deadlines and for her instrumental role in helping to include one of the chapters; D. Ricci (Artech House) for editing the material; M. Walsh (Artech House) for keeping after me to get the book done and L. Case (Artech House) for helping in the production of the book.

Others who helped through editing, discussions, or in other ways were M.T. Borkowski (Raytheon), Dr. M. Michelson (Raytheon), F. Daum (Raytheon), R. Logan (Texas Instruments), S. Drabowitch (Thomson-CSF), H. Schrank (Westinghouse), and R. Klem and Dr. W.D. Wirth (both of Forschungs institut für Funk und Mathematik). Joanne Portsch and Ester Stevelman (both of the Raytheon Equipment Division Library) helped locate and provide many of the references used in preparing Chapter 2.

Many thanks are due Barbara Rolinski (formerly of Raytheon), who transcribed many of the tapes from the lecture series as well as several revisions of Chapter 2 before having her baby the morning after her last day at Raytheon; S. Peters (Raytheon) for taking over for Barbara and doing many additional revisions of Chapter 2 and for always smiling and having a pleasent disposition while doing them, helping to make the task more enjoyable; T. Spence (Raytheon) for transcribing the lecture tapes that led to Chapter 4; D. George (Raytheon) for editing Chapter 1; C. Temple (Raytheon) for helping with the artwork of Chapter 2; S. Ruscitti (Raytheon) who did the artwork for Chapter 1 as well as some of Chapter 2.

Finally, my greatest indebtedness and appreciation are due my wife, Ethel, who has given continuing encouragement for endeavors such as this one. Without her understanding, patience, tolerance, and encouragement, this book would not have been possible.

Eli Brookner
June 1988
Lexington, MA

Dedication

TO ETHEL—32 years my wife, my friend, my love—we are still together, but a little apart. You listen when I talk, you feed me when I am hungry, and you are there when I need you. Here is to you. I care.

AND TO OUR SONS, LARRY AND RICHARD.

Chapter 1
RADAR—AN OVERVIEW

by R.T. Hill

This chapter might have been titled, "Radar—Off to a Good Start!" to reflect the author's commitment to two things. The first concerns the field of radar as a whole: an engineer entering into radar work must have a grasp of the fundamentals governing all radars, regardless of his or her leaning toward one or another specialty within radar, reflecting more fervent interests or the intentions of employers. The second commitment is to this book: *Radar—Past, Present, and Future;* it, too, deserves to get "off to a good start"! So, whether you are on the threshold of a new career in engineering or merely on the threshold of reading this book, we invite your attention to this overview of the "whole radar," hoping that it will provide a foundation upon which to build more specialized knowledge.

We will treat first the origins of radar, introducing the fundamental idea and common types of radar, then move to the radar equations that permit us to quantify radar performance. This second step will require an introduction to detection theory and the character of targets as reflectors or scatterers of radio waves. We will next explore each of the four subsystems of radar: the transmitter, the antenna, the receiver and signal processor, and the control and interface apparatus. The chapter concludes with a discussion of the challenges facing the modern radar system designer and the trends of future developments in radar.

1.1 RADAR ORIGINS

1.1.1 Radio Waves

While *electromagnetic waves* were somewhat understood (amazingly well, considering the limited human experience at the time!) from the work of Maxwell, Hertz, and others of the 19th century, and while several pioneers (Hulsmeyer, Marconi) in the early 20th century thought that the reflective character of objects could be very useful, the experiments of the 1920s and early 1930s, conducted quite independently by several nations, brought the needed evidence to induce further investment, with which came the British preparation for defense in the late 1930s and British-American collaboration immediately prior to World War II. Figure 1.1 shows the experience typical of the origins of radar; *multipath* interference in radio transmission was a specific way that scientists at the Naval Research Laboratory (NRL) in Washington, DC, encountered the radar (not then so-named, of course) phenomenon. Such continuous wave (CW) interference is depicted in Figure 1.2, which is a *vector diagram* that represents the strength and phase of the *direct path* signal (obviously, the intended signal in the illustrated radio set-up) and the "bothersome signal" *reflected* by the ship moving in the river channel. A rotating vector can indeed be thought of as generating a sine wave through *projection* (i.e., by considering the values of one component of the

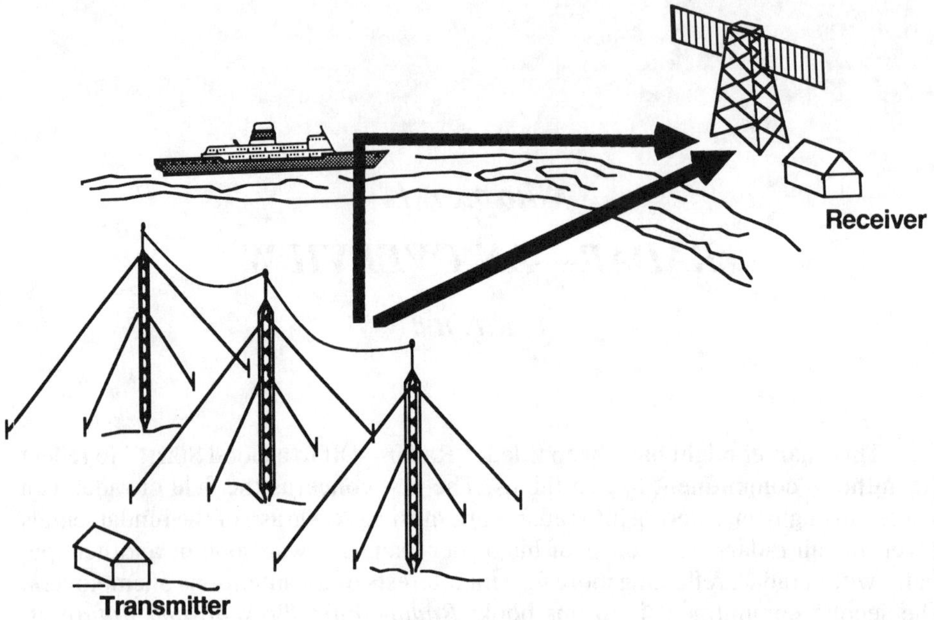

Figure 1.1 Early Naval Research Laboratory (NRL) Radar Experience.

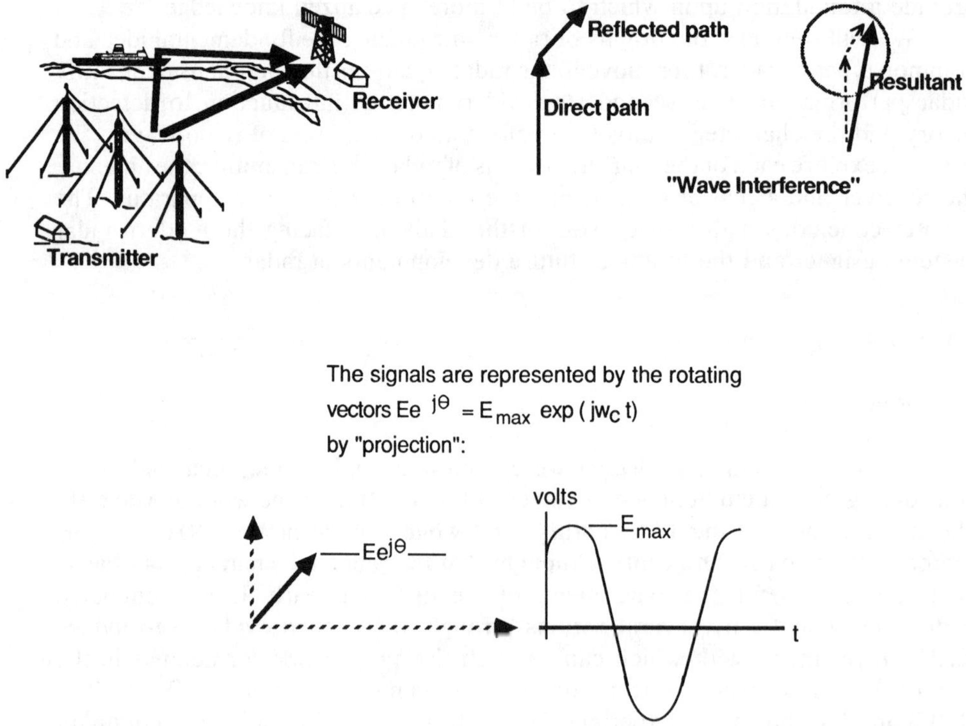

Figure 1.2 Beat Signal due to Target Multipath.

vector, the y component, as a function of time). We can thus represent the electric-field strength (one constituent of the radiating *electromagnetic field*) arriving at a *field point* (e.g., the receiver location) by such a vector, $E_t = E_{\max} e^{j\omega c t}$, in general form, as illustrated. The *two signals* (by the direct and reflected paths) interfere (as shown in the upper right), with a phase relationship dependent on the path-length difference involved. For many cycles of the radio-frequency (RF) signal being used as a carrier (f_c), this phase relationship is nearly fixed; however, as the ship moves, the phase relationship slowly changes, and the received signal is alternately stronger and weaker, in much the same way a television picture ''beats'' as an interfering (reflecting) airplane flies near the transmission path. It is the detection of this beat signal that indicates the presence of the target.

Such vector (or *phasor*) diagrams serve well in understanding several matters associated with radar signals and apparatus, and the prudent student will want a confident understanding of this useful tool.

1.1.2 Pulsed Radar

In Figure 1.3, we see the fundamental idea of *pulsed* radar: the use of pulses as modulation applied to the carrier frequency signal was an outgrowth of the early experimentation, permitting straightforward measurement of range to a target when receiver and transmitter are located together (thus making a *monostatic* radar). The time diagram shows the *envelope* of the transmitted signal, but does not graph every cycle of the carrier within each pulse. Shown in simplified form is the echo from a target (e.g., the aircraft sketched in the beam of the directive antenna), which is about 100 nmi away, leading to the echo time of 1220 μs (1.22 ms). The *transmitter, antenna, receiver and signal processor,* and the *control and interface apparatus* (which generally includes some form of video display terminal (VDT), or *cathode ray tube* (CRT) display) are introduced here as the principal subsystems of the radar. The parameters of interest are the *peak,* or *pulse,* power of the radar, *P*, the *pulse repetition frequency* (PRF = 400 pulses per second, leading to the 2500 μs total listening period, or *pulse repetition interval,* PRI), and the pulsewidth (τ = 1 μs). The radar shown here has a *duty cycle* (or, more properly, *duty factor*) of .0004, which is the fraction of the time that the radar actually transmits. Average power in radar is the product of peak power and duty factor.

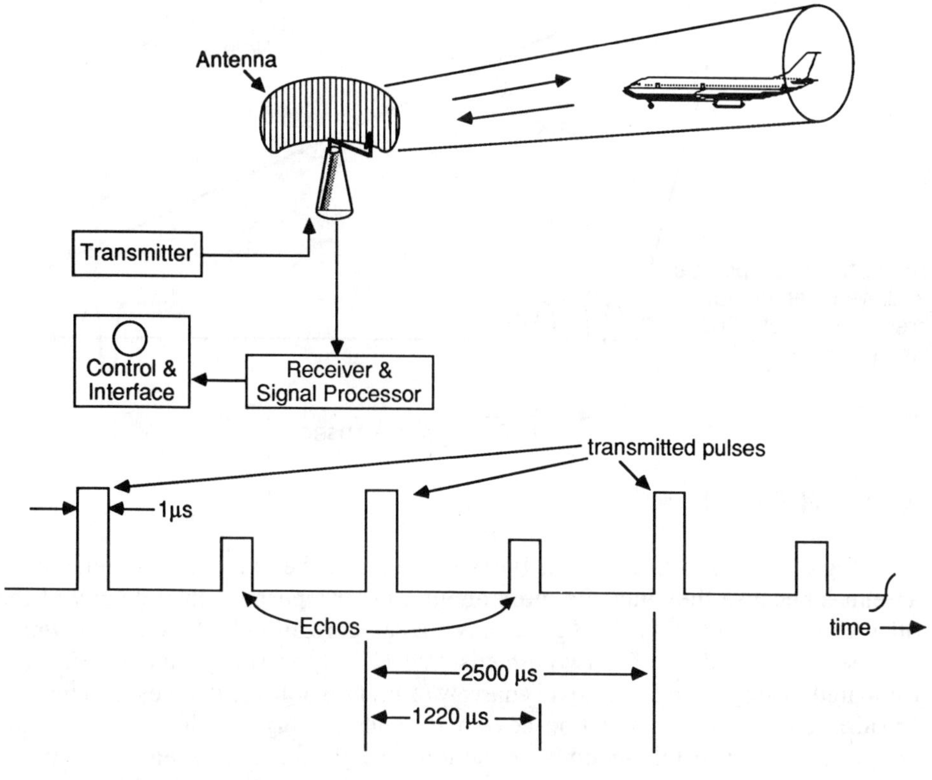

Figure 1.3 Pulsed Radar.

Figure 1.4 illustrates the pulse itself, and introduces a few more typical characteristics and principles involved in the "basic idea" of radar. Radio waves travel at approximately the speed of light, c, and common radar carrier frequencies are in the *microwave* region of the radio spectrum, which is from 1000 MHz (one gigahertz = 1 GHz = 1000 MHz) to 10 GHz, with the radar frequencies of L-band, S-band, C-band, and X-band being approximately 1, 3, 5, and 10 GHz, respectively. As in all propagating waves, the product of the frequency and the wavelength ($f \times \lambda$) is the speed of propagation:

$$c = \mathbf{3} \times \mathbf{10^8}\ \text{m/s} \doteq \mathbf{1}\ \text{ft/ns}$$

Figure 1.5 shows that the radar designer must somehow account for the many things to which the signal is subject during its journey to the target and back: we see rain or other weather effects as *clutter* (although some radars are designed to "see" such *targets*); surface reflections, also as clutter; surface then reflecting some of the signal to give a multipath interference, which alters total signal strength, as discussed above; and an interfering radio signal as might be sent to us by a *jammer,* represented by the aircraft on the far right of the figure. Some of the details shown in the figure within each subsystem will be discussed below. Certainly, the radar's components tend to distort the signal, or at least to weaken it, causing further signal path aggravation, which the designer must not ignore.

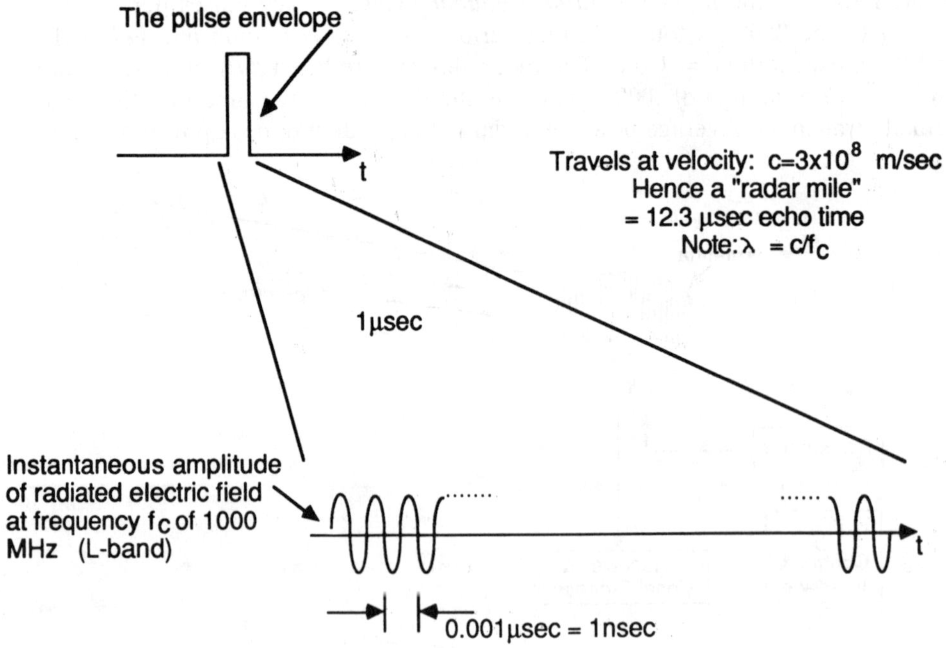

Figure 1.4 Radar Pulse.

Figure 1.6 sketches several types of radars. (The sketches are actually of antennas because they indicate more about a radar's purpose than do any of the other three subsystems.) The figure shows a categorization by function, in which we see search radars of a two-dimensional (2D) type (range and bearing are estimated), using a *fan beam* that is narrow (e.g., two or three degrees) in azimuth, broad in elevation (for elevation coverage), and rotating in azimuth for periods of typically four to ten seconds. A nodding height-finder is sketched, but the functions of a 2D radar and a height-finder have been combined in the contemporary three-dimensional (3D) search radars (range, bearing, and elevation are estimated on each detection), where a single beam, narrow in both azimuth and

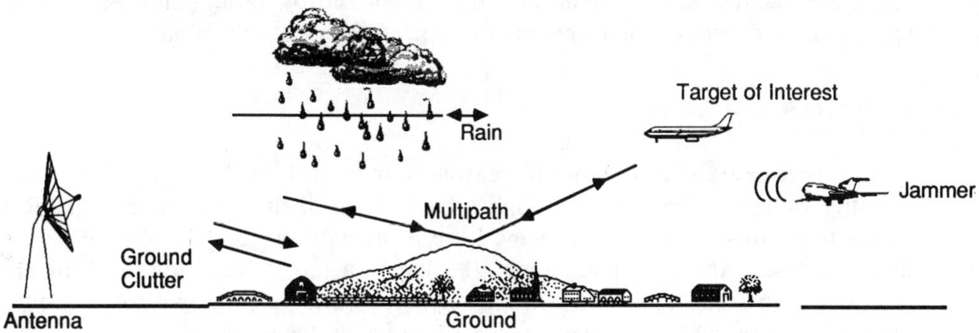

Figure 1.5 Environment: Atmospheric Attenuation, Backscatter, Refraction; Multipath; Ground Clutter; Jamming; Sky and Ground Noise.

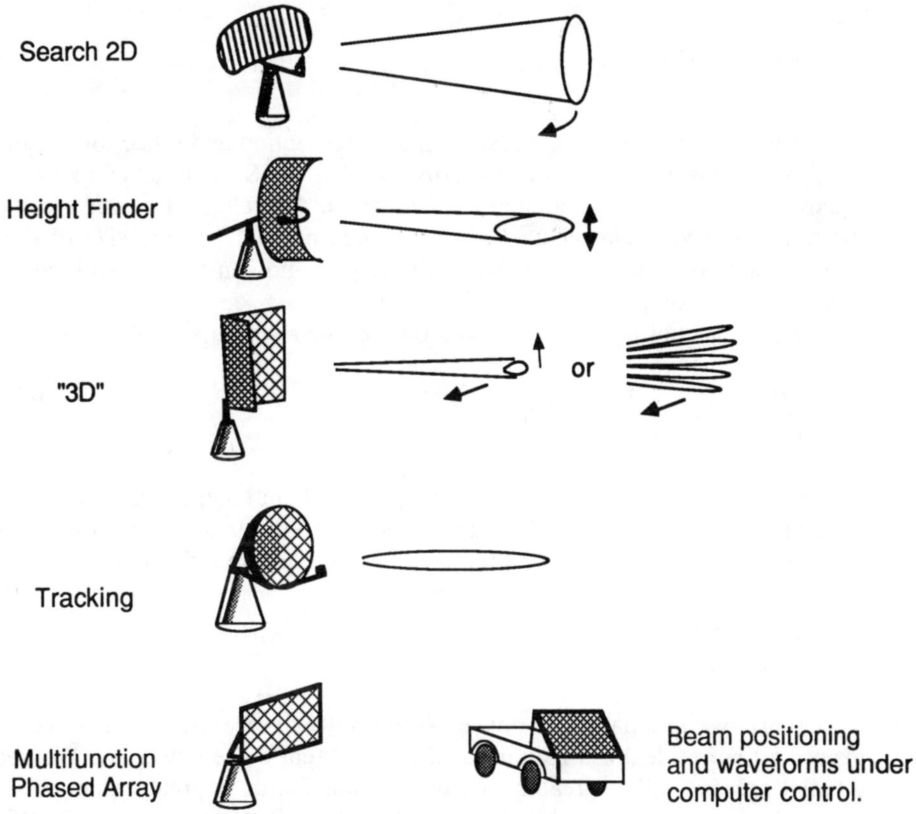

Figure 1.6 Radar Types.

elevation, or a *stack* of simultaneous beams (implying some sort of receiver channeling) scans rapidly in elevation as the antenna rotates in azimuth. These search radars may feed a *track-while-scan* (TWS) computer to associate the detections from scan-to-scan to form target *tracks*. We more properly call these *surveillance* radars when this potential exists. Next, the familiar "dish" antenna illustrates a precision *tracking* radar of the searchlight or single-target type, associated with specific fire control systems or test-range instrumentation. Such radars require a *designation* from a search radar, or some other external source, in order to *acquire* the intended target and begin to track it continuously. Finally, Figure 1.6 shows the versatile, computer-controlled phased-array radar in which several functions can be "time-shared" without mechanical positioning or rotation of the antenna. Chapter 2 gives many examples of recently developed 2D, 3D, phase-phase steered, and other radars. Radars could also be categorized by siting (land, sea, air, or space) or by nature of their application (e.g., military, commercial).

1.2 RADAR EQUATION

It is important to be able to reason your way through expressions which quantify radar performance, through a basic form of the radar range equation, rather than solely relying on looking it up or memorizing a form of it. We want first to express the power received, P_r, due to a target reflection, and then to compare it to some criterion. The power received is proportional to the power incident to a target, which we assume to be in the beam of a directive antenna of gain G_t, when the transmitter produces the power P_t to be sent to the antenna through a system with loss L_t (expressed here as a number greater than unity so that $1/L_t$ of the power is radiated). This incident *spatial power density,* in units of watts per square meter (W/m^2) at a distance of R meters, then, is

$$\frac{P_t G_t}{L_t 4\pi R^2} \quad (\mathrm{W/m^2}) \tag{1.1}$$

The presence of a target results in an interception and reflection of some of the power, that given by its *radar cross section* (RCS, denoted σ, expressed in square meters), which is defined to account for the reflected power received at the radar receiver. The value σ then depends not only on the size of the real target, but also on the complexity of the target, and can vary considerably as a function of its composition and the orientation.

This reflected power is received by the radar with a spatial density of

$$\frac{P_t G_t}{L_t 4\pi R^2} \cdot \sigma \cdot \frac{1}{4\pi R^2} \quad (\mathrm{W/m^2}) \tag{1.2}$$

where an antenna of effective receiving aperture A_e intercepts some of the power, and, with receiving losses of L_r (again, expressed as a number greater than unity), provides the power available to the receiver, due to the reflection:

$$P_r = \frac{P_t G_t \sigma A_e}{L_t L_r (4\pi)^2 R^4} \quad (\mathrm{W}) \tag{1.3}$$

Next, we compare this power to the power present in the receiver, here expressed in an idealized manner as a total system noise power N_r. We declare a target present if P_r is greater than N_r by some factor W greater than one. This comparison allows us to solve, then, for the range R at which a target of RCS = σ produces such a signal level for which we can claim desirable statistics of declaring the target to exist. In this discussion, we represent the total system noise power as $N_r = kT_sB$, where k is Boltzmann's constant, T_s is an overall, effective system noise temperature (which includes ambient noise external to the antenna, the losses previously designated as L_r, and the receiver's noise represented by its noise figure) and B is the receiver's bandwidth, which is sufficient to embrace the signal's principal modulation sidebands or adjusted to the spectral width of the transmitted pulse envelope, assuming that the receiver is well matched to deliberate modulation within the envelope (see the following comments on *pulse compression*). The criterion for detection is

$$P_r \geq WkT_sB \tag{1.4a}$$

or

$$\frac{P_t G_t \sigma A_e}{L_t (4\pi)^2 R^4} > WkT_s B \tag{1.4b}$$

and, solving the limit case (equality) for R^4, we have

$$R^4 = \frac{P_t G_t \sigma A_e}{L(4\pi)^2 WkT_s B} \tag{1.5}$$

Here, we write just an L without index for losses which include those previously identified as transmission losses L_t as well as other system losses not specifically accounted in the overall noise temperature T_s (*beam shape* loss on receive, for example). The expression (1.5) is one form of the radar range equation for a single pulse propagating in free space.

A number of variations to the radar range equation are commonly used and important for us to note. First, many radars transmit several pulses, perhaps from 10 to 40, while the antenna scans just one beamwidth. The postdetection integration of these pulses (as on a CRT display) can be credited only with an *integration efficiency,* and it can be as low as about 50% for 40 pulses integrated. In the numerator of the R^4 expression we account for this integration by the factor $nE(n)$, when n pulses are integrated, where $E(n)$ is an efficiency factor. For example, for $n = 40$, $E(n) \doteq 0.5$ as mentioned.

Another variation recognizes that the gain of the antenna on receive is sometimes more convenient to use than its effective aperture. The expression:

$$G_r = 4\pi \frac{A_e}{\lambda^2} \tag{1.6}$$

thus accommodates us. Here, λ is the wavelength of the carrier frequency, and we see frequency brought into the equation for the first time (since $f = c/\lambda$). Frequency, however, is not *fundamental* to radar sensitivity in the same way that receive *aperture* is. Finally, an *energy* form of the equation results from representation of the receiver bandwidth B by the reciprocal of the *transmitted* pulsewidth τ. We can do this only if the receiver is well matched to deliberate modulations of the signal (introduced for range *resolution* purposes, the object of pulse compression on receive) *within* τ. We note from this that *sensitivity* in radar is a function of *energy* transmitted ($P_t\tau$) and the spectral density of the noise power in the receiver (kT_s). These simple expressions assume that the noise power spectral density is uniform over the actual spectrum occupied by signal and to which the receiver is matched. (Note that its units are of energy as well, watts per cycle per second.) From these several variations we have

$$R^4 = \frac{P_t \tau G_t G_r \lambda^2 \sigma n E(n)}{L(4\pi)^3 WkT_s} \tag{1.7}$$

which shows the range achieved in multiple-pulse operation, in free space (i.e., no earth-multipath or path-atmosphere attenuation effects included).

In addition to receiver noise, we may have other *competing signals* with which to contend: reflection from rain or chaff, surface clutter, or an external source of power in the radar's passband. Hence, an expression developed in much the same way as for the desired reflection would be written in the denominator, added to the receiver noise power (or in place of it, if sufficiently dominant). Another factor is required in the numerator of the equation to account for multipath signal reinforcement (or reduction). Accounting for atmospheric attenuation requires another loss factor to be put in the denominator. (Note that this loss may depend on the target range being calculated, requiring iterative computation of a final value of range.) The reader will find more detailed coverage of the radar equation in the works of Blake [1,2].

The factor W represents a margin, perhaps 20 to 1, by which the reflected signal power must exceed that of the competing signal (its receiver noise, at least) to be sufficiently strong that we can declare the presence of a target. To understand this better, we must review detection fundamentals based on the statistics of receiver output. In Figure 1.7, we see a *probability density function* (pdf), representing the relative frequency of occurrence (the ordinate labeled "probability") of detector output, having value v (the abscissa in volts) when noise alone is present. For the case in which noise and a target signal are both present (called the *null* hypothesis, the *alternate* hypothesis being that only noise is present), the second pdf prevails. The *threshold test,* which must be made when viewing a display driven by the detector output v (or, as in an automatic detector, in which a computer notes, resolution cell by resolution cell, the succession of values of v) is illustrated in Figure 1.8, and the probabilities of error are graphed in Figure 1.9. Had the detector been viewed after many pulses were integrated (as in a *range correlator;* CRT displays in *plan-position-indication,* PPI, format inherently provide such a medium), then the test would be based on the mean of several samples of v and the two hypotheses are represented by the much more peaked pdf values of Figure 1.10. For actual pdf values governing these two hypotheses, $W = 20$ (or a 13 dB *signal-to-noise ratio* in our equation) typically results in error sizes of about 0.5 and 10^{-6} for *miss* and *false-alarm* probabilities, respectively, sizes quite commonly cited in range calculations. The echo strength is itself statistical in nature, and so we will now examine target scattering.

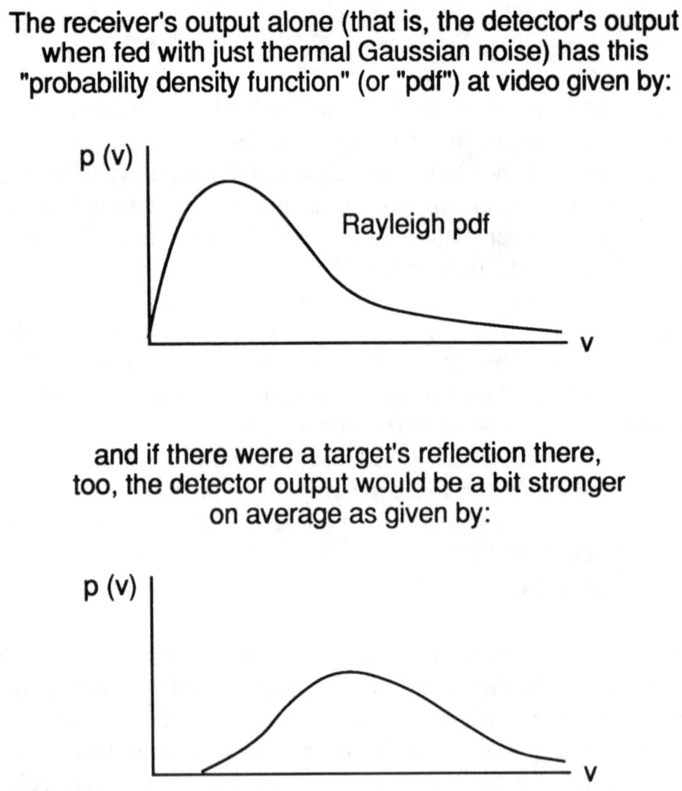

Figure 1.7 Receiver Output Probability Density Functions (pdf).

1.2.1 Radar Target

Figure 1.11 shows the reflecting *size* (σ) of an elementary reflector, a sphere, as a function of the wavelength, λ (inverse to frequency, f) of the electromagnetic

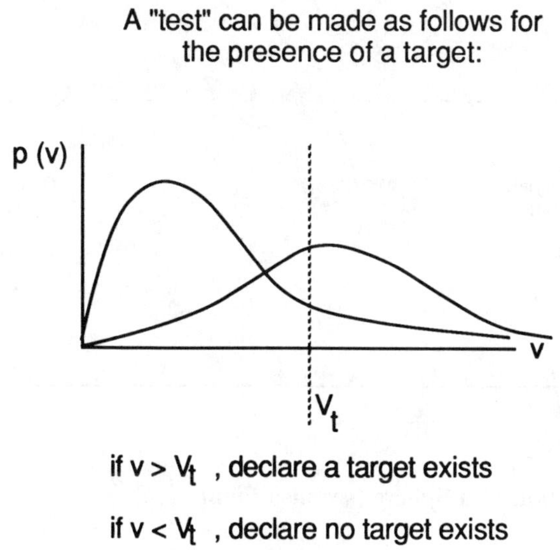

Figure 1.8 Test for Presence of Echo Signal.

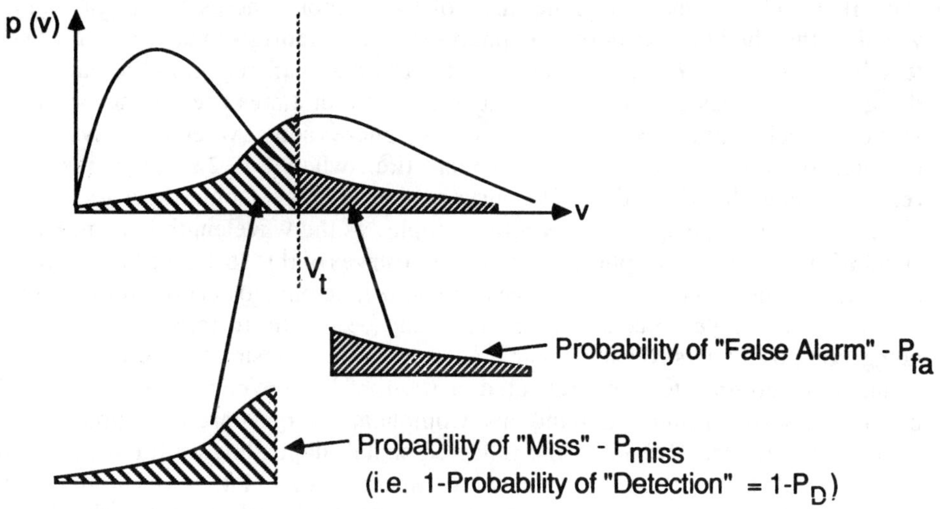

Figure 1.9 Probability of Wrong Decision.

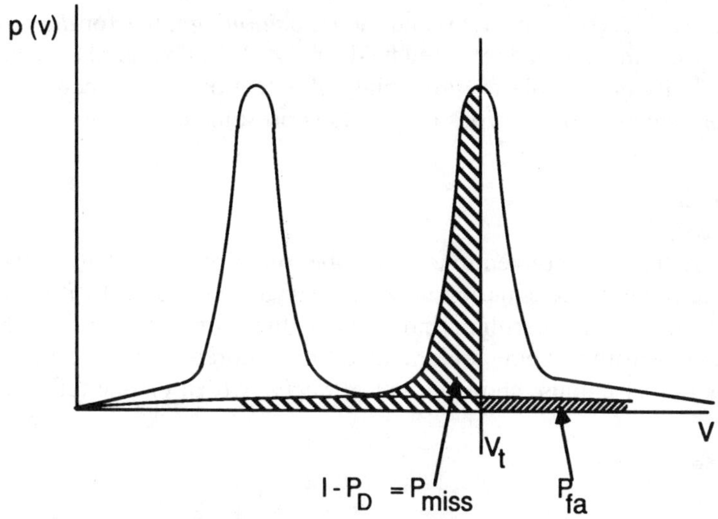

Figure 1.10 Detection Situation When a Number of Pulses Have Been Integrated.

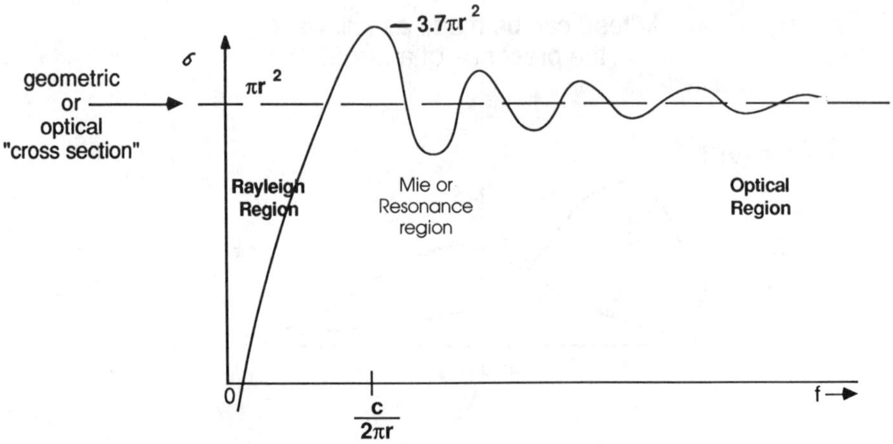

Figure 1.11 **Cross Section of a Sphere (see also Figure 8.6).**

wave. This is the classical curve posed by Lord Rayleigh in his investigations of why the sky appears to be blue, among other things. At the left of the curve is the sloping fourth-power dependent region (appropriately called the *Rayleigh region*), which relates an echoing area of tiny vapor spheres to optical (light) wavelengths (the blue-sky case), or spheres such as raindrops to microwave wavelengths (of more interest in the radar case). The *resonant* region in the middle of the graph produces a peak value that is nearly four times the geometric cross section (specifically, $\sigma = 3.7\pi r^2$, where r = sphere radius) when the wavelength is equal to the circumference of the sphere (i.e., when $\lambda = 2\pi r$). This resonant region is called the *Mie region*. The curve proceeds to the geometric cross section (i.e., $\sigma \doteq \pi r^2$) in the *optical region* to the right, as the wavelength becomes very small with respect to the sphere's actual size. Curves of this sort have been derived for other geometries (cylinders, ogives, *et cetera*), helping to account for the radar reflectivity of single objects. However, it may be helpful to think of a target as being composed of several similar scatterers (such as small spheres), and to examine the composite signal reflected or *scattered* by the ensemble. Figure 1.12 provides a basis for understanding this. Complicated targets present a great range of possible *resultant* values to the reflected signal, depending on the orientation of individual scatterers and the viewing aspect. Figure 1.13 gives four useful categories of targets thus composed, these being the *Swerling* cases (after Peter Swerling) based on two properties: the nature of the local collection of scatterers of which the target is comprised (all uniform in cross section or composed of one dominant scatterer; see Figure 1.13) and the *repositioning* times (or *decorrelation* times of the resultant summed reflected field strength). Of course, the exact values for the probability of false alarm, probability of detection, and detection criterion W depend on which of these types of targets is presumed.

1.2.2 The Decibel Convenience

Let us review the conventions of decibel notation. In radar, as the introduction of the radar range equation may have suggested, we deal with very large factors, and the use of logarithmic notation is therefore quite convenient, with the addition of logarithms being preferable to "brute-force" multiplications. Table 1.1 provides the essentials, showing that in *decibels* (dB, or one-tenth of a *bel,*

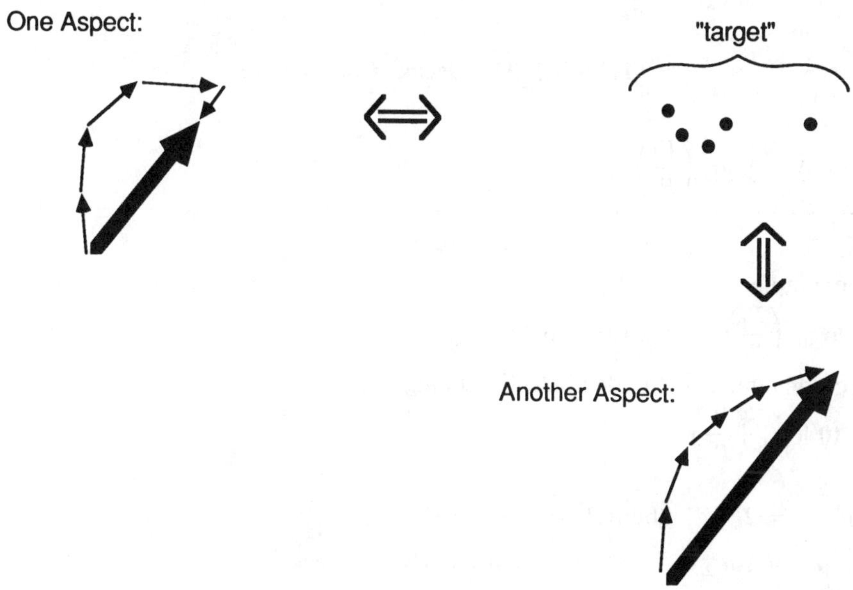

Figure 1.12 Multiple-Scatterer Targets.

	Uniform field of scatterers:	One dominant scatter in field of many small equal scatterers:
Slowly reposition themselves ("decorrelated" only scan to scan)	I	III
Rapidly do so ("decorrelated" pulse to pulse)	II	IV

Figure 1.13 Four Cases of Fluctuating Targets (Swerling Cases I–IV).

B) a power ratio of 2 is 3 dB, 10 is 10 dB, and 100 is 20 dB, so that a ratio of 20 (2×10) is 13 dB (3 dB + 10 dB), or a ratio of 4000 is 36 dB, *et cetera*. We see, too, that ratios of less than one are expressed as negatives in decibels, or so many dB *down* (e.g., $-n$ dB $= n$ dB down). The dB is an indispensable tool in radar work, and the student must become comfortable with these manipulations.

Table 1.1 The Decibel Convenience

$$1 \text{ bel (B)} = \log_{10}\left(\frac{P_1}{P_2}\right)$$

For example, if $P_1 = 2P_2$, then

$$\log_{10}\left(\frac{P_1}{P_2}\right) = 0.30103 \approx 0.3$$

More convenient is the decibel (dB), being

$$10 \log_{10}\left(\frac{P_1}{P_2}\right)$$

or, if $P_1 = 2P_2$	then $P_1 = 3 \text{ dB} > P_2$
$P_1 = 10P_2$	$P_1 = 10 \text{ dB} > P_2$
$P_1 = 20P_2$	$P_1 = 13 \text{ dB} > P_2$
$P_1 = 100P_2$	$P_1 = 20 \text{ dB} > P_2$
$P_1 = 0.5P_2$	$P_1 = 3 \text{ dB} < P_2$
$P_1 = 0.1P_2$	$P_1 = 10 \text{ dB} < P_2$
⋮	⋮

Important:

1 dB represents a factor of approximately 1.25.

−1 dB represents a factor of approximately 0.8.

These are all useful and close approximations. Of course, the *decades* (e.g., 10 dB is a factor of 10) are exact.

1.3 RADAR SUBSYSTEMS

We now proceed to the principal subsystems of radar, and, following the path of a signal, by analogy, we treat them in this order:

- Transmitter (Section 1.3.1);
- Antenna (Section 1.3.2);
- Receiver and Signal Processor (1.3.3);
- Control and Interface Apparatus (1.3.4).

1.3.1 Transmitter

In a brief overview such as this, we must resist the temptation to go too far in component-level examination, but we will give one further level of subdivision. The transmitter comprises three major components: the *power supply,* the *modulator,* and the *power amplifiers* (or, in some radars, the *oscillator;* in any case, we refer to the "tubes"). The overall purpose of the transmitter is to convert utility power (such as from a motor-generator set or a power distribution system that is not a part of the radar proper) into the desired RF signals. This is a formidable task, and we ought to appreciate the vast difference between the nature

HILL
CHAPTER 1

and dimensions of the input and those of the output. We show these three components working together in Figure 1.14, which gives a sketch of a *master-oscillator, power-amplifier* (MOPA) type of transmitter, with some more description of their functions. A second type of transmitter involves only a single high-power oscillator that is switched on by the modulator, but the MOPA is generally considered to be the more versatile and "modern" of the two. Finally, we must note that, despite the transmitter's intended purpose (conversion of power from main to RF), the chief product is the undesired *heat* stemming from the only modest efficiencies of the process, 10 to 30% overall for transmitters is typical, requiring the designer to pay attention to the removal of heat (called *dissipation*).

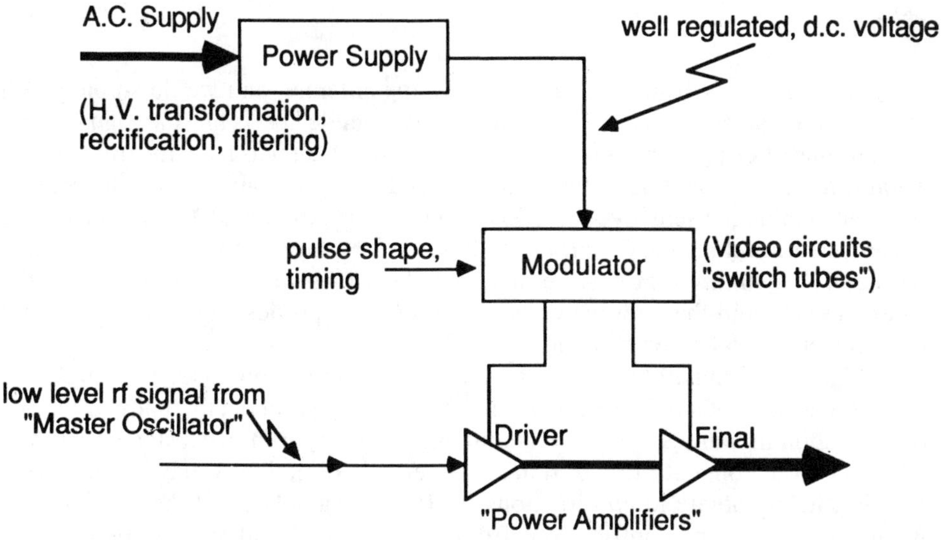

Figure 1.14 The Transmitter: Power Supply, Modulator, Power Amplifiers or Oscillator.

While tubes are only one component of a transmitter, they are the most frequently discussed, and for many they are the most challenging and interesting. The two principal kinds of tubes are *crossed-field devices* and *beam power tubes.* In crossed-field devices, the applied (generally by the modulator) high voltage and current move electrons within the vacuum in the tube, in the form of a cloud-like structure, following the electron-ballistic physics of charges moving in crossed electric and magnetic fields. In the beam power tubes, a beam of electrons is formed in a focused cathode-to-anode path, responding to the voltage gradient (but not that of a magnetic field). The *magnetron oscillator* and the *crossed-field amplifier* (CFA) are crossed-field devices (i.e., with cloud-like mobile electron regions), while *klystrons* and *traveling-wave tubes* (TWTs) are beam power tubes.

In principle, all tubes convert applied electrical power to RF power by moving electrons and then having them interact with RF resonant structures (in these microwave tubes, the interaction occurs within the vacuum envelope) to produce an RF output at the expense of the electrons' total motive energy (called *electromotive force,* EMF). Incidentally, the magnetron is almost invariably an oscillator. The beam power tubes can also be made to oscillate, but they have their major value in radar as amplifiers.

Modern radar demands that our signals be *coherent,* which, in turn, demands that the transmitter designer address both *amplitude* and *phase* properties of the signal's path. Efficiency and the handling of high voltages are major design concerns, with efficiency having profound reliability implications because of the heat, and with power handling having not only arc-induced damage involved, but also matters of human safety being of utmost importance.

We will give only brief attention to the important field of *solid-state* amplification of microwave energy, wherein banks of *modules* are assembled or arranged in some manner (such as a "corporate" structure) to achieve the total high powers needed by most radars [3]. *Avalanche-transit-time devices* and *transistors* make worthwhile subjects for the student or worker to pursue. See Chapters 2 (Section 2.6.5) and 4 for discussions of contemporary developments in solid-state discrete and monolithic microwave integrated circuitry (MMIC) and bipolar transistor technology.

1.3.2 Antenna

We move now from the transmitter to the antenna, and we do so on a path that involves a *transmission line,* often a waveguide, carrying the signal, with perhaps line filtering to trap out undesired "spurious" signals from the transmitter, a *duplexer* to permit this antenna to be used both to transmit and to receive (without ruining the sensitive receiver) as quickly as possible after the transmitted pulse is sent, and a rotary joint at the base of the antenna to accommodate any mechanical *scanning* that may be involved. This brings us to a discussion of the antenna itself, and the microwave *beam-forming* properties with which the radar engineer ought to be comfortable.

Figure 1.15 illustrates a most fundamental antenna composed of a *feed horn* at the focal point of a reflector, which is a section of a parabola (actually, a surface of revolution in many such physical antennas). The figure further illustrates that, while the axial concentration of reflected energy in direction A may be intuitively familiar, fields radiate in *all* directions (to B, for example), and the net field is a matter of phasing the contributions from every excited unit area of the reflector as received in direction B. Phasing at the generalized field point is further illustrated with the modest array of discrete radiators shown in Figure 1.16, where we note that the summed field is less at B than at A. At some larger angle, the phasor diagram for point B closes on itself without any resultant value. This is the angle of the *first null* in this antenna's *pattern*. We also note that had the array been larger (wider), this null would have occurred at point B, as shown, illustrating the important principle that large antennas (large in dimensions as measured in wavelengths, not physical size) form *narrow beams,* and small (or narrow) antennas form *broad beams*. Figure 1.17 shows the relationship between the incremental phase angles of the successive contributions to the total field and the dimensions of the array. This figure also suggests that if the radar could *control the excitation phase* of each element, the direction in the field at which a resultant maximum could be achieved would itself be controlled. This, then, is the principle behind beam steering in a *phased array,* without involving physical positioning of the antenna.

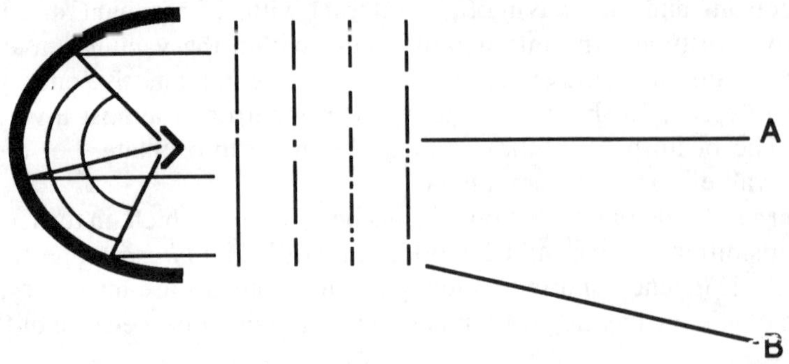

Figure 1.15 Reflector Antenna.

HILL
CHAPTER 1

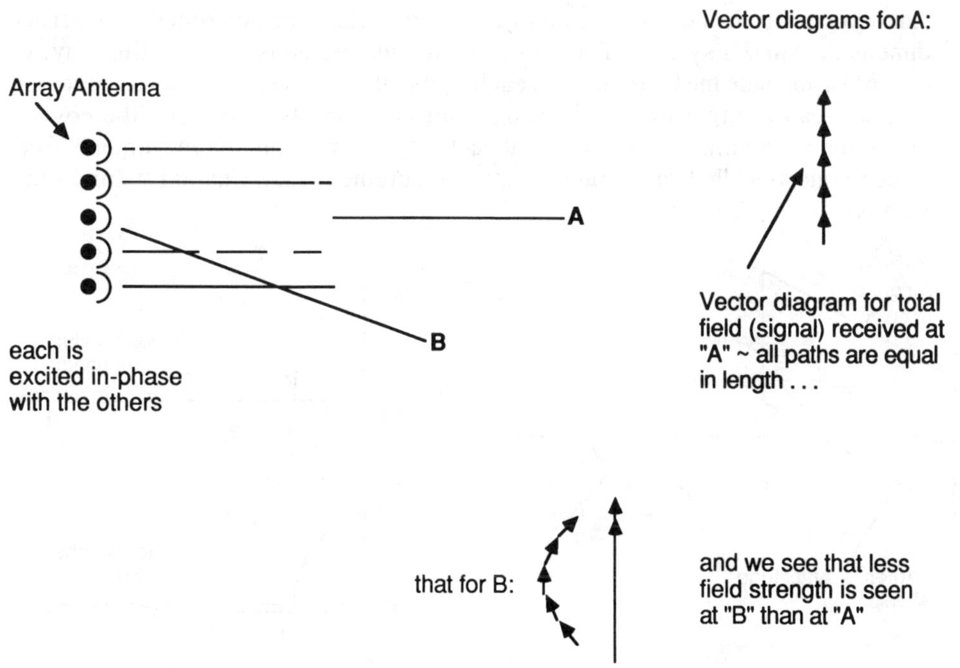

Figure 1.16 Antenna Beam Formation.

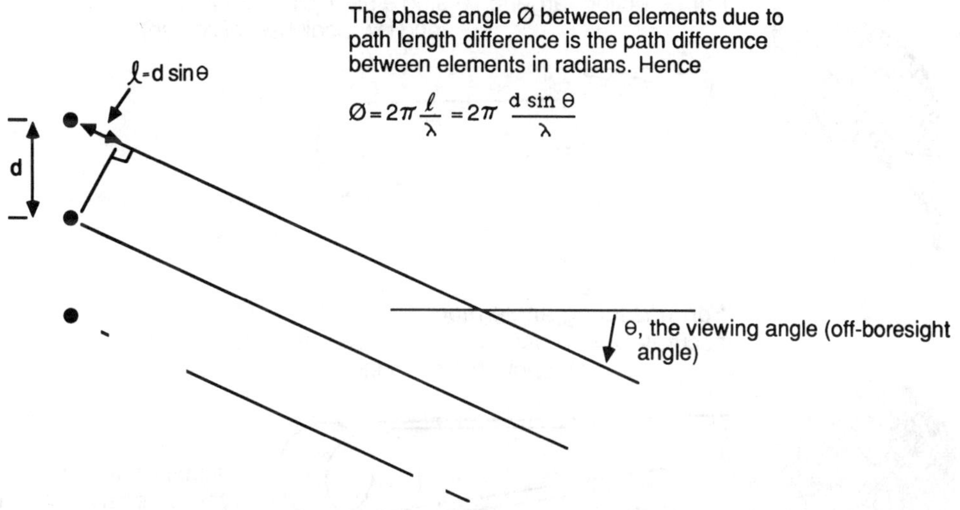

Figure 1.17 Antenna Beam Formation and Basis for Phase Steering.

Figure 1.18 illustrates that the successive values of the resultant field strength (those values being squared for the *relative power* values) form a pattern when plotted against the viewing angle θ. This pattern is dependent on angle alone, provided that the field domain is sufficiently distant from the antenna so that the paths to the individual elements of the antenna are essentially parallel. The figure further shows that this pattern may be shaped by *weighting,* in amplitude, the excitation (or illumination) function across the antenna. However, applying a weighting function that is other than uniform (which unfortunately has fairly high *sidelobes*) can only be done (such as to achieve lower sidelobes) at some expense in main-beam gain and broadening.

Our antenna overview is completed with Figure 1.19, which shows the previously mentioned fan beam, often used in 2D search radars, and also the more precise angle measurements on targets that can be achieved if several beams are simultaneously formed (the beams being narrow in both horizontal and vertical dimensions) and a system of signal-strength comparison is used. In this way, we can make an angle measurement on each transmitted pulse (which does not require that we scan a target to see changing values of successive pulses), thereby significantly improving the measurement accuracy (even on fluctuating targets). Such a technique is called *monopulse* angle measurement (see Chapter 5 for further details).

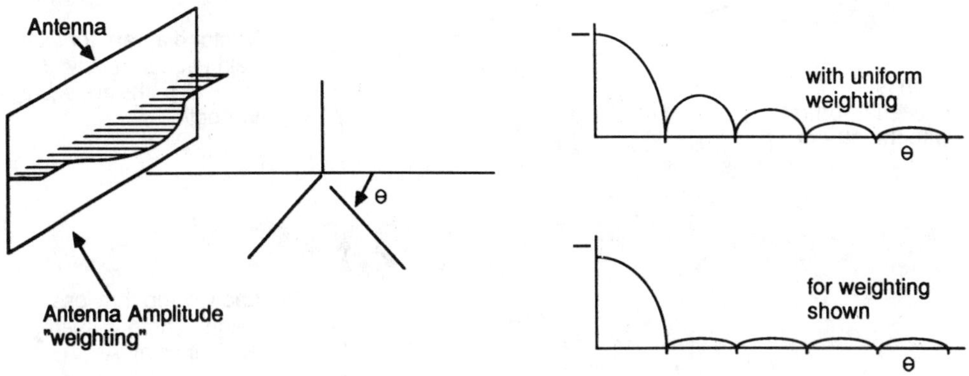

Figure 1.18 Antenna Pattern.

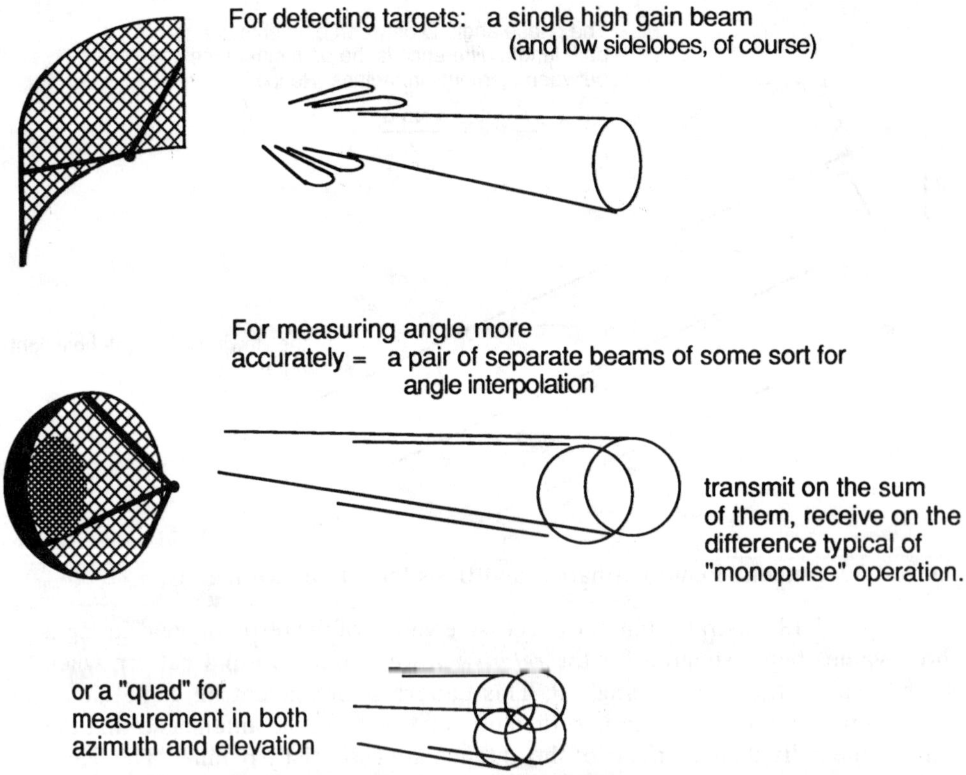

Figure 1.19 Types of Antenna Beams.

The antenna serves the twofold purpose of causing radiation into space in a well matched way (the antenna is indeed an impedance transformer) and causing a desirable spatial concentration of the energy into a main beam. The same antenna gathers all signals incident to it during the listening periods between transmitted pulses, whereupon signals are routed to the third major subsystem of the radar, the receiver and signal processor.

1.3.3 Receiver and Signal Processor

The receiver and signal processor subsystem comprises several components: RF *preamplifiers,* a *local oscillator* (LO) and *mixer* combination, and a *signal processor.* We will treat the latter subject in more detail. Note that RF preamplification is not always used in radar; when it is used, it is often done with solid-state amplifiers of low noise figure (*NF*) and considerable gain (*G*), located as close as possible to the antenna assembly. (Typically $NF = 2$ or 3 dB and $G =$ 20 or 30 dB, so that the receiver's total noise figure is dominated by this stage.) The local oscillator and mixer *down-convert* the signals received to a convenient, fixed *intermediate frequency* (IF, where 18 and 30 MHz are common in radar). A process of *dual* (up and down) *conversion* is sometimes used to reduce the effects of the *image frequency* associated with single (up or down) conversion. The signal processing then performed on the coherent IF signal is the next subject to be discussed.

For convenience, we may think of signal processing as being applied to each individual pulse and to a succession of pulses, and also as involving only the amplitude or envelope of the received signal (*noncoherent processing*), or involving both the amplitude and phase (*coherent processing*) of the signal. These four kinds of signal processing are listed in Table 1.2 with examples given in a matrix representation of the process.

Table 1.2 Examples of Four Types of Signal Processing

	Single-Pulse Processing	*Multipulse Processing*
Noncoherent Processing	Crystal Video Detection	Range Correlation (Video Integration on a Scope, or in the Receiver Processor)
Coherent Processing	Pulse Compression	Doppler Processing Canceler Moving-Target-Indication (MTI) Pulsed Doppler (PD) Filtering

Coherent processing of a single pulse, called *pulse compression,* deserves further attention. Both sensitivity (requiring a lot of energy in a pulse, suggesting long pulses) and resolution (seemingly requiring very short pulses) in radar signals can be jointly achieved. This is achieved by using the long pulse required for high sensitivity, but modulating the pulse, in frequency or phase, so that it has the broad bandwidth of the very short pulse required for fine resolution. This fine resolution is achieved by processing correctly the broad *bandwidth* of the long pulse. A receiver which achieves this "correct processing" of the modulated pulse is called a *matched filter* receiver. Its output is *not* a replica of the signal being received, but rather that of the desired very short pulse. The pulsewidth of this response will be approximately the reciprocal of the bandwidth of the modulated pulse, much narrower in time (thus giving good range resolution, if the receiver remains "linear" or unsaturated for multiple closely spaced inputs) than the long pulse envelope. In fact, the ratio of the width, in time, of this response to the original long pulsewidth transmitted is called the *pulse compression ratio,* and it is often on the order of several hundred to one. Actual pulse compression achieved will be affected by distortions to the signal caused by amplification in the transmitter and all the apparatus, the medium, and the target characteristics that the signal experiences before arriving at the matched filter processor. The matched filter also has the property of providing a signal-to-noise ratio at the output that is the maximum possible for the received, modulated signal. Hence, this filter is optimal for signal detection. For further reading on pulse compression, see [4].

A second kind of coherent signal processing is in the multiple-pulse category, specifically doppler processing. Most people are familiar with the doppler principle of frequency shift experienced when a source of sound is moving toward or away from a listener—such as the familiar analogy of when a train whistle or police siren passes by. In radar, we deal with signals reflected by objects, of course, but in many cases the dimensions are such that the shift is so slight (even for very fast moving targets) that the received signal is more like an unchanged carrier frequency incurring a phase change (relative to the signal that had been transmitted) from one pulse to the next. So, instead of building a receiver *at a different frequency* to accommodate and measure doppler shift, the radar engineer builds pulse-to-pulse phase-comparison circuits called *cancelers* to distinguish returns so shifted (moving targets) from unshifted returns (nonmoving clutter). If a number of pulses were processed, the total phase shift between the first and last pulse would be large and a bank of narrow doppler filters could be used. The doppler phenomenon is summarized in Figure 1.20.

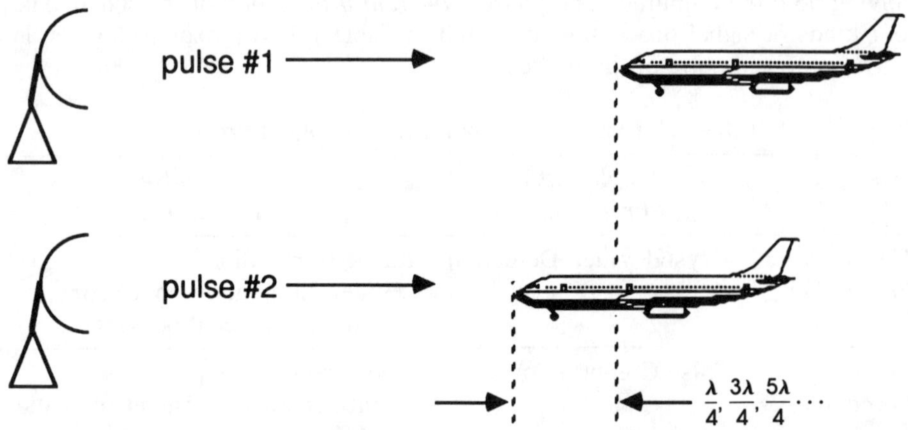

Figure 1.20 Doppler Processing Showing 180° Phase Shift between Received Pulses (as "Optimal" Velocity).

An interesting relationship (very significant in modern radar) exists between antennas and doppler processing receivers, particularly in airborne radar, as illustrated in Figure 1.21. If we concede that ground clutter (undesired) would have a doppler shift associated with one's own aircraft velocity, we might suggest that an adjustment in the phase reference apparatus could be made to *zero out* the clutter return, and the adjustment would depend *only* on the direction the antenna is pointing, relative to the aircraft path (shown on the left-hand side of Figure 1.21). However, if we account for the antenna sidelobes, the single adjustment mentioned will be inadequate. There will seem to be clutter moving "at other speeds" (as shown on the right-hand side of the figure), and this return will *mask* targets, even in the main beam, which are moving relative to the clutter. Such targets *should* have been seen because of the relative doppler shift. The only way to restore sensitivity to these moving targets is to reduce markedly the sidelobes of the antenna itself, and thereby contain the clutter spectrum in the main-beam-only doppler-shifted position. Systems of this nature have prompted the achievement of *ultra-low-sidelobe antennas* (ULSA), made possible by computer-aided design and high-precision manufacturing techniques.

It is important to note that, as a function of the radar's PRF, the whole structure of clutter response *folds over,* or is "aliased" in doppler sense, obscuring signals of even greater than forward-flight clutter doppler (denoted by f_{d_1} on the

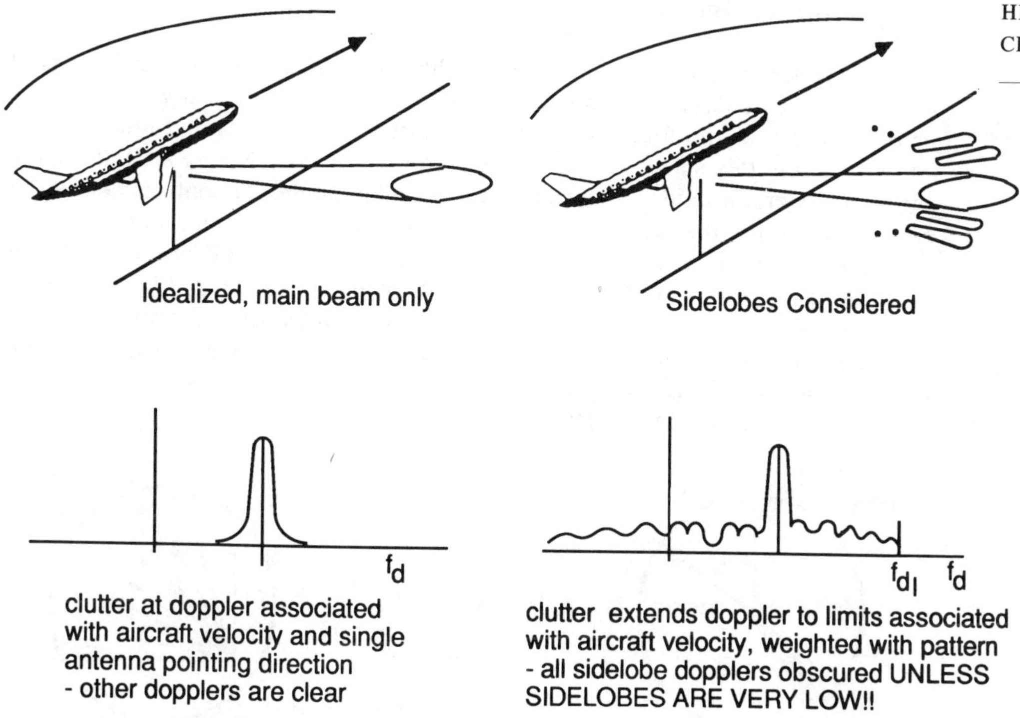

Figure 1.21 Considerations in Airborne Radar: Clutter, Antenna Sidelobes, and Doppler Processing.

sketch). The advantage of higher PRFs in such radars is that more of a clutter-clear region is maintained between this obscuring structure and the first doppler ambiguity. This region is then limited only by instabilities in the transmitted signal and the processing itself.

Coherent signal processing can be brought about by modern digital processing using paired synchronous detection of complex (or "vector") radar signals. Here, amplitude *and* phase information is retained in two scalar "channels," the *in-phase* (I) and *quadrature* (Q) channels. This important technique is illustrated in Figure 1.22. The advantages of computer-like circuitry then accrue.

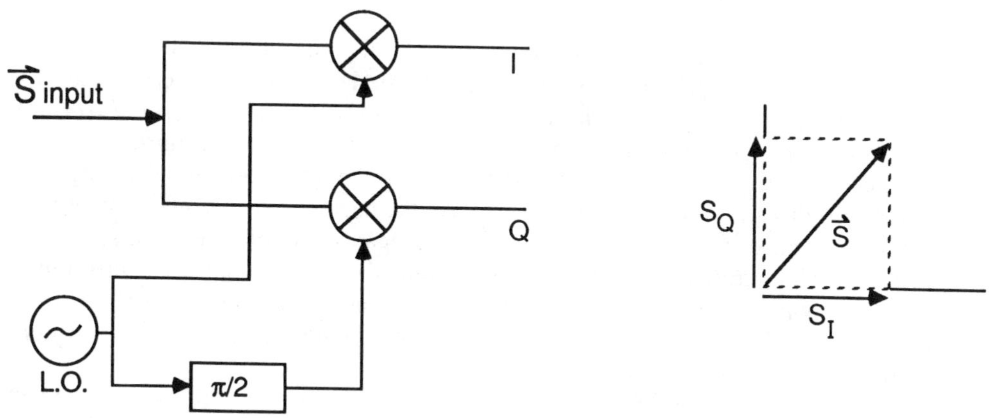

Figure 1.22 Digital Coherent Processing.

HILL
CHAPTER 1

1.3.4 Control and Interface Apparatus

We proceed now as if a radio detector had been applied to the signal received and processed as described above, and that we are about to deal with a voltage or *video* signal in the fourth principal subsystem, the control and interface apparatus. Often, this subsystem is an "operator's console," where a CRT display shows the operator evidence of radar echoes and a number of controls permit the operator to adjust the radar, call for various modes of operation, and report his interpretation of radar data to some user system that the radar is intended to serve. A few of the many formats for CRT displays are shown in Figure 1.23, of which the *plan-position indicator* (PPI) is by far the most common and generally useful. The operator can call for various calibration marks (range *rings*, or bearing *strobes*) and even local maps to be superimposed on such video displays as an aid in interpretation and reporting.

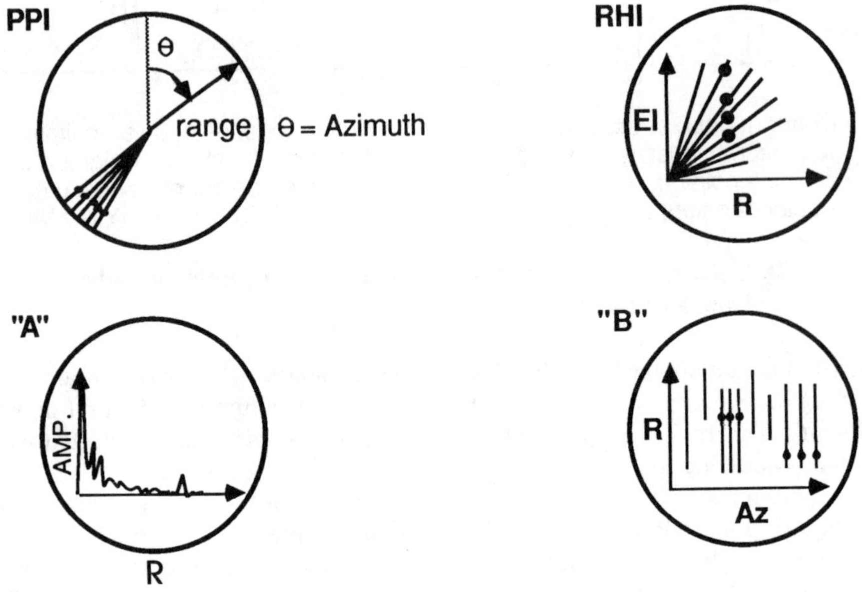

Figure 1.23 **Control and Interface Apparatus (RHI = Range-Height Indicator; PPI = Plan-Position Indicator).**

A competent operator can keep up with interpretation and reporting of only a few new or changing events every few seconds. In more severe situations, computers can be of great help in performing *automatic detection and tracking* functions, and such automation has been successfully developed since the 1960s. Figure 1.24 illustrates how an automatic detector might work, first dealing with a variable threshold test during each interpulse period, and second dealing with the several pulse periods associated with one beamwidth of azimuthal scan. The techniques illustrated are the *mean level detector* (MLD) and a "centroiding" function, which, when parameters are properly adjusted, can provide *constant false-alarm rate* (CFAR) detection in dense target environments, even with non-uniform background conditions.

The detections or "plots" produced on successive scans of the antenna can be associated by computer processing and formed into tracks. Considering that the plots have a certain inaccuracy (at a minimum, what would be called a "measurement" inaccuracy, producing "measurement noise," made worse by such things as jamming or target fading), it is well to filter the succession of plots in processes such as *exponential filtering* illustrated in Figure 1.25 (shown for a single dimensional space). In some radar systems, and more certainly in networks

HILL
CHAPTER 1

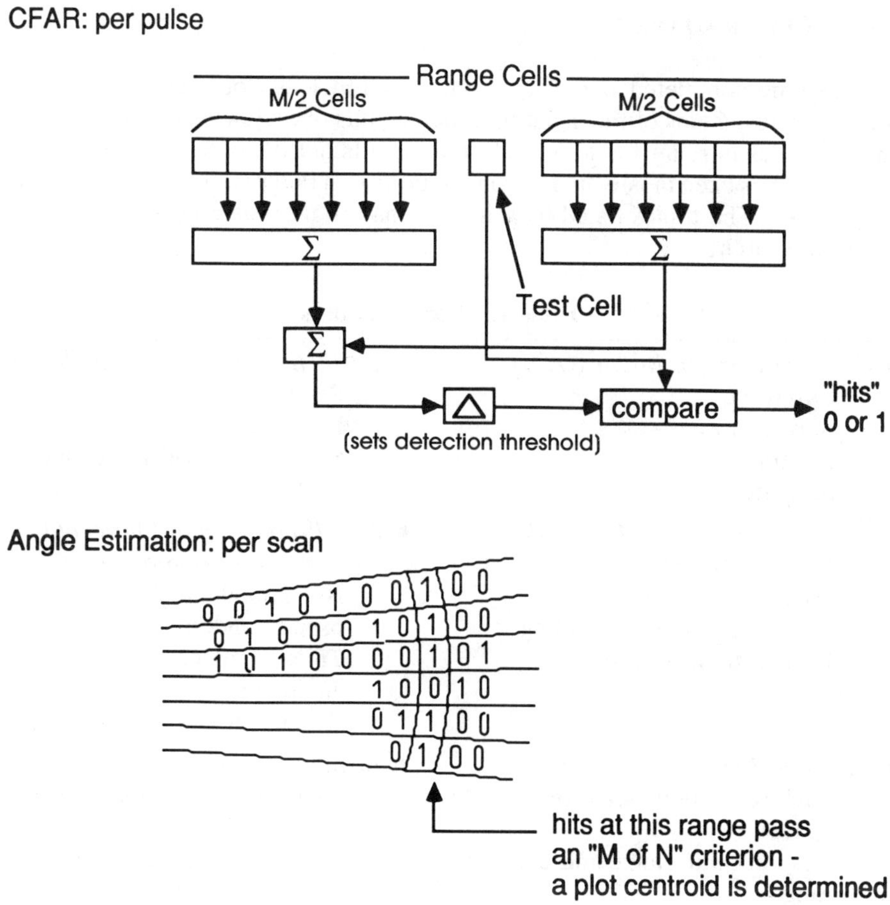

Figure 1.24 Automatic Detection and Plot Extraction.

an "α filter: $\quad \bar{X}_n = X_{pn} + \alpha (X_n - X_{pn})$

where $\quad X_{pn} = \bar{X}_{n-1}$

or an "α -β filter": $\quad \bar{X}_n$ as above

$$X_{pn} = \bar{X}_{n-1} + \bar{\dot{X}}_{n-1} T$$

where $\quad \bar{\dot{X}}_n = \bar{\dot{X}}_{n-1} + \frac{\beta}{T}(X_n - X_{pn})$

Figure 1.25 Automatic Tracking. (*Note*: Detections at the output of the automatic detector can be fed, scan by scan, to a track-filtering computer; exponential track filters are typically used for "smoothing" the tracks. The weights (α and β) can be varied, depending on the state of one's radar, or of different radars contributing "plots," or on experienced error, as in Kalman filtering.)

of radars sharing plots and tracks, this kind of postdetection processing is called radar "data processing," whereas the term "signal processing" is reserved for the predetection treatment within the radar's receiver.

1.4 SPECIAL RADARS

Having completed our discussion, which at least touches on each of the principal subsystems of radar, let us conclude this overview by citing a few other special topics in radar for which full treatment is not possible here. Our context heretofore has been mostly in the monostatic search radar at "regular" microwave frequencies. The topics listed in Table 1.3 may trigger the reader's desire to do further research.

Table 1.3 Special Radars

• *Over-the-Horizon Radar (OTH)* Sky wave Surface wave Ducting Satellites	• *Synthetic Aperture Radar (SAR)* Resolution Focused operation Sampling, ambiguities, and limits
• *Millimeter-Wave (MMW) Radar* Fine resolution Small components Power generation, handling Propagation, medium	• *Low Probability of Intercept (LPI)* Communication origins Noise modulation; pseudorandom $TB \gg 1$: matched filter advantage Other LPI matters, approaches
• *Imaging Radar* High resolution (scatterer space) Resolution in range and cross-range Inverse SAR (SAR^{-1} or ISAR) Other work in radar signature	• *Bistatic Radar* Ovals of Cassini—sensitivity Ellipses of arrival—range Performance relative to monostatic radar

Over-the-horizon (OTH) bistatic radars are discussed briefly in Chapter 2, and millimeter-wave (MMW) radar is the topic of Chapter 7. Synthetic aperture radar (SAR) is discussed in detail by Brookner, in Part 4 of [4]. Inverse SAR (ISAR) is discussed in [5,6].

1.5 FUTURE CHALLENGES

Finally, the challenges in modern radar design are somewhat different than those of power and sensitivity, which guided radar work for so many of its first 50 years. The topics of high-information-content signals and processing, the exploitation of flexibility in modern radar apparatus, subject to adaptive control by computers, and the extraction of information ("estimation") for a user system that is very much tailored to its need are worthy of the radar designer's attention. When developed, these considerations will provide radars with properties of:

- *Tenacity*—suggesting that a radar can adjust immediately to secure track, even on tenuous initial detections;
- *Adaptivity*—suggesting that a radar can operate with tenacity for a wide variety of circumstances or environments;
- *Capacity*—suggesting that a radar can achieve tenacity and adaptivity in a dense target environment.

Such properties will make radars remarkably more valuable to the overall system they may serve.

1.6 CONCLUSION

The student of modern radar has ample opportunity to profit by the experience of his or her colleagues. The open literature is plentiful, in both textbooks and the professional journals. Notable among the journal literature are the various *Transactions* of the Institute of Electrical and Electronics Engineers (IEEE), the leading professional association in the field. IEEE activities also include periodic conferences throughout the world, where the younger engineer can associate with many others of varied experiences. The IEEE, furthermore, distinguishes itself by such activities as the lecture series run so well by the Boston section, which led to this book. It has been this writer's pleasure and honor to participate in the book with this overview of radar.

So, then, what is it that is "off to a good start"? Perhaps, it is the reader in this book itself. Perhaps, it is the student in his or her radar engineering career. Or, perhaps, it is the very field of radar as a whole, which, in its first half century of demonstrated success, has seen but a beginning.

REFERENCES

1. Blake, L.V., *Radar Range-Performance Analysis,* Artech House, Norwood, MA, 1986.
2. Blake, L.V., Chapter 2 in *Radar Handbook,* M.I. Skolnik (ed.), McGraw-Hill, New York, 1970.
3. Ostroff, E.O., *et al., Solid-State Radar Transmitters,* Artech House, Norwood, MA, 1984.
4. Brookner, E. (ed.), *Radar Technology,* Artech House, Norwood, MA, 1977.
5. Brookner, E., "Radar Imaging for Arms Control," Chapter 11 in *Arms Control Verification, the Technologies that Make it Possible,* K. Tsipis, D.W. Hafemeister, and P. Janeway (eds.), Pergamon-Brassey's, London, 1986.
6. Wehner, D.R., *High Resolution Radar,* Artech House, Norwood, MA, 1987.

Chapter 2
TRENDS IN RADAR SYSTEMS AND TECHNOLOGY TO THE YEAR 2000 AND BEYOND

by Eli Brookner

2.1 INTRODUCTION

Radar has recently gone through many dramatic advances—in phased array radars (2D scan and 1D scan); in the trend toward solid-state transmitters; in the development of over-the-horizon (OTH) radars; in ultra-low-sidelobe array antennas (near-in sidelobes 40 to 50 dB down); in adaptive array processing, which gives large (50 dB) rejection of interfering signals; in microstrip antennas; in the use of more optimum clutter rejection techniques; in digital radar signal and data processors; and in the development of *monolithic microwave integrated circuitry* (MMIC), which has the potential of reducing the cost of microwave circuitry, just as integrated circuits did for the computer industry. This chapter discusses these recent advances and the future trends. We also cover the future effect of *very large scale integration* (VLSI), *very high speed integrated circuits* (VHSIC, US militarized VLSI), and *very high performance integrated circuits* (VHPIC, the British version of VHSIC) on digital radar signal processors and data processors.

2.2 ARRAY RADARS

Since the mid-1970s, we have seen phase-phase steered antennas, which are phased arrays that electronically steer their beam in elevation and azimuth, come into their own. This has been exhibited by the construction of numerous large phased arrays and the initiation of production runs for many other phase-phase steered arrays. Phase-phase steered arrays are needed to track simultaneously a large number of dynamic targets, when the target dynamics do not permit the use of a mechanically steered antenna to achieve the required update rate. This trend toward phase-phase steered arrays can be expected to continue into the 1990s and beyond. The dramatic trend toward phase-phase steered arrays in recent years is illustrated by the examples in the following subsection and Section 2.2.3.

2.2.1 Phase-Phase Steered Array Systems

2.2.1.1 Cobra Dane (AN/FPS-108) [1–6,321,346,360]

The COBRA DANE radar of Figure 2.1 represents an excellent example of the significant advances that have been made in phase-phase steered array radars in the past decade. This is a 95-ft diameter, 15.4-MW peak power, and 0.92-MW

BROOKNER
CHAPTER 2

Figure 2.1 COBRA DANE (AN/FPS-108) (from Brookner [1]).

average power L-band radar deployed in 1977 on the next to last island in the Aleutian chain of Alaska, providing close proximity to the USSR for observing the conformance of test firings of Soviet intercontinental ballistic missile (ICBM) systems as agreed by the SALT I treaty. The COBRA DANE can simultaneously track hundreds of targets. In addition to its function of intelligence-gathering on Soviet missile systems undergoing test firing, COBRA DANE is used for space track support and ICBM early warning.

A significant feature of this radar is its ability to provide high range resolution observations of targets at ranges on the order of 1000 nmi. These wideband measurements are used to obtain target size and shape data. A 200 MHz waveform, which can provide a 2.5-ft range resolution, is used to give the high range resolution measurements. Normally, it is not possible to obtain such a high range resolution measurement from a 95-ft aperture antenna because of the time dispersion across the antenna. For example, when pointing up to a target which is 22° off boresite in elevation, the signal from the lower part of the aperture arrives at the target 95 ft × sin 22° = 35.6 ft after the signal from the top of the array arrives at the target. As a result, the 2.5-ft pulse is spread out over 35.6 ft, making it impossible to achieve the desired high range resolution. To circumvent this problem, the 95-ft diameter array is broken up into smaller subarrays. These subarrays are sized so that the time dispersion across any one of them is small compared to the desired range resolution. Furthermore, the signal to each subarray is delayed by the amount necessary to make the signals from all subarrays arrive simultaneously at the target. Thus, for the example given above, the top subarray unit receives a time delay of about 36 ns relative to the bottom subarray unit. This process is called *time delay steering*. Phase steering is used over each subarray. The 95-ft aperture is divided into 96 subarrays, each having 160 active radiating elements that are phase steered.

Figure 2.2 is a simplified block diagram of the system, which shows the *time delay steering unit* (TDU) used for each subarray and the phase steering within the subarray. There are a total of 15,360 (160 × 96) active elements in the array face. Each subarray is fed by a 160-kW peak power TWT (Raytheon ring bar QKW 1723).

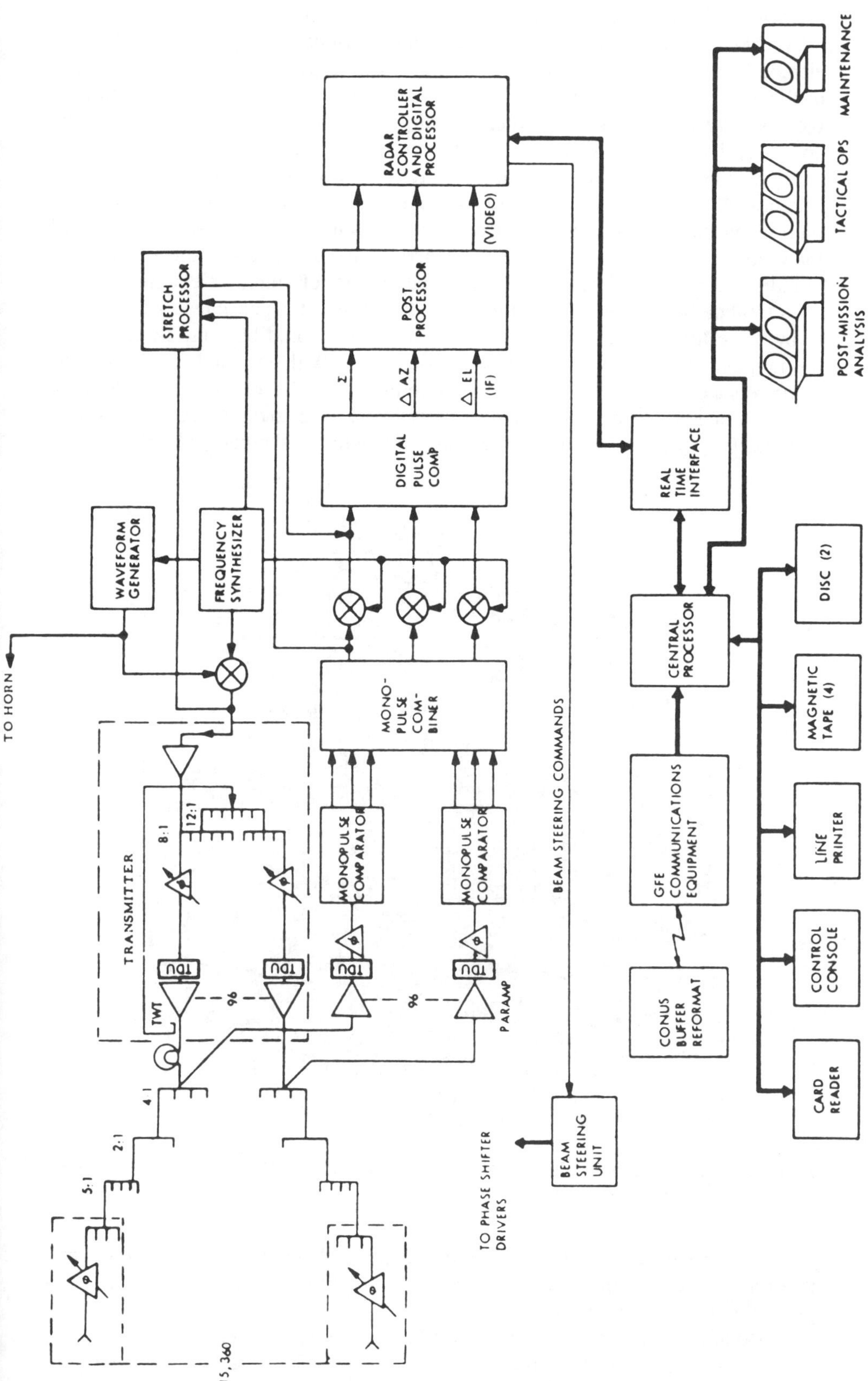

Figure 2.2 COBRA DANE Block Diagram (from Filer and Hart [4] © 1976 IEEE).

Another interesting feature of the radar is the use of array thinning [7]. Figure 2.3 illustrates the concept of *array thinning*. The black dots in the figure indicate positions in the array at which active elements are placed. Where there are no black dots, nonradiating elements, called *dummy elements*, are used. These elements are terminated into a dummy load with impedance which matches that of the active element loads. Notice that most of the elements near the center of the aperture are active, with their density decreasing toward the edge of the aperture. This density variation effectively provides amplitude weighting of the aperture illumination on transmission. As a result, low near-in sidelobes are achieved. The density tapering was designed to provide a 35 dB tapered-Taylor weighting on transmission. Array thinning also resulted in a narrower main-lobe beamwidth than would have been possible if a full array having the same number of active radiating elements had been used. The total number of dummy elements is 19,403 for a total of 34,769 elements in the array. If desired, it is possible at a future date to increase the system's sensitivity by 9 dB. This would be done by increasing the number of active elements by a factor of two, with 15,360 dummy elements being replaced by active elements, and another 96 TWTs feeding these additional active elements. Figure 2.4 shows a row of 1-to-160 subarray power divider units as viewed approximately along the major diameter of the backside of the array face.

Figure 2.3 Illustration of Thinned Circular Array (not COBRA DANE) (from Skolnik [8] © 1964 IEEE).

Dual-polarized circular waveguide radiating elements are used in the array; these are currently connected for vertical polarization [9]. Growth featuring dual polarization is a system option.

Detailed parameters for this radar, as well as many other radars to follow, are given by Table 1 of Chapter 1 in [1]. An update of that table, to include more recent radars, is given in Appendix 2A. Additional data are also given in this chapter and [10].

Figure 2.4 View of 1:160 Power Divider Units Along Backside of Array.

2.2.1.2 PAVE PAWS Radar (AN/FPS-115) [1, 11, 12, 321, 346, 360]

The impressive advancement made in large UHF phase-phase steered array radars over the past decade is exemplified by the PAVE PAWS system. Figure 2.5 shows the PAVE PAWS radar built at Otis Air Force Base in Massachusetts. The first large phased array radar was the AN/FPS-85, which also operated at UHF [1]. For this radar, it was less expensive to use separate antennas for transmitting and receiving than to compensate for the duplexer insertion loss and pay for the cost of the duplexer [358]. The power generation was supplied by tetrode tubes driving each transmit radiating element. In contrast, the PAVE PAWS system transmits and receives on the same antenna. In addition, the transmitter is all solid-state, using bipolar silicon UHF transistor power amplifiers, with each active element of the array driven by a solid-state transceiver (T/R) module (which contains the transmitter and receiver as well as the phase shifter needed for each radiating element to have electronic steering of the antenna beam). These modules receive the exciter signal, which is first passed through a phase shifter on transmission and then through a solid-state UHF power amplifier. On reception, the signal comes from a radiating element and passes through a low-noise receiver in the transceiver module, after which it is switched through the same phase shifter before going to a beamformer. PAVE PAWS has two array antenna faces instead of one. The two faces provide continuous coverage for two 120° azimuth regions, for a total coverage of 240° in azimuth.

Figure 2.6 shows a simplified block diagram of the PAVE PAWS radar. As in the case of the COBRA DANE, the array is divided into subarrays, each consisting of 32 active modules to feed 32 radiating elements. There are 56 such subarrays, which are, in turn, driven by modules identical to the array element modules, except that they do not use the unneeded receiver portion. A predriver drives the 1-to-56 divider, which supplies the subarray signals. This predriver also uses a module identical to the array element module, except that, again, it does

Figure 2.5 Multifunction UHF Phased-Array Radar AN/FPS-115 (PAVE PAWS) (from Brookner [1]).

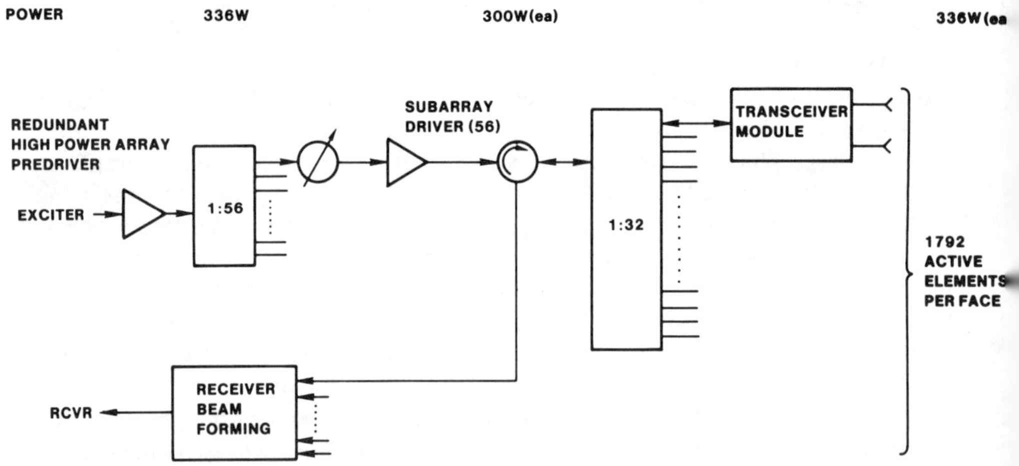

Figure 2.6 Simplified Block Diagram of PAVE PAWS System (from Campbell and Hoft [11]).

not use the receiver portion. Thus, to build the PAVE PAWS array, a large number of the UHF modules proliferate. Once it is known how to build these UHF modules, these radars can be developed with relative ease.

Figure 2.7 shows a view of the backside of the array, comparable to that of Figure 2.4 for the COBRA DANE system. A subarray driver module is visible on the left, mounted on the girder. The units standing in the middle of the floor are the power supply units for the array modules. What is striking, when comparing Figure 2.7 with Figure 2.4, is how much less hardware (especially cabling) is needed for the PAVE PAWS solid-state array than for the TWT COBRA DANE array. The large 1-to-160 subarray divider units of the COBRA DANE system, and the large three-inch high-power coaxial cable are not needed. The high-voltage modulators and power supplies (42 kV), not shown, can also be eliminated.

Figure 2.7 Backside of PAVE PAWS Along Major Diameter of the Array.

Figure 2.8 shows the modules in the backside of the array face, near the center of the array. The subarray driver modules are mounted on the girders. Array thinning is used, similar to COBRA DANE. There are 885 dummy elements and 1792 active elements over a 72.5-ft diameter. Provision has been made to allow the active area of the antenna to grow to 102 ft, if desired at a future time, permitting the system sensitivity to increase by about 10 dB. The number of active elements would be increased by slightly more than a factor of two in this case. The system is intended for early warning and attack characterization of submarine-launched missiles. A secondary mission is the support of the Air Force Space Track program. This 420–450 MHz radar has a peak power of about 600 kW and average power of about 150 kW per face. The radar's range is 3000 nmi for a 10-m^2 target. Figure 2.9 shows the eight-inch high, tilted, crossed-dipole radiating elements of the array. These radiating elements are bent downward to alter electromagnetic interactions of adjacent elements, which otherwise can, at certain steering angles, prevent energy from radiating (where the energy instead goes back into the radar), with the result that the target cannot be seen at those steering angles. The radar is thus called *blind* at these angles. This is referred to as the *phased array radar blindness phenomenon* [1]. The metal stubs interspersed among the antenna elements aid in producing a circularly polarized beam, a characteristic that helps to maximize the system's sensitivity.

Three other PAVE PAWS have been built, one at Beale Air Force Base in California, in 1980, another at Robins Air Force Base in Georgia, in 1986, and a third near Goodfellow Air Force Base in Texas, in 1987 [13–17].

The US Air Force plans to have the PAVE PAWS radar at Beale grow by 6 dB in sensitivity. A sensitivity growth of 10 dB is also planned for the Robins site to perform the Space Track function, and thereby replace the aging AN/FPS-85 radar system at Eglin Air Force Base in Florida. For further discussion on the PAVE PAWS, see Chapter 4.

Figure 2.8 View of Backside of PAVE PAWS Array Near Center of the Array. Modules and driver modules shown.

Figure 2.9 View of Front Face of PAVE PAWS Array Showing Bent Dipoles.

2.2.1.3 BMEWS Radar Upgrade (*AN/FPS-123*(*V*)*5*) [14]

The aging BMEWS, in Thule, Greenland, has been upgraded with a solid-state, two-faced, UHF phase-phase steered array radar (designed and built by Raytheon), using modules which are identical to those of the PAVE PAWS radar (see Figure 2.10 [14,18,388]). Each face has an 84-ft diameter array, containing 2560 active elements and 1024 dummy (inactive) elements. Similar upgrades are planned for Fylingdales Moor, England, and Clear, Alaska [13,16]. The Fylingdales system is to have three faces for 360° coverage [19].

Figure 2.10 UHF Solid-State BMEWS System in Thule, Greenland. (Courtesy of Raytheon Co.)

2.2.1.4 COBRA JUDY (AN/SPQ-11) [20]

The COBRA JUDY radar has a 22.5-ft, 12,288-element, phase-phase steered planar array placed aboard a ship (see Figures 2.11 and 2.12). The radar's transmitter consists of 16 TWTs (see Figure 2.13). The COBRA JUDY is used to obtain data for Strategic Arms Limitation Treaty (SALT). The array is mounted on a trainable turret.

2.2.1.5 PATRIOT Radar [1]

The PATRIOT system, well into its production phase as of 1988, illustrates the ability of a single multifunction phased array radar to replace a system that uses many dish radars. The multifunction phased array PATRIOT tactical air defense system (Figure 2.14) replaces older systems, such as the Nike-Hercules, which typically have four radars, and even then are limited in the number of simultaneous engagements that can be performed with the several tracking-illuminator radars dedicated to each battery control center because each tracking-illuminator radar must be assigned to a single missile for the entire missile flight. The PATRIOT radar is space-fed and uses about 5000 ferrite phase shifters [1]. Raytheon's PATRIOT radar is being deployed by the Federal Republic of Germany, Japan, and the Netherlands.

Figure 2.11 COBRA JUDY on Observation Island Ship. (Courtesy of Raytheon Co.)

Figure 2.12 Close-up of COBRA JUDY Array Face.

2.2.1.6 B-1B X-Band Phased Array Radar [10,21,22]

A production run of 100 B-1B X-Band phase-phase steered array radars has been completed by Westinghouse (see Figures 2.15 and 2.16 [22]). This radar, designated the AN/APQ-164, is a derivative of Westinghouse's Electronically Agile Radar (EAR) and F-16 radar (The AN/APG-68) [23–30]. The B-1B consists

Figure 2.13 View of Some of the COBRA JUDY TWTs. The Hoses Are for Water Cooling of the TWTs. (Courtesy of *Aviation Week and Space Technology*.)

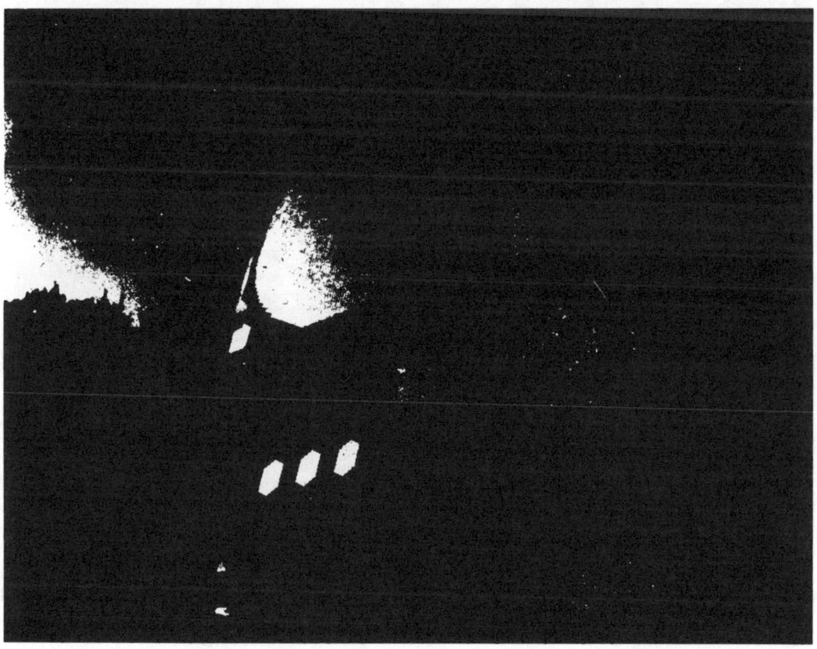

Figure 2.14 PATRIOT (formerly SAM-D) System (from Brookner [1]).

of a flat, oval-shaped antenna, tilted back about 30° to provide a lower cross section than the three antennas of its B-1A prototype [29]. The antenna can be rolled to one of three different positions: toward either side of the aircraft or forward, the latter being its normal mode and the one shown in Figure 2.15 [10]. When rolled to one side, the B-1B's field of coverage is from nose-on to about

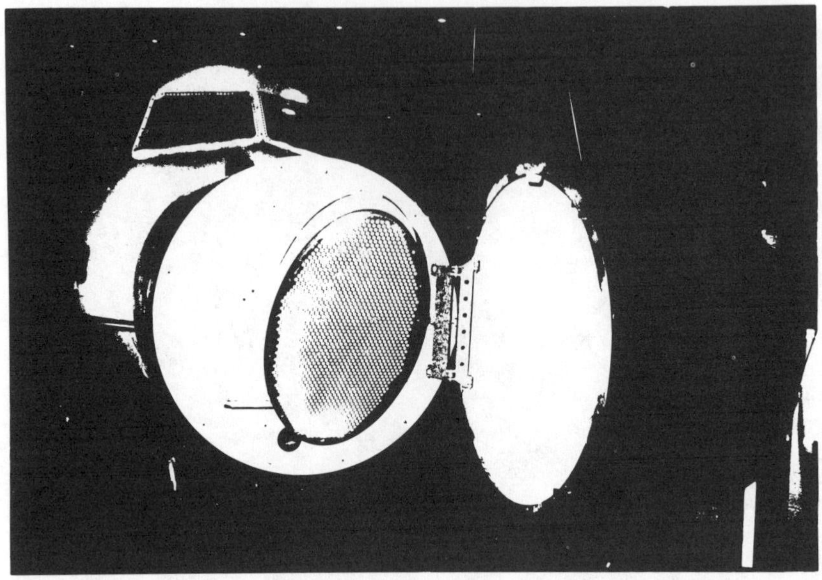

Figure 2.15 Westinghouse B1-B Antenna in BAC-111. (Courtesy of Westinghouse Co.)

Figure 2.16 Westinghouse B1-B Antenna Phase Control Modules and Radiating Elements. (Courtesy of Westinghouse Co.)

105° back. This permits the aircraft to look off to the side without changing its heading. The antenna consists of 1526 radiating elements and ferrite phase shifters (provided by Raytheon and Microwave Applications Group (MAG) [28,29]). The beam switching takes 150 μs [29]. This multimode radar is used for navigation, penetration, weapon delivery, and air refueling [10]. Navigation data is obtained by using a high-resolution *synthetic aperture radar* (SAR) mode [1], a real beam ground-mapping mode, weather detection, and the display of ground beacon returns over radar images [10]. In the SAR mode, the radar can provide high-resolution maps of regions from 18.5 km^2 down to 1.1 km^2 [24,29]. Weapons delivery is aided through the use of a ground *moving-target detection* (MTD) mode, an altimeter mode, and a velocity update mode. Penetration is aided through the use of an automatic terrain-following and terrain-avoidance mode.

2.2.1.7 AEGIS AN/SPY-1A

RCA's AEGIS AN/SPY-1, four-faced (for 360° coverage), shipboard S-band phase-phase steered array [31] (see Figure 2.17) is also well into production with 16 Ticonderoga-class cruisers contracted [32]. The AEGIS AN/SPY-1A on the *USS Ticonderoga* has performed successfully during sea trials in the Eastern Mediterranean [33]. See Chapter 3 for a detailed discussion of the AEGIS AN/SPY-1A and AN/SPY-1.

Figure 2.17 AEGIS System on *USS Valley Forge*. (Courtesy of Raytheon Co.)

2.2.1.8 Firefinder AN/TPQ-37 [34] *and VSTAR* [10,411]

The AN/TPW-37 of Figure 2.18 represents another phase-phase steered array radar in production, although phase steering in the elevation plane is quite limited. An interesting feature of this radar is the use of the moving-target detector (MTD) technique for suppressing clutter [2,34,35,323]. This S-band artillery locating radar scans ±45° in azimuth and a few degrees in elevation (about 10°) [34]. Targets in track are updated a few times per second. Table 2.1 summarizes some of its characteristics (see also Table 2A.1(e) of Appendix 2A).

The Variable Search and Track Air Defense Radar (VSTAR) is a derivative of the AN/TPQ-37 (Hughes). VSTAR uses a 7 × 11-m, S-band, phase-phase steered planar array, which is continuously rotated mechanically to provide 360° coverage (see Figure 2.19). Two versions of this radar have been built: one has a modified AN/TPQ-37 antenna mounted vertically to provide ±45° of azimuth scan, with up to 30° of elevation scan; the other has its antenna turned on its side

Figure 2.18 **Firefinder AN/TPQ-37 Artillery Locator Radar. (Courtesy of Hughes Aircraft Co.).**

Table 2.1 3D AN/TPQ-37 Artillery Locator [34]

Frequency	S-Band
Power (Prime)	45 kW
Transmitter	TWT (Shadow Grid)
Scan: Type	Phase-Phase (*pin* Diodes)
AZ	±45° (Trainable)
EL	Few Degrees
Waveform	Nonlinear FM
Monopulse	Time Multiplexed
Range	>30km (Artillery)
Processing	Digital Moving Target Detector (MTD): MTI/Doppler Processor
Other Features	Kalman Filter
	Automatic First-Round Location (Usually)
	Track and Search Simultaneously
	Stationary Antenna, Thus Difficult to Detect
	Digital Map Interface (Height Correction)

(Figure 2.19) to provide ±45° scan in elevation, with limited scan (up to 30°) in azimuth [10,36]. The *precision tracker* (PT) variant, the VSTAR-PT, is turned on its side.

Figure 2.19 Highly Mobile VSTAR. (Courtesy of Hughes Aircraft Co.)

The azimuth scan of these radars permits electronic scanning behind and ahead of mechanical boresite for immediate target verification and track initiation, *multiple target tracking* (MTT) in a high traffic density, variable track update rates, a long dwell for noncooperative identification, and *electronic countermeasures* (ECM) burn-through. The VSTAR can track 400 targets to a range of 250 km and an altitude of 30 km. These radars are said to have very low antenna sidelobe emission. The VSTAR-PT searches the horizon region every 3 s in azimuth scan, while higher elevation angles are revisited at 6, 9, or 12 s intervals. The VSTAR is configured in a number of forms, ranging from static to transportable to mobile on a tracked vehicle [10].

2.2.1.9 Multiple-Target Instrumentation Radars

Instrumentation radars are now taking advantage of the high accuracy, multiple target tracking capability afforded by phased array radars. One such radar is the MIR developed (by Raytheon) for the Naval Air Systems Command (NASC). (See Figure 2.20 [37].) This 12-ft diameter, C-band, space-fed 8973-element array system can simultaneously track up to 16 targets. The system not only reduces the number of tracking radars required, but also eliminates the radar-to-radar bias tracking errors that exist when multiple tracking radars are used to track simultaneously many targets. The MIR system has a relative tracking error of 0.05 mr, absolute tracking error of 0.1 mr, and range error of 2.0 yd rms. Table 2.2 gives a more detailed listing of its parameters.

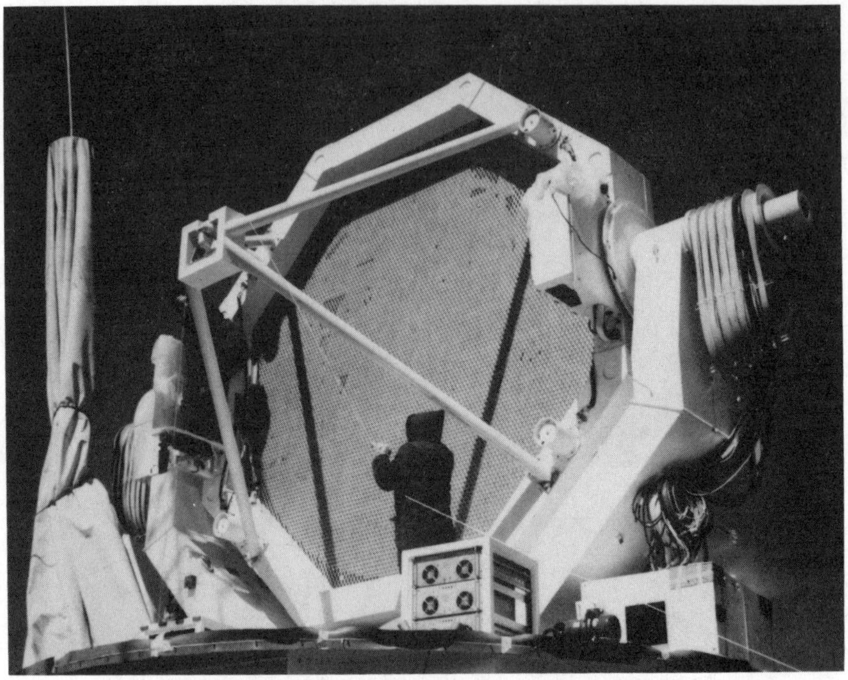

Figure 2.20 Multiple-Target Instrumentation Radar (MIR). (Courtesy of Raytheon Co.)

In the early 1980s RCA initiated the design of a C-band, phased-array, instrumentation radar [38,395], called the Multiple Target Tracking Instrumentation Radar (MTTIR), formerly called Multiple Object Tracking Radar (MOTR) [395]. Figure 2.21 shows an engineering model of the MTTIR. Table 2.3 gives some of its parameters, comparing them to two existing C-band, single-target tracking, reflector radars, the AN/FPS-16 and the AN/MPS-36. (A more complete listing is given in Table 2A.1(e) of Appendix 2A.) The MTTIR has a very large azimuth acceleration capability, permitting it to track a small missile during proximity fly-by or lift-off at close range. The MTTIR is claimed to have 10 dB lower near-in sidelobes than the AN/FPS-16 or AN/MPS-36 parabolic reflector antenna, which yields a 20 dB improvement in signal-to-clutter (*S/C*) ratio and a 3-to-1 reduction in tracking error due to multipath.

2.2.1.10 Thomson-CSF Prototype Multimode Phased Array System [39]

Figure 2.22 shows the Thomson-CSF multimode phased array system (MPAS) under development for the French Navy. This antenna system is intended for shipboard air surveillance, tracking, and weapons guidance in a multiple target environment. To obtain 360° of azimuth coverage, like the VSTAR and VSTAR-PT, this phase-phase steered array is mechanically rotated in azimuth. The MPAS has full azimuth and elevation phase-phase scanning, and the array uses diode phase shifters and spiral radiators.

2.2.1.11 Flexible Adaptive Radar (FLEXAR) [10,40]

The FLEXAR (Hughes) is an X-band, phase-phase steered, planar array radar, intended for surface-ship fire control (see Figure 2.23). The antenna (about 40 inches in diameter) is rotated at about 60 rpm to provide hemispherical coverage. The FLEXAR adapts its waveform to the situation in real time. For example, a doppler mode is used for search on the horizon in clutter, while another

Table 2.2 Raytheon C-Band Multiple-Target Instrumentation Radar (MIR)

SYSTEM		*TRACK*	
Sensitivity	SNR 25 dB 1m^2 at 100 nmi	Number of Targets	16 or 32 High Dynamic Skin or Beacon
Resolution	0.1 m^2 Target 75 ft from 100 m^2 Target		
TRANSMITTER		*TRACKING ACCURACY*	
Type	Coherent TWT	Absolute Angle	0.1 mr rms
Frequency	5.4–5.9 GHz	Relative Angle	0.05 mr rms
Frequency Diversity	8 Frequencies	Range	2.0 yd rms
Average Power	5.28 kW	*ANTENNA*	
Equivalent Peak Power (Chirp)	132 MW	Type	Pedestal Mounted Phased Array with Trainable Radome
Duty Cycle	0.032	Size	12 ft
Power Programming	10, 20, or 40 Hz Target	Feed	Optical
		Gain	42 dB
Pulsewidth	100.0 μs Chirp (0.12 μs) 0.25 and 2.0 μs	Beamwidth	1° (Broadside)
		Electronic Scan	$\pm$35° Azimuth $\pm$30° Elevation
Pulse Coding	IRIG 2-Pulse Beacon Codes	Array Elements	8973
RECEIVER			
Range Sidelobes (Chirp)	-39 dB		
Type	3-Channel Monopulse		

Table 2.3 Performance Comparison (from Hammond and Wedge [38])

Characteristic	*AN/FPS-16*	*AN/MPS-36*	*MTTIR*
Absolute Angle Accuracy* (20 dB *S/N*)	<0.2 mil	<0.4 mil	<0.2 mil
Relative Angle Accuracy* (20 dB *S/N*)	0.3 mil	0.6 mil	0.15 mil
Range Accuracy*	5 yd	3 yd	<1 yd
Range (on 1 m^2 target)	150 nmi	190 nmi	310 nmi (10 μs pW) 460 NM (50 μs pW)
Azimuth			
Rate mil/sec	750	1,000	2,000
Acceleration mil/s^2	1,020	400	25,000
Elevation			
Rate mil/sec	400	500	1,500
Acceleration mil/s^2	1,020	400	25,000
Range			
Rate yd/s	10,000	20,000	20,000
Acceleration yd/s^2	1,400	5,000	5,000

*All values are rms.

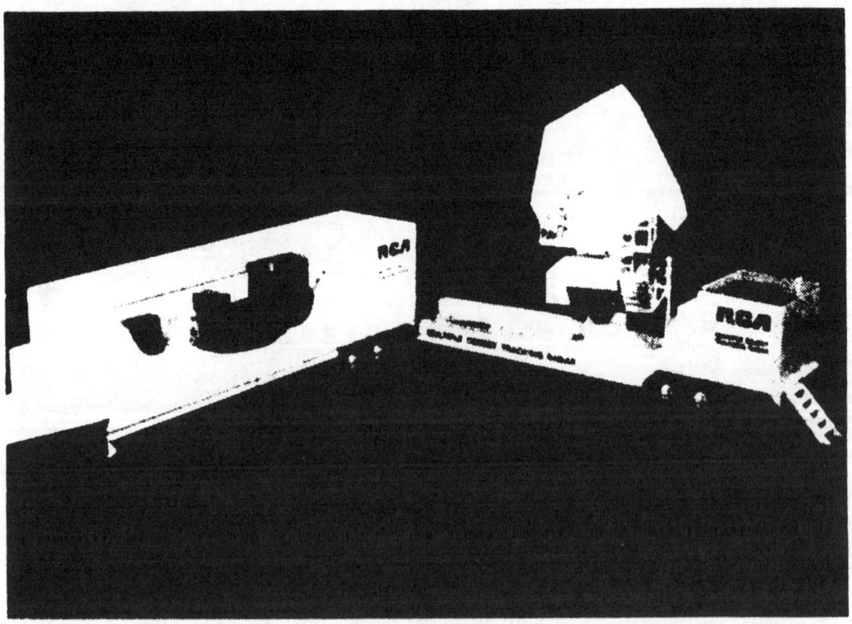

Figure 2.21 Engineering Model of the AN/MPS-39 Multiple-Target-Tracking Instrumentation Radar (MTTIR) (from Stegall [395]).

waveform is used for track. Its aperture design permits multiple, simultaneous beams to be independently steered, while the radar is in a target-illumination mode for guidance of semiactive missiles.

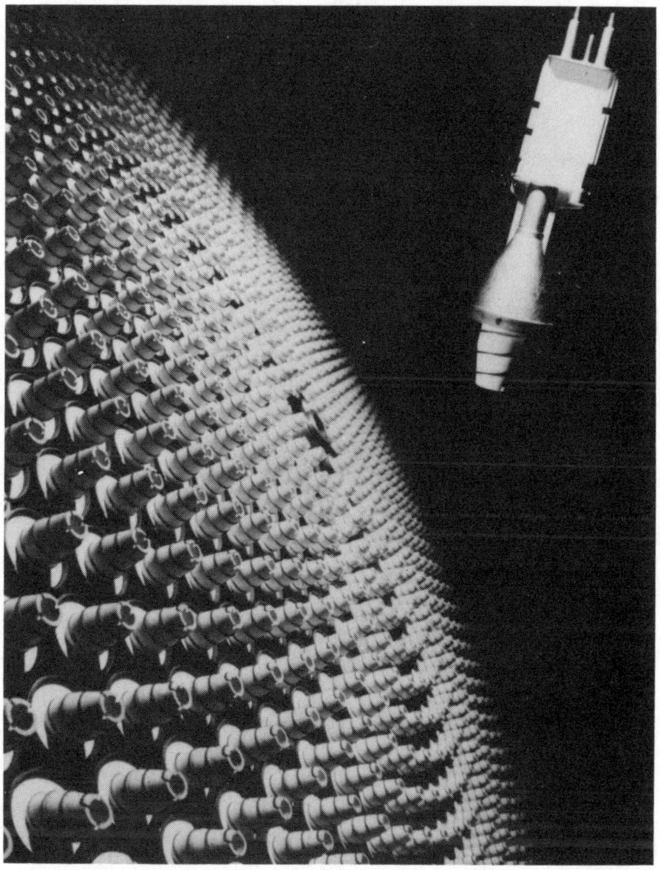

Figure 2.22 Multimode Phased Array. (Courtesy of Thomson-CSF.)

2.2.1.12 Other Phase-Phase Array Systems

Examples of other phase-phase steered array systems are given here as follows: the Siemens 243-dipole-element, S-band, VM 256 surface-to-air missile (SAM) multifunction adaptive array [42] (see Figure 2.24); the 2200-element, X-band, ORSET, which was under development by Siemens, has provisions for being rotated continually to provide 360° coverage, and has adaptive sidelobe canceling and sidelobe blanking for jammer suppression [10]; the Forschungsinstitut fuer Funk und Mathematik experimental phased array system (ELRA), which uses sequential detection and adaptive processing (see Figure 2.25 [43,44, 45]); the Toshiba and Mitsubishi phased array radars under development [46] (*Note*: Whether these are actually phase-phase steered array radars has not been explicitly indicated in the open literature. Most likely, they are one-dimensional electronically scanning systems, like those of Section 2.2.4. More recently, it was reported that Mitsubishi is starting to deploy an active phased array air defense radar at 28 sites in Japan, operating in the L- and S-bands [47].); the Thomson-CSF experimental Louxor, 2500-dipole, phased array, C-band system, intended for short-range air defense [10] (see Figure 2.26); the space-based radar system for which technology is being developed (see Figure 2.27 [18,48,49,166,345]). Although the program was canceled by the Air Force during the development phase, Sperry's Advanced Tactical Radar (ATR), consisting of a cylindrical phase-phase steered array antenna (Figure 2.28) is worth mentioning. The ATR was to

Figure 2.23 US Navy's Flexible Adaptive Radar (FLEXAR) System. (Courtesy of Hughes Aircraft Co.)

Figure 2.24 Siemens VM 256 Multifunctional Radar System. (Courtesy of Siemens.)

Figure 2.25 **Experimental Phased Array Radar System (ELRA). (Courtesy of Forschungsinstitut fuer Funk und Mathematik.)**

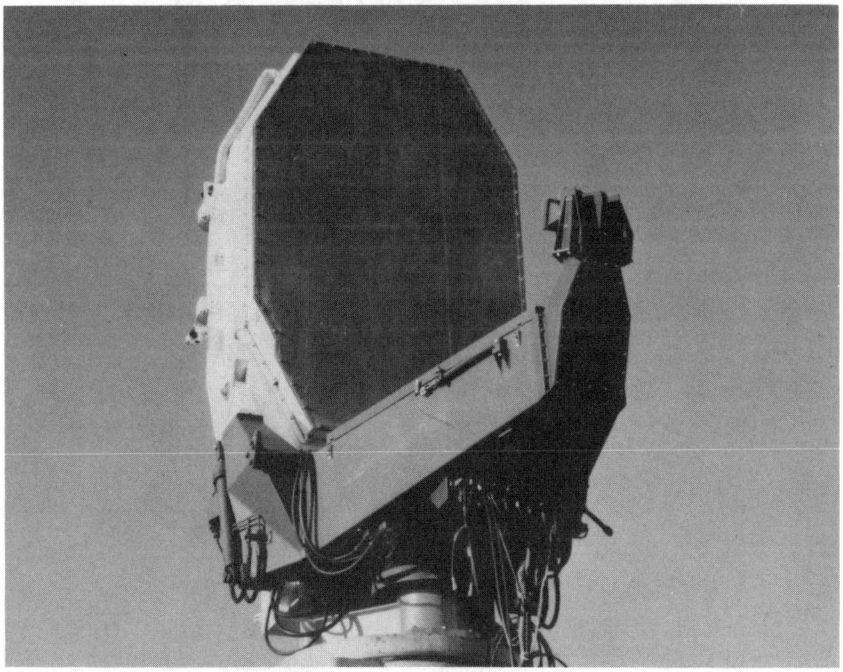

Figure 2.26 **Louxor Experimental Electronically Scanned Radar for Short-Range Air Defense Applications. (Courtesy of Thomson-CSF.)**

use spread-spectrum techniques, a very-low-sidelobe antenna (−45 dB *versus* −25 dB for previous systems), and adaptive search, track, and target identification capabilities [50]. The ATR was intended to replace the AN/TPS-43 surveillance radar. Reconfigurable, cylindrical, array antennas with electronic scanning are

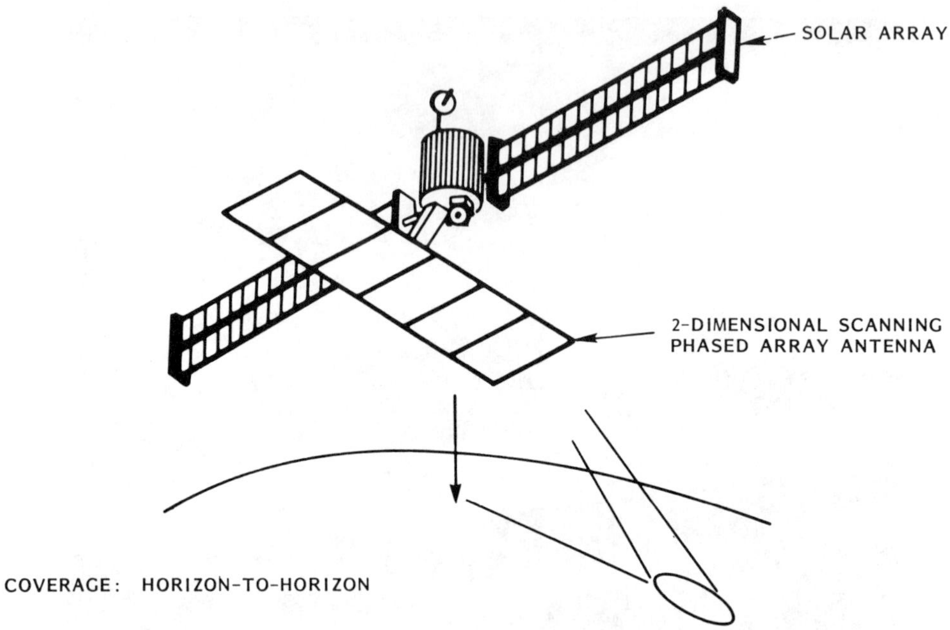

Figure 2.27 Phase-Phase Steered Space-Based Radar (SBR) (from Brookner and Mahoney [49, 166]).

Figure 2.28 Advanced Tactical Radar (ATR). (Courtesy of Sperry-Unisys.)

used on Sperry's unattended, short-range, two-dimensional, gap-filler radars for the Defense Early Warning (DEW) radars. There will be 39 such radars [51].

The UK (Plessey) is developing an early warning, S-band, phase-phase steered array radar system called the Multi-Function Electronic Scanning Adaptive Radar (MESAR) [105,382]. This radar is primarily intended for use on warships where one array face of about 4000 elements covers a single quadrant, with four elements needed to cover 360°, as in the AEGIS system. *Microwave integrated circuits* (MICs), manufactured in gallium arsenide (GaAs) technology, are to be used for the transceiver modules, with one module for each radiating element. The transceiver module output is 2 W peak, and the transmitter chain has an efficiency of 22% [383]. The noise figure of the T/R module is 3.5 dB. A 400 μs coded pulse at a PRF of 500 Hz would be used for early warning search, with shorter pulses (e.g., 36 μs) being used for other modes. The MESAR uses an optical fiber feed for sending the azimuth-elevation commands to each module, which calculates its own phase shift command. The optical fiber feed is used for weight reduction and resistance to *radio frequency interference* (RFI). Adaptive nulling of jammers will be used to within 2° of the main beam. The initial demonstration unit is intended to be able to track 50 targets. The MESAR is expected to be put into service by 1990.

Marconi Radar, Selenia, and Thomson-CSF are jointly developing the C-band European Multifunction Phased Array Radar (EMPAR) for shipboard use [384]. This radar is initially being configured as two back-to-back phase-phase steered arrays, rotated at 45 rpm. Each face provides ±60° scanning in elevation and ±45° in azimuth for a possible data rate of 3 Hz. This electronic scan is also used for electronic compensation of ship's roll and pitch. Each array face consists of 2000 radiating elements, fed by four-bit *pin*-diode phase shifters from a waveguide and stripline constrained feed. A Blass type of feed is used, providing one sum (Σ) and two difference (Δ) beams for monopulse measurements (see Chapter 5). The RF power is supplied by a high-power TWT. The system is planned to be modular, allowing it to be reconfigured as a one-face radar rotating at 60 rpm, or a nonrotating radar of four (or even five) faces, like the AEGIS. This allows the radar to be configured for the ship's size and requirements. The modules consist of the antenna, transmitter, receiver, signal processor, radar management computer, and software.

The USSR has recently developed two very impressive, huge, four-faced phase-phase steered array antiballistic missile radars, providing 360° coverage [52]. The Radar at Pushkino is housed in a four-faced structure, 120-ft high and 500-ft wide (see Figure 2.29), while a radar at Abalakovo is even larger. The USSR is also constructing a large radar (30-story receiver, 11-story transmitter) at Krasnoyursk. This high-power ($\approx$20 MW [352]) radar has an estimated carrier frequency of 180 MHz.

As indicated previously, this trend towards phase-phase steered array radars will continue into the 1990s and beyond. By the end of the 1990s, a solid-state phase-phase steered tactical radar is expected to replace the AN/TPS-43 and AN/TPS-70. Such a radar would use transceiver modules similar to PAVE PAWS (see Section 2.2.1.2), but they would be made with GaAs monolithic microwave integrated circuit (MMIC) MESFET technology (see Section 2.6.5.2 and Chapter 4), and they would not be thinned. The next generation of *advanced tactical fighter* (ATF) will likely be a solid-state, active, phased array multifunction radar.

2.2.2 Phase-Frequency Two-Dimensional Electronic Scanning Systems [10,34,36,53]

For many applications, only a few degrees of coverage are needed in elevation. So, a simple, low-cost, frequency-scanning technique is sometimes used,

BROOKNER
CHAPTER 2

Figure 2.29 USSR's New Antiballistic Missile Radar Located at Pushkino (Artist's Concept).

instead of a phase scanner. This is the case for the Firefinder AN/TPQ-36 mortar locator (see Figure 2.30 and Table 2.4). Like the AN/TPQ-37, the AN/TPQ-36 is a weapon locating radar (artillery, rockets, and mortars). The AN/TPQ-36 is similar in operation to the AN/TPQ-37, in that a fixed array is employed for electronically scanning the horizon, and phase scanning is used in azimuth to achieve $\pm 45°$ coverage [34]. Unlike the AN/TPQ-37, however, the AN/TPQ-36 employs frequency scanning in elevation with a slot array antenna (Figure 2.31). Because of the frequency dispersive character of an end-fed slotted waveguide linear array, the scan angle of the beam formed by the array will depend on the frequency of the pulse applied to the array. Typically, the higher is the carrier frequency, the higher in elevation the beam will point. The phase scanning in azimuth is achieved by using a ferrite phase shifter at the input to each column of the slotted waveguide array.

The AN/TPQ-36 operates in X-band. The surveillance, tracking, and back-tracking functions are nearly identical to those of the AN/TPQ-37. The AN/TPQ-36 is a medium-PRF radar, operating ambiguously in range and doppler. The principle differences between the AN/TPQ-36 and AN/TPQ-37 involve the design point-location ranges. The AN/TPQ-36 has about half the range of the AN/TPQ-37, with somewhat less accuracy, considerably less transmitter power, and employs sequential lobing for tracking. The lower power and simpler design of the AN/TPQ-36 results in an air-cooled, lighter, and less costly system, with a resulting improvement in mobility. Note that both the AN/TPQ-36 and AN/TPQ-37 are fully automatic.

The low altitude surveillance radar (LASR), developed by Hughes, evolved from the AN/TPQ-36 X-band radar (see Figure 2.32). The LASR is used for detection of low flying aircraft and missiles, and has a low-sidelobe antenna [10]. A large number of these radars are being built for a surface-to-air missile system [36]. The LASR uses an antenna similar to that of the AN/TPQ-36, turned 90° onto its side and mounted on a rotating pedestal to provide 360° of azimuth

Figure 2.30 3D Frequency-Phase Steered AN/TPQ-36 Artillery Locator. (Courtesy of Hughes Aircraft Co.)

coverage (once every two seconds) as well as full phase scanning in elevation. The frequency scan provides limited look-forward and look-back in azimuth [36, 53]. The LASR uses *moving-target indication* (MTI) and *range-gated pulsed doppler* (RGPD) processing for clutter rejection, which permits target detection while looking down on clutter from a high vantage position. The LASR is claimed to track aircraft doppler velocities less than 40 knots, and to track hovering helicopters, by using special doppler analysis techniques to detect rotor movement. The instrumentation measured radar range is to 40 nmi.

The AN/TPN-22 (ITT-Gilfillan) precision approach and landing radar is also a phase-frequency scanning radar. This X-band radar scans a 46° azimuth sector by 8° elevation sector [10].

2.2.3 Limited Scan Phased Array Systems

A variety of techniques for limited scanning have been developed to reduce the cost of phase-phase steered array radars. These techniques are useful where coverage over small angles is required, such as ±10°, instead of ±45° to ±60°. Limited scan techniques significantly reduce the number of phase shifters required for the system, in some cases by an order of magnitude.

The X-band AN/TPN-25 *precision approach radar* (PAR) of the AN-TPN-19 system (built by Raytheon) is one example (see Figure 2.33). This system guides the aircraft during final approach, and only requires a limited coverage sector, 15° in elevation by 20° in azimuth [1]. To achieve this limited angle coverage, a phased array is used to illuminate a reflector having an offset hyperbolic

Table 2.4 3D AN/TPQ-36 Artillery Locator [34]

Frequency	X-Band
Power (Prime)	10 kW
Transmitter Output Tube	TWT (Shadow Grid)
Scanning	Electronic: AZ: Phase (Ferrite) ±45° Trainable EL: Frequency
Sequential Lobing	
Range	≈12 km (Artillery)
Processing	Digital Moving Target Detector (MTD): MTI/Doppler Processor
Other Features	Kalman Filter Automatic First-Round Location (Usually) Track and Search Simultaneously Stationary Antenna, Thus Difficult to Detect Digital Map Interface (Height Correction)

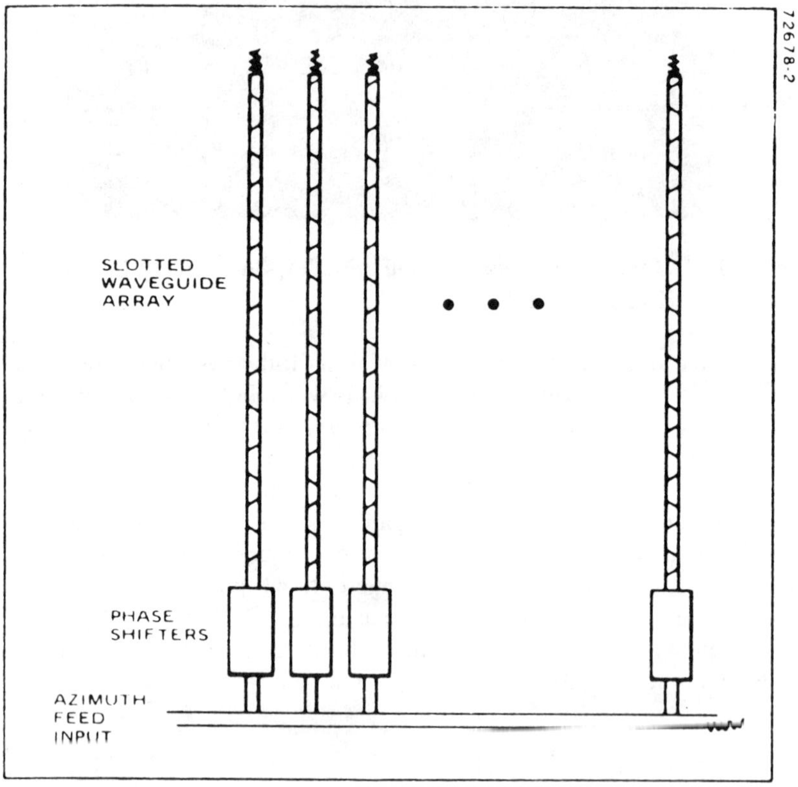

Figure 2.31 3D Frequency-Phase Steered AN/TPQ-36 Antenna Feed (from Ethington [34] © 1977 IEEE).

surface. This phased-array feed is fed by a horn, as shown in Figure 2.34. By changing the phase setting of the phase shifters in the array, the horn feed image appears to be in a different position (see Figure 2.34). Moving the horn feed image effectively alters the position of the beam formed by the main reflector. The number of phase shifters needed in the array feed is 824.

A fixed, low-cost version of the AN/TPN-25 radar is the AN/GPN-22, shown in Figure 2.35. Its coverage in elevation is 8°, instead of 15° [1]. Only 443 shifters

Figure 2.32 Hughes LASR Low-Level Air Defense Radar. (Courtesy of Hughes Aircraft Co.)

Figure 2.33 AN/TPN-25 Precision Approach Radar (PAR) of AN/TPN-19 System (from Brookner [1]).

are needed in the AN/GPN-22 array feed.

The two PAR systems can simultaneously track up to six aircraft and maintain performance in heavy rain. Specifically, the PAR system is capable of tracking a 1-m^2 target, such as a small jet aircraft, at a range of 10 nmi through a 5-nmi

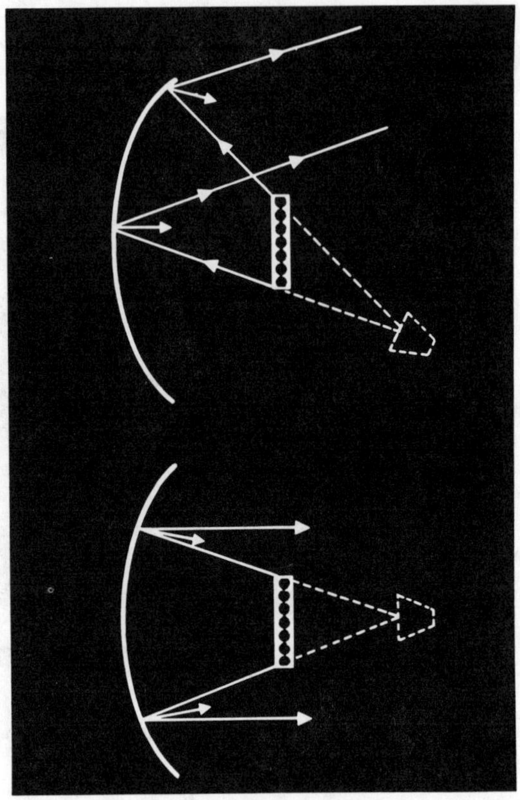

Figure 2.34 Principle of Limited Scan Antenna System (from Howell [54]).

cell of rain falling at a rate of 2 in/hr [55]. To achieve this performance, the system turns on a high-power *crossed-field amplifier* (CFA) tube [1,402] at the output of the TWT during heavy rain. The CFA tube in the on-state provides a 15 dB increase in the system output power; in the off-state, the CFA passes the signal through with a small loss of about 1 dB. Such a CFA is said to be transparent in the off-state, being transparent to the system. The system also uses circular polarization which reduces the rain backscatter by 20 dB. Noncoherent MTI is also used for suppressing the rain clutter. Finally, the system's range resolution is improved to about 5 ft during track, to help reduce the clutter return relative to the target return.

Eleven AN/TPN-25 radars have been produced for the US Air Force, and more than 50 AN/GPN-22 radars have been produced for the Air Force and overseas customers. The AN/TPN-25 was the first phased array radar to be put into production.

Another limited scan phased array is the Thomson-CSF Artois, shown in Figure 2.36. This C-band system electronically scans in a cone having a vertex angle of at least 10° [10]. The 3-m diameter antenna can be mechanically trained to any direction.

A third limited-scan phased array is the 400-MHz *Stradivarius* satellite tracking radar (Thomson-CSF), using a 13-m × 20-m main cylindrical reflector (see Figure 2.37 [56,57]). Its sweep is ±20° in azimuth and 5.5° in elevation (from 12° to 17.5°).

2.2.4 Array Radars Using One-Dimensional Electronic Scanning

Lower cost, one-dimensional (1D), electronic scanning systems have seen great popularity. They are increasingly using such features as pulse compression,

Figure 2.35 AN/GPN-22 Fixed Based PAR System (from Brookner [1]).

coherent processing, solid-state transmitters, digital processing, CFAR, frequency agility, low-sidelobe planar arrays, and improved *built-in test equipment* (BITE). One-dimensional electronic scanning is used extensively in three-dimensional (3D) radars, where a pencil beam is electronically scanned in elevation, while the antenna is rotated 360° to provide full coverage in elevation and azimuth. The elevation scan can be obtained by using either frequency scanning or phase scanning. Three-dimensional radars have the advantage of providing three-dimensional target measurements: range, azimuth, and elevation (or, equivalently, height). The phase-phase steered, phase-frequency, and limited scan radars of Sections 2.2.1 to 2.2.3 are also 3D radars. The stacked beam radars discussed in Section 2.2.5 are also 3D radars, with elevation coverage provided by simultaneous and contiguous multiple receiving beams. Three-dimensional radars make possible the use of one radar, instead of two, to provide 3D information. The two-radars approach uses a two-dimensional (2D) search radar to provide range and azimuth, with a separate height-finder radar for the third dimension. The 2D radar typically is a vertical fan beam radar that rotates 360° in azimuth.

The nonphase-phase steered 3D radars are not as capable as the phase-phase steered radars, but neither are they as expensive. Hence, where the threat does not necessitate fast track update rates for large numbers of targets, which forces the use of phase-phased steered array radars, the less expensive nonphase-phase steered 3D radars can be used.

Figure 2.36 Artois Limited-Scan Multitarget Tracking Radar. (Courtesy of Thomson-CSF.)

2.2.4.1 Frequency Scanning in Elevation

The S-band radar of Figure 2.38 (ITT-Gilfillan Series 320), which has low antenna sidelobes, achieves frequency scanning in elevation by using a vertically mounted frequency dispersive feed. The feed outputs excite successive rows of the array. (These rows are made of horizontally oriented slotted waveguide arrays.) The feed is made very dispersive in frequency by using a serpentine waveguide with slots that feed into the antenna rows [58]. This serpentine waveguide feed makes possible the achievement of large scan angles (in this case, 28°) with reasonable changes in the pulse carrier frequency. By changing the pulse carrier frequency, the beam points at different angles in elevation. Five beams are actually formed simultaneously by contiguously transmitting five pulses, with each at a successively higher carrier frequency (see Figure 2.39). Five simultaneous beams are necessary to achieve the 28° × 360° coverage with a reasonable scan frame time. Five receivers tuned to the five different frequencies are used to process the signals from an equal number of beams. Table 2.5 summarizes the Series 320 system parameters, Table 2.6 lists those of the transmitter, and Table 2.7 provides those of the antenna [59]. Figure 2.40 gives its coverage [59].

In contrast to the Series 320, the AR-3D (Figure 2.41) scans all elevation angles within a single pulse. This is done by transmitting a nonlinear frequency modulated pulse, which causes the beam to scan 30° in elevation within a single pulsewidth. The target elevation angle is determined by measuring the carrier frequency of the echo pulse [2,60,61].

Plessey is upgrading its three-dimensional elevation, frequency scanning AR-3D radar [1] to the AR320, which uses phase scanning in elevation as well as

Figure 2.37 Stradivarius 400-MHz Limited-Scan Satellite Tracking Radar. (Courtesy of Thomson-CSF.)

Figure 2.38 Frequency Scanned 3D Series 320 Radar [59]. (Courtesy of ITT-Gilfillan.)

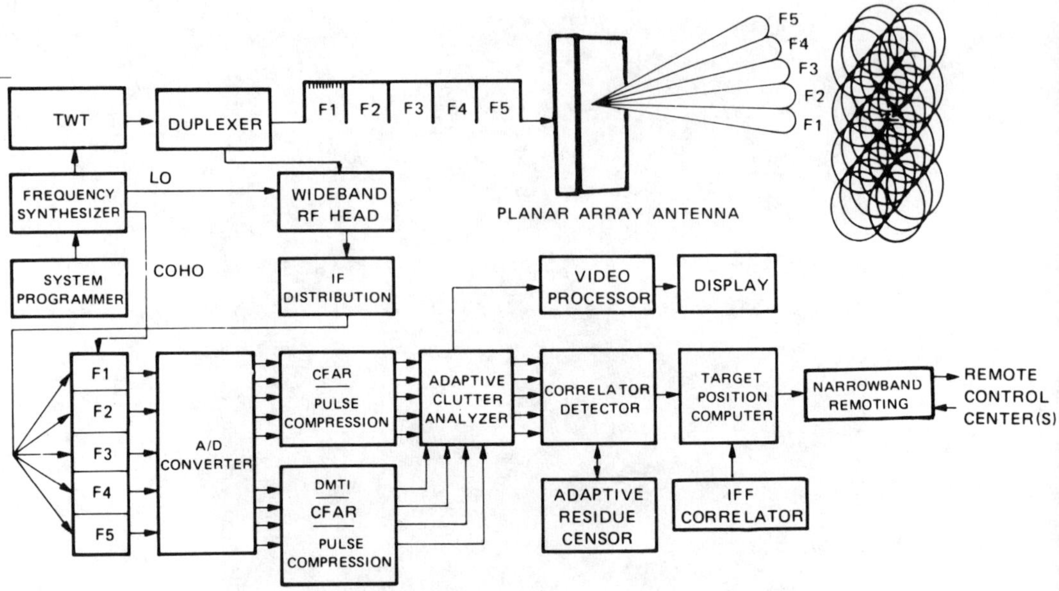

Figure 2.39 ITT Series 320 S-Band Frequency-Scanned Radar Block Diagram (from Pfister [59]).

Table 2.5 3D Frequency-Scanned Series 320 (after Pfister [59])

Frequency	S-Band (2.9–3.1 GHz)
Instrumental Coverage:	
Range (normal)	300 km
Range (MTI)	100 km
Azimuth	360°
Elevation	28°
Height	25 km
Data Rate	10 s
Detection Range for 1 m² Target (Swerling I):	
P_D = 0.5 (1° elevation)	250 km
P_D = 0.9 (1° elevation)	185 km
Digital Plot Output Data:	
Plot Output Rate	80/s
False Plot Rate	<(1)/s
Accuracy at 185 km:	
Range	20 m
Azimuth	600 m
Height	1100 m
Resolution at 185 km:	
Range	125 m
Azimuth	3°
Elevation	5°
Ground Clutter Performance in MTI:	
Improvement Factor	>35 dB
First Blind Speed	Mach 2
Processing	Digital Pulse Compression, MTI (3-Pulse), and CFAR

Table 2.6 S-Band Series 320 Transmitter Characteristics (from Pfister [59])

Transmitter Type	Crystal Controlled Master Oscillator Power Amplifier
Operating Frequency Band	2900–3100 MHz
RF Power Output	125 kW Peak, 5 kW Average
RF Pulsewidth	65 μs at Low Angles with 5 Frequency Subpulses 31.5 μs at High Angles with 5 Frequency Subpulses
Subpulse Coding	31-bit Pseudorandom Phase Coding at Low Angles 15-bit Pseudorandom Phase Coding at High Angles
Interpulse Period	550 to 3000 μs Variable with Elevation Angle
Stability	>40 dB
RF Tube Type	Gridded TWT PPM Focused, Liquid Cooled
Predicted MTBF	1800 hr
Predicted MTTR	1 hr

Table 2.7 S-Band Series 320 Array Characteristics (from Pfister [59])

Physical Characteristics:	
Overall Width	510 cm
Height	430 cm
Antenna Tilt Back	10°
Number of Arrays	76
Number of Slots per Array	80
Electrical Characteristics:	
Frequency Band	2900–3100 MHz
Polarization	Horizontal
Gain	41 dB
Azimuth Beamwidth	1.4°
Elevation Beamwidth	1.5°
Sidelobe Level	−25 dB max.

frequency scanning [62]. Figure 2.42 gives a sketch of the frequency-phase scanning antenna concept, and Table 2.8 lists some of its characteristics. As in the AR-3D, the AR320 uses a serpentine waveguide feed, which permits a large phase-scanning angle in elevation for a small change in the radar carrier frequency [58]. Slots along the serpentine waveguide feed into rows of slotted waveguides, which form a rectangular antenna with a pencil beam. Each row of the slotted waveguide, however, has a phase shifter at its input. These phase shifters permit the elevation scan angle of the beam to be independent of the radar carrier frequency, thus making the system less vulnerable to a jammer. The standard antenna for frequency-scanning in elevation without the use of phase shifters in each row is vulnerable to jamming because the elevation angle is one-to-one, relative to the radar carrier frequency. As a result, if the enemy determines what elevation angle the defense radar has to scan to see him, he knows precisely what jamming frequency to use and does not have to resort to wideband barrage jamming. (The

BROOKNER
CHAPTER 2

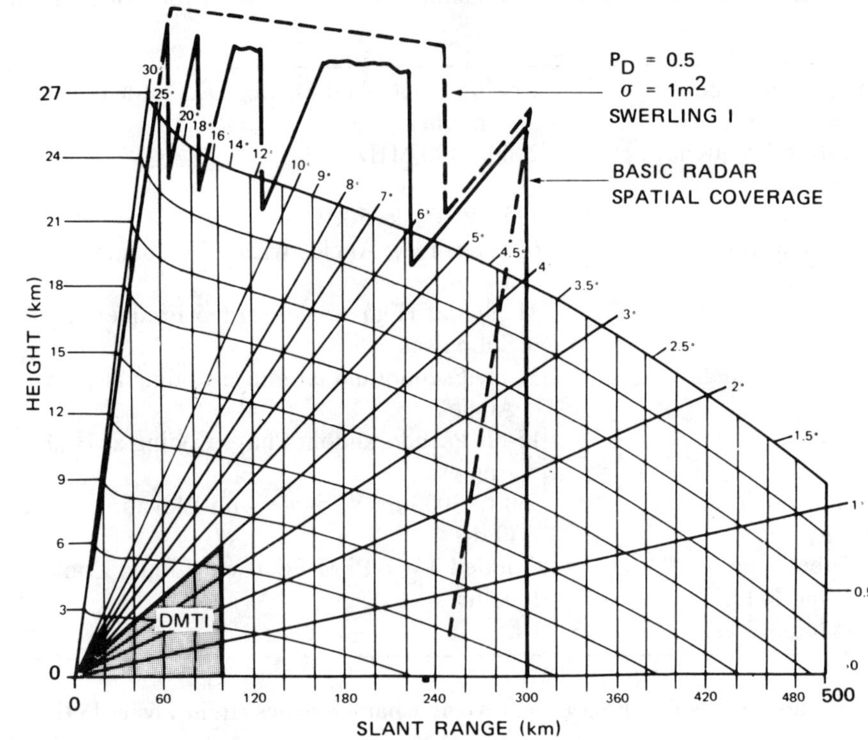

Figure 2.40 Coverage of Series 320 S-Band Frequency-Scanned Radar (from Pfister [59]).

Figure 2.41 Plessey S-Band AR-3D 3D Frequency-Scanned Radar (from Brookner [1]).

antenna, shown in Figure 2.43, is being built for Plessey by ITT Gilfillan.) This type of combined frequency-phase scanning in elevation is used on the Japanese

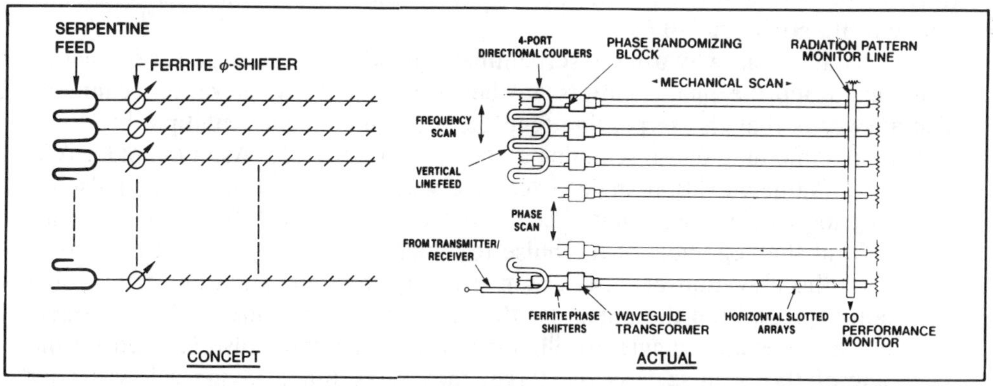

Figure 2.42 **Plessey AR320 3D Frequency-Phase Scanning Air Defense Radar (after Mathias and Cunningham [62]).**

Table 2.8 **Plessey AR320 S-Band 3D Radar [62]**

• Waveform:	Similar to AR-3D plus Fixed Frequency and Pulse Burst Agile
• Transmitter:	CFA (Transparent) Output, 24 kW Average, Permits Pulse-to-Pulse Power Control TWT Driver
• Antenna:	5 × 5M Planar Array (Provided by ITT, Looks Similar to ITT Series 320)
• Features:	Sidelobe Blanking (SLB), Secondary Surveillance Radar (SSR), CFAR, MTI, ECCM Processing
• Status:	In Production

Figure 2.43 **Plessey-ITT Gilfillan AR320 3D Radar. (Courtesy of Plessey.)**

(NEC) mobile three-dimensional radar systems, the NPM-510, and its nonmobile version, the NPG-880 [10,63].

We might ask, why use the serpentine feed, which is intended for frequency scanning, when the phase shifters are being used to allow elevation scanning? The answer is that the serpentine feed permits simultaneous multiple beams in elevation, something phase scanning alone does not provide. As in the older AR-3D radar, a large elevation angle is scanned within a single transmitted pulse by sweeping the carrier frequency over the pulse. As it sweeps from a low carrier frequency at the beginning of the pulse to a high carrier frequency at the end of the pulse, the elevation beam is scanned from a low elevation angle to a high elevation angle, within the pulse, by the dispersive serpentine feed [1]. Targets having a low elevation angle are illuminated by a narrow pulse (caused by the sweeping of the beam across the target in a short time) having a low carrier frequency, while targets at a high elevation angle are illuminated by a narrow pulse, at a high carrier frequency. The elevation angle of the target can then be determined by measuring the carrier frequency of the echo pulse, using a bank of receivers to cover the carrier frequencies that correspond to the different transmitted frequencies. We can think of the system as having multiple beams in elevation, with each receiver corresponding to a different beam. As mentioned previously, this frequency scanning within the pulse is the same as used in the older AR-3D radar [1]. The addition of phase scanning permits changing the frequency band needed to sweep the required elevation angle, thus providing less vulnerability to jamming.

2.2.4.2 Phase Scanning in Elevation

An example of recently developed 3D radars that use phase scanning in elevation is the solid-state L-band radar (GE 592) of Figure 2.44 (which evolved from the USMC AN/TPQ-59 long-range air search radar) [64,65]. Twelve of the AN/FPS-117 version of this radar are used for modernization of the Alaskan air defense radar network, and thirteen of the AN/FPS-117s will replace the outdated (1950s) technology of the radars in the Defense Early Warning (DEW) network [10,51,66]. Sixteen AN/FPS-117 systems are to be installed in Saudi Arabia [10]. A version of the AN/FPS-117 is being proposed for the US Federal Aviation Agency Radar Replacement (FARR) project [10] (see Section 2.2.5).

The Selenia S-Band, three-beam, elevation-scanning radar of Figure 2.45 uses simultaneous phase-scanning of three independent beams in elevation, each having a different, randomly selected carrier frequency [18,67,68,69]. The radar's parameters are given in Table 2.9, and Figure 2.46 shows its coverage in elevation. The coverage of the lowest elevation beam is given by the coverage of beam group 1 in Figure 2.46; hence, that of the second beam is given by beam group 2, *et cetera*. Figure 2.47 shows its elevation angle beam pattern. To eliminate clutter on the horizon, a null is placed in the antenna pattern along the horizon, independent of the scanning of the main beam (see Figure 2.47 [70–72]). The achievement of low antenna sidelobes and the simultaneous use of multiple frequencies provide the RAT-31S system with good characteristics for *electronic counter-countermeasures* (ECCM). More than 25 RAT-31S radars have been ordered (as of 1988) [10,73]. An improved version, the RAT-31SL, is being developed, which has better antijam features (e.g., a sidelobe blanker channel, ECM receiver, and jammer strobing) [71]. The RAT-31SL also has a narrower elevation beamwidth of 1.3°.

Other examples of modern 3D radars are the coherent S-band air defense TRS 2215 (Figure 2.48 and Table 2.10) and TRS 2230 mobile and fixed radars (Thomson-CSF); the TRS 2215D, the ultra-low-sidelobe antenna version of the TRS 2215, discussed in Section 2.6.2.1 (there is also an ultra-low-sidelobe version of the TRS 2230 [396]); the AWACS airborne surveillance system [74]; and the

Figure 2.44 GE 592 Solid-State L-Band 3D Radar. (Courtesy of General Electric.)

Hughes Air Defense Radar (HADR) S-band, 4.6-m × 6-m nominal size, planar array (see Figure 2.49), which is a derivative of the Hughes VSTAR and Navy AN/SPS-52C surveillance radar [1,411]. The AN/SPS-52C uses a planar array antenna, the AN/SPA-72B, which is made of rows of slotted waveguides that are end-fed with a serpentine feed to allow frequency scanning in elevation (as done with the Series 320; see Section 2.2.4.1). To allow frequency agility with its better performance against jammers, the dispersive serpentine feed was replaced by a column of high-power, ferrite phase shifters, thus permitting phase scanning in elevation. Two AN/SPA-72B antennas are stacked, one on top of the other, to form the HADR 4.6-m × 6-m nominal size antenna. Various array sizes exist for different applications.

The HADR rotates with a period of 10 or 12 s, depending on the array size, and has a typical detection range of 170 nmi on a 1-m^2 target, coverage to 100,000-ft altitude up to a 25° elevation angle. The HADR searches out the elevation coverage before the antenna moves one beamwidth in azimuth [10,75]. To achieve the increased long-range capability over the VSTAR, a CFA [1,402] has been added to the TWT output [411]. This CFA has a 10 dB power gain. For short-to-medium range search-in-clutter modes, the CFA is turned off and the output of the TWT passes through with a small loss, as is done on Raytheon's PAR systems (see Section 2.2.3 [411]).

The HR-3000, the newest generation of the HADR, will be used in Portugal and integrated with the Hughes-built NATO Air Defense Ground Environment (NADGE) in Greece, Turkey, and Italy [362]. The HR-3000 can detect and precisely locate fighter-sized aircraft at more than 250 nmi. Multiple modes incorporating various clutter rejection capabilities are adaptively adjusted to the

Figure 2.45 S-Band 3D RAT-31S System. (Courtesy of Selenia.)

environment. The planar array antenna is designed for extremely low sidelobes to minimize jamming and clutter effects. Elevation monopulse allows single transmission height reports on all detected targets. Radar and beacon radar reports are automatically correlated and assessed priority for transmission to multiple users.

Thomson-CSF's TRS 2215 and 2230 both have three contiguous beams in elevation, which are scanned together in phase by using four-bit diode phase shifters. The three beams are formed by using a dispersive feed (which looks like a waveguide serpentine feed; see Figure 2.48), just as is done on the Plessey AR320 (Figures 2.42 and 2.43). Three pulses, having three contiguous carrier frequencies, are transmitted, as is done in the Series 320 with five pulses (see Figure 2.39). The antijam characteristic is the same as obtained with the frequency-phase scanning AR320 and NEC's mobile three-dimensional radar system, described in Section 2.2.4.1.

The prototype 3D Telefunken Radar Mobile Search (TRMS) [10,76] is another example of a system which uses 1D elevation scanning and mechanical scanning in azimuth. This mobile, C-band, array search radar (shown in Figure 2.50) has the useful features of polarization agility (horizontal, vertical, and circular), low sidelobes, the capability of continuously varying the antenna beamwidth in elevation, if desired (by a factor of up to eight), and an antenna "pop-up" capability, which gives good low-altitude target coverage capability. When using circular polarization, the integrated cancelation ratio for rain clutter rejection is about −20 dB (approximately 0.8 dB axial ratio). Polarization agility is achieved by using two sets of azimuth corporate-feed structures per row of the array, with each corporate-feed structure having its own ferrite phase shifters. The outputs of one pair of azimuth corporate-feed structures feed a row of dual-polarized

Table 2.9 Selenia S-Band or 3D RAT-31S [69]

KEY CHARACTERISTICS

- A multiple simultaneous pencil beam antenna to provide a suitable dwell time for appropriate doppler processing and for good azimuthal accuracy.
- A true monopulse system compatible with frequency agility and hard-limited processing.
- An adaptive anticlutter processing compatible with pulse-to-pulse frequency agility.
- An agile coded waveform to resist sophisticated jammers.
- An automatic frequency selection to reduce to the minimum the interference in the receiver bandwidth.
- A low peak power for ARM detection before lock-on.
- A strict control of false alarms by hard-limiting, CFAR threshold and postprocessing filtering.
- A sophisticated BITE for an instantaneous on-line fault indication.
- Self-erecting devices for easy deployment of the complete radar system.

PARAMETERS:

Frequency	3100–3500 MHz
Power	P_{peak} = 170 kW; P_{av} = 4.2 kW
Transmitter Type	Coherent Amplifier Chain
PRF	450 Hz ± 10%
Pulsewidth	Uncompressed: 2.6 μs —All Beam Positions Compressed to 0.864 μs —8-Pulse Burst
Antenna type	Rotating Planar Array, Rear Feed, Electronic Scan in Elevation
MTBF	450 hr
MTTR	35 min
Antenna Size	4.7 m high × 4.0 m wide × 1.6 m deep
Radome	Polyurethane
Antenna Gain	37.2 dB
Azimuth Beamwidth	1.5°
Elevation Beamwidth	1.5° (First 4 Beams) 3° (Remaining Beams)

PARAMETERS (cont'd):

Polarization	Linear Horizontal
Signal Processing Type	Digital: MTI (Double Canceler) plus 8 FFT Doppler Filters and Pulse Compression
Prime Power	220/330 V, 3 Phase, 50 Hz, 90 kW
Coverage	Instrumental Range to 330 km Elevation— 0–21° (30° Optional) up to 30 km in Altitude Azimuth—360°
Detection Performance, σ_{tgt} = 2.0 m²	Range—280 km, P_d = 90% P_{fa} = 10^{-6}
Target Speed Range	Up to Mach 3
Accuracy	Range = 38 m Azimuth = 0.15° } at 280 km Height = 500 m } at 280 km
Resolution	Range = 130 m Azimuth = 1.5° Elevation = 1.5° (Low Beams)
Scan Rate	10/5 rpm; 12/6 Optional
Availability	0.998

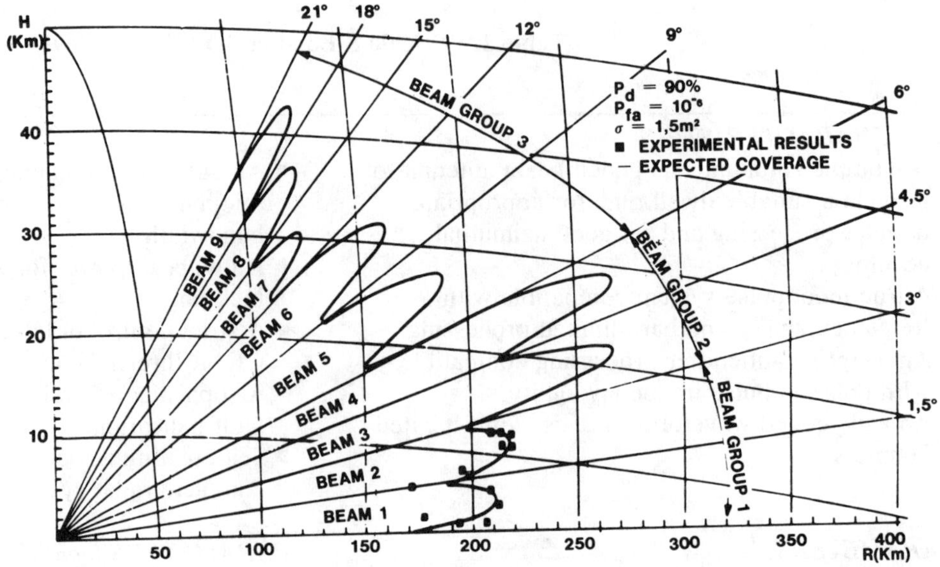

Figure 2.46 Selenia 3D RAT-31S Coverage (from DiLazzaro and Giaccari [69]).

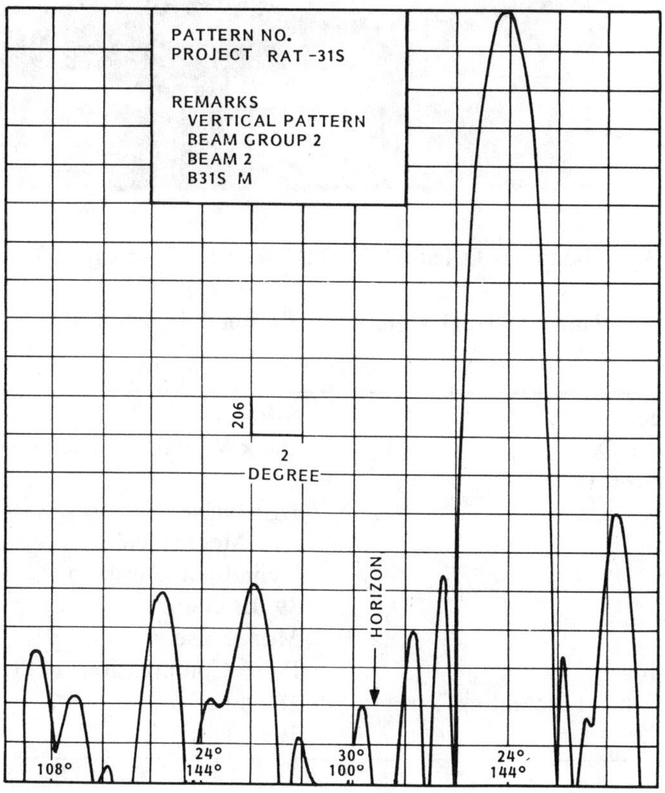

Figure 2.47 **Selenia 3D RAT-31S Elevation Antenna Pattern (after DiLazzaro and Giaccari [69]).**

Figure 2.48 3D S-Band Air Defense TRS 2215 Radar. (Courtesy of Thomson-CSF.)

Table 2.10 3D Thomson-CSF Mobile* TRS 2215

Frequency	S-Band
RF Power Out	700 kW Peak, 10 kW Average
Transmitter Tube	CFA
Scanning	EL: Phase ($-3°$ to $+30°$)
	AZ: Mechanical
Antenna	Cylindrical-Parabola
	39 dB Gain*
	Monopulse
Processing	Pulse Compression, MTI, CFAR
Range (2 m^2, Fluctuating Target)	180 nmi*
Ceiling	100,000 ft

*TRS 2230 is fixed version, which has 3 dB higher gain and 230 nmi range.

radiating elements (see Figure 2.51). Adjusting the relative phase between a pair of azimuth corporate-feed structures provides the desired polarization control.

Westinghouse is developing a solid-state 3D radar, which uses 1D electronic scanning in elevation, the W-2000 [77,78]. Figure 2.52 shows the array antenna (an ultra-low-sidelobe antenna), Figure 2.53 gives its predicted coverage, and Table 2.11 summarizes its parameters. For this system, the solid-state transmitter is not distributed over the array antenna (as is done with the GE 592 systems)

Figure 2.49 Hughes Air Defense Radar (HADR). (Courtesy of Hughes Aircraft Co.)

but instead the transmitter is housed off of the antenna. Specifically, all the solid-state transmitter modules are summed into a single port, and the power is then fed to the antenna through a rotary joint. Such a transmitter is called a solid-state "bottle" because transistors replace the usual high-power tube transmitters [78]. A distributed solid-state transmitter was originally proposed by Westinghouse for this L-band system [77]. However, the developer has since decided on the solid-state bottle approach. The advantages of this approach are (1) the transmitter is easier to cool, by air or water flow, if it is not on the antenna; (2) the antenna is colder, and thus does not provide a large infrared signal that the enemy could exploit with a heat-seeking missile; (3) elevation sidelobes are not degraded due to transmitter module failures; (4) maintenance of the transmitter is easier. The main disadvantage of the bottle approach is that the transmitter suffers a large microwave loss, resulting from the solid-state combiner, the long microwave line from the transmitter to the antenna, and the elevation-angle phase shifters, which operate after power amplification. The loss will also be higher on reception because all these losses are also suffered in the receiver. This system employs three simultaneous beams in elevation to achieve fast coverage of the search volume.

Hollandse Signaalapparaten is developing a 3D multibeam radar called SMART (*Signaal* Multibeam Acquisition Radar for Targeting). It is designed to search and track small Mach 3+ sea-skimming or diving missiles [360]. Its coverage goes from the horizon to zenith [10]. SMART uses 1D phase scanning in elevation on transmission [360,425]. On reception, SMART uses a stack of 12 simultaneous beams to cover from the horizon to beyond the zenith [10]. The beams are formed digitally from the 16 row outputs of the ultra-low-sidelobe stripline planar array

BROOKNER
CHAPTER 2

Figure 2.50 AEG-Telefunken Radar Mobile Search (TRMS). (Courtesy of AEG-Telefunken.)

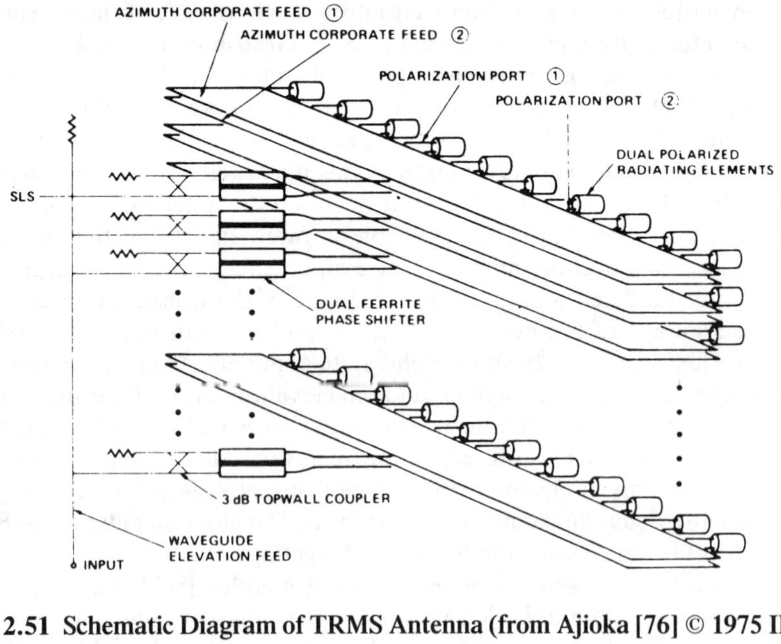

Figure 2.51 Schematic Diagram of TRMS Antenna (from Ajioka [76] © 1975 IEEE).

Figure 2.52 L-Band Solid-State 3D W2000 Antenna. (Courtesy of Westinghouse.)

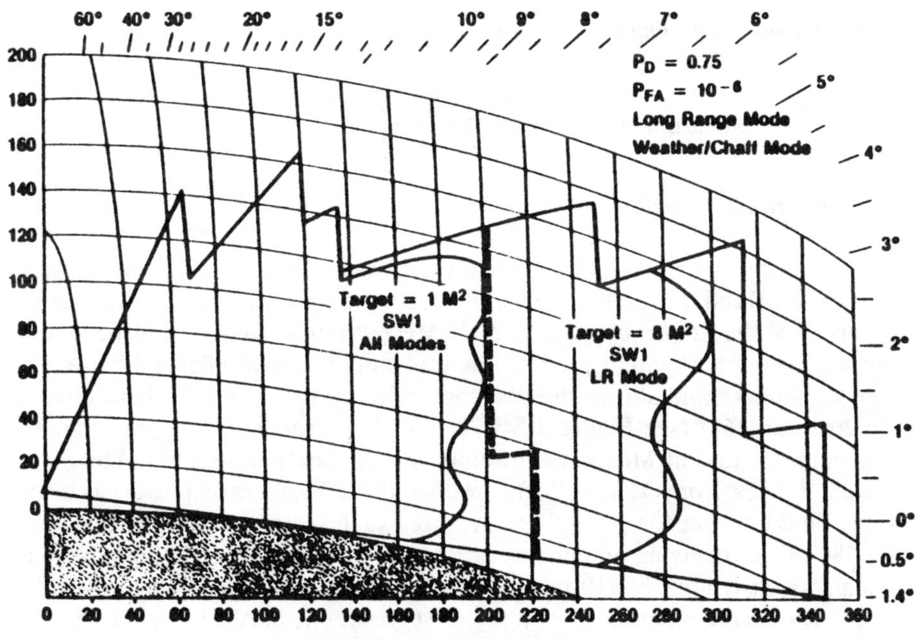

Figure 2.53 L-Band Solid-State 3D W2000 Radar Coverage (from Foster [78]).

antenna used [10,360]. The antenna is squintless (see Section 2.6.2.1) [360]. This radar is being considered for use as an *airborne early warning* (AEW) radar. Detailed parameters for SMART are given in Table 2A.1(c) of Appendix 2A.

Table 2.11 L-Band Solid-State 3D W2000 Parameters (from Foster [78])

Coverage	270 nm Range, 360° Azimuth to 20° Elevation, to 100,000 ft Altitude	Resolution	Two 15 M^2 Swerling I Targets; at nominal range of 135 nmi Probability of Resolution = 80%
		Range	Separation 0.16 nmi
		Azimuth	Separation 1.3°
Detection Capability	210 nm, $P_d > 80\%$, Target Type = 1 m^2 Swerling I	MTI Improvement Factor	>50 dB
False Alarm Control	Adaptive Range Cell-Averaging CFAR		Frequency Agility Pulse to Pulse Burst to Burst (Fully Compatible with Doppler Processing)
Accuracy	1 m^2 Swerling I Target to 135 nm Range	Ultra-Low-Sidelobe Antenna Sidelobe Blanking	
Range	<0.05 nm rms	Jamming Analyzer	
Azimuth	<0.13° rms		
Height	Relative <1800 ft rms		

2.2.4.3 Phase Scanning in Azimuth

There is also a class of radars that uses phase scanning in azimuth and no scanning in elevation. Several radars of this class are described below.

Over-the-Horizon Radar

A one-dimensional, electronically scanned, HF, over-the-horizon backscatter (OTH/B) radar is built by General Electric in Maine [16,66,79,80]. The system operates in the frequency range of 5–28 MHz. At these frequencies the ionosphere acts like a mirror, and as a result the transmitted signal reflects downward from the ionosphere and allows the detection of aircraft beyond the horizon, such as to 2000 mi [363] (see Figure 2.54). The system uses a 3630-ft long transmitting antenna, located at Moscow-Caratunk, Maine (see Figure 2.55). The antenna's height ranges from 35 to 135 ft. Twelve 100-kW (average power) transmitters provide the power for one 60° sector; three such sectors provide a total coverage of 180°. The receiving antenna consists of a 4980-ft long, 64-ft high array of 246 elements. It is located 110 mi southeast at Columbia Falls, Maine. Three other OTH/B sites are planned, one on the west coast, one in Alaska, and one in central US, looking south and possibly east and west [66,362].

The Raytheon Company is developing a relocatable over-the-horizon radar (ROTHR) for the US Navy. The ROTHR is a tactical land-based, bistatic, ionospheric backscatter radar system. Its mission is to provide wide-area over-the-horizon surveillance of both aircraft and ships in support of tactical forces in locations of national interest. ROTHR extends the range of horizon-limited microwave radars, and provides the Navy with early warning of approaching naval and airborne threats at any altitude.

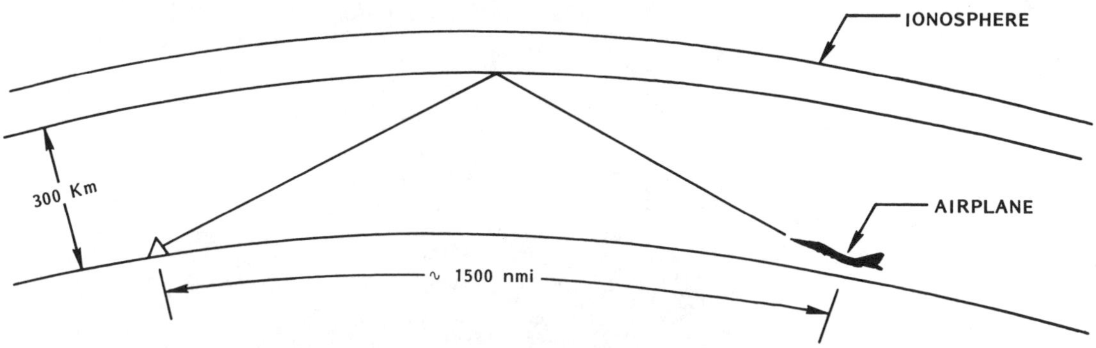

Figure 2.54 Over-the-Horizon Backscatter (OTH/B).

Figure 2.55 AN/FPS-118 Over-the-Horizon Backscatter (OTH/B) Transmitting Antenna. (Courtesy of General Electric Company).

In contrast to the OTH/B of Figure 2.55, the ROTHR transmitter is all solid-state for better reliability and ease of maintenance. Figures 2.56 and 2.57 show respectively the transmitting and receiving antennas, which are 50 to 100 nmi apart.

Operationally, ROTHR can automatically detect and track both aircraft and ships. At the operator's option, the system can be used for overall surveillance and tracking within the coverage area, spotlighting specific regions to handle targets of special interest, or assessment of the size of the attacking raid. All target information is forwarded to the user command center.

ROTHR is designed to be relocated to prepared sites in support of tactical surveillance missions. This cost-effective feature preserves the high value assets and eliminates the need for on-station personnel at nonoperational sites. Because ROTHR can be relocated, it must adapt to changing ionospheric conditions throughout the world. This requirement is met by eliminating the need for external sounder data inputs. ROTHR

Figure 2.56 Relocatable Over-the-Horizon Radar (ROTHR) Transmitting Site. (Courtesy of Raytheon Co.)

Figure 2.57 ROTHR Receiving Site. (Courtesy of Raytheon Co.)

uses a collocated *quasivertical incidence* (QVI) sounder and a radar backscatter sounder for *propagation management assessment* (PMA). This unique approach automatically adapts to the ionospheric conditions of the operational sites.

There is claimed to be a renewed interest in HF because of its effectiveness against future "stealth" aircraft [426]. It is said to be impractical to use *radar absorbent material* (RAM) at these frequencies [426]. There is even thought to be consideration for using VHF and UHF to cope with stealth [426,429].

Airborne Solid-State Surveillance Radar

Consideration is being given to replacing the rotating antenna of the E-2C type of airborne surveillance radar systems with phased array radars located either on pods or conformal to the fuselage and wings [1,81]. The Naval Air Development Center (NADC) has been involved in the development of a nonuniform lattice array for this application. The array (see [1], Fig. 4, p. 326) uses density tapering along its length to provide array weighting, as is done in the COBRA DANE and PAVE PAWS systems. Additional weighting is provided by the elliptical shape of the antenna and by the use of 40-dB tapered-Taylor weighting on reception. For increased reliability, a solid-state transmitter is used with a transceiver module behind each radiating element. With this sytem, phase scanning is used in azimuth, with the elevation coverage achieved by using a fan beam in elevation. The antenna scans $\pm 45°$ in azimuth. To achieve 360° coverage, a number of array antennas would be used around the aircraft. Bent dipoles on printed circuit boards are used for the radiating elements. The use of bent dipoles permits a $\lambda/4$ spacing at the center of the array so that the desired large power-aperture product can be achieved with a smaller aperture area. Measurement results obtained indicate that it is possible to achieve the 60-dB two-way antenna sidelobe level required for the intended application. Work is also under way to replace the E-3A AWACS rotodome antenna with solid-state, S-band, phased arrays (see Section 2.6.5.1 and [162]).

Israel Aircraft Industries is developing an AEW, L-band, solid-state, phased array radar for the new Boeing 707 jet aircraft [430]. It is intended as an alternative to the USAF-Boeing E-3 AWACS and the Navy-Grumman E-2 Hawkeye (see Table 2A.4 for a description of these two systems). This solid-state system, called Phalcon, is to use six antennas to provide 360° coverage—two forward and two aft of the wings on each side of the aircraft, one in the nose, and one under the tail. These antennas are electronically scanned, having several hundred elements per array with one transceiver module per element. This arrangement avoids fuselage, wing, or tail blockage, and provides the improved search, tracking, and reliability advantages of an adaptive, electronically scanned, distributed, solid-state array. L-band was selected because of its capability in bad weather, its accuracy, and the practical size of antenna that can be used.

2.2.5 Stacked-Beam Three-Dimensional (3D) Systems

A three-dimensional stacked-beam radar system transmits with a vertically oriented fan beam, which simultaneously illuminates all elevation angles to be covered. This beam is mechanically scanned 360°, as is done with conventional 2D radars to provide 360° coverage. For reception, the 2D radars also use a fan beam (usually the same as the transmitting beam). As a result, these radars only provide 2D data on the target—range and azimuth. In contrast to 2D radars, 3D stacked-beam radars use a stack of narrow contiguous elevation beams on reception, with each beam having its own receiver channel. Correlation of receive beams with elevation angle thus provides the target elevation angle. The Martello L-band S713 radar, shown in Figure 2.58, is such a stacked-beam system [82–84].

Figure 2.58 Marconi-Martello S713 Tube 3D Stacked Beam L-Band Antenna. (Courtesy of J. Forrest, Marconi.)

It uses eight simultaneous receive beams. Figure 2.59 shows the coverage and some parameters of the S713 radar [84]. Additional parameters are given in Table 2.12 (see also Table 2A.1(b) of Appendix 2A). An informative discussion of the trade-offs which led to the selection of L-band for this system is given in [83]. A planar array antenna is employed, which provides ultra-low sidelobes in azimuth. The antenna is formed from 60 rows of linear dipole arrays, each of which uses a center-fed squintless feed network.

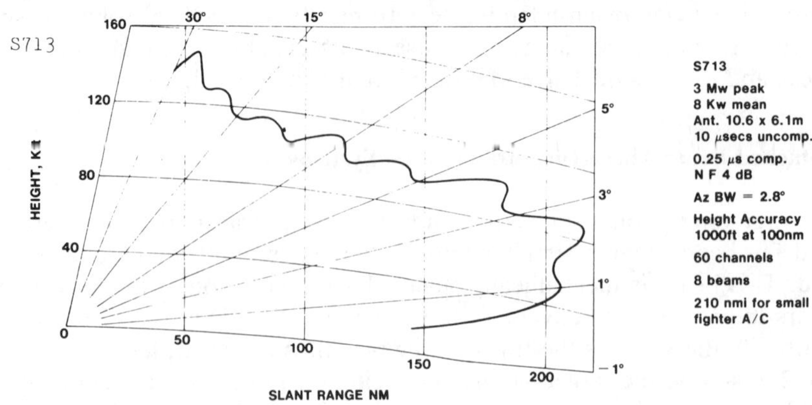

Figure 2.59 Coverage of Marconi-Martello S713 Tube System (after Latham [84]).

The Marconi-Martello S713 is a tube radar (see Table 2.12). Marconi has developed a solid-state version of this tube system, the S723, which has about the same coverage [84]. Figures 2.60 and 2.61 show front and back views of the antenna for this system, while Figure 2.62 shows its coverage and some of its parameters. Additional parameters are given in Table 2.13 and Table 2A.1(b). For this system, the solid-state transmitter is mounted in the vertical spine behind the array (see Figure 2.61). Using this approach results in a lower system loss, as indicated previously. The S713 has a higher receiver noise figure than the S723 (4 dB *versus* 2.5 dB). (See Appendix 2A, Table 2A.1(b).) As a result, the S713 is less vulnerable to jamming by 1.5 dB [84]. There are two versions of the S723, the S723A with eight receiving beams in elevation, and the S723C with six beams in elevation [10]. A derivative of the Martello S723 is being bid by Raytheon and Marconi for the CONUS Federal Aviation Administration Radar Replacement (FARR) network of 48 minimally attended radars for the US Joint Surveillance System (JSS). This fourth-generation air route surveillance radar (ARSR-4) contract is valued at $500 million. The other competitors are General Electric and Westinghouse, as of this writing.

Figure 2.60 Marconi-Martello S723 Solid-State 3D Stacked Beam L-Band Antenna: Front Side.

The Westinghouse AN/TPS-70 ultra-low-sidelobe antenna, stacked-beam system is also now in production (see Section 2.6.2 [85,86]). The AN/TPS-70 can tag and track up to 1000 targets [420]. It has been ordered by the United Arab Emirates for surveillance of the critical Persian Gulf and surrounding areas [364]. (See Table 2A.1(b) in Appendix 2A for parameters of AN/TPS-70.) Westinghouse is developing a stacked-beam, L-band radar for the FARR/ARSR-4 program. This radar uses a solid-state bottle and a reflector antenna with an array feed, consisting of dual-polarized radiating elements on a curved surface. The antenna achieves −35 dB and −30 dB azimuth and elevation sidelobes, respectively, and nearly 28 dB of circular polarization integrated cancelation for detection of aircraft in heavy rain.

Table 2.12 Additional Parameters For Twystron Tube 3D Marconi Martello S713 [82–84]

Frequency	1.3 GHz
Wavelength	23 cm
Peak Power	3.3 mW*
Output Tube	Twystron (Capable of 7 MW Peak)
Antenna	35 ft High
	20 ft Wide
	60 Row Array
	61 Receivers
	Triplate Glass-Fiber
Horizontal Beamwidth	2.8°
Vertical Beamwidth	1.5°
Polarization	Linear
Gain:	
Transmit	32 dB
Receive	38 dB
Sidelobes:	
Horizontal (1st Sidelobe)	30 dB
Horizontal (15° Off Main Beam)	45 dB
Vertical	30 dB
Vertical Coverage	0–30° (Heights to 24°)
Rotation Rate	6 rpm
Pulsewidth (Effective)	0.25 μs
Receiver Noise Figure	4 dB
Cancellation (Ground Clutter)	40–55 dB
Instrumented Cover up to	200,000 ft
Height Accuracy (Single Look)	1,000 ft rms at 100 nmi
Time into Action	5 hr
ECCM Features	Frequency Agility
	Random PRF Stagger
	Antispoof Jitter
	Pulse Compression
	Pulse Doppler Processing

*Derated to provide 10,000 hr life.

Stacked-beam radars and the other 3D radars are becoming increasingly popular because, as indicated previously, they eliminate the need for a separate height-finding radar in conjunction with a search radar. With the expected progressive reduction in cost of receiver electronics, stacked-beam radars may well become even more popular. The future development of MMICs (see Section 2.6.5.2 and Chapter 4) will help reduce the cost of microwave circuitry for such radars.

2.3 OTHER RADARS UNDER DEVELOPMENT OR COMPLETED DEVELOPMENT

2.3.1 Two-Dimensional Radars

At the same time major advances are being made in phase-phase steered array radars, 3-D radars, OTH radars and bistatic radars, major advances are

Figure 2.61 Marconi-Martello S723 Solid-State 3D Stacked Beam L-Band Antenna: Back Side.

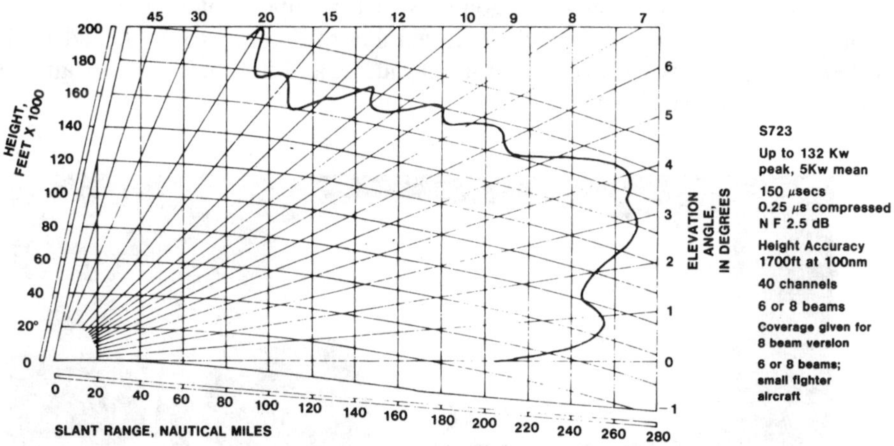

Figure 2.62 Coverage of Marconi-Martello S723 Solid-State System (after Latham [84]).

being made in simple 2-dimensional radars, i.e., mechanically rotating (360°) fan beam radars that provide range and azimuth measurements of the target. These advances are in terms of:

(1) the use of better clutter rejection techniques, such as the moving target detector (Lincoln Laboratory), and the use of dual receiving beams;

(2) better reliability, through the use of solid-state transmitters (see Section 2.6.5 and discussion of RAMP radar to be given shortly in this section), integrated receiver circuitry, more reliable signal processors (see Section 2.6.6), and the use of built-in test equipment;

(3) better automatic target detection and tracking techniques, which also help in the rejection of clutter;

Table 2.13 Additional Parameters For Solid-State 3D Marconi Martello S723 [84]

Frequency	L-Band (23 cm)
Antenna	7.1 × 12.2 m, 40 rows, squintless in AZ.
Azimuth Resolution	1.5° (0.2 nmi at 100 nmi)
Sidelobes Off Axis	−45 dB
Range: Instrumental	14–256 nmi
Accuracy	525 ft (160 m) rms
Resolution	607 ft (185 m); 0.25 μs Comp. Pulse
Elevation Coverage	0–20°; 200 kft Altitude
Detection Range	270 nmi on Small Fighter Aircraft
Transmitter	Solid-State with up to 40 Modules
Frequency Agility Band	10%
Signal Processing	One Processor per Beam 4-Pulse DMTI with Doppler Compensation, Time Varying Weights, and Pseudorandom Stagger

(4) better electronic counter-countermeasure (ECCM) techniques, such as improved adaptive nulling of jammers (see Section 2.6.4), and the use of ultra-low-sidelobe antennas (see Section 2.6.2);

(5) the use of the main reflector for the IFF and beacon interrogator antenna, instead of a separate antenna, which leads to lower cost and better performance (due to less ground multipath).

The cost of 2D radars has gone down for some applications. A dramatic example is the 24-inch commercial marine radar of Figure 2.63, which costs about \$2200. These trends will continue for 2D radars, and their use will sustain where they perform satisfactorily.

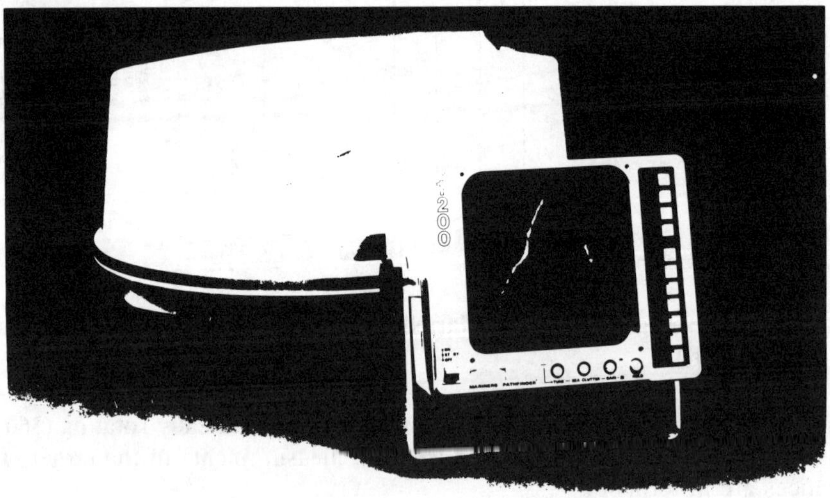

Figure 2.63 \$2200 Commercial Marine 2D Radar. (Courtesy of Raytheon Co.)

Another excellent example of a 2D radar, currently under development, which incorporates a solid-state transmitter and advanced clutter rejection features, is the radar of the Canadian Radar Modernization Program (RAMP) [87–89]. It uses a solid-state bottle transmitter consisting of 14 parallel solid-state modules.

Other interesting features are the use of a four-pulse moving target detector and dual-beam antenna to eliminate ground clutter, circular polarization to eliminate rain clutter, and an orthogonal channel for observing weather. One of the L-band, solid-state, air traffic control radars will be located at each of the 22 major airports across Canada [87]. Table 2.14 gives some of the parameters of this radar. Figure 2.64 gives its coverage. Figure 2.65 shows the antenna. The Canadian Government has placed the contract to build the RAMP system with Raytheon Canada. Still another example of a 2D solid state bottle radar is the S-band AN/TPS-73 being developed by Unisys and Selenia; see Table 2A.1(a) in the appendix of this chapter.

Table 2.14 Ramp L-Band Solid-State Radar Parameters [87]

Frequency	L-Band
Peak Power	20.8 kW
Transmitted Pulsewidths	1 μs (CW)
	+ 100 μs (Nonlinear Chirp)
Pulse Repetition Frequency (Average)	676 Hz
Antenna Gain	33.5 dB
Beamwidth (AZ)	1.5° (Average)
Rotation Rate	12 rpm
Noise Figure	1.5 dB
Pulse Compression Ratio	100:1
A/D	12 bits
Doppler Filters	4
Clutter Rejection	4-Pulse MTD
	Circular Polarization
	Dual-Beam Antenna
Weather Channel	

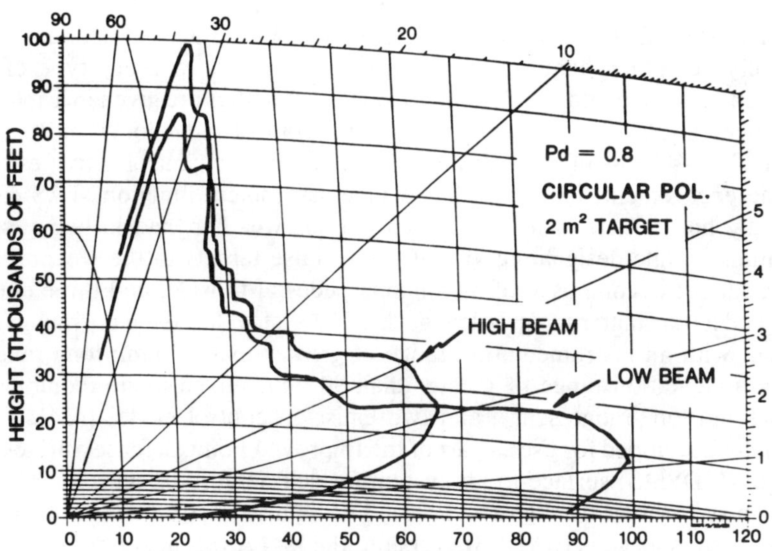

Figure 2.64 L-Band Solid-State RAMP Radar Coverage (from *Microwave Journal*, December 1984 [87]).

A 2D radar designed to cope with the sea-skimming "pop-up" high-altitude diving threat, as well as other targets, is the Hughes Target Acquisition System

Figure 2.65 RAMP Antenna. (Courtesy of Raytheon Co.)

(TAS), Mk23 [411]. This is a modern pulsed doppler radar that provides hemispherical surveillance from 0° to 75° in elevation and 360° in azimuth for automatically detecting, acquiring and tracking all targets in every type of natural at-sea environment and ECM. (The TAS characteristics are given in Table 2A.1(a) of Appendix 2A. TAS ensures that the fire control radar locks on to the right threat after assignment and coordinates the missile fire control activities throughout the engagement, including the survive assessment function. TAS can automatically find targets in heavy ground clutter, weather, and in heavy sea-state conditions, while designating virtually zero false targets to the weapon system. The radar set is composed of an antenna pedestal topside, and three equipment racks and a transmitter group below decks. The system has three operating modes—normal, with an instrumentation range of greater than 20 nmi; long range, with an instrumentation range of greater than 90 nmi; mixed mode (combination of normal and long-range scan); and two emission control modes [424]. TAS was originally developed for use as part of the Improved Point Defense Surface Missile System (IPDSMS) surface-to-air system for self-defense of ships against the antiship missile threat through the 1990s. TAS is also used to direct other weapon systems, such as guns [10]. A 3D version, the M-TAS, is under development [10, 411]. M-TAS will use one-dimensional phase steering in elevation [411].

2.3.2 Weather Radar Systems

The Federal Aviation Administration (FAA) is planning to deploy a national weather-forecasting network consisting of long-range, S-band, doppler radars for the continental US, Alaska, Hawaii, the Caribbean, and Western Europe. These radars will be used for predicting the formation of tornadoes and hurricanes, and for tracking weather in general [90]. Unisys (formerly Sperry) has been contracted, for $60 million, for the first 10 systems [403]. With 185 additional radars planned, the program is ultimately valued at about $400 million. The FAA is also looking at the addition of doppler C-band radars at airport terminals for use in the detection of microburst wind shear, which was responsible for the crash of a Delta Airlines L-1011 at the Dallas-Ft. Worth airport in 1985.

2.3.3 Airborne 3D Solid-State Radars

Westinghouse has been selected to develop an airborne, solid-state, "ultra-reliable radar" using an active-element phased array antenna for the Air Force Avionics Laboratory. The objective is to achieve a 500-hour mean-time-between-failure (MTBF), which represents a tenfold increase in reliability over that obtained with existing systems. This airborne radar, which will have about 2000 elements, will use the Texas Instruments solid-state X-band transceiver modules and array developed under the Solid State Phased Array (SSPA) Program [91] (see Section 2.6.5.1).

2.3.4 JSTARS

Grumman Aerospace Corporation and United Technologies-Norden have been selected to build the USAF-Army Joint Surveillance Target Attack Radar System (JSTARS or Joint STARS) [92,93]. Norden will build the airborne radar for the system. This multimode radar is capable of SAR (see [1]) operation to detect fixed surface targets and doppler processing to spot slow-moving vehicles in a time-shared mode ([92–94; see also [95]).

2.3.5 ASAR

The new Advanced Synthetic Aperture Radar System (ASARS-2), developed by Hughes, is capable of spotting military ground targets and creating radar maps with a resolution equal to that of an infrared sensor. ASARS-2 is to be flown on the US Air Force TR-1 reconnaissance aircraft [96]. The radar has a "squinted" antenna and provides a spotlight mode [1] with real-time imagery to a ground station in all weather [96]. A pod-mounted version is also being developed [97]. This latter version would significantly improve SAR imaging capabilities for multirole fighter-reconnaissance aircraft. A discussion of the development of space-based SAR systems is given in Section 2.5.

2.3.6 Netted Radars

The netting of radars has been done for air traffic control systems [98], and such netting has a number of advantages. When an overlap in coverage exists for the region shadowed by one radar, it overcomes the problem of hill shadowing and provides better target track. This feature should be even more important for a tactical radar system, as illustrated in [428]. With netting, the tracking in clutter and through blind velocities would potentially be improved [98]. There is expected to be an increased trend toward netting of tactical radars in the future.

2.4 BISTATIC RADAR SYSTEMS

For a bistatic radar, such as the OTH/B, the transmitter and receiver are at separate locations. The types of bistatic radar systems to be addressed in this

BROOKNER
CHAPTER 2

section are the tactical ones located along the battle front. For such tactical applications, a bistatic radar has the advantage that the receiver can be covert because it does not radiate. As a result it has decreased vulnerability to jamming because the enemy is less likely to know its location. The receiver can also be placed in a more vulnerable forward position with the transmitter located behind the enemy lines. Another advantage is decreased vulnerability of the receiver to anti-radiation missiles (ARMs). For best performance, such a system must use a stack of narrow azimuth beams for the receiver. The transmitter would either be one designed for the system, or use could be made of someone else's transmitter, even that of the enemy if its parameters are known and it can be exploited. With the increased feasibility for building stacked-beam radars, such systems are becoming more feasible.

Attempts have been made in the past to construct bistatic radar systems, but, to the author's knowledge, none have become operational [99]. An interesting system was built by University College of London [100–103]. Researchers there demonstrated that it is possible to build an inexpensive bistatic radar system using someone else's transmitter. Specifically, the Marconi S650 [1] 2D air traffic surveillance radar at Heathrow Airport was used. This radar operates at a carrier frequency near 600 MHz [1]. The researchers obtained excellent results on a low budget with the transmitter at Heathrow being used in an uncooperative mode. Figure 2.66 shows the bistatic video display obtained at University College of London. The display was corrected to provide true range and angle at the receiving site [100]. The feasibility of bistatic SAR using two aircraft has been demonstrated by the Environmental Research Institute of Michigan (ERIM) [104]. This result has potential for more survivable reconnaissance and weapon delivery systems [94].

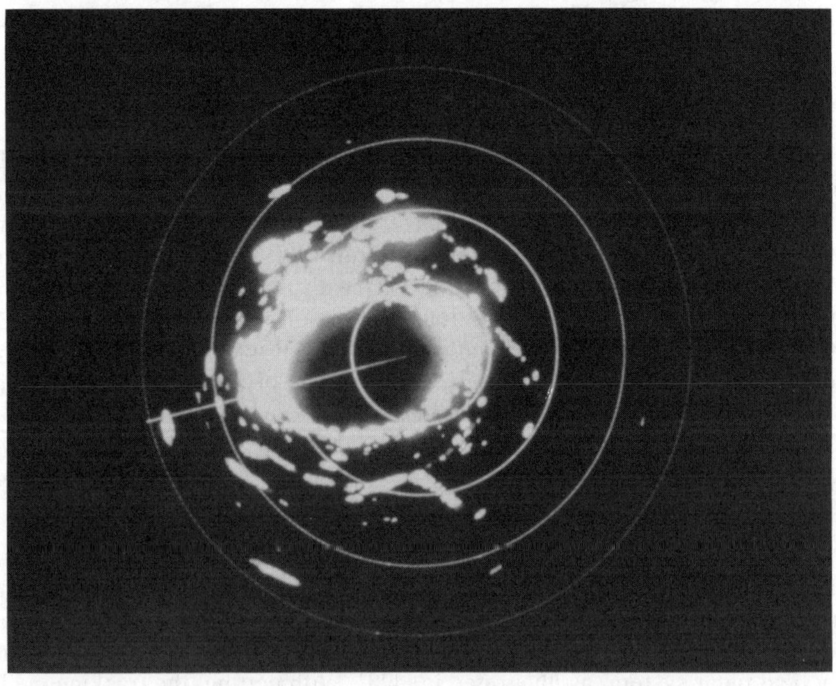

Figure 2.66 Bistatic Radar Display. (Courtesy of J. Forrest, Marconi.)

In the UK, Plessey, Marconi, and the Royal Signals and Radar Establishment (RSRE) are working together to develop an S-band bistatic radar intended for air defense applications. They are building an experimental system having an illuminator with a 1° pencil beam that scans at 4 rpm. The receiver would consist of

20 contiguous beams in azimuth. An IF resistive matrix would be used to form these beams, as done on the Marconi-Martello system. Pulse synchronization of the transmitter and receiver is to be achieved by referencing each of these to an atomic clock, with the transmitter and receiver clocks synchronized to each other once per day [105].

2.5 SPACE-BASED SYNTHETIC APERTURE RADAR (SAR) SYSTEMS [1,92–97,104,106–122]

The last decade has seen the deployment of three SARs in space, the SEASAT in June 1978, the Shuttle Imaging Radar-A (SIR-A) in November 1981 on the second space shuttle *Columbia* flight, and the SIR-B on a later shuttle flight in 1984 [106,107,109]. Such satellite imaging radars have the advantage of being able to provide maps of regions of the earth that are covered by clouds most of the time or are otherwise hard to access. Also, these satellite radars can be used to help monitor crop growth and city growth, aid in the search for mineral and oil resources, monitor iceberg movement to assist in navigation through the icy waters near the poles, monitor the world's oceans and storms, provide soil moisture mapping, give geological exploration data, and produce stereo imaging for topographical maps. Figure 2.66 shows a digitally processed map of Los Angeles obtained with the SEASAT. The resolution is 25 × 25 m. To achieve such a resolution in the along-track direction with a real antenna at the SEASAT altitude of 432 nmi would require an antenna of 4.4 nmi in length (or about the length of 21 Empire State Buildings). Obviously, it is not possible to deploy such a long antenna. The actual SEASAT antenna is only 10.7 m in length. The long antenna needed to achieve the resolution obtained for Figure 2.66 is synthesized by coherently processing the echo signal received by the satellite as it travels a distance equivalent to the length that a real antenna would need to achieve the desired resolution. (Actually, an effective length of only half that of a real antenna is needed for the synthetic aperture. Hence, the synthetic antenna length needed for the above example is 2.2 nmi, or that of about 10.5 Empire State Buildings [1,120].) It took 600 × real-time to obtain the digitally processed map shown in Figure 2.67 [121]. However, real-time or near-real-time digital processing of satellite SAR data can be done with the Jet Propulsion Laboratory's (JPL) Advanced Digitally SAR Processor (ADSP) [116,122].

Compared with visible and infrared satellite mapping instruments, the satellite radar imager has the advantage of being able to do mapping during rain and nighttime conditions as well as when it is cloudy. The SIR-A demonstrated the radar's ability to obtain images below the sand of the Sahara Desert. Figure 2.68 shows a LANDSAT photograph obtained of the Sahara Desert. Figure 2.69 shows a strip map obtained with the SIR-A for part of the same region. Whereas the LANDSAT, which operates in the visible and near-visible regimes, gives images of only the sand surface, the SIR-A penetrates the sand surface to show pictures of ancient river beds up to 5 m below the sand. The SIR-B provided stereo and 3D imagery from space [109,110].

Many future space-based SARS are being planned. The next major step forward is the joint US-Germany-Italy SIR-C/X-SAR experiment, which will simultaneously provide images at L-, C-, and X-band with all polarizations at L- and C-band (see Table 2.15 [109]). The X-band X-SAR is a joint German-Italian project [111]. The short-duration (one week), shuttle-attached SIR systems are to be followed by many long-duration, free-flyer satellite systems, such as the European ERS-1 [112, 113] and Japanese ERS-1 approved flights, which are expected to have lifetimes of two to three years [109] (see Table 2.15). The ERS-1 systems will provide only localized coverage because of their limited swath width (about 100 km). As a result, the JPL is looking at the feasibility of a global radar imager

Figure 2.67 25-m High Resolution SAR Map of Los Angeles, Obtained August 12, 1978 with SEASAT. (Courtesy of C. Elachi, JPL.)

(GLORI) (see Table 2.15). This sensor would have a dual 400-km swath. Consideration has been given to a UHF system to achieve a penetration factor of three better than the SIR-A achieved over the Sahara Desert and to many kilometers of depth into the Antarctic ice. Canada is looking at a free-flyer for monitoring

Table 2.15 Past and Future Spaceborne SAR Systems (after Elachi [109])

	Shuttle SIR-A/B 1981, 1984	*Shuttle SIR-C/D 1991–1992*	*EOS SIR-E 1997*
Frequency (GHz)	1.25 (L-Band)	1.25 (L-Band) 5.3 (C-Band) 9.6 (X-Band)*	1.25 5.3 9.6*
Polarization	HH	HH, VV, HV, VH	HH, VV, HV, VH
Resolution (m)	20–40	10–60	10–20
Swath Width (km)	30–50	15–80	50–300
Illumination Control	Mechanical	Electronic	Electronic
Architecture	Conventional	Distributed	Distributed

*Germany, Italy.

Table 2.15 (cont'd)

	ESA/ERS-1 1990	*Japan/JERS-1 1991*	*Canada/RADARSAT 1994*	*US/EOS 1997*	*GLOBAL EOS RADAR IMAGER (GLORI) 1997*
Frequency (GHz)	5.3	1.2	5.3	1.2 5.3 9.6*	5.3
Resolution (m)	30 or 100	18	10–100	10–20	100
Swath Width (km)	80	75	55–500	50–300	700–800
Illumination Angle	23°	42°	20–45°	10–55°**	—
Principal Application	Oceanography	Geology	Polar Ice	Multidisciplinary	Global Coverage Every Two Days

*Germany, Italy.
**Controllable.

Figure 2.68 LANDSAT Picture of Sahara Desert in Egypt. (Courtesy of C. Elachi, JPL.)

iceberg movement (see Table 2.15 [115,117]). Other SAR systems, as well as the ones discussed above, are summarized in Table 2.15.

NASA plans to fly a SAR around the planet Venus in 1990 for the Magellan mission. The radar is to operate at a carrier frequency of 2385 MHz and provide a resolution down to 250 m [109]. This SAR will provide the first high-resolution maps of the surface of Venus, which is continually covered by clouds. (The Pioneer 12 spacecraft provided radar maps of the surface of Venus from 1978 to 1980 with a resolution of only 30 km.) By the late 1990s the Cassini spacecraft may be put in an orbit around Saturn and image its moon, Titan, at L- and K-band on flybys [109].

2.6 TECHNOLOGY AND TECHNIQUES

2.6.1 Scanning Techniques

2.6.1.1 Hemispherical and Omnidirectional Coverage

Hemispherical coverage is needed for many applications. Often a three- or four-faced phased array system is used to achieve such coverage, with each face being a phase-phase steered array. An alternative approach is to use a single phase-phase steered array mechanically rotated 360°, as was done for the Thomson-CSF multifunction array radar of Figure 2.22. Still another approach is to use the dome antenna concept [1,123,124], illustrated in Figure 2.70. In this approach a phase-phase steered planar array is laid flat on its back, with its boresite pointing straight up. A *dielectric* lens is placed over this array. When the planar array scans the beam to 60° away from vertical, the dielectric lens bends the rays

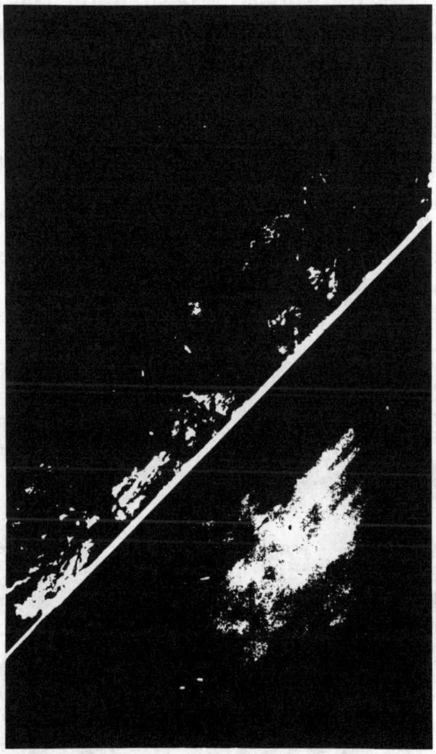

Figure 2.69 SAR Picture, Obtained with the Space Shuttle SIR-A, of Strip through the Same Region as in Figure 2.68. (Courtesy of C. Elachi, JPL.)

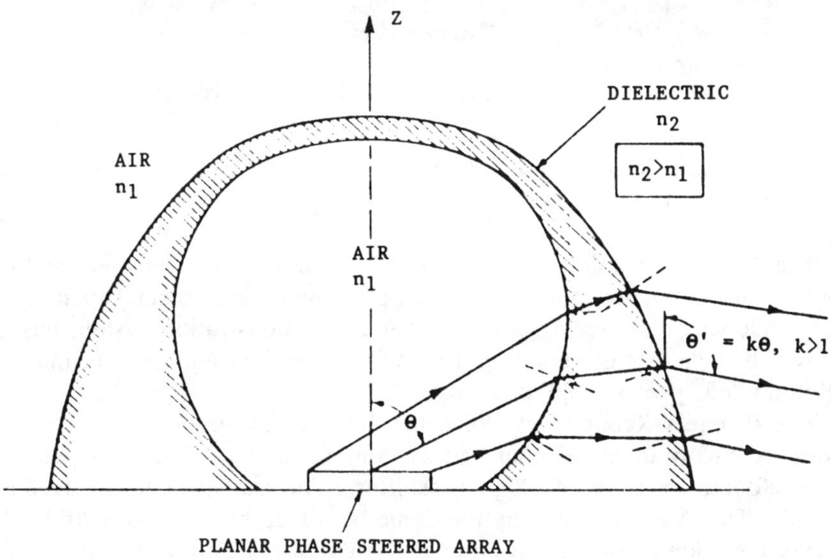

Figure 2.70 Principle by which Dome Lens Provides Greater than Hemispherical Coverage for a Planar Array; n_1 and n_2 represent indices of refraction of air and dielectric, respectively (from Brookner [1]).

down to 90° away from the vertical, so as to point horizontally. When the planar array points the beam vertically, the lens does not bend the ray so that the beam is still pointed vertically. For all scan angles of the planar array between 0° and

60°, the beam is scanned after it goes through the dielectric lens from 0° to 90°, thus achieving hemispherical coverage. If desired, the lens can be made stronger so that the beam is bent to 20° or 30° below the horizon when the planar array scans the beam to 60° away from boresite. Such increased coverage is needed for shipboard applications to compensate for ship's roll and pitch. The dielectric lens is usually implemented with a dome in which fixed phase shifters are embedded to cause the dome to act like a dielectric lens, as described above [1]. Figure 2.71 shows a dome antenna (built by Sperry) intended for tactical application. Figure 2.72 shows the dome fixed phase shifters, while Figure 2.73 shows how they are clustered to form the dome antenna. Part of the steerable phase-phase steered planar array is shown in Figure 2.74.

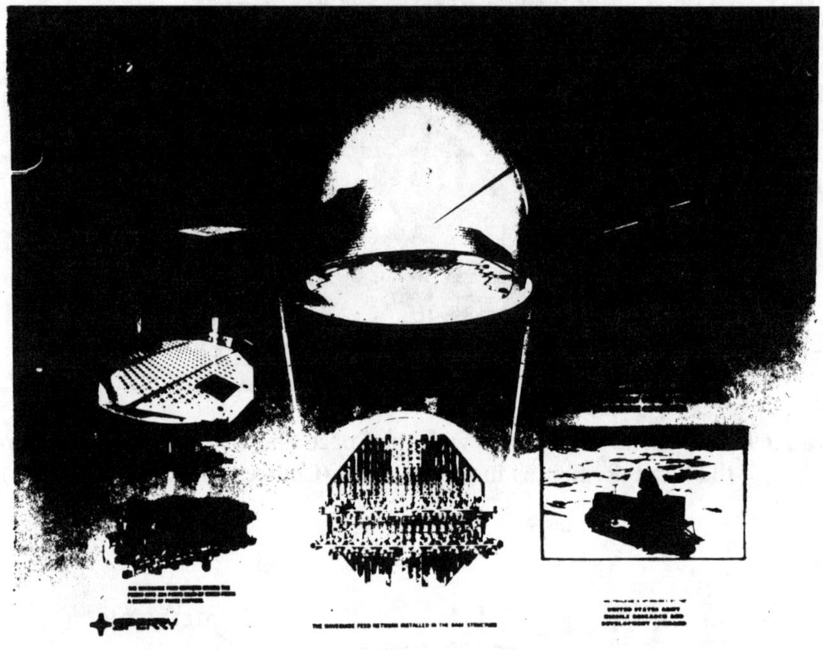

Figure 2.71 Recent Sperry Dome Antenna and System. (Courtesy of Sperry-Unisys Corp.)

The dome antenna has been combined with the Rotman lens-fed multibeam array to realize an octave bandwidth, multiple simultaneous beam capability, and better sidelobes [125] (see Figure 2.75). A 2D demonstration system has been built, which is capable of operating from 8 to 15 GHz. Twenty-nine beamports in the Rotman lens provide 180° coverage [126].

The People's Republic of China has built an experimental C-band dome antenna. A 0.85-m diameter planar array containing 297 radiating elements was used. The dome consisted of 300 phase shifters. The scan angle for the feed array was ±35°. This was magnified by the dome by a factor of two to ±70° [127].

No operational dome radar antenna system has been built to date. A major question is whether the dome antenna is more economical than using a four-faced phase array system. For many radar applications, the dome and array antennas are about equal in cost.

An antenna which provides nearly omnidirectional coverage is the developmental "crow's nest" antenna. This is a 3D array of antenna loops distributed randomly over a spherical volume, as shown in Figure 2.76 [128]. There is no ground plane for this antenna system so that omnidirectional coverage is achievable in principle. The coverage is of course limited by the element pattern coverage. For the antenna systems shown in Figure 2.76, open-loop elements are

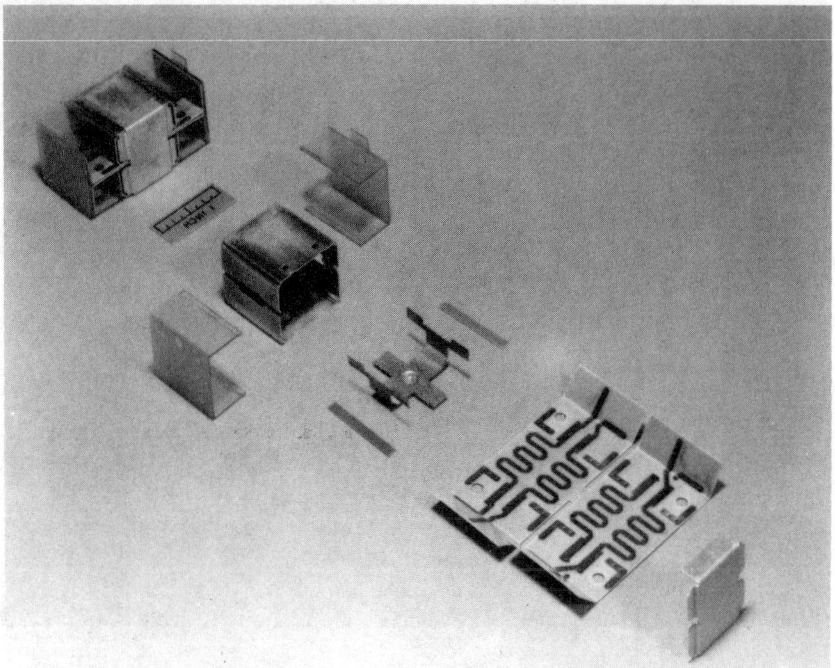

Figure 2.72 **Sperry Dome Module Consisting of Fixed Phase Shifters. (Courtesy of Sperry-Unisys Corp.)**

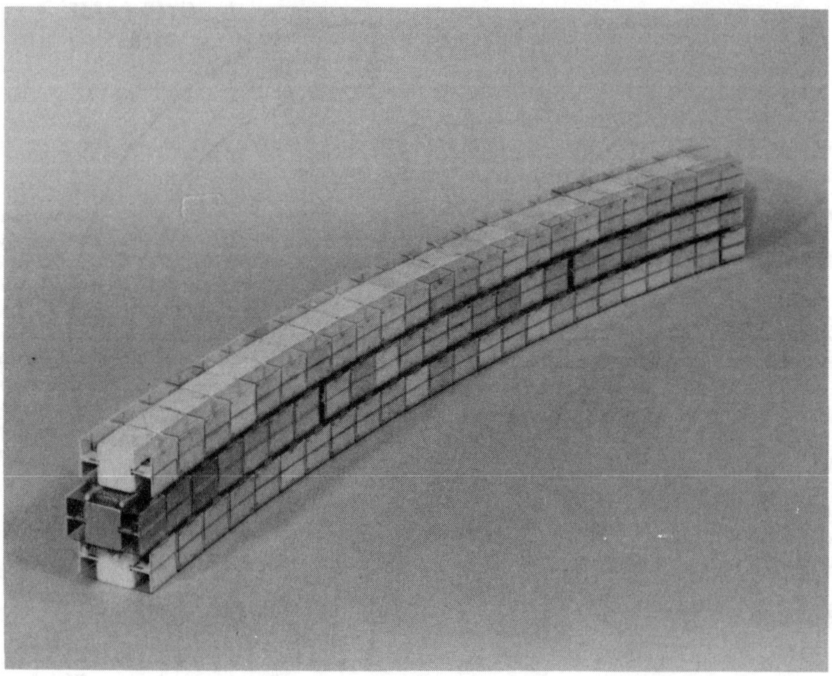

Figure 2.73 **Sperry Dome Lens Module Cluster. (Courtesy of Sperry-Unisys Corp.)**

used to provide isotropic coverage in azimuth and a cosine coverage in elevation with the peak of the element gain being placed at the horizon. The array shown (which was built by Forschungsinstitut fuer Funk und Mathematik) consists of 39 elements having an average spacing of 1.2λ. The radius of the sphere over which the elements are placed is 3λ. Figure 2.77 shows measured antenna patterns

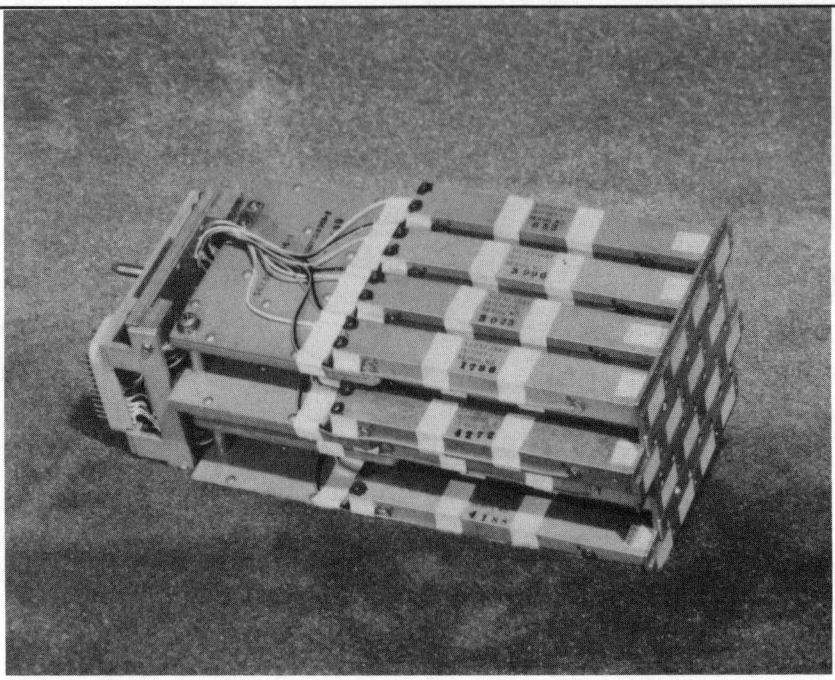

Figure 2.74 Part of Sperry Dome Antenna Phase-Phase Steered Array; Specifically, Phase Shifter Subarray Module. (Courtesy of Sperry-Unisys Corp.)

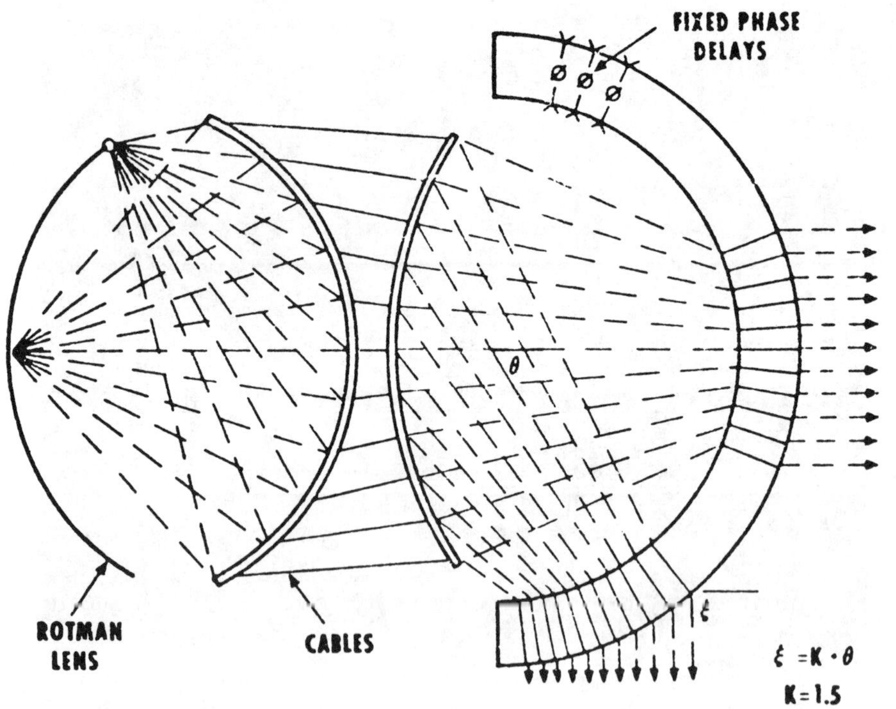

Figure 2.75 Wide-Angle Array-Fed Lens (WAAFL) (from Thomas [125] © 1979 IEEE).

that have been obtained. Whether this type of antenna is practical to manufacture and to use remains to be seen. If used, it would probably be for some special application.

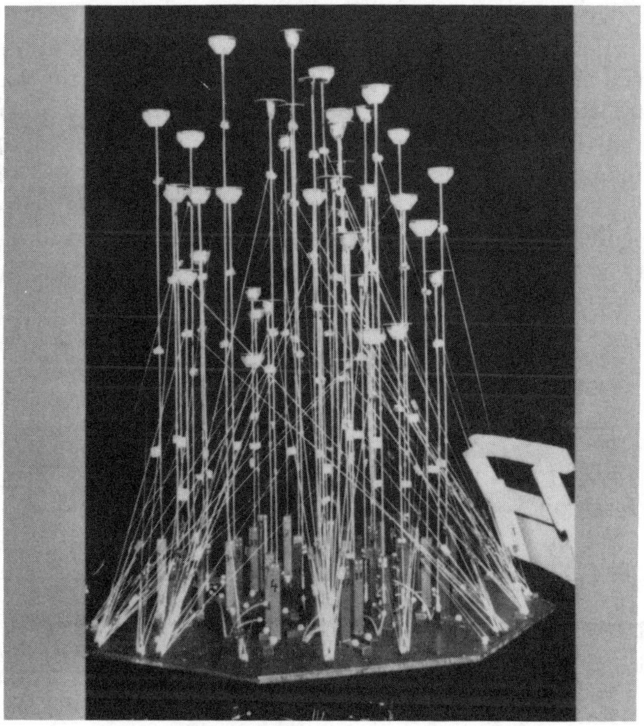

Figure 2.76 The Crow's Nest Antenna. (Courtesy of Wilden [128].)

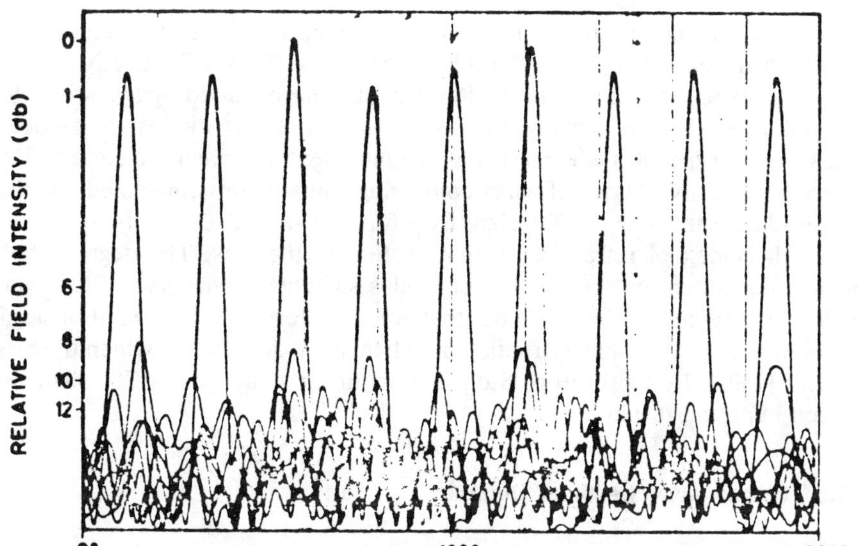

Figure 2.77 Crow's Nest Azimuth Antenna Pattern when the Antenna Is Focused to Different Azimuth Directions (from Wilden [128] © 1979 IEEE).

2.6.1.2 Continuous Aperture Scanning

Several efforts have been made to produce continuous-aperture scanning and to reduce the number of controlled elements in an array [1, p. 330]. Among these efforts, one promising approach involves the use of a series ferrite scan technique (see Figure 2.78 [1,129]). Two-dimensional scanning of a pencil beam can be achieved in the following way. N rows of the series ferrite waveguides of Figure 2.78 are used to form a planar array having a pencil beam. An additional series ferrite waveguide is then mounted vertically along the side of the array with each radiator port of this waveguide feeding a different row of the array. For this array, $N+1$ controls are needed to provide 2D scanning, instead of N^2, as conventionally needed. This technique, however, is limited to narrow bandwidth signals and to small-sized apertures.

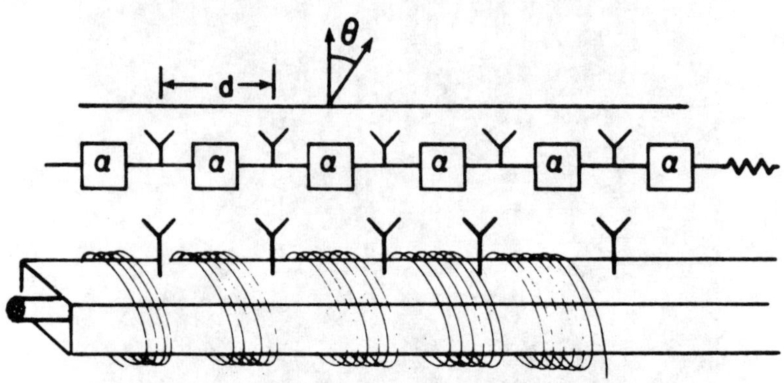

Figure 2.78 Series Ferrite Scan Technique. ***Note:*** **Phase scanning is obtained by controlling the propagation velocity down the ferrite-loaded nonmagnetic waveguide. This control is achieved by varying the current in the solenoid coil wrapped around the waveguide (from Brookner [1]).**

Another technique under development, which provides 2D scanning with $2N$ controls instead of N^2 is the RADANT technique. This technique is hoped to replace the expensive diodes of standard diode phase-shifted arrays with inexpensive diodes. However, many more of these inexpensive diodes are needed in the RADANT type of system. The technique uses an artificial dielectric lens, consisting primarily of grids of wires containing many diodes connected together. Changing diodes in a section of the lens from forward-bias to reverse-bias condition changes the index of refraction in that section of the lens. The degree of this change is dependent on the number of diodes switched from forward to reverse bias. By varying the number of diodes that are in the forward-bias condition across the lens, we change the lens refraction, and hence the direction in which the beam is pointed [130,131]. A prototype S-band 10-m^2 array, consisting of 75,000 diodes, has been built and tested.

2.6.1.3 Thinned Arrays without Dummies

It is now possible to design large, thin, phased arrays that do not need to use dummy elements, as done in the COBRA DANE and PAVE PAWS radars [132,133]. Eliminating the costly dummy elements, however, causes a loss in the antenna gain. Consequently, the saving resulting from the elimination of the dummy elements must be traded off against the cost of compensating for the loss in system sensitivity. For some applications the elimination of the dummy elements may be cost effective for future systems.

2.6.2 Ultra-Low-Sidelobe Antennas

In order to cope with the ever increasing jammer threat, the radar antiradiation missile, sidelobe clutter and sidelobe interference it is essential to have antennas with very low sidelobes. The last decade and a half has seen the development of antennas having what are called ultra-low sidelobes. Ultra-low sidelobes are those which are either down 40 dB from the main-lobe peak or 20 dB below that of an isotropic antenna; that is, its sidelobes are at a level of −20 dBi (see [134] and Chapter 6 of this book). Specifically, an antenna would have its sidelobes at a level of −20 dBi, if the signals coming from the sidelobes were 20 dB below that which would be radiated from an isotropic antenna with the same power applied at its input as that applied to the real antenna. Ultra-low sidelobes have been achieved with array antennas that do not scan electronically and with fixed-beam reflector antennas (see Table 2.16). For these antennas, the scanning is achieved by mechanically rotating or moving the antenna. Achieving ultra-low or near-ultra-low sidelobes with nonelectronically scanned arrays is described in Section 2.6.2.1. Section 2.6.2.2 discusses the achievement of low sidelobes with electronically scanned antennas, and Section 2.6.2.3 covers their achievement with reflectors. Section 2.6.2.4 discusses the sources of errors, and the effects of environmental factors are discussed in Section 2.6.2.5.

Table 2.16 Types of Antennas Providing Ultra-Low and Near-Ultra-Low Antenna Sidelobes

FIXED BEAM ARRAY ANTENNAS
Radiating Element:
• Slotted Waveguide
• Dipoles
• Helix
• Slotted Square Coaxial Cable
REFLECTOR ANTENNAS

2.6.2.1 Nonelectronically Scanned Array Antennas

Nonelectronically scanned array antennas have been effectively used for achieving ultra-low antenna sidelobes because the illumination function across the antenna can be accurately controlled for such antennas. For these antennas the scanning is achieved by mechanically rotating or moving the antenna. The ability to achieve ultra-low sidelobes is dependent on the ability to control the amplitude and phase illumination across the antenna; the "heavier" is the antenna weighting, the lower are the antenna sidelobes (see, for example, [1, page 381]. Another advantage of array antennas is that they do not have sidelobes generated by spillover or blockage, such as occurs with reflector antennas [135]. Table 2.16 lists the types of array radiating elements used to achieve near-ultra-low to ultra-low sidelobes. These arrays are discussed in the following paragraphs.

Slotted-Waveguide Array Antennas

An array antenna comprised of rows of edge-slotted waveguides, as shown in Figure 2.79, has been reported to be able to achieve sidelobes of 40 to 50 dB down from the main beam [136]. An array of edge-slotted waveguides can achieve ultra-low sidelobes because the slots can be accurately machined by using *numerically controlled* (NC) machining. Also, for such array antennas, Westinghouse

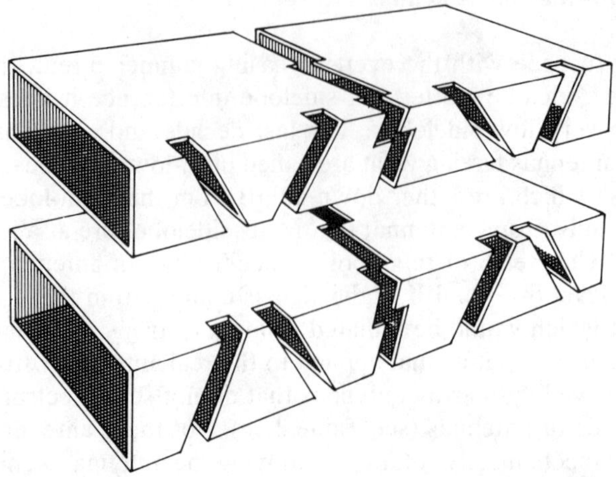

Figure 2.79 Edge-Slotted Waveguide Array for Low Sidelobes (from Burgess [136]).

has been able to develop the theory to make possible the accurate prediction of the effects of the mutual coupling between the slotted waveguide's radiating elements, which is essential for the achievement of ultra-low sidelobes. Westinghouse was the first to build an ultra-low-sidelobe slotted waveguide array. An example of such a slotted array is the AWACS antenna (see Figures 2.80 and 2.81). This antenna is put in a rotodome above the aircraft, as indicated in Figure 2.80. The rotodome rotates mechanically in order to provide 360° of azimuth. The antenna actually uses 1D electronic scanning in elevation by using a single phase shifter per row of the array. Because the antenna uses electronic scanning in elevation, the elevation sidelobes will not be so low as they are in azimuth, where no electronic scanning is used.

Figure 2.80 E3A (AWACS) in Flight, Showing Rotodome Antenna Above the Aircraft. (Courtesy of Westinghouse.)

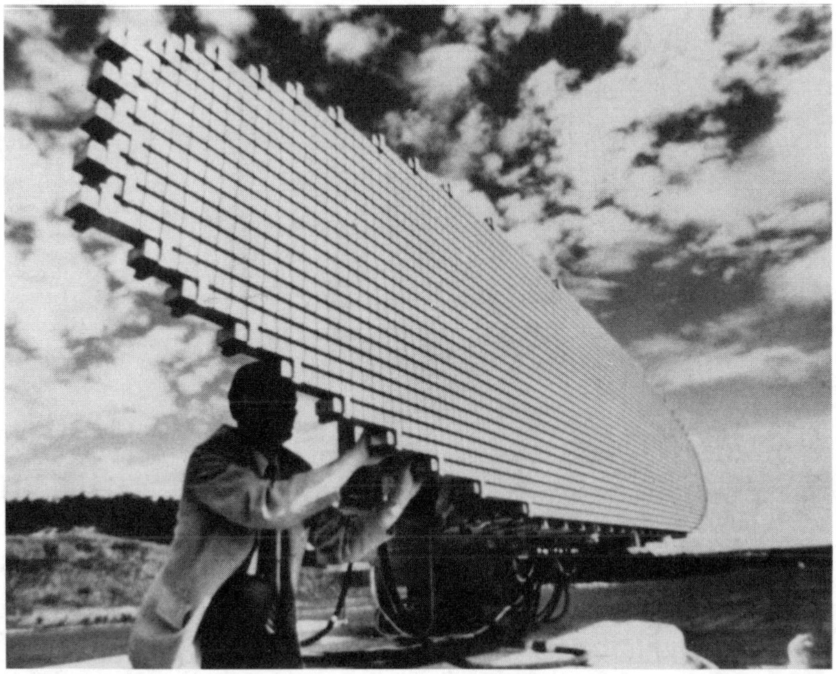

Figure 2.81 AWACS (AN/APY-1 E3A Radar) Low-Sidelobe S-Band Antenna. (Courtesy of Westinghouse.)

Westinghouse has built a slotted-waveguide ultra-low-sidelobe antenna (ULSA) as a replacement for the reflector antenna of the S-band, 3D, AN/TPS-43 of Figure 2.82 [137]. With this antenna, the radar is designated as the AN/TPS-70 [10,134] (see Figure 2.83). This system is now in production [85,86]. The fixed-beam array antenna of the F-16 uses a broad-wall, rather than narrow-wall, slotted-waveguide antenna (see Chapter 6).

Dipole Array Antennas

It is possible to build an ultra-low-sidelobe planar array by using dipole radiating elements. Such an array was built by Westinghouse (see Figure 2.84 [134,138]). Figure 2.85 shows the very impressive sidelobes achieved with this L-band corporate-fed, planar array consisting of 512 elements. This antenna provided low sidelobes over almost an octave band, from 800 to 1400 MHz. The planar array used an air-dielectric (suspended) stripline manifold to achieve precise control of the ampliude and phase across the antenna.

Raytheon has also built an ultra-low-sidelobe, L-band, dipole array antenna (Figure 2.86 [139]). It consists of 12 rows, with each row having 32 dipoles. This 15 × 6-ft planar array has a gain of about 30 dB. The antenna provides ultra-low sidelobes over a wide (15%) bandwidth from 1.2 to 1.4 GHz (see Figure 2.87). This figure indicates that the peak sidelobes are 40 dB or more below the mainlobe peak, while the far-out rms sidelobes are down to −75 dB (or, equivalently, about −45 dBi) in some regions. This dipole array has a number of interesting features. First, the 16 dipole radiating elements, making up half of one row of the array, together with the 1-to-16 corporate-feed network for these 16 elements (see Figure 2.88), are all photoetched out of a single sheet of copper. This has the advantage of providing an inexpensive way of mass producing the antenna with tight control of the amplitude and phase errors across the aperture. Two of these

BROOKNER
CHAPTER 2

Figure 2.82 S-Band 3D Stacked-Beam AN/TPS-43. (Courtesy of Westinghouse.)

Figure 2.83 AN/TPS-70 3D Stacked Beam Radar with Ultra-Low-Sidelobe Antenna. (Courtesy of Westinghouse.)

16-element linear arrays are combined to form one 32-element row. The feed of Figure 2.88 has equal path lengths to all the dipole radiating elements in a given row. This provides the antenna with a wide instantaneous bandwidth, or equivalently with no scanning when the signal frequency is changed, as it would be if a linear FM chirp waveform were used (i.e., the antenna's main beam is squintless).

Figure 2.84 Nonelectronic Scanning (Fixed Beam) L-Band, Wideband, Ultra-Low-Sidelobe Corporate-Fed Dipole Array. (Courtesy of H. Schrank [134], Westinghouse.)

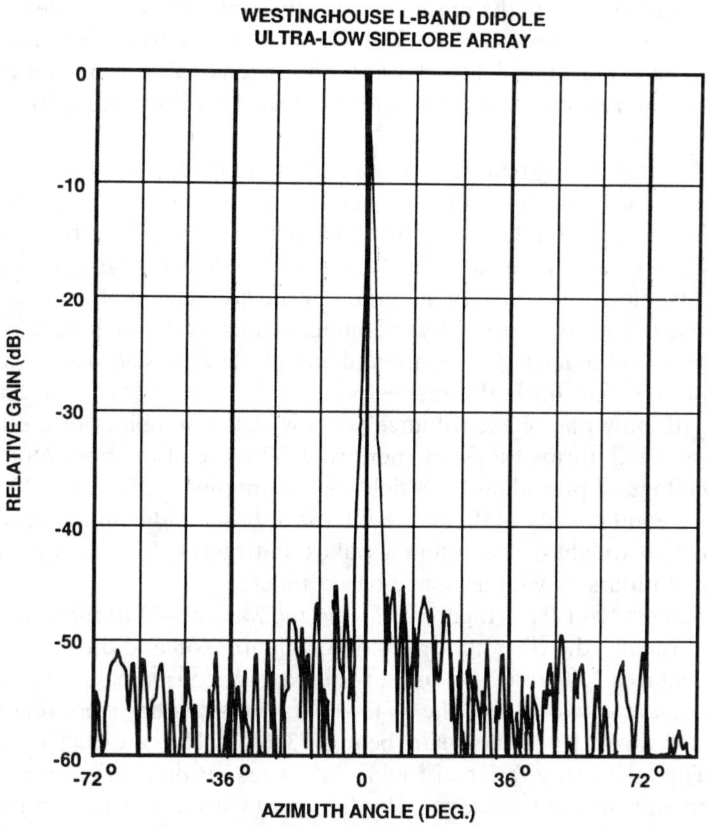

Figure 2.85 Typical Pattern of Ultra-Low-Sidelobe L-Band Dipole Array. (Courtesy of H. Schrank [134], Westinghouse.)

Figure 2.86 L-Band Ultra-Low-Sidelobe Planar Array Using Stripline Corporate Feed and Dipole Radiating Elements which Are Etched Together. (Courtesy of Raytheon Co.)

The use of equal path lengths in the feed for this antenna is equivalent to the use of time delay steering in the COBRA DANE radar, as discussed in Section 2.2.1.1, except that here the time delay steering only needs to be provided at boresite because the antenna does not steer electronically, as does the COBRA DANE system.

Another interesting feature of the antenna in Figure 2.86 is the use of a very low dielectric constant (ϵ) material (hence called ''low-epsilon'') for the stripline of the corporate array feed (see Figures 2.88 and 2.89). Specifically, a honeycomb strip transmission line is used, which has a low dielectric constant of $\epsilon = 1.13$. With such low dielectric constant material, the introduction of phase and amplitude errors across the array, caused by inhomogeneities in the dielectric constant of the medium, is minimized. The phase and amplitude errors across the array were found to be 0.8° and 0.07 dB, respectively [139]. These very low errors were achieved with only one phase trimmer per row at the combiner of the two 16-to-1 combiners which forms the 32-element row. The use of the honeycomb has the added advantage of providing for a lightweight antenna, only 1.3 lbs/ft^2 (approximately the weight density of the SEASAT space-based radar antenna; see Section 2.6.3). The light weight of the antenna makes it attractive for use in ground-based transportable radars as well as space-based radars.

The Selenia RAT-31S (Figure 2.45) and the Marconi-Martello stacked-beam, 3D, L-band radars (the S713 tube version of Figure 2.58 and the S723 solid-state version of Figures 2.60 and 2.61) use a planar array, consisting of rows of dipoles fed by a stripline network. For the S713 radar, the sidelobes were reported to be −45 dB, 15° away from the main beam [83,140]. The Westinghouse W-2000 L-band, 3D planar array radar of Figure 2.52 uses air-dielectric stripline to help achieve very low azimuth sidelobes. Printed-circuit stripline usually provides poor sidelobes because of dielectric variations [77]. The W-2000 achieves 1D scanning in elevation by using one phase shifter per row. The antenna for this radar is probably a derivative of the antenna in Figure 2.84.

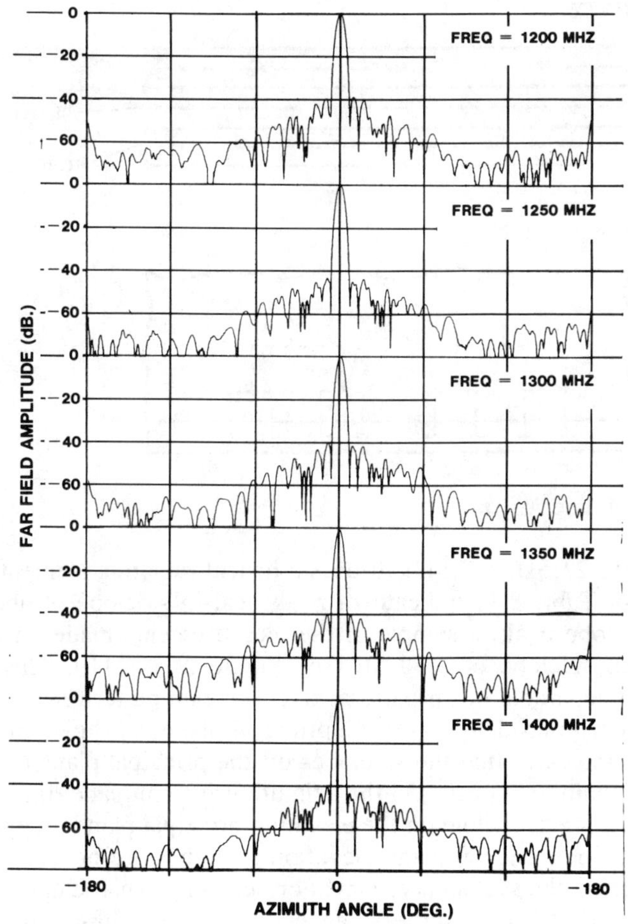

Figure 2.87 **Measured Antenna Patterns for Ultra-Low-Sidelobe Planar Array (after Dahl *et al.* [139]). (Pattern Data for Raytheon ULSA Planar Array.)**

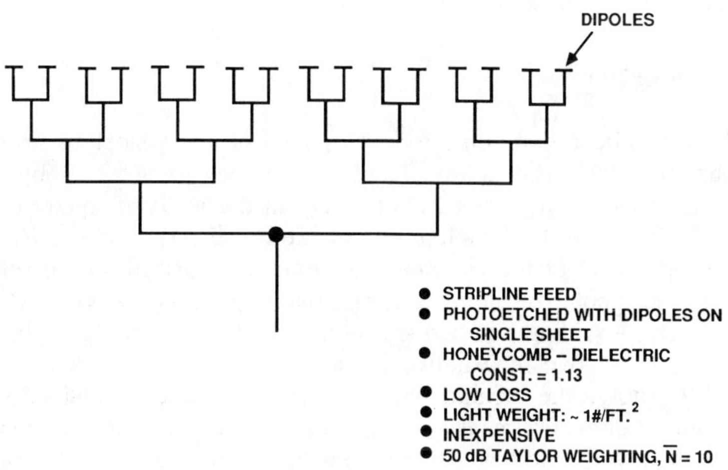

Figure 2.88 **Nondispersive Stripline Feed; Equal Path Length, Hence Wideband (15%, i.e., 200 MHz) and Squintless.**

Helical Radiating Elements

Thomson-CFS has developed an ultra-low-sidelobe planar array antenna for the TRS 2215 S-band, 3D radar of Figure 2.48 (Section 2.2.4.2) and has redesig-

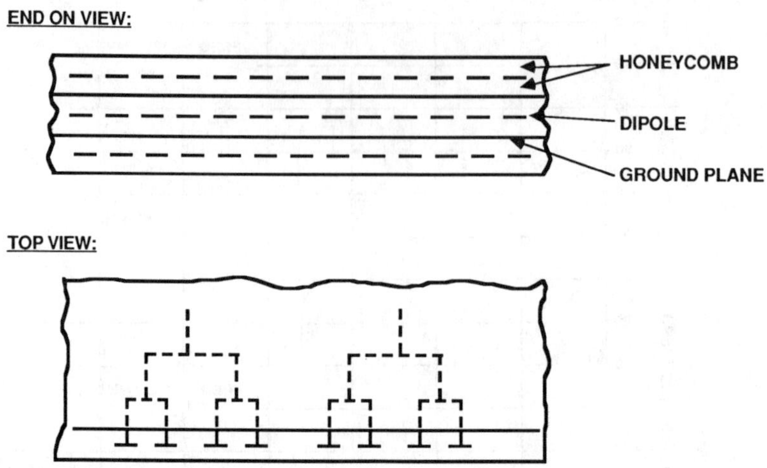

Figure 2.89 Stripline Antenna.

nated it the TRS 2215D [141,142]. It uses a helical radiating element (see Figure 2.90). Although Table 2.17 indicates a peak near-in sidelobe of about −30 dB from the main-lobe peak, a sample of the measurements made on this antenna indicate a peak sidelobe of −40 dB (see Figure 2.91 [143]). This rectangular antenna uses a separable beam pattern. Such a beam pattern has the advantage of providing very low sidelobes off the principal plane, as shown in Figure 2.92. Thomson-CSF indicates that the sidelobes off the principal plane are −70 dB or more from the main-lobe peak. On the azimuth principal plane, the sidelobes are indicated to be −45 dB, while on the elevation principal plane they are indicated to be −35 dB. The principal plane elevation sidelobes are poorer than they are in azimuth because the system uses elevation scanning. This scanning is achieved by using four-bit diode phase shifters, one per row. These phase shifters introduce extra amplitude and phase errors to degrade the elevation sidelobes. The sidelobes shown in Figure 2.87 for the stripline dipole array (Raytheon) of Figure 2.86 are along the azimuth principal plane, while off the principal plane the sidelobes can be expected to be further down.

Polarization Agility

Work has been done on the development of a low-sidelobe array that also has complete polarization agility. The low sidelobes are achieved by forming the array from low-cost, dual-slot radiators cut in the walls of square coaxial lines (see Figure 2.93 [144]). Complete polarization agility is achieved by having the even and odd rows of the antenna orthogonally polarized (alternately plus and minus 45° linear polarization) and using separate feed networks for the even and odd rows. By providing a network which can adjust the relative phase and amplitude into these two feed networks, arbitrary polarizations can be obtained. Figure 2.94 shows the azimuth sidelobe antenna pattern obtained for vertical polarization. Preliminary measurements of several experimental antennas indicate that sidelobe levels below 40 dB can be achieved over a 15% bandwidth.

2.6.2.2 Phase-Phase Steered Arrays

It is more difficult to achieve ultra-low sidelobes for an electronically scanned array than for an array that does not scan electronically, such as those discussed in Section 2.6.2.1. This is because the phase shifters added to the array to allow it to electronically scan (see Figure 2.95) can introduce considerable phase errors.

Figure 2.90 Thomson-CSF Mobile TRS-2215D Tactical Defense Radar Employing Low-Sidelobe-Level Planar Array. (Courtesy of Thomson-CSF.)

Table 2.17 Thomson-CSF TRS 2215D Planar Array System—Typical Values (from Drabowitch [141])

Frequency Band	15% S-Band
Structure	50 Horizontal Subarrays
Number of Feeds	60 Feeds per Array
Total	3000 Feeds
Dimensions:	
Width	4.80 m
Height	3.20 m
Feeding:	
Azimuth	Nondispersive
Elevation	Dispersive; 4-bit Phase-Shifters
Elevation Patterns	Monopulse (Sum and Difference)
Near Sidelobe Level	−30 dB
Outside Main Lobe Planes	<−70 dB
In Main Lobe Plane	Elevation −35 dB Azimuth −45 dB
Electronic Scan	−5° + 30°

For example, for a three-bit phase shifter, the least significant bit has a step size of 45°, which results in an rms phase error of $45°/\sqrt{12} = 13°$ due to the phase quantization error alone, not including any other errors, such as inaccuracies in the bit sizes.

The lower limit in antenna sidelobe gain resulting from random element amplitude and phase errors and element failures is approximately given by

$$G_S \doteq (\overline{\delta^2} + \overline{\Delta^2} + P_f)G_e \tag{2.1}$$

where δ is the element rms phase error in radians, Δ is the rms fractional amplitude error, P_f is the number of radiating elements not functioning, and G_e is the gain of a single array radiating element [7]. For $\delta = 13°$ (the rms phase quantization error of a three-bit phase shifter), $\Delta = 0$, $P_f = 0$, and $G_e = 6$ dB (a typical value for the gain of an array radiating element), $G_S = -6.9$ dB. This value results from only the three-bit phase shifter's quantization error, and does not include any other errors. Figure 2.96 gives a plot of G_S for the assumption that $G_E = 1$ and $P_f = 0$. The main-lobe gain of an array antenna is approximately given by

$$G_M \doteq \eta G_e N_e \tag{2.2}$$

where N_e is the number of elements in the array and η is the antenna efficiency, which includes the weighting loss. The antenna sidelobe level—the gain of the antenna sidelobes relative to the gain of the antenna's main lobe—is found from (2.1) and (2.2) to be given by

$$\text{SLL} = G_s/G_M \doteq (\overline{\delta^2} + \overline{\Delta^2} + P_f)/\eta N_e \tag{2.3}$$

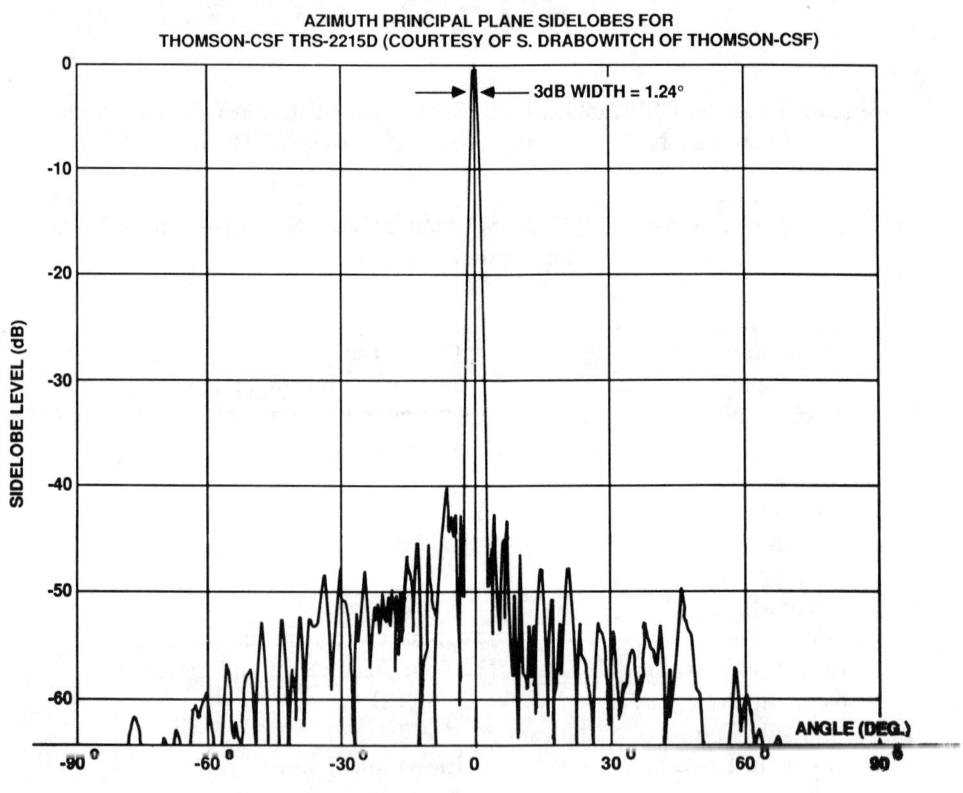

Figure 2.91 Azimuth Principal Plane Sidelobes for Thomson-CSF TRS-221D. (Courtesy of S. Drabowitch, Thomson-CSF.)

The array of Figure 2.95 is called a *passive* array. This is because it has phase shifters at only each element of the array, and no *power amplifiers* (PA) or receiver *low-noise amplifiers* (LNA). An array that includes the phase shifters shown and either power amplifiers or low-noise amplifiers, or both, as shown in Figure 2.97, has still greater difficulty in achieving ultra-low sidelobes than the passive array of Figure 2.95. This is because of the additional amplitude and phase errors introduced by the PAs or LNAs. An array such as that of Figure 2.97 is called an *active* array. The unit containing the phase shifter, PA, and LNA is the

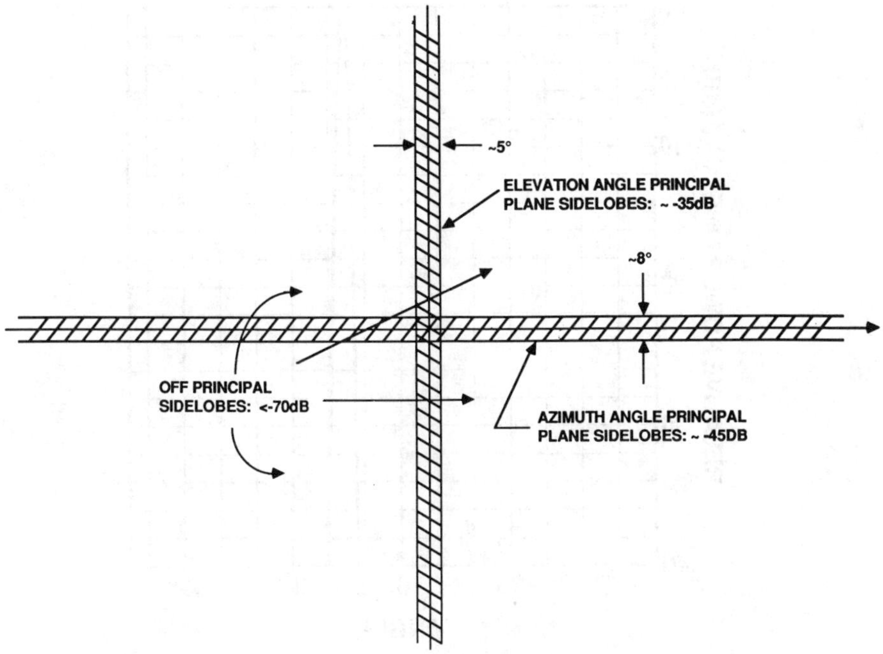

Figure 2.92 Sidelobes for Thomson-CSF TRS-2215D.

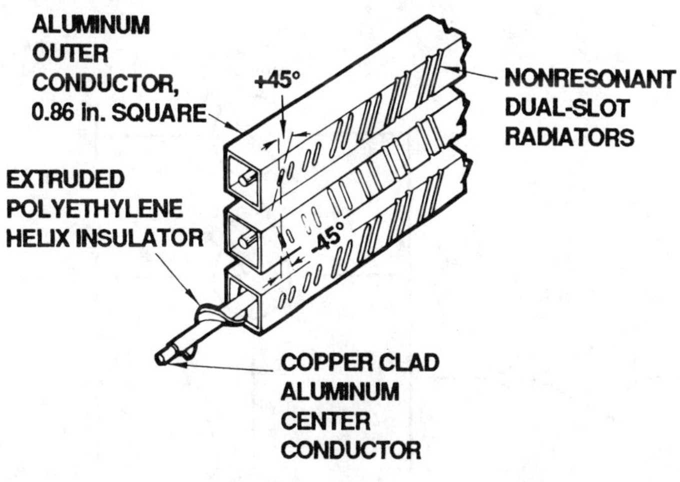

Figure 2.93 Construction of Dual-Slotted Square Coaxial Line Array (from Wolfson and Cho [144] © 1980 IEEE).

transceiver (T/R) module. Consideration is being given to real-time calibration and correction of the T/R module errors [147]. The PAVE PAWS radar antenna of Section 2.2.1.2 is an example of an active array, and the PATRIOT radar antenna of Section 2.2.1.5 is an example of a passive array.

Low-sidelobe technology has been applied to the Westinghouse X-band, phase-phase scanned, Electronically Agile Radar (EAR) [134] (see also Chapter 6). This array uses dielectrically filled circular-waveguide radiators with a dual-mode ferrite phase shifter [134]. The B-1 bomber's radar array (see Figures 2.15 and 2.16) evolved from the EAR array.

The problem of attaining low sidelobes in a tactical phase-phase steered array has been studied by Patton [148]. In this study the arrays were assumed to be subarrayed with column beam-forming being done first, followed by azimuth beam-forming. Patton indicated that better than 50 dB rms sidelobes can be achieved

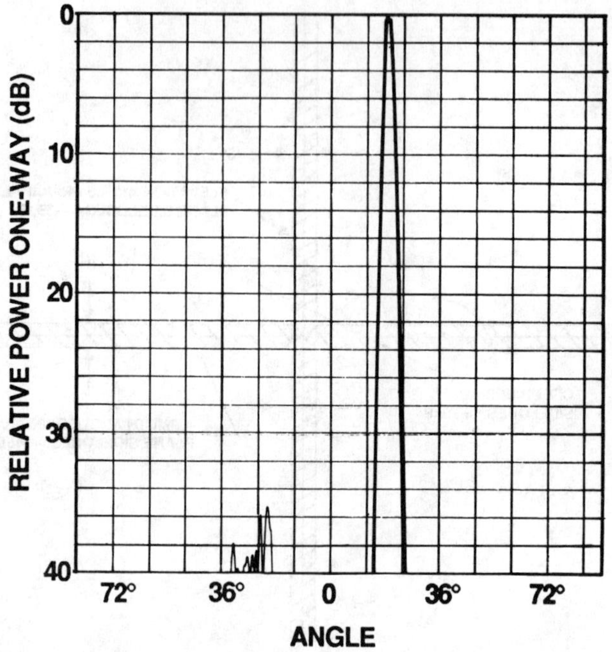

Figure 2.94 Azimuth Pattern for 16-Element Experimental Dual-Slotted Square Coaxial Line Array; Vertical Polarization (from Wolfson and Cho [144] © 1980 IEEE).

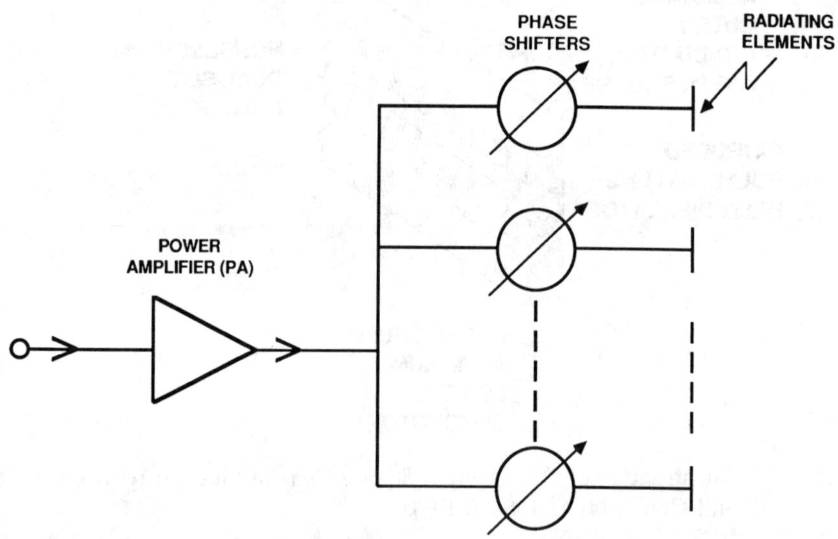

Figure 2.95 Passive Phase-Steered Array.

for this antenna, over all space, if the array errors were: element-to-element, 4.8% in amplitude and 11.0° in phase; subarray-to-subarray, 4% and 7.8°; column-to-column, 2.8% and 3.4°. Note that a five-bit phase shifter was assumed.

By the year 2000, we expect that phase-phase steered arrays having ultra-low sidelobes will have been developed. These arrays probably would use real-time calibration measurements of the array with built-in provisions for storing the calibration errors and correcting them so that ultra-low sidelobes could be achieved. For example, each phase shifter would have associated shift registers for storing the calibration data (as a function of phase shift command, frequency, beam-steering direction, and temperature). These error corrections would be added to the normal phase-steering commands.

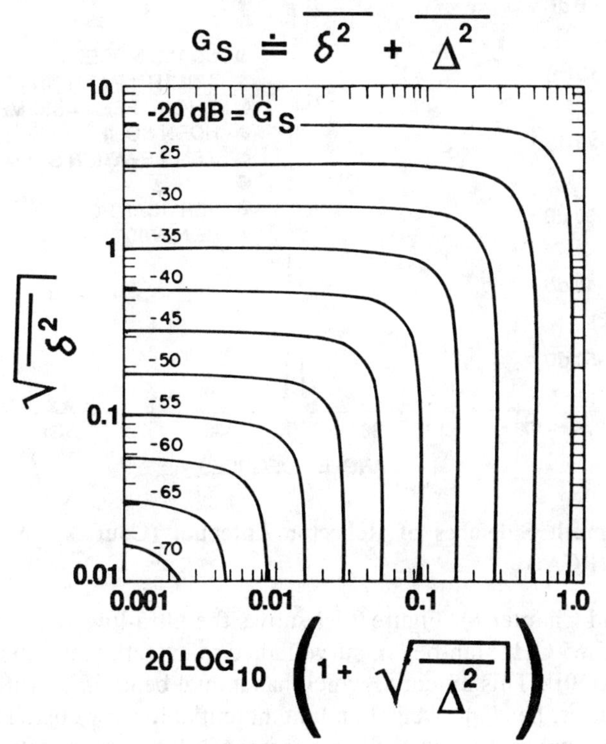

Figure 2.96 Array Antenna Sidelobe Gain due to rms Random Amplitude and Phase Errors (σ_A and σ_ϕ, Respectively); Element Gain Assumed Equal to Unity (after Mayhan [146]).

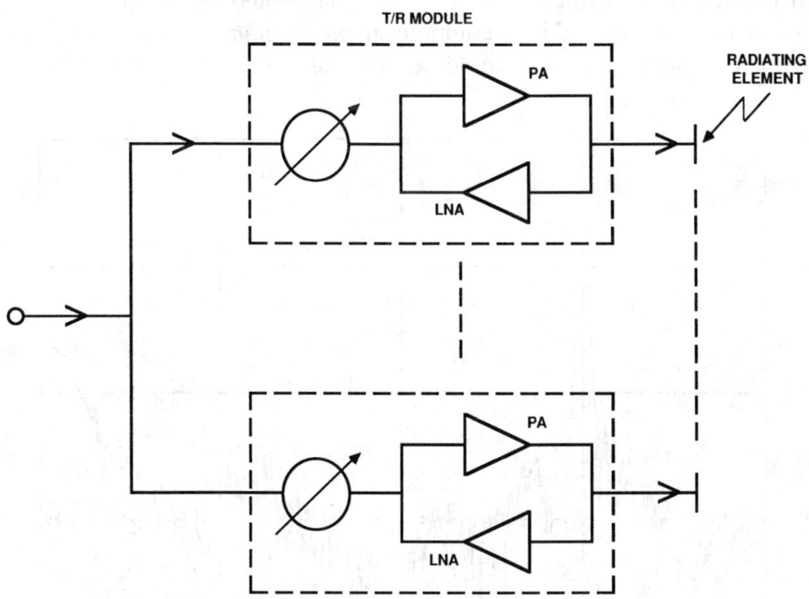

Figure 2.97 Active Phase-Steered Array.

2.6.2.3 Reflector Antennas

Considerable progress has been made in achieving ultra-low sidelobes by using reflector antennas. Figure 2.98 shows the superb azimuth sidelobes obtained with a scale model of a reflector, the measurements being made at 35 GHz ([149],

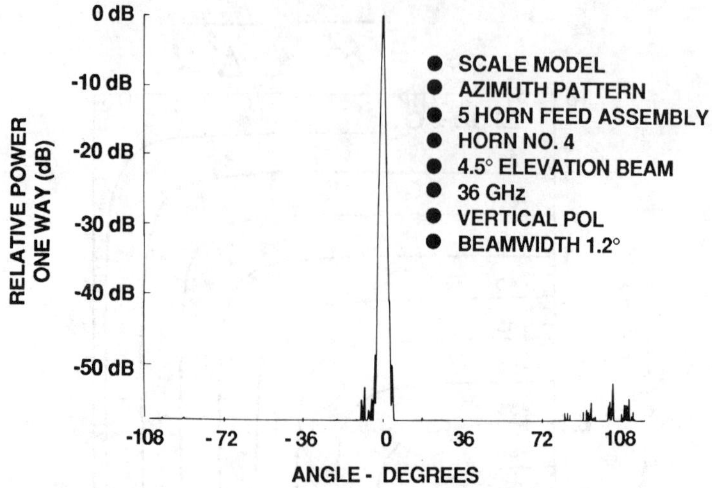

Figure 2.98 Azimuth Sidelobes of Reflector Antenna. (Courtesy of W. Patton, GE/RCA.)

see also [138] and Chapter 6). Figure 2.99 shows the ultra-low azimuth sidelobes obtained with a 5.65 GHz, fan-beam, surveillance radar reflector antenna (shown in Figure 2.100 [150]). This antenna, which has a dual beam in elevation to eliminate close-in clutter, has light weight and the potential for low cost. The reflector consists of a sandwich of two glass-fiber-reinforced plastic skins with an alumina honeycomb core [150,397]. The front skin reflective surface consists of an integrated metallic mesh. A trade-off between an aluminum and a carbon-fiber-reinforced plastic (CFRP) tube framework for support of the reflector showed that a reduction in weight of about 20 kg was obtained by using the CFRP framework. A mold is used to form the reflector. This mold would be machined with NC inputs from a tape generated by a computer using *computer-aided manufacturing* (CAM) for its design. The mold could be made of either wood or metal.

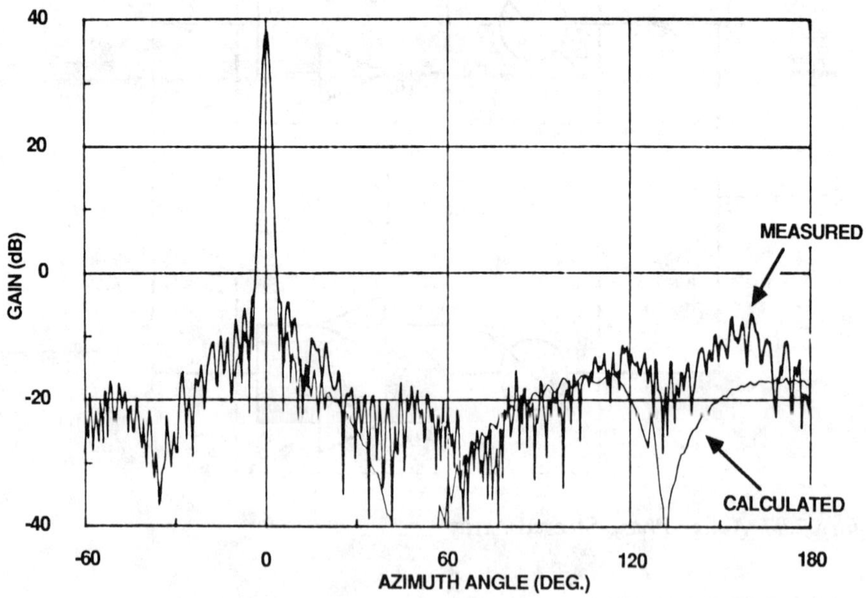

Figure 2.99 Calculated and Measured Azimuth Pattern of Dual-Beam Reflector Antenna (from Carlsson, *et al.* [150]).

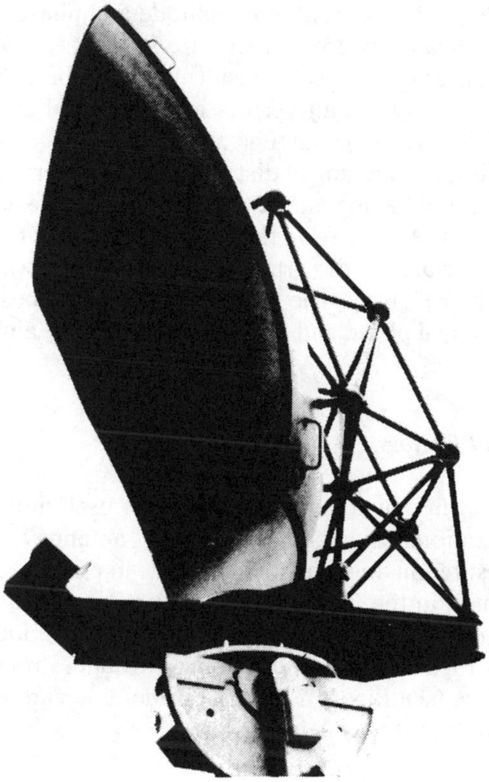

Figure 2.100 Dual-Beam 5.65 GHz Antenna (from Carlsson, *et al.* [150]).

ERA Technology, Ltd. and ASWE, both of the UK, also have designed and built a reflector having ultra-low sidelobes [151]. The elliptical antenna used an offset four-horn feed, with the offset chosen to avoid blockage. The four horns provided two beams in elevation had dual polarization. ERA and ASWE have built and tested a scale model of the antenna. They also use a construction which lends itself to low cost and light weight. Specifically, they use a half-inch honeycomb in a core with a carbon-fiber-cloth outer skin. The tolerances were ±0.15 mm. The reflector was formed from an aluminum mold made by using high-precision NC machining. More recently, ERA demonstrated through simulation the ability of obtaining −40 dB sidelobes (essentially, a 40-dB Taylor weighting) with a shaped Gregorian reflector [398]. These sidelobes were obtained in the azimuth, elevation, and diagonal planes for vertical and horizontal polarization over a 10% frequency band. The antenna efficiency was 60% (spillover, phase errors, and diffraction losses included). A good difference pattern was realized and the cross-polar pattern was −40 dB. A Gregorian was selected over a Cassegrain because the subreflector could be made larger without introducing aperture blockage, thus minimizing the diffraction effects of a smaller subreflector. The Gregorian also permits a wide feed-cone angle, which eases the design of a broadband primary feed. Finally, the Gregorian layout is more favorable for introducing RF absorbing shields around the subreflector to minimize spillover from the primary feed. Perhaps, even better sidelobes than demonstrated can be achieved.

For many communication systems, shrouds around reflector antennas and horn antennas have been effectively used to achieve very low sidelobes (see Chapter 6).

2.6.2.4 Errors

Equations (2.1) and (2.3) gave the array antenna's sidelobe gain and level relative to the main lobe due to random amplitude and phase errors and element failures. In addition to random errors there are systematic and mutual coupling errors. The systematic errors typically result in main-lobe or near-main-lobe errors. For example, if the systematic errors are sinusoidal and have a period of about one cycle or less across the antenna array, these errors will result in lobing errors such as main-lobe broadening or distortion. If the errors consist of sinusoidal variations across the array having one or more periods, these errors will introduce sidelobes at angles $\pm m\lambda/D$ away from the main-lobe peak, where m is the number of periods for this error across the array, λ is the antenna wavelength, and D is the size of the antenna [7,145]. The level of these sidelobes will be $\overline{\delta_{pk}^2}/2$ and $\overline{\Delta_{pk}^2}/2$ for a peak sinusoidal phase and amplitude error of δ_{pk} and Δ_{pk}, respectively, across the array.

2.6.2.5 Environmental Factors

The environment in which an antenna is to be used must be factored when determining its operational sidelobe levels. The antenna may have excellent sidelobes when measured in an ideal environment, like an anechoic chamber, but when actually operated in the field may have much poorer sidelobes. A number of factors can contribute to the sidelobe degradation. One factor is the use of a radome as shown in Figure 2.101. The radome introduces reflections, which can result in high sidelobes (see Figure 2.101). Considerable care must be used in the design of the radome to reduce such sidelobes.

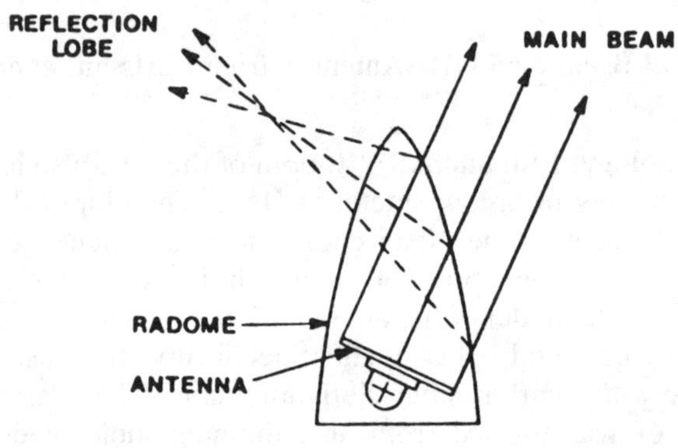

Figure 2.101 Generation of High Sidelobe due to Main-Lobe Reflection from Radome (from Roulston [152]).

Such things as the superstructure of a ship at sea can result in poorer sidelobes. Figure 2.102 depicts the situation where the ship's superstructure provides blockage for the antenna at certain viewing angles.

The subject of ultra-low sidelobe array antennas is discussed in greater detail in Chapter 6.

2.6.3 Microstrip Antennas [153,154]

Considerable progress has been made in the design and development of low-cost, lightweight, microstrip array antennas [153,154]. For such arrays, the radiating elements (patches), the corporate-feed network, and the phase shifter's microwave and dc power supply circuitry (not including the diode switches themselves) in some instances can all be simultaneously etched onto a single surface.

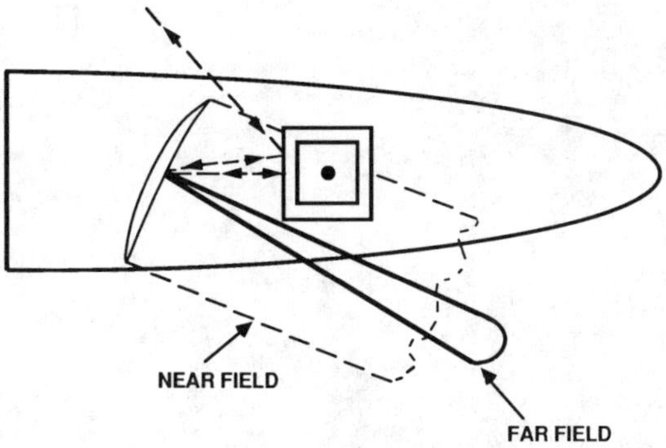

Figure 2.102 Sidelobe Degradation due to Scattering from Ship's Superstructure.

Such an array is called a *monolithic microstrip array*. Figure 2.103 shows one such microstrip antenna, used on an aircraft for communication to the ATS-6 satellite. This antenna illustrates the ability to make such arrays conformal to the aircraft body. The SEASAT system used a microstrip antenna because of its compactness and light weight, weighing only 1 lb/ft^2 and being only $\frac{1}{4}$-inch thick (see Figure 2.104 [153]).

Microstrip antennas will find increased use for fixed and transportable radar systems, where the bandwidth and power requirements permit their use.

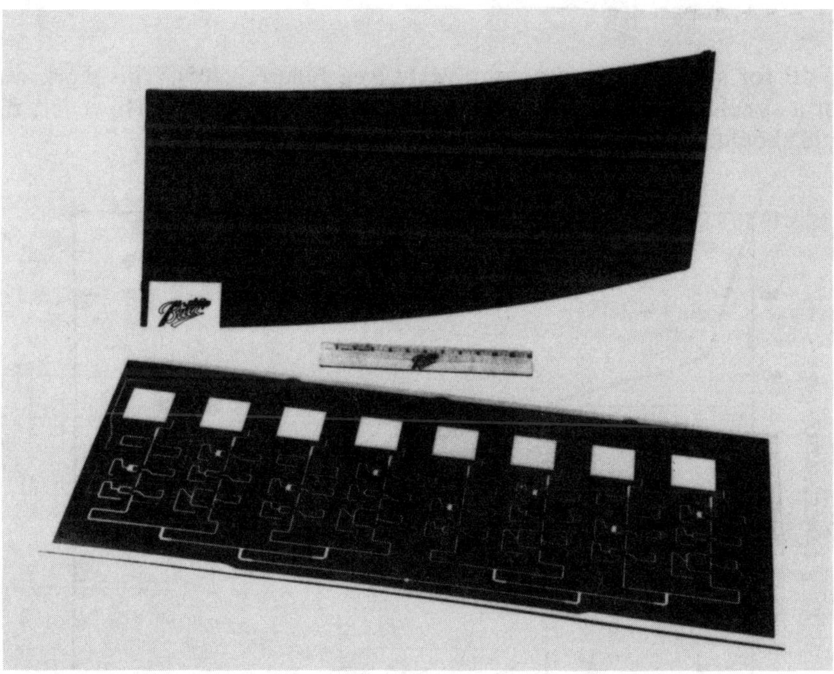

Figure 2.103 Conformal Monolithic Microstrip Antenna Used on an Aircraft. (Courtesy of Munson, Ball Aerospace Co.)

2.6.4 Adaptive Array Processing

Significant advances have been made in the area of adaptive array processing for eliminating sidelobe clutter and interference (either intentional or unintentional). A very impressive performance was achieved by Mayhan [155]. Using an adaptive array, he experimentally obtained cancelations of interfering signals of

Figure 2.104 SEASAT Microstrip Phased Array Antenna. (Courtesy of Munson, Ball Aerospace Co.)

over 50 dB for signal bandwidths of 5 MHz (see Figure 2.105). This work was done for a synchronous altitude communication antenna system. However, the technique should be applicable to radar systems.

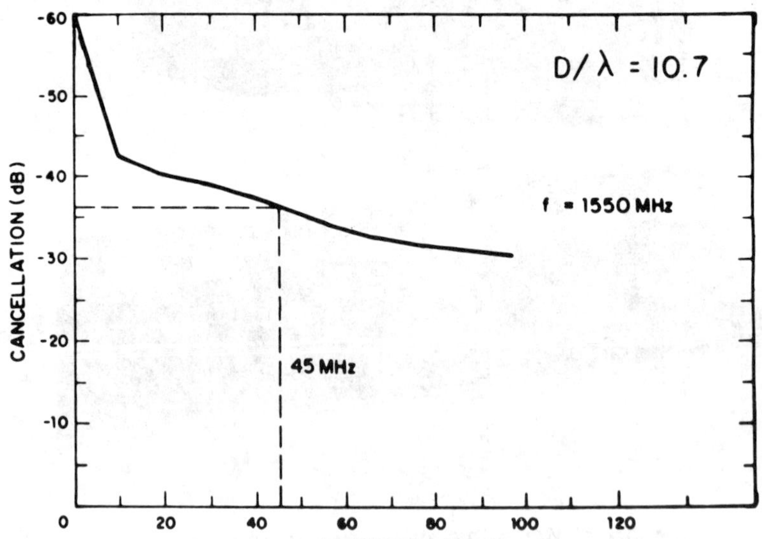

Figure 2.105 Cancelation Ratio *versus* Signal Bandwidth for 8-Element Adaptive Array Antenna System; Jammer at 4° in Azimuth and 2° in Elevation (after Mayhan [146]; see also [155] © 1979 IEEE).

A multichannel adaptive array radar processor was under development at Lockheed, which reportedly would provide 45 dB interference suppression [156]. Microwave integrated circuitry was to be used to ensure good phase tracking.

The technique of breaking the signal up into contiguous narrowband channels and applying sidelobe canceling to each of these channels will permit improved sidelobe cancelation in the future. Another technique that may find greater use in the future is the Gram-Schmidt orthogonalization technique, which provides good cancelation and a fast transient response. The increased signal processing throughput capability offered by VLSI/VHSIC/VHPIC will make these algorithms more feasible in the future. Systolic array processors may become feasible for adaptive nulling. This type of architecture lends itself to VLSI circuitry (see Section 2.6.6.9).

A new technique called *adaptive-adaptive array processing* has been developed, which provides the jammer-cancelation advantages of a fully adaptive array without its many disadvantages, such as excessive computations, poor sidelobes (see Figure 2.106) in directions other than the jammer locations, and poor transient response [157]. The advantages are achieved at the expense of hardware complexity. This technique involves transforming a large array of N elements into an equivalent small array of $J+1$ elements, where J is the number of jammers present (see Figure 2.107). The technique also involves estimating the number and locations of the jammers by a *discrete Fourier transform* (DFT) of the array element outputs, the use of standard *maximum entropy methods* (MEM), other super-resolution techniques, or just searching out the sky with the array main beam. Once the number and locations of the jammers have been determined, beams are formed in the direction of the jammers by using the whole array. The outputs of these jammer beams, together with the output of the main signal beam, now consist of $J + 1$ ports, instead of N ports. The standard *sample matrix inversion* (SMI) or the Applebaum algorithm can be applied to the $J+1$ ports of the equivalent adaptive-adaptive array. Whereas N may be very large (e.g., 10,000 for large arrays), $J + 1$ for the equivalent array can be very small. For example, if there are only 10 jammers, $J + 1 = 11$. The total number of multiplications needed for adaptive array processing reduces from something on the order of $2N^3 = 2(10{,}000^3) = 2 \times 10^{12}$ to approximately 10^5, a reduction in the computation complexity by seven orders of magnitude. In addition, the settling time for the adaptive-adaptive array is much faster. For the above example, the settling time for the full array is about 20,000 samples, whereas it is only 22 time samples for the adaptive-adaptive array, an improvement of three orders of magnitude.

2.6.5 Solid-State Components

There has been a strong trend toward radars using solid-state transmitters, for 3D, 2D, and 1D electronically scanned arrays and for mechanically scanned 2D radars, which mechanically rotate a vertically oriented fan beam 360° in azimuth about the local vertical. It is expected that this trend will continue in the future. Solid-state transmitters and radars offer the advantages of better system reliability and easier maintenance and repair. Better reliability results because the failure of a number of transistors still allows the system to function without a significant reduction in performance. Easier maintenance results from the low voltages involved and the typically small size of the solid-state modules to be replaced or repaired. The trend toward solid-state systems is exhibited by many of the new radars that are developed or under development, which were discussed in the previous sections of this chapter. Examples of these were the PAVE PAWS and BMEWS phase-phase steered array radars, Marconi-Martello S723 stacked-beam radar, RAMP 2D radar, HF ROTHR, MESAR, W2000, AN/TPS-117, AN/TPS-73, SEASAT, and airborne "ultra-reliable radar."

Two types of solid-state components are in use, *discrete* and *monolithic*. Discrete technology, which is what has been used to date for solid-state radars, consists of microwave components, composed of single transistors, diodes, resistors, capacitors, and inductors, with no integrated circuitry. The T/R module

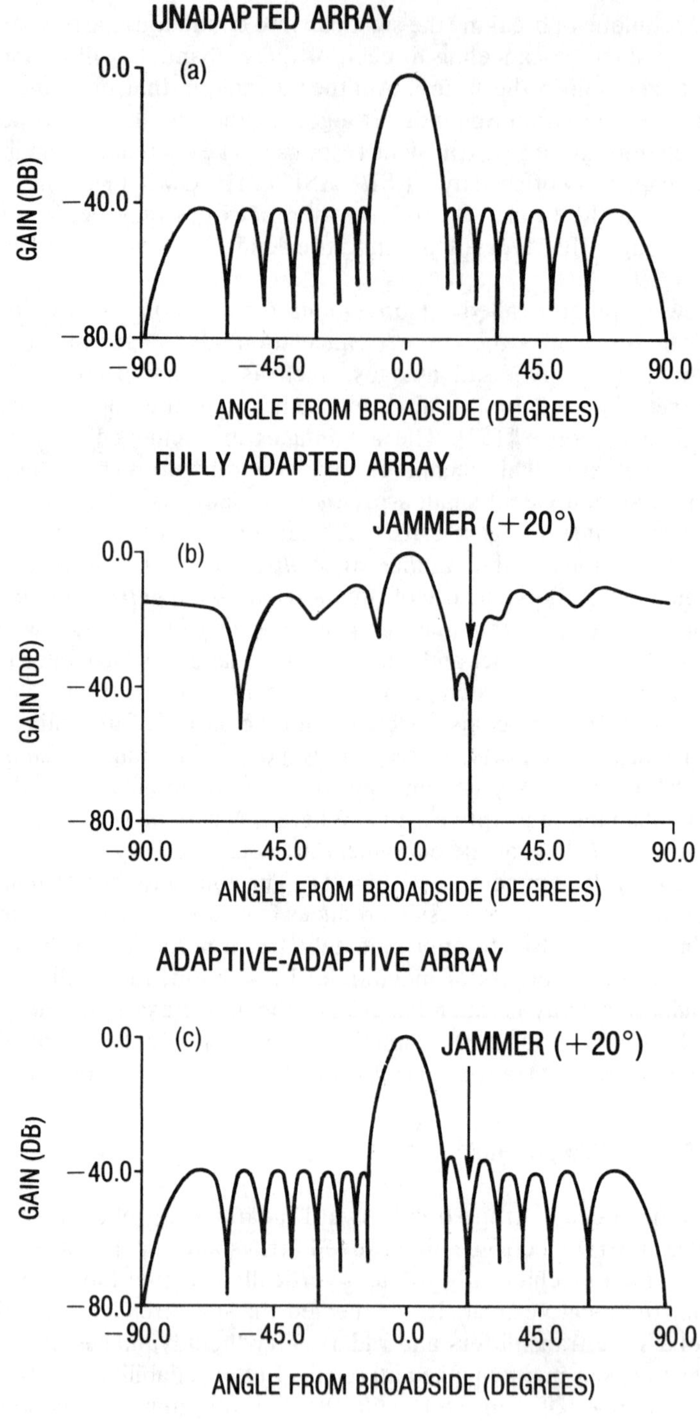

Figure 2.106 N = **16-Element Array with 40-dB Antenna Sidelobes (Chebyschev Weighting). Jammer at 20° (Peak of Second Sidelobe; from Brookner and Howell [157]).**

using this technology typically consists of hundreds of parts. Monolithic circuitry (MMIC) consists of microwave integrated circuits, and as a result a T/R module using this technology will consist of a few parts. MMIC is the technology of the future. Discrete technology, however, will still be needed in the future for the high-power circuits, typically transistors providing outputs of 10 W or more. Section 2.6.5.1, which follows, will elaborate on the recent developments in discrete solid-state technology, while Section 2.6.5.2 will cover the current advances in MMIC solid-state technology.

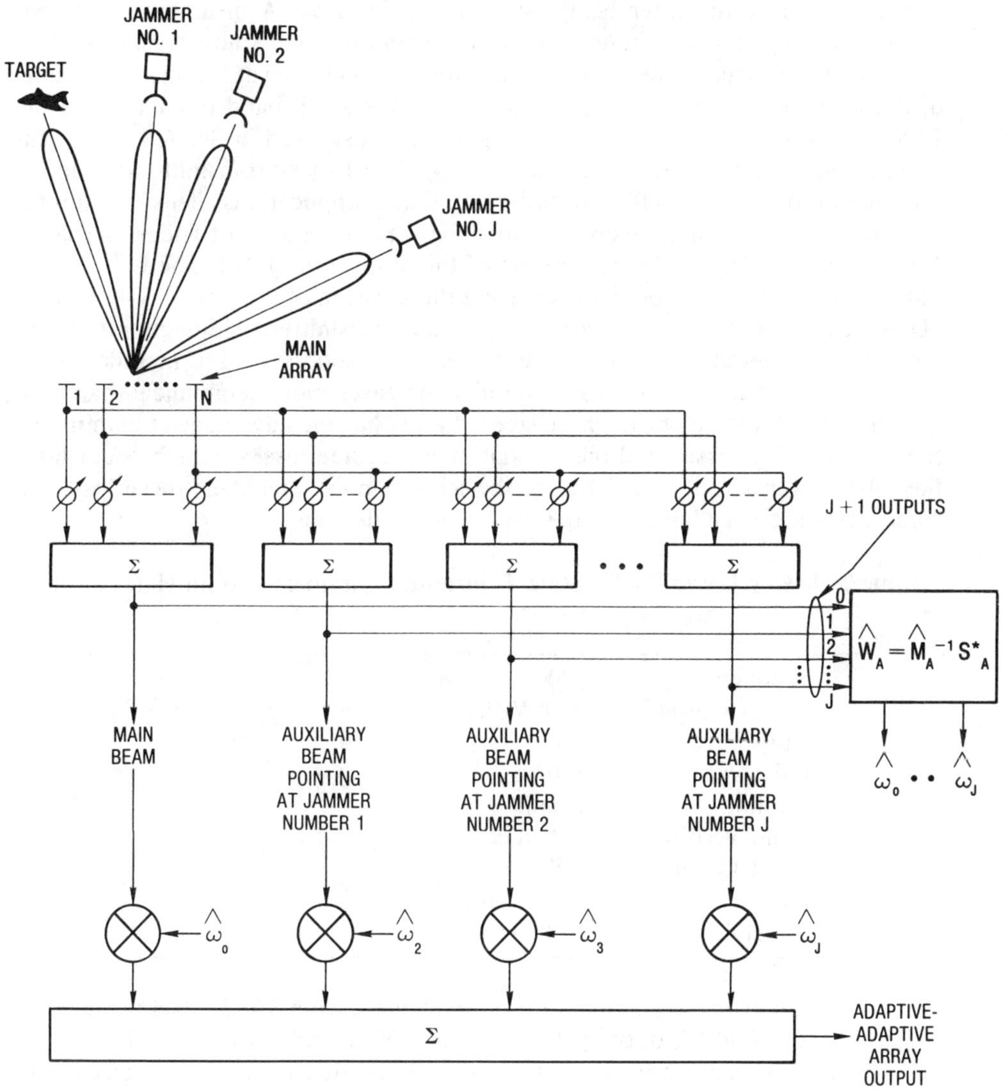

Figure 2.107 Adaptive-Adaptive Array Configuration; *N*-Element Array Is Reduced to *J* Plus One-Element Array, where *J* Equals Number of Jammers (from Brookner and Howell [157]).

2.6.5.1 Discrete Technology

Current advances in UHF T/R modules for active phased arrays are well illustrated by the PAVE PAWS radar, which uses thousands of 336 W peak-power modules. There is a trend also to replace the tube transmitters of older radars with all-solid-state transmitters. One dramatic example is, as mentioned before, the construction by Westinghouse of a 250 kW peak-power-output solid-state transmitter to replace the existing klystron tube transmitter of the Navy AN/SPS-40 surveillance radar. The solid-state replacement transmitter combines one hundred twelve 2500 W peak-power modules in parallel to produce the 250 kW peak-power output. These 2500 W peak-power modules use eight 400 W peak-power transistors in parallel in the output stage. The transmitter occupies three racks, has an instantaneous bandwidth of 50 MHz (400 to 450 MHz), and an average power of 4 kW [158]. These solid-state replacements for tube transmitters are called *solid-state "bottle" transmitters*. The Navy UHF AN/SPS-40 with the 250 kW solid-state bottle transmitter replacement has successfully undergone six months of sea trials, with the transmitter being 100% available [159]. A mean-time-between-failure (MTBF) of 10,000 hours is expected with this solid-state transmitter.

The bottle transmitter replacements are also occurring at L-band, at the tens of kilowatts level. An example of a bottle solid-state L-band transmitter is the 20.8 kW transmitter used in Raytheon's RAMP, discussed in Section 2.3.1. Another example is the M/A-COM 80 kW peak-power L-band transmitter developed for shipboard and air traffic control surveillance applications. This transmitter was originally developed to replace a tube for one of the radars of The Netherlands Navy. Table 2.18 gives the parameters of this transmitter [160]. The US Navy is now looking at the possibility of replacing the klystron of the AN/SPS-49 L-band 2D search radar (Raytheon) with a solid-state transmitter. Although an L-band phase-phase steered array has not been built using solid-state T/R modules, such systems may become competitive and in some cases more economical than tube systems. A solid-state phase-phase steered array has the advantage of eliminating the phase-shifter losses and the coaxial or waveguide losses, which occur after the tube power amplifier and before the RF low-noise amplifier; these losses in some cases are very large, up to 3 dB or more one-way.

Table 2.18 Key L-Band Solid-State Transmitter Parameters (from Hoft [160])

Frequency	1200–1415 MHz
Power Output	80 kW Peak
Duty Factor	7%
Pulsewidth	130 μs
Cooling	Liquid
Load VSWR	3:1 max.
Input Level	5 W
Pulse Droop	0.6 dB max.
Reliability	10,000 hr with Graceful Degradation

Bipolar power transistors are now available for the 3.1–3.5 GHz band with a peak power of 50 W, duty cycle of 5%, and 50 μs pulsewidth; for the 2.9–3.1 GHz band with a peak power of 100 W, duty cycle of 1%, and 10 μs pulsewidth; for 2.7–2.9 GHz band with a peak power of 100 W, duty cycle of 10%, and pulsewidth of 100 μs [161]. Eighteen-watt field-effect transistors (FETs) are available at C-band (5.9–6.4 GHz) [359]. Solid state bottle transmitters are being developed at S-band, the 10 kW AN/TPS-73 being a good example of one; see Section 2.3.1 and Table 2A.1(a).

The USAF Rome Air Development Center at Hanscom Air Force Base, Massachusetts, has funded development of a 64-element, active-aperture, S-band radar array as the precursor to the successor of the E-3A [162]. Ericsson Radio of Sweden is developing an AEW S-band, solid-state, phased array radar for Fairchild's small Metro 3 commuter aircraft [10,431,432]. The developers claim it will cost one-tenth of the more than $100 million for an AWACS, with probably 50 to 75% of its capability. Furthermore, it is claimed that operating cost for the Metro is $300 hr to $400/hr *versus* $6000/hr for the AWACS. It is intended initially for use as a highly mobile gap-filler radar with the command and control elements on the ground. At a later date, an autonomous version would be developed. The antenna is a 9.51-m long by 0.79-m high planar array, mounted vertically above the fuselage along the length of the aircraft. This antenna would use about 200 T/R modules. Its range would be about 160 nmi on a fighter-sized target. S-band was selected because it was thought to provide the best performance in an intense EW environment.

Building an efficient, high-power, solid-state amplifier at X-band for the development of an airborne phased array is now possible. A 9–10 GHz, GaAs

FET amplifier has been developed for the NRL, providing 4 W, 20% efficiency (power added), and 24 dB gain [163,164]. NRL had awarded Texas Instruments (TI) a contract to produce (by 1984) 50 complete X-band T/R modules having 3.5 W peak power, expected noise figure of about 4.3 dB, and module efficiency of about 11%, although 15% is believed possible. The bottom of Figure 2.108 shows the TI X-band, 3.5 W, airborne array radar T/R module. The top module in the figure is the TI MERA module, while shown in the middle is the RASSR dual module [1].

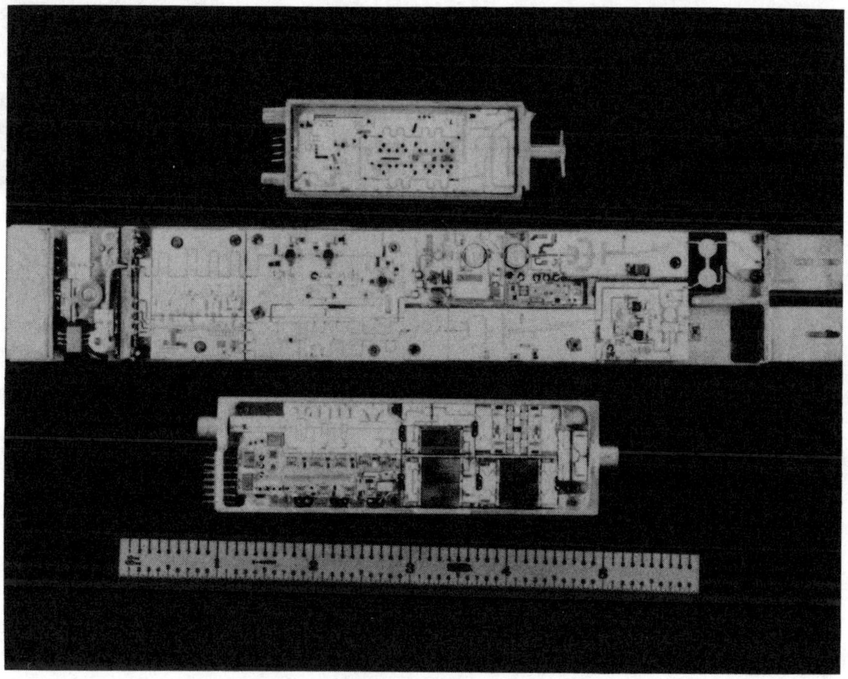

Figure 2.108 Starting from the Top, Texas Instruments MERA, RASSR, and Recent NRL 3.5-W Peak Power Module. (Courtesy of Texas Instruments.)

Hughes Aircraft Corporation has completed a contract from Wright-Patterson Avionics Laboratory to produce about 100 complete X-band T/R modules intended for an airborne array radar. Hughes expected these modules to have 2.4 W peak power with an overall module efficiency of 12 to 16%. The MTBF goal is 60,000 hours [165]. The module arrangement in an airborne phased array is illustrated in Figure 2.109. The Avionics Laboratory awarded TI a contract for the construction of an X-band array antenna, using solid-state, T/R modules, having about 2 W peak power [165,385]. The module noise figure is 4.6 dB, the rms phase errors are 4° and 1.5° on transmission and reception, respectively. Over 1000 of these modules were built. With the 60,000-hour MTBF per module, such an array would lose only 2% of its modules after 1000 hours of use [165]. This array, called the *solid-state phased array* (SSPA), is to be used for the "ultra-reliable radar" discussed in Section 2.3.3. (See also Chapter 4, Section 4.4.)

As mentioned in Section 2.2.4.3, Raytheon is building a Relocatable Over-the-Horizon Radar (ROTHR, which uses a solid-state transmitter). M/A-COM has indicated that there are plans in several countries for long-range OTH surveillance radars, and they have developed a 20 kW solid-state transmitter for these radars [160]. These modern OTH systems would typically use transmitters with 200 kW CW or more of power delivered to an array of 10 to 20 antenna radiators.

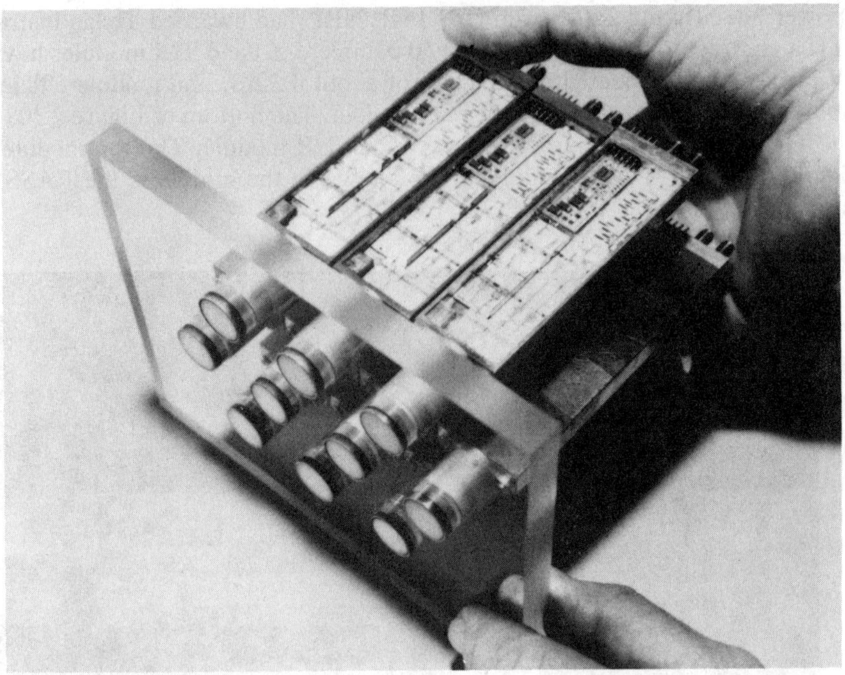

Figure 2.109 Hughes Aircraft Modules as They Would Be Arranged in an Airborne Phased Array. (Courtesy of Hughes Aircraft Corp.)

2.6.5.2 Monolithic Microwave Integrated Circuitry (MMIC)

MMIC is to microwave circuitry what VLSI, VHSIC, and VHPIC are to digital circuitry. MMIC is the development of integrated circuitry for microwave circuits; that is, the placement on one chip of many transistors, diodes, resistors, capacitors, and inductors to develop one or more PAs, LNAs, phase shifters, and T/R switches. Such MMIC chips can be combined to form a T/R module. Just a few such chips are needed, instead of hundreds, as previously required for hybrid T/R modules. In some applications, it may be possible to form the T/R module with a single MMIC chip. The reduction in the number of parts required for a single T/R module by one or more orders of magnitude will lead to a significant reduction in the cost of T/R modules, and in their size and weight. The goal is for a $500 T/R module [367]. In contrast, the hybrid, X-band, 2.4 W T/R module that TI has built for the "ultra-reliable radar" (see Sections 2.6.5.1 and 2.3.3) is projected to cost between $1500 and $2000 each, in quantities of millions [385]. This is based on actual cost experience gained with a few thousand modules. The reduction in the number of parts would also result in a significant increase in reliability [365]. These improvements are important for ground-based transportable and fixed radar systems as well as airborne and space systems.

Considerable progress has been made in the area of MMIC. Figure 2.110 shows a 10 W peak-power, S-band, GaAs T/R module, comprised of six MMIC chips (built by Raytheon) [166]. Figures 2.111, 2.112, and 2.113 show three L-band MMIC chips that can be used to build an L-band T/R module [166,167]. These consist of a 1 W PA chip, LNA chip, and five-bit phase shifter chip (all built by Raytheon). Raytheon has indicated that the L-band PA and LNA can be put on the same chip, with the LNA having 3 dB noise figure and 25 dB gain, and the three-stage PA providing a 1 W peak-power output with a gain of 33 dB and 16% power efficiency [168].

Texas Instruments has developed an X-band, single-chip, monolithic GaAs T/R module that incorporates a four-bit phase shifter, LNA and PA [169,366].

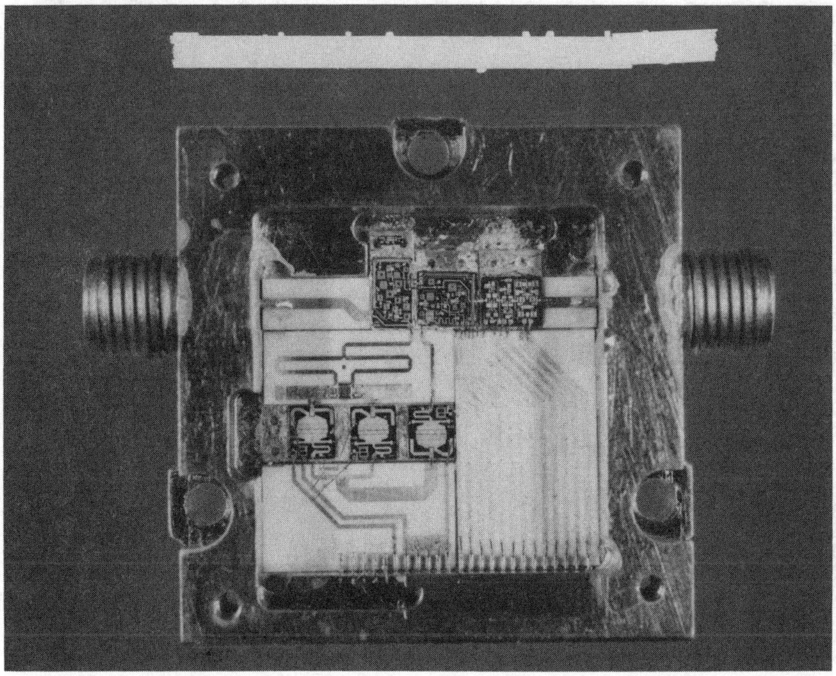

Figure 2.110 MMIC S-Band GaAs 10-W Transceiver Module. (Courtesy of Raytheon Co.)

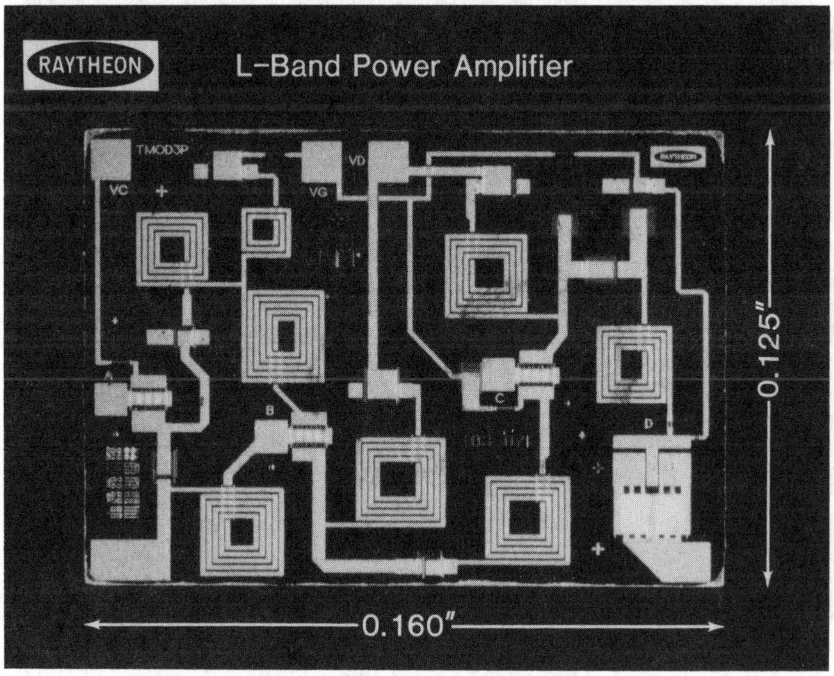

Figure 2.111 Monolithic L-Band GaAs Power Amplifier. (Courtesy of Raytheon Co.)

The module provides an output power of 0.5 W [169,366]. A 12.5% efficiency and 5.5 dB noise figure were realized on their best chip [170,366]. This module has potential applications for a space-based radar [169]. Plessey (UK) announced plans to build about 100 multi-MMIC chip, S-band, T/R radar modules, intended for a phased array radar [171]. The module is expected to have a noise figure of about

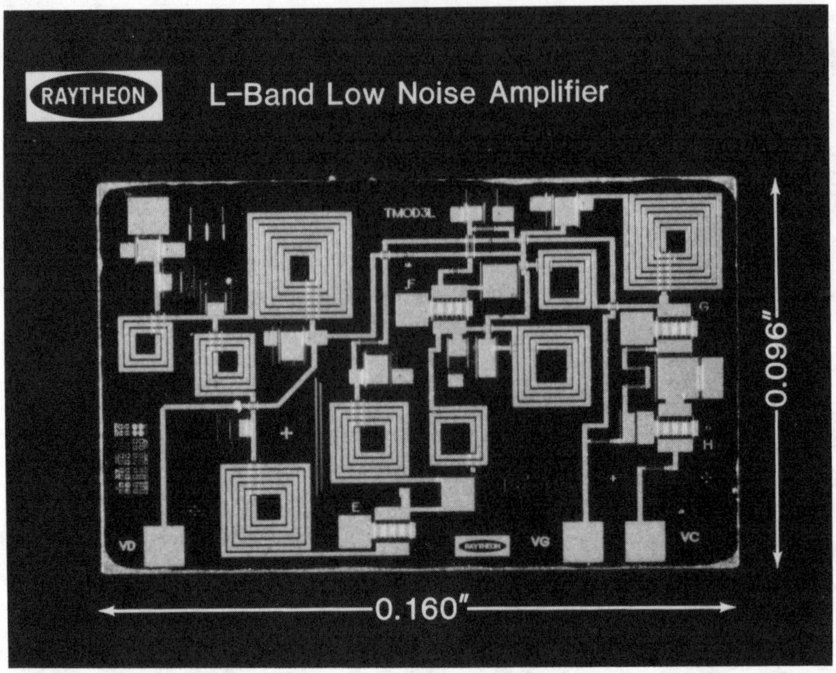

Figure 2.112 L-Band Low-Noise Amplifier. (Courtesy of Raytheon Co.)

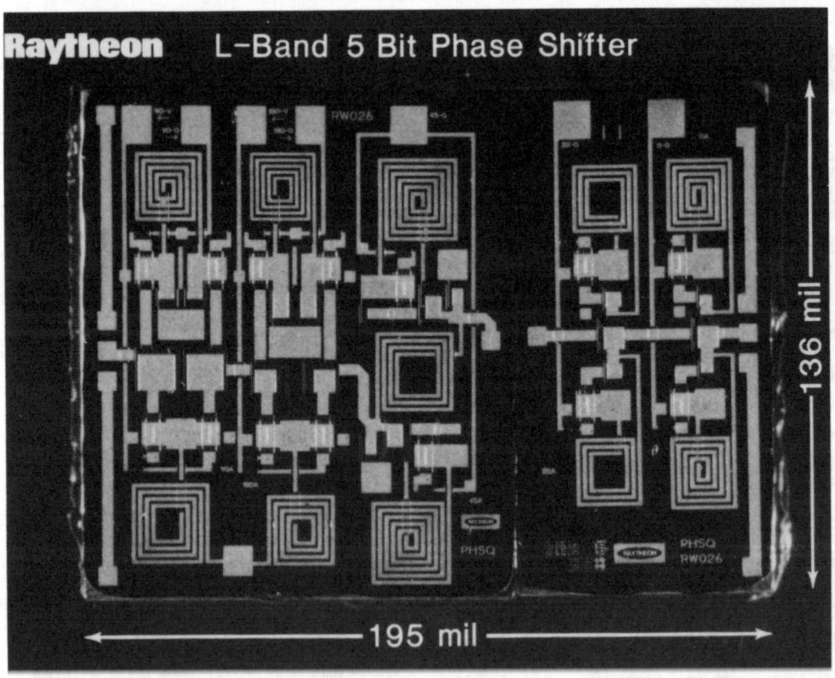

Figure 2.113 Monolithic L-Band GaAs 5-Bit Phase Shifter. (Courtesy of Raytheon Co.)

3.5 dB with a receiver gain of 15 dB, a power output of 2.5 W with a 28 dB transmitter gain. Plessey claims a yield of 30 to 40%, peaking at 60% on two-inch-diameter GaAs wafers [172]. By 1990, Plessey expects the yield to be 85%, and will be using three-inch wafers. General Electric Company is developing a dual-frequency, wideband, T/R module for the US Air Force (Rome Air Development Center) [173]. This module, intended for phase-array radar applications, would operate at S-band (3 GHz) and C-band (5 GHz) with a peak output power of 5 W. The module is to provide a 40% bandwidth at C-band so that wideband frequency agility can be used to frustrate enemy jammers. Raytheon has developed a 6–18 GHz MMIC amplifier providing 3.1 W [174]. Raytheon has also developed a phase shifter consisting of two MMIC chips that provide 360° of continuous phase shift and more than 20 dB of gain control [368]. Texas Instruments has developed a broadband, MMIC, 0.0032-in^2 chip, which covers the band from 0.1 to 2.0 GHz with an exceptional gain flatness of ±0.25 dB, 18 dB gain, and 2.4 dB noise figure [175]. The 3-dB band is from 0.1 to 3.5 GHz, over five octaves. The output power at the 1-dB gain compression is +15 dBm. This chip will find applications at IF as well as RF. COMSAT has reported a K_a-band (28 GHz) MMIC amplifier providing 0.5 W [175].

The first (to the author's knowledge) phased array using MMIC chips has been built by Raytheon [176–178]. This is a 128-element, SHF (7.25–7.75 GHz), phase steered array consisting of receive-only modules in a conformal microstrip antenna; see Section 2.6.3 for a discussion of microstrip antennas. The antenna is intended for use on board an aircraft for reception from a satellite. The receive-only modules consist of three MMIC chips—a limiter, LNA, and three-bit phase shifter. Figure 2.114 shows the receiver module, Figure 2.115 shows the 128-element conformal patch radiating element array, and Figure 2.116 shows half of the corporate combiner. Table 2.19 summarizes the parameters and yield of the MMIC chips.

Figure 2.114 SHF (7.25–7.75 GHz) Receive-Only Module Made of GaAs MMIC Chips. (Courtesy of Raytheon; see also [176–178].)

The US Department of Defense (DoD) originally approved $135 million for a GaAs MMIC program [179,180,181], now programmed for in excess of $500 million [182,391]. This program, called the Microwave/Millimeter-Wave Monolithic Integrated Circuits (MIMIC) is expected to do for analog GaAs MMICs what VHSIC has done for silicon VLSI circuitry. One application intended for the program is a phase-phase steered array radar using solid-state MMIC T/R modules. A contract may be let for the fabrication of a subarray consisting of hundreds of modules, intended for the Advanced Tactical Radar (ATR), which

Figure 2.115 SHF (7.25–7.75 GHz) 128-Element Microstrip Receiving Antenna. (Courtesy of Raytheon; see also [176–178].)

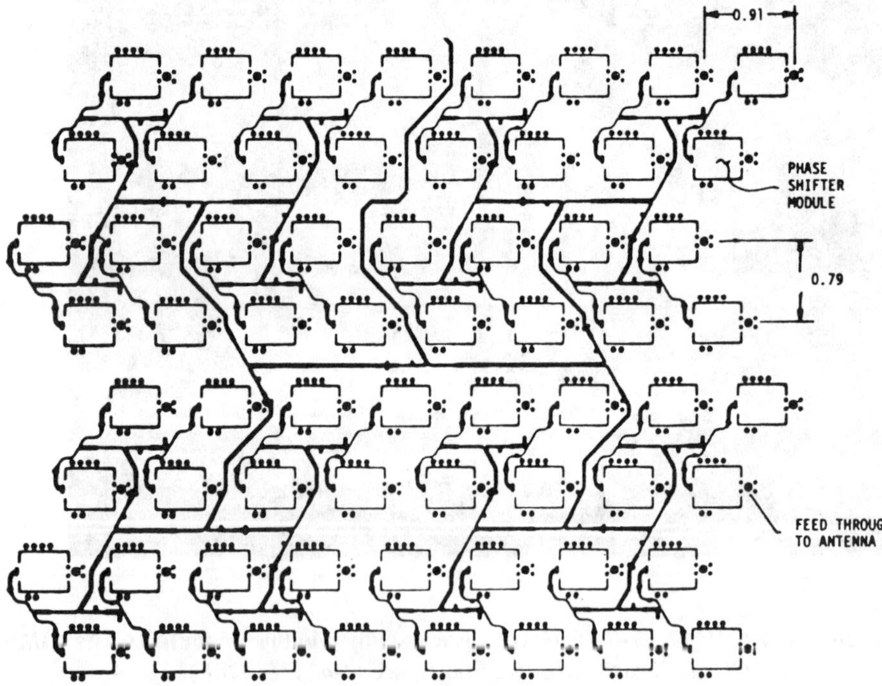

Figure 2.116 64:1 Corporate Combiner (from Considine [178]).

may in turn consist of thousands of modules. This effort will establish the eventual price of such modules.

The following statement made by Dr. Robert Pucel in the preface of his published anthology gives some feel for how fast things have moved and are expected to move ([183]; see also [184]):

It is my belief that the influence of MMICs on the design of future microwave systems and on consumer applications of microwave circuits will be so profound that we are not yet able to fully appreciate it. From a modest beginning only six years ago, with circuits designed for the S to X bands, the range of applicability of GaAs MMICs has been extended to frequencies as low as 50 MHz, and as high as 100 GHz.

By the end of the century, we expect that array radars using T/R modules comprised of one or more MMIC chips will have been developed and will be in or near production.

Table 2.19 SHF (7.25–7.75 GHz) MMIC Chip Parameter and Yield [176–178]

Chip	*Parameters*	*Yield (%)*
Limiter	0.5 dB max insertion loss, VSWR 1.24:1, max input 10W CW	—
LNA	Gain 7 dB, noise figure 3.5 dB, 1 μm FET gate	19
3 bit phase shifter	Accuracy $\pm 5°$ per bit, loss 3.8 dB − 4.4 dB	16*
Total module	VSWR $\leq$ 2:1, insertion loss 0 $\pm$ 1 dB, phase error per bit $\leq$ 10°	—

*Yield per bit.

2.6.5.3 GaAs HEMT

A technology that bears close watching for both analog and digital circuitry is HEMT (high electron mobility transistor), which is also called MODFET (modulation-doped FET) by The University of Illinois and Rockwell International, TEGFET (two-dimension electron GaAs FET) by Thomson-CSF, and SDHT (selectively doped heterojunction transistor) by Bell Labs [185–193]. Gould, Inc. has come out with a commercially available HEMT LNA that challenges GaAs FETs [192]. The HEMT LNA has a lower noise figure than a comparable GaAs FET. For example, its minimum noise figure at 8 GHz is 0.8 dB with a gain of 12 dB, while the comparable GaAs FET has a minimum noise figure of 1.4 dB and associated gain of 11.5 dB. Computer predictions indicate that a 2 dB noise figure at 60 GHz is achievable with an associated gain of 10 dB [185]. General Electric has produced a HEMT device that achieves a 2.1 dB noise figure at 40 GHz with an associated gain of 7 dB [186]. When this device is cooled to 12 K, the noise figure was measured to be 0.05 dB at 3.3 GHz. This device is to be used by the JPL's radio telescopes for receiving signals transmitted to the earth from deep-space exploration vehicles. The telescopes now use masers, which are bulky, expensive, and difficult to cool adequately. Varian has developed a 2–20 GHz HEMT distributed amplifier, having a 3 dB noise figure at 10 GHz, rising to 5 dB at the band edges [194]. This represents improvement by a factor of two in noise figure over MESFET, with a factor of four improved stage gain.

These devices also have potential as PAs. A power amplification efficiency approaching 40% in Class-A operation has been achieved [185]. At 30 GHz, a power-added efficiency of 26% has been achieved, making HEMT a strong contender for space-based systems [189].

2.6.6 Digital Technology

The radar signal processing of the 1990s will be primarily digital. Digital processing clearly has the economic momentum and the technical potential to achieve an increasing predominance and ultimately to perform almost all radar processing functions. At the same time, we will see a radar system with more intelligence, versatility, reliability, stability, dynamic range, and performance than the systems of the 1980s.

The future increase in the use of digital *programmable signal processors* (PSPs) will enable the same signal processor to be used for many different radars (as was done in the past with a signal processor used for a pulsed doppler processing radar and a CW radar). This, in turn, will lead to a reduction in field maintenance, logistics, and life-cycle costs for the radar system; this is especially true for distributed signal processors, consisting of many, identical processing elements. Scheduling risk for developing a signal processor for a new radar would in many cases be removed, since an existing programmable signal processor could be used. Such signal processors would also prevent their early obsolescence because they could be programmed to allow for the use of new, improved waveforms, and the signal processors could be programmed for improved signal processing algorithms as they are developed.

2.6.6.1 Effect of VLSI, VHSIC and VHPIC on Radar Processors and Data Processors [195,196]

Very high speed integrated circuits (VHSIC) technology is the militarized version of very large scale integration (VLSI) technology. Very high performance integrated circuitry (VHPIC) is the UK's version of VHSIC. Table 2.20 gives some typical radar signal processors [195,196]. Figure 2.117 dramatically indicates the effect that the *VHSIC Phase 1* program, which employs 1.25 μm features sizes, is expected to eventually have on the F-18 airborne signal processor (Hughes) of Table 2.20. The figure shows that an increase of about two orders of magnitude in throughput can be achieved with a VHSIC Phase 1 signal processor of the same size and prime power consumption.

The trend toward the VHSIC Phase 1 performance of Figure 2.117 is demonstrated by the replacement (in 1987) of the F-15 APG-63 radar processor of Table 2.20 with a new PSP having 3 to 5 times the speed and 10 times the memory [197]. The radar is redesignated the APG-70. It provides a high-resolution ground map with 8.5-ft resolution at 20 nmi. This new F-15 signal processor, which weighs only 65 pounds, has the capability of 200,000 desk-top personal computers, which would weigh roughly 3.4 million pounds without any add-on memory cards, disk drives, or monitors [369]. The new F-16 APG-68 PSP incorporating VHSIC chips is expected to provide a throughput of 40 million complex operations per second (MCOPS),* more than double the 16 MCOPS of the non-VHSIC one, in less than half the size, and for about one-third the power (800 W *versus* 2000 W) [198].

Data processors will also be affected by VLSI circuitry. This will be exhibited by the increased use of small microprocessors to do tasks that are normally done by large computers [195,196]. For further details on the effects of VLSI and

---

*A complex operation as used here is a complex radix-two-butterfly, which consists of one complex multiplication and two complex additions. This, in turn, is equivalent to four real multiplications and six real additions. Hence, for every complex operation, there are 10 real operations. The number N_c of complex radix-two butterflies to do an N-point complex fast Fourier transform (FFT) is $N_c = (N/2) \log_2 N$. Even if the FFT were done by the DSP using some other radix FFT (e.g., four), the throughput is measured in terms of the equivalent number of complex radix-two butterflies needed to do the computation. This provides a common measure for comparing the throughput capability of different DSPs when doing the same complex FFT.

Table 2.20 Radar Signal Processors* (after Brookner [196])

	Architecture	Throughput (MCOPS)**	Chip Technology	Applications	Comments
Current Radar Programmable Signal Processors (PSP)					
PSP (Raytheon, Wayland, MA)	Two arithmetic elements (AE) in parallel. One for FFT has pipeline butterfly; one for non-FFT has 3 ALUs in parallel, hardware multiplication, and magnitude units.	10	Schottky TTL, very small amount of ECL.	Ground radars, pulsed doppler, FFT, CFAR, threshold detection.	User-oriented, higher-order language; increased throughput by using more butterfly or ALU units. Block floating-point FFT; standard 6-ft cabinet. 1.5 kW, 200 6 × 8 cards (late 1970 technology).
General-Purpose Signal Processor (GPSP) (Raytheon, Bedford, MA)	Pipeline ALU with 4 real multipliers and several adders (to do butterfly).	5 (future: 20 possible)		Ground radars, PAVE PAWS (4 used).	Parallel ALUs or GPSPs for increased throughput. 3 kW and 6.5 × 2.5 × 2.5 ft. cabinet per GPSP.
PSP (Hughes, El Segundo, CA)	Complex (12 + j12) multiplier, pipelined.	7.1	Schottky and ECL, AMD 2505 MSI, 4 × 2-bit chips used for multiplier.	F-15 (AN/APG-63); F-18 (AN/APG-65); air-to-air and air-to-ground modes.	1300 W, 3000 ICs, 1 ft^3; in production.
PSP (Westinghouse, Baltimore, MD)		16		F-16 (AN/APG-66)	≈2000 W, 1 ft^3 (with radar computer).
PSP (Hughes, El Segundo, CA)	Complex (12 + j12) multiplier, pipelined.	18	ECL-10K, SSI, MSI, commercial parts.	Radar signal processing, ground-based.	2 kW, 3000 ICs; wire wrap boards; in small relay rack.

*Condensed from [195] which contains survey of 28 radars and related digital signal processors under development, developed, or in production.
**MCOPS ≡ Million complex radix two butterflies per second. The butterfly is a subalgorithm of the fast Fourier transform algorithm. It consists of four real multiplications and six real additions.

BROOKNER
CHAPTER 2

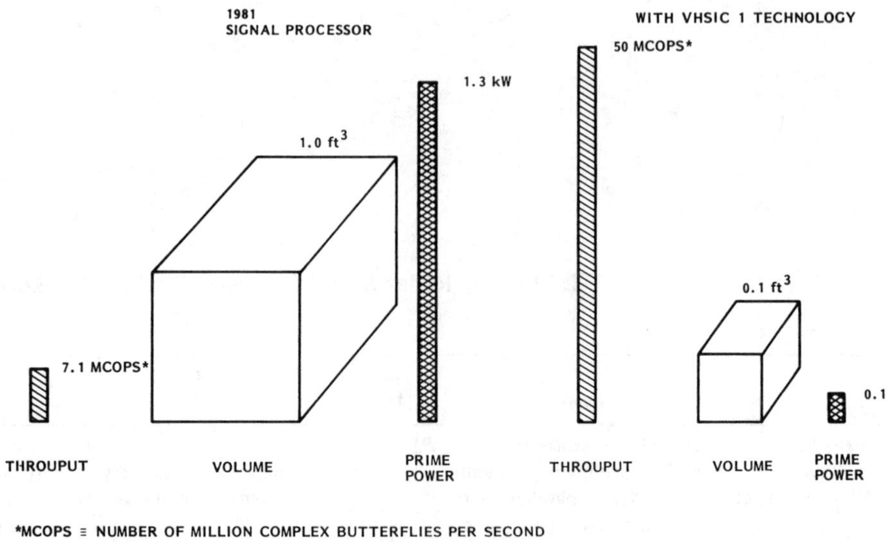

Figure 2.117 Effect of VHSIC on F-18 (AN/APG-65) Airborne Programmable Signal Processor (after Brookner [199]).

VHSIC on signal and data processors, the reader is referred to [195] and [196]. The latter reference cited also indicates useful alternatives before VLSI and VHSIC technology become available.

By the end of the 20th century, digital radar signal processors and data processors should be extensively using these technologies. These technologies would make available more powerful, programmable signal processors, which should provide increased capability for (1) more optimum clutter rejection, which would adapt to the environment; (2) more optimum jammer-rejection algorithms, as described previously; (3) target identification capability based on the use of various target signatures, such as wideband range-profile measurements of the target. This technology would enable the reduction of the number of parts of a signal processor having a given throughput by about two orders of magnitude, which in turn would increase the signal processor's reliability by about two orders of magnitude.

All the VHSIC Phase 1 contractors [195] have supplied working chips, and a number of them are providing commercial VHSIC chips to qualified buyers [200–203]. Texas Instruments' VHSIC 72K, n-MOS, *static random-access memory* (SRAM) chip sells for only $2.25, while its bipolar 1750A processor costs $3,200, but it is expected to reduce to $500 by 1990 with higher volume [204]. Efforts are underway to insert this technology into the F-16 AN/APG-68 Airborne Fire-Control Radar's programmable signal processor [168,205], the AN/APG-65 [206], the Firefinder radars [1,207], the Raytheon-RADC (Rome Air Development Center) distributed Advanced On-board Signal Processor (AOSP) (applicable to space-based radars, communication, and electro-optics systems) ([195,208]; see also Section 2.6.6.9), and other systems [206]. IBM indicates that the yields of VHSIC Phase 1 signal processing chips have increased from 1 or 2% to 34% [204]. IBM, TRW, and Honeywell are the VHSIC Phase 2 contractors [207]. The goal for VHSIC Phase 2 is to attain a size of 0.5 μm (*versus* 1.25 μm for Phase 1) with a 100-MHz clock speed, a factor of four times that of the Phase 1 chips. Many of these chips will contain a half million transistors. TRW had elected to develop "superchips" with up to 34.7 million devices [209,210]. These had three factors that improve reliability and manufacturing yields: self-contained spare components, built-in testing, and software reconfiguration. Yields of 20 to 50% were expected, about the best achieved with commercial LSI chips [209]. In-orbit lifetimes of about 50 years are expected, due to self-repair features [211]. Unfortunately, TRW has retreated somewhat from these ambitious goals.

The total DoD funding of VHSIC is now expected to be about $1 billion [182]. Industry and government spending exceeded $1 billion in its first seven years [204].

2.6.6.2 Silicon and GaAs Compilers

We all are familiar with high-order languages, such as FORTRAN, which enable us to program computers without having to use assembly language. There is now an even higher order programming language for which we need only define the logic that we want a particular chip to have, and this higher-order language transcribes the logic equations into the chip mask designs. This is called the *silicon compiler language* [211]. Compilers are now available for VHSIC Phase 1 technology that would produce chips containing over 100,000 transistors [212]. It is claimed that chips designed with these silicon compilers are only 15 to 20% larger than chips designed manually. Honeywell has teamed up with Silicon Compiler, Inc. to obtain such a compiler. Several other manufacturers are using Silicon Compiler's product [213]. One of these is General Electric, which is also offering semiconductor foundry services for custom microcircuits with 1.25 μm or 2 μm featured sizes [214]. General Electric has developed a compiler for internal use, which goes from functional description to layout in three working days [215]. One chip designed contained 35,000 transistors and used 1.25 μm CMOS [215], where only 100 lines of compiler code were required. A silicon compiler is now available for an IBM PC-AT compatible personal computer [216]. Silicon compilers are also becoming available for high-speed GaAs logic (see Section 2.6.6.7).

2.6.6.3 Memory Chips

Toshiba has had its one-megabit (1-Mb) CMOS (dynamic RAMS, or DRAMS) in production for some time [217,218]. This chip has 2.2 million transistors on a 4.5 × 12.3 mm chip. United States companies that are providing 1-Mb DRAMS are IBM, AT&T, and TI [219]. Toshiba uses 12 of these chips in its IBM-compatible "lap-top" personal computer [218]. IBM Japan is using its 1-Mb DRAM in several of its computers [218]. Hitachi, Mitsubishi, and NEC are also mass producing 1-Mb *static RAMs* (SRAMs) [220–222]. The 4-Mb DRAM is soon to follow, with IBM, TI, Siemens, AG, NEC, Mitsubishi, Hitachi, and Toshiba, among others, having produced such chips ([223–226,389,434] [227]; see [228–231]). Production of these is expected in 1989, with preproduction samples in 1988 [226,227,232,440]. Siemens and Philips are jointly working to develop 1-Mb and 4-Mb DRAMS ([225]; see also [446]). This joint project, called *Megaproject,* is expected to cost the two companies $1.5 billion [233]. Work is underway for a 16-Mb DRAM. IBM announced a technique for a 16-Mb RAM using 0.5 μm featured sizes [232,234,235]. Nippon Telegraph and Telephone (NTT) described a prototype 16-Mb DRAM at the 1987 International Solid-State Circuits Conference (ISSCC) [232,235,236]. NEC and Matsushita Electric are also each working on a 16-Mb RAM [386,393]. Matsushita workers gave a paper at ISSCC 1988 on their 16-Mb DRAM [441,446]. Matsushita indicated at the December 1987 SEMI Technology Symposium of Japan that they have developed a krypton fluoride *excimer* laser (providing 0.4 μm rules) for the fabrication of their 16-Mb chips [445]. It is also expected by Nippon Kogaku K.K. (Tokyo) that excimer laser steppers will be used for producing 64-Mb chips [445].

A 1-Mb expandable programmable ROM (EPROM) using 1.5 μm is available from Advanced Micro Devices, Inc. [219,237,238], and others have followed [239].

Toshiba has announced a 4-Mb EPROM [434]. GigaBit Logic, Inc. has produced a one-kilobit (1-kb) GaAs RAM having a 1 ns access time and requiring a 1.8 W dissipation power [239]. McDonnell-Douglas has produced 1-kb static RAM GaAs chips with yields of 47% for a DARPA-funded contract [240,241]. NTT (Japan) has produced a 16-kb GaAs SRAM having a 4.1 ns access time, consuming 2.52 W, containing 100,000 1-μm gate FETs on a 7.18 × 6.24 mm chip [242]. Rockwell International has built 16-kb GaAs SRAMS, also under DARPA funding [450]. A yield of 1% was achieved, but DARPA expects to achieve yields of 3 to 10%.

2.6.6.4 Analog-to-Digital (A/D) and Digital-to-Analog (D/A) Converters

Considerable progress has been made over the last decade in A/D and D/A converters. Analog Devices now provides a single-chip (monolithic), eight-bit, flash A/D converter, which handles signals having a 115-MHz bandwidth. This converter consumes only 0.75 W [243]. The sampling rate is in excess of 150 megasamples/s. Honeywell and Siemens respectively supply eight-bit single-chip converters with 250 and 100 megasamples/s [244,442]. Sony provides an eight-bit ECL flash converter chip, which has a 300-MHz sampling rate, handles a 180-MHz input bandwidth, consumes 1.3 W through the use of ECL III 1-μm technology [442]. TRW has a monolithic 10-bit flash A/D chip, which has a conversion speed of 25 megasamples/s, uses 1-μm technology, consumes 4.6 W, and has a signal-to-noise ratio of 54 dB for a 3.58-MHz input signal [245]. Honeywell provides a 10-bit flash A/D chip that samples at a 50-MHz rate, and handles signals having bandwidths up to 25 MHz [256,442,447]. Honeywell's chip consumes 3.7 W and is derived from their VHSIC Phase 1 program using a 1.2 μm process. Using the TRW monolithic eight-bit flash converter, TRW and Hybrid Systems, Inc. are offering an eight-bit, 20 megasample/s, hybrid A/D converter in a 0.8 × 1.3-inch package which does not require analog support circuitry, since the converter has the necessary voltage reference, input amplifier, output register, and conversion state [247]. Datel, Inc. provides a hybrid 12-bit A/D and 1.3-MHz sample-and-hold (S/H) in a 46-pin DIP package [248], which consumes 2.4 W. Crystal Semiconductor has developed a monolithic 12-bit A/D having a 1.5-MHz throughput, typical conversion time of 1 μs (2.5 μs maximum) [447,448]. The sample-and-hold circuit is on the chip. Crystal Semiconductor provides a monolithic 16-bit A/D having a conversion time of 16.25 μs [449]. Elsin, Inc. offers a 14-bit, 5.1-MHz A/D, occupying a 3.5 × 7 × 11.5-inch module, weighing eight pounds and consuming 87 W. Using two 4-bit flash *A/D converters* (ADC) operating at 200 MHz, LeCroy has produced a 200 megasample/s, eight-bit ADC. LeCroy also has a 1.3 gigasample/s, eight-bit, modular waveform recorder. Sony Tektronix has a waveform digitizer instrument, which operates at 200 megasample/s with 10-bit resolution [249]. Micro Networks provides a 20-bit floating-point ADC having a 0.32-MHz conversion rate. The ADC occupies a 3 × 4 × $\frac{5}{8}$-inch card. DARPA is funding a GaAs ADC. C. House of Hewlett-Packard predicts that today's high-priced, six-bit, 1-GHz A/D converters will evolve to tomorrow's low-cost, 10-GHz devices, probably by 1994 [250]. TriQuint Semiconductor, Inc. offers a monolithic 1-GHz, eight-bit D/A converter [251], which consumes 3.5 W.

2.6.6.5 Multipliers and Dividers

TRW offers a 32 × 32-bit *multiplier-accumulator chip* (MAC), which does a 32 × 32 multiplication in 50 ns, while consuming only 0.65 W [252]. This MAC uses 1 μm CMOS. The chip can also do 64-bit multiplication, division, and swap operations, useful for FFTs. Analog Devices offers a monolithic chip that does a 16 × 16 multiplication in 15 ns (maximum), while consuming 3.5 W [253]. A supplier of a similar type of chip is Bipolar Integrated Technology (BIT), Inc. [254]. Floating-point MACs are available from Advanced Micro Devices (AMD), LSI Logic, TI, and Weitek, ranging from 24 to 32 bits [254]. The chips typically have on-chip multiplying speeds of less than 100 ns and power dissipations that are ≤1 W. BIT offers an *emitter-coupled logic* (ECL), floating-point multiplier chip, which does 32-bit (single-precision) and 64-bit (double-precision) floating-point multiplications in typically 27 and 35 ns, respectively. A single-precision division is done in 81 ns. The ECL-compatible version of this chip consumes 7.8 W. Other suppliers of multiplier chips are Cypress Semiconductor, Fairchild, Integrated Device Technology, Intersil, Logic Devices, Marconi Electronic Devices, Micro Power Systems, NCR Micro Electronics, Stanford Telecommunicating, and Waferscale Integration [254,255,394].

GaAs offers the potential for high-speed, low-power multiplier chips. Rockwell had built, some time ago, a monolithic GaAs chip, which did an 8 × 8 multiplication in 6 ns, while consuming only 0.3 W [256]. GaAs logic is finally commercially available (see Section 2.6.6.7). TRW offers the first "stand-alone" VLSI divider chip [257]. This 1 μm CMOS chip does a 32-bit floating-point division in 400 ns.

2.6.6.6 Digital Signal Processor (DSP) Chips

Now chips are being offered which have a hardwired multiplier (or multipliers), possibly an *arithmetic logic unit* (ALU) and memory for storing instructions and data. These are called *digital signal processor* (DSP) chips [258–261, 370]. These chips have features that make them very suitable for fast "number crunching" radar signal processing tasks, such as pulse compression using the fast convolver algorithm [196], or for pulsed doppler processing using the FFT algorithm. DSP chips are also suitable for implementing *finite impulse response* (FIR) and *infinite impulse response* (IIR) digital filters.

There are basically two types of DSPs [262]. One is the general-purpose DSP consisting of a hardwired multiplier, ALU, and memory. This type of chip offers a great amount of flexibility, as does the general-purpose microprocessor (μP) chip. General-purpose DSPs have been referred to as a result by some as the "μP-like DSP chips" [262]. They differ from the general-purpose μP chips, however, in that they are optimized for multiply-and-accumulate operations needed for a radar signal processor. Because of their flexibility and speed, some of these chips are used instead of general-purpose μP chips. The other type of DSP chips are more specialized, and as a result less flexible; however, they offer a higher throughput. Typically, they have many multipliers on a single chip with an architecture suitable for synthesis of FIR filters as well as for doing the discrete Fourier transform (DFT) algorithm. These DSPs are referred to by some as the "non-μP-like DSP chips" [262]. These two types of DSP chips are discussed in the following paragraphs.

A large number of companies are providing μP-like DSP chips. Among them are AT&T, Analog Devices, Fujitsu, ITT Semiconductors, Motorola, NEC Electronics, Oki Semiconductor, Philips, TI, and Thomson Components-Mostek [262, 370]. Texas Instruments provides a third-generation μP-like DSP chip, the TMS320C30, which provides a 32-bit floating-point multiplication in about 60 ns [262–264]. This chip uses 1 μm CMOS. The previous two generations of this chip were slower and provided only 16-bit fixed-point computation [262]. The third-generation chip can do a 1024-point complex FFT in 9.65 ms for a throughput of 0.53 MCOPS, or, equivalently, 5.3 million real operations per second (MROPS; see footnote in Section 2.6.6.1) [195,196]. Other companies supplying floating-point μP-like DSP chips are AT&T, Fujitsu, NEC, and Oki [261,265–268]. The AT&T floating-point μP-like DSP chip is claimed to be able to do a 1024-point complex FFT in 4.4 ms for a throughput of 1.2 MCOPS [268].

As indicated, the μP-like DSP has features that accommodate high-speed calculations needed for radar signal processing tasks. The chips typically have a Harvard type of architecture, instead of the classic von Neumann architecture [259]. With the classic von Neumann architecture, the instructions and data are both stored in the same memory. As a result, the instructions and data have to be sequentially accessed. With the Harvard architecture, the instructions and data are stored in separate memories, and thus can be accessed simultaneously, resulting in an increased speed. In addition, the use of "pipelining" and three busses allows the user to take maximum advantage of the Harvard architecture; two of the busses are used to fetch two operands simultaneously, with the third bus used for the result [259]. Some have bit-reversal addressing, which aids in performing the FFT computation [262]. For higher throughput, multiple DSP chips can be used [371].

The AT&T 32-bit, floating-point DSP, which uses 0.75 μm CMOS technology, employs the von Neumann architecture, instead of the Harvard architecture, because the former is easier to program [269]. To achieve high throughput, its bus supports four memory accesses per instruction cycle. This allows it to fetch two operands from memory, do a multiply-and-accumulate operation, and put the result in memory during a single 80-ns instruction cycle.

The non-μP-like DSP, as indicated above, typically has more than one multiplier on the chip. As a result, they allow for a higher throughput, in some cases, by as much as one order of magnitude, than obtainable with the μP-like DSP chips. Examples of non-μP-like DSP chips that are efficiently geared to do FIR or IIR are the Zoran 33XXX, Calmos-Intersil 29C128, NCR 45CF8, INMOS A-100, RCA CDSP-100, Gould-AMI 614381, Motorola 56200, and Kurzweil KSC 2408 [262]. By way of example, the Zoran ZR33891 *digital filter processor* (DFP) chip has eight 9 $\times$ 9 multipliers and 26-bit accumulators organized as a programmable digital tap delay-line filter, having a finite impulse response (i.e., as a digital FIR) [270]. Each multiplier is capable of a 20-MHz rate. One such chip is thus capable of implementing an eight-tap FIR filter operating at a 20-MHz rate. A number of these chips can be cascaded to synthesize a higher-order filter. FIRs can be synthesized having over 1000 taps without any overflow. Also, they can be paralleled to increase the precision. Furthermore, a single chip can synthesize many more taps or provide greater precision by operating at a slower rate than 20 MHz. Alternately, increased speed (e.g., a factor of two), can be achieved by paralleling chips. The Zoran DFP chip can be used for doing a DFT. To do the FFT algorithm, Zoran has another chip called the *vector signal processor* (VSP) chip to be described below.

Another member of the non-μP-like DSP chips having an FIR architecture is the INMOS IMS A-100 CMOS chip. This chip contains thirty-two 16 $\times$ M multipliers, each with its own 36-bit accumulator, where M is the number of bits for the FIR filter weights, with 16 being the number of bits for the input data. M can be either 4, 8, 12, or 16, depending on the throughput rate, which is 10 MHz for M = 4 bits and 2.5 MHz for M = 16 bits. Motorola's DSP 56200 non-μP-like DSP chip implements an adaptive least mean square (LMS) FIR filter [259]. The LMS algorithm is implemented in the silicon, making it unnecessary to write the software to implement the LMS algorithm.

The Zoran VSP chips are very close to the μP-like DSP chips. Their architecture is optimized to provide very high performance DSP operation. In return for high performance in one area, the VSP chips give up some flexibility in other areas. Their instruction set is powerful because it is geared for signal processing tasks, such as doing an FFT. For example, to do a 1024-point complex FFT requires only three instructions—load, execute the FFT, and finally, store the result. Using 16-bit fixed-point precision, the Zoran VSP can do a 1024-point complex FFT in 2.6 ms for a throughput of 2.0 MCOPS. The VSP chip can also do this FFT by using 16-bit block floating-point arithmetic in 3.3 ms. Using 32

VSPs, it is possible to do a 1024-point complex FFT in 100 μs for a throughput of 51.2 MCOPS. The VSP chip is also capable of doing a "zoom" FFT; that is, it can apply the FFT over a small part of the signal spectrum. This is done by first bandpass filtering the data to the band of interest before doing the FFT over the filtered band. The bandpass filtering would be done with the same VSP. A 32-bit floating-point version of the VSP chip will also be available [349,372,378]. In addition to efficiently doing FFTs (1024-point complex in 1.7 ms), the new VSP chips can efficiently do peak detection, comparisons, and thresholding. The VSP is called a "vector signal processor" because it processes blocks of data, as with the FFT operation. Some of the other instructions, besides the FFT instruction, that the chip has are magnitude, demodulate, complex conjugate, magnitude square and accumulate, and modulate.

A great deal of support is available from the DSP suppliers as well as others in the design, simulation, and debugging of the DSP chips [260,348,370]. Simulators are available to check the programs developed by the user for the DSP chips. Emulators are also available, and these take the place of the DSP. They plug into the DSP socket, and allow the DSP software and supporting hardware to be run in real time. If the program or system fails, most of the emulators can trace through the system to see how it failed. Finally, many filter design packages are available, which allow the user to try different filter designs in quick fashion. Various signals can be input to these filters to see the output responses.

2.6.6.7 *Gigahertz and GaAs Logic*

Considerable progress has been made with respect to *gigahertz GaAs logic*. Gigahertz GaAs logic is now available from companies such as Ford Microelectronics, Gain Electronics, Gigabit Logic, Gigahertz Technology, Harris, Honeywell, TI, Triquint Semiconductor, NEC, and Vitesse Semiconductor ([241,271–273, 353,354,373–375]; see also [1] and [274,275,376,377]). Harris Corp. has developed a GaAs 300-gate array chip having a clock speed of up to 3 GHz [272]. The cost is $65,000 for a design plus the first 25 chips. Gain Electronics has a 3500-equivalent-gate GaAs chip [276]. It provides double the speed of equally complex ECL arrays at one-quarter the power. A 14-by-16 crosspoint switch built with the chip can operate at 1 GHz while consuming 1.1 W. Unloaded, its gate delay is 100 ps. TriQuint Semiconductor has developed a 3000-gate GaAs array offering twice the speed of silicon while requiring half the power [277]. Vitesse achieves a yield of 28 to 32% per GaAs wafer (with 6000 to 8000 devices per chip) [278]. Chips with complexities of 10,000 to 20,000 transistors are to be released shortly [278]. Vitesse predicts 10,000-gate GaAs array capability by 1990 [275]. Vitesse is offering a GaAs version of the AM2900 (Advanced Micro Devices, Inc.) four-bit-slice μP, which is functionally compatible with the AMD ECL equivalent, but two to four times faster. A typical GaAs 2900-bit-slice processor goes for about $50, double the cost of the silicon version [353,354]. Vitesse also offers a 3-ns 1-kb SRAM [354]. Ford Microelectronics and TI will be offering a 2.5-ns 1-kb SRAM [353]. Vitesse claims a 5000 wafer per month capability [354]. Vitesse will provide a 4-ns RAM, which is a plug-in replacement for an existing Fairchild Semiconductor RAM, but with two to three times the speed (4 ns access) ([374]; see also [443] and [451]). ITT Defense Technology has developed a GaAs 940 ps 4 × 4-bit multiplier chip with a yield exceeding 50% [377].

Seymour Cray, father of the Cray supercomputers, is working on two GaAs computers, the Cray 3 and 4 [241,279,356,376,434]. The Cray 3 uses fully customized GaAs 300 to 500-gate chips and some 1-kb GaAs memory chips [356]. It has a tenfold higher throughput than the Cray 2 [376]. The next generation, Cray 4, will provide an even higher throughput of 100 billion floating point operations per second (Gflops) provided by a 1-ns cycle time and 64 parallel processors, *versus* 16 for the Cray 3 and 4 for the Cray 2 [376]. McDonnell Douglas Astronautics is developing a fault-tolerant GaAs signal processor intended for space systems in the 1990s and beyond [380]. When performing a 1024-point radix 4 FFT (the type of algorithm for which it is optimized) this vector signal processor is capable of a 382 Mflop rate when configured with nine parallel *processing elements* (PEs) [380].

DARPA is funding a 32-bit GaAs microprocessor having a 200-MHz clock speed [280,281]. DARPA is funding AT&T to provide the Department of Defense radiation-resistant GaAs 3000 to 5000-gate array chips, operating from 200 to 400 MHz ([434]; see also [282]). DARPA is also funding Rockwell-Honeywell and McDonnell Douglas for 50-MHz chips. Texas Instruments has built a GaAs 4000-gate array chip for DARPA ([281]; see also [1]).

A technique has been developed for placing GaAs circuits on top of a silicon base, thus allowing the marriage of both technologies on a single chip [283,355,392]. Sample commercial GaAs-on-silicon 2 to 6-inch wafers are available from Kopin Corp. [434].

HEMT (discussed in Section 2.6.5.3), is not only very attractive for analog microwave circuits, but also for fast digital circuits [185,191,284]. A HEMT ring oscillator has a 12-ps delay at room temperature [284]. Large scale integration has been achieved with HEMT. Specifically, a 4-kb static RAM having 27,000 HEMTs with 1.5-μm gate length has been built [185]. It has a 2-ns access time at 77 K. A 1-kb HEMT SRAM has been built by Rockwell International and it has a 0.6-ns access time at room temperature [285].

Seattle Silicon Corp. and Gigabit Logic, Inc. have teamed up to develop the first GaAs logic compiler [286]. Its availability will boost the acceptance of GaAs among designers. The compiler was used to design Gigabit Logic's 10G181 four-bit arithmetic logic unit (ALU), which operates at a 1.3-GHz rate, three times faster than the equivalent 10 K or 100 K emitter coupled logic (ELC) part, while requiring two thirds the power. The area required was about the same as if done by hand (handcrafted). It requires two-thirds the area of a GaAs gate array design while operating four times faster and consuming 30% less power.

2.6.6.8 CMOS and ECL Gate Arrays [287–290]

Fujitsu, NEC, and Honeywell announced a CMOS 20,000-gate array chip [287]. The Fujitsu chip has gate delays of less than 1 ns and maximum power consumption of 500 mW. Honeywell's is based on its 1.25-μm VHSIC process [288]. Performance Semiconductor is producing for ETA systems a CMOS 20,000 gate array having a 500-ps loaded gate delay [289]. Thomson-Mostek announced a CMOS 128,000 gate array chip of which 50,000 to 90,000 could be used, the latter being achievable if the designer works interactively with the layout and routing tools [290]. The internal propagation delays are typically less than 200 ps (maximum 400 ps), close to ECL delays at CMOS densities and power levels. The transistor gate lengths are 1.2 μm.

Raytheon announced plans for an ECL 20,000 gate array having the lowest speed-power product of any bipolar logic arrays—less than 0.1 pJ per gate [291]. Typical loaded equivalent gate delay is 300 ps. Consumption is typically 5 W when 5000 equivalent gates and 60 outputs are used for their 12,500 equivalent gate chip.

2.6.6.9 New Architectures

A new architecture that is worth watching for systems of the future is the *systolic array architecture*. With it the signal processor is made up of many identical (or nearly identical) processing elements in a one-dimensional or two-dimensional array [292,293]. This architecture is ideal for VLSI and VHSIC circuits, where it is desired to minimize the number of circuit interconnections. Systolic array architecture together with the use of VLSI and VHSIC, which reduce the number of chips needed for a system, can eliminate the wiring problem illustrated in Figure 2.118.

Figure 2.118 Cyber-205 Supercomputer Complex Interconnections (Courtesy of Control Data Corp.).

Mitre has developed an interesting residue-arithmetic, systolic-array, finite-impulse-response (FIR) filter [294,295]. Mitre's technique can be used to do radar pulse compression and pulsed doppler processing through the implementation of the fast fourier transform (FFT).

Studies have been made on the use of systolic arrays for adaptive nulling [296–299], the FFT [300], and matched filtering [300]. One of the VHSIC Phase 2 superchips (see Section 2.6.6.1) TRW has proposed is an FFT chip using a systolic architecture [211]. This chip would be able to do a 1024-point FFT in 41 μs. The Standard Telecommunications Laboratories in Harlow, England are developing a systolic, adaptive array antenna, test-bed system [301,302].

Another important type of signal processing architecture worth watching is the fault tolerant programmable one, consisting of many parallel processing elements (PEs), some of which are standby spares that can be automatically switched in when an operating PE fails [195]. The McDonnell Douglas GaAs signal processor, mentioned in Section 2.6.6.7 is such a processor [380], another is the Advanced On-board Signal Processor (AOSP) [381].

2.6.6.10 Radiation Hardened Chips

Great strides have been made in obtaining radiation-hardened chips [303–311]. Sandia National Laboratories have developed a CMOS eight-bit microprocessor, which is hardened to 10^7 Rads Si [303–304]. This chip is pin-compatible to Intel's 8085 microprocessor, containing 20,000 transistors and using 3-μm design features. Harris Corp. sells commercially a version of this microprocessor chip which is hardened to 10^6 Rads Si [305]. Harris has developed the first hardened static CMOS 16-bit processor chip, the 80C86 [312]. It is hardened to 10^5 Rads and is free of upsets to 100 M Rads/s of transient gamma radiation. Total dose capabilities greater than 10^5 Rads will also be available. Intel has developed a version of its 80386 μP hardened to 10^5 Rads [313]. Harris and GE/RCA sell a whole line of hardened CMOS chips. GE/RCA Solid State Division is mass producing hardened silicon-on-sapphire (SOS) CMOS (CMOS/SOS) [314]. A 3 μm design rule is used, but 2 μm design rule speeds are achieved. The chips will be generally specified to have a total dose hardness of greater than 10^5 Rads. A hardness level of 10^7 Rads was achieved with their HCTS240 octal buffer device. They will also supply a hardened 16-kB static RAM (SRAM). Honeywell's Solid-State Electronics Division is offering a 16-kB hardened SRAM chip based on 1.25-μM featured sizes [315]. The worst-case access time for the chip is 160 ns. Universal Semiconductor offers CMOS gate array chips having 100 to 2400 gates per chip, which are hardened to 10^6 Rads Si [306]. Fairchild has CMOS and ECL gate arrays hardened to 10^6 Rads Si [307]. Honeywell has developed a 10,000-gate array chip hardened to 10^6 Rads Si that operates at a clock speed of over 50 MHz [387]. It is based on their 1.25-μm VHSIC bipolar technology. It consumes about half the power of an equivalent ECL-based gate array.

Sandia National Laboratories are working on having 65-kb static RAMs, 256-kb EEPROMs, and a 32-bit microprocessor (compatible with National Semiconductor's 32032), all hardened to 10^7 Rads Si. Sandia expects to progress to hardness approaching 10^8 Rads Si [311]. AT&T and Sandia are jointly working on radiation-hardened 1-μm CMOS chips as of mid-1988 [347]. (Most present CMOS hardened chips currently use 2-μm sizes.) AT&T and Sandia expect to reach a hardness of 10^7 Rads.

2.6.6.11 Digital Beam Forming

Considerable work has been done toward the goal of achieving digital beam forming. Digital beam forming has the advantages of being able to correct for array errors adaptively, to change at will the designed antenna beam pattern and sidelobe level, and to provide very accurate phase and amplitude weighting across the antenna [316–320]. It is not expected, however, that digital beam forming will be used extensively for some time to come, except for applications where the signal bandwidth is narrow (much less than 1 MHz) and where only one-dimensional electronic scanning is required and where it is desired that multiple simultaneous beams be available (as is the case for the SMART system under development; in Section 2.2.4.2). It has been applied to the University College London bistatic radar, discussed in Section 2.4 [103].

2.6.6.12 Beam-Steering Computer

The future phased array radars are expected to use distributed beam-steering computers, instead of centralized beam-steering computers, which require thousands of wires in some cases to carry the phase shifter's signal commands from the centralized beam-steering computer to the phase shifters [321]. For the distributed beam-steering computer, each phase shifter will have its own computer associated with it, consisting of a small microprocessor [322,323,344]. All the microprocessors in the array will receive the same information as to where the beam is to be scanned in azimuth and elevation on the next transmission. Each microprocessor would then calculate the phase shift needed for its associated phase shifter. This is a simple calculation involving only a few multiplications and additions.

2.6.6.13 Superconductivity [324–326,357]

With the revolutionary achievement of *superconductivity* above 93 K, beyond liquid nitrogen (77 K), there has been an immensely revived interest around the world in superconductors [324,325,390]. Nobel Prize physicist John Bardeen, who received a Nobel Prize in 1972 for his work on superconductivity, indicates that there is no theoretical reason for not achieving superconductivity at room temperature [324], which was reported by three groups [357,433]. However, the earlier reported results have been questioned [357]. Practical use of superconductivity at temperatures above liquid nitrogen for radar signal processing will probably wait until the 21st century. However, superconductivity holds the future promise of providing high throughput at low powers for signal processors and data processors. Superconductivity also has the potential for providing low-noise, high-Q, microwave circuits and wideband chirp networks. Superconductive Components, Inc. (SCI) offers kits ($50 and $200) for demonstrating high-temperature superconductivity. The lower cost kit provides the superconducting disk, the higher priced one provides the raw material for making your own [439]. Liquid nitrogen must be obtained by the user.

2.6.6.14 Fast-Turnaround Gate Arrays [435–438]

Nonrecurring costs and time to put a unit into production have been major problems in getting custom, semicustom, and even gate arrays into a product or system. By the time the chip is ready, the market could disappear or a competitor could be there [196]. A potential solution to this problem is the development of the fast-turnaround gate arrays using laser technology. Now, a completely packaged chip can be developed in minutes to hours with in-house capability, or in days using a fabrication service, instead of 6 to 12 weeks. The technology uses a laser either to cut prefabricated interconnections (Elron Electronic Industries Ltd., Haifa, Israel) or to write them (Lasa Industries, Inc., Santa Clara, CA). Elron and Lasa will sell complete turn-key systems. Elron's system will cost about $500,000 and provide a packaged 2500-gate array chip in about 40 minutes, and a 10 K chip in about 90 minutes using a skilled operator. Lasa's system will cost about $3 million and provide a packaged chip in minutes to hours, depending on complexity, using an untrained operator. The Elron system requires specially prepared blank gate array chips (VLSI Technology and Seiko are potential vendors). The Lasa system can use standard blank gate array chips. Both will be eventually capable of fabricating chips having up to 50,000 gates. Both are self-contained units, about the size of two telephone booths in volume. No clean room is necessary. These units make it possible to have an in-house capability at a modest cost. Laserpath Corp. (Sunnyvale, CA) offers a fabrication service. Lasarray Corp. (Irvine, CA) and Bid (Switzerland) offer turn-key and fabrication services.

2.6.7 Analog Signal Processing [1,196]

Despite the rapid advances in integrated digital circuitry and GHz circuitry, analog processing will still find a place in radar signal processing through the end of this century. This is especially true for the processing of wideband waveforms (that is, waveforms having a bandwidth of greater than 30 MHz). For such applications, *surface acoustic wave* (SAW) devices will be used because of their ability for easy pulse compression of chirp and other coded waveforms having up to 500 MHz bandwidth, using small packages and little power. One example of this is the SAW pulse compressor used for TI's AN/APS-116 periscope radar, which uses a 500-MHz bandwidth chirp waveform [406]. The SAW device for this processor is made by using E-beam lithography. This SAW processor replaces a folded taper meander line (FTML), which had a size two orders of magnitude larger and was less reliable.

Even for some applications where narrow bandwidths (less than 10 MHz) are used, SAW and IMCON would be used if the flexibility of a digital signal processor were not needed [196], where, for example, the pulse compression of only one waveform has to be done.

A programmable analog signal processor is now available which ought to see increased use. This is the monolithic SAW convolver. This device is capable of handling waveforms having a bandwidth up to about 100 MHz, and simultaneously handling uncompressed pulsewidths up to about 20 μs with no restriction on the waveform modulation in amplitude, frequency, or phase. The device has the capability of being programmable in real time so that a noise waveform can be pulse compressed with the waveform changing from pulse to pulse. This makes the system much less vulnerable to jamming. When a chirp waveform is used, the device can correct for quadratic and cubic phase errors in the transmitted waveform by programming the device in real time to match the errors in the transmitted waveform. This is done by coupling some of the transmitted signal into the receiver processor, where it is A/D converted and stored for application to the monolithic SAW convolver when the signal echo is received. (This type of correction can also be employed when digital signal processing is used [323]). Dynamic ranges of 60 dB or more are reported to be available with this device. It has been reported that Plessey is using a monolithic SAW convolver in one of their production radars.

Charged coupled devices (CCDs), *bucket brigade devices* (BBDs) and *charged transfer devices* (CTDs) have not found any widespread use as yet for radar signal processing (see [1] for an explanation of these devices). The hope for the CCDs, BBDs, and CTDs was in their potential for providing a high throughput in a small package requiring low power. However, the very rapid advances in digital integrated circuitry and the lack of a high dynamic range with these analog devices have held back their wider use. They still hold promise, however, for applications where a very high dynamic range is not required, and space, weight, and power are very limited, as in an RPV or missile application. Promising chips are being developed by MIT Lincoln Laboratory [196,327–329]. One of these devices, clocked at 10 MHz, performs a 16-point cosine transform in 100 ns and dissipates only 700 mW. Another is a programmable tapped delay line filter having 32 taps that can be clocked at 26 MHz, has six-bit tap weights, and a 60-dB linear dynamic range [328]. Still another of these chips can do pulsed doppler processing of 32 range cells simultaneously for as many as 42 pulses [327]. A dynamic range of 50 dB was achieved at an input sampling rate of 10 MHz. The device can be designed to handle up to 100 doppler cells. It is capable of giving the signal spectrum to a much higher precision than the Fourier transform resolution limit of $1/T_c$, where T_c is the coherent integration time. Spectral outputs at one tenth the resolution were demonstrated (i.e., every $\frac{1}{10}T_c$) over a limited spectral region of interest. This capability is the result of a discrete Fourier transform (DFT) algorithm being implemented. The FFT algorithm can also be implemented in CCD hardware [329].

One area where CCDs, BBDs, and CTDs are proving useful is in marine radars for increasing the brightness of the PPI display in the short-range mode (on the order of a one mile maximum range). The increased brightness is achieved by using a CCD, BBD, or CTD to stretch the time before displaying the short-range data. This is achieved by storing the short-range echo in one of these devices and then reading it out at a rate one or two orders of magnitude slower so that the PPI trace is swept in much slower than real time, giving a much brighter display.

VLSI expert Carver Mead, the man who "wrote the book" on VLSI designs [292], suggests that it is time for digital designers to refocus their attention on analog circuitry [330]. He contends that instead of trying to do every function digitally we should examine semiconductors to see if these same functions can be carried out by virtue of the physical properties of semiconductors.

2.6.8 Tubes

Although the technology of solid-state transmitters has progressed rapidly during the 1980s, tube technology still has a place in many systems because of its high-power and high-efficiency capabilities. Figures 2.119, 2.120, and 2.121 show the current CW, peak, and average power of various tubes. Although gyrotrons have achieved high peak powers (like the Soviet oscillator gyrotron tube, which produced 1 MW at 100 GHz with a power efficiency of 40% for 100-ns pulses, and the Varian amplifier gyrotron tube, which produced 212 kW peak at 28 GHz), they are too noisy for radar applications and it is not known when they will be useful for radar. They are being used primarily in thermonuclear fusion work. Figure 2.121 also shows the current average-power capabilities for solid-state single and paralleled devices (solid-state bottles). This figure indicates the regions where the tube dominates over the solid-state devices, where the reverse is true, and where they are competitive with each other. The original figures did not include spatially combined solid-state systems like the PAVE PAWS phased array radar or a potential solid-state version of the COBRA DANE, only solid-state bottle transmitters being considered. Both PAVE PAWS and the potential solid-state version of COBRA DANE are added to the figure, indicating the region of competition for solid-state phased array systems is beyond that for bottle transmitters. Figure 2.122 indicates the specific weights of various tubes.

An interesting trade-off was made by Hallberg *et al.* [399] between an airborne, S-band, solid-state and tube phased array system. They concluded that, for a system requiring an average radiated RF power of 2 to 3 kW, the solid-state system is a good solution. However, if higher RF power is required, the tube solution is favored. Their results were based on the availability of 30 to 40 W transmitters, having a 5-dB gain and 30% collector efficiency. Better S-band transistors are now available—50 W peak power, 7 dB gain, and 35% collector efficiency (MA/COM-PHI). Such transistors will move up the average RF power output point at which a solid-state phase array is favored.

2.6.9 Fiber Optics

Fiber optics may soon find application as a replacement for the corporate-feed structure of a large array radar because of light weight, assuming dynamic range and signal level problems are resolved [331–337,339]. The attenuation of fiber optics has gone from 1000 dB/km in 1970 to 0.2 dB/km today [1]. Furthermore, there is the potential for this attenuation decreasing to 0.01–0.001 dB/km [338,339]. Fiber optics is already being used in radar as a replacement for coaxial and microwave lines between the radar's antenna and control unit [340,341]. The UK is using an optical fiber feed for the phase shifter commands on the MESAR S-band phase-phase steered array now under development [105,382] (see Section 2.2.1.12).

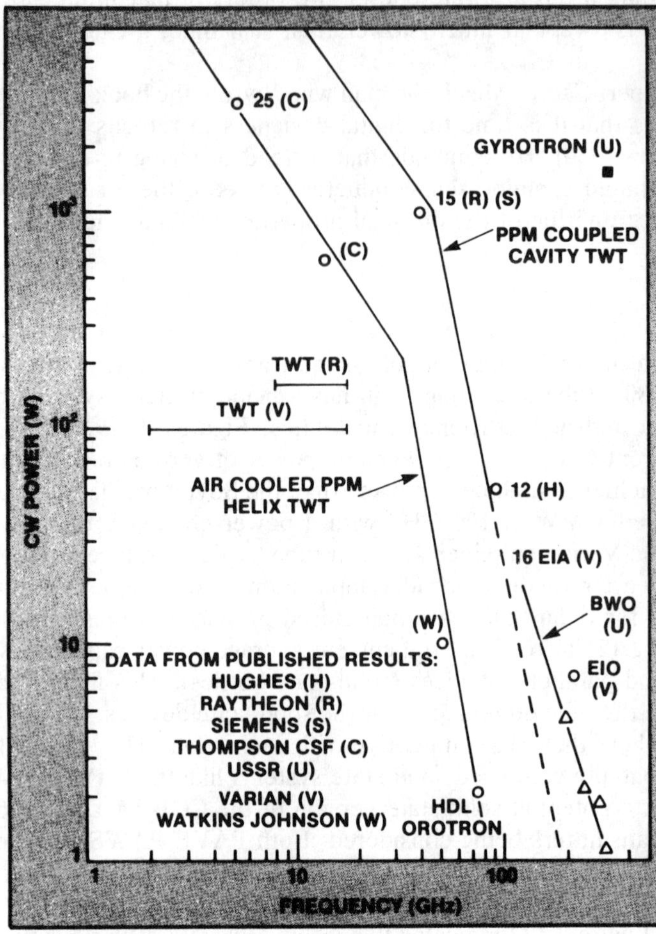

Figure 2.119 Performance for Various Types of Tubes. Numbers by Data Points Represent Efficiency in Percent; TWT = Traveling Wave Tube; PPM = Periodic Permanent Magnet; EIA = Extended Interaction Amplifier; BWO = Backward Wave Oscillator; EIO = Extended Interaction Oscillator (from Hieslmair *et al.* [401]).

2.6.10 Flat Displays

In the not too distant future the high-voltage CRT will be replaced by flat low-voltage displays. Planar Systems (Beaverton, OR) has an electroluminescent (EL) display, 0.55-inch thick, with 640 × 200 pixels, operating on 12 V, and weighing only one pound [342]. Planar Systems has developed two-color, thin-film EL (TFEL) prototype displays and is working on three-color prototypes [343]. Considerable progress has been made by Cherry Corp. in introducing thick-film, direct-current EL technology, which shows promise of lower cost than contemporary TFEL displays.

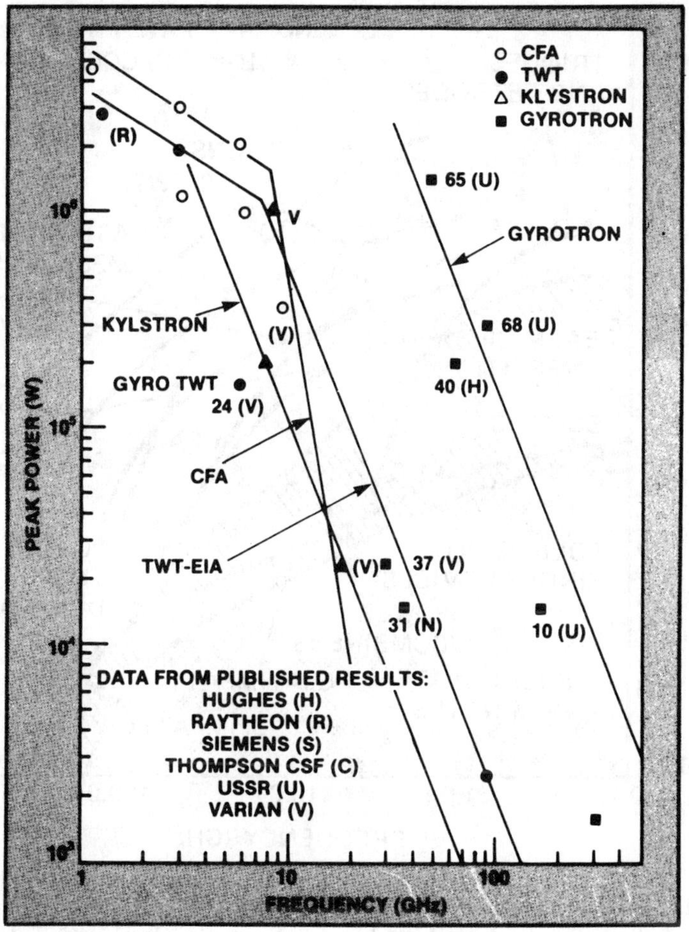

Figure 2.120 **CFA, TWT, Klystron, and Gyrotron Performance Data. Numbers by Data Points Represent Efficiency in Percent; CFA = Crossed Field Amplifier (from Hieslmair *et al.* [401]).**

2.7 CONCLUSIONS

It is apparent that radar is a many faceted field, which is still evolving in some fascinating ways. We see it developing into more complex radars having increased capabilities, while it is also developing into simple low-cost radars. A dramatic example of this is the $2200 2D mechanically scanning commercial marine radar of Figure 2.63 (see Section 2.3.1). With the advent of speech-recognition hardware, we can expect to be able to talk to the radar, which may possibly talk back to the radar operator.

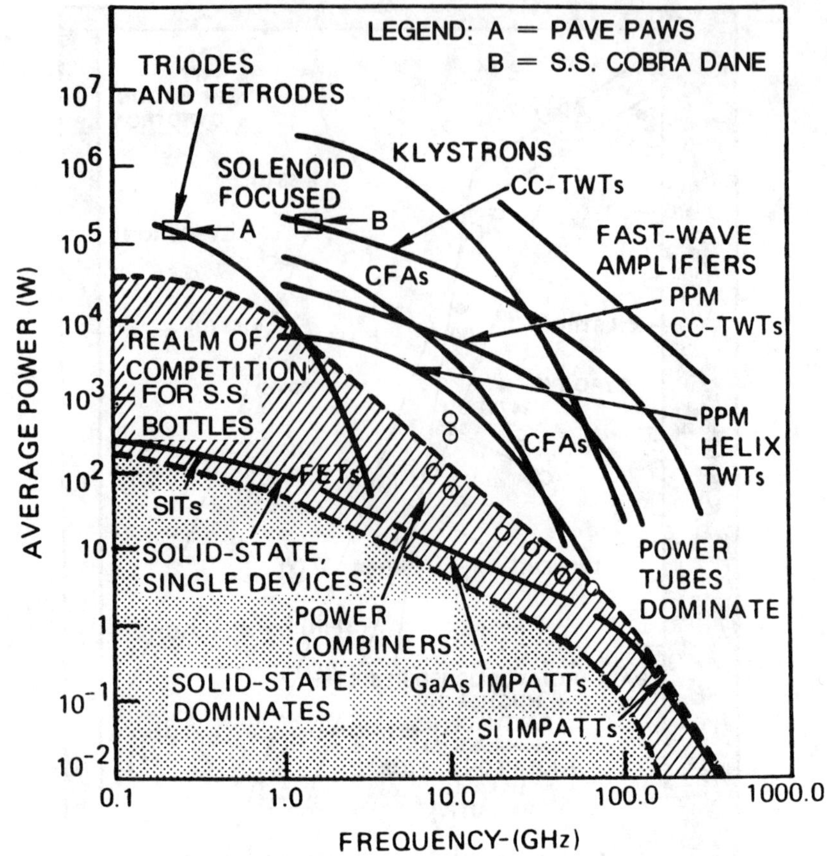

Figure 2.121 Average Power of Single and Parallel Devices, Solid-State (S.S.) Bottles. Points A and B Are for Spatially Combined Tube Systems; FET = Field-Effect Transistor; CC = Cavity-Coupled; SIT = Silicon (Bipolar) Transistor (after Woods [400]).

REFERENCES

1. Brookner, E. (ed.), *Radar Technology*, Artech House, Norwood, MA 1977.
2. Brookner, E., "Radar Technology," IEEE Radar Lecture Notes, Boston, 1984.
3. Daley, E.J., and F. Steudel, "Modern Electronically-Scanned Array Antennas," *Electronic Progress* (Raytheon), Winter 1974, pp.10–17.
4. Filer, E., and J. Hartt, "COBRA DANE Wideband Pulse Compression System," IEEE EASCON, Paper No. 61, 1976.
5. Coraine, R.T., "COBRA DANE (AN/FPS-108) Radar System," *Signal*, May-June 1977.
6. Johnson, R.C., and H. Jasik (eds.), *Antenna Engineering Handbook*, McGraw-Hill, New York, 1984.
7. Brookner, E., "Practical Phased-Array Systems," *Microwave Journal Intensive Course Notes*, Lectures 2 and 3, April 1975.
8. Skolnik, M.I., in R.E. Collin and F.J. Zucker (eds.), "Antenna Theory," Pt. 1, McGraw-Hill, New York, 1969; Skolnik, M.I., J.W. Sherman, III,

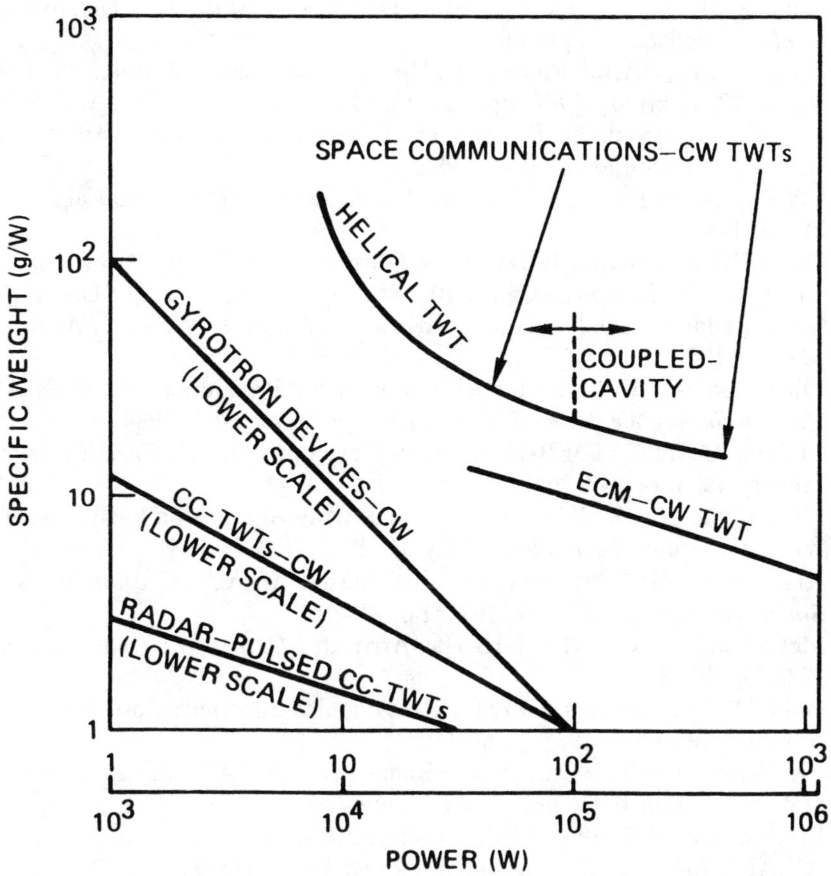

Figure 2.122 Specific Weights of High Power Tubes (from Woods [400]).

and F.C. Ogg, Jr., "Statistically Designed Density-Tapered Arrays," *IEEE Trans. Antennas and Propagation*, Vol. AP-12, No. 7, July 1964, pp. 408–417.

9. Hanfling, J.D., and F. Beltran, "Elements for Array Antennas," *Electronic Progress* (Raytheon), Winter 1974, pp. 33–44.
10. Petty, R.T. (ed.), *Jane's Weapon Systems, 1987–88*, Jane's Publishing, London, 1987.
11. Campbell, R.B., and D.J. Hoft, "Solid-State Transmitters for Radar Applications," *Int. Radar Conf.*, Paris, December 1978, pp. 487–496.
12. Stein, K.J., "PAVE PAWS Phased Array Radar Nears Acceptance," *Aviation Week*, April 9, 1979.
13. "PAVE PAWS BMEWS Radar Site Updates will Broaden Missile Threat Coverage," *Aviation Week and Space Technology*, December 9, 1985, pp. 52–54.
14. Stein, K.J., "BMEWS Update Program Progresses at Thule Site," *Aviation Week and Space Technology*, August 20, 1984, pp. 87–91.
15. Filter Center, "Raytheon Will Build Third PAVE PAWS," *Aviation Week and Space Technology*, November 28, 1983, p. 155.
16. Correll, J.T., "Harvest and Seedtime in C^3I," *Air Force Magazine*, June 1985, pp. 54–63.

BROOKNER
CHAPTER 2

17. Gumble, B., "Air Force Upgrading Defenses at NORAD," *Defense Electronics*, August 1985, pp. 86–108.
18. Brookner, E., "Array Radars: An Update," *Microwave Journal*, Pt. 1, Vol. 30, No. 2, February 1987, pp. 117–138; Pt. 2, Vol. 30, No. 3, March 1987, pp. 167–174; Brookner, E., "Radar of the '80s and Beyond—An Update," IEEE Electro-86, May 13–15, 1986.
19. "Phased Array Radar for North Yorkshire Moors," *Electronic Engineering*, March 1987, p. 17.
20. Stein, K.J., "COBRA JUDY Phased-Array Radar Tested," *Aviation Week and Space Technology*, August 10, 1981, pp. 70–73; see also "USAF Shipborne Radar Detailed," *Aviation Week and Space Technology*, August 17, 1981, p. 105.
21. Filter Center, "Contract for $553 Million for Production of 92 AN/APQ-164," *Aviation Week and Space Technology*, August 6, 1984, p. 79.
22. "Boeing Gears Up For B-1B Avionics Development," *Defense Electronics*, January 1983, pp. 47–56.
23. Filter Center, "USAF Will Use a Phase-Array Antenna on B-1B," *Aviation Week and Space Technology*, May 31, 1982, p. 69.
24. Klass, P.J., "B-1 Pioneers Airborne Phased Array," *Aviation Week and Space Technology*, April 9, 1984, pp. 84–89.
25. McClelland, L.A., "The B-1B Has Arrived," *Defense Electronics*, August 1985, pp. 72–84.
26. Gold, D., "It's Bombs Away for B-1 Avionics Subcontractors," *Electronic Business*, March 1, 1985, p. 66–69.
27. "Rockwell Installs Westinghouse Radar Set in B-1 Avionics Integrator Aircraft," *Aviation Week and Space Technology*, June 11, 1984, p. 17.
28. Filter Center, "Raytheon Will Produce 50,000 Phase-Control Modules for AN/APQ-164 Radar," *Aviation Week and Space Technology*, February 11, 1985, p. 83.
29. Adams, J.A., "Technology '85: Aerospace and Military," *IEEE Spectrum*, Vol. 22, No. 1, January 1985, pp. 85–88.
30. Skolnik, M.I., "Fifty Years of Radar," *Proc. IEEE*, Vol. 73, No. 2, February 1985, pp. 182–197.
31. Patton, W.T., "Compact, Constrained Feed Phased Array for AN/SPY-1," *Microwave Journal Intensive Course Notes, Practical Phased-Array Systems*, Lecture 8, April 1975.
32. "Government Programs Are More Diverse," *Raytheon News*, February 1986, p. 5.
33. Threston, J.T., and G.R. Meinig, "AEGIS and the AN/SPY-1A Radar—A Description and a Quick Look Report on the First Deployment," *Military Microwaves Conf. Proc.*, London, October 24–26, 1984, p. 216.
34. Ethington, D.A., "The AN/TPQ-36 and AN/TPQ-37 Firefinder Radar Systems," *EASCON-77 Conf. Rec.*, 1977, pp. 4-3A to 4-3F.
35. O'Donnell, R.M., and C.E. Muehe, "Automated Tracking for Aircraft Surveillance Radar Systems," *IEEE Trans. Antennas and Propagation*, Vol. 27, No. 7, July 1979, pp. 508–517.
36. Ethington, D.A., P.J. Kahrilas, and G.D. Wright, "Multifunction Rotating Electronically Scanned Radar (RESR) for Air Surveillance," *Proc. IEEE*, Vol. 73, No. 2, February 1985, pp. 340–354.
37. "Multiple-Target Instrumentation Radar," Raytheon Co. brochure, Wayland, MA.
38. Hammond, V.W., and K.H. Wedge, "The Evolution and Application of Multiple Target Tracking Instrumentation Radar," *Rec. First Chinese Int. Conf. Radar*, Nanjing, November 4–7, 1986, pp. 44–49.
39. Stein, K.J., "French Emphasizing System Concepts," *Aviation Week*, December 11, 1978, pp. 75–81.

40. Smith, B.A., "Prototype Adaptive Radar Evaluated," *Aviation Week and Space Technology*, February 2, 1981, pp. 47–49.
41. "More Fire Power Projected for Improved Hawk," *Aviation Week and Space Technology,* August 3, 1981, p. 54.
42. Rauh, K., "Siemens Radar Developments in the Course of the Next Decade: Adaptation to More Demanding Tasks," *IEEE Int. Radar Conf.,* 1975, pp. 28–30.
43. Wirth, W.D., "Fast and Efficient Target Search with Phased Array Radars," *IEEE Int. Radar Conf.,* 1975, pp. 198–203.
44. Wirth, W.D., "Radar Signal Processing with an Active Receiving Array," *IEE Radar-77,* October 1977, London, pp. 218–221.
45. Fleskes, W., and G. Van Keuk, "Adaptive Control and Tracking with the ELRA Phased Array Radar Experimental System," *IEEE Int. Radar Conf.,* 1980, pp. 8–13.
46. Fawcett, J., "Microwaves in Japan," *Microwave Systems News (MSN),* Vol. 26, No. 10, February 1980, p. 44.
47. Filter Center, "Japan's Mitsubishi Electric Corporation Has Delivered its First Advanced Air Defense Radar System," *Aviation Week and Space Technology,* October 27, 1986, p. 71.
48. "Prospects for Radar Satellites Hinge On Emerging Technologies," *MSN,* April 1979, pp. 19–23; "Competition Intensifies in Monolithic Gallium Arsenide," April 1979, pp. 35–48, and September 1978, p. 17.
49. Brookner, E., and T.F. Mahoney, "Derivation of a Satellite Radar Architecture for Air Surveillance," IEEE EASCON-83 *Conf. Rec.,* September 19–21, 1983, pp. 465–475; see also [166].
50. "New-Generation Radar," *Microwaves and RF,* Vol. 23, No. 2, February 1984, p. 44.
51. "Sperry Wins Short-Range Radar Contract," *Aviation Week and Space Technology,* August 27, 1984, p. 79.
52. Mohr, C., "US and Soviets Discuss Whether Moscow Violated Terms of Two Arms Pacts," *The New York Times,* Wednesday, October 5, 1983, p. A8.
53. Witt, M., and D. Reeder, "A New Generation Air Defense Radar," *Microwave Journal,* Vol. 27, No. 2, February 1984, pp. 160–164.
54. Howell, J.M., "Limited Scanning Arrays and Hybrids," *Electronic Progress* (Raytheon), pp. 18–23, Winter 1974.
55. Ward, H.R., "AN/TPN-25 & AN/GPN-22 Precision Approach Radars," *IEEE Int. Radar Conf.,* 1980, pp. 26–31.
56. "Increasingly Agile Radars," *TELONDE* (Thomson-CSF), 1978, pp. 10–13.
57. Drabowitch, S., and F. Gautier, "Phased Arrays," *Revue Technique Thomson-CSF,* March 1980, pp. 124–141.
58. Hammer, I.W., "Frequency-Scanned Arrays," Chapter 13 in Skolnik, M.I. (ed.), *Radar Handbook,* McGraw-Hill, New York, 1970.
59. Pfister, G., "Series 320 Radar Family—3D Air Surveillance for the 1980s," *Int. Conf. Radar,* Paris, December 4–8, 1978, pp. 83–92.
60. White, I., "Signal Processing Simulation of a Non-Linear FM Frequency Scanned Pulse Compression Radar," *IEE Radar-77,* London, October 1977, pp. 463–467.
61. Cochran, J.G., "Design of Air Defense Ground Radar Systems," *IEE Radar-77,* London, October 1977, pp. 16–19.
62. Mathias, R.F., and N.A. Cunningham, "AR320 Long-Range Defense Radar," *Military Microwave Conf. Proc.,* October 24–26, 1984, pp. 67–72.
63. Toyoda, R., and I. Matsukasa, "Concise Three Dimensional Surveillance Radar," *IEEE Int. Radar Conf.,* 1975, pp. 48–51.
64. King, I.E., *et al.,* "The AN/TPS-59 and GE592—A Family of Solid-State Radars," *IEE Radar-77,* October 1977, pp. 12–15.

BROOKNER
CHAPTER 2

65. Klass, P.J., "Solid-State 3D Radar for NATO Tested," *Aviation Week,* May 21, 1979.
66. "New Radar Installations Promise 360° Air Defense Perimeter," *Aviation Week and Space Technology,* December 9, 1985, pp. 56–63.
67. Palumbo, B., and A. Cucci, "Design Criteria and Test Results of an Array Antenna with Multiple Scanned Beams for a Monopulse 3D Radar," *1978 Military Microwaves Conf. Proc.,* London, October 1978, pp. 208–219.
68. Bega, L.J., "System Design Philosophy of a 3D Antenna for Radar Applications," *Int. Conf. Radar Applications,*" Paris, December 4–8, 1978, pp. 351–356.
69. DiLazzaro, M., and E. Giaccari, "Description and Field Evaluation of a New Family of Three Dimensional, Medium Range Military Surveillance Radars," *Int. Conf. Radar,* Paris, May 21–24, 1984, pp. 511–516.
70. Palumbo, B., and A. Cucci, "Multibeam Monopulse Array Antenna with Independent Elevation Beam Scanning," *AGARD Conf. Proc.,* No. 197, The Hague, The Netherlands, June 1976.
71. Giaccari, E., and S. Strappaveccia, "RAT 31-SL: A Multiple Beam Radar for 3D Long Range Detection," *Military Microwaves Conf. Proc.,* London, June 24–26, 1986, pp. 75–81.
72. Steyskal, H., "Synthesis of Antenna Patterns with Prescribed Nulls, "*IEEE Trans. Antennas and Propagation,* Vol. AP-30, No. 2, March 1982.
73. "Selenia Three-Dimensional Radar Being Produced," *Aviation Week,* January 1, 1979, p. 49.
74. Skillman, W.A., "Utilization of the E-3A Radar in Europe," *Military Electronics Defense Expo-78, Interavia,* Geneva, October 3–5, 1978.
75. "Radar System Begins Operating in Germany," *Aviation Week and Space Technology,* December 27, 1982, p. 12.
76. Ajioka, J.S., "TRMS Antenna," *IEEE Int. Radar Conf.,* 1975, pp. 382–384.
77. Miller, C.J., and M.J. Hodges, "W2000—An Advanced Long-Range 3-D Radar," *Microwave Journal,* Vol. 26, No. 10, October 1983, pp. 103–120.
78. Foster, T.M., "The Westinghouse W2000—An Advanced Solid-State Long-Range Radar," *Military Microwaves Conf. Proc.,* London, October 24–26, 1984, pp. 73–78.
79. Filter Center, "USAF's Electronic Systems Division is Studying Concepts for Siting Over-the-Horizon/Backscatter (OTH/B) High-Frequency Radars," *Aviation Week and Space Technology,* August 27, 1984, p. 79.
80. Wolfe, A., "Radar Enters High-Tech '80s" *Electronics Week,* March 25, 1985, pp. 36–38.
81. Smith, J.K., "Space Density Tapered Arrays for Airborne Surveillance Radar," *IEEE Int. Radar Conf.,* 1980, pp. 263–270.
82. Wood, D., "Martello—A New 3-D Radar for the 1980s," *Int. Defense Review,* No. 6, 1978.
83. Radford, M.F., and A.D. Stevens, "A New 3-Dimensional Surveillance Radar," *Military Microwaves Conf. Proc.,* London, October 25–27, 1978, pp. 193–200.
84. Latham, C., "Martello—A Long-Range 3-D Radar," *Military Microwaves Conf. Proc.,* London, October 24–26, 1984, pp. 79–84.
85. "ULSA Antennas in Production," *Microwaves and RF,* January 1985, p. 19.
86. Washington Window, "Antenna Makes Enemy's Detection of Radar Difficult," *Microwave Systems News and Circuit Technology (MSN & CT),* January 1985, p. 43.
87. Dyck, J.D., and H.R. Ward, "RAMP's New Primary Surveillance Radar," *Microwave Journal,* Vol. 27, No. 12, December 1984, pp. 105–113.
88. Ward, H.R., "The RAMP PSR, A Solid-State Surveillance Radar," *IEE Int. Radar Conf., Radar-87,* London, October 1987.

89. MacDonald, J.C., and H.R. Ward, "RAMP-Canada's New ATC Surveillance System," *J. ATC,* September 1986.
90. Kachman, M., "Turbulent Times for NEXRAD," *Microwaves and RF,* January 1986, pp. 34–40.
91. Filter Center, "Westinghouse Has Been Selected to Develop an 'Ultra-Reliable Radar' Using a Solid-State Active-Element Phased-Array Antenna," *Aviation Week and Space Technology,* March 11, 1985, p. 75.
92. "Joint USAF/Army JSTARS System Set for Airborne Test in 1988," *Aviation Week and Space Technology,* December 9, 1985, pp. 91–99.
93. "Grumman Team to Develop Joint STARS," *Defense Electronics,* November 1985, pp. 35–37.
94. Scott, W.B., "Side-Looking Radars Provide Realistic Images Under Adverse Weather Conditions," *Aviation Week and Space Technology,* September 7, 1987, pp. 93–97.
95. Lenorovitz, J.M., and K.F. Mordoff, "Europeans Seek Interoperable Battle Reconnaissance Systems," *Aviation Week and Space Technology,* September 7, 1987, pp. 77–85.
96. *Hughes' Science/Scope,* Hughes Aircraft Company, Los Angeles, CA.
97. "Hughes Developing Pod-Mounted Derivative of ASARS-2 Radar," *Aviation Week and Space Technology,* September 7, 1987, pp. 98–99.
98. Farina, A., "Survey of Data Processing Systems in Netted Radars," *Proc. Int. Radar Symp., India-83,* October 9–12, 1983, Bangalore, pp. 161–167.
99. Fleming, F.L., and N.J. Willis, "Sanctuary Radar," *Military Microwaves Conf. Proc.,* London, October 22–24, 1980, pp. 103–108.
100. Schoenenberger, J.G., and J.R. Forrest, "Principles of Independent Receivers for Use with Co-Operative Radar Transmitters," *The Radio and Electronic Engineer,* Vol. 52, No. 2, February 1982, pp. 93–101.
101. Forrest, J.R., H.D. Griffiths, J.G., Schoenenberger, and D.A. Williams, "Bistatic Radar," *University College London Annual Review,* Department of Electronic and Electrical Engineering, October 1982, pp. 26–27.
102. Schoenenberger, J.G., J.R. Forrest, and C. Pell, "Active Array Receiver Studies for Bistatic/Multistatic Radar," *IEE Int. Conf. RADAR-82,* London, 1982, pp. 174–178.
103. Griffiths, H.D., J.R. Forrest, A.D. Williams, and C. Pell, "Digital Beamforming for Bistatic Radar Receiver," *IEE Int. Conf. Antennas and Propagation,* 1983.
104. Auterman, J.L., "Phase Stability Requirements for a Bistatic SAR," *Proc. IEEE National Radar Conf.,* March 13–14, 1984, pp. 48–52.
105. J. Clark, "The Evolution of Early Warning Radar in the United Kingdom," *Rec. First Chinese Int. Conf. Radar,* Nanjing, November 4–7, 1986, pp. 7–20.
106. Elachi, C., and J. Granger, "Spaceborne Imaging Radars Probe 'In Depth'," *IEEE Spectrum,* Vol. 19, No. 11, November 1982, pp. 24–29.
107. Elachi, C., T. Bicknell, R.L. Jordan, and C. Wu, "Spaceborne Synthetic-Aperture Imaging Radars: Applications, Techniques, and Technology," *Proc. IEEE,* Vol. 70, No. 10, October 1982, pp. 1174–1209.
108. *Spaceborne Imaging Radar Symposium,* January 17–20, 1983, Jet Propulsion Laboratory, JPL Publication 83-11, July 1, 1983, pp. 96–97.
109. Elachi, C., "Spaceborne Imaging Radar Research in the '90s," *Second Spaceborne Imaging Radar Symp.,* April 28–30, 1986, Jet Propulsion Laboratory, JPL Publication 86-26, December 1, 1986, pp. 45–55.
110. Cimino, J., C. Elachi, and M. Settle, "SIR-B—The Second Shuttle Imaging Radar Experiment," *IEEE Trans. Geoscience and Remote Sensing,* Vol. GE-24, No. 4, July 1986, pp. 445–482.
111. Ottl, H., "The X-SAR System," *Second Spaceborne Imaging Radar Symp.,*

April 28–30, 1986, Jet Propulsion Laboratory, JPL Publication 86-26, December 1, 1986, pp. 218–221.

112. Schussler, H., and R. Wagner, "Features and Technologies of ERS-1 (ESA) and X-SAR Antennas," *Second Spaceborne Imaging Radar Symp.,* April 28–30, 1986, Jet Propulsion Laboratory, JPL Publication 86-26, December 1, 1986, pp. 125–129.
113. Gredel, J., W. Markwitz, W. Noack, and G. Schreier, "Off-Line Processing of ERS-1 Synthetic Aperture Radar Data with High Precision and High Throughput," *Second Spaceborne Imaging Radar Symp.,* April 28–30, 1986, Jet Propulsion Laboratory, JPL Publication 86-26, December 1, 1986, pp. 130–141.
114. Ishiwada, Y., and Y. Nemmoto, "Present Status of Japanese ERS-1 Project," *Second Spaceborne Imaging Radar Symp.,* April 28–30, 1986, Jet Propulsion Laboratory, JPL Publication 86-26, December 1, 1986, pp. 79–83.
115. Raney, R. K., "An Alternative Multi-Mode SAR for RADARSAT," *Second Spaceborne Imaging Radar Symp.,* April 28–30, 1986, Jet Propulsion Laboratory, JPL Publication 86-26, December 1, 1986, pp. 74–78.
116. Curlander, J. C., "The SIR-C Ground Data System: Digital Processor, Data Products, Information Flow," *Second Spaceborne Imaging Radar Symp.,* April 28–30, 1986, Jet Propulsion Laboratory, JPL Publication 86-26, December 1, 1986, pp. 149–153.
117. Foley, T. M., "Canada Approves Development of Scaled-Back Radarsat," *Aviation Week and Space Technology,* July 13, 1987, pp. 51–53.
118. *Aviation Week and Space Technology,* special issue on Reconnaissance, September 7, 1987, pp. 68–99.
119. Klass, P. J., "Inverse Synthetic Aperture Technology Aids Radar Identification of Ships," *Aviation Week and Space Technology,* September 7, 1987, pp. 88–89.
120. Brookner, E., "Radar Imaging for Arms Control," Chapter 11 in *Arms Control Verification, the Technologies that Make it Possible,* K. Tsipis, D. W. Hafemeister, and P. Janeway, (eds.), Pergamon-Brassey's, London, 1986.
121. Brookner, E., and B. R. Hunt, "Synthetic Aperture Radar Processing," *Trends and Perspectives in Signal Processing* (Prof. Alan V. Oppenheim, Editor, MIT, Cambridge, MA 02139), Vol. 2, No. 1, January 1982, pp. 2–4.
122. Bicknell, T., "Real-Time Digital Processing of SAR Data," *Spaceborne Imaging Radar Symp.,* January 17–20, 1983, Jet Propulsion Laboratory, JPL Publication 83-11, July 1, 1983, pp. 96–97.
123. Schwartzmann, L., and P. M. Liebman, "A Report on the Sperry Dome Radar," *Military Microwaves Conf. Proc.,* London, October 1978, pp. 167–176.
124. Schwartzmann, L., and P. M. Liebman, "A Report on the Sperry Dome Radar," *Microwave Journal,* Vol. 22, No. 3, March 1979, pp. 65–69.
125. Thomas, D. T., "Design Studies of Wide Angle Array Fed Lens," *IEEE Int. Symp. Antennas and Propagation,* 1979, pp. 340–343.
126. Thomas, D. T., and S. D. Bixler, "Hardware Demonstration of a 2-Dimensional Wide Angle Array Fed Lens," *IEEE Int. Symp. Antennas and Propagation,* 1979, pp. 344–347.
127. Jigun, X., M. Guilan, and C. Linmei, "A C-Band Phased Array Dome Antenna," *Rec. First Chinese Int. Conf. Radar,* Nanjing, November 4–7, 1986, pp. 571–576.
128. Wilden, H., "The Crow's-Nest Radar—An Omnidirectional Phased Array System," *IEEE Int. Radar Conf.,* 1980, pp. 253–258.
129. Davis, R. T., "Inexpensive Phased Array," *Microwaves,* August 1975, pp. 14–16.
130. Chekroun, C., D. Herrick, Y. Michel, R. Pauchard, and P. Vidal, "RADANT: New Method of Electronic Scanning," *Microwave Journal,* February 1981, pp. 45–53.

131. Michel, Y., R. Pauchard, and P. Vidal, "RADANT: New Electronic Scanning Process," *L'Onde Electrique,* Vol. 59, No. 12, 1979.
132. Beltran, F., "Elimination of the Dummy Elements in Thinned Phased Arrays," Final Report on Contract DASG60-78-C-0124 for BMDATC, Huntsville, AL, August 23, 1979, AD B043885L.
133. Beltran, F., and F. King, "Elimination of the Dummy Elements in Thinned Phased Arrays," *IEEE APS Symp.,* Los Angeles, June 15–19, 1981.
134. Schrank, H. E., "Low Sidelobe Phased Array Antennas," *IEEE Antennas and Propagation Society Newsletter,* Vol. 25, No. 2, April 1983, pp. 5–9.
135. Skolnik, M. I., "Antenna Options in Radar System Design," *Microwave Journal,* Vol. 26, No. 12, December 1983, pp. 75–82.
136. Burgess, J. S., "Recent Advances in Radar Technology," *IEE Conf. Radar-77,* London, pp. 8–11.
137. "Dish Outperformed by Slot Array; Loses Place in AF Tactical Radar," *Microwaves,* April 1980, p. 20.
138. Evans, G. E., and H. E. Schrank, "Low-Sidelobe Radar Antennas," *Microwave Journal,* Vol. 26, No. 7, July 1983, pp. 109–117.
139. Dahl, C. G., C. E. Fogelstrom, W. W. Gansz, and P. R. Merrill, "Low-Sidelobe Tactical Radar Antenna," *Proc. Antenna Applications Symp.,* University of Illinois, Urbana, IL, September 26–28, 1979.
140. "Martello—First of a New Generation of Long Range 3D Radars," *Radar Systems Int.* (Marconi Newsletter), No. 24, 1978, pp. 1–3.
141. Drabowitch, S., "Electronic Scanning Antennas," *Proc. Int. Radar Symp. India-83,* Bangalore, October 9–12, 1983, pp. 368–382.
142. Rogers, J., "Defense Radar Antennas at Thomson-CSF Planar Arrays and Recent Studies," *Proc. Int. Radar Symp. India-83,* Bangalore, October 9–12, 1983, pp. 88–92.
143. Drabowitch, S., private communication, Thomson-CSF, Bagneaux, France.
144. Wolfson, R. I., and C. F. Cho, "A Wideband, Low-sidelobe, Polarization-Agile Array Antenna," *IEEE Int. Radar Conf.,* 1980, pp. 284–287.
145. Skolnik, M. I. (ed.), *Radar Handbook,* McGraw-Hill, New York, 1970.
146. Mayhan, J. T., "Bandwidth Limitations on Achievable Cancellation for Adaptive Nulling Systems," MIT Lincoln Lab. Tech. Note 1978-1, February 17, 1978.
147. Forrest, J., *et al.,* Rome Air Development Center (RADC), Rep. No. RADC-TR-83-218, November 1983.
148. Patton, W. T., "Low-Sidelobe Antennas for Tactical Radars," *IEEE Int. Radar Conf.,* 1980, pp. 284–287.
149. Patton, W., private communication, RCA, Moorestown, NJ.
150. Carlsson, E., A. Derneryd, E. R. Karlsson, and J. O. Larsson, "Search Radar Reflector Antenna with Extremely Low Sidelobes," *Military Microwaves Conf. Proc.,* London, October 20–22, 1982, pp. 500–505.
151. Williams, N., P. Varnish, and D. J. Browning, "Reduced Cost Low Sidelobe Reflector Antenna Systems," *IEE Radar-82,* IEE Conf. Pub. No. 219, October 18–20, 1982, pp. 351–354.
152. Roulston, J. F., "Multimode Airborne Radar—A Review," *Military Microwaves Conf. Proc.,* London, October 20–22, 1984, pp. 137–147.
153. Mailloux, R. J., J. F. McIlvenna, and N. P. Kernweis, "Microstrip Array Technology," *IEEE Trans. Antennas and Propagation,* Vol. AP-29, No. 1, January 1981, pp. 25–37.
154. Carver, K. R., and J. W. Mink, "Microstrip Antenna Technology," *IEEE Trans. Antennas and Propagation,* Vol. AP-29, No. 1, January 1981, pp. 2–24.
155. Mayhan, J. T., "Some Techniques for Evaluating the Bandwidth Characteristics of Adaptive Nulling Systems," *IEEE PGAP,* May 1979, pp. 363–373.
156. "Adaptive Array Radar Looks Away from Jammers," *Microwaves,* March 1980, p. 18.

157. Brookner, E., and J. M. Howell, "Adaptive-Adaptive Array Processing," *Phased Arrays Symp. Proc.*, The Mitre Corp., Bedford, MA, October 15–18, 1985, pp. 133–146. See also: RADC Rep. No. RADC-TR-85-171, Electromagnetics Science Div., RADC, Hanscom AFB, Bedford, MA, Air Force Systems Command, August 1985; *IEE Int. Conf. Radar-87,* London, October 19–21, 1987; *Proc. IEEE,* Vol. 74, No. 4, April 1986, pp. 602–604.
158. Lee, K. J., C. Corson, and G. Mols, "A 250 kW Solid-State AN/SPS-40 Radar Transmitter," *Microwave Journal,* Vol. 26, No. 7, July 1983, pp. 93–105.
159. Mols, G., C. Corson, and K. Lee, "The Solid-State Transmitter for the AN/SPS-40 Radar System," *IEEE Int. Radar Conf.,* 1985, pp. 76–81.
160. Hoft, D. J., and F. Agi, "Solid State Transmitters for Modern Radar Applications," *Rec. First Chinese Int. Conf. Radar,* Nanjing, November 4–7, 1986, pp. 775–781.
161. M/A-COM PHI, Torrance, CA.
162. Filter Center, *Aviation Week and Space Technology,* August 22, 1983, p. 135.
163. Cohen, E. D., *Microwave Journal,* Vol. 21, No. 8, August 1978.
164. Cohen, E. D., *Microwave Journal,* Vol. 22, No. 2, February 1979, pp. 30–31.
165. Klass, P. J., "New Array to Impede Radar Failures," *Aviation Week and Space Technology,* January 4, 1982, pp. 60–61.
166. Brookner, E., and T. S. Mahoney, "Derivation of a Satellite Radar Architecture for Air Surveillance," *Microwave Journal,* Vol. 29, No. 2, February 1986, pp. 173–191; see also M.I. Skolnik (ed.), *Radar Applications*, IEEE Press, New York, 1988.
167. Bernstein, M., "Yield the Key Topic at GaAs IC Symposium," *Microwave and RF,* September 1985, pp. 41–47.
168. "Gallium-Arsenide Microcircuit Technology Meeting Planned," *Aviation Week and Space Technology,* September 30, 1985, pp. 99–101.
169. "Filter Center," "Monolithic Gallium-Arsenide Transmit-Receive Module that Includes Phase-Shifter, which Operates at X-Band with an Output Power of 0.5 W," *Aviation Week and Space Technology,* October 28, 1985, p. 93.
170. J. Van Nostrand, "New Chip Design Methodology, Advanced ICs Shine at GOMAC," *Electronic Engineering Times,* November 24, 1986, p. 26; see also *Government Microcircuit Applications Conf. (GOMAC) Proc.,* 1986.
171. Pengelly, R. S., T. B. Banbridge, A. Lane, J. R. Suffolk, J. B. Swift, T. J. Simmons, A. M. Howard, and P. N. Rigby, "An Integrated Transmit/Receive Module Employing GaAs Monolithic Integrated Circuits," *Military Microwaves Conf. Proc.,* London, October 24–26, 1984, pp. 449–456.
172. Mitchell, B., "Plessey Rushes to Put GaAs MMICs in Production," *Microwaves and RF,* November 1985, pp. 47–51.
173. Filter Center, "General Electric Will Develop Both Dual-Frequency and Wideband Transmit/Receive Modules Using Monolithic Microwave Integrated Circuit (MMIC)," *Aviation Week and Space Technology,* March 2, 1987, p. 89.
174. Filter Center, "Raytheon Electromagnetic Systems Div. Has Demonstrated a MIMIC Amplifier with 3.1-W Radio-Frequency Output Power at 6–18 GHz," *Aviation Week and Space Technology,* June 29, 1987, p. 105.
175. Jones, M., D. Seymour, S. Thompson, D. Lacinski, and L. Cook, "High-Gain MMIC Boosts Broadband IF Applications," *Microwaves and RF,* September 1987, pp. 198–208.
176. Considine, B. C., and D. L. Wandrei, "SHF SATCOM Microstrip Phased Array Antenna," *Antenna Applications Symp.,* Monticello, IL, September 22–24, 1982.
177. Considine, B. C., and D. L. Wandrei, "X-Band Receive Module Using Monolithic GaAs MMIC's," *IEEE GaAs Integrated Circuits Conf.,* Phoenix, AZ, October 25–27, 1983.

178. Considine, B. C., "An Active Microstrip Phased Array for SHF SATCOM Applications," *Phased Arrays Symp. Proc.,* The MITRE Corp., Bedford, MA, October 15–18, 1985, pp. 91–107; RADC-TR-85-171, August 1985.
179. Kachmar, M., "DoD Gets Serious About GaAs MMICs," *Microwaves and RF,* November 1985, pp. 35–36.
180. Rhea, J., "MIMIC Begins in Earnest," *Microwaves and RF,* March 1987, pp. 35–41.
181. Klass, P. J., "Pentagon Program Seeks Advanced Analog Chips," *High Technology,* July 1987, pp. 48–50.
182. Van Nostrand, J., "DoD May Up the Ante in MIMIC Program to $500 M," *Electronic Engineering Times,* November 17, 1986, p. 10.
183. Pucel, R. A., *Monolithic Microwave Integrated Circuits,* IEEE Press, New York, 1985.
184. Bookmark, *Microwaves and RF,* February 1986, p. 118.
185. Liechti, C.A., "Heterostructure Transistor Technology—A New Frontier in Microwave Electronics," *15th European Microwave Conf.,* Paris, September 9–13, 1985, pp. 21–29.
186. "General Electric Brings Latest HEMT to Life," *Microwaves and RF,* November 1985, p. 124.
187. Berenz, J.J., "HEMT Technology Gains on MM-Waves," *Microwaves and RF,* November 1985, p. 124.
188. Berenze, J.J., *et al.,* "HEMT 60-GHz Amplifier," *Electronics Letters,* Vol. 21, No. 22, October 24, 1985, p. 1028.
189. Communication News, "GE Develops HEMTs for Satellite Communications and Radar," *MSN & CT,* December 1985, p. 30.
190. Derewonko, H., and D. Delangebeaudeuf, "Thomson Continues Work on 18-to-40-GHz 'TEGFETs'," *Microwaves and RF,* 1985, pp. 126, 159.
191. Wilson D., "High-Performance ICs Fuel Up With GaAs," *Digital Design,* October 1985, pp. 41, 46.
192. Swanson, A., *et al.,* "First Commercial HEMT Challenges GaAs FETs," *Microwaves and RF,* November 1985, pp. 107–118.
193. Browne, J., "Looking Forward to the Promise of the HEMT," *Microwaves and RF,* March 1985, p. 17.
194. "HEMTs Fire Varian Amp," *Microwaves and RF,* May 1987, p. 31.
195. Brookner, E., "Developments in Digital Radar Processing," *Trends and Perspectives in Signal Processing* (Prof. Alan V. Oppenheim (ed.), MIT, Cambridge, MA 02139), Vol. 2, No. 1, January 1982, pp. 7–12; Brookner, E., "A Guide to Radar Systems and Processors," *Trends and Perspectives in Signal Processing,* Vol. 2, No. 1, January 1982, pp. 13–23.
196. Brookner, E., "Trends in Radar Signal Processing," *Microwave Journal,* Vol. 25, No. 10, October 1982, pp. 20–39.
197. Filter Center, "Hughes Radar Systems Group Produced the Last APG-63 Radar for the USAF/McDonnell Douglas F-15 in Oct.," *Aviation Week and Space Technology,* November 3, 1986, p. 151.
198. "VHSIC Demonstrates Significant Benefits In Performing Signal Processing Functions," *Aviation Week and Space Technology,* September 7, 1987, pp. 98–99.
199. Brookner, E., "Radar of the '80s and Beyond," *IEEE Electro-84,* Session 4, Boston, MA, May 15–17, 1984.
200. Special Issue on VHSIC, *Aviation Week and Space Technology,* July 30, 1984.
201. Filter Center, "The Honeywell-developed Very High Speed Integrated Circuits Phase 1 Chipset of Three Microcircuits, which Can Be Used for Real-Time Image Processing," *Aviation Week and Space Technology,* July 22, 1985, p. 181.
202. Iversen, W.R., "IBM Goes to Market With Six VHSIC Chips," *Electronics Week,* November 19, 1984, pp. 18–19.

203. Baker, S., "CMA To Sell Gate Arrays Using TRW Process," *Electronics Engineering Times,* July 22, 1985, p. 37.
204. Waller, L., "VHSIC Finally Builds a Head of Steam," *Electronics,* April 16, 1987, pp. 84–86.
205. Ishiwada, Y., and Y. Nemmoto, "Present Status of Japanese ERS-1 Project," *Second Spaceborne Imaging Radar Symp.,* April 28–30, 1986, Jet Propulsion Laboratory, JPL Pub. 86-26, December 1, 1986, pp. 79–83.
206. Patterson, D.O., and E.D. Maynard, "VHSIC Today and Tomorrow," *NATO AGARD Conf. Proc.,* "The Impact of Very High Performance Integrated Circuits on Radar, Guidance and Avionics Systems," No. 380, Paper No. 1, May 1985.
207. Adam, J.A., "Fast Chips Start to Fly," *IEEE Spectrum,* Vol. 23, No. 1, January 1986, p. 79.
208. "TRW to Provide VHSIC Technology for Signal Processors," *Defense Electronics,* February 1987, p. 27.
209. Lyman, J., "VHSIC Moves Headlong into the Submicron Stage," *Electronics,* April 16, 1987, pp. 91–93.
210. Waller, L., "Update TRW's Superchip Project Forges Ahead," *Electronics,* July 23, 1987, p. 73.
211. Guteri, F., *et al.,* "In Pursuit of the One-Month Chip," *IEEE Spectrum,* September 1984, Vol. 21, No. 9, pp. 28–49.
212. "Silicon Compilation Aids VHSIC Insertion," *Aviation Week and Space Technology,* January 13, 1986, p. 157.
213. Merrifield, J.T., "General Dynamics Signs Pact for Silicon Compilation Expertise," *Aviation Week and Space Technology,* November 4, 1985, pp. 80–81.
214. Filter Center, "General Electric is Offering Semiconductor 'Foundry' Service for Custom Microcircuits with 1.25-micron or 2-micron Feature Sizes," *Aviation Week and Space Technology,* October 21, 1985, p. 167.
215. "Compiler Goes from Functional Description to Layout in 3 Days," *Electronic Engineering Times,* March 16, 1987, p. 51.
216. McLeod, J., "Update: Round Two for a Desktop Silicon Compiler," *Electronics,* June 25, 1987, p. 100.
217. "Toshiba Begins Production of 1M DRAM," *Semiconductor International,* February 1986, pp. 24–26.
218. "1-Mbit DRAMs Get Designed In, but Still Costly," *Electronic Engineering Times,* October 27, 1986, p. 41.
219. Fischetti, M. A., "Memories on Parade," *IEEE Spectrum,* Vol. 23, No. 1, January 1986, p. 55.
220. "Philips' 1-Mbit SRAM Challenges Japan," *Electronics,* July 23, 1987, p. 36.
221. "Mbit SRAM Module Samples," *Electronic Engineering Times,* February 2, 1987, p. 15.
222. Lammers, D., "NEC Goes After Toshiba in 1-Mbit DRAM Stakes," *Electronic Engineering Times,* January 26, 1987, p. 14.
223. Andrews, W., "4-Mbit DRAMs To Highlight '86 ISCC," *Electronic Engineering Times,* January 20, 1986, pp. 45, 60.
224. "Siemens Joins the Select 4-Mb DRAM Club," *Electronics,* March 19, 1987, p. 53.
225. Woolnough, R., "Siemens Aims 4 Mbit For Logic Line, Plans ASIC Role," *Electronic Engineering Times,* June 8, 1987, pp. 32–34.
226. Baker, S., and M. Gold, "TI: 4-Mbit DRAM In '88," *Electronic Engineering Times,* September 14, 1987, pp. 1, 8.
227. Bursky, D., "Megabit Dynamic RAMs Target System Solutions," *Electronic Design,* July 9, 1987, pp. 35–38.
228. Cushman, R. H., "ISSCC '86 Speakers Predict Future Developments in Solid-State Technology," *Electronic Design,* February 6, 1986, pp. 77–80.

229. "Toshiba Develops Technology for 4M DRAM," *Semiconductor International,* February 1986, p. 22.
230. Wright, M., "Dynamic RAMs," *Electronic Design,* February 20, 1986, pp. 134–148.
231. ISSCC, Electronic Design Reports, *Electronic Design,* February 20, 1986, pp. 99–123.
232. Brown, C., and S. Baker, "IBM, NTT Provide DRAMatics at ISSCC," *Electronic Engineering Times,* March 2, 1987, pp. 1, 14.
233. "4M EPROM," *Electronic Engineering Times,* December 15, 1986.
234. "IBM Cashes in on its New Superchips," *Science Digest,* August 1985, p. 65.
235. Innovations, "16-Megabit Memory Chips Are Near," *High Technology,* July 1987, p. 6.
236. "16-Mbit Dynamic RAM Tops ISSCC Memory Line Up," *Electronic Design,* March 5, 1987, p. 23.
237. "Fast-Accessing 1-Mbit CMOS EPROM Makes its Debut," *Electronic Design,* May 2, 1985, p. 27.
238. "One-Megabit EPROMs Invade Disk Territory," *Electronics Week,* April 22, 1985, pp. 52–54.
239. Costlow, T., "GaAs NMOS, and Bipolar Parts Add Zip to High-Speed Tasks," *Electronic Design,* September 19, 1985, p. 225.
240. Brown, C., "Process Achieves Record Yields on GaAs SRAM," *Electronic Engineering Times,* January 31, 1986, p. 21.
241. Brown, C., "GaAs ICs: Niches, Not Commodities," *Electronic Engineering Times,* January 13, 1986, pp. 1, 32.
242. Bernstein, M., "The Sixth IEEE GaAs IC Symposium Spotlights a Discipline Ready to Roll," *Microwaves and RF,* January 1985, pp. 37–40.
243. Logan, M., "Flash ADC Achieves High Speed at Low Power," *RF Design,* May 1987, p. 27.
244. "Honeywell Ups Performance, Not Price, on A/D Flash Converter," *Electronic Products,* July 15, 1987, pp. 51, 54.
245. "TRW's 10-Bit ADC Leads the Pack," *Electronics,* August 20, 1987, pp. 33–34.
246. Goodenough, F., "Single-Chip 10-bit Flash ADC Samples at a 50-MHz Rate," *Electronic Design,* September 3, 1987, pp. 49–52.
247. Weiss, R., "Ruggedized 8-bit ADC Samples at 20 MHz," *Electronic Design,* March 5, 1987, pp. 53–54.
248. "ADC Samples, Holds at 1.3 MHz," *Electronic Design,* March 19, 1987, p. 98.
249. "Sony Tektronix Waveform Digitizer Claims A-D Conversion Record," Products Newsletter, *Electronics,* March 19, 1987, p. 25.
250. "10 Views from the Top," *Electronic Design,* January 10, 1985, pp. 261–281.
251. Goodenough, F., "GaAs Fuels 1-GHz, 8-bit Video DAC," *Electronic Design,* April 30, 1987, pp. 57–58.
252. "TI's 8836 32-bit MAC in Production," *Electronic Products,* September 1, 1987, pp. 67, 71.
253. "'Speediest' Multipliers Claimed," *Electronic Engineering Times,* March 23, 1987, p. 56.
254. Titus, J., "Chips," *Electronic Design,* June 11, 1987, pp. 115–126.
255. News Briefs, "Fairchild Bows 16-Bit MAC, First VLSI FACT Device," *Electronic Engineering Times,* November 17, 1986, p. 2.
256. *Electronic Design,* April 12, 1980.
257. "TRW Has First Dividing Stand-Alone VLSI Chip," *Electronic Engineering Times,* May 25, 1987, p. 2.
258. Titus, J., "Special Report, DSP ICs," *Electronic Design,* October 16, 1986, pp. 162–176.

BROOKNER
CHAPTER 2

259. Wiegand, J., "Digital Signal Processing Enters the Mainstream," *Electronic Design,* August 6, 1987, pp. 111–118.
260. Shear, D., "Wide Choice of Tools Breaks Down Barriers to DSP Design," *Electronic Design,* August 20, 1987, pp. 183–188.
261. Shear, D., "Design and Build a Transponder Using DSP Tools," *Electronic Design,* September 3, 1987, pp. 137–150.
262. Cushman, R. H., "μP-like DSP Chips," *Electronic Design,* September 3, 1987, pp. 155–186.
263. Simar, R., Jr., and M. Hames, "CMOS DSP Chip Packs Punch of a Supercomputer," *Electronic Design,* March 19, 1987, pp. 103–110.
264. Baker, S., "TI DSP Hits 33 Mflops, Packs FPU," *Electronic Engineering Times,* March 23, 1987, pp. 1, 16.
265. Bindra, A., and R. Doherty, "Next Generation of DSPs Faster, More Powerful," *Electronic Engineering Times,* April 27, 1987, pp. 41–42.
266. Morrow, J., "CMOS Floating-Point DSP Chip Features 80-nsec Cycle Time," *Electronic Design,* August 20, 1987, p. 21.
267. Bursky, D., "On-Chip Cache Lets DSP IC Cycle at 75 ns," *Electronic Design,* March 5, 1987, p. 144.
268. Gold, M., "CMOS DSP Hits 25 MFLOPS," *Electronic Engineering Times,* July 20, 1987, pp. 1, 16.
269. Ferro, F., and K. Ulery, "25-MFLOPS DSP Looks for New Worlds to Conquer," *Electronic Products,* September 15, 1987, p. 37.
270. "A New Class of DSP Chip: The Vector Signal Processor," *Electronics,* July 24, 1986, pp. 59–66.
271. Doherty, R., "ECL-Compatible GaAs IC," *Electronic Engineering Times,* February 17, 1986, p. 14.
272. Panasuk, C., "3-GHz GaAs Logic Array Fits Discrete ECL Designs into a 300-Gate Package," *Electronic Design,* May 30, 1985, p. 157.
273. Smith, D., "Gallium Arsenide Challenges Silicon in High-Speed-Array Applications," *Electronic Design,* June 27, 1985, pp. 75–78.
274. "The Impact of Very High Performance Integrated Circuits on Radar, Guidance and Avionics Systems," *NATO AGARD Conf. Proc.,* No. 380, May 1985.
275. "GaAs Wedges Deeper into Digital LSI Market," *Electronic Products,* April 15, 1987, pp. 15, 20.
276. "3456-Gate Array Debuts in GaAs," *Electronic Design,* March 19, 1987, p. 24.
277. Filter Center, "A 3,000-gate Configurable-Cell Gallium-Arsenide Array, which Reportedly Offers a Twofold Higher Speed than Silicon," *Aviation Week and Space Technology,* October 13, 1986, p. 135.
278. Design Alert, "GaAs Product Yields Let Prices Come Down," *Electronic Design,* September 3, 1987, p. 32.
279. Filter Center, "First Details on the Gallium-Arsenide Microcircuits to Be Used in the Next Generation Cray-3," *Aviation Week and Space Technology,* September 14, 1987, p. 75.
280. Filter Center, *Aviation Week and Space Technology,* July 23, 1984, p. 155.
281. Filter Center, "TI Produces 4K Gallium Arsenide Chip," *Aviation Week and Space Technology,* June 10, 1985, pp. 125–127.
282. Filter Center, "Defense Advanced Research Projects Agency Plans to Create a Third Industrial Gallium-Arsenide Microcircuit Manufacturing Facility," *Aviation Week and Space Technology,* December 16, 1985, p. 77.
283. Peterson, I., "A High-Speed Match Made in Silicon," *Science News,* Vol. 129, p. 102.
284. Morkoc, H., and P. M. Solomon, "The HEMT: A Superfast Transistor," *IEEE Spectrum,* February 1984, pp. 28–35.
285. Filter Center, *Aviation Week and Space Technology,* October 6, 1986, p. 99; see also GaAs IC Symposium, Grenelefe, FL, October 28–30, 1986.

286. McLeod, J., "The First GaAs Compiler is Already Producing Chips," *Electronics,* April 30, 1987, pp. 58–62.
287. Bindra, A., "CMOS Device Offers 20,000 Gates," *Electronic Engineering Times,* October 14, 1985, p. 36.
288. Baker, S., "Honeywell Plans 20K-Gate Array with CMOS-III," *Electronic Engineering Times,* September 2, 1985, p. 28.
289. "20,000-Gate CMOS Array Achieves 500 ps Leaded Delays," *Computer Design,* January 1, 1987, p. 9.
290. Bursky, D., "Arrays Packing 50,000 Gates Help Build Faster Systems," *Electronic Design,* August 20, 1987, pp. 60–61.
291. Wirbel, L., "Raytheon Rolling Out ECL Arrays On Bipolar Integrated Tech's Fab," *Electronic Engineering Times,* July 6, 1987, p. 46.
292. Mead, C., and L. Conway, *Introduction to VLSI Systems,* Addison-Wesley, Reading, MA, 1980.
293. Kung, H. T., "On the Implementation and Use of Systolic Array Processors," *IEEE Int. Conf. Computer Design, VLSI in Computers,* 1983, pp. 370–373.
294. Haggarty, R. D., B. L. Johnson, and E. A. Palo, "A Revolutionary Design Method for VLSI Signal Processors," *NATO AGARD Conf. Proc.* No. 380, "The Impact of Very High Performance Integrated Circuits on Radar, Guidance and Avionics Systems," May 1985, Paper No. 25.
295. Vaccaro, J. J., B. L. Johnson, and C. L. Nowacki, "A Systolic Discrete Fourier Transform Using Residue Number Systems Over the Ring of Gaussian Integers," *IEEE ICASSP Conf.,* May 1986.
296. Ward, C. R., A. J. Robson, and P. J. Hargrave, "The Application of a Systolic Least Squares Processing Array to Adaptive Beamforming," *IEEE ICASSP Conf.,* 1984, pp. 34A.3-1–4.
297. McWhirter, J. G., "Recursive Least-Squares Minimization Using a Systolic Array," *Proc. SPIE,* Vol. 431, *Real Time Signal Processing,* Vol. 1, 1983, pp. 105–112.
298. Griffiths, J. W. R., "Adaptive Array Processing—A Tutorial," *IEE Proc.,* Vol. 130, Pts. F, H, No. 1, February 1983, pp. 3–10.
299. Ward, C. R., and P. J. Hargrave, "A Systolic Array for High Performance Adaptive Beamforming," *NATO AGARD Conf. Proc.* No. 380, "The Impact of Very High Performance Integrated Circuits on Radar, Guidance and Avionics Systems," May 1985, Paper No. 19.
300. Willey, T., T. S. Durrani, and R. Chapman, "An FFT Systolic Processor and its Applications," *IEEE ICASSP Conf.,* 1984, pp. 34A.4-1–4.
301. Ward, C. R., P. J. Hargrave, and J. G. McWhirter, "A Novel Algorithm and Architecture for Adaptive Digital Beamforming," *IEEE Trans. Antennas and Propagation,* Vol. AP-34, No. 3, March 1986, pp. 338–346.
302. McWhirter, J. G., "A Brief Review of Adaptive Null Steering Techniques," Royal Signals and Radar Establishment, RSRE Memorandum No. 3939, February 1986.
303. "Sandia-Made Microprocessor Resists High Radiation Doses," *Electronic Design,* September 9, 1985, pp. 27–28.
304. Baker, S., "Sandia's 8-bit Microprocessor Handles Ten Times More Radiation," *Electronic Engineering Times,* July 22, 1985, p. 39.
305. "Rad-Hardened CMOS Version of 8085 μP Cuts Power Consumption by as Much as 90%," *Electronic Design,* April 11, 1985, p. 183.
306. Universal Semiconductor, "RAD Hard CMOS Arrays," *Electronic Engineering Times,* January 20, 1986, p. 29.
307. "Gate Arrays Absorb High Radiation Dose," *Electronic Engineering Times,* November 11, 1985, p. 8.
308. Merrifield, J. T., "Sandia Extending Radiation-Hardening Technology to Complex Microcircuits," *Aviation Week and Space Technology,* August 5, 1985, pp. 89–91.

309. Filter Center, "RCA will Receive a Defense Nuclear Agency Contract to Develop a High-Speed Radiation-Resistant 64K Static Random Access Memory (SRAM) Microchip," *Aviation Week and Space Technology,* July 8, 1985, p. 129.
310. Filter Center, "Radiation Hardened Gate Array that Provides the Equivalent of 1,000 Bipolar Gates, with Internal Gate Speeds of 0.8 nanosec. After Radiation Exposure," *Aviation Week and Space Technology,* April 15, 1985, p. 85.
311. Fischetti, M. A., "A Leap in Radiation Hardness," *IEEE Spectrum,* Vol. 23, No. 1, January 1986, p. 56.
312. Microsystems News, "Harris' Rad Hard CMOS Breakthrough," *DS&E News,* July 1986, p. 64.
313. Filter Center, "Samples of the New Intel 32-bit 80386 Microprocessor, which Are Radiation Resistant to Total Dose of More Than 10^5," *Aviation Week and Space Technology,* April 21, 1986, p. 89.
314. Baker, S., "RCA Puts Out CMOS SOS Rad-Hard Parts—Head for Volume Production," *Electronic Engineering Times,* July 13, 1987, p. 1, 6.
315. Filter Center, "Radiation-Hardened 16K Static Random Access Memory (SRAM) Chip that Uses 1.25-micron Feature Sizes and Offers Worst-Case Access Time of 160 nanosec.," *Aviation Week and Space Technology,* March 30, 1987, p. 101.
316. Petri, U., "Results From an Experimental Receiving Array Antenna with Digital Beamforming," *IEE Radar-82 Conf. Proc.,* pp. 399–402.
317. Sanders, W., "Beamforming with Phased Array Antennas," *IEE Radar-82 Conf. Proc.,* pp. 403–407.
318. Wardrop, B., "Digital Beamforming in Radar Systems—A Review," *Military Microwaves Conf. Proc.,* London, October 24–26, 1984, pp. 319–323.
319. Old, J. C., D. J. Day, and G. N. Harvey, "'GASP'—The GEC Array-Signal-Processing Test Bed," *IEEE Int. Radar Conf.,* 1985, pp. 220–225.
320. Litva, J., "Processing Requirements for Radar Digital Beamforming," *NATO AGARD Conf. Proc.* No. 380, "The Impact of Very High Performance Integrated Circuits on Radar, Guidance and Avionics Systems," May 1985, Paper No. 20.
321. Brookner, E., "Phased-Array Radars," *Scientific American,* Vol. 252, No. 2, February 1985, pp. 94–102.
322. Works, G., "Distributed Beam Steering Computer," US Patent 4,445,119.
323. Rabinowitz, S. J., C. H. Gager, E. Brookner, C. E. Muehe, and C. M. Johnson, "Applications of Digital Technology to Radar," *Proc. IEEE,* Vol. 73, No. 2, February 1985, pp. 325–339.
324. Lemonick, M.D., "Superconductivity Heats Up," *Time,* March 2, 1987, p. 62.
325. Gleick, J., "In the Trenches of Science," *The New York Times Magazine,* August 16, 1987, pp. 28*ff*.
326. "Thin-Film Deposition of Superconductors Researched," *Electronic Engineering Times,* June 29, 1987, pp. 1, 8.
327. Chiang, A. M., "A Real Time CCD Parallel Radar Processor," *SPIE Real Time Signal Processing Symp.,* July 1987.
328. Chiang, A. M., and B. E. Burke, "A High Speed CCD Digitally Programmable Transversal Filter," *IEEE J. Solid-State Circuits,* 1984, Vol. SC-18, p. 745.
329. Chiang, A. M., and C. M. Rader, "High Speed Single Chip Signal Processing Devices for Radar and Sonar," *GOMAC Digest,* 1982, pp. 222–225.
330. Johnson, R. C., and C. Brown, "Once Again, Designers' Thoughts Turn to Analog," *Engineering Electronic Times,* July 13, 1987, pp. 41, 53.
331. Montgomery, J., J. R. Klein, and J. Glasco, "Fiber Optics Takes Hold for Analog Applications," *Microwaves and RF,* February 1986, pp. 47–48.

332. Special Issue on Fiber Optics, *Microwave Journal,* April 1985.
333. Forrest, J. R., "Opto-Electronics in Practice and Promise," *GEC J. Research,* Vol. 2, No. 2, 1984, pp. 62–65.
334. Forrest, J. R., F. P. Richards, A. A. Salles, and P. Varnish, "Optical Fiber Networks for Signal Distribution and Control in Phase Array Radars," *IEE Int. Conf. Proc. Radar-82,* pp. 408–412, 1982.
335. Forrest, J., F. Richards, D. Esdale, and D. Wight, "Optical Techniques for Signal Distribution and Control in Phased Array Radars," *IEEE Int. Radar Conf. Proc.,* 1985, pp. 410–415.
336. Billam, E. R., "New Technologies in Phased Array Radar," *15th European Microwave Conf.,* Paris, September 9–13, 1985, pp. 57–66.
337. Daryoush, A., P. Herczfeld, V. Contarino, A. Rosen, Z. Turski, and P. Wahi, "Optical Beam Control of mm-Wave Phased Array Antennas for Communications," *Microwave Journal,* March 1987, pp. 97–104.
338. Unger, H. G., "Recent Developments and Trends in Optical Communication," *15th European Microwave Conf.,* Paris, September 9–13, 1985, pp. 48–56.
339. Lay, R. W., "Fiber Optics Advance Cited," *Electronic Engineering Times,* August 26, 1985, pp. 1, 14.
340. *Electronic Warfare/Defense Electronics,* April 1979, p. 90.
341. "Surface Radar at Stockholm Airport Uses Fiber-Optic Links," *Aviation Week and Space Technology,* August 12, 1985, pp. 72–73.
342. Hawkins, W. J., "Electronics Newsfront," *Popular Science,* February 1986, p. 24.
343. Manuel, T., "The Picture Brightens In Flat-Panel Technology," *Electronics,* May 28, 1987, pp. 55–64.
344. Waldron, T. P., S. K. Chin, and R. J. Naster, "Distributed Beamsteering Control of Phased Array Radars," *Microwave Journal,* September 1986, pp. 133–146.
345. Tsandoulas, G. N., "Space-Based Radar," *Science,* July 17, 1987, Vol. 237, pp. 257–262.
346. Brookner, E., "A Review of Array Radars," *Microwave Journal,* Vol. 24, No. 10, October 1981, pp. 25–42, 61, 114.
347. News Briefs, "AT&T, Sandia Cooperating on Rad-Hard CMOS ICs," *Electronic Engineering Times,* April 6, 1987, p. 15.
348. Bursky, D., "Software Tool Kit Eases Design of DSP Algorithms," *Electronic Design,* June 25, 1987, pp. 37–40.
349. Products Newsletter, "Zoran's 32-Bit Floating-Point Processor Is Loaded with New Features," *Electronics,* September 17, 1987, p. 25.
350. Broad, W. J., "Soviet Radar on Display," *New York Times,* September 9, 1987.
351. Mann, P., "Administration Disputes Findings of U.S. Visit to Soviet Radar," *Aviation Week and Space Technology,* September 14, 1987, pp. 26–28.
352. Broad, W. J., *New York Times,* private communication.
353. Waller, L., "Commercial Quantities of LSI GaAs Are Finally Here," *Electronics,* September 17, 1987, pp. 48–49.
354. Manuel, T., "Update: GaAs LSI from Vitesse Is on Track," *Electronics,* September 17, 1987, p. 100.
355. Iversen, W. R., "Laser Milestone: GaAs on Silicon," *Electronics,* September 17, 1987, p. 32.
356. Iversen, W. R., "Cray Is Still Using GaAs—but not Its Own," *Electronics,* September 17, 1987, p. 49.
357. Sakurai, M., "Scientists Play Down Superconductivity Discoveries," *Electronic Engineering Times,* September 14, 1987, p. 30.
358. Reed, J. E., "The AN/FPS-85 Radar System," *Proc. IEEE,* Vol. 57, No. 3, March 1969, pp. 324–335.

BROOKNER
CHAPTER 2

359. Fujitsu Microelectronics, Sanda Clara, CA.
360. Brookner, E., "Radar Trends to the Year 2000," *Interavia,* May 1987, pp. 481–486.
361. "Interavia Surface Radar Catalogue," *Interavia,* April 1987, pp. 366–378; "Interavia Review of Western Airborne Attack Radar Systems," *Interavia,* March 1981, pp. 232–241.
362. Hughes Aircraft, "The HR-3000, A New Generation Version of the Hughes Air Defense Radar (HADR)," *Hughes Science Scope,* September 1987.
363. Washington Window, "Antenna Sites Chosen for Alaskan Radar System," *MSN & CT,* September 1987, p. 102.
364. International Report, "US Radars Ordered to Guard Persian Gulf," *Microwave Journal,* September 1987, p. 55.
365. Ho, P., T. Andrade, and E. Johnson, "GaAs MMIC Reliability Analysis and Its Impact on Microwave Systems, Part II," *MSN & CT,* September 1987, pp. 130–133.
366. Wisseman, W. R., L. C. Witkowski, G. E. Brehm, R. P. Coats, D. D. Heston, R. D. Hudgens, R. E. Lehmann, H. M. Macksey, and H. Q. Tseng, "X-Band GaAs Single-Chip T/R Radar Module," *Microwave Journal,* September 1987, pp. 167–173.
367. Chilton, R. H., "MMIC T/R Modules and Applications," *Microwave Journal,* September 1987, pp. 131–146.
368. Selin, J. R., "Continuously Variable L-Band Monolithic GaAs Phase Shifter," *Microwave Journal,* September 1987, pp. 211–218.
369. Hughes Aircraft, "A Processor Utilizing Advanced Microchips Will Offer Processing Power," *Hughes' Science Scope,* September 1987.
370. Martin, S. L., "Wave of Advances Carry DSPs to New Horizons," *Computer Design,* September 15, 1987, pp. 69–83.
371. Contolini, R. E., C. J. Mason, and G. A. Sterling, "Multiple DSPs Provide Speed for Digital Servo Control," *Computer Design,* September 15, 1987, pp. 87–92.
372. Bursky, D., "Floating-Point Signal Processor Packs High-Level Instructions," *Electronic Design,* September 17, 1987, pp. 55–58.
373. Bond, J., "GaAs, ECL and CMOS Gate Arrays Crash through Speed Barriers," *Computer Design,* September 15, 1987, pp. 20–30.
374. Gold, M., "Vitesse Points GaAs at Commodity IC Markets," *Electronic Engineering Times,* September 21, 1987, p. 54.
375. Woolnough, R., "Esprit's GaAs Advances to Highlight Esscirc '87," *Electronic Engineering Times,* September 21, 1987, p. 56.
376. Costlow, T., "Plan for Next Cray Revealed," *Electronic Engineering Times,* September 21, 1987, pp. 1, 8.
377. Filter Center, "ITT Defense Technology Corp. Has Developed a Gallium Arsenide 4 × 4 bit Array High-Speed Multiplier Integrated Circuit," *Aviation Week and Space Technology,* September 21, 1987, p. 99.
378. Wirbel, L., "Zoran Details 32-bit Floating Point Processor," *Electronic Engineering Times,* September 21, 1987, p. 8.
379. Pan, J. J., "Integrated Optic Components for Microwave Coherent Lightwave System," *MSN & CT,* September 1987, pp. 44–59.
380. Geideman, W. A., T. L. Rasset, T. A. Misko, and J. W. Wine, "An All GaAs Signal Processing Architecture," *Microwave Journal,* Vol. 30, No. 9, September 1987, pp. 105–127.
381. Dechant, D. J., and F. A. Horrigan, "AOSP—Multiprocessor Architecture and System Considerations," *AIAA Computers in Aerospace VI Conf.,* Wakefield, MA, October 7–9, 1987.
382. Billam, E. R., and D. H. Harvey, "MESAR—An Advanced Experimental Phased Array Radar," *IEE Int. Conf. Proc. Radar-87,* London, October 19–21, 1987, pp. 37–40.

383. Claridge, P. A., C. R. Green, L. I. Gregory, M. D. R. Tench, A. A. Lane, and D. M. Manners, "Affordable GaAs Tx/Rx Modules for Phased Array Radar," *IEE Int. Conf. Proc. Radar-87,* London, October 19–21, 1987, pp. 41–45.
384. Thompson, G. R. G., "A Modular Approach to Multi-Function Radar Design for Naval Application," *IEE Int. Conf. Proc. Radar-87,* London, October 19–21, 1987, pp. 32–36.
385. Logan, R., "Airborne Solid State Phased Arrays: A System Engineering Perspective," *IEE Int. Conf. Proc. Radar-87,* London, October 19–21, 1987, pp. 26–31.
386. Cole, B. C., "Selective Growth and New DRAM Cells Share the IEDM Spotlight," *Electronics,* November 26, 1987, pp. 109–110.
387. "A Rad-Hard Array Family Evolves from VHSIC," *Electronics,* November 26, 1987, pp. 139–140.
388. Washington Window, "Raytheon Upgrades Greenland Radar Station," *MSN & CT,* November 1987, p. 66.
389. Weiss, R., "Toshiba Opens 4-Mbit Front," *Electronic Engineering Times,* November 23, 1987, pp. 1, 8.
390. Brown, C., "High-Temp Superconductors," *Electronic Engineering Times,* November 30, 1987, pp. 45–46.
391. Cohen, E. D., "MIMIC: Improving Military Electronic Systems for the 1990s," *Defense Electronics,* November 1987, pp. 87–98.
392. Brown, C., "GaAs-On-Silicon Technology Scores Advances," *Electronic Engineering Times,* November 2, 1987, pp. 41, 47.
393. "Excimer Laser Stepper Aimed at 16M DRAMS," *Electronic Engineering Times,* November 2, 1987, p. 47.
394. Rogers, K., "CMOS Multipliers Sprint to New Speed Records," *Electronic Engineering Times,* November 2, 1987, p. 7.
395. Stegall, R. L., "Multiple Object Tracking Radar (MOTR)," *Military Microwaves Conf. Proc.,* London, June 24–26, 1986, pp. 69–74.
396. Grave, E., "Thomson CSF Planar Array 3D Radar," *Military Microwaves Conf. Proc.,* London, June 24–26, 1986, pp. 94–99.
397. Dahlsjo, O., B. Ljungstrom, and H. Magnusson, "Plastic and Composite Materials for Antennas; an Evolution Towards More Demanding Applications," *Military Microwaves Conf. Proc.,* London, June 24–26, 1986, pp. 237–242.
398. Adatia, N., and S. Tun, "Low Sidelobe Antennas Using Shaped Reflector Techniques," *Military Microwaves Conf. Proc.,* London, June 24–26, 1986, pp. 139–145.
399. Hallberg, L., J. E. Nilsson, and S. Wallander, "Tube versus Solid State Trade-Off for S-band Phased Array Radar," *Military Microwaves Conf. Proc.,* London, June 24–26, 1986, pp. 87–93.
400. Woods, R. L., "Microwave Power Source Overview," *Military Microwaves Conf. Proc.,* London, June 24–26, 1986, pp. 349–353.
401. Hieslmair, H., C. DeSantis, and N. J. Wilson, "State of the Art of Solid-State and Tube Transmitters," *Microwave Journal,* Vol. 26, No. 10, October 1983, pp. 46–48.
402. Weil, T. A., "Transmitters," Chapter 7 in Skolnik, M. I. (ed.), *Radar Handbook,* McGraw-Hill, New York, 1970.
403. "Unisys Wins $60-Million Contract for Nexrad Doppler Weather Radar," *Aviation Week and Space Technology,* December 7, 1987, p. 35.
404. Long, W. H., III, and K. A. Harriger, "Medium PRF for the AN/APG-66 Radar," *Proc. IEEE,* Vol. 73, No. 2, February 1985, pp. 301–311.
405. Smith, J. M., and R. H. Logan, "AN/APS-116 Periscope-Detecting Radar," *IEEE Trans. Aerospace and Electronic Systems,* Vol. AES-16, No. 1, January 1980, pp. 66–73.

BROOKNER
CHAPTER 2

406. Lipka, M., "Pulse Compression Filter and Wideband Evaluation," *IEEE Int. Radar Conf.,* 1975, pp. 283–287.
407. Clarke, J., "Airborne Early Warning Radar," *Proc. IEEE,* Vol. 73, No. 2, February 1985, pp. 312–324.
408. Staudaher, F. M., "Airborne MTI," Chapter 18 in *Radar Handbook,* Skolnik, M. I. (ed.), McGraw-Hill, New York, 1970.
409. Petty, R. T. (ed.), "Jane's Weapon Systems," Jane's Publishing, London, 1977, pp. 609–611.
410. Skillman, W. A., "Utilization of the E-3A Radar in Europe," *Military Electronics Defence Expo '78,* October 3–5, 1978, pp. 334–343.
411. Ethington, D., "Military Radar System Evolution Extends Service Life and Fulfills New Missions," *MSN,* February 1984, pp. 54–64.
412. Taylor, J. W., Jr., and G. Brunins, "Design of a New Airport Surveillance Radar (ASR-9)," *Proc. IEEE,* Vol. 73, No. 2, February 1985, pp. 284–289.
413. Cole, E. L., M. J. Hodges, R. G. Oliver, and A. C. Sullivan, "Novel Accuracy and Resolution Algorithms for the Third Generation MTD," *IEEE National Radar Conf. Proc. Radar-86,* 1986, pp. 41–47.
414. Taylor, J. W., Jr., and G. Brunins, "Design of a New Airport Surveillance Radar (ASR-9)," *Rec. First Chinese Int. Conf. Radar,* Nanjing, November 4–7, 1986, pp. 33–38.
415. "Westinghouse Ground Radar Systems," Westinghouse Co. brochure.
416. "Know-How Built It—Operations Refined It—The AN/TPS-43 Radar Set," Westinghouse Co. brochure.
417. "Air-Defense Radar to Protect Marines Against Low Altitude Attacks," *Microwaves,* March 1978, p. 14.
418. "The AN/TPS-63 Radar Set," Westinghouse Co. brochure.
419. Taylor, J. W., Jr., and H. J. Blinchikoff, "The Quadriphase Code—A Radar Pulse Compression Signal with Unique Characteristics," *IEE Int. Conf. Proc. Radar-87,* London, October 19–21, 1987, pp. 315–319; see also *IEEE Trans. Aerospace and Electronic Systems,* Vol. AES-24, No. 3, March 1988, pp. 156–170.
420. Westinghouse Co. flier.
421. Pfister, G., "The Tactical Three-Dimensional Air Surveillance Radar," *Electrical Communication* (ITT-Gilfillan Magazine), Vol. 53, No. 1, 1982.
422. King, I. E., A. A. Albanese, and H. A. Burris, "The AN/TPS-59 and GE592—A Family of Solid-State Radars," *IEEE Radar-77 Symp.,* October 1977.
423. "Solid State L-Band Radars," General Electric Co. brochure, Electronic Systems Div., Syracuse, New York.
424. Ethington, D., Hughes Aircraft Corp., private communication.
425. "Dutch Firm Completes Tests of 3-D Radar," *Aviation Week and Space Technology,* September 1, 1986, p. 238.
426. Pengelley, R., "Ground Air Defense Sensors for the 1990s; Technologies Enter Gradually," *International Defense Review,* December 1987, pp. 1619–1625.
427. Farina, A., and F. A. Studer, *Radar Data Processing:* Vol. I, *Introduction and Tracking* and Vol. II, *Advanced Topics and Applications,* John Wiley and Sons and Research Studies Press, New York, 1985 (Vol. I), 1986 (Vol. II).
428. Mirkin, M. I., C. E. Schwartz, and S. Spoerri, "Automated Tracking with Netted Ground Surveillance Radars," *IEEE Int. Radar Conf. Proc.,* April 28–30, 1980, pp. 371–379.
429. Farina, A., and G. Galati, "Surveillance Radars: State of the Art, Research and Perspectives," *Alta Frequenza,* No. 4, Vol. LIV, 1985, pp. 261–277; see also Farina, A. (ed.), *Optimized Radar Processors,* Peter Peregrinus and the Institution of Electrical Engineers, London, 1987, pp. 161–178.

430. "Israel Developing 707-Based Surveillance Platform," *Aviation Week and Space Technology,* June 15, 1987, p. 175.
431. Shifrin, C. A., "Fairchild Delivers Metro 3 AEW Testbed to Sweden," *Aviation Week and Space Technology,* October 19, 1987, pp. 127–130.
432. "Radar Designed for Intense EW Environment," *Aviation Week and Space Technology,* October 19, 1987, p. 130.
433. Filter Center, "Georgia Tech Has Developed Copper Oxide Compounds that Physicists Said Remain Superconductive at Temperatures up to 500 K (440°F) and Seem to Be Stable and Reproducible," *Aviation Week and Space Technology,* January 4, 1988, p. 57.
434. Wollard, K., "Solid State, Technology '88," *IEEE Spectrum,* Vol. 25, No. 1, January 1988, pp. 44–46.
435. Cole, B. C., "Gate Arrays' Big Problem: They Take Too Long to Build," *Electronics,* November 12, 1987, pp. 69–71.
436. Cole, B. C., "Lasa's Mobile ASIC Fab May Do the Job in Minutes," *Electronics,* November 12, 1987, pp. 72–74.
437. Cole, B. C., "How a Worry Turned into a Company," *Electronics,* November 12, 1987, p. 75.
438. Cole, B. C., "Laser Micromachining Brings Quick Prototypes," *Electronics,* November 12, 1987, pp. 76–78.
439. Wittenberg, R. C., "'Do-It-Yourself' Kit Spreads Gospel of Superconductivity to the Masses," *Electronic Engineering Times,* December 28, 1987, p. 21.
440. "4-Mbit DRAM Scheduled for Fall '88 Bow," *Electronic Engineering Times,* December 14, 1987, p. 76.
441. Gold, M., and A. Bindra, "At ISSCC '87: The Shape of Chips to Come," *Electronic Engineering Times,* December 21, 1987, p. 24.
442. Bindra, A., "New Advances in 10-bit Parts—Flash Converters Lead Speed-Driven ADC World," *Electronic Engineering Times,* December 21, 1987, p. 30.
443. Weiss, R., "Vitesse GaAs 1-kbit SRAM Is ECL Compatible," *Electronic Engineering Times,* December 21, 1987, p. 32.
444. Wanstall, B., *Interavia,* private communication, December 1987.
445. Baker, S., "0.5-micron Production Approaches on Horizon," *Electronic Engineering Times,* January 11, 1988, pp. 66, 72.
446. Cole, B. C., "Look for Hot Developments in Every Area at the ISSCC," *Electronics,* December 17, 1987, pp. 103–104.
447. Curran, L., "These Chips Are Breaking the Linear Bottleneck," *Electronics,* December 17, 1987, pp. 61–64.
448. Cole, B. C., "Crystal Builds the First 12-bit Flash Converter Chip," *Electronics,* December 17, 1987, pp. 67–69.
449. Riezenman, M., "Special Report: High-Resolution Arrives for ADC Chips and Hybrids," *Electronics,* January 7, 1988, pp. 133–138.
450. Military/Aerospace Newsletter, "Rockwell Breaks the 1% Yield Barrier for GaAs 16-kbit SRAMs," *Electronics,* January 7, 1988, p. 162.
451. Manuel, T., "Vitesse's 1-kbit Static RAM Is Twice as Fast as Competition," *Electronics,* December 17, 1987, p. 133.

Appendix 2A
Catalogue of Radar Parameters

This Appendix provides a catalogue of ground and airborne radar parameters. Table 2A.1 gives an update of the ground radars covered in Brookner ([1] Table 1 of Chapter 1). Radars developed since Table 1 [1] was compiled are primarily given. The abbreviations used in Table 2A.1 are defined at the end of this appendix, where we also define the US AN/XYZ designation. Table 2A.2 is a reprint of a ground radar listing given in *Interavia* magazine (April 1987, pp. 366–378). A few of the radars listed in Table 2A.1 [1] are also listed in Table 2A.2 with, in some cases, additional information being obtained by looking up both entries. Table 2A.2 also includes secondary radars. Several secondary radars not included in Table 2A.2 are summarized in Table 2A.1(f). Table 2A.3 gives a reprint of the listing of Western airborne attack radar systems published in *Interavia* (March 1981, pp. 232–242). This table is updated with Table 2A.4, which includes some nonattack radar systems as well. The abbreviations used for Table 2A.4 are also explained at the end of this appendix.

Table of Contents for Catalogue of Radar Parameters

Table 2A.1(a) 2D Surveilance and Search Radars and Fan Beam Precision Approach Radars (PARs)

Radar	ADS (Air Defense System) [415]		
Type	Complete air defense and air traffic control system		
Manufacturer	Westinghouse		
Location	Ground. Static or transportable.		
Frequency Band	L (2D Search Radar)	and	S (Nodding-Beam HFR[b])
Frequency (MHz)			
Power, Peak (MW)	5*		
Power, Av (kW)	3.3*		
Antenna Size (H × W) (m) (or Type)			
Antenna Gain (dB)	34.5 (low); 33 (high)		
Beam Shape	Fan beam		
Type of Scanner	Mechanical in AZ		
No. of Beams	Dual on receive in EL		
Beamwidth, EL (deg)			
Beamwidth, AZ (deg)	2		
Angular Coverage, EL (deg)	to 40°, 60kft*		
Angular Coverage, AZ (deg)	360°		
Scan Rate, EL	N.A.		
Scan Rate, AZ	5 rpm		
PRF	310 to 364 Hz*		
PRF Stagger	8 PRFs*		
Pulsewidth, Transmit (μs)	2 [a]		
Pulsewidth, Compressed (μs)	0.5 [a]		
Pulse Compression Ratio	4		
Type of Tube			
Noise Figure (dB)			
Range	240 nmi diplex (2m^2, P_D = 80%, P_{fa} = 10^{-6})		
MTI	50 IF		
Polarization	Linear/Circular*		
Frequency Agility	used; 39 freq.*		
ECCM Features	CPACS, JATS		
Misc. Comments	Appears to be militarized version of ARSR-3. Has better resolution of 250 ft instead of 1000 ft.		Height acc. = ± 1000 ft at 100 nmi
Fig. No.	Fig. 43 in [1]		

*These are ARSR-3 values. Assumed same for ADS. ASR-30 is an ASR version of ARSR-3 with a range to 120 nmi for 2m^2 target [10].

[a] Data extrapolated by author from available information.

[b] See Glossary for definitions of abbreviations used in Tables 2A.1 and 2A.4.

Table 2A.1(a) (cont'd)

Radar	AN/TPS-61
Type	2D medium and short range gap-filler radar
Manufacturer	Westinghouse
Location	Ground. Tactical or fixed site.
Frequency Band	L
Frequency (MHz)	
Power, Peak (MW)	
Power, Av (kW)	
Antenna Size (H × W) (m) (or Type)	
Antenna Gain (dB)	28.5
Beam Shape	
Type of Scanner	
No. of Beams	1
Beamwidth, EL (deg)	
Beamwidth, AZ (deg)	2.7
Angular Coverage, EL (deg)	
Angular Coverage, AZ (deg)	
Scan Rate, EL	
Scan Rate, AZ	6, 12, 15 rpm
PRF	
PRF Stagger	Used.
Pulsewidth, Transmit (μs)	
Pulsewidth, Compressed (μs)	0.2
Pulse Compression Ratio	
Type of Tube	
Noise Figure (dB)	
Range	100 nmi (1.75m^2, P_D = 75%, P_{fa} = 10^{-6})
MTI	50 IF
Polarization	
Frequency Agility	Used.
ECCM Features	CPACS, JATS, FA, PRF stagger [b].
Misc. Comments	Used also for range instrumentation.
Fig. No.	

Table 2A.1(a) (cont'd)

Radar	AN/TPS-63 [10, 415, 417, 418, 420]
Type	2D Marine Corps. air surveillance radar. Also balloon-borne low-altitude surveillance radar (LASR).*
Manufacturer	Westinghouse
Location	Ground. Transportable or mobile. Tethered balloon [10].
Frequency Band	L (51 selectable frequencies)
Frequency (MHz)	1250–1350
Power, Peak (MW)	0.10
Power, Av (kW)	3
Antenna Size (H × W) (m) (or Type)	6.7 × 5.5 Vertically oriented par. cycl.
Antenna Gain (dB)	32.5
Beam Shape	Fan beam
Type of Scanner	Mechanical in AZ
No. of Beams	1Σ
Beamwidth, EL (deg)	
Beamwidth, AZ (deg)	2.7
Angular Coverage, EL (deg)	40 kft, 0° to 40°
Angular Coverage, AZ (deg)	360°
Scan Rate, EL	N.A.
Scan Rate, AZ	6, 12, 15 rpm
PRF	774 Hz (ave)
PRF Stagger	
Pulsewidth, Transmit (μs)	13 and 26 subpulses, 39 total
Pulsewidth, Compressed (μs)	1.2
Pulse Compression Ratio	11, 22 (32.5 with single 39μs pulse)
Type of Tube	CFA
Noise Figure (dB)	3.5
Range	80 nmi (1m^2, P_D = 90%, P_{fa} = 10^{-7}, S-1)
MTI	60 and 35 dB IF for ground and rain, respectively.
Polarization	Vertical
Frequency Agility	Subpulses transmit at 2 frequencies.
ECCM Features	CPACS, PRF stagger
Misc. Comments	Digital 4- and 3-pule MTI for ground and rain clutter. Quadriphase CPACS [b, 419]. Deployment in one hr. Over 70 delivered to U.S. Marines. Some 34 to be deployed in Egypt [10].
Fig. No.	

*Also, low-altitude surveillance system (LASS).
[b] See Glossary for definitions of abbreviations used in Tables 2A.1 and 2A.4.

Table 2A.1(a) (cont'd)

Radar	AN/TPS-65 [10, 415]
Type	2D Marine airfield surveillance radar (ASR)
Manufacturer	Westinghouse
Location	Ground. Tactical or fixed site.
Frequency Band	L (51 selectable frequencies)
Frequency (MHz)	1250–1350
Power, Peak (MW)	0.1
Power, Av (kW)	3
Antenna Size (H × W) (m) (or Type)	4.9 × 5.5 Vertically oriented par. cycl.
Antenna Gain (dB)	
Beam Shape	Fan beam
Type of Scanner	Mechanical in AZ
No. of Beams	1Σ
Beamwidth, EL (deg)	
Beamwidth, AZ (deg)	2.7
Angular Coverage, EL (deg)	40 kft, 0° to 40°
Angular Coverage, AZ (deg)	360°
Scan Rate, EL	N.A.
Scan Rate, AZ	6, 12, 15 rpm
PRF	Parameters of this radar appear to be basically the same as for the AN/TPS-63. A different sized antenna may be used in some configurations.
PRF Stagger	
Pulsewidth, Transmit (μs)	
Pulsewidth, Compressed (μs)	
Pulse Compression Ratio	
Type of Tube	
Noise Figure (dB)	
Range	
MTI	
Polarization	
Frequency Agility	
ECCM Features	
Misc. Comments	Initial production completed.
Fig. No.	

Table 2A.1(a) (cont'd)

Radar	AN/TPS-73
Type	US Marine Corps and Air Force 2D tactical airfield surveillance radar (ASR)
Manufacturer	Unisys and Selenia
Location	Ground. Transportable.
Frequency Band	S
Frequency (MHz)	2700–2900
Power, Peak (MW)	0.01
Power, Av (kW)	1.1
Antenna Size (H × W) (m) (or Type)	Reflector
Antenna Gain (dB)	
Beam Shape	Fan beam
Type of Scanner	Mechanical in AZ
No. of Beams	1
Beamwidth, EL (deg)	5
Beamwidth, AZ (deg)	1.45
Angular Coverage, EL (deg)	0.2°–50°
Angular Coverage, AZ (deg)	360°
Scan Rate, EL	N.A.
Scan Rate, AZ	10, 12, 15 rpm
PRF	Fixed or staggered.
PRF Stagger	
Pulsewidth, Transmit (μs)	
Pulsewidth, Compressed (μs)	
Pulse Compression Ratio	
Type of Tube	Solid-state bottle
Noise Figure (dB)	1.5
Range	60 nmi (1m^2, P_D = 90%, P_{fa} = 10^{-6}, S-1)
MTI	Adaptive MTD.
Polarization	Horizontal, circular (left or right).
Frequency Agility	Yes.
ECCM Features	Automatic frequency selection (AFS).
Misc. Comments	Integrated radar-beacon antenna.
Fig. No.	Monopulse beacon system. Mode-S compatibility.

Table 2A.1(a) (cont'd)

Radar	AN/UPS-3
Type	Tactical battlefield radar.
Manufacturer	GEC Astronics Corp.
Location	Ground. Transportable (≈210 lbs)
Frequency Band	
Frequency (MHz)	Frequency-selectable
Power, Peak (MW)	230W
Power, Av (kW)	20W
Antenna Size (H × W) (m) (or Type)	Slotted waveguide planar array
Antenna Gain (dB)	22
Beam Shape	
Type of Scanner	
No. of Beams	
Beamwidth, EL (deg)	17°
Beamwidth, AZ (deg)	8°
Angular Coverage, EL (deg)	
Angular Coverage, AZ (deg)	
Scan Rate, EL	
Scan Rate, AZ	60° or 90° per second
PRF	
PRF Stagger	
Pulsewidth, Transmit (μs)	
Pulsewidth, Compressed (μs)	
Pulse Compression Ratio	
Type of Tube	
Noise Figure (dB)	
Range	20km IR
MTI	55db IF
Polarization	
Frequency Agility	
ECCM Features	
Misc. Comments	DSP. Reduced sidelobe antenna.
Fig. No.	

Table 2A.1(a) (cont'd)

Radar	AR-I [444]
Type	Precision approach radar (PAR).
Manufacturer	Hindustan Aeronautics Ltd.
Location	Ground.
Frequency Band	X
Frequency (MHz)	10,000
Power, Peak (MW)	0.225
Power, Av (kW)	2.2
Antenna Size (H × W) (m) (or Type)	
Antenna Gain (dB)	
Beam Shape	
Type of Scanner	
No. of Beams	
Beamwidth, EL (deg)	
Beamwidth, AZ (deg)	0.62
Angular Coverage, EL (deg)	
Angular Coverage, AZ (deg)	
Scan Rate, EL	
Scan Rate, AZ	80.4 rpm
PRF	2,170 Hz
PRF Stagger	
Pulsewidth, Transmit (μs)	
Pulsewidth, Compressed (μs)	
Pulse Compression Ratio	
Type of Tube	
Noise Figure (dB)	
Range	20 nmi ($3m^2$)
MTI	DMTI
Polarization	Linear elliptical or circular.
Frequency Agility	
ECCM Features	
Misc. Comments	
Fig. No.	

Table 2A.1(a) (cont'd)

Radar	ASR-9 [412–414, 420]
Type	ATC 2D airport surveillance radar.
Manufacturer	Westinghouse
Location	Ground. Static.
Frequency Band	S
Frequency (MHz)	2,700–2,900
Power, Peak (MW)	1.12
Power, Av (kW)	1.5
Antenna Size (H × W) (m) (or Type)	
Antenna Gain (dB)	33.5 (low), 32.5 (high)
Beam Shape	Fan beam
Type of Scanner	Mechanical in AZ
No. of Beams	Dual in EL on receive.
Beamwidth, EL (deg)	25kft, 4.8°
Beamwidth, AZ (deg)	1.4°
Angular Coverage, EL (deg)	
Angular Coverage, AZ (deg)	360°
Scan Rate, EL	
Scan Rate, AZ	12.5 rpm
PRF	980, 1250 Hz
PRF Stagger	
Pulsewidth, Transmit (μs)	1
Pulsewidth, Compressed (μs)	N.A.
Pulse Compression Ratio	N.A.
Type of Tube	Klystron
Noise Figure (dB)	4.5 (System)
Range	≈ 50 nmi (1m^2, S-1, P_D = 80%, P_{fa} = 10^{-6}, simplex)
MTI	MTD; 10- and 8-pulse CPI; 42 dB SCV [b].
Polarization	Linear or circular (selectable)
Frequency Agility	
ECCM Features	
Misc. Comments	Capacity: 700 a/c plus 300 non-a/c; 44 dB CR nominally, 52 dB for mountain clutter. Over 100 to be installed in US.
Fig. No.	

[b] See Glossary for definitions of abbreviations used in Tables 2A.1 and 2A.4.

Table 2A.1(a) (cont'd)

Radar	RAMP [87–89]	
Type	Air traffic surveillance radar system for Canada.	
Manufacturer	Raytheon (primary radar)	Cossor (secondary)
Location	Ground. Static.	Ground. Static.
Frequency Band	L	L
Frequency (MHz)	1250–1350	1030 (trans.); 1090 (rec.)
Power, Peak (MW)	0.021	
Power, Av (kW)	1.4[a]	0.002
Antenna Size (H × W) (m) (or Type)	6.7 × 10.1 (Reflector)	1.5 × 7.9 (Vertical array antenna)
Antenna Gain (dB)	33.5	
Beam Shape	Fan beam	Fan beam
Type of Scanner	Mechanical in AZ	Mechanical in AZ
No. of Beams	Dual in EL on receive.	Σ, ΔAZ, Omni (in AZ).
Beamwidth, EL (deg)	See Fig. 2.64.	
Beamwidth, AZ (deg)	1.5	
Angular Coverage, EL (deg)	0° to 40°	
Angular Coverage, AZ (deg)	360°	360°
Scan Rate, EL	N.A.	N.A.
Scan Rate, AZ	12 rpm	12 rpm
PRF	676 (ave) Hz	
PRF Stagger		
Pulsewidth, Transmit (μs)	1100 (nonlinear chirp)	
Pulsewidth, Compressed (μs)	1	
Pulse Compression Ratio	100	
Type of Tube	Solid-state "bottle"	Solid-state "bottle"
Noise Figure (dB)	1.5	
Range	95 nmi (2m^2, P_D = 0.8)	250
MTI	(4-pulse MTD, 40 dB SVC)	
Polarization	Circular	
Frequency Agility		
ECCM Features		
Misc. Comments	See Secs. 2.3.1 and 2.6.5.1	Mounted atop primary antenna at 22 terminal sites, stand-alone at 17 en-route sites.
Fig. No.	2.65 (see also 2.64)	

[b] See Glossary for definitions of abbreviations used in Tables 2A.1 and 2A.4.

Table 2A.1(a) (cont'd)

Radar	RSR-1A [444]
Type	ATC en-route surveillance.
Manufacturer	Hindustan Aeronautics Ltd.
Location	Ground.
Frequency Band	L
Frequency (MHz)	1300
Power, Peak (MW)	2.2 or 5.0
Power, Av (kW)	2.2
Antenna Size (H × W) (m) (or Type)	
Antenna Gain (dB)	
Beam Shape	
Type of Scanner	
No. of Beams	
Beamwidth, EL (deg)	
Beamwidth, AZ (deg)	1.25 (main)
Angular Coverage, EL (deg)	
Angular Coverage, AZ (deg)	360°
Scan Rate, EL	
Scan Rate, AZ	5, 10 rpm
PRF	360
PRF Stagger	
Pulsewidth, Transmit (μs)	
Pulsewidth, Compressed (μs)	
Pulse Compression Ratio	
Type of Tube	
Noise Figure (dB)	
Range	160 nmi (with 2.2 MW peak)
MTI	
Polarization	DMTI
Frequency Agility	
ECCM Features	Log/FTC, Dicke-fix
Misc. Comments	Two ordered by Indian Civil Aviation Dept.
Fig. No.	

Table 2A.1(a) (cont'd)

Radar	TAS MK23 (IPD/TAS MK23) [411, 424]
Type	2D point defense designation, medium-range surveillance
Manufacturer	Hughes
Location	Shipboard
Frequency Band	L
Frequency (MHz)	Multiple
Power, Peak (MW)	
Power, Av (kW)	
Antenna Size (H × W) (m) (or Type)	
Antenna Gain (dB)	
Beam Shape	Fan beam
Type of Scanner	Mechanical in AZ
No. of Beams	1
Beamwidth, EL (deg)	
Beamwidth, AZ (deg)	
Angular Coverage, EL (deg)	0° to 75°
Angular Coverage, AZ (deg)	360°
Scan Rate, EL	N.A.
Scan Rate, AZ	15, 30 rpm.
PRF	Multiple
PRF Stagger	
Pulsewidth, Transmit (μs)	
Pulsewidth, Compressed (μs)	
Pulse Compression Ratio	
Type of Tube	
Noise Figure (dB)	
Range	90 nmi (instrumental long range), > 20 nmi (instrumental normal range).
MTI	DMTI; PD
Polarization	
Frequency Agility	
ECCM Features	
Misc. Comments	See Sec. 2.3.1
Fig. No.	

Table 2A.1(a) (cont'd)

Radar	TCR-4T [444]
Type	Terminal area control.
Manufacturer	Hindustan Aeronautics Ltd.
Location	Ground.
Frequency Band	L
Frequency (MHz)	1300
Power, Peak (MW)	0.5
Power, Av (kW)	0.5
Antenna Size (H × W) (m) (or Type)	
Antenna Gain (dB)	
Beam Shape	
Type of Scanner	
No. of Beams	
Beamwidth, EL (deg)	1.25°
Beamwidth, AZ (deg)	
Angular Coverage, EL (deg)	
Angular Coverage, AZ (deg)	360°
Scan Rate, EL	
Scan Rate, AZ	15, 7.5 rpm
PRF	1000 Hz
PRF Stagger	
Pulsewidth, Transmit (μs)	
Pulsewidth, Compressed (μs)	
Pulse Compression Ratio	
Type of Tube	
Noise Figure (dB)	
Range	70 nmi
MTI	
Polarization	
Frequency Agility	
ECCM Features	
Misc. Comments	Dual-diversity option available.
Fig. No.	

Table 2A.1(b) 3D Radars (Stacked-Beam)

Radar	AN/TPS-43E [10, 415, 416]
Type	3D stacked-beam surveillance radar.
Manufacturer	Westinghouse
Location	Ground. Air transportable.
Frequency Band	S
Frequency (MHz)	2900–3100
Power, Peak (MW)	4.0
Power, Av (kW)	6.7 (42, prime)
Antenna Size (H × W) (m) (or Type)	4.3 × 6.2
Antenna Gain (dB)	40.6
Beam Shape	Pencil beam. Broadened at high EL.
Type of Scanner	Mechanical in AZ.
No. of Beams	6 (stacked) in EL.
Beamwidth, EL (deg)	1.55 to 8.1 (20 tot.)
Beamwidth, AZ (deg)	1.1
Angular Coverage, EL (deg)	10kft, 20°
Angular Coverage, AZ (deg)	
Scan Rate, EL	
Scan Rate, AZ	6 rpm
PRF	250 (ave) Hz
PRF Stagger	6 PRFs (±10% ave).
Pulsewidth, Transmit (μs)	6.5
Pulsewidth, Compressed (μs)	0.5
Pulse Compression Ratio	13
Type of Tube	Linear-beam twystron.
Noise Figure (dB)	4.5
Range	220 nmi (2.5m^2, P_D = 80%, P_{fa} = 10^{-6})
MTI	Digital 4-pulse cancelation (30 dB IF).
Polarization	Vertical
Frequency Agility	16 frequencies
ECCM Features	CPACS, JATS, SLB, PRF stagger, auto ARM detection and shut-off [b].
Misc. Comments	MTBF: 150 hr; MTTR: <1 hr; 1 hr assumed. Over 170 delivered. Some retrofited with ULSA of AN/TPS-70(V)-1.
Fig. No.	2.82

[b] See Glossary for definitions of abbreviations used in Tables 2A.1 and 2A.4.

Table 2A.1(b) (cont'd)

Radar	AN/TPS-70(V)-1 [10, 85, 86, 420]
Type	Basically, AN/TPS-43E with ultra-low-sidelobe antenna (ULSA).
Manufacturer	Westinghouse
Location	Ground. Mobile or static.
Frequency Band	S
Frequency (MHz)	2900–3100
Power, Peak (MW)	3.5
Power, Av (kW)	6.2
Antenna Size (H × W) (m) (or Type)	2.5 × 5.5 (see Sec. 2.6.2.1)
Antenna Gain (dB)	36
Beam Shape	Pencil beam.
Type of Scanner	Mechanical in AZ
No. of Beams	6 in EL on receive.
Beamwidth, EL (deg)	
Beamwidth, AZ (deg)	1.6°
Angular Coverage, EL (deg)	0° to 20°
Angular Coverage, AZ (deg)	360°
Scan Rate, EL	N.A.
Scan Rate, AZ	6.38 ± 10%; 10, 20 rpm
PRF	250 Hz*
PRF Stagger	Yes.
Pulsewidth, Transmit (μs)	6.5
Pulsewidth, Compressed (μs)	0.5*, CPACS.
Pulse Compression Ratio	
Type of Tube	Twystron
Noise Figure (dB)	
Range	240 IR. 180 nmi (1.7 m^2, P_D = 75%, P_{fA} = 10^{-6})
MTI	4-pulse; 50 dB IF (all beams).
Polarization	
Frequency Agility	
ECCM Features	ULSA (< 35 dB), CPACS, FA, JATS, ARM detection and shut-off [b]
Misc. Comments	(See Secs. 2.2.5 amd 2.6.2.1.) IFF. Static version, IPS-700, has same performance.
Fig. No.	2.83

*AN/TPS-43E value; believed to be same for AN/TPS-70.
[b] See Glossary for definitions of abbreviations used in Tables 2A.1 and 2A.4.

Table 2A.1(b) (cont'd)

Radar	Martello S713 [10, 82–84]
Type	3D stacked-beam air surveillance radar.
Manufacturer	Marconi
Location	Ground. Transportable.
Frequency Band	L
Frequency (MHz)	1300 (10% BW)
Power, Peak (MW)	3
Power, Av (kW)	8
Antenna Size (H × W) (m) (or Type)	10.7 × 6.1 planar array.
Antenna Gain (dB)	32 on transmit; 38 on receive.
Beam Shape	Fan beam on transmit.
Type of Scanner	Mechanical in AZ.
No. of Beams	8
Beamwidth, EL (deg)	1.5 (on boresight).
Beamwidth, AZ (deg)	2.8
Angular Coverage, EL (deg)	0° to 30°; 200 kft.
Angular Coverage, AZ (deg)	
Scan Rate, EL	Stacked-beam total coverage 30°.
Scan Rate, AZ	6 rpm
PRF	
PRF Stagger	Random
Pulsewidth, Transmit (μs)	10
Pulsewidth, Compressed (μs)	0.25
Pulse Compression Ratio	40
Type of Tube	Twystron
Noise Figure (dB)	4
Range	210 nmi (small fighter).
MTI	40–55 dB of CC.
Polarization	Horizontal.
Frequency Agility	Yes.
ECCM Features	LSA, FA, anti-spoofer jitter.
Misc. Comments	In production. (See Secs. 2.2.5 and 2.6.2.1).
Fig. No.	2.58 (see also 2.59).

Table 2A.1(b) (cont'd)

Radar	Martello S 723 [10,84]
Type	3D stacked-beam air surveillance radar.
Manufacturer	Marconi
Location	Ground. Transportable.
Frequency Band	L
Frequency (MHz)	1300 (10% BW)
Power, Peak (MW)	to 0.132
Power, Av (kW)	5
Antenna Size (H × W) (m) (or Type)	7.1 × 12.2 planar array.
Antenna Gain (dB)	
Beam Shape	Fan beam on transmit.
Type of Scanner	Mechanical in AZ.
No. of Beams	6 or 8 (on receive).
Beamwidth, EL (deg)	
Beamwidth, AZ (deg)	1.5°
Angular Coverage, EL (deg)	0° to 20°; 200 kft.
Angular Coverage, AZ (deg)	360°
Scan Rate, EL	Stacked-beam total coverage 20°
Scan Rate, AZ	6 rpm
PRF	
PRF Stagger	
Pulsewidth, Transmit (μs)	150
Pulsewidth, Compressed (μs)	0.25
Pulse Compression Ratio	600
Type of Tube	Solid-state
Noise Figure (dB)	2.5
Range	260 nmi (small fighter).
MTI	
Polarization	Horizontal.
Frequency Agility	Yes.
ECCM Features	SLB. Frequency agility.
Misc. Comments	See Secs. 2.2.5 and 2.6.2.1. In production.
Fig. No.	2.60, 2.61 (see also 2.62)

Table 2A.1(c) 3D Radars (1D Electronic Scan in Elevation)

Radar	AN/FPS-117 [10]
Type	2D air defense radar. Derivative of AN/TPS-59 [1] and GE 592.
Manufacturer	General Electric
Location	Ground. Fixed.
Frequency Band	L
Frequency (MHz)	1215–1400
Power, Peak (MW)	0.025
Power, Av (kW)	
Antenna Size (H × W) (m) (or Type)	7.3 × 7.3 planar array.
Antenna Gain (dB)	
Beam Shape	Pencil beam.
Type of Scanner	Mechanical in AZ; phase in EL.
No. of Beams	
Beamwidth, EL (deg)	
Beamwidth, AZ (deg)	
Angular Coverage, EL (deg)	−6° to +20°
Angular Coverage, AZ (deg)	360°
Scan Rate, EL	
Scan Rate, AZ	
PRF	
PRF Stagger	
Pulsewidth, Transmit (μs)	
Pulsewidth, Compressed (μs)	
Pulse Compression Ratio	
Type of Tube	Distributed solid-state modules; 100-W (peak) per module.
Noise Figure (dB)	
Range	200 nmi
MTI	MTI; PD.
Polarization	
Frequency Agility	20 frequencies (pseudorandom, selectable)
ECCM Features	
Misc. Comments	Sidelobe nulling used to eliminate ground clutter for high beam positions.
Fig. No.	(See 2.44)

Table 2A.1(c) (cont'd)

Radar	AR 320 [10, 62]
Type	3D air defense; upgrade of AR-3D.
Manufacturer	Plessey/ITT-Gilfillan
Location	Ground.
Frequency Band	S
Frequency (MHz)	
Power, Peak (MW)	
Power, Av (kW)	24
Antenna Size (H × W) (m) (or Type)	5 × 5 planar array.
Antenna Gain (dB)	43 [a]
Beam Shape	Pencil beam.
Type of Scanner	Mechanical in AZ, frequency in EL.
No. of Beams	1 on transmit, stacked on receive.
Beamwidth, EL (deg)	1.2 [a]
Beamwidth, AZ (deg)	1.3 [a]
Angular Coverage, EL (deg)	
Angular Coverage, AZ (deg)	360°
Scan Rate, EL	
Scan Rate, AZ	
PRF	(Some parameters not shown are estimated to be the same as for AR-3D [1].)
PRF Stagger	
Pulsewidth, Transmit (μs)	
Pulsewidth, Compressed (μs)	
Pulse Compression Ratio	
Type of Tube	CFA (transparent).
Noise Figure (dB)	
Range	
MTI	Digital.
Polarization	Horizontal. [a]
Frequency Agility	Used.
ECCM Features	SLB, ECMR, FA, LSA.
Misc. Comments	NLFM (as in AR-3D) for frequency scan in EL. (See Secs 2.2.4.1 and Table 2.8. In production.)
Fig. No.	2.43 (see also, 2.42)

[a] Data extrapolated by author from available information.

Table 2A.1(c) (cont'd)

Radar	GE 592 [422, 423]
Type	2D air defense radar; Derivative of AN/TPS-59 [1].
Manufacturer	General Electric
Location	Ground. Fixed.
Frequency Band	L (20 selectable frequencies).
Frequency (MHz)	1215–1400
Power, Peak (MW)	
Power, Av (kW)	35 (prime)
Antenna Size (H × W) (m) (or Type)	7.3 × 7.3 planar array.
Antenna Gain (dB)	
Beam Shape	Pencil beam.
Type of Scanner	Mechanical in AZ; phase in EL.
No. of Beams	
Beamwidth, EL (deg)	2.0°
Beamwidth, AZ (deg)	2.1°
Angular Coverage, EL (deg)	100 kft
Angular Coverage, AZ (deg)	360°
Scan Rate, EL	
Scan Rate, AZ	6, 12 rpm.
PRF	
PRF Stagger	
Pulsewidth, Transmit (μs)	
Pulsewidth, Compressed (μs)	0.4, 1.6 (LFM) [a]
Pulse Compression Ratio	
Type of Tube	Distributed solid-state modules; 100-W (peak) per module.
Noise Figure (dB)	
Range	220 nmi (2m^2, P_D = 80%, 5 FA/scan)
MTI	55dB and 37 dB CR for ground and rain, respectively.
Polarization	Horizontal
Frequency Agility	
ECCM Features	
Misc. Comments	Has 44 rows in antenna; rms height error of 2000 ft at 150 nmi.
Fig. No.	2.44

Table 2A.1(c) (cont'd)

Radar	HADR [10,411]
Type	3D air defense radar; derivative of VSTAR.
Manufacturer	Hughes
Location	Ground. Static or transportable.
Frequency Band	S
Frequency (MHz)	(12% BW)
Power, Peak (MW)	
Power, Av (kW)	
Antenna Size (H × W) (m) (or Type)	6 × 4.6 planar array.
Antenna Gain (dB)	
Beam Shape	Pencil beam. Broadened at high EL.
Type of Scanner	360° Mechanical in AZ, phase in EL.
No. of Beams	1Σ (probably one ΔAZ and ΔEL).
Beamwidth, EL (deg)	1.1° (broadened at high EL angles).
Beamwidth, AZ (deg)	1.7°
Angular Coverage, EL (deg)	to 24°, 100 kft.
Angular Coverage, AZ (deg)	360°
Scan Rate, EL	24°/47 ms.
Scan Rate, AZ	5 or 6 rpm.
PRF	
PRF Stagger	
Pulsewidth, Transmit (μs)	
Pulsewidth, Compressed (μs)	
Pulse Compression Ratio	
Type of Tube	CFA driven by TWT.
Noise Figure (dB)	
Range	170 nmi (1 m^2)
MTI	2 and 3 pulses respectively to 60 nmi and 100 nmi. 7-pulse PD for rain and chaff.
Polarization	
Frequency Agility	
ECCM Features	FA, coded pulses, multiple pulsewidths and PRFs, LSA, SLB, BT, JS [b].
Misc. Comments	In operation.
Fig. No.	2.49

[b] See Glossary for definitions of abbreviations used in Tables 2A.1 and 2A.4.

Table 2A.1(c) (cont'd)

Radar	HR-3000 [10, 362]
Type	New generation HADR.
Manufacturer	Hughes
Location	Ground.
Frequency Band	Parameters probably the same as for HADR, except for improvements consisting of better ECCM capabilities, wider receiver BW, better clutter-rejection waveforms, and faster antenna scan rate.
Frequency (MHz)	
Power, Peak (MW)	
Power, Av (kW)	
Antenna Size (H × W) (m) (or Type)	
Antenna Gain (dB)	
Beam Shape	
Type of Scanner	
No. of Beams	
Beamwidth, EL (deg)	
Beamwidth, AZ (deg)	
Angular Coverage, EL (deg)	
Angular Coverage, AZ (deg)	
Scan Rate, EL	
Scan Rate, AZ	
PRF	
PRF Stagger	
Pulsewidth, Transmit (μs)	
Pulsewidth, Compressed (μs)	
Pulse Compression Ratio	
Type of Tube	
Noise Figure (dB)	
Range	250 nmi (fighter-sized aircraft).
MTI	
Polarization	
Frequency Agility	
ECCM Features	
Misc. Comments	In full production. See Sec. 2.2.4.2.
Fig. No.	

Table 2A.1(c) (cont'd)

Radar	RAT-31S [10, 69]
Type	3D air defense, tactical air, and civil or military ATC.
Manufacturer	Selenia
Location	Ground. Air and land transportable.
Frequency Band	S
Frequency (MHz)	3100–3500
Power, Peak (MW)	0.17
Power, Av (kW)	4.2
Antenna Size (H × W) (m) (or Type)	4.7 × 4.0 Dipole planar array.
Antenna Gain (dB)	37.2
Beam Shape	Pencil beam.
Type of Scanner	Mechanical in AZ; Phase in EL.
No. of Beams	3 pairs of Σ/ΔEL beams in EL.
Beamwidth, EL (deg)	1.5° (1st 4); 3 (remaining).
Beamwidth, AZ (deg)	1.5°
Angular Coverage, EL (deg)	0°–21° (30° optional); to 30 km.
Angular Coverage, AZ (deg)	360°
Scan Rate, EL	
Scan Rate, AZ	10/5 rpm; 12/6 optional.
PRF	450 Hz ± 10%
PRF Stagger	
Pulsewidth, Transmit (μs)	2.6
Pulsewidth, Compressed (μs)	0.864
Pulse Compression Ratio	8
Type of Tube	TWT
Noise Figure (dB)	
Range	150 nmi (2 m^2, P_D = 90%, P_{fa} = 10^{-6})
MTI	3-pulse; digital.
Polarization	Horizontal.
Frequency Agility	Used.
ECCM Features	LSA, FA, simultaneous 3 carrier frequencies.
Misc. Comments	See Sec. 2.2.4.2 and Table 2.9. In production. More than 25 ordered.
Fig. No.	2.45 (see also 2.46 and 2.47)

Table 2A.1(c) (cont'd)

Radar	RAT-31SL [71]
Type	3D; upgrade of RAT-31S; better ECCM.
Manufacturer	Selenia
Location	Ground. Transportable.
Frequency Band	S
Frequency (MHz)	3100–3500
Power, Peak (MW)	0.15
Power, Av (kW)	13.5
Antenna Size (H × W) (m) (or Type)	Dipole planar array.
Antenna Gain (dB)	
Beam Shape	Pencil beam
Type of Scanner	Mechanical in AZ, Phase in EL.
No. of Beams	3 pairs of Σ/ΔEL beams in EL.
Beamwidth, EL (deg)	1.3° (less than RAT-31S)
Beamwidth, AZ (deg)	1.5°
Angular Coverage, EL (deg)	
Angular Coverage, AZ (deg)	360°
Scan Rate, EL	
Scan Rate, AZ	
PRF	Low for long range. High for medium range.
PRF Stagger	
Pulsewidth, Transmit (μs)	Binary phase coded or FM.
Pulsewidth, Compressed (μs)	
Pulse Compression Ratio	
Type of Tube	PPM grided TWT.
Noise Figure (dB)	
Range	250 nmi instrumental range.
MTI	Ground, rain, chaff simultaneously suppressed.
Polarization	Horizontal.
Frequency Agility	Used.
ECCM Features	Better antijam than RAT-31S. SLB, LSA, FA, ECMR, and JS.
Misc. Comments	
Fig. No.	

Table 2A.1(c) (cont'd)

Radar	SERIES 320 [59, 421]
Type	3D air defense radar.
Manufacturer	ITT-Gilfillan
Location	Ground. Transportable.
Frequency Band	S
Frequency (MHz)	2900–3100
Power, Peak (MW)	0.125
Power, Av (kW)	
Antenna Size (H × W) (m) (or Type)	4.3 × 5.1 planar array
Antenna Gain (dB)	41
Beam Shape	Pencil beam [a].
Type of Scanner	Mechanical in AZ, frequency scan in EL.
No. of Beams	5 in EL.
Beamwidth, EL (deg)	1.5°
Beamwidth, AZ (deg)	1.4°
Angular Coverage, EL (deg)	28°; 80 kft.
Angular Coverage, AZ (deg)	360°
Scan Rate, EL	28°/3.5 ms
Scan Rate, AZ	6 rpm
PRF	333–1818 Hz
PRF Stagger	
Pulsewidth, Transmit (μs)	31.5 and 65 (5 Frequency Subpulses)
Pulsewidth, Compressed (μs)	0.42
Pulse Compression Ratio	15 to 31[a] (digital).
Type of Tube	Gridded TWT (PPM).
Noise Figure (dB)	4
Range	100 nmi (1 m^2, S-1, P_D = 90%, 1° EL).
MTI	>35 dB IF.
Polarization	Horizontal.
Frequency Agility	
ECCM Features	Hard-limited non-MTI channel; LSA.
Misc. Comments	See Sec. 2.2.4.1 and Tables 2.5 to 2.7.
Fig. No.	2.38 (see also 2.39, 2.40)

[a] Data extrapolated by author from available information.

Table 2A.1(c) (cont'd)

Radar	SMART [10, 360, 425]
Type	3D stacked-beam naval surveillance radar; used for weapon system control.
Manufacturer	Hollandse Signaalapparaten.
Location	All ships—fast patrol boats upward.
Frequency Band	S
Frequency (MHz)	
Power, Peak (MW)	
Power, Av (kW)	
Antenna Size (H × W) (m) (or Type)	2.5 × 3.6 planar array
Antenna Gain (dB)	approximately equal to 39[a]
Beam Shape	
Type of Scanner	Mechanical in AZ; phase in EL on transmit.
No. of Beams	1 phase-scanned on transmit; stack of 12 on receive
Beamwidth, EL (deg)	12 on receive
Beamwidth, AZ (deg)	2
Angular Coverage, EL (deg)	to zenith.
Angular Coverage, AZ (deg)	360°
Scan Rate, EL	Stacked on receive.
Scan Rate, AZ	27 rpm.
PRF	3700 Hz
PRF Stagger	
Pulsewidth, Transmit (μs)	
Pulsewidth, Compressed (μs)	
Pulse Compression Ratio	Used.
Type of Tube	TWT
Noise Figure (dB)	
Range	Designed for 0.1 m^2 target [10].
MTI	PD [b]
Polarization	
Frequency Agility	
ECCM Features	ULSA, PRFA, FA [b]
Misc. Comments	11 being produced [10]. A C-band version is available. See Sec. 2.2.4.2.
Fig. No.	

[a] Data extrapolated by author from available information.
[b] See Glossary for definitions of abbreviations used in Tables 2A.1 and 2A.4

Table 2A.1(c) (cont'd)

Radar	TRS 2215 [10,141,142,420]
Type	3D air defense. New version is TRS 2215D.
Manufacturer	Thomson-CSF
Location	Ground. Transportable (ground or helicopter).
Frequency Band	S
Frequency (MHz)	(200 BW)
Power, Peak (MW)	0.7
Power, Av (kW)	10
Antenna Size (H × W) (m) (or Type)	3.2 × 4.8 cylindrical parabola.
Antenna Gain (dB)	39
Beam Shape	Pencil beam.
Type of Scanner	Mechanical in AZ, phase in EL.
No. of Beams	3 pairs of Σ/ΔEL beams in EL.
Beamwidth, EL (deg)	1.9° to 4°
Beamwidth, AZ (deg)	1.5°
Angular Coverage, EL (deg)	−2° to +26° (>100 kft)
Angular Coverage, AZ (deg)	360°
Scan Rate, EL	
Scan Rate, AZ	6 rpm.
PRF	375 Hz (mean).
PRF Stagger	
Pulsewidth, Transmit (μs)	3 × 13 = 39 (3 subpulses at 3 adjustable frequencies).
Pulsewidth, Compressed (μs)	0.2
Pulse Compression Ratio	65
Type of Tube	CFA (eff. >70%)
Noise Figure (dB)	2
Range	180 nmi (2m^2, fluctuating target).
MTI	35 dB IF.
Polarization	Circular.
Frequency Agility	Pseudorandom.
ECCM Features	FA, IPFD, BT, ECCM console, PR-PRF [b].
Misc. Comments	Deployed in <1 hr. See Sec. 2.2.4.2. In production. Helical radiating feed elements [a].
Fig. No.	2.48

[a] Data extrapolated by author from available information.
[b] See Glossary for definitions of abbreviations used in Tables 2A.1 and 2A.4.

Table 2A.1(c) (cont'd)

Radar	TRS 2215D [10, 141, 142, 396, 420]
Type	3D air defense. TRS 2215 with ULSA. TRS 22XX belongs to this family of radars.
Manufacturer	Thomson-CSF
Location	Ground. Transportable.
Frequency Band	S
Frequency (MHz)	(200 BW) [a]
Power, Peak (MW)	0.7 [a]
Power, Av (kW)	10 [a]
Antenna Size (H × W) (m) (or Type)	4.5 × 5 planar array.
Antenna Gain (dB)	39
Beam Shape	Pencil beam.
Type of Scanner	Mechanical in AZ, phase in EL.
No. of Beams	3Σ and 3Δ in EL [a].
Beamwidth, EL (deg)	2° to 4°
Beamwidth, AZ (deg)	1.5°
Angular Coverage, EL (deg)	−5° to +30°; >100 kft.
Angular Coverage, AZ (deg)	360°
Scan Rate, EL	
Scan Rate, AZ	6 rpm
PRF	375 Hz (ave).
PRF Stagger	
Pulsewidth, Transmit (μs)	39 (3 subpulses at 3 adjustable frequencies) [a].
Pulsewidth, Compressed (μs)	0.2
Pulse Compression Ratio	65
Type of Tube	CFA (transparent).
Noise Figure (dB)	2
Range	180 nmi ($2m^2$, fluctuating target)
MTI	≥35 dB IF.
Polarization	Circular
Frequency Agility	Pseudorandom
ECCM Features	FA, IPFD, BT, ECCM console, ULSA [b]. See Sec. 2.6.2.1 and Table 2.17.
Misc. Comments	
Fig. No.	2.90 (see also 2.91 and 2.92)

[a] Data extrapolated by author from available information.
[b] See Glossary for definitions of abbreviations used in Tables 2A.1 and 2A.4.

Table 2A.1(c) (cont'd)

Radar	TRS 2230 [10, 396, 420]
Type	3D air defense. Same as TRS 2215 with 3 dB higher gain antenna and nontransportable.
Manufacturer	Thomson-CSF
Location	Ground. Static.
Frequency Band	S
Frequency (MHz)	(200 BW)
Power, Peak (MW)	0.7
Power, Av (kW)	10
Antenna Size (H × W) (m) (or Type)	Cylindrical parabola*.
Antenna Gain (dB)	42 [a]
Beam Shape	Pencil beam.
Type of Scanner	Mechanical in AZ; phase in EL.
No. of Beams	3Σ and 3Δ in EL.
Beamwidth, EL (deg)	1.1 to 2.5
Beamwidth, AZ (deg)	1.3
Angular Coverage, EL (deg)	−2° to +26° (>100 kft)
Angular Coverage, AZ (deg)	360°
Scan Rate, EL	
Scan Rate, AZ	6 rpm
PRF	380 Hz (mean).
PRF Stagger	
Pulsewidth, Transmit (μs)	39 (3 subpulses at 3 adjacent frequencies).
Pulsewidth, Compressed (μs)	0.2
Pulse Compression Ratio	65
Type of Tube	CFA
Noise Figure (dB)	2
Range	230 nmi (2 m^2, fluctuating target) [a].
MTI	35 IF
Polarization	Circular
Frequency Agility	Pseudorandom
ECCM Features	FA, IPFD, BT, ECCM console [b].
Misc. Comments	See Sec. 2.2.4.2. In production. Helical radiating element in feed [a].
Fig. No.	(See 2.48)

*New version of system with ultra-low-sidelobe planar array, similar to that of TRS 2215D now in production.
[a] Data extrapolated by author from available information.
[b] See Glossary for definitions of abbreviations used in Tables 2A.1 and 2A.4.

Table 2A.1(c) (cont'd)

Radar	TRS 22XX
Type	3D air defensive radar.
Manufacturer	Thomson—CSF
Location	
Frequency Band	S
Frequency (MHz)	15%
Power, Peak (MW)	
Power, Av (kW)	20
Antenna Size (H × W) (m) (or Type)	4.5 × 6 (planar)
Antenna Gain (dB)	40
Beam Shape	Pencil beam.
Type of Scanner	Mechanical in AZ; phase in EL.
No. of Beams	1
Beamwidth, EL (deg)	1.6
Beamwidth, AZ (deg)	1.2
Angular Coverage, EL (deg)	20°
Angular Coverage, AZ (deg)	360°
Scan Rate, EL	20° (100 kft)
Scan Rate, AZ	6 rpm
PRF	
PRF Stagger	
Pulsewidth, Transmit (μs)	85, 20
Pulsewidth, Compressed (μs)	≈0.8 [a]
Pulse Compression Ratio	25, 110 [a]
Type of Tube	
Noise Figure (dB)	2.2
Range	190 nmi (very small fighter).
MTI	PD.
Polarization	Linear
Frequency Agility	Yes.
ECCM Features	ULSA, FA, JS, ECMR, JATS.
Misc. Comments	Sidelobes 60 dB down in intercardinal plane.
Fig. No.	2.83

[a] Data extrapolated by author from available information.

Table 2A.1(c) (cont'd)

Radar	W-2000 [78]
Type	3D air defense and civil surveillance radar. Under development.
Manufacturer	Westinghouse
Location	Ground.
Frequency Band	L
Frequency (MHz)	1300
Power, Peak (MW)	0.18
Power, Av (kW)	3.2
Antenna Size (H × W) (m) (or Type)	7.3 × 9.1 (planar array) [10].
Antenna Gain (dB)	≈39 (receive) [a].
Beam Shape	Pencil beam.
Type of Scanner	Mechanical in AZ; phase in EL.
No. of Beams	3Σ, 3ΔAZ (Independent).
Beamwidth, EL (deg)	2° to 4° (low to high EL).
Beamwidth, AZ (deg)	≈1.5°
Angular Coverage, EL (deg)	0° to 20°; >100 kft.
Angular Coverage, AZ (deg)	360°
Scan Rate, EL	
Scan Rate, AZ	6 rpm.
PRF	Medium: >300 Hz; high: >475 Hz.
PRF Stagger	Variable interpulse period.
Pulsewidth, Transmit (μs)	3(16) = 48 (3 subpulses at 3 independent frequencies).
Pulsewidth, Compressed (μs)	0.25 (LFM)
Pulse Compression Ratio	64
Type Tube	Solid-state bottle.
Noise Figure (dB)	
Range	210 nmi (1 m^2, P_D = 80°, S − 1)
MTI	>50 dB; 4-pulse. PD.
Polarization	Horizontal [a]
Frequency Agility	Over 10% band.
ECCM Features	ULSA, SLB, ESMR.
Misc. Comments	Uses 120 1.5 KW L-band modules. Rejects ground, rain, chaff clutter simultaneously. See Secs. 2.2.4.2 and 2.6.2.
Fig. No.	2.52 (2.53 and Table 2.11)

[a] Data extrapolated by author from available information.

Table 2A.1(d) 3D Radars (Phase-Frequency Scanned)

Radar	Firefinder AN/TPQ-36 [10, 34, 411]
Type	3D weapon (mortar, artillery, rocket) locator. Used with AN/TPQ-37.
Manufacturer	Hughes
Location	Ground. Transportable.
Frequency Band	X [10, 53]
Frequency (MHz)	
Power, Peak (MW)	
Power, Av (kW)	(<10, prime)
Antenna Size (H × W) (m) (or Type)	Planar array, slotted waveguide.
Antenna Gain (dB)	
Beam Shape	Pencil beam.
Type of Scanner	Phase in AZ, frequency in EL.
No. of Beams	1
Beamwidth, EL (deg)	
Beamwidth, AZ (deg)	
Angular Coverage, EL (deg)	Few degrees
Angular Coverage, AZ (deg)	±45° (trainable, in seconds).
Scan Rate, EL	
Scan Rate, AZ	Few times each second on horizon.
PRF	Medium [411]
PRF Stagger	
Pulsewidth, Transmit (μs)	
Pulsewidth, Compressed (μs)	
Pulse Compression Ratio	
Type of Tube	TWT (shadow grid)
Noise Figure (dB)	
Range	≈7 nmi.
MTI	MTD.
Polarization	
Frequency Agility	
ECCM Features	
Misc. Comments	See Sec. 2.2.2 and Table 2.4. Almost 300 delivered [10].
Fig. No.	2.30, 2.31

Table 2A.1(d) (cont'd)

Radar	LASR (AN/TPQ-36(A)) [10, 36, 53, 411]
Type	3D low-altitude air surveillance radar.
Manufacturer	Hughes
Location	Ground. Transportable.
Frequency Band	X [53]
Frequency (MHz)	
Power, Peak (MW)	
Power, Av (kW)	
Antenna Size (H × W) (m) (or Type)	Same as AN/TPQ-36, on side.
Antenna Gain (dB)	
Beam Shape	Pencil beam.
Type of Scanner	360° mechanical and a few degrees frequency in AZ; phase in EL.
No. of Beams	1Σ [a].
Beamwidth, EL (deg)	
Beamwidth, AZ (deg)	
Angular Coverage, EL (deg)	90°
Angular Coverage, AZ (deg)	
Scan Rate, EL	360° mechanical, few degrees electronic.
Scan Rate, AZ	30 rpm
PRF	
PRF Stagger	
Pulsewidth, Transmit (μs)	
Pulsewidth, Compressed (μs)	
Pulse Compression Ratio	
Type of Tube	
Noise Figure (dB)	
Range	≈25 nmi [53].
MTI	
Polarization	
Frequency Agility	
ECCM Features	Low-sidelobe antenna.
Misc. Comments	See Sec. 2.2.1.8.
Fig. No.	2.32

[a] Data extrapolated by author from available information.

Table 2A.1(e) 3D Radars (Phase-Phase Scanned)

Radar	Firefinder AN/TPQ-37 [34, 411]
Type	3D weapon (artillery, rocket) locator. Used with AN/TPQ-36.
Manufacturer	Hughes
Location	Ground. Transportable.
Frequency Band	S [34]
Frequency (MHz)	
Power, Peak (MW)	
Power, Av (kW)	(45, prime)
Antenna Size (H × W) (m) (or Type)	11 × 7 planar array.
Antenna Gain (dB)	
Beam Shape	Pencil beam.
Type of Scanner	Phase-phase steered.
No. of Beams	1 Σ, ΔAZ, ΔEL.
Beamwidth, EL (deg)	0.6° [a]
Beamwidth, AZ (deg)	0.9° [a]
Angular Coverage, EL (deg)	Few degrees (≈ 10°).
Angular Coverage, AZ (deg)	±45° (trainable, in seconds).
Scan Rate, EL	
Scan Rate, AZ	Horizontal, few times each second.
PRF	Medium [411]
PRF Stagger	
Pulsewidth, Transmit (μs)	Nonlinear FM.
Pulsewidth, Compressed (μs)	
Pulse Compression Ratio	
Type of Tube	TWT (shadow grid).
Noise Figure (dB)	
Range	16 nmi
MTI	MTD.
Polarization	
Frequency Agility	
ECCM Features	
Misc. Comments	See Sec. 2.2.1.8. 72 delivered to US Army [10].
Fig. No.	2.18

[a] Data extrapolated by author from available information.

Table 2A.1(e) (cont'd)

Radar	AN/FPS-123(V)5 BMEWS Upgrade [13, 14]
Type	Ballistic missle early-warning system (BMEWS).
Manufacturer	Raytheon
Location	Ground. Static. (Thule, Greenland)
Frequency Band	UHF
Frequency (MHz)	
Power, Peak (MW)	
Power, Av (kW)	
Antenna Size (H × W) (m) (or Type)	25.6 diameter (2 faces)
Antenna Gain (dB)	
Beam Shape	Pencil beam.
Type of Scanner	Phase-phase.
No. of Beams	
Beamwidth, EL (deg)	
Beamwidth, AZ (deg)	
Angular Coverage, EL (deg)	
Angular Coverage, AZ (deg)	
Scan Rate, EL	
Scan Rate, AZ	
PRF	
PRF Stagger	
Pulsewidth, Transmit (μs)	
Pulsewidth, Compressed (μs)	
Pulse Compression Ratio	
Type of Tube	Distributed solid-state T/R modules, identical to those of PAVE PAWS.
Noise Figure (dB)	
Range	
MTI	
Polarization	
Frequency Agility	
ECCM Features	
Misc. Comments	2560 active elements per face, 1024 passive. Also does Space Track. See Sec. 2.2.1.3. Same radiating element as PAVE PAWS.
Fig. No.	2.10

Table 2A.1(e) (cont'd)

Radar	AN/SPQ-11 Cobra Judy [20]
Type	3D phase-phase steered radar. Used to obtain data for SALT.
Manufacturer	Raytheon
Location	On-board ship. (Observation Island)
Frequency Band	
Frequency (MHz)	
Power, Peak (MW)	
Power, Av (kW)	
Antenna Size (H × W) (m) (or Type)	6.9 diameter.
Antenna Gain (dB)	
Beam Shape	
Type of Scanner	Phase-phase. (Array mount on trainable turret.)
No. of Beams	
Beamwidth, EL (deg)	
Beamwidth, AZ (deg)	
Angular Coverage, EL (deg)	
Angular Coverage, AZ (deg)	
Scan Rate, EL	
Scan Rate, AZ	
PRF	
PRF Stagger	
Pulsewidth, Transmit (μs)	
Pulsewidth, Compressed (μs)	
Pulse Compression Ratio	
Type of Tube	16 TWTs
Noise Figure (dB)	
Range	
MTI	
Polarization	
Frequency Agility	
ECCM Features	
Misc. Comments	Has 12,228 radiating elements.
Fig. No.	2.11, 2.12, 2.13

Table 2A.1(e) (cont'd)

Radar	MTTIR [38,395]
Type	Multiple-target-tracking instrumentation radar. (Formerly called MOTR.)
Manufacturer	RCA
Location	Ground. Mobile.
Frequency Band	C
Frequency (MHz)	5.4 to 5.9
Power, Peak (MW)	1
Power, Av (kW)	5
Antenna Size (H × W) (m) (or Type)	3.7 diameter (planar SFL).
Antenna Gain (dB)	45.9 (directive).
Beam Shape	Pencil beam.
Type of Scanner	Phase-phase.
No. of Beams	1 Σ, ΔAZ, ΔEL.
Beamwidth, EL (deg)	1.0°
Beamwidth, AZ (deg)	1.0°
Angular Coverage, EL (deg)	±60° (On EL-over-AZ pedestal.)
Angular Coverage, AZ (deg)	±60° (On EL-over-AZ pedestal.)
Scan Rate, EL	1500 mils/s.
Scan Rate, AZ	2000 mils/s.
PRF	20, 40, 80, 160, 320, 1280.
PRF Stagger	
Pulsewidth, Transmit (μs)	CW: 0.25, 0.5, 1; Chirp: 3.125, 12.5, 50.
Pulsewidth, Compressed (μs)	0.25 (LFM)
Pulse Compression Ratio	12.5, 50, 200
Type of Tube	CFA, TWT (0.005 dc) [b].
Noise Figure (dB)	1.5 (GaAs FET).
Range	310 nmi (10 μs pulsewidth); 460 nmi (50 μs pulsewidth).
MTI	
Polarization	
Frequency Agility	
ECCM Features	
Misc. Comments	8359 elements in array. 3-bit diode phase shifter. Track 10 targets simultaneously.
Fig. No.	2.21

[b] See Glossary for definitions of abbreviations used in Tables 2A.1 and 2A.4.

Table 2A.1(e) (cont'd)

Radar	Multiple-Target Instrumentation Radar (MIR)
Type	3D phase-phase steered instrumentation radar.
Manufacturer	Raytheon
Location	Ground.
Frequency Band	C
Frequency (MHz)	5400–5900.
Power, Peak (MW)	0.165
Power, Av (kW)	5.28
Antenna Size (H × W) (m) (or Type)	3.7, space-fed.
Antenna Gain (dB)	42
Beam Shape	Pencil beam.
Type of Scanner	Phase-phase
No. of Beams	1 Σ, 3-channel monopulse.
Beamwidth, EL (deg)	1°
Beamwidth, AZ (deg)	1°
Angular Coverage, EL (deg)	±30° (mechanically trainable).
Angular Coverage, AZ (deg)	±35° (mechanically trainable).
Scan Rate, EL	
Scan Rate, AZ	
PRF	320 [a]
PRF Stagger	
Pulsewidth, Transmit (μs)	100 (LFM).
Pulsewidth, Compressed (μs)	0.12, 0.25, 2.0.
Pulse Compression Ratio	50, 400, 833
Type of Tube	TWT
Noise Figure (dB)	
Range	100 nmi (1 m^2).
MTI	
Polarization	
Frequency Agility	8 frequencies.
ECCM Features	
Misc. Comments	8973 elements; see Sec. 2.1.9 and Table 2.2.
Fig. No.	2.20

Table 2A.1(e) (cont'd)

Radar	AN/FPS-115 PAVE PAWS [1,11,12]
Type	3D SLBM detection and warning system radar. Also used to support Space Track.
Manufacturer	Raytheon
Location	Ground. Four in USA. See Sec. 2.2.1.2.
Frequency Band	UHF
Frequency (MHz)	420–450
Power, Peak (MW)	0.6 per face.
Power, Av (kW)	≈150 per face.
Antenna Size (H × W) (m) (or Type)	22.1 diameter (31.1 for growth)
Antenna Gain (dB)	approximately equal to 38.5 [a] (1792 active elements)
Beam Shape	Pencil beam.
Type of Scanner	Phase-phase.
No. of Beams	
Beamwidth, EL (deg)	≈2.2 [a]
Beamwidth, AZ (deg)	≈2.2 [a]
Angular Coverage, EL (deg)	to 60°
Angular Coverage, AZ (deg)	±60° per face (240 total)
Scan Rate, EL	
Scan Rate, AZ	
PRF	
PRF Stagger	
Pulsewidth, Transmit (μs)	16,000
Pulsewidth, Compressed (μs)	
Pulse Compression Ratio	
Type of Tube	Distributed solid-state T/R modules.
Noise Figure (dB)	2.9 dB rms max.
Range	3000 (10 m^2)
MTI	
Polarization	Circular
Frequency Agility	
ECCM Features	
Misc. Comments	≈ 335 W peak per T/R module. Has two faces. See Sec. 2.2.1.2.
Fig. No.	2.5 (see also 2.6, 2.7, 2.8, 2.9)

[a] Data extrapolated by author from available information.

Table 2A.1(e) (cont'd)

Radar	VSTAR [10,36,411]	
Type	3D air defense radar.	
	VSTAR	VSTAR-PT
Manufacturer	Hughes	
Location	Ground. Static. Transportable or mobile.	
Frequency Band	S [10]	
Frequency (MHz)		
Power, Peak (MW)		
Power, Av (kW)		
Antenna Size (H × W) (m) (or Type)	7 × 11 planar array	11 × 7 planar array
Antenna Gain (dB)		
Beam Shape	Pencil beam.	
Type of Scanner	Phase-phase	
No. of Beams	1Σ, ΔAZ, ΔEL [a].	
Beamwidth, EL (deg)		
Beamwidth, AZ (deg)		
Angular Coverage, EL (deg)	30°	±45°
Angular Coverage, AZ (deg)	±45°*	30°*
Scan Rate, EL		(3 s for horizontal).
Scan Rate, AZ		20 rpm
PRF		
PRF Stagger		
Pulsewidth, Transmit (μs)		
Pulsewidth, Compressed (μs)		
Pulse Compression Ratio		
Type of Tube		
Noise Figure (dB)		
Range	Track to 135 nmi	
MTI		
Polarization		
Frequency Agility		
ECCM Features	BT, very low-sidelobe antenna [b].	See Sec. 2.2.1.8. Production contracts received.
Misc. Comments	Derivative of AN/TPQ-37.	Derivative of AN/TPQ-37.
Fig. No.	2.19	

*VSTAR has electronic coverage and rotates mechanically about vertical to provide 360° coverage.
[a] Data extrapolated by author from available information.
[b] See Glossary for definitions of abbreviations used in Tables 2A.1 and 2A.4.

Table 2A.1(f) Secondary Surveillance Radars (SSR)

Radar	IRS-10 Secondary Radar
Type	Dual-SSR for ATC or military applications.
Manufacturer	Ceselsa
Location	Ground. Static.
Frequency Band	L
Frequency (MHz)	1030 (transmit); 1090 (receive)
Power, Peak (MW)	1200 W
Power, Av (kW)	
Antenna Size (H times W) (m) (or Type)	7.93 wide
Antenna Gain (dB)	>21
Beam Shape	
Type of Scanner	
No. of Beams	Σ, Δ, Ω
Beamwidth, EL (deg)	
Beamwidth, AZ (deg)	2.35°
Angular Coverage, EL (deg)	
Angular Coverage, AZ (deg)	360°
Scan Rate, EL	
Scan Rate, AZ	7.5 rpm.
PRF	
PRF Stagger	
Pulsewidth, Transmit (μs)	
Pulsewidth, Compressed (μs)	
Pulse Compression Ratio	
Type of Tube	Solid-state.
Noise Figure (dB)	
Range	
MTI	
Polarization	
Frequency Agility	
ECCM Features	
Misc. Comments	Military version includes IFF MARK X and MARK XII and full-mode 4.
Fig. No.	

Table 2A.1(f) (cont'd)

Radar	IRS-20 MP Secondary Monopulse Radar
Type	Civil ATC and Air defense.
Manufacturer	Ceselsa
Location	Ground. Static.
Frequency Band	L
Frequency (MHz)	1030 (transmit); 1090 (receive)
Power, Peak (MW)	2
Power, Av (kW)	
Antenna Size (H × W) (m) (or Type)	Planar array (340 radiating elements).
Antenna Gain (dB)	>23 TMA, >29 en route [b].
Beam Shape	
Type of Scanner	
No. of Beams	Σ, Δ, Ω
Beamwidth, EL (deg)	35°
Beamwidth, AZ (deg)	2.35°
Angular Coverage, EL (deg)	
Angular Coverage, AZ (deg)	
Scan Rate, EL	
Scan Rate, AZ	7.5 rpm.
PRF	
PRF Stagger	
Pulsewidth, Transmit (μs)	
Pulsewidth, Compressed (μs)	
Pulse Compression Ratio	
Type of Tube	
Noise Figure (dB)	
Range	250 nmi IR.
MTI	
Polarization	
Frequency Agility	
ECCM Features	
Misc. Comments	18 units ordered by Spanish Civil Aviation Administration. Mode S compatible. MARK X and MARK XII IFF.
Fig. No.	

[b] See Glossary for definitions of abbreviations used in Tables 2A.1 and 2A.4.

Table 2A.1(f) (cont'd)

Radar	Messenger
Type	Secondary surveillance radar.
Manufacturer	Marconi
Location	Ground. Static.
Frequency Band	L
Frequency (MHz)	1030 (transmit); 1090 (receive)
Power, Peak (MW)	0.5 to 2.0 (4 modules).
Power, Av (kW)	
Antenna Size (H × W) (m) (or Type)	Planar dipole LVA [b].
Antenna Gain (dB)	29.1
Beam Shape	
Type of Scanner	
No. of Beams	Σ, Δ, control.
Beamwidth, EL (deg)	
Beamwidth, AZ (deg)	2.45°
Angular Coverage, EL (deg)	
Angular Coverage, AZ (deg)	360°
Scan Rate, EL	
Scan Rate, AZ	7.5 rpm.
PRF	
PRF Stagger	
Pulsewidth, Transmit (μs)	
Pulsewidth, Compressed (μs)	
Pulse Compression Ratio	
Type of Tube	Solid-state.
Noise Figure (dB)	
Range	250 nmi IR.
MTI	
Polarization	
Frequency Agility	
ECCM Features	
Misc. Comments	ISLS or IISLS. Can be updated to full mode S system [b].
Fig. No.	

[b] See Glossary for definitions of abbreviations used in Tables 2A.1 and 2A.4.

Table 2A.1(f) (cont'd)

Radar	Mode-S/S-M [420] (for ASR-9)
Type	Secondary surveillance radar (SSR).
Manufacturer	Westinghouse
Location	Ground. Static or transportable.
Frequency Band	L
Frequency (MHz)	
Power, Peak (MW)	
Power, Av (kW)	
Antenna Size (H × W) (m) (or Type)	
Antenna Gain (dB)	
Beam Shape	
Type of Scanner	
No. of Beams	
Beamwidth, EL (deg)	
Beamwidth, AZ (deg)	2.4 [a]
Angular Coverage, EL (deg)	
Angular Coverage, AZ (deg)	360°
Scan Rate, EL	
Scan Rate, AZ	12.5 rpm.
PRF	
PRF Stagger	
Pulsewidth, Transmit (μs)	
Pulsewidth, Compressed (μs)	
Pulse Compression Ratio	
Type of Tube	
Noise Figure (dB)	
Range	255 nmi.
MTI	
Polarization	
Frequency Agility	
ECCM Features	
Misc. Comments	
Fig. No.	

[a] Data extrapolated by author from available information.

Table 2A.2 Interavia Ground Radar Catalogue (From *Interavia Aerospace Review*, April 1987, pp. 366–378. Reprinted with permission.)

All but one of the manufacturers have updated their product information in this catalogue. The categories covered are surveillance, military 3-D, secondary radars, airport surface movement radars, precision approach radars and specialised military radars.
The following familiar expressions are employed: ATC–Air Traffic Control; IFF–Interrogation Friend or Foe; SIF–Selected Identification Feature; PAR–Precision Approach Radar; GCA–Ground-Controlled Approach; MTI–Moving Target Indicator; MTD–Moving Target Detection; CFAR–Constant False Alarm Rate; ECCM–Electronic Counter-Counter-Measures; FFT–Fast Fourier Transform; FTC–Fast Time Constant; R/ISLS–Receiver/Interrogation Side-Lobe Suppression. Note also: NA–not applicable; ND–not disclosed; S–secret, either for military or commercial security reasons.

Manufacturer Designation	Application	Wave-length (cm)	Pulse power (kW peak/mean)	Pulse repetition frequency per sec	Max range for 3 m² target (nmi)	Antenna rotation speed (rev/min)	Horizontal beamwidth	*Remarks* (Clutter suppression facilities, commercial history, etc.)
Surveillance radars: ATC or military 2-D								
AEG								
SRE-M5	ATC	23	2,500/2.5	400	200	5	1.05°	Circular polarisation, CFAR, clutter-locked MTI; first introduced 1979
Elta Electronics Industries								
L/M-2215	ATC (airport surveillance)	10	450 (peak)	900-1,050	75 (2 m²)	12.5/6	1.4° Az. 5° El.	Dual-channel diversity, digital MTI (33 dB improvement factor), linear/circular polarisation
L/M-2220	Mobile air surveillance	10	425/ND	1,000/750	80	ND	ND	Three versions available. Quadrature digital MTI. IFF. High clutter suppression
Hindustan Aeronautics Ltd								
ARSR-1A	ATC en-route surveillance	ND	ND	ND	ND	ND	ND	Company did not reply
ATCR-4T	Terminal area control	ND	ND	ND	ND	ND	ND	As above
Hollandse Signaalapparaten								
LAR II	ATC en-route surveillance and approach control	23	150/5.1	Up to 500/600	200	6 or 12	1.45°	MTI or MTD, pulse compression, circular polarisation. First installation 1979. Second generation 1983
STAR III	ATC approach control	10	145/4.9	Up to 1,000	100	15	1°	MTI, circular polarisation or MTD with weather extractor, pulse compression
Marconi								
S511	Approach control	10	2×650/ 0.55	Adjustable to 1,000	87 (64 instru-mented)	15	1.5°	Adaptive signal processing. Sub-weather clutter visibility. Weather processing channel
S511C	Approach/terminal control	10	2×1,000/ 1.0	1,000 (adjustable)	100	15	1.5°	Adaptive signal processor, long life coaxial magnetron. Weather processing channel
S512	Terminal area control	10	2×160/ 1.9	825 or 625	112 (100 instru-mented)	15	1.5°	Driven TWT with pulse compression. Adaptive signal processor. Weather processing channel
Plessey								
Watchman	Military suvreillance and ATC	10	58/1.3	1,100 or 840	100 or 110	15 or 12	1.5°	Adaptive MTD, air-cooled TWT. Frequency diversity in single channel. Air transportable Frequency agility, ECCM facilities
Guardsman	Military surveillance low cover	5	S	S	55	15, 30 or 60	1.5°	Air-cooled TWT. Adaptive MTD. Frequency agility, ECCM facilities. Plot extraction, remote operation. Air transportable

▼ Marconi S511 approach control radar.

▼ Bright colour display of the Racal LAR 200 local area radar.

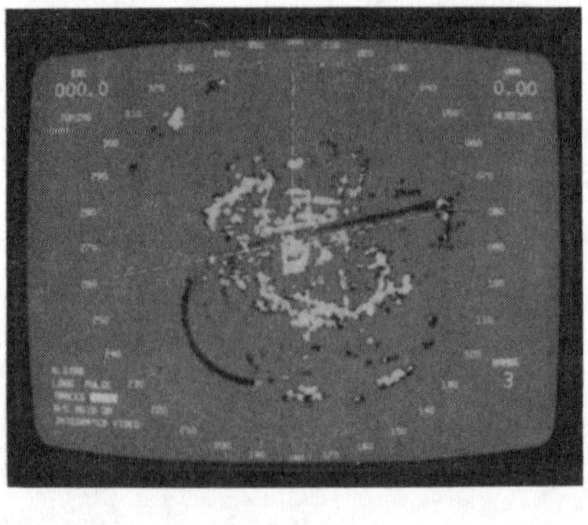

▼ Toshiba TW1253 air route surveillance radar inside its dome.

Table 2A.2 (cont'd)

Manufacturer Designation	Application	Wave-length (cm)	Pulse power (kW peak/ mean)	Pulse repetition frequency per sec	Max range for 3 m² target (nmi)	Antenna rotation speed (rev/min)	Horizontal beamwidth	*Remarks* (Clutter suppression facilities, commercial history, etc.)
AR-15/2C	ATC and military surveillance	10	625/0.425	700 or 1,000	60	15	1.5°	Digital MTI and circular polarisation; static diversity operation
ACR-430	ATC approach guidance	3	55 nominal	1,000 & 2,000	22	20 & 40	0.55°	Static
Racal Avionics								
LAR 200	Air detection and tracking	10	30/–	650 & 1,300	15	24 (4-48)	2°	High-definition colour displays. Low-cost, lightweight mobile. Easy to install. Military or civil
Raytheon								
AN/TPN-24	ASR-2D Part of TPN-19 PAR/GCA system	10	500×2/ 1.0	1,000	100	15	30°	Circular polarisation, coherent and non-coherent digital MTI. First introduced 1971; fixed or mobile
RAMP	ATC ASR	23	21/1.4	676	110	12	1.5°	41 dB SCV. Solid-state transmitter unattended operation. Prototype on test
Raytheon Canada								
ASR-8000 Series	ATC ASR-8300 ASR-8400 ASR-8800	23 23 10	500 nominal/ ND	600-1,000 350-600 800-1,000	60-120 120-200 60-80	7.5/15	1.3°/1.5°	MTS and fully adaptive clutter map. Dual beam antenna. Doppler thresholding. Circular polarisation. CFAR
ASR-9000	ATC	23	ND	ND	80	6/12	ND	22 sites on order for Canadian Radar Modernisation program
Selenia								
Argos 10	Military surveillance (Early Warning)	23	S	S	260	5-6	2°	Frequency agility, digital MTI, adaptive clutter attenuator, etc. First installed 1977; at least 10 sold
Pluto	Military surveillance (low coverage)	10	135	S (prefixed or random stagger)	65	15	1.5°	MTI (45 dB improvement), jam strobe (optional). First introduced 1978 at least 20 sold; highly mobile (air transportable)
ATCR-22D	En-route ATC	23	1,800/ 2.25	300-800 (6 staggered repetition periods)	250	5-10, 6-12, 7.5-15 (standard)	1.2° (main beam)	Adaptive linear processing, 4-pulse, movable notch MTI, improvement factor (sub-clutter visibility) 47 dB-37 dB; Jincar/circular polarisation; log FTC. First introduced 1974. Last version 1985
ATCR-44D	Terminal ATC	23	500/0.5	1,000 (typical)	100	5-10, 6-12, 7.5-15	1.25° (main beam)	Adaptive linear processing, 4-pulse movable notch MTI, improvement factor (sub-clutter visibility) 47 dB-37 dB; various types of polarisation; log FTC. First introduced 1974. Last version 1985
ATCR-33D	Approach ATC	10	500/0.5	1,000 (typical)	80	Dual, various	1.45° (both beams)	Adaptive linear processing, 4-pulse, movable notch MTI improvement factor 47 dB-37 dB; various types of polarisation; log FTC. First introduced 1974. Last version 1985
Thomson-CSF								
TA 10M	ATC (terminal and approach)	10	650/0.5	750	60	15	1.4°	Adaptive digital processing (MTD). Building or shelter installation
TA 10K	ATC (terminal and approach)	10	1,500/1.9	900	80	15	1.4°	Coherent; adaptive target & weather processing (MTD). Fixed & transportable
TR 23MA	ATC (terminal and approach)	23	500/0.5	750	80	15	1.7°	Adaptive digital processing (MTD). Integral plot extractor
TR 23MR	ATC (med. range)	23	2,200/2.5	500	140	12	1.7°	As above
TR 23K	ATC (med. range)	23	3,500/4.2	750	160	15	1.7°	Coherent high-performance radar (MTD)
LP 23M	ATC (en-route)	23	2,200/2.5	340	>200	5	1.2°	Adaptive digital processing (MTD). Integral plot extractor
LP 23K	ATC (en-route)	23	3,500/ 4.2	340	215	5	1.2°	Coherent high-performance radar (MTD)
TRS 2100 (Tiger S)	Low-cover tactical	10	70/1.0	1,100/2,200	75	12	1.3°	Pulse compression–ECCM. Frequency agility. Fixed-echo and weather clutter suppression
TRS-2105 (Tiger G)	Low-cover tactical	5	70/1.5	1,250/2,000	50	5/10	1.3°	Pulse compression. Frequency agility

Table 2A.2 (cont'd)

Manufacturer Designation	Application	Wave-length (cm)	Pulse power (kW peak/mean)	Pulse repetition frequency per sec	Max range for 3 m² target (nmi)	Antenna rotation speed (rev/min)	Horizontal beamwidth	*Remarks* (Clutter suppression facilities, commercial history, etc.)
Toshiba								
TW1374	Airport surveillance radar	10	500/–	880	ND	15	ND	Dual beam, digital signal processor, CFAR, MTI
TW1446	Airport surveillance radar	10	500-1,000/–	750-880	ND	15	ND	As above
TW1378	Airport surveillance radar	10	1,200-3,000/–	880	ND	15	ND	As above
TW1253	Air route surveillance	23	3,000/–	346-349	ND	6	ND	Dual-beam/centre-co-fed (ARSR/SSR), MTI, 200 or 256 mi range
Westinghouse								
ASR-30	ATC	23	2,000/3.65	524	140 (1 m²)	12	1.25-2.0°	50 dB MTI improvement, PRF stagger; unmanned and remote controlled. Separate weather channel. Built-in digitiser. BIT/FIT
ASR-30A	ATC	23	2,000/3.65	1,000	80 (2 m²)	12	1.25°	50 dB MTI, MTD weather contours, unmanned remote control and maintenance. Built-in digitiser
ASR-9	ATC	10	1,200/1.5	1,200 7/9 staggers per CPI	60/80/100 (2 m²)	12.5	1.3°	MTD, RMM. Separate weather channel, CFAR digital output, unmanned with remote control and built-in fault detection and isolation
ARSR-3	ATC	23	5,000/3.5	310-365	240 (2 m²)	5	1.25°	50 dB MTI improvement. PRF stagger. Separate weather channel. Unmanned and remote controlled. Transportable MERF version available. Built-in digitiser. BIT/FIT
AN/TPS-63 (V)	Military surveillance	23	100/3	774	120	6, 12, 15	2.7°	60 dB MTI improvement factor. Pulse compression, DTE data
SPS-58/65 (V)	Military surveillance	23	12/0.26	3,000	25	15	5.8°	60 dB MTI improvement factor, IFF compatible, DTE data
W-630	ATC	23	100/3.21	774	80 (1 m²)	6, 12, 15	2.7°	60 dB MTI improvement. 35 dB movable notch weather suppression. CPACS, PRF stagger. Built-in digitiser. Mobile. Civil version of TPS-63
SPS-65ER	2-D air surveillance	23	25/1.25	625-300 variable	64	20/10	3°	60 dB MTI improvement. Coded compressed pulse. Instrumented to 100 nmi
Military 3-D radars								
AEG								
TRM-AD	Military surveillance	C-band	S	S	33	S	S	Coherent transmitter. Information renewal rate 2/sec. Altitude 20,000 ft. In production
TRM-S	Military surveillance	5	S	S	110	Variable	S	Circular polarisation, step-scan MTI. CFAR. Frequency diversity. First installation 1979
General Electric								
AN/TPS-59	3-D surveillance tactical	23	46/8.3 (av)	Variable	300	ND	3.4°	Solid-state transmitter. Extensive ECCM
AN/FPS-117		23	24.75/ND	Variable	200	ND	ND	USAF SEEK IGLOO radar and chosen for PEACE SHIELD program in Saudi Arabia
GE592		23	25/4(av)	Variable	250	ND	2.2°	Bought by UK and Belgium
GE592 Transportable		23	25/4(av)	Variable	250	ND	2.2°	MTBF for all of this family stated to be over 1,000 hours
Hughes Aircraft								
HADR	Military surveillance (3-D)	10	1,200/30	300-400	225+	5 or 6	1.7°	Digital MTI, CFAR, range-gated pulse Doppler, linear polarisation. Bought by Germany, Malaysia and Norway
HR-3000	Military surveillance (3-D)				225+	5 or 6		Digital MTI, CFAR, range-gated pulse. Doppler, linear polarisation. 8 ordered by Italy, Greece, Turkey and Norway

Table 2A.2 (cont'd)

▼ Hughes Aircraft LASR mobile short-range air defence radar for the Norwegian Adapted Hawk system.

▼ Plessey AR 325 Commander military 3-D surveillance radar.

Manufacturer Designation	Application	Wave-length (cm)	Pulse power (kW peak/ mean)	Pulse repetition frequency per sec	Max range for 3 m² target (nmi)	Antenna rotation speed (rev/min)	Horizontal beamwidth	*Remarks* (Clutter suppression facilities, commercial history, etc.)
VSTAR	Medium-range air defence, mobile					15-30		Based on AN/TPQ-37. Phase-phase planar array. Several fixed-site systems in operation
LASR	Short-range air defence, mobile					15-30		Based on AN/TPQ-36. Automatic digital data interface. 24 on order by Norway
ITT Gilfillan								
AN/TPS-32	Tactical military surveillance (also mobile version AN/TPS-64)	10	2,200/S	1,090-3,772	300 (for 1 m² target)	6	2.15°	MTI, CFAR. First installation 1970
320-300 Series	Mobile, long-range air defence	10	1,100/24	300-1,400	250	6	Az 2.1° El 0.84°	Clutter-gated 4-pulse MTI, pulse compression designed for automated system
Marconi								
Martello S713	Long-range 3-D	23	3,300/8.5	250	220	6	2.8°	Mobile. Adaptive signal processing. Twystron with pulse compression. Low sidelobe planar array
Martello S723	Long-range 3-D	23	62/2.5	240	—	6	1.4°	Mobile. Solid-state transmitter. Pulse compression with adaptive signal processing. Very low sidelobe planar array
Plessey								
AR-3D	Military surveillance	10	1,110/10	250	190	6	Az 1.5° El 1.5°	Mobile, air transportable
AR320	Military surveillance	10	S	S	S	S	S	Mobile, air transportable
AR325	Military surveillance	10	S	S	S	S	S	Mobile, air transportable
Selenia								
RAT-31S	Military surveillance	10	153/3.78 (minimum values)	450 average (prefixed or random stagger)	Over 160	5-10	1.5°	Linear horizontal polarisation, CFAR, adaptive clutter attenuator; coherent limiter, moving window azimuth correlator, etc. First installed 1978; at least 20 sold
RAT-31Sd	Military surveillance	10			Over 210	6		As above, with better antenna performance
Thomson-CSF								
TRS 2215/ 2230 Family	Surveillance radar, static or mobile	10	700/10	360	200-225	6	1.5°	Pulse compression, frequency agility, MTI improvement over 40 dB. Planar array antenna

Table 2A.2 (cont'd)

Manufacturer Designation	Application	Wave-length (cm)	Pulse power (kW peak/ mean)	Pulse repetition frequency per sec	Max range for 3 m² target (nmi)	Antenna rotation speed (rev/min)	Horizontal beamwidth	*Remarks* (Clutter suppression facilities, commercial history, etc.)
Westinghouse								
AN/TPS-43	Military surveillance, mobile, 3-D	10	3,000/4.9	250	220 ($2.5\,m^2$)	6	1.1°	50 dB improvement fixed-echo suppression; CFACS, JATS, frequency agility, PRF stagger. Major enhancements in production, off-the-shelf delivery
AN/TPS-70	Military surveillance; mobile & fixed, 3-D	10	3,000/5.4	250-275	240	6	1.6°	Total volume search and track, air or truck transportable. 50 dB MTI improvement
Vigilant Series	Military/ATC surveillance, mobile 3-D	10	2,000/5	1,200 nominal	60	20/10	1.6°	49 dB MTI improvement factor. Pulse Doppler processing. Auto calibration and alignment. Monitoring and fault isolation
W-2000A	Military/ATC surveillance	23	2,000/3.6	275	Approx. 220	5	1.34°	50 dB MTI improvement factor; unattended & remote controlled; weather channel, CFAR, ECCM. Adaptive Doppler processing
W-160	3-D Fire control and search	3	16/0.2	Variable 1,300-12,500	32	15/20 & 25 search	1.8°	52 dB MTI improvement factor. 16 search & track modes. In production
W-160A	3-D Fire control and search	3	16/0.2	Variable 1,300-12,500	48	6/15 search	1.35°	52 dB MTI improvement factor. 16 search & track modes. Missile illumination capability

Secondary radars (SSR/IFF)

Eaton Corp AIL Division						
AN/TPX-42	Ground-based mil/civil SSR/SIF	1, 2, 3/A, B, C, D	Any usual interlace	150-400	2,000	Processing/display systems, extractors. Digital scan-converted Brite displays, digital mappers
VLATME/SPARTAN	Ground-based mil/civil SSR/SIF	1, 2, 3/A, C	Any one of six	200-400	1,000	Lightweight systems with plasma panel displays, ISLS, beam sharpening, GTC and RSLS
AN/MSR-()	Military IFF	1, 2, 3/A, C	Any one of six	200-400	1,000	Military IFF. Tracking system

◄ Westinghouse AN/TPS-70 mobile military surveillance radar.

Table 2A.2 (cont'd)

Manufacturer Designation	Application	Modes	Mode interlace facility	Pulse repetition frequency/sec	Transmitter pulse power (peak) W	Associated equipment/Remarks
ISC Cardion Electronics						
AN/UPX-27	IFF/SIF interrogator	1, 2, 3/A, 4 (Mk XII) (B+D optional)	4 sequences: 1, 2, 3/A, C, etc. 1, 4, 2, 4, 3/A, 4, C, 4, etc.	150-450	2,000 (min)	Mode 4 continuous. Normally controlled remotely by Decoder Group and Interrogator Control C 8430
AN/GPX-504	ATC Interrogator system with Dual AN/UPX502 Interrogators and Automatic Switching Unit C-5483/UPX	1, 2, 3/ A, B, C, D	Four patterns (PRI, SEC, TER, QUAT), each selectable for any mode	150-450, 4-step stagger available	Conventional SLS 2,000 (min) Improved ISLS P1-P2 2,000 (min) or P1 1,000 P2 2,000	Group also contains Decoder Test Set, Interrogator Test Set, Control Monitor, Remote Control Monitor, and Interference Blanker
CMSSR-201	Dual monopulse interrogator system with automatic switching	1, 2, 3/A, B, C, D	PRI, SEC, TER, QUAT patterns selectable for any mode. Mode S compatible	150-450 7-step stagger available	Conventional SLS 2,000 (min) Improved SLS P1+P2 2,000 or P1 1,000 P2 2,000	Phase processing for optimum accuracy. 4-channel reply processing for optimum defruiting, degarbling & resolution. Internal self-test. Remote maintenance monitoring available
Cossor Electronics						
IFF 870 Series	Miniature general purpose IFF equipment	1, 2, 3/A	No	150-450	27 dBW ±1 dB	Supplied with defruiter/decoder. Cossor antennas CRS 379 or CRS 389 suitable
IFF 880 Series	"On-weapon" IFF for air defence guns and missile systems	1, 2, 3/A	No	150-450	13 dBW (P_1P_3) 18 dBW (P_2)	One version supplied for Bofors RBS 70 weapon system. Variants for other weapons including Blowpipe and Javelin available
CEL 850 System	General purpose SSR/IFF system for static, mobile and naval applications	1, 2, 3/A, B, C, D, External	Single, Dual, Triple	150-1,000	33 dBW ±0.5 dB adjustable by −3 dB and −6 dB	Wide range of antennas, processing and display units. Single and dual-channel configurations with auto-changeover
SSR 950 Monopulse SSR System	ATC Systems	1, 2, 3/A, B, C, D, External	Single, Dual, Triple, External	50-450	a) Conventional ISLS 33 dBW ±0.5 dB b) Improved ISLS P_1P_3-1 kW P_2-2 kW	Single or dual channel with automatic changeover. Remote control and monitoring. To be superseded by Condor 9600. Solid-state SSR 955 integrated for Transport Canada by Raytheon Canada
Condor 9600	Monopulse SSR System	ND	ND	ND	ND	Supersedes SSR 950. Single or dual channel with automatic changeover. Plot extractor, remote control and monitoring. Totally solid-state transmitter with improved system configuration and automatic fault diagnosis
Hazeltine Corp						
AN/TPX-54 (V)	All-purpose civil/military SSR/IFF for ground shipboard use. Redesignated AN/OX-58/TPQ-43 for USAF SEEK SCORE system	1 or B, 2 or D 3/A, C, 4	Up to 5 modes, any combination	450 max accepts radar triggers to 4,050 PRF with 16 cycle stagger	2,000/4,000, stepped lower values available	Internal options include defruiting, RSLS/beam sharpening, automatic Mode 4 decoding, Target extractor digital interfaces. For use with AN/UPA-59 decoder. Antennas: AN/GPA 123 or 128, or AS-2167/TPX-46
AN/TPX-54 (V) USAF Variant	IFF for TPS-43E. Replaces UPX-14/21 in USAF	1 or B, 2 or D 3/A, C, 4	Up to 5 modes any combination	450 max accepts radar triggers to 4,050 PRF with 16 cycle stagger	4,000 (max)	Defruiting, RSLS/beam sharpening, automatic Mode 4 evaluator, automatic BIT to LRU
Model 2679	Tactical IFF for short-range air defence	1, 2, 3/A, 4	One SIF mode plus mode 4	Radar PRF counted down to 450	200	Memory stored codes, manual control or digital interfaces for computerised fire control systems, ISLS, RSLS, beam sharpening
OX-60/FPS-117	IFF for minimally attended radar	1, 2, 3/A, C, 4	3, 2, C or 3, 2, 3, C	300-350	2,000	Target position information over data bus, ISLS, RSLS; used with AN/FPS-117 radar

Table 2A.2 (cont'd)

Manufacturer Designation	Application	Modes	Mode interlace facility	Pulse repetition frequency/sec	Transmitter pulse power (peak) W	Associated equipment/Remarks
AN/TPX-46 (V)	Used with PATRIOT, HAWK, HERCULES air defence systems and radars	1, 2, 3/A, 4	Various combinations	Radar PRF counted down to 450	2,000	RSLS, ISLS, beam sharpening, MIL qualified replacement. AS-2167 antenna group. Solid-state transmitter available
AN/TPX-50	IFF used with tactical surveillance radar	1, 2, 3/A, 4	Normally single mode	375-400	165, 82 or 41 selectable	RSLS, ISLS, beam sharpening, internal defruiting
Italtel						
MM/UPX 708	IFF for ground and naval applications full Mode 4 with internal friend evaluator	1, 2, 3/A, C, 4 (Mk XII)	Up to 5 modes programmable for priority or repetitions	400-450 100-1,000 with autostagger locking	Up to 1,300	Dual channel. RF for sum/difference or directional/omni antennas. Used with up to 6 AN/UPA-59A decoders. VLSI technology
SIT 422	Short/medium-range SIF/IFF interrogator for armoured vehicle applications	1, 2, 3/A, C, 4 (Mk XII)	1 SIF Mode + Mode 4, two patterns available	442/663 PPS jittered internal trigger for independant operation	200	Sum/difference antenna RF connections. Computer interface for remote control and data processing (serial RS 422). Stored code facility for SIF/Mode 4 operation. Control box available for independant operation. Internal plot extractor and data generator for target identification
Selenia						
SIR-R	Civil/Military SSR/IFF	1, 2, 3 (4) A, B, C (D)	Any combination	150-450	2kW (4kW option)	Fully solid-state. Both Tx and Rx fully programmable. Built-in extractor and BITE
SIR-M	Civil/Military SSR/IFF	1, 2, 3 (4) A, B, C (D) up gradeable to S	Any combination	150-450	2kW (4kW option)	Monopulse. Fully solid-state. Both Rx and Tx fully programmable. Built-in extractor and BITE
Siemens						
DII 2 (MSR 400)	Short/medium-range IFF/SIF interrogator, primarily for air defence	1, 2, 3/A/C, Emergency Mk XII (Mk X-A + Mode 4)	2 modes	300-600 Hz jitter, av 450 Hz	400 (adjustable)	Interrogation and reply beam sidelobe suppression. Sum and difference beam sharpening. Interrogation pulse jittering
DII 111 (MSR 200)	Short/medium-range interrogator with control box incl Mode 4 capability	1, 2, 3/A Emergency Mk XII	2 modes	As above	Up to 600 (adjustable)	Interrogation and reply beam sidelobe suppression. Sum/difference sharpening. Interrogation pulse jittering
DII 511 (MSR 2000)	All-purpose IFF (ATC, air defence systems incl shipborne)	1, 2, 3/A, 4, B, C, D, Emergency Mk XII	3 modes, priority controlled	150-400, adjustable	Up to 2,000	As above; used with UPA 59 and/or SIF/Mode 4 extractor
Thomson-CSF						
RS 870	ATC SSR	1, 2, 3/A, B, C, D	Up to 3 modes	450	500 (solid-state) 2,500 (amplified)	RSLS facility. AS 809 elevation scan antenna available
RSM 870	ATC SSR monopulse	1, 2, 3/A, B, C, D	Up to 4 modes	450	3,000 (solid-state)	RSLS facility. BITE. Mode S compatible. Large vertical aperture antenna. Integral extraction and tracking
Toshiba						
SSR	SSR/IFF	1, 2, 3/A, C	2 or 3 modes	ND	ND	Centre co-fed on main radar antenna or separate open planar array. Conforms to ICAO Annex 10. Monopulse and Mode S offered
Westinghouse						
ATCBI-30	SSR/IFF	1, 2, 3/A, C, Mil. 4	S	S	2,000	Solid-state
MODE S	SSR/IFF	2, 3/A, C and S	3 plus S	Up to 450	2.4kW primary 9.4kW secondary	Monopulse. Solid-state, remote control and monitoring, built-in, computer-aided fault detection and isolation
MODE S/SSR	SSR/IFF	1, 2, 3/A, B, C, D	3	Up to 450	2.4kW	Monopulse SSR which can be upgraded to full Mode S

Manufacturer Designation	Application	Wave-length (cm)	Pulse power (kW peak/mean)	Pulse repetition frequency per sec	Max range for 3 m² target (nmi)	Antenna rotation speed (rev/min)	Horizontal beamwidth	*Remarks* (Clutter suppression facilities, commercial history, etc.)
Airport surface movement radars								
Hollandse Signaalapparaten								
ASDE		33/ 1.8	60/0.012	4,096/8,192	5.5	60	0.25°/ 0.4°	Circular or linear polarisation. Digital scan conversion. Introduced 1977
Racal Avionics								
ASMI 18X		3	15/–	4,096	1.5	60	0.4°	Clutter suppression, range suppression blanking, pan/zoom. Operating at Gatwick, Manchester, Arlanda, Kimpo. Ordered for Luton, Sydney, Amsterdam, Frankfurt, Madrid

Table 2A.2 (cont'd)

Manufacturer Designation	Application	Wave-length (cm)	Pulse power (kW peak/ mean)	Pulse repetition frequency per sec	Max range for 3 m² target (nmi)	Antenna rotation speed (rev/min)	Horizontal beamwidth	*Remarks* (Clutter suppression facilities, commercial history, etc.)
Thomson-CSF								
ASTRE		1.9	30/0.012	10,000	8	60	0.3°	Good visibility in heavy rain. Fully solid-state. Digital scan-converted bright display. Fully solid-state.

Precision and Ground-Controlled Approach Radars (PAR/GCA)

Manufacturer Designation	Application	Wave-length (cm)	Pulse power (kW peak/ mean)	Pulse repetition frequency per sec	Max range for 3 m² target (nmi)	Antenna rotation speed (rev/min)	Horizontal beamwidth	Remarks
Hindustan Aeronautics Ltd								
PAR-1	Precision approach	ND	ND	ND	ND	ND	ND	Company did not reply
ITT Gilfillan								
Quadradar MTI	ATC, PAR, surface movement & heightfinding	3	180/NA	1,925 or 3,300	30	15 (Az.)	Az. 0.95° El. 2.5°	Digital MTI, GAGC. Circular polarisation, log amplification. Highly mobile
AN/TPN-18 and -18A	ATC, PAR & heightfinding	3	200/NA	1,200	35 (1 m²)	15 (Az.)	Az. 1.1° El. 3.2°	Circular polarisation, log amplification. TPN-18 first introduced 1967; -18A introduced 1980
AN/TPN-22	Automatic landing, tactical system	3	120/0.18	4,200 to 6,000	10 (1 m²)	Electronic sector scan	0.98°	MTI, CFAR, GAGC. 3 operational modes including full autoland. Introduced 1977; 17 sold
Series 52 (AN/MPN-23V, AN/FPN-63)	Precision approach	3	100/NA	2,750 or 3,300	24 (1.12 m²)	ND	Az. 0.92° El. 3.84°	Digital MTI and circular polarisation; sector scan (2 looks per sec). Introduced 1976; Mobile and static versions
PAR-80	Precision approach	3	150/NA	3,450	30 (1 m²)	ND	1.1°	Clutter sensor, CFAR optional. Dual-channel modification of Quadradar. Turntable operation
Raytheon								
AN/TPN-25	PAR part of TPN-19 system	3	325/1.0	3,000	20+	2 scans per sec	1.4× 0.75°	Agile beam; 15×20° raster. Circular polarisation, coherent and non-coherent digital MTI. Introduced 1971
AN/GPN-22	Precision approach (fixed or mobile)	3	325/1.0	3,000	20+	2 scans per sec	1.4× 0.75°	Agile beam, 8×20° raster. Circular polarisation coherent and non-coherent digital MTI. Introduced 1975
AN/FPN-62	Precision approach (fixed)	3	50/0.2	4,000	15	120 each	0.5× 2.2°	Digital echo MTI, circular polarisation. Introduced 1976
Thomson-CSF								
TRS 2300 Perceval	Surveillance and helicopter recovery	3	ND	ND	>40	15	1.2°	Surveillance, landing and combined use
TRS 2310	Precision approach GCA & TA-1002 Tiger	3	30/0.1	Variable	20	2 scans per sec	Az 1.1° El 0.55°	Fixed-echo and weather-clutter suppression and ECCM facilities. Mobile and static versions
Toshiba								
PAR	Precision approach	3	50-80/–	3,400/2,400	20	ND	0.8°	MTI, rain echo canceller, Log-CFAR, digital signal processor

Specialised military radars

Manufacturer Designation	Application	Wave-length (cm)	Pulse power (kW peak/ mean)	Pulse repetition frequency per sec	Max range for 3 m² target (nmi)	Antenna rotation speed (rev/min)	Horizontal beamwidth	Remarks
Electronique Serge Dassault								
Rodeo 2	Short-range air defence	10	0.2/S	S	6.5	40	6°	Simultaneous multiple target tracking. Helicopter identification. Frequency agile. Shipborne version available
Rodeo 2S	Short-range air defence	10	0.45/S	S	10	30	6°	—
Elta Electronics Industries								
EL/M-2106	Mobile early warning	23	0.1/0.01	8,000-16,000	16 (2 m²)	ND	10°	Man-portable. Can be remoted up to 100 m. Programmed sector scan available
EL/M-2106 H	Point defence alert radar	23	0.15/0.01	153	8.6 (2 m²)	ND	5°	Vehicle mounted or man transportable. Detects hovering helicopters and low-flying aircraft. Made under licence and improved as the TDAR by Lear Siegler

Table 2A.2 (cont'd)

Manufacturer Designation	Application	Wave-length (cm)	Pulse power (kW peak/mean)	Pulse repetition frequency per sec	Max range for 3 m² target (nmi)	Antenna rotation speed (rev/min)	Horizontal beamwidth	Remarks (Clutter suppression facilities, commercial history, etc.)
Ericsson Radio Systems								
Giraffe (Basic version)	Military surveillance (2-D); short-range air defence systems. Mobile	5	15/0.3	6,000/3,000	15	60	2°	55 dB fixed-echo and weather improvement factor. Fully coherent. Auto-frequency agility and target tracking. IFF as option
Giraffe (Super version)	As above, for medium-range SAM systems. Mobile	5	60/1.1	3,000	22	60	2°	As above
UAR 1021	Combined search/fire control. Part of Skyguard air defence system	3	25/0.2	8,500	10	60 (search)	Search: 1.7° (cosec²) Track: 2.4°	Also 1.4° fan beam. 45 dB fixed-echo and weather improvement factor. IFF option
HARD (Helicopter and Aircraft Radar Detection)	3-D local search air defence radar	3	ND	ND	9	40	2.5°	All solid state 3-D radar. Automatic target detection and tracking. Helicopter detection function. Adaptive MTI. Frequency agility
Eagle	Tracking air defence radar	0.8	ND	ND	10	ND	0.6°	Fully coherent transmitter. Simultaneous frequency agility (pulse-to-pulse) and MTI. Adaptive MTI
Hollandse Signaalapparaten								
AA Tank Radar	a) 2-D surveillance with IFF	3	220 (peak)	6,400-8,000	10	60	1.5°	Mobile. Introduced 1972
	b) Tracking with 2 35 mm guns	3 and 0.8	120 and 12 (peak)	6,400-8,000	10	ND	4.3° 0.9°	
Flycatcher	a) 2-D surveillance with IFF	3	220 (peak)	4,800-6,000	11	40	1.1°	Mobile. Introduced 1974
	b) Tracking	3 and 0.8	120 and 15 (peak)	4,800-6,000	11	ND	2.4° 0.6°	
Reporter	a) 2-D surveillance with IFF	3	220 (peak)	2,400-3,000	22	48	1.4°	Mobile. Introduced 1981
Marconi								
S711	Tactical/gap-filler	10	160/2	800-1,000	80	15	1.5°	Elevating antenna for use against low level targets. Adaptive signal processing. TWT with pulse compression. Mobile
S712	Battlefield/point defence	10	160/2	800-1,000	64	15	3	Highly mobile. Adaptive signal processing. TWT with pulse compression
Siemens								
MPDR 30	Low-level air defence	23	20/0.2	S	S	16	4°	> 60 dB fixed-echo clutter improvement factor. First installed 1967. Mobile
DR 151 (MPDR 45/E)	Low-level air defence	23	20/0.2	S	24	16	4°	> 50 dB fixed-echo clutter improvement factor. First installed 1978
DR 161 MPDR 60/E	Low-level air defence	23	20/0.2	S	33	16	4°	> 50 dB fixed-echo clutter improvement factor. First installed 1979
DR 171 (MPDR 90/E)	Low-level air defence	23	20/S	S	48	16	2.9°	> 50 dB fixed-echo clutter improvement factor. First installed 1969
MPDR 12	Low-level air defence	12.5	4/0.1	S	7.5	60	6.3°	> 57 dB fixed-echo clutter improvement factor. First installed 1969
DR 621 (MPDR 18/X)	Low-level air defence	3.3	200/0.26	Statistical	10	60	1.6°	> 35 dB fixed-echo clutter improvement factor. First installed 1977. Highly mobile
DR 631 (MPDR 30/X)	Low-level air defence	3.3	200/0.26	Statistical	16	40	1°	> 33 dB fixed-echo clutter improvement factor. First installed 1981
MPDR 3002/S	Low-level air defence	12.5	S	S	16	S	S	First installed 1980. Highly mobile

Table 2A.2 (cont'd)

Manufacturer Designation	Application	Wave-length (cm)	Pulse power (kW peak/ mean)	Pulse repetition frequency per sec	Max range for 3 m² target (nmi)	Antenna rotation speed (rev/min)	Horizontal beamwidth	Remarks (Clutter suppression facilities, commercial history, etc.)
Thomson-CSF								
TRS 2120 (Œil Vert)	Surveillance and target acquisition	17	0.1/ 0.0125	Several modes	>8 >11	60	10°	2 antennas available
TRS 2620/ 2621 Gerfaut	Surveillance and acquisition for antiaircraft weapon systems	9	0.1/0.02	~10	>7	40	Several modes	Antenna under radome
TRS 2106	Low altitude surveillance for ground to air defence systems	5	7/0.12	Several	>25	30	1.7° (typical)	Planar antenna optional
TRS 2150	Surveillance and acquisition for Shorads	12	4/0.12	~8	>10	60	3°	—
TRS 2630 Grifon	Surveillance and acquisition for Shorads and spread-out sub-Shorads	9	0.1/0.02	ND	>10	40	5°	Planar antenna
TRS 3250	Fire control for AA guns	0.8	35/0.01	Stagger	6	ND	1°	—

Table 2A.3 Interavia Review of Western Airborne Attack Radar Systems (From *Interavia Aerospace Review*, March 1981, pp. 232–241. Reprinted with permission.)

UNITED STATES

Designation AN/APQ-153/-157 ASTAR

Manufacturer Emerson Electric Co, Electronics and Space Division, 8100 West Florissant Avenue, St. Louis, Missouri 63136, USA

Basic function Airborne search and range tracking radar

Technical characteristics

Operating modes	Stabilized Search, Boresight Missile Aiming (head-down), Gun Aiming (head-up)
Frequency range	I band (8-10 GHz)
Peak power	80 kW
Antenna	Parabolic dish, 30.5×40.6 cm (12×16 in)
Coverage	90° elevation, 90° azimuth * (beamwidth 7° in elevation)
Range	(Search) 37 km (20 nmi), lock-on at 19 km (10 nmi) for missiles
System weight	-153 : 50 kg (110 lb); -157 : 64 kg (143 lb)

Associated display systems 12.7 cm direct view storage tube, sector B-scan (-157)

Application F-5E, armed with Sidewinder AIM-9 or Matra Magic missiles, etc
AN/APQ-157 radar is virtually identical, with duplicated indicator display systems for two cockpits of F-5F aircraft.

* Coverage data given in single degree figures imply symmetrical coverage about the line of flight (e.g. 90° = ±45°). Exceptions are given thus: +80°−40°.

Designation EC-153 Hunter (developed version of AN/APQ-153)

Manufacturer Emerson Electric Co, Electronics and Space Division, 8100 West Florissant Avenue, St. Louis Missouri 63136, USA

Basic function Pulse Doppler air-to-air search and track radar for lightweight fighters

Technical characteristics

Operating modes	Search: Look-up (pulse); Look-down (pulse Doppler); Automatic Acquisition and Tracking: (Optional modes: Air-to-ground ranging; Missile Illumination; Terrain Avoidance)
Frequency range	I/J band (8-20 GHz)
Transmitter type	Solid state (GaAs FET)
Peak power	10 kW
Antenna type	Flat plate
Range	Max: 150 km (80 nmi) Search and track: (look-up) 37 km (20 nmi) (look-down) 24 km (13 nmi)
Search coverage	Azimuth and elevation: 90°

Associated display systems High-brightness CRT

Applications In development. Compatible with AGM-65A Maverick missile (?)

Designation AN/APQ-159/(V)

Manufacturer Emerson Electric Co, Electronics and Space Division, 8100 West Florissant Avenue, St Louis, Missouri 63136, USA

Basic function Multimode air-to-air search and tracking fire control radar

Technical characteristics

Operating modes	Search; Manual/Automatic Acquisition and Tracking (on or off boresight); Gun Fire Control (Dogfight and AA1/AA2); optional Electro-Optical modes for display of imagery from TV-guided missiles (in air-to-ground role?)
Frequency range	I/J band (8-20 GHz)
Transmitter type	Frequency-agile (100 MHz excursion)
Peak power	90 kW
Antenna type	Planar, 48×31 cm (19×12 in), vertical polarization, stabilized
Range	(Search) up to 74 km (40 nmi)
Coverage	Elevation: 11° beam scanning ± 45° Azimuth: 90°

Associated display systems High-brightness CRT with electro-optical raster image ((V)-1/2 only). Duplicate display for rear cockpit with (V)-2/4

Total system weights (V)-1: 63 kg (139 lb)
(V)-2: 71 kg (157 lb)
(V)-3: 56 kg (123 lb)
(V)-4: 71 kg (157 lb)

Applications F-5E ((V)-1/3); F-5F ((V)-2/4)
Compatible with AIM-9 series Sindwinder and AGM-65 Maverick missiles ((V)-1/2 versions only)

Designation AN/APQ-137

Manufacturer Emerson Electric Co, Electronics and Space Division, 8100 West Florissant Avenue, St. Louis, Missouri 63136, USA

Basic function Pulse-Doppler ground-target detection radar

Technical characteristics

Operating modes	Search, Target Ranging
Frequency range	K band (34.5 GHz)
Peak power	25 kW
Range	18 km (9.7 nmi)
Coverage	40° sector during search
Special capabilities	Moving-Target Indication (3-30 km/h)

Associated display systems Storage tube, sector B-scan

Application AH-1G HueyCobra helicopter (pod-mounted)

Table 2A.3 (cont'd)

Designation AN/APQ-113, -114, -144, -161 (-163)

Manufacturer General Electric Company, Aircraft Equipment Division, French Road, Utica, New York 13503, USA

Basic function Multimode air-to-air, air-to-ground and navigational radar

Technical characteristics

Operating modes Position Fixing, all-weather Weapon Delivery, Air-to-ground Ranging; Track-while-scan; Beacon (-114/-144)
Frequency range J band (16-16.4 GHz)
Antenna Phased array
Scan angle 20° or 90°; max. 120° (-163)
Detection range up to 300 km (160 nmi)

Application -113: USAF TAC's F-111As, F-111Es, Australian F-111Cs
-114: USAF SAC's FB-111As
-144: F-111Fs
-161: F-111Fs (PAVE TACK)
Air-to-air mode compatible with AIM-9 Sidewinder missiles

Note: AN/APQ-163 prototype version developed for Rockwell International B-1 strategic bomber; three assemblies supplied; nuclear-hardened; Doppler beam sharpening

---

Designation AN/AWG-9

Manufacturer Hughes Aircraft Company, El Segundo, California 90245, USA

Basic function Long range all-weather detection, target illumination and fire control

Entry into service 1972

Technical characteristics

Type Pulse Doppler, normal pulse and continuous wave
Operating modes Track-while-scan; range-while-search; single target track; weapons delivery; air-to-surface mapping and ranging; gun director; and several dogfight modes.
Antenna type Slotted-plate planar array
Range In excess of 115 nautical miles (213 km)
Special capabilities Look-down/shoot-down; simultaneous tracking of up to 24 targets and simultaneous control of up to six Phoenix missiles, with mid-course guidance. System is currently undergoing digital modification to incorporate programmable signal processor and digital display.

Associated display systems 2 CRTs for multimode raw and processed data, latter with alphanumeric/symbolic identification labels (including data link inputs), and target assignments.

Application F-14s of U.S. Navy with AIM-54A & C Phoenix, AIM-7E/F Sparrow, AIM-9L Sidewinder missiles and M-61 Vulcan 20 mm cannon.

Designation AN/APG-63

Manufacturer Hughes Aircraft Company, El Segundo, California 90245, USA

Basic function Multimode air-to-air/air-to-surface digital pulse Doppler radar with programmable signal processor (incorporation began into radars for F-15Cs & Ds in October 1979).

Entry into service 1974

Technical characteristics

Operating modes Track-while-Scan; Range-while-Search (high/medium PRF); Velocity Search (high PRF); Boresight, Super-search (HUD frame) and Vertical Acquisition modes with automatic/manual lock-on; Single Target Track; Weapons Delivery; Real Beam Ground Map; Doppler Beam Sharpening Ground Map; Air-to-surface Ranging; Precision Velocity Update; fixed and moving Ground Target Track; Beacon; Sniff-passive/active; Flood (dogfight).
Transmitter Gridded TWT
Antenna type Planar array, gimballed in 3 axes
System weight 494 lb (224 kg) (total)
Range In excess of 185 km (100 nmi)
Special capabilities Look-down/shoot-down; synthetic array radar high-resolution ground map mode flight demonstrated November, 1980, to be ready for production in 1984

Associated display systems HUD (lock-on, range & steering data), vertical situation display CRT (with alphanumerics and symbols), Sector-B scan, PPI

Application USAF F-15: AIM-7F Sparrow and AIM-9L Sidewinder missiles, M-61 Vulcan 20 mm cannon.

Table 2A.3 (cont'd)

Designation AN/APG-65

Manufacturer Hughes Aircraft Company, El Segundo, California 90245, USA

Basic function Multimode air-to-air/air-to-surface digital pulse Doppler radar, with programmable signal processor

Entry into service 1982

Technical characteristics

Operating modes Velocity Search (high PRF); Range-while-Search (high/medium PRF); Track-while-Scan (maintains 10, display 8); Single Target Track; Gun Director; Raid Assessment; HUD, Boresight and Vertical Acquisition modes with automatic lock-on; Terrain Avoidance; Precision Velocity Update; fixed and moving Ground Target Track; Sea Surface Search; Air-to-Surface Ranging; long range Real Beam Ground Mapping; Doppler Beam-Sharpened Patch; Doppler Beam-Sharpened Sector

Transmitter Software programmable gridded TWT

Antenna type Low sidelobe, planar array, fully balanced, with direct electric drive

Special capabilities Look-down/shoot-down

Associated display systems HUD and multifunction CRT displays

System weight 340 lb (154 kg), excluding rack

Application US Navy & Marine Corps F/A-18, Canadian Armed Forces CF-18 with AIM-7F Sparrow and AIM-9L Sidewinder missiles and M-61 Vulcan 20 mm cannon

---

Designation AN/APQ-148/156

Manufacturer Norden Systems (United Technologies Corp), Helen Walk, Norwalk, Connecticut 06856, USA

Basic function Multimode air-to-air and air-to-ground radar

Entry into service 1971 (APQ-148)

Technical characteristics

Operating modes Search, Ground Mapping, Tracking and Ranging, Terrain Avoidance and Following, Beacon Detection, Track-while-scan

Frequency range J Band (10-20 GHz)

Antenna type Paraboloid, about 100 cm (40 in) wide

Special capabilities APQ-156 incorporates forward-looking infra-red/laser target recognition and attack multi-sensor (TRAM)

Associated display systems Storage tube and terrain data display (pilot), direct-view CRT (navigator)

Application US Navy A-6E

Designation APQ-140

Manufacturer Raytheon Company, Missile Systems Division, Hartwell Road, Bedford, Massachusetts 01730, USA

Basic function Multimode attack radar

Technical characteristics

Frequency range J Band (10-20 GHz)

Antenna Phased array, 72 cm (28 in) diam, electronic scanning and beam steering, circular and linear polarization

Application Experimental (flown on KC-135 aircraft)

Designation Tornado IDS radar

Manufacturer Texas Instruments Inc, PO Box 226015, M/S 228, Dallas, Texas 75222, USA (in collaboration with Panavia)

Basic function Multimode forward-looking ground-mapping radar, optimized for air-to-ground use on all-weather penetration missions

Entry into service 1980

Technical characteristics

Operating modes Ground Mapping (wide/narrow, fast/slow scanning, pencil and "spoiled beam" – latter for general mapping); Contour Mapping (on boresight clearance plane); Height Finding; Air-to-Ground Ranging; Lock-on; Air-to-Air Tracking; Automatic Manual Terrain Following; plus Standby and Test modes.

Transmitter type Broadband TWT

Antenna type Planar array

ECCM capabilities Pulse compression, random frequency agility (possibly others)

Associated display systems HUD, moving map display with superimposed scan-converted radar data, two TV tabulator displays (navigator)

Application Tornado IDS variant (UK RAF, Italian and West German Air Forces, West German Navy), with various weapons systems
Note: Attack/ground mapping radar is used in conjuntion with separate Terrain-Following Radar (also by Texas Instruments). Computer for terrain-following radar is shared with ground-mapping radar.

Table 2A.3 (cont'd)

Designation AN/APQ-120
Manufacturer Westinghouse Electric Corp, PO Box 1897, Baltimore, Maryland 21203, USA
Basic function Fire control radar

Technical characteristics
Operating modes Search, Track
Frequency range I band (8-10 GHz)
Peak power 165 kW
Antenna typ- Paraboloid, 81 cm (32 in) diam
Range 56 km (30 nmi) against 1 m^2 target (?)
Total system weight 290 kg (639 lb)
Application F-4E/Gs in service with USAF and in Iran, Israel, Greece, Turkey, West Germany, Korea, Japan

Designation AN/APG-66
Manufacturer Westinghouse Defense and Electronic Systems Center, PO Box 1897, Baltimore, Maryland 21203, USA
Basic function Multimode digital coherent pulse-Doppler radar: air-to-air, air-to-ground, mapping and sea-search
Entry into service 1978 (over 500 systems delivered)

Technical characteristics
Frequency range I/J band (8-20 GHz; probably around 10 GHz)
Operating modes (Air-to-air) Look-up/look-down Search and Track; Air Combat Search (Dogfight); Automatic Track (Air-to-ground) Real-beam Mapping, with expanded beam and scan freeze; Ranging; Beacon; two Sea-Surface search modes, for sea states up to and above 4
Transmitter type TWT
Antenna type Planar array, 74 × 48 cm (29 × 19 in)
Search angles Up to 120° in azimuth and elevation
Search range Up to 148 km (80 nmi)
Special capabilities Doppler beam sharpening for mapping mode
Frequency agility for sea search mode
ECCM capability Frequency agility in certain modes
Associated display systems CRT, HUD
Applications F-16, with AIM-9 Sidewinder and AIM-7F Sparrow

Designation Advanced AN/APG-66 fire control radar
Manufacturer Westinghouse Defense and Electronic Systems Center, PO Box 1897, Baltimore, Maryland 21203, USA
Basic function Multimode air-to-air/air-to-ground digital pulse Doppler radar, with programmable signal processor and dual mode transmitter
Status In full scale development; entry into service 1984

Technical characteristics
Operating modes *Air-to-air:* Range-while-Search, Track-while-Scan (10 targets), Velocity Search (cued track-while-scan, high/med PRF), med PRF Uplook Search, Raid Cluster Assessment, multiple auto-acquisition Air Combat Mode, additional growth mode compatibility
Air-to-ground: Real Beam Mapping with Doppler beam sharpening and scan freeze, Sea Surface Search, Beacon, Ground Moving Target Indication/Track, Fixed Target Track, Terrain Following/Avoidance (growth), high-resolution Spotlight Map (growth)
Frequency range I/J band (8-20 GHz)
Transmitter Gridded, multiple peak power TWT
Antenna Planar array, 74 × 48 cm (29 × 19 in)
Search range Up to 296 km (160 nmi)
Special capabilities Extensive software reprogrammability and growth potential, multiple target awareness (multiple missile launch), rapid-reaction ECCM, high-resolution ground mapping, beyond visual range identification
Associated display systems CRT, HUD
Total system weight 153 kg (336 lb)
Application USAF F-16, with AIM-9 Sidewinder, AIM-7F Sparrow, and AMRAAM compatibility

Designation WX series
Manufacturer Westinghouse Defense and Electronic Systems Center, Baltimore, Maryland 21203, USA
Basic function Pulse-Doppler fire control radars. Family of modular radars differing chiefly in range performance:
WX-50: Range 28 km (15 nmi); non-coherent 30-40 GHz; modes: Ground Mapping, Terrain Avoidance, Ground Moving Target Indicator, Air-to-Ground Range; Sensor Cueing (TV). Weight 61 kg (135 lb)
WX-150: Range 37 km (20 nmi); reduced-power version of WX-200 (see below)
WX-200: (baseline version): Range: 70 km (38 nmi) Pulse Doppler Air-to-Air Combat radar, with TWT transmitter and Cassegrain antenna; frequency-agile for ECCM. Weight 160 kg (350 lb)
WX-300: Doppler-beam-sharpening version of -200 for high-resolution Ground Mapping. Weight 190 kg (420 lb)
WX-400: Long-range model, 185 km (100 nmi) No air-to-ground facilities. Weight 240 kg (530 lb)
Applications WX-50 and -200 versions in flight test. F-16 radar based on -200/400 designs

Designation AN/AWG-10/-10A/-11/-12
Manufacturer Westinghouse Electric Corporation, Aerospace Division, Baltimore-Washington International Airport, P.O. Box 746, Baltimore, Maryland 21203, USA
Basic function Multimode weapons control system
Entry into service 1967

Technical characteristics
Operating modes Air-to-air Search and Track (pulsed): Look-down (pulse Doppler); Target Illumination; Mapping (high/low); Air-to-ground Ranging
Frequency range I/J band (8-20 GHz)
Antenna type Paraboloid, 81 cm (32 in)? wide
ECCM capabilities Present, but unspecified
Applications US Navy F-4J Phantoms (AN/AWG-11/-12: UK RAF and RN F-4K/Ms)

Table 2A.3 (cont'd)

FRANCE

Designation Aida II

Manufacturer Electronique Marcel Dassault, 55 quai Carnot, 92214 St-Cloud, France

Basic function Automatic fire control and ranging: air-to-air, air-to-ground, air-to-surface

Entry into service 1971

Technical characteristics

Frequency range	I band (8-10 GHz)
Transmitter type	Magnetron
Peak power	80 kW
Antenna	Fixed lens, 18 cm (7 in) diameter
Beamwidth	16° azimuth and elevation (fixed)
Range	Air-to-air (1m² target): 0.25-8.0 km, depending on altitude Air-to-surface: 35 km

Associated display systems Gyro gunsight (no CRT needed)

System weight 30 kg (66 lb)

Application Mirage 5 and F1, for use with guns, rockets, infra-red missiles and bombs. Over 300 manufactured.

Designation AGAVE Radar

Manufacturer (Joint venture) Thomson-CSF, Avionics Division, 178 boulevard Gabriel-Péri, 92240 Malakoff, France
Electronique Marcel Dassault, 55 quai Carnot, 92214 St-Cloud, France

Basic function Lightweight multirole radar optimized for naval applications, with interception and ground-attack capabilities

Entry into service 1977

Technical characteristics

Operating modes	Air-to-Surface/Air-to-Air-Search; Target Designation; Automatic Tracking (Air-to-Air, Air-to-Surface); Ground Mapping; Air-to-Air/Air-to-Ground or Surface Ranging. Options: Contour Mapping; Blind Penetration; Beacon Homing
Frequency range	I band (8-10 GHz); Frequency agility option
Peak power	70 kW
Antenna type	Inverted Cassegrain, size optional ("standard" 46 cm (18 in)); 2-axis stabilized
Coverage	(Search) Azimuth 140°, elevation 60°
Range *	(Large aircraft) 46 km (25 nmi) (Patrol boat) 40 km (22 nmi) (* According to published data but not released by manufacturer)
Special capabilities	Monopulse; 2 pulse widths and 3 PRFs; beam sharpening

Associated display systems HUD, CRT Head-down display, moving map (COMED)

Total system weight 65 kg (144 lb)

Applications Super Etendard (French Navy); Jaguar International; Mirage 5 and 50

Designation RDI – Radar Doppler à Impulsions

Manufacturer (Joint venture) Thomson-CSF, Avionics Division, 178 boulevard Gabriel-Péri, 92240 Malakoff, France
Electronique Marcel Dassault, 55 quai Carnot, 92214 St-Cloud, France

Basic function Air-to-air and air-to-ground high-PRF pulse Doppler radar

Entry into service 1985 (projected)

Technical characteristics

Operating modes	Air-to-air: Search, Track-while-Scan, Continuous Track, automatic Lock-on Air-to-ground: Ground Mapping, Terrain Avoidance, Ranging
Frequency band	I band (8-10 GHz)
Transmitter type	Coherent TWT
Antenna type	Flat-plate, slotted
Coverage (search)	120° conical
Range	100 km (54 nmi) for 5 m² airborne target
Associated display systems	Multimode trichromatic head-down CRT display; head-up CRT display

Application Mirage 2000 Air Defence version

Table 2A.3 (cont'd)

Designation Antilope V

Manufacturer Electronique Marcel Dassault, 55 quai Carnot, 92214 St-Cloud, France in cooperation with Thomson-CSF, Avionics Division

Basic function Pulse-Doppler multimode radar

Entry into service 1986 (planned)

Technical characteristics

Operating modes Terrain-Following, Navigation (Ground Mapping and Updating)
Air-to-air: Search, Continuous Track, automatic Lock-on
Air-to-ground: Air-to-Ground Ranging
Air-to-sea: Search and Track

Frequency range J band (12-20 GHz)

Transmitter type Coherent, with TWT

Antenna type Flat plate, slotted

ECCM performance Facilities provided, no details available

Associated display systems Head-up display and multimode trichromatic head-down display

Application For penetration version of Mirage 2000

Designation Cyrano series radar (IV/IV M/IV MR/IV MS)

Manufacturer Thomson-CSF, Avionics Division, 178 boulevard Gabriel-Péri, 92240 Malakoff, France

Basic function Monopulse forward-looking radars for air-to-air and air-to-ground/sea search and tracking

Frequency range All models: I band (8-10 GHz)

Transmitter type All models: Coaxial magnetron

Peak Power All models: 200 Kw

Antenna All models: Inverted Cassegrain, 57 cm (22.4 in) diameter

Coverage All models: Elevation 60°, Azimuth 120°

	Cyrano IV MR	Cyrano IV MS	Cyrano IV	Cyrano IV M
Entry into service	1982	1982	1972	1980
Technical characteristics				
Operating modes	*Air-to-air:* Search Track-while-Scan Continuous Track Auto. Lock-On	*Air-to-air:* Search Track-while-Scan Continuous Track Auto. Lock-On	*Air-to-air:* Search Continuous Track Auto. Lock-On	*Air-to-air:* Search Track-while-Scan Continuous Track Auto. Lock-On
	Air-to-ground: Ground Map/Waypoint Nav. Reset/Update Terrain Avoidance Blind Let-Down Ranging	*Air-to-ground:* Ground Map/Waypoint *(Options)* Terrain Avoidance Contour Mapping Blind Let-Down Ranging	*Air-to-ground:* Ground Mapping Contour Mapping Blind Let-Down Terrain Avoidance Ranging	*Air-to-ground:* Ground Mapping *(Options)* Contour Mapping Terrain Avoidance Blind Penetration Ranging
		Air-to-sea: Search Track		*Air-to-sea:* Search Track

	Cyrano IV	Cyrano IV M	Cyrano IV MR	Cyrano IV MS
Range (2 m² airborne target, no clutter)				
Detection	60 km (32 nmi)	60 km (32 nmi)	60 km (32 nmi)	50 km (27 nmi)
Tracking	45 km (24 nmi)	45 km (24 nmi)	45 km (24 nmi)	45 km (24 nmi)
ECCM features	Variable pulse width Homing on jammer	PRF coding and variable pulse width Homing on jammer	PRF coding and variable pulse width Homing on jammer	PRF coding and variable pulse width Homing on jammer
Associated display systems	Storage tube Head-up display	Storage tube or head down display Head-up display or gunsight	Storage tube or multimode head-down display Head-up display	Storage tube or multimode head-down display Head-up display
Application	Mirage F1	Mirage F1	Mirage F1	Mirage 50 and Mirage III retrofit

Table 2A.3 (cont'd)

Designation	Agrion 15
Manufacturer	Thomson-CSF, Avionics Division, 178 boulevard Gabriel-Péri, 92240 Malakoff, France
Basic function	Sea search, missile guidance and navigation radar
Entry into service	from 1982
Technical characteristics	
Operating modes	Search, Track-while-scan, Over-horizon targeting
Frequency range	I band (8-10 GHz)
Transmitter type	Pulse compression
Antenna	Parabolic
Coverage	360° or sector scan
Special capabilities	Differential tracking of missile and target; can pass track data to ships, aircraft, etc. Effective range (90 % hit probability on fast patrol boat) believed 15 km (8 nmi); data not released by manufacturer
ECCM capabilities	Frequency agility
Associated display systems	CRT
Applications	Anti-ship and ASW aircraft, with AS. 15 TT missiles; SA 365 N Dauphin

Designation	RDM – Radar Doppler Multifunction
Manufacturer	Thomson-CSF, Avionics Division, 178 boulevard Gabriel-Péri, 92240 Malakoff, France
Basic function	Multipurpose low PRF Doppler radar: Air-to-Air, Air-to-Ground, Air-to-Sea
Entry into service	1983 (projected)
Technical characteristics	
Operating modes	Air-to-air: Search, Track-while-Scan, Continuous Track, automatic Lock-On Air-to-ground: Ground Mapping, Contour Mapping, Blind Let-Down, Terrain Avoidance, Ground Moving Target Indicator, Ranging Air-to-sea: Search, Track
Frequency range	I band (8-10 GHz)
Transmitter type	Coherent TWT
Antenna type	Inverted Cassegrain
Coverage (search)	120° conical
Range	100 km (54 nmi) for 5 m² target, under clutter-free conditions
Special capabilities	Continuous-wave illuminator optional. Growth potential: Beacon mode, Doppler beam sharpening, Sniff, IFF
ECCM capabilities	Frequency agility, PRF coding, pulse compression, variable pulse width
Associated display systems	Multimode trichromatic look-down CRT display; Head-up CRT display
Applications	Designed for Mirage 2000 multirole version; proposed for Mirage 4000

Designation	ORB 31D
Manufacturer	Omera-Segid SA, 49 rue Ferdinand Berthoud, 95100 Argenteuil, France
Basic function	Target designation radar for detection, tracking and location of surface vessels (part of Heracles I system). Also used for navigation and weather mapping.
Technical characteristics	
Frequency range	I band (9.375 GHz)
Peak power	40 kW
PRF	400/1,600 Hz
Antenna	Parabolic, 72×36 cm (46×31 cm also available), pitch and roll stabilization
Coverage	Azimuth 180° Elevation 30°
Range	80 km for 1,000 m² target (in sea state 4. 4 mm/h rain)
Special capabilities	Automatic target tracking
Associated display systems	2 PPIs, 1 B-scope
System weight	79 kg
Application	Used on SA.321 Super Frelon, with Exocet missile system

Designation	ORB 32 W (WAS, WTD)
Manufacturer	Omera-Segid SA, 49 rue Ferdinand Berthous, 95100 Argenteuil, France
Basic function	Airborne surveillance and target designation and tracking. Target coordinates automatically transmitted to semi-active missiles launched by carrier aircraft or other platforms. (Part of Heracles II system)
Technical characteristics	
Frequency range	I band (8-10 GHz)
Peak power	≏ 80 kW
PRF	300, 600 and 1,200 Hz
Antenna	Parabolic, 72×36 cm (28×14 in)
Coverage	360° scan (alternatively 60°, 120° or 180° sector scan)
Special capabilities	Track-while-scan facility Target coordinates permanently displayed. Sea-clutter elimination
System weight	44 kg (32 W), 129 kg (32 WAS), 140 kg (32 WTD)
ECCM performance	Coded jump-frequency and frequency-agile (1,000 Hz expansion) transceivers available
Associated display systems	PPI & B-scopes
System weight	32 W: 44 kg; 32 WAS: 129 kg; 32 WTD: 572 kg (last two moving parts only)
Application	SA.321 Super Frelon, SA.332 Super Puma

Table 2A.3 (cont'd)

ISRAEL

Designation EL/M-2001B

Manufacturer Elta Electronics Industries Ltd, Ashdod, Israel

Basic function Ranging only

Technical characteristics:

Operating modes Air-to-Ground, Air-to-Air (automatic acquisition)
Frequency range I or J band (8-20 GHz) (?)
Antenna Fixed, aperture 19 cm (7.5 in)

Associated display systems HUD

System weight less than 50 kg (110 lb)

Application Kfir (Israel)

Designation EL/M-2021

Manufacturer Elta Electronics Industries Ltd, Ashdod, Israel

Basic function Air-to-air and air-to-ground multimode combat radar

Technical characteristics

Operating modes Air-to-Air: Search and Automatic Track (Track-while-scan); Fire Control for Interception and Dogfight (Scan-while-track)
Air-to-Ground: Ranging; Bombing and Gunnery Fire Control; Terrain Mapping and Navigation; Terrain Following and Avoidance
Frequency range I/J band (8-20 GHz)?
Peak power 2kW (200 W average)
Antenna Inverted Cassegrain
Search angles Tracking: Elevation +80° to −40°, Azimuth 140°
Search/air-intercept: Elevation 10°, Azimuth 90°
Dogfight: Elevation 0° to +60°, Azimuth 20°
Special capabilities Look-up, look-down; continuously computed impact and release point modes for air-to-ground weapons delivery. Doppler beam sharpening available for mapping mode. Beacon display.

ECCM capability May use spread-spectrum techniques

Associated display systems HUD, Head-down

System weight 120 kg (265 lb)

Applications Suitable for F-4 retrofit

ITALY

Designation MM/APQ-706

Manufacturer SMA-Segnalamento Marittimo ed Aereo, PO Box 200, Villa S. Martino a Soffiano, 50100 Florence, Italy

Basic function Maritime search and attack

Technical characteristics

Operating modes Surface Search, Navigation, Target Acquisition and Missile Guidance
Frequency range I band (8-10 GHz)
Peak power 75 kW
Antennas Back-to-back parabolic

ECCM capabilities Frequency agility on one antenna channel

Application SH3-D helicopters (Italian Navy), for use with Marte air-to-surface missile

Table 2A.3 (cont'd)

SWEDEN

Designation	PS-46/A
Manufacturer	LM Ericsson, PO Box 1001, S-431 26 Mölndal, Sweden
Basic function	Multimode pulse-Doppler interception radar
Technical characteristics	
Operating modes	Search, Track-while-scan, Continuous Track, Dogfight, Target Illumination, Air-to-Ground Ranging (with optional mapping)
Frequency range	I band (8.6-9.6 GHz)
Transmitter	TWT-Klystron
Peak power	50 kW (illuminator 200 W CW)
Antenna	Cassegrain, 70 cm (28 in) diam.
Coverage	120° elevation and azimuth
Special capabilities	Look-down; target data extraction by Kalman filtering
Associated display systems	HUD, Head-down
System weight	300 kg (661 lb) including radar rack
Application	JA37 Viggen, armed with Skyflash or AIM-9 Sidewinder missiles, rockets, gun

Designation	PS-37/A, PS-371/A
Manufacturer	LM Ericsson, PO Box 1001, S-431 26 Mölndal, Sweden
Basic function	Pulse ground attack, search and surveillance radar
Technical characteristics	
Operating modes	Ground Mapping, Navigation Update, Air-to-Ground Ranging, Sea Surveillance and Target Designation, Air-to-Air Ranging
Frequency range	I/J band (8-20 GHz)
Peak power	200 kW
Antenna	Cassegrain, 70 cm (28 in)
Coverage	130° elevation and azimuth
Associated display systems	Sector PPI or B-scope
System weight	225 kg (496 lb) including radar rack
Application	AJ/JA37 Viggen armed with AIM-9 Sidewinder and AGM-65 Maverick missiles, rockets, bombs, gun

Designation	PS-011/A
Manufacturer	LM Ericsson PO Box 1001, S-431 26 Mölndal, Sweden
Basic function	Interception search and track pulse radar
Technical characteristics	
Operating modes	Search, Continuous Track
Frequency range	I/J band (8-≃12 GHz)
Peak power	200 kW
Antenna	Cassegrain, 45 cm (18 in) diam, 3 axis pedestal
Coverage	Elevation 65°, azimuth 120°
Special capabilities	Integrated IR search and track
System weight	230 kg (507 lb)
Application	J35F Draken (Sweden), armed with AIM-4 Falcon and AIM-9 Sidewinder missiles, rockets, guns

UNITED KINGDOM

Designation	Blue Kestrel
Manufacturer	Ferranti Ltd, Radar Systems Department, Ferry Road, Edinburgh EH5 2XS, UK
Basic function	Multirole anti-submarine and anti-ship search and tracking radar
Status	In development
Technical characteristics	
Operating modes	Surveillance (long-range detection). Track-while-Scan
Transmitter type	TWT
Antenna type	Large aperture, high gain
Coverage	360°
Special capabilities	Frequency agility, pulse compression. Interfacing with other avionics through MIL-1553 data bus
Application	For EH 101 (Agusta/Westland) helicopter, in development for Italian and UK navies

Table 2A.3 (cont'd)

Designation ARI 5979 Seaspray

Manufacturer Ferranti Ltd., Radar Systems Dept, Ferry Road, Edinburgh EH5 2XS, UK

Basic function Surface ship detection and tracking, missile guidance

Entry into service 1978-9

Technical characteristics

Operating modes Search and Track, Transponder Beacon, Navigation
Frequency range I band (8-10 GHz)
Antenna Paraboloid, 69 × 25 cm (27 × 10 in), monopulse
Coverage Wide-angle azimuth (180°) or limited scan; elevation 10° steerable

Display systems Scan-converted bright TV display, alphanumeric range and bearing labels

ECCM capabilities Frequency agility

System weight 64 kg (141 lb)

Application Westland Lynx helicopter (UK, Netherlands, Brazil, Argentina, Denmark, Germany, Norway); provision for controlling Sea Skua anti-shipping missile

---

Designation Blue Fox

Manufacturer Ferranti Ltd, Radar Systems Department, Ferry Road, Edinburgh EH5 2XS, UK

Basic function Air Interception and air-to-surface search and strike (developed from Seaspray)

Entry into service In production and delivered 1980

Technical characteristics

Operating modes Search, Attack (tracking and weapon-aiming), Boresight, Transponder. Radar Navigation update
Frequency range I band (8-10 GHz)
Antenna Flat-aperture (two-plane monopulse), 55 cm (22 in) diameter, 3-axis stabilized
Search angle (azimuth) 110°

ECCM capabilities Frequency agility

Associated display systems PPI, Sector B-scan, HUD, raster TV

Total system weight 85 kg (187 lb)

Application Sea Harrier

Designation Foxhunter AI (Airborne Interception) radar

Manufacturer Marconi Avionics, Elstree Way, Borehamwood, Herts WD6 1RX, UK; Ferranti Ltd, Radar Systems Dept, Ferry Road, Edinburgh EH5 2XS, UK (subcontractor responsible for scanner mechanism and transmitter mechanism)

Basic function Multimode airborne interception radar

Technical characteristics

Type High PRF pulse Doppler
Frequency range I band (8-10 GHz)
Operating modes Air-to-air Search and Track, Track-while-scan, Vis. Ident., Target illumination, Air Combat Search and Acquisition, Air-to-Ground Ranging, Ground Mapping
Range (detection) Up to 185 km (100 nmi)
Antenna Twist-reflecting Cassegrain
Special capabilities Look-down; Signal processor employing Fast Fourier Transform signal processing

ECCM capabilities "Extensive", unspecified

Associated display systems HUD, Head-down display, Radar warning display, 2 TV displays (navigator)

Application ADV Tornado F Mk2 (UK RAF)

Table 2A.4(a) Airborne Attack Radar Systems

Radar	F-16 AN/APG-66 [10, 404]
Type	Multimode; air-to-air; look-up, look-down; search, track; air-to-ground; real-beam mapping; doppler beam sharpening (8:1); sea search, ranging, beacon homing.
Manufacturer	Westinghouse
Location	Nose of F-16.
Frequency Band	X
Frequency (MHz)	
Power, Peak (MW)	
Power, Av (kW)	
Antenna Size (H × W) (m) (or Type)	7.4 × 4.8 planar array.
Antenna Gain (dB)	
Beam Shape	
Type of Scanner	Mechanical, except for electronic sequential lobing on receive.
No. of Beams	
Beamwidth, EL (deg)	
Beamwidth, AZ (deg)	
Angular Coverage, EL (deg)	120°
Angular Coverage, AZ (deg)	120°
Scan Rate, EL	
Scan Rate, AZ	
PRF	Medium.
PRF Stagger	
Pulsewidth, Transmit (μs)	
Pulsewidth, Compressed (μs)	
Pulse Compression Ratio	
Type of Tube	Wideband TWT.
Noise Figure (dB)	
Range	20 and 40 nmi for look-down and look-up, respectively.
MTI	
Polarization	
Frequency Agility	
ECCM Features	Frequency agility, low-sidelobe antenna.
Misc. Comments	More than 2000 built as of 1986.
Fig. No.	

Table 2A.4(a) (cont'd)

Radar	F-20 AN/APG-67 [10, 30]
Type	Multimode; air-to-air and air-to-surface. Pulsed doppler.
Manufacturer	General Electric
Location	Nose of F-20 Tigershark tactical fighter.
Frequency Band	200 MHz bandwidth at X.
Frequency (MHz)	
Power, Peak (MW)	
Power, Av (kW)	0.20
Antenna Size (H × W) (m) (or Type)	0.32 × 0.51 flat plate slotted array with low sidelobes.
Antenna Gain (dB)	
Beam Shape	
Type of Scanner	Mechanical.
No. of Beams	3(Σ, ΔAZ, ΔEL) on receive.
Beamwidth, EL (deg)	
Beamwidth, AZ (deg)	
Angular Coverage, EL (deg)	±60°
Angular Coverage, AZ (deg)	±60°
Scan Rate, EL	
Scan Rate, AZ	
PRF	Low, medium, and high
PRF Stagger	
Pulsewidth, Transmit (μs)	Variable.
Pulsewidth, Compressed (μs)	
Pulse Compression Ratio	Used.
Type of Tube	Coherent TWT.
Noise Figure (dB)	
Range	30 and 22 nmi for look-up and look-down (fighter target), respectively; 35 and 30 nmi in sea-states 1 and 4, respectively (50 m^2 target).
MTI	Used for ship detection in high sea states.
Polarization	
Frequency Agility	Used for ship detection.
ECCM Features	Low sidelobes.
Misc. Comments	Main missions: supersonic intercept, combat air patrol, and air superiority in air-to-air mode; close air support and sea detection in air-to-surface mode.
Fig. No.	

Table 2A.4(a) (cont'd)

Radar	F-16 AN/APG-68 [10]
Type	Improved AN/APG-66.*
Manufacturer	Westinghouse
Location	Nose of F-16.
Frequency Band	
Frequency (MHz)	
Power, Peak (MW)	
Power, Av (kW)	
Antenna Size (H × W) (m) (or Type)	
Antenna Gain (dB)	
Beam Shape	
Type of Scanner	
No. of Beams	
Beamwidth, EL (deg)	
Beamwidth, AZ (deg)	
Angular Coverage, EL (deg)	
Angular Coverage, AZ (deg)	
Scan Rate, EL	
Scan Rate, AZ	
PRF	Low PRF for air-to-surface engagements, medium and high for long-range interception.
PRF Stagger	
Pulsewidth, Transmit (μs)	
Pulsewidth, Compressed (μs)	
Pulse Compression Ratio	
Type of Tube	Dual-mode transmitter.
Noise Figure (dB)	
Range	
MTI	
Polarization	Vertical [a].
Frequency Agility	
ECCM Features	
Misc. Comments	Future growth available to very high resolution SAR because of new VHSIC signal processor. (See Sec. 2.6.6.1.)
Fig. No.	

*Allows launch for AIM-20 AMRAAM beyond visual range. Also provides improved ground-mapping resolution by a factor of 8.
[a] Data extrapolated by author from available information.

Table 2A.4(a) (cont'd)

Radar	F-5 AN/APG-69(V) [10]
Type	Multimode; pulsed doppler fire control radar. Air-to-air and air-to-surface modes.
Manufacturer	Emerson
Location	Nose of F-5
Frequency Band	X
Frequency (MHz)	
Power, Peak (MW)	0.008
Power, Av (kW)	
Antenna Size (H × W) (m) (or Type)	Flat plate slotted array.
Antenna Gain (dB)	
Beam Shape	
Type of Scanner	Mechanical
No. of Beams	Σ and Δ on receive.
Beamwidth, EL (deg)	
Beamwidth, AZ (deg)	
Angular Coverage, EL (deg)	
Angular Coverage, AZ (deg)	
Scan Rate, EL	
Scan Rate, AZ	
PRF	Low for unambiguous range, higher PRF also available.
PRF Stagger	
Pulsewidth, Transmit (μs)	
Pulsewidth, Compressed (μs)	
Pulse Compression Ratio	Used with low-PRF waveform.
Type of Tube	TWT
Noise Figure (dB)	GaAS FET LNA.
Range	
MTI	
Polarization	Vertical [a].
Frequency Agility	
ECCM Features	
Misc. Comments	Real-beam ground mapping available. Frequency diversity used for ship detection. Under development.
Fig. No.	

[a] Data extrapolated by author from available information.

Table 2A.4(a) (cont'd)

Radar	F-15E AN/APG-70 [10]
Type	Upgrade of AN/APG-63 multimode radar.
Manufacturer	Hughes
Location	Nose of F-15E
Frequency Band	X
Frequency (MHz)	
Power, Peak (MW)	
Power, Av (kW)	
Antenna Size (H × W) (m) (or Type)	Same as AN/APG-63.
Antenna Gain (dB)	
Beam Shape	
Type of Scanner	Mechanical.
No. of Beams	
Beamwidth, EL (deg)	
Beamwidth, AZ (deg)	
Angular Coverage, EL (deg)	
Angular Coverage, AZ (deg)	
Scan Rate, EL	
Scan Rate, AZ	
PRF	Multiple PRFs.*
PRF Stagger	
Pulsewidth, Transmit (μs)	
Pulsewidth, Compressed (μs)	
Pulse Compression Ratio	
Type of Tube	High average power gridded TWT.*
Noise Figure (dB)	
Range	Improved sensitivity.*
MTI	
Polarization	
Frequency Agility	
ECCM Features	Improved tracking in ECCM.*
Misc. Comments	Has programmable signal processor.*
Fig. No.	

*Improvements over AN/APG-63.

Table 2A.4(a) (cont'd)

Radar	F-111F AN/APQ-146 [10]
Type	Multimode, forward-looking radar. Modes: terrain following; terrain avoidance; ground mapping.
Manufacturer	Texas Instruments
Location	Nose of F-111F
Frequency Band	
Frequency (MHz)	
Power, Peak (MW)	
Power, Av (kW)	
Antenna Size (H × W) (m) (or Type)	
Antenna Gain (dB)	
Beam Shape	
Type of Scanner	
No. of Beams	Elevation monopulse used.
Beamwidth, EL (deg)	
Beamwidth, AZ (deg)	
Angular Coverage, EL (deg)	
Angular Coverage, AZ (deg)	
Scan Rate, EL	
Scan Rate, AZ	
PRF	
PRF Stagger	
Pulsewidth, Transmit (μs)	
Pulsewidth, Compressed (μs)	
Pulse Compression Ratio	
Type of Tube	
Noise Figure (dB)	
Range	
MTI	
Polarization	
Frequency Agility	
ECCM Features	
Misc. Comments	After future modifications to be designated AN/APQ-171.
Fig. No.	

Table 2A.4(a) (cont'd)

Radar	F-5E Radar
Type	Multimode Radar.
Manufacturer	General Electric
Location	Nose of F-5E
Frequency Band	X
Frequency (MHz)	
Power, Peak (MW)	
Power, Av (kW)	
Antenna Size (H × W) (m) (or Type)	2.8 × 4.0 planar slotted array.
Antenna Gain (dB)	
Beam Shape	
Type of Scanner	
No. of Beams	
Beamwidth, EL (deg)	
Beamwidth, AZ (deg)	
Angular Coverage, EL (deg)	
Angular Coverage, AZ (deg)	to ±40°.
Scan Rate, EL	
Scan Rate, AZ	
PRF	
PRF Stagger	
Pulsewidth, Transmit (μs)	
Pulsewidth, Compressed (μs)	45-m range resolution in air-to-ground mode.
Pulse Compression Ratio	
Type of Tube	
Noise Figure (dB)	
Range	30 nmi (look-up), 19 nmi (look-down) for fighter in air-to-air mode. 20 nmi in air-to-ground doppler beam-sharpening mode. 40 nmi in ground track MTI mode. 80 nmi in ground-mapping mode. 31 nmi for stationary or moving ships in low sea state.
MTI	
Polarization	Vertical [a].
Frequency Agility	
ECCM Features	Low antenna sidelobes.
Misc. Comments	Coherent MTI used for high sea states to detect targets having dopplers over 8 kts. In production.
Fig. No.	

Table 2A.4(b) Air Surveillance and Bomber Radars

Radar	AN/APS-116 [405, 406]
Type	Periscope-detection radar; also detects ships.
Manufacturer	Texas Instruments
Location	Placed in US Navy S-3A Viking CB A/C [b].
Frequency Band	X
Frequency (MHz)	9,500–10,000
Power, Peak (MW)	0.50
Power, Av (kW)	0.50
Antenna Size (H × W) (m) (or Type)	1.1 × 0.61
Antenna Gain (dB)	35
Beam Shape	
Type of Scanner	Mechanical.
No. of Beams	
Beamwidth, EL (deg)	4.0°
Beamwidth, AZ (deg)	2.4°
Angular Coverage, EL (deg)	
Angular Coverage, AZ (deg)	
Scan Rate, EL	
Scan Rate, AZ	300 rpm.
PRF	500–2000 Hz
PRF Stagger	
Pulsewidth, Transmit (μs)	0.5 (LFM chirp)
Pulsewidth, Compressed (μs)	0.003 (1.5 ft)
Pulse Compression Ratio	250
Type of Tube	TWT, CFA.
Noise Figure (dB)	7.5
Range	15 nmi (1 m^2 periscope, sea-state 3). 76 nmi (destroyer, sea-state 3). 42 nmi instrumental range.
MTI	
Polarization	
Frequency Agility	
ECCM Features	
Misc. Comments	SAW pulse compressor used. Built by using E-beam lithography [406].
Fig. No.	

[b] See Glossary for definitions of abbreviations used in Tables 2A.1 and 2A.4.

Table 2A.4(b) (cont'd)

Radar	AN/APS-137(V) [10, 30]
Type	Upgrade of AN/APS-116 to include ISAR for long-range ship classification.
Manufacturer	Texas Instruments
Location	Placed on S-3B Viking CB A/C.*
Frequency Band	X
Frequency (MHz)	
Power, Peak (MW)	
Power, Av (kW)	
Antenna Size (H × W) (m) (or Type)	
Antenna Gain (dB)	
Beam Shape	
Type of Scanner	
No. of Beams	
Beamwidth, EL (deg)	
Beamwidth, AZ (deg)	
Angular Coverage, EL (deg)	
Angular Coverage, AZ (deg)	
Scan Rate, EL	
Scan Rate, AZ	
PRF	
PRF Stagger	
Pulsewidth, Transmit (μs)	
Pulsewidth, Compressed (μs)	
Pulse Compression Ratio	
Type of Tube	
Noise Figure (dB)	
Range	
MTI	
Polarization	
Frequency Agility	
ECCM Features	
Misc. Comments	See AN/APS-116.
Fig. No.	

*Also aboard US Navy P-3C and US Coast Guard C-130.

Table 2A.4(b) (cont'd)

Radar	E-2C AN/APS-138/125 [10, 407]
Type	Airborne early warning. Works over land and sea. Evolved from AN/APS-96, -111, -120. To be upgraded to AN-APS-139 and -145.
Manufacturer	General Electric
Location	In rotodome above carrier-based E-2C Hawkeye.
Frequency Band	Low UHF
Frequency (MHz)	
Power, Peak (MW)	1.0
Power, Av (kW)	
Antenna Size (H × W) (m) (or Type)	2 Yagi arrays having 8 elements. (7.4 × 0.77 m elliptical cross section for rotodome.)
Antenna Gain (dB)	
Beam Shape	
Type of Scanner	Mechanical.
No. of Beams	Σ and Δ in AZ on receive to provide DPCA processing [408].
Beamwidth, EL (deg)	
Beamwidth, AZ (deg)	
Angular Coverage, EL (deg)	
Angular Coverage, AZ (deg)	360°
Scan Rate, EL	
Scan Rate, AZ	6 rpm.
PRF	Low-PRF pulsed doppler.
PRF Stagger	
Pulsewidth, Transmit (μs)	
Pulsewidth, Compressed (μs)	sub-μs.
Pulse Compression Ratio	Used.
Type of Tube	Coaxial transmitter.
Noise Figure (dB)	7.5
Range	
MTI	Double-delay line.
Polarization	Horizontal.
Frequency Agility	
ECCM Features	Sidelobe canceler.
Misc. Comments	Bank of 16 filters for doppler processing. Digital CFAR. Height of target measured by using difference in arrival times between direct and earth multipath signals.
Fig. No.	

Table 2A.4(b) (cont'd)

Radar	B-1B AN/APQ-164 [10]
Type	Multifunction phased array for B1-B.
Manufacturer	Westinghouse
Location	Nose of B1-B bomber.
Frequency Band	X
Frequency (MHz)	
Power, Peak (MW)	(Same as F-16 AN/APG-68.)
Power, Av (kW)	
Antenna Size (H × W) (m) (or Type)	1.1 × 5.6 (1526 phase shifters).
Antenna Gain (dB)	approximately equal to 38 [a]
Beam Shape	
Type of Scanner	Phase-phase.
No. of Beams	
Beamwidth, EL (deg)	0.3°
Beamwidth, AZ (deg)	1.7°
Angular Coverage, EL (deg)	±60° (about boresight).
Angular Coverage, AZ (deg)	±60° (about boresight).
Scan Rate, EL	
Scan Rate, AZ	
PRF	
PRF Stagger	
Pulsewidth, Transmit (μs)	
Pulsewidth, Compressed (μs)	
Pulse Compression Ratio	
Type of Tube	Gridded multiple TWTs.
Noise Figure (dB)	
Range	
MTI	
Polarization	
Frequency Agility	
ECCM Features	
Misc. Comments	See Sec. 2.2.1.6.
Fig. No.	2.15, 2.16

[a] Data extrapolated by author from available information.

Table 2A.4(b) (cont'd)

Radar	E-3A AWACS AN/APY-2 [10, 407, 409, 410]
Type	Airborne early warning. Air and sea-surface surveillance. Formerly AN/APY-1.
Manufacturer	Westinghouse
Location	In rotodome atop KC-135.
Frequency Band	S
Frequency (MHz)	
Power, Peak (MW)	
Power, Av (kW)	
Antenna Size (H × W) (m) (or Type)	Planar array consisting of slotted waveguide; see Sec. 2.6.2.1 and 7.3.1.
Antenna Gain (dB)	
Beam Shape	
Type of Scanner	Mechanical in AZ; phase in EL.
No. of Beams	
Beamwidth, EL (deg)	
Beamwidth, AZ (deg)	
Angular Coverage, EL (deg)	
Angular Coverage, AZ (deg)	360°
Scan Rate, EL	
Scan Rate, AZ	360° in 10 s.
PRF	High for operation against ground clutter. Low for long range, beyond the horizon.
PRF Stagger	
Pulsewidth, Transmit (μs)	
Pulsewidth, Compressed (μs)	
Pulse Compression Ratio	Used for long range.
Type of Tube	Wideband klystron.
Noise Figure (dB)	
Range	
MTI	
Polarization	Horizontal.
Frequency Agility	
ECCM Features	ULSA; jammer strobing in AZ, used to locate jammer by triangulation; BT.
Misc. Comments	28 high-power phase shifters used for EL scan.
Fig. No.	2.80 and 2.81

NOTES AND GLOSSARY

Notes for Tables 2A.1 and 2A.4

[a] Data extrapolated by author from available information.

[b] See Glossary for definitions of abbreviations used in Tables 2A.1 and 2A.4.

For bibliographic citations, see References at the end of Chapter 2.

Throughout this appendix, the author has used the US AN/XYZ designation, where

X = A = airborne;
= C = air transportable (inactivated);
= F = fixed, static;
= G = ground;
= M = mobile, installed on vehicle dedicated to transporting radar;
= S = shipborne;
= T = ground, transportable;
= U = general utility (two or more platforms: airborne, shipboard, and ground);
Y = P = radar;
Z = G = fire control or search light directing;
= N = navigational aids (landing, altimeters, beacons, *et cetera*);
= Q = special, or combination of purposes;
= S = detecting or range and angle measurement system;
= Y = surveillance (search, detect, multiple target tracking) and control (both fire control and air control).

Glossary

Abbreviation	*Definition*
ARM	antiradiation missile
BT	burnthrough
BW	bandwidth
CC	clutter cancelation
CPACS	coded pulse anticlutter system
CPI	coherent processing interval
DC	duty cycle
DMTI	digital MTI
DSP	digital signal processor
ECCM	electronic counter-countermeasures
ECM	electronic countermeasures
ECMR	electronic countermeasure (ECM) receiver
ESM	electronic support measures
ESMR	electronic support measure (ESM) receiver
FA	frequency agility
FET	field-effect transistor
FM	frequency modulation
FTC	fast time control (to help combat CW jamming)
GaAs	gallium arsenide
HFA	height-finder radar

IISLS	improved interrogation sidelobe suppression
IPFD	intrapulse frequency diversity
IR	instrumental range
ISLS	interrogation sidelobe suppression
JATS	jammer analysis and transmission selection
JS	jammer strobing
LFM	linear frequency modulation
LSA	low-sidelobe antenna
LVA	large vertical aperture
MTBF	mean time before failure
MTD	moving target detector [323]
MTTR	mean time to repair
NLFM	nonlinear frequency modulation
PAR	precision approach radar
P_D	probability of detection
P_{fa}	false-alarm probability
PPM	periodic permanent magnet [1]
PRF	pulse repetition frequency
PRFA	PRF agility
PRI	pulse repetition interval
PR-PRF	pseudorandom PRF
rpm	revolutions per minute
SFL	space-fed lens
S-I	Swerling I target model [1]
SLB	sidelobe blanker
SLBM	submarine launched ballistic missle
SSR	secondary surveillance radar
TMA	terminal array
T/R	transceiver module (see Sec. 2.2.1.2)
TWT	traveling wave tube
ULSA	ultra-low-sidelobe antenna
Δ	difference pattern
Σ	sum pattern
Φ	phase
Ω	omnidirectional pattern

Chapter 3
THE AEGIS SYSTEM

by J. Sensi, Jr.

Since the advent of the cruise missile, the deployment of the phased array AN/SPY-1A radar and its associated AEGIS weapon system has produced the most significant change in the arena of naval warfare.

This naval combat system (shown in Figure 3.1), the most sophisticated in the world today, is installed aboard the *USS Ticonderoga* (CG 47) class of cruisers. Future generations of the AN/SPY-1A will be the AN/SPY-1B and AN/SPY-1D. These radar systems also will be deployed aboard the *USS Ticonderoga* (CG 47) class of cruisers, starting with hull number CG 65 and *USS Arleigh Burke* (DDG 51) class of destroyers, respectively, in the late 1980s. The AEGIS combat system is currently installed and used at sea on the reliable *Spruance* class of destroyer hull. The CG 47 class cruisers closely resemble the *Spruance* from the main deck and below, but the ship's environment is bristling AEGIS weaponry. Utilizing an existing hull increases dependability, eases maintenance, and facilitates logistics. Upgrades of the electrical and supporting auxillary systems were required to support the increased electrical load needed for the new combat system.

AEGIS ships are fast and highly maneuverable vessels. For propulsion, CG 47 and DDG 51 utilize four gas turbine jet engines adapted for marine use. The LM2500 gas turbines develop over 80,000 shaft horsepower, and they power two controllable, reversible pitch propellers. The ships reach a top speed of greater than 30 knots in a little more than a minute and measure their stopping distance in hundreds of feet. The unique starting and stopping ability is obtained through the controllable, reversible pitch propellers. The shafts maintain a single direction of rotation, and the pitch of the screw itself varies to provide either ahead or astern propulsion. *USS Yorktown* (CG 48) conducted a humorous "road test" for *Car and Driver* magazine's February 1986 issue. The road test included everything from a standing quarter-mile time to measured stopping distance at top speed.

These impressive ships are expensive, $930 million dollars for CG 47, $820 million dollars for CG48, with cost reductions to additional ships in the class [1].

The mission of these ships is to destroy enemy aircraft, cruise missiles, submarines, and surface combatants, thereby providing protection to a carrier battle group. These are standard assignments for a cruiser or destroyer. The *Ticonderoga* class performs all these missions well, but the quantum improvement is in air defense and associated command and control.

Conventional warships employing rotating types of radars cannot approach the performance delivered by a phased array radar system. If 100% detection rates were achieved, rotating array radar ships could, at best, detect a target once on each rotation. Considering the long-range requirements of an air search radar, this rate equates to perhaps four to ten detection opportunities per minute. In today's cruise missile environment, missiles travel on the order of 8–20 miles per minute. Against low flying cruise missiles, the warship's radar horizons are typically in the 12–20 nmi range, depending on the ship's antennas' height and cruise

SENSI
CHAPTER 3

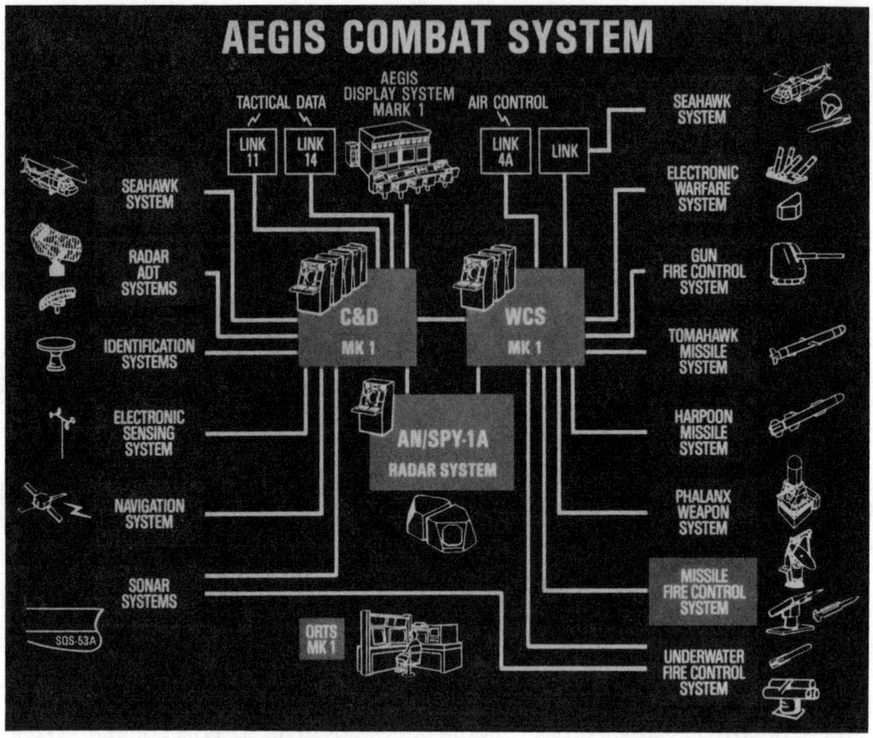

Figure 3.1 Schematic Diagram for AEGIS Combat System. Sensor and Weapon Systems Are Linked Together via Computer to Provide a Coherent Tactical Naval Picture. (Diagram Courtesy of RCA Electronic Systems Department.)

missile attack altitude. We can see the inherent problem of antiship missile defense with only single-digit detection opportunities before impact. A further complication is dealing with saturation attacks by many cruise missiles. Conventional ship air-search radar utilizes a long-range, slow scan rate, low-frequency, and low pulse repetition frequency (PRF) radar to make initial detections. These detections are often made by a human radar operator, and the number of targets that the operator can track efficiently is limited, approaching six tracks every two seconds. When a detection is made, radars of this type are not accurate enough to provide fire-control quality data. The search radar must pass the target to a high-frequency, high-PRF radar, or a continuous wave (CW) radar fire-control system. A normal cruiser combat control center suite has on the order of three or four fire control systems, and the normal destroyer complement is two or three. An obviously unpleasant dilemma occurs for the unfortunate vessel simultaneously targeted by more than four cruise missiles.

The AEGIS weapon system AN/SPY-1A counters this cruise missile threat by exploiting the inherent advantages of the phased array radar system. The AN/SPY-1A does not rotate, but rather a four-faced, fixed, phased array octagonal antenna searches the airspace from horizon to zenith at a heretofore unmatched speed. Figure 3.2 shows one such array face. The figure shows the AN/SPY-1A superstructure. On top of the superstructure is the dome housing the AN/SPQ-9 surface-to-surface fire-control radar antenna. To the left and right of the dome are two MK 99 illuminators.

Each flat antenna face covers slightly more than 90° of azimuth. Together, the four faces essentially provide continuous 360° coverage. Instead of mechanically rotating a single antenna, each array face contains more than 4000 individual

Figure 3.2 Forecastle: Superstructure with AN/SPY-1 Octagonal Antenna Clearly Visible. Foreground: 5″, 54-Caliber Rifled Naval Gun. Background: MK26 Guided Missile Launcher. (Photo Courtesy of RCA Electronic Systems Department.)

radiating elements, which are electronically steered. This electronic steering generates phase shifts, which allow the radar to steer pencil beams in all directions. The developed transmissions are called *dwells*. The AN/SPY-1A is computer controlled and the pencil beams of energy are scheduled to cover the predetermined search envelope. While conducting the search, areas close to the ship, in which low-flying cruise missiles may first appear, are searched more frequently than the air space hundreds of miles away, where time is not critical. When a detection occurs, computer-scheduled tracking dwells then search the airspace containing the target at an increased rate. This allows a target position to be updated at a rate of several times per second. The value of this feature becomes more apparent when we consider that updates can occur simultaneously for over 200 contacts without any human intervention. The AN/SPY-1A also is a powerful radar. There are few rotating antennas as large or with as high an output. Each radar face is almost 12 feet across, and the output power is on the order of four to six megawatts [2]. Many rotating pulsed radars have maximum outputs well below one megawatt (MW).

The AN/SPY-1A positional data, acquired from the high update rate previously mentioned, provide an important element of tactical performance. The positional data eliminate the need for independent fire-control systems, circumventing the cumbersome process of making the transition from a search radar to a fire-control system. The dynamic target parameters obtained by the AN/SPY-1A are the most accurate ever realized in naval warfare. Target positions are measured within feet in all three dimensions, as opposed to hundreds of yards in previous systems. The improved accuracy allows AN/SPY-1A radar to act as both a search and fire-control radar. The obvious advantage is having antiair missile fire-control capability available simultaneously on more than 200 targets. A second benefit lies in the radar's command and control mission. A battle group antiair commander not only can direct missile employment from an AEGIS ship, but can also provide extremely accurate interception vectors for friendly fighter aircraft.

Through the Naval Tactical Data System (NTDS), other missile ships can actually fire their missiles utilizing AN/SPY-1A fire-control data from the AEGIS-equipped ship. NTDS is a digital computer link between ships and aircraft, which allows sensor information to be exchanged. There are two procedures for using remote AN/SPY-1A tracking data. The first procedure, *remote track launch on search* (RTLOS) involves launching a missile in a radar emission controlled state, and then using on-board fire control systems to acquire the target from fire-control data provided by AN/SPY-1A through the NTDS computer-to-computer encrypted radio link. The second procedure involves using an on-board fire-control system to acquire an AN/SPY-1A remote Link-11 NTDS track and then consummating a missile engagement.

Both systems have extensive tactical implications. Imagine a traditional long-range pulsed radar that has been completely neutralized by jamming. The ship then employs a remote AN/SPY-1A radar display to keep its missile battery in action. Other scenarios could envision a missile ship using electronic emission silence to remain undetected until an aircraft flies into the ship's missile range. In addition to using the Link-11 NTDS encrypted radio link for missile firings, other ships in the battle group can exploit this versatile display for command and control functions.

The AEGIS weapon system utilizes three sets of computer banks to link together the radar display, weapons control, and decision making. Each set of computer banks utilizes four general-purpose AN/UYK-7 computers. The *Navy Times* stated that, "The program for the central computer contains 32 million bytes, or pieces of information" [3].

When a detection is observed with the AN/SPY-1A, its computers relay that information to the *command and decision* computer (C & D) (see Figure 3.1). C & D is not only responsible for such things as display, but also compares inputs from other sources such as *identification friend or foe* (IFF) to provide operators with as much information as possible. Naval officers additionally preprogram C & D with what basically comprises tactical software, nicknamed "doctrine." Doctrine encompasses many categories, including identification, alert, and firing criteria. C & D performs other functions, such as recommending optimum weapon selection, calculating closest points of approach between passing ships, and recommending which way to maneuver the ship to optimize helicopter launchings or landing winds.

A typical engagement scenario might be as follows. A target meets preprogrammed engagement parameters. Command and decision (C & D) interacts with the *weapons control system* (WCS) computers, which drive the fire-control system. WCS and C & D interact to ensure that highest threat targets are engaged first. WCS proceeds to load MK26 launchers on CG 47 through CG 51, or alerts MK41 vertical launching group on CG 52 and above. The main battery for this battle cruiser is the Standard Missile Two (SM2). *Jane's Fighting Ships* states the Standard Missile Two has a range of more than 40 nmi [4]. Improvements in missiles of the same type are accomplished through production "block" variants. This 40-nmi range missile is a *block one* variant and its successor, SM2, is *block two,* and utilizes a new propulsion motor, which extends the missile's range from medium to long. These nearly one-ton missiles can be hurled toward their targets so rapidly that we could imagine observing the ship's navigational draft changing during a rapid engagement. Various sources differ on how many targets an AEGIS ship can engage simultaneously and the actual performance is, of course, classified. The *Navy Times* reported the number as 16 [5], and the *National Journal* reported it as 18 [6]. In either case it represents at minimum a fourfold increase in firepower over existing US and NATO cruisers. The AEGIS weapon system has a "hands-off" mode, called *auto special*. In this configuration there is no human intervention, with all detection, missile firing, and kill evaluation performed automatically by the system. This type of automation is required for a modern warship to be able to screen a high-value target (normally an aircraft carrier) effectively against air, sea, and subsurface launched cruise missiles.

Given the awesome firepower assigned to these AEGIS ships, we might question how to find aviators of one's own force who will fly through this awesome air defense shield to land aboard the carrier. This system actually provides a margin of safety never before obtained. As aircraft leave the catapult, AEGIS develops a course and its associated track information. A four-digit track number is used to access a particular computer address, which stores such data as a contact's course, speed, altitude, identity, location, and IFF responses. When an aircraft responds to a mode-four IFF challenge (i.e., secure encrypted mode), the plane's track immediately assumes a friendly identity and upon receipt of additional IFF responses (modes one, two, and three), an operator knows exactly which particular friendly aircraft has taken off.

Although the AEGIS identification process appears to be trivially simple, this process has hitherto required scores of operators, who can never achieve the same accuracy as with the AEGIS system.

The shipborne commander no longer looks over a sailor's shoulder at a cathode ray tube, nor squints at plastic vertical plotting boards with sailors writing backwards on the status boards. Instead, a system called the AEGIS Display System (ADS) with four 42-inch blue screens, resembling large-screen televisions, provide the commander with a crystal clear display, with a 2000-mile range. ADS can superimpose sensor track data over a map so that a geographical tactical picture can be observed. This is of utmost importance in areas such as the Mediterranean, where coastlines, national borders, air traffic control routes, air defense zones, and other points of interest (harbors and airfields) are all displayed within clear eyesight of the battle group decision-maker. It is extremely important for the shipborne commander to have a firm grasp on how the geographic setting relates to the tactical situation. The vehicular track identities of ships, aircraft, missiles, and submarines moving on this display can easily be discerned. Friendly tracks appear as portions of circles, with aircraft using the top half of the circle, submarines using the bottom half, and surface ships using a full circle. Similarly, hostile units appear as portions of a diamond, tilted such that the corners face cardinal headings. Unknown contacts are displayed as portions of a box. Operation specialists attach alphanumeric tags to vehicular or stationary tracks to expand known identities. These tags are as simple as "Bear" or "Tomcat," or may indicate such things as airport identities. Tactical zones, such as missile engagement zones (MEZ), or fighter engagement zones (FEZ) are displayed to provide ease of weapon assignment and coordination. This is usually referred to as *combat air patrol* (CAP), *surface-to-air missile* (SAM) coordination.

The AEGIS weapon system is highly computer-intensive, and we may question its operational (on-line) reliability. The AEGIS weapon system is monitored by a system called *operational readiness test system* (ORTS). In addition to performing many test functions, ORTS provides computerized fault isolation, pinpointing the exact locations of problems for technicians to replace faulty components. The system has additional redundancy, whereby any bank of four AN/UYK-7 computers can perform at nearly 100% efficiency by utilizing three of the four computers. This allows technicians to replace a suspect major component, such as an entire computer bay, quickly, without taking the tactical system off-line. When the suspect bay has been replaced and the system regains normal capability, the faulty bay is inserted into the *maintenance availability computer* (MAC), which is an AN/UYK-7 optimized for fault isolation. Tactical computer programs for the AEGIS system reside in a shadow-writing tactical two-disk-drive system. In the event of a disk-drive failure, all of the current tactical data are available on the back-up disk.

The main battery for the AEGIS weapon system is the SM2 (see Figures 3.3, 3.4, 3.5, 3.6, and 3.7. SM2 is a semiactive, AEGIS command guided missile, which is 14.7-ft long and weighs 1386 pounds [7]. SM2 travels at high mach-plus speed, out to a maximum range well over the horizon. When launched, the SM2 receives midcourse command guidance from the AN/SPY-1A radar through an uplink. Utilizing AN/SPY-1A radar guidance, and not relying on semiactive target reflections for the entire time of flight, allow the missile to follow an optimized, longer range flight profile. When target interception is a few seconds hence, the SM2 switches from command guidance to semiactive guidance, provided by four time-shared MK99 CW illuminators. These illuminators receive scheduling orders from WCS and their own AN/UYK-20 general-purpose computer. To maximize the number of missiles in flight at any given moment, computers schedule the time-shared illuminators. Two illuminators forward and two aft provide redundant 360° coverage (see Figures 3.2 and 3.8).

Figure 3.3 Standard Missile (SM2) Is Loaded into the MK26 Magazine Aboard *USS Yorktown* (CG 48). (Photo Courtesy of RCA Electronic Systems Department.)

Figure 3.4 *USS Yorktown* **(CG 48) Fires a Standard (SM2) Surface-to-Air Missile from the Ship's Aft MK26 Missile Launcher. (Photo Courtesy of RCA Electronic Systems Department.)**

Figure 3.5 *USS Yorktown* **(CG 48) Launches a Standard (SM2) Surface-to-Air Missile from the Ship's Forward MK26 Missile Launcher. (Photo Courtesy of RCA Electronic Systems Department.)**

Figure 3.6 Forecastle: Foreground MK26 Missile Launcher Utilized to Fire Standard Missiles (SAM) and ASROC Antisubmarine Weapons. Background: Forward AN/SPY-1A Phased Array Antenna. (Photo Courtesy of RCA Electronic Systems Department.)

If successive missile engagements should fail, the AEGIS weapon system would employ the 20-mm *close-in weapon system* (CIWS), a six-barreled Gatling gun, which is a self-contained, close-in air defense system (see Figure 3.9). When integrated with AEGIS, CIWS receives target assignments from WCS. These would normally be SM2 targets that the computerized WCS kill assessment had determined to be still a threat and which were within minimum SM2 range. Once a target designation is obtained, CIWS utilizes its own high-frequency, high-PRF, self-contained radar to achieve a track and to engage. The six 20-mm barrels fire depleted-uranium bullets designed to penetrate hard targets like the cruise missile at a rate of several thousand rounds per minute.

The CIWS also has an autonomous mode of operation, whereby it can be programmed to cover the critical airspace near its ship with its own radar and engage hostile air tracks, independently of the AEGIS weapon system.

Having alluded to the AN/SPY-1A's prowess as a tracking radar, we must additionally consider its ability to operate in an electronically dense and hostile environment. The inherent characteristics of phased array radars complicate most radar jamming techniques. Conventional jammers look for predictable frequency bandwidth, scan rate, PRF, and pulsewidth parameters, which, however, vary on a phased array radar. In addition to the inherent characteristics of a phased array radar, the AN/SPY-1A uses a number of techniques to counter jamming.

Figure 3.7 Aft MK41 Vertical Launching System on USS Bunker Hill (CG52). These Launchers Are Placed on CG52 (and above) Systems and Capable of Firing Standard (SM2), Tomahawk, and Vertically Launched ASROC Missiles. (A SH60 LAMPS MKIII Helicopter Is Partly Visible in the Background Behind Edward Mahoney (ESD, Hanscom AFB, MA) in the Left Foreground.)

Complementing the AN/SPY-1A radar is the AN/SLQ-32 *electronic warfare* (EW) system. This system acts as the AEGIS cruiser's *electronic surveillance system* (ESS). Additionally, the AN/SLQ-32 can employ active electronic jamming and it can control the deployment of chaff. Chaff is normally aluminum cut to a specific radar wavelength and designed to give deceptive radar returns.

The AEGIS cruiser and all of its computerized equipment must be able to withstand the hazards of modern-day warfare. To ensure proper performance under battle conditions, the first or second of a new class of ships is usually subjected to shock testing. *USS Yorktown* (CG 48) underwent such testing for the *USS Ticonderoga* (CG 47) class. The test consists of detonating tons of high explosives at close proximity to the ship, with the crew embarked (this author included) and the AEGIS combat system fully operational. An extensive network of accelerometers and shock-sensing devices were installed to measure the forces to which the ship would be subjected in actual battle. These shock tests were the most successful in naval history, and proved that the AEGIS combat systems has the ability to sustain punishment and to continue performing.

Aboard AEGIS cruisers, the AEGIS weapon system encompasses the AN/SPY-1A and most of the antiair equipment and weaponry. The AEGIS combat system expands to include the other sensors and weapons utilized by the AEGIS cruiser for *antisurface warfare* (ASUW) and *antisubmarine warfare* (ASW) (see Figure 3.1).

AEGIS ships employ the surface-to-surface, medium-range Harpoon missile for ASUW (see Figure 3.10). Harpoon is an active, self-contained, "launch and forget" missile. *USS Bunker Hill* (CG 52) and subsequent cruisers will also have the Tomahawk missile (see Figure 3.7). Tomahawk comes in several variants, including antiship and antishore types, with varying warhead payloads. Tomahawk

SENSI
CHAPTER 3

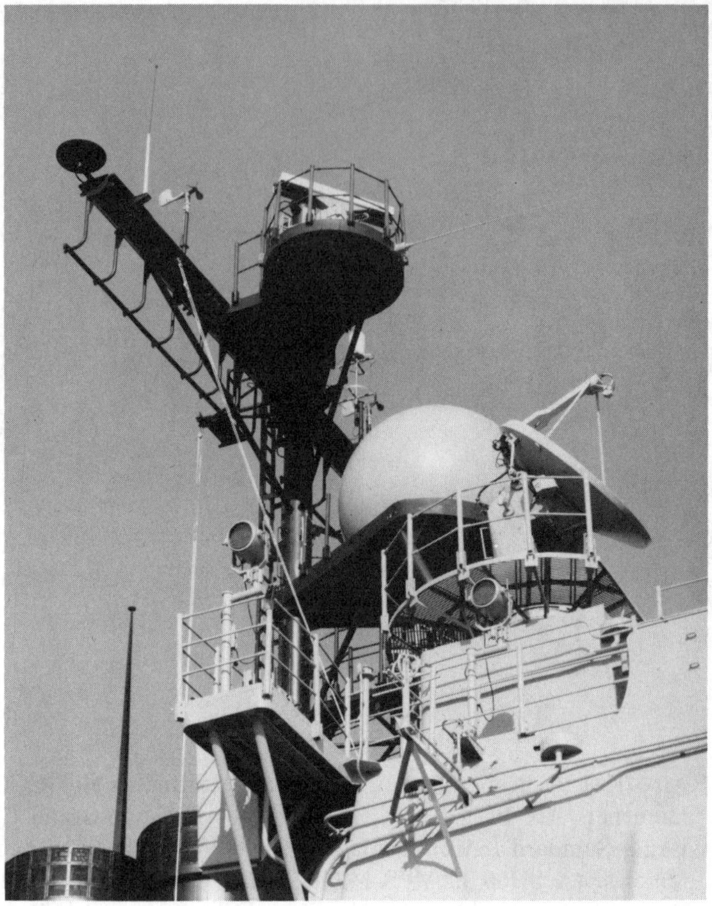

Figure 3.8 Forward Mast of CG 47 Class Cruiser. To the Right of the Dome Is One of Two Forward MK 99 Illuminators. The Radar Dome Houses the AN/SPQ-9 Surface-to-Surface Fire Control Radar Antenna. (Photo Courtesy of RCA Electronic Systems Department.)

brings added dimensions of warfare to the AEGIS class cruiser, since the missile enables the ship to strike either opposing ships or land targets from hundreds of miles away.

Over-the-horizon (OTH) targeting for surface missile engagements can be provided from numerous sensors. SH60 LAMPS MKIII helicopters can link OTH radar pictures back to engagement planners on-board. Acoustic or electronic interceptions provide sufficient information for cruise missile targeting.

For close-in surface engagements or shore bombardment, AEGIS ships employ the common five-inch, 54-caliber, lightweight automated gun; (see Figure 3.2). The gun receives guidance from the MK86 gunfire control system. The MK86 utilizes the AN/SPQ-9A track-while-scan (TWS) radar and is capable of multiple surface target engagement.

AEGIS ships couple their unique antiair capacity with the proven *USS Spruance* (DD963) class' antisubmarine prowess. The CG 47 class of cruisers comprise the most acoustically silent capital ships ever deployed. A system called *Prairie-Masker* provides a belt of noise-absorbing air bubbles between propellers, machinery rooms, and the sound transmitting medium of the sea. Virtually all moving machinery is isolated from the hull through sound-deadening mounts, normally constructed of rubber-like materials. The above characteristics develop a low acoustic sound source level, minimizing the detection opportunities for opposing submarines. A low source level additionally enhances the passive in-

Figure 3.9 20-mm Close-in Weapon System. (*Note*: This Close-in Weapon System Scans the Horizon for Low-Flying Aircraft and Cruise Missiles. The Self-Contained System Can Receive Designations from the AEGIS Weapon System to Engage Close-in Targets.) Self-Contained Gatling Gun Fires Thousands of Rounds per Minute; White Radar Dome Encloses Sensing Equipment; Magazine Is Located in the Base of the Mount. (Photo Courtesy of RCA Electronic Systems Department.)

terception capability of the AEGIS cruiser. ASW sensors include the hull-mounted, low-frequency, AN/SQS-53A sonar and sophisticated sonobuoy receivers. Active and passive sonobuoys can be deployed by the ship or its embarked LAMPS I or LAMPS III helicopters. Later ships of the AEGIS class will be outfitted with a tactical towed array, which will supplement the hull-mounted sonar.

Antisubmarine weapon control is integrated into the AEGIS WCS (see Figure 3.1). MK26 launchers are equipped to fire not only SM2 antiair missiles, but also ASROC rocket-thrown torpedoes or depth charges (see Figure 3.6). The torpedo payload, the MK46, is a homing torpedo, which also can be deployed from embarked helicopters or over-the-side triple-torpedo tube mounts for close-in engagements.

To defeat incoming torpedoes, the ship employs the AN/SLQ-25 torpedo countermeasure system. This countermeasure, towed astern of the ship, is an acoustical jammer. Acoustical environmental conditions and subsequent sensor or weapon range predictions are obtained by sampling the sea temperature to determine the sound velocity profile by using expendable bathythermograph probes. The gathered data are then analyzed by an on-board computer system, called SIMAS, allowing tactical decisions to be based on current conditions.

Figure 3.10 **Fantail Surface-to-Surface Antiship Harpoon Missiles. Harpoons Are Stored and Launched from Cannisters. Two of Possible Eight Harpoon Missiles Are Shown. (*Note*: One of the AN/SPY-1A Octagonal Antennas Is Shown in Background at Junction of the Cannisters.) (Photo Courtesy of RCA Electronic Systems Department.)**

In conclusion, ships of the CG 47 class of cruiser and the nearly operational DDG 51 destroyer bring new dimensions of performance to the complex naval warfare arena. The ability to accomplish multiple warfare missions in hostile environments with heretofore unprecedented proficiency makes these vessels the world's most desirable naval escorts. A technologically superior design, coupled with the planned ability to adapt to new innovations in weapon and sensor systems, provide these warships with tremendous future growth potential. Ships of the CG 47 and DDG 51 classes should prove to be the mainstay of air, surface, and subsurface defense for carrier and surface battle groups well into the 21st century.

SENSI
CHAPTER 3

REFERENCES

[1] *Jane's Fighting Ships 1986–1987,* Capt. J. Moore, R.N. (UK), ed., (Jane's Publishing, London, 1986), pp. 724–725.
[2] Elliot, F., "An Inside Look at Aegis, Shield for the Fleet," *Navy Times,*, August 25, 1986, pp. 35–36.
[3] Elliot, *loc. cit.,* p. 35.
[4] Moore, *loc. cit.,* pp. 684–687.
[5] Elliot, *loc. cit.,* p. 35.
[6] Gordon, M., "AEGIS: A High Priority Naval Defense Project . . . But Will it Work When the Chips Are Down?" *National Journal,* July 24, 1982, pp. 207–208.
[7] *Aviation Week,* March 10, 1986, p. 178.

Suggestions for Further Reading

[8] Grier, P., "AEGIS to Put Swagger in the Navy's Step," *Christian Science Monitor,* August 21, 1986, pp. 3–4.
[9] Jeannes, W., "A Journey to Pascagoula," *Car and Driver,* February 1986, pp. 95–100.
[10] Patton, W.T., "Microwave Design for Reliability-Availability—The AN/SPY-1 Radar System," *Military Microwaves Conf. Proc.,* London, October 1974, pp. 51–55.
[11] Scudder, R.M., and W. H. Sheppard, "AN/SPY-1 Phased-Array Antenna," *Microwave Journal,* Vol. 27, No. 5, May 1974, pp. 51–55.

Appendix 3A
Selected Readings

This appendix comprises two reprinted papers to supplement the total system discussion of the AEGIS given in this chapter with detailed coverage of the AN/SPY-1 and AN/SPY-2 radars.

We present herein the following papers:

Patton, W.T., ''Microwave Design for Reliability-Availability—The AN/SPY-1 Radar System,'' *Military Microwaves Conf. Proc.,* London, October 1978, pp. 177–192. (Reprinted with permission.)

Scudder, R.M., and W.H. Sheppard, ''AN/SPY-1 Phased Array Antenna, *Microwave Journal,* May 1974, pp. 51–55. (Reprinted with permission.)

For additional reading and research on the topics of radar, antennas, and electromagnetics at sea, we suggest the books by Preston E. Law, Jr., *Shipboard Antennas,* 2nd Ed. (Artech House, 1986), and *Shipboard Electromagnetics* (Artech House, 1987).

3A.1 MICROWAVE DESIGN FOR RELIABILITY/AVAILABILITY—THE AN/SPY-1 RADAR SYSTEM

by W.T. Patton

The AN/SPY-1 radar, a major element of the AEGIS weapon system, is the latest tactical multifunction phased array radar system designed for use at sea. Since coming on-line in 1973 at the Land Based Test Site and in subsequent operations at sea in *USS Norton Sound,* it has achieved an outstanding record of availability due to both system design and component reliability. This paper reviews that record from the point of view of that component of the AN/SPY-1 radar of greatest interest to the microwave community—the phased array antenna.

The acquisition of the AEGIS Weapon System is now proceeding with the construction of the lead AEGIS destroyer DDG-47. A brief history of the AEGIS program to date will be reviewed. Again the developments centering around the antenna will be emphasized as they progressed from early R&D through AN/SPY-1 for AEGIS.

Design of the AN/SPY-1 Array

The design of the AN/SPY-1 radar system began in 1970. This design was based upon concepts and critical component developments that evolved over the preceding six years. The design goals for component reliability were such that major repairs to the array would be limited to a shipyard availability at 18-month intervals. The experience has been that, after five years operation (*circa* 1978), a major repair is still not required. This is due in part to component reliability that substantially exceeded goals.

The phase shifter (phaser) and driver are the elements of the phased array that determine its reliability. The experience with these units in the AN/SPY-1 design during a four-year period from 1 January 1974 through 31 December 1977 is given in Table 3.A.1-1. Here it is noted that the phase shifter driver bettered its failure allocation by more than three times, while the phase shifter itself was nearly 100 times better than its allocation.

Table 3.A.1-1 AN/SPY-1 Array Reliability (1974–1977 Record)

Component	*Device Hours*	*Total Events*	*Event Rate/ 10^6 hours*	*Allocation/ 10^6 hours*
Phase Shifter	132,106,240	7	0.0529	5
Phase Shifter Driver	33,026,560	238	7.2	25

The array that achieved these results consists of 4480 phase shifters and 1120 quad drivers, all arranged in 140 modules of 32 elements each. Two array modules are combined to form a receiving subarray, and two receiving subarrays are combined to form a transmitting subarray. The subarrays are grouped to form the array as shown in Figure 3.A.1-1. The receive subarrays are arranged in 10 columns; each subarray is connected to a column beamformer as shown in Figure 3.A.1-2.

CHAPTER 3

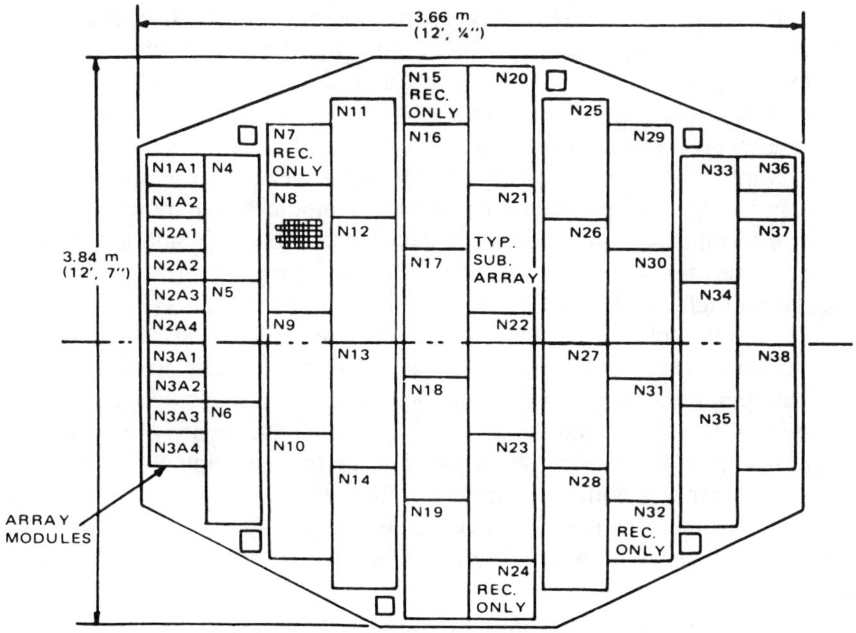

Figure 3.A.1-1. Rear View of Subarray Arrangement.

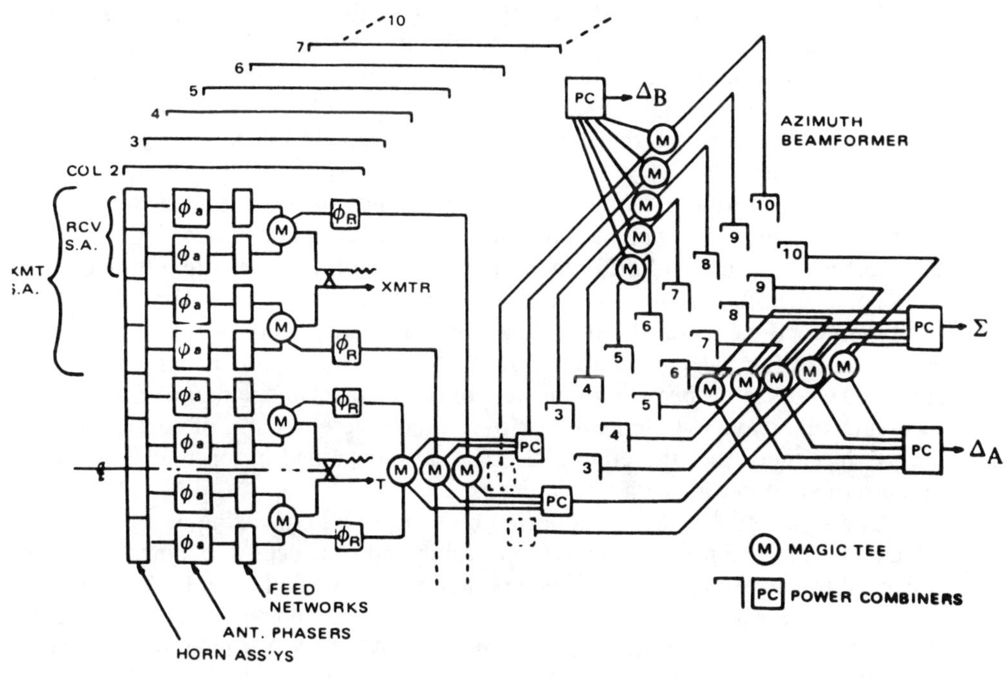

Figure 3.A.1-2 Schematic Drawing of Microwave System.

The initial step in beam formation, applicable to both the transmitting and receiving functions, combines the array modules into 32 transmitting and 68 receiving subarrays, with separate transmitting and receiving terminals. Hybrid tee junctions combine pairs of array modules, where the E-arm of each junction is defined as the receiver subarray terminal. H-arms from two junctions are combined via a quadrature hybrid, such that its third terminal becomes the input terminal for the transmit subarray encompassing four array modules. The remaining quadrature hybrid terminal is connected to an RF load capable of handling the power of the associated transmitter HPA. This arrangement basically provides the function of isolating high array face mismatch (such as that which may be caused by heavy rain) from the transmitter amplifiers, while the magic tees isolate the transmitter power from the receiver terminals. This, of course, is made possible through the use of nonreciprocal phase shifters. During the transmitting mode, the antenna phase shifters in the array modules must be set to accommodate the quadrature phase relationship of the transmitter hybrid. During the receiving mode, the phasing between the array modules is altered to an antiphased condition, such that received signals appear at the E-arm of the magic tee hybrids, and from there proceed to the 68 receiver phase shifter components and to the next level of receiver beam formation.

In the beam formation sequence, receiving subarrays in each column are first combined to obtain the sum of the received signals from all subarrays in a column, as well as the difference between the signals in the upper and lower halves of that column. For those columns having odd numbers of receiving subarrays, the central subarray is omitted in the generation of the difference function.

In the second step, the 10 sum outputs from the 10 columns are summed to produce the complete reference pattern; the 10 column difference outputs generated in the first step are also summed to produce the complete array elevation difference pattern. In addition, the five sum outputs of one side of the array are subtracted from those of the opposite side to produce the total array azimuth difference pattern.

This beam-forming network utilizes waveguide of reduced height as a compromise between space-weight considerations and RF loss characteristics. Magic tees perform the sum and difference functions, amplitude weighting utilizes networks of directional couplers to maintain adequate control of phase, and amplitude excitation coefficients on the receiving subarrays. Relatively tight control of the excitations of the receive subarrays by the beamformer components is required, since any errors in the beam-forming networks are correlated, either over complete receiving subarrays (64 elements) or over a complete column of receiving subarrays. In either case, the effects of errors are magnified in terms of this effect on pattern sidelobes near the main beam.

The AN/SPY-1 array is supported by a structure consisting of a perforated 3.2 cm (1-$\frac{1}{4}$ mil) front plate welded to a frame and four deep T-beams. The total weight of the array is 7700 kg (17,000 pounds), including all mounted equipment.

After extensive testing of the array in the Automated Antenna Pattern Data Acquisition System (Figure 3.A.1-3) and of the AN/SPY-1 Radar System in the Land Based Test Site (Figure 3.A.1-4), the radar was installed in *USS Norton Sound* for an extensive period of at-sea operation and tests beginning 20 May 1974.

Figure 3.A.1-3 Array Mounted in Test Station at Automated Antenna Pattern Data Acquisition System Site.

Figure 3.A.1-4 Land-Based Test Site.

System Description

The AEGIS weapon system [1] is a fast-reaction, high-performance antiair weapon system that directs the standard missile against air and surface targets. It is an integrated shipboard detection, control, and engagement system, and is the first Navy system capable of fully automatic reaction to an intense air attack.

Based upon the excellent performance of the AN/SPY-1 array and lessons learned in its production and testing, a decision was made in 1974 to develop a modified design for the array to be used in the AN/SPY-1A radar. This design would reduce cost and weight, and would further improve reliability to the point where shipboard spares could be eliminated and overhaul cycles could be increased from the initial goal of 18 months for AN/SPY-1 to more than 48 months.

Radar system AN/SPY-1A (Figure 3.A.1-5) is the primary air and surface radar sensor of the entire combat system for an operational AEGIS ship. It is a multifunction phased-array radar capable of search and automatic detection, transition to track, and track of air and surface targets located in preselected coverage volumes. Digital control, high power output, and advanced signal processing techniques are used to provide adaptive search and multitarget tracking. The radar, capable of operating in both heavy clutter and ECM environments, has a high track capacity and is effective against a wide spectrum of target characteristics.

In addition to tracking surface and air targets, the AN/SPY-1A provides a communication link with the SM2 missile throughout its flight for exchange of midcourse guidance commands and missile status messages.

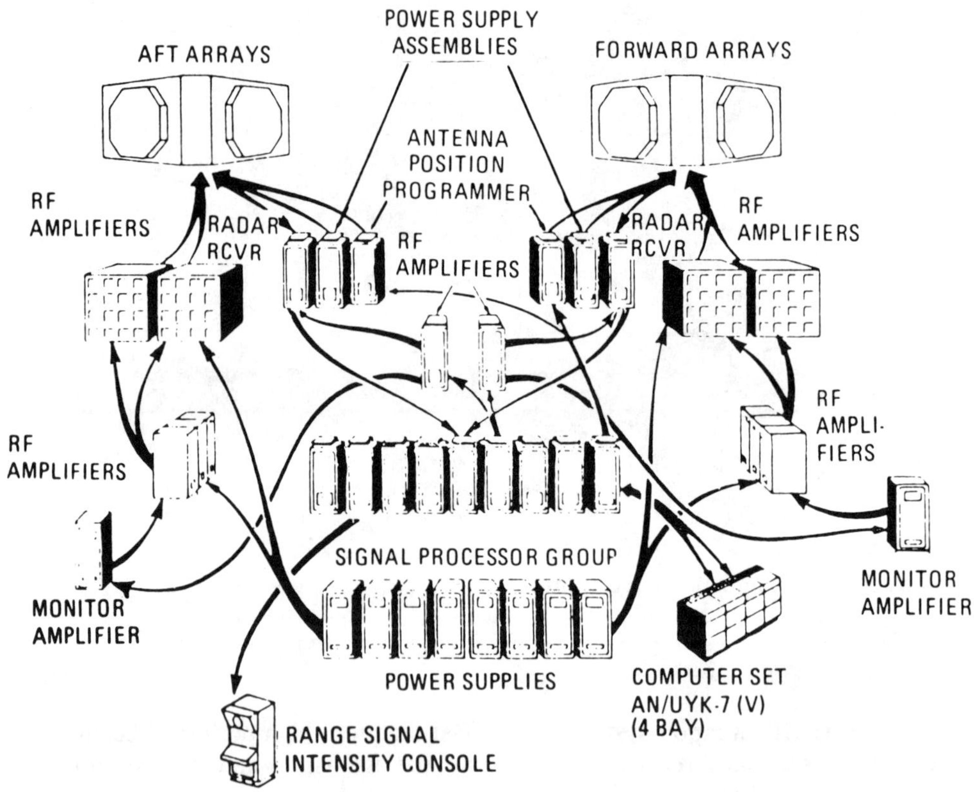

Figure 3.A.1-5 AN/SPY-1A Radar System Pictorial Diagram.

The AN/SPY-1A operation is summarized as follows. Under the direction of its AN/UYK-7 digital computer, the signal processor generates the appropriate search or track RF waveform. The RF signal, amplified by transmitter high-power RF amplifiers, is fed to the selected array face. A beam-steering command is converted by the antenna position programmer to array phase shifter commands that cause the RF output of the transmitter to be formed by the antenna face into a beam at the proper spatial angle. Received signals from the array sum channel and two difference channels are first amplified in RF preamplifiers (in the array), down-converted and amplified in the radar receiver, and then sent to the signal processor. The signal processor performs constant false-alarm rate and pulse compression processing, and angle and range estimating on the received signals. The data are used by the AN/SPY-1A computer to develop radar tracks. If additional clutter rejection is required, MTI processing is available for use in both search and track.

The AN/SPY-1A consists of the antenna group, transmitter group, signal processor group, control group, and auxiliary equipment. Only the antenna group is described here. The antenna group transfers, forms, and steers the high-power transmitting beam. It also receives and amplifies the received signals and converts them to IF. It consists of the following items:

Antenna—The antenna consists of four electronically steerable phased array faces providing 360° of bearing, and up to the zenith radar coverage. RF preamplifiers associated with each antenna face provide initial amplification of the received signals.

Power Supply Assembly—Two power supply assemblies (one forward and one aft) provide the required voltages to the phase shifter drivers in the arrays.

Radar Receiver—Two units (one forward and one aft) process the radar returns from the RF preamplifiers and convert the RF to IF.

Antenna Position Programmer—Two special processors (one forward and one aft) generate the required transmit and receive phase shifter commands for the associated array faces.

A high availability for the AN/SPY-1A radar is ensured by means of on-line redundancy and rapid computer-controlled reconfiguration. The radar must operate continuously in spite of individual component malfunctions with limited numbers of operational and maintenance personnel required. The Operational Readiness Test System (ORTS) conducts continuous system availability testing and automatically isolates detected faults to the level of a line replaceable unit.

AN/SPY-1A Design

One rather significant change was in the method of beam-steering control. The AN/SPY-1 uses subarray phase control. The phase command used for each element in a typical subarray is computed in the beam-steering controller and supplied to corresponding elements in all subarrays. Phase alignment between the subarrays is provided by a separate fine-steering phase shifter in the receiving and transmitting paths to the subarrays. The AN/SPY-1A uses row-column phase control with the row phase increment and the column phase increment each presented to the individual element drivers as an offset in the time of arrival of a pulse train. This type of control provides a simple addition modulo 2 of the row and column phase increments for each radiating element as shown in Figure 3.A.1-6.

CHAPTER 3

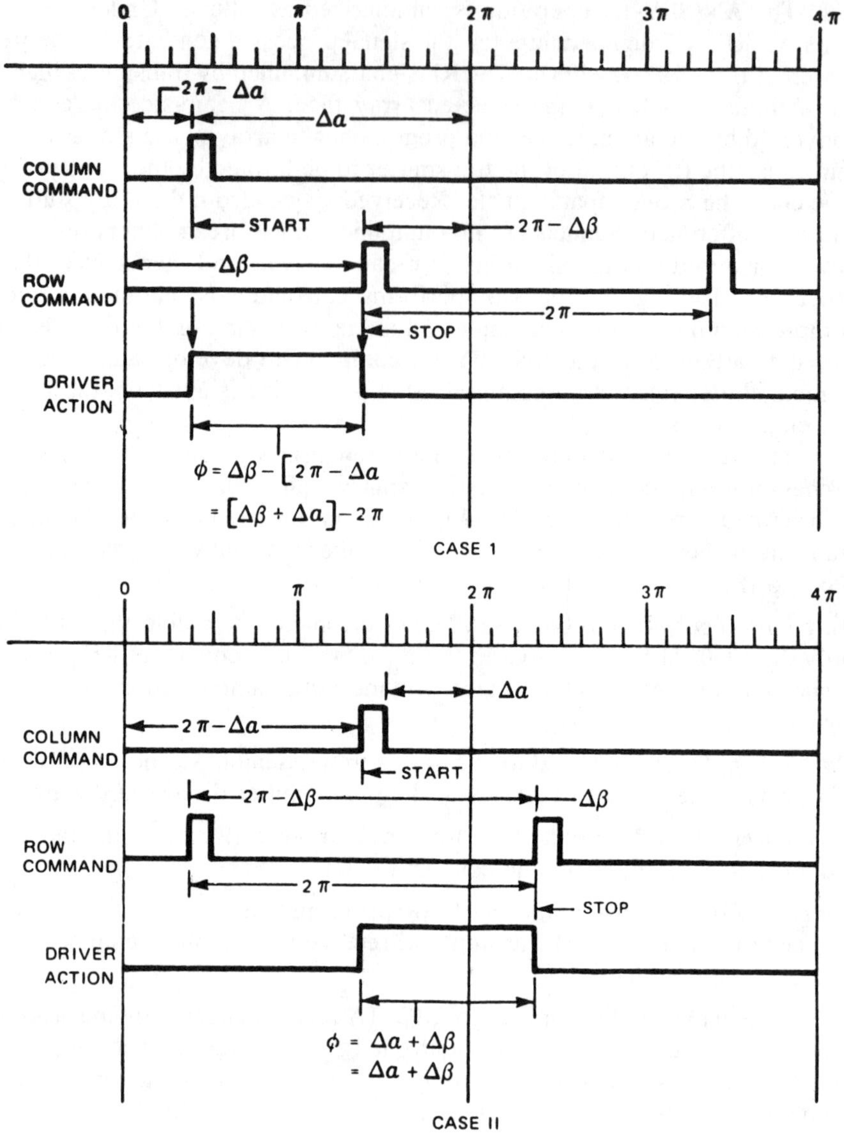

Figure 3.A.1-6 Time Analog Flux Drive Profile.

This development allowed the elimination of the fine phase shifters in the receiving and transmitting beamformers, reducing RF loss, weight, and cost. Elimination of the fine phase shifters relieved the component error budget and reduced the number of steering bits required in the antenna phase shifters. This permitted the antenna phase shifter length to be reduced by 5 cm (2 in), further reducing loss, weight and cost.

These changes, together with the detailed changes discussed below, have resulted in an array (Figure 3.A.1-7) that is 30% lighter and which can be produced at a 30% cost savings. At the same time, performance and reliability have been increased from their previous high levels [2].

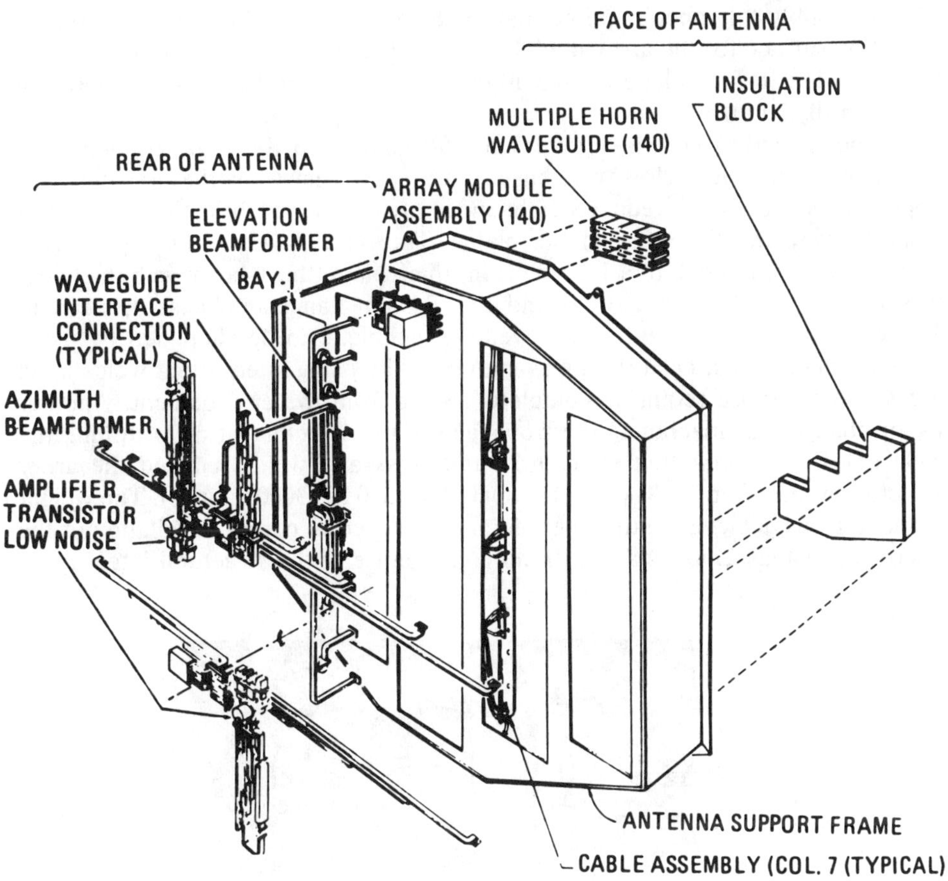

Figure 3.A.1-7 Phased Array Assembly, Exploded View.

Array Structure

The AEGIS phased array makes use of a large (4.06 × 3.89 × 0.813 cm) (160 × 153 × 32 in) welded aluminum structure with critical tolerance control, including flatness, parallelism, element spacing relationships, and plane-to-plane parallelism. The basic aluminum structure consists of a base plate containing 4480 holes welded to an octagonal outside solid frame with four equally spaced inside T-beam sections 81.3-cm (32 in) deep.

The first system utilized a 3.18-cm (1-$\frac{1}{4}$ in) thick front face, 1.27-cm ($\frac{1}{2}$ in) outside sheets forming the octagonal shape, T-beams of 1.27-cm ($\frac{1}{2}$ in) web thickness, and 2.54-cm (1 in) flange thickness. The fabrication sequences and welding techniques were developed to minimize distortion. The different material thickness required that specific weld joints be designed and sample-tested prior to application to the final structure. Of particular interest was the welding of the 1.27-cm ($\frac{1}{2}$ in) T-beam webs to the 3.18-cm (1-$\frac{1}{4}$ in) thick face plate. The face plate was machined from raw stock 7.62-cm (3 in) thick to the octagonal shape with 1.27-cm ($\frac{1}{2}$ in) thick upright webs. This permitted the welding of the T-section to the face plate with metal of the same thickness.

The T-beams were then welded with continuous welds to the face plate, starting with the T-sections on the central part of the face plate and continuing to the outer edges. The welding sequence was carefully adjusted to control warpage over the entire face plate. Upon completion of the T-section welding, the other skirts were welded to complete the enclosure of the structure. Subsequent to welding, the structure was inspected and optically checked for flatness and parallelism prior to drilling the 4480 holes in the face plate.

Assembly tolerance required that the front surface of the face plate and the mounting surface for the array modules be parallel within 0.102 cm (0.040 in) and the mounting surface plane for the 140 array modules be in a plane, flat within 0.051 cm (0.020 in).

The second structure (Figure 3.A.1-8) maintained the same configuration and number of holes. However, based on the experience of the first unit, the approach was revised to reduce weight and improve ease of fabrication. The major changes were the thickness of the plates. The face plate was machined from 1.91-cm ($\frac{3}{4}$ in) raw stock to a 1.27-cm ($\frac{1}{2}$ in) thick plate, the T-beam web thickness was reduced to 0.953 cm ($\frac{3}{8}$ in) and the T-beam flange thickness reduced to 1.27 cm ($\frac{1}{2}$ in). This permitted the welding technique to be simplified by using standard fillet welds. Distortion was minimized by using intermittent welds in all areas, except those requiring complete sealing from the environment. Another major change was machining the front face of the structure prior to drilling the 4480 holes. The reduction in material thickness and welds reduced the array structure weight from 2270 kg (5000 pounds) to 1360 kg (3000 pounds). The weight reduction was achieved within the tolerances needed to permit installation of the microwave subassemblies and with no compromise of the structural integrity.

Figure 3.A.1-8 Array Structure.

Radiating Element

The radiating elements (Figure 3.A.1-9) are rectangular horns arranged in a "brick-wall" pattern forming a quasirectangular module of 32 elements. The spacing has been established on the basis of the required coverage for each array face. This lattice makes allowance for ship motion and provides 360° coverage by the four AEGIS arrays (two forward and two aft).

The basic approach for both systems was a dip-brazed modular assembly containing 32 elements, with each element containing a soldered beryllium oxide window covered with silicone rubber.

The first array, AN/SPY-1, required oven-brazing for window soldering using a spring-loaded fixture to seal all 32 windows and touch-up of solder voids afterwards. The AN/SPY-1A array incorporated a hot plate approach, heating the assembly sufficiently to allow hand soldering, thus eliminating the touch-up operation and spring loaded fixture for window sealing.

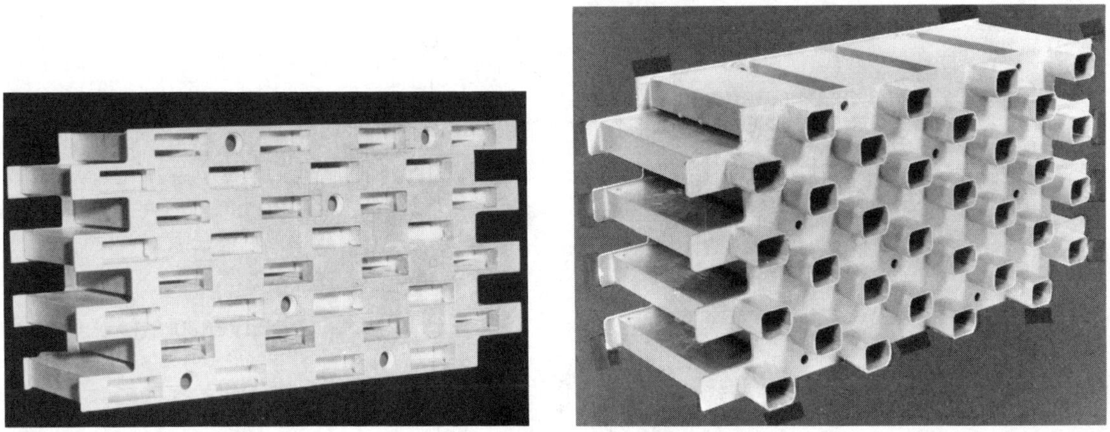

(a)

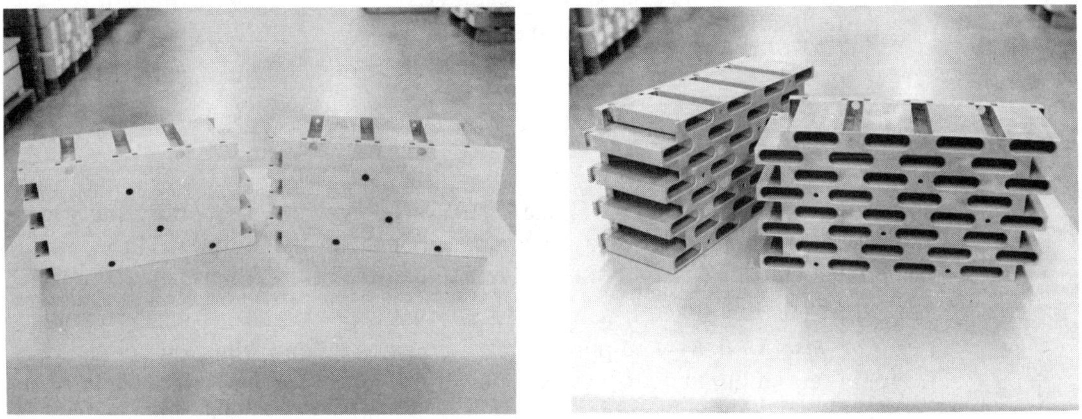

(b)

Figure 3.A.1-9 Radiating Elements (Horn Modules): (a) AN/SPY-1 Design; (b) AN/SPY-1A Design.

Both systems apply a layer of 0.030 ± 0.008 cm (0.012 ± .003 in) thickness of silicone rubber for environmentally sealing the front face. The AN/SPY-1 required a fast cure cycle by using a humidity chamber for a 24-hour period; the AN/SPY-1A permitted the option of allowing the unit to be stored in a humid area for seven days. Phase shifter insertion on the AN/SPY-1 was accomplished by a funnel-type casting added to each horn element. This required a larger hole in the structure to accept each casting. In addition, since a flat surface was required, the fabricator had to machine around the 32 funnels. In the AN/SPY-1A, the size of the hole was reduced, and a lead-in chamber was added in the face plate hole. This permitted utilization of a simple face cut on the horn assembly to achieve the required flatness. The size of the hole in the face plate in the AN/SPY-1A was selected after electrical tests to determine optimum hole size for minimum RF leakage.

Array Module

An array module assembly, shown in Figure 3.A.1-10, consists of a 32:1 power divider, 32 phase shifters, an electronic nest containing eight driver boards,

CHAPTER 3

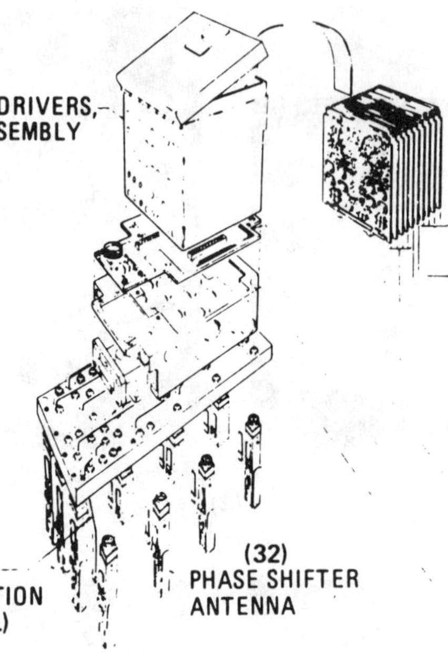

Figure 3.A.1-10 Array Module Assembly.

a line receiver board, and a voltage regulator board with interconnecting wiring and support brackets. The array is composed of 140 array modules. Three critical components in this assembly are the ferrite phase shifters, their drivers, and the 32:1 power divider.

Ferrite Phase Shifter—The phase shifter provides an end-launched coaxial connector at one end to mate to a coaxial power divider and a blind connection, using spring contacts, at the other end to mate to a radiating horn, where no mechanical fasteners are permitted.

For the AN/SPY-1, the phase shifter consisted of three 5.08-cm (2 in) hollow rectangular segments of garnet material. These were bonded into an essentially monolithic element with the widths held to within .003 cm (0.001 in) through its 15.2-cm (6 in) length. With a latching wire threaded through its center, and input and output transformers at its ends, the garnet was assembled into its close-fitting waveguide case. For the AN/SPY-1A, the garnet construction was changed to a one-piece unit 10.2-cm (4 in) long. This reduction in length from 15.2 cm (6 in) to 10.2 cm (4 in) reduced the material cost of the garnet core by 30% and eliminated the labor required to bond together three 5.08 cm (2 in) pieces.

This change in manufacturing technique reduced phase shifter costs and RF losses (by 0.2 dB). The negative effect was an increase in phase error when the phase shifter would be required to operate at maximum temperature (on hottest day at maximum duty). The results, however, are well within the required phase tolerances.

The requirement of installing the array module by making 32 blind connections simultaneously was accomplished by mounting the phase shifter to the power divider with a spring clamp, which allows up to $\pm 0.5°$ misalignment from perpendicularity, and springs at the horn end of the phase shifter. This permitted self-alignment and self-connection between the array module assembly and the horn assembly as shown in Figure 3.A.1-11.

Phase Shifter Driver—The phase shifter driver controls the latched flux level in the garnet toroids (the phase shifter) to a commanded value, establishing the relative phase shift through the device. To prevent the hysteresis memory effect of past toroidal magnetic states affecting the

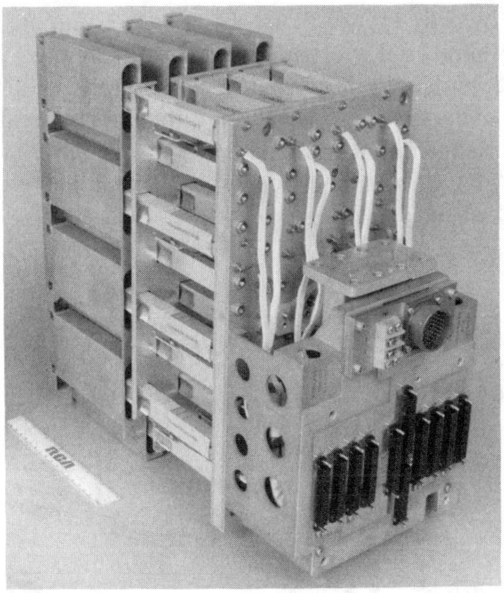

Figure 3.A.1-11 Array Module and Horn Module Assemblies.

current state, the phase shift is reset before setting. The reset-set operation is reversed during transition between transmitting and receiving modes.

For the AN/SPY-1, the drivers employed analog control to a voltage level supplied by the beam-steering controller by a flux-feedback technique to control accurately the differential phase in the antenna elements.

In the AN/SPY-1A, the elimination of the fine-steering phase shifters allowed a larger error budget for the drivers. A simpler open-loop driver was designed with high reliability (military standards) digital integrated circuits, replacing the custom analog components of the flux-feedback driver. Fewer power supply voltages are required and the multilayer printed circuit boards are eliminated, reducing manufacturing complexity and cost, while further enhancing reliability.

Power Divider—The 32:1 power divider is a unique device in that the outer conductor is a square cross section fabricated from punched plates (fine press blanking) with a circular cross section center conductor. The unpressurized power divider is required to handle substantial RF input power, split equally among each of the 32 outputs.

For the AN/SPY-1, the center conductor was fabricated from 90 precision pieces fixtured and silver-soldered to obtain a total assembly. For the AN/SPY-1A, the center conductor tolerances were released, requiring only 11 diecast parts with minimal machining to fit pieces together. The reduction in cost was 80%, and the effect on electrical performance was a slight change in maximum VSWR over the total operating frequency band from 1.35 to 1.45.

Receiving Beamformer

The AN/SPY-1 required insertion of a fine-steering phase shifter in each of 68 waveguide runs. Each of these runs had to be phased matched to each other within 4° over the total frequency band. The required trimming necessitated a minimum of three 0.95-cm ($\frac{3}{8}$ in) thick spacers per run.

AN/SPY-1A eliminated the fine-steering phase shifter and the spacers in the elevation beamformer shown in Figure 3.A.1-12. Phase trimming was accomplished by means of thick flanges. By eliminating phase trimming capabilities in the eight longest runs, the net length of waveguide in all paths within the column beamformers was substantially reduced. In addition, the implementation of bench trimming instead of *in situ* adjustment reduced test and assembly effort by 50%, and allowed the AN/SPY-1A to be more closely phase-matched than its predecessor.

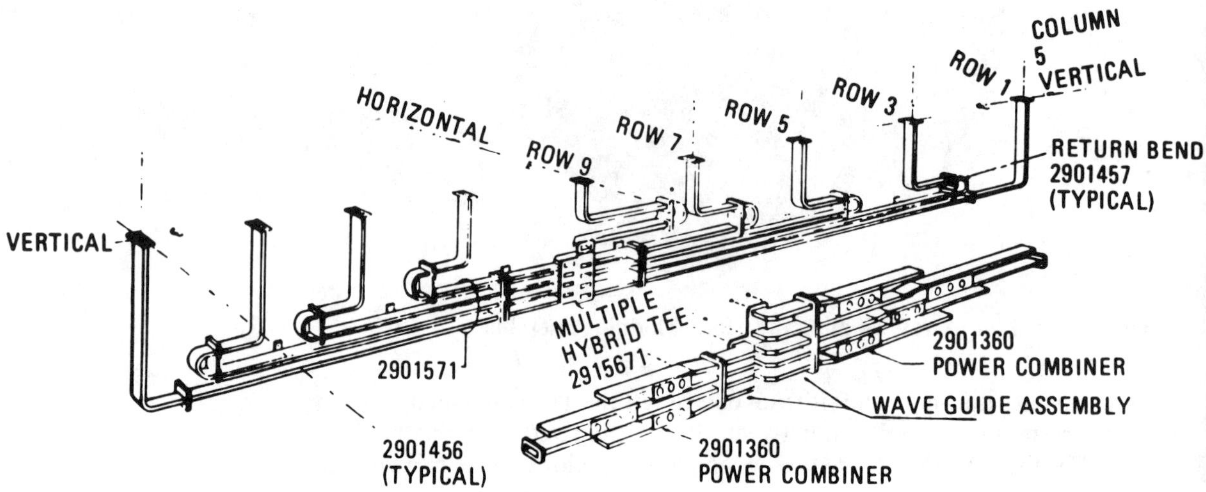

Figure 3.A.1-12 Beam-Forming Network.

Low-Noise Amplifiers

In the AN/SPY-1 design, the low-noise amplifiers (LNAs) take the form of a two-stage parametric-amplifier–transistor-amplifier, mounted off-array in the radar receiver cabinet. This LNA is an 11-channel receiver covering the three monopulse channels of the main array, a sidelobe blanking channel, and seven auxiliary channels.

The AN/SPY-1A design makes use of advances in the transistor amplifier art, providing the required system noise temperature with smaller transistor amplifiers. These amplifiers use a GaAs FET input stage with a bipolar transistor output stage. A noise figure of 2.5 dB has been achieved with this configuration.

The smaller size and weight of the AN/SPY-1A RF amplifier permits it to be mounted on the array at the waveguide beamformer ports, thus minimizing loss. In addition to the reliability improvement inherent in the simple transistor amplifier, availability is further enhanced by a cross-switching arrangement, which will allow switching at least one of the auxiliary channel amplifiers into a main array or sidelobe blanking channel should the primary amplifier there enter a fault mode.

Reliability

The early reliability experience with the AN/SPY-1A antenna group equipment (depicted in Figure 3.A.1-13) is measured against the allocation of 27.53 events per thousand hours. For comparison, the event allocation for the AN/SPY-1 antenna group is 55 events per thousand hours of operation. A detailed record of one of the principal contributors to this event rate is shown in Figure 3.A.1-14. The phase shifter driver event rate for the AN/SPY-1 array had settled to the allocation level by the end of the initial test and evaluation period at the factory, and went on to better that rate by a wide margin in service in *USS Norton Sound*. The phase shifter driver in the AN/SPY-1A array has already bettered its lower allocation substantially, while still in the initial test and evaluation period.

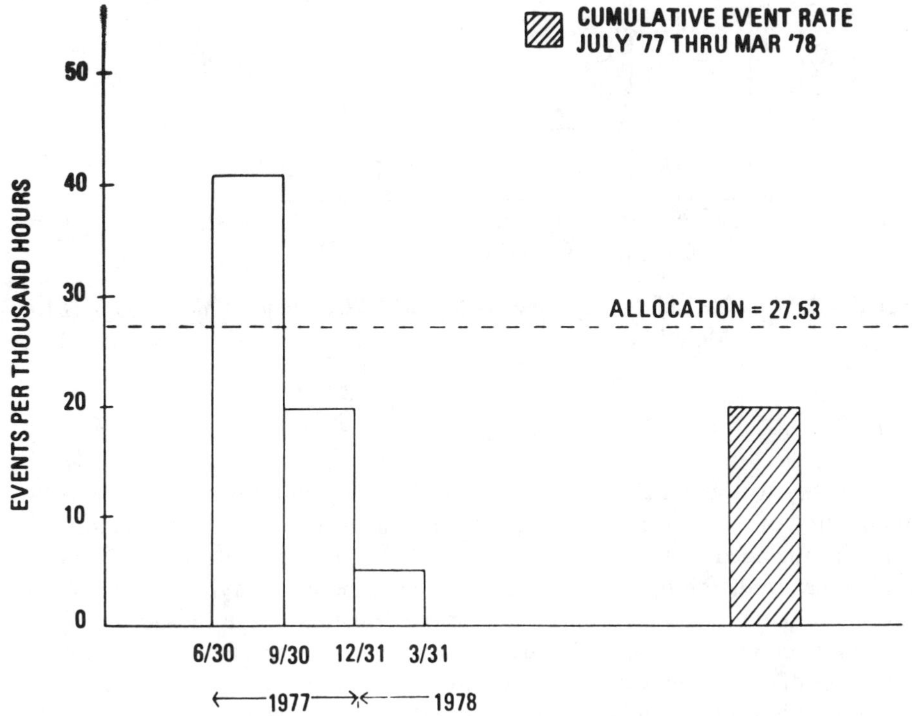

Figure 3.A.1-13 Reliability Experience—AN/SPY-1A Antenna Group.

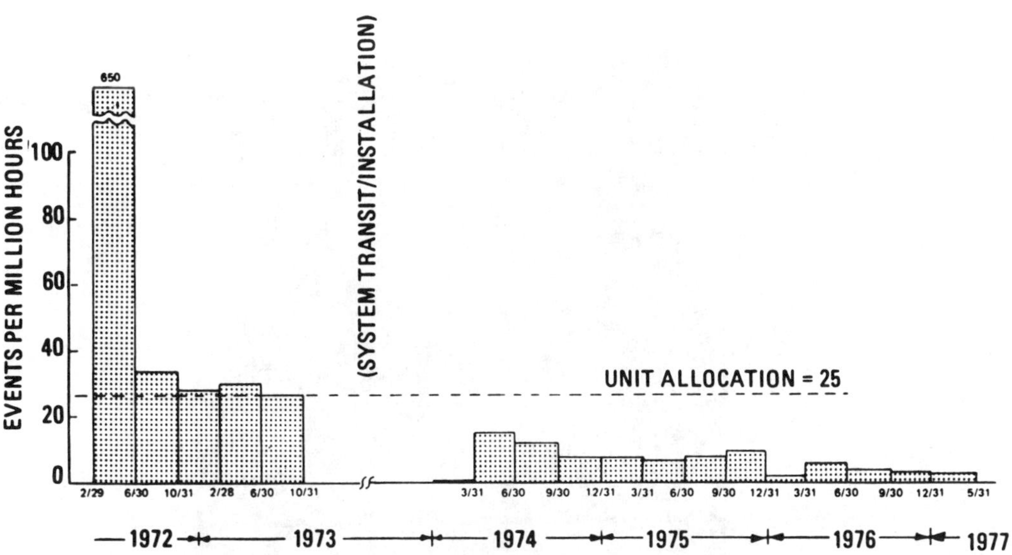

Figure 3.A.1-14(a) Reliability Experience—AN/SPY-1 Phase Shifter Driver Modules.

CHAPTER 3

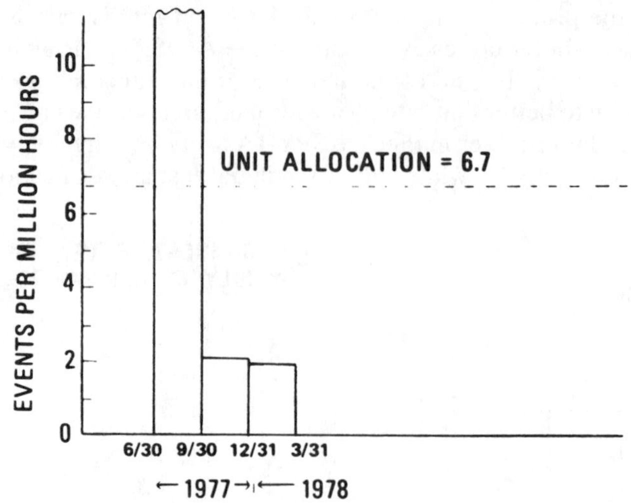

Figure 3.A.1-14(b) Reliability Experience—AN/SPY-1A Antenna Phase Shifter Driver Modules.

Conclusions

The key to the excellent record of availability achieved by the AN/SPY-1 radar has first of all been careful design of components to achieve high reliability—minimizing the need for repair action. Second, ORTS quickly detects failure modes and identifies faulty components. Third, those components having greatest maintenance frequency are made easily accessible (Figures 3.A.1-15 and 3.A.1-16). These three, combined with the highly parallel nature of phased array components and the use of redundancy switching where critical components are in sequence, are responsible for the availability demonstrated for AN/SPY-1 and enhanced by a factor of at least two in the AN/SPY-1A design. In addition, performance was improved by reducing receiving loss by 1.2 dB. At the same time, both weight and cost have been reduced in the new design by 30%.

Figure 3.A.1-15 Rear View of AN/SPY-1 Phased Array.

CHAPTER 3

Figure 3.A.1-16 Rear View of AN/SPY-1A Phased Array.

References

[1] R. Adm. Wayne E. Meyer, USN, "AEGIS—The Record," *1977 Electro Conference Record,* Section 8, p. 8–2.

[2] R.F. DiFelice and J. Drenik, "Mechanical Design and Integration of Phased Arrays," *The Record of the IEEE Mechanical Engineering in Radar Symposium,* IEEE Pub. No. 77CH 1250-0 AES, 1977, pp. 230–235.

3A.2 AN/SPY-1 PHASED-ARRAY ANTENNA

by R.M. Scudder and W.H. Sheppard

General Functional Requirements

The phased array described in this paper is a major unit of the AN/SPY-1 radar, which in turn is a segment of the AEGIS system. It operates in conjunction with other major units including the transmitter, signal processor, and the beam-steering control unit, which make up the remainder of the AN/SPY-1 segment. Primary interfaces of the array antenna also include monitoring functions of the AEGIS Operational Readiness Test System (ORTS).

The phased array must support the functional requirements of the AEGIS system, including: search, both active and passive; monopulse angle tracking; and ECCM.

In addition, the phased array must be designed to provide highly reliable operation in the stringent shipboard environment, which includes: vibration; underwater and Nuclear Shock; icing conditions to −20°F, 70-knot relative wind; and limited weight, volume, and prime power.

Functional Description of the Array

Figure 3.A.2-1 illustrates the organization of the array aperture, including the subarray layout for both the transmitting and receiving functions of the main radar channels and the auxiliary functions. Transmitting and receiving functions utilize subarrays of different sizes. The receiving function also uses slightly more total elements than does the transmitting function, primarily to achieve a symmetrical arrangement for best monopulse pattern characteristics and peak sidelobe control. These additional elements are contained in the four "receive-only" subarrays shown. The illumination function in the transmitting mode is uniform; that for receive is a modified circular Taylor [1] distribution for the monopulse reference channel, and utilizes modified Bayliss [2] distributions for the azimuth and elevation error channels.

As shown in Figure 3.A.2-1, the basic building blocks of the array face are termed array modules; each array module is associated with a horn-module assembly. Two array and two horn modules constitute a receiving subarray, of which there are 68 for the main array function and two others which are separately utilized for the "pre-look" and sidelobe-blanking functions. Four modules (two receiving subarrays) constitute a transmitting subarray, of which there are 32, corresponding to the number of parallel high-power amplifier stages in the radar transmitter subsystem.

The relationship of the major components of the array is illustrated in Figure 3.A.2-2. Each of the ten columns is identical except for the number of elements. The horn-module assembly is mounted to the external surface of the array structure and consists of 32 horns with the required impedance matching network and outer covering. The array-module assemblies consisting of 32 phase shifters integrated with a 32:1 power divider and eight phase-shifter driver boards, plug into each of the horn assemblies from inside the array structure. Four array modules are combined via a network of the three RF hybrids shown such that transmitter-receiver isolation (self-duplexing) is obtained via phase control of the nonreciprocal antenna phase shifters. This hybrid network further implements receiving subarrays, which are half the size of transmitting subarrays, and provides isolation of array face mismatch from the transmitter high-power amplifiers.

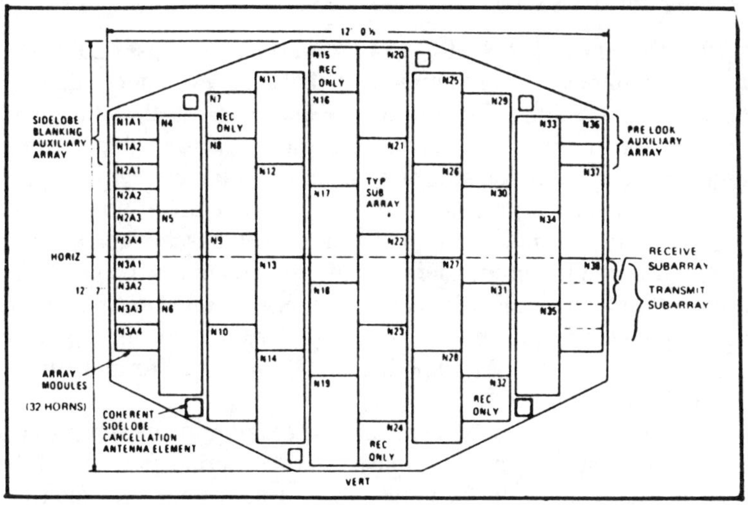

Figure 3.A.2-1 Subarray Arrangement, Rear View.

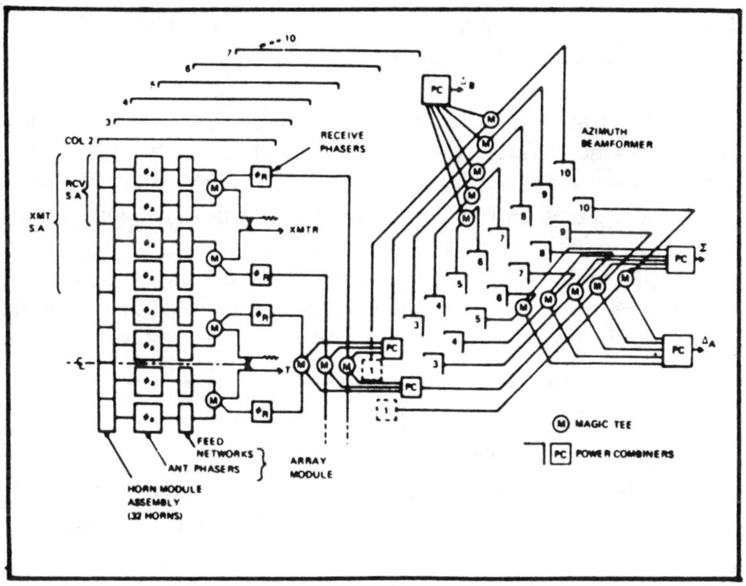

Figure 3.A.2-2 Schematic of Microwave System.

In effect, all subarrays are given identical phase-steering commands, thus allowing a simplified beam-steering unit, and a minimum of interface cabling between that unit and the array face. Relative phasing of the subarrays in the transmit mode is accomplished within the transmitter unit at the low-power input of the 32 final amplifiers. Relative phasing of the subarrays in the receiving mode utilizes the receiver phase shifters shown.

Receiving-beam formation is accomplished in two steps. The receiving subarrays in each of the 10 columns are combined to produce reference sum and monopulse-elevation difference outputs; in the second step, these outputs are combined to produce the complete array-reference sum beam and the complete monopulse-elevation difference beam. In addition, the monopulse-azimuth beam is derived in the second step.

CHAPTER 3

PHYSICAL DESCRIPTION

- Overall Dimensions 159.3 × 153 × 51 in
- Array Weight 17,137 lbs
- No. of Elements
  Transmit: 4096
  Receive: 4352
  Auxiliary: 128

Further Details on AN/SPY-1

Phase Shifter-Driver. The AN/SPY-1 phased-array radar uses subarray steering to minimize the number of phase commands that must be generated. This requires three phase-shifter designs. Transmitting phasers are used at the inputs to the final CFA tubes; receiving phasers are used at the inputs to the receive beamformer and antenna phasers are associated with each radiating element.

All three designs [4,5] can be described as waveguide, nonreciprocal, flux-drive, latching garnet configurations. Each phaser has a rectangular toroid of garnet mounted in the longitudinal center of a rectangular waveguide. The toroid slot contains two magnetizing wires and is otherwise filled with a ceramic which has a relative dielectric constant of 50, except for the transmitting phaser where the dielectric constant is 16. An impedance matching transformer is used at each end of the toroid.

The transmitting and receiving phase shifters require greater phase accuracy than the antenna phaser. To accomplish this, the precision units have sturdy housings with waveguide interfaces and have a mated driver which is adjusted during production for optimum performance.

Data measured on production units of the three phaser and driver designs show that differential phase shift is linear with command within an rms error of less than 2.5°. Over a ±40°F temperature range, insertion phase changes ±5.0° while differential phase at the maximum bit command changes ±6.0°. Differential phase errors, as a function of operating frequency, are small and are included in the above figures.

Insertion loss for the phasers is typically 0.90 dB when averaged over frequency and command settings SWR is low in all designs and power handling capability is tailored to the various requirements. Table 1 lists the specification requirements and typical performance of the three designs. Differential and insertion phase errors include the effects of ambient temperature, RF power, dc power supply drift, and an estimate for aging. Phase errors for the antenna phaser are considerably below the specification values. Switching energy includes that which is delivered to the toroid and switching wires as well as driver efficiency.

TABLE 1

PHASE SHIFTER CHARACTERISTICS

	ANTENNA PHASER		RECEIVE PHASER		TRANSMIT PHASER	
	SPEC	TYP	SPEC	TYP	SPEC	TYP
SWR, MAX	1.35	1.25	1.25	1.20	1.25	1.15
INSERT LOSS, AVG (dB)	1.1	0.9	0.9	0.9	1.2	0.9
INSERT LOSS, PK (dB)	1.5	1.15	1.2	1.10	1.5	1.15
# BITS	5	5	5	5	5	5
ERROR °rms	9.5	5.0	4.5	4.0	4.5	4.0
INSERT PHASE ERROR °rms	14.2	6.0	4.0	3.0	4.0	3.0
POWER, PK SURVIVE (kW)	6.0	16.0	10.0	16.0	40	80
TEMP, OPERATING (°F)	80 TO 160	80 TO 160	80 TO 105	80 TO 105	80 TO 105	80 TO 105
TEMP, SURVIVE (°F)	-80 TO 167	-80 TO 167	-80 TO 167	-80 TO 167	-80 TO 167	-80 TO 167
Sw SPEED, RESET (μs)	10	4	7	5	13	6
Sw SPEED, SET (μs)	COMBINED	2	6	3	10	4
Sw RATE, AVG (Hz)	800	800	800	800	800	800
Sw ENERGY, PER PRF (μ joule)	-	600	-	600	-	800
Sw CURRENT, P	13±1	13	-	13	-	18
MTBF (h)	88,700	-	83,333	-	83,333	-
SHOCK	MIL-STD-202D		MIL-STD-202D		MIL-STD-202D	
VIBRATION	MIL-STD-167		MIL-STD-167		MIL-STD-167	
HUMIDITY, DEWPOINT (°F)	50		50		50	

Beam-Forming Networks. As indicated in Figure 3.A.2-2, the initial step in beam formation, applicable to both the transmitting and receiving functions, combines the array modules to form transmit and receive subarrays, each with separate terminals (as described by Patton).

Temperature Control System. To meet the array pointing-accuracy specification, the temperature profile of the phase shifters across the entire face and the temperature of the receive waveguide components must be maintained within prescribed limits. Cooling protection for the driver cards and other electronic components must also be provided. In addition, the temperature control system provides de-icing for the outer covering of the radiating aperture.

An important design constraint is that no liquid cooling means be utilized within the confines of the antenna structure. Ships services, i.e., power and chilled water, are also restricted to remain within the total budget. This evolved a thermal control system design which utilizes the waste heat from the components for de-icing purposes.

A simplified temperature-control system was configured utilizing two blowers (one is standby for reliability) each having a capacity of 1700 CFM, air-to-water heat exchangers external to the array, heaters for de-icing and the necessary thermostats or control valves to regulate the system. The temperature control system was modeled using computer simulation and provides the required performance within the system constraints.

Antenna Structure. An aluminum structure, consisting of a $1\frac{1}{2}$ inch thick base plate welded to an octagonal outside frame with four inside "T"-beams, provides the mechanical support for all array components including the 140 array modules and associated horn modules. Location tolerances of the modules are controlled to ±.020 inch. The design of the structure is based on computer simulation, which produced a configuration with a natural resonant frequency well in excess of the required 25 Hz, which was verified by static load testing.

CHAPTER 3

ARRAY COMPONENTS
- Radiating Elements
- Array Modules
  Corporate Structure
  Power Divider-Combiner
  RF Loss: 0.2dB max.
  Phase Tracking (port to port, matched): 3° rms; (module to module): 3° rms.
  SWR = 1.35
- Phase Shifters & Drivers
- Beam Forming Networks
- Temperature Control System
- Antenna Structure

Representative Array Performance

The antenna pattern data presented in this report were measured using the Automatic Antenna Pattern Data Acquisition System (AAPDAS), an automated test range located at the RCA Moorestown Plant. Figure 3.A.2-3 shows the AN/SPY-1 array mounted on the test positioner at the test range. The facility provides automatic test capability by means of computer control of the following:

- Selection of array beam position via the AN/SPY-1 Beam Steering Controller (BSC),
- Two-axis angular positioning of the AN/SPY-1 array antenna for pattern measurements,
- Selection of test frequency via remote-controlled microwave frequency synthesizer,
- Data acquisition and magnetic tape data recording,
- Monitoring of the array and BSC self-test equipment (ORTS).

A total of more than 900 patterns of various types, plus several other additional tests, were recorded for the AN/SPY-1 antenna utilizing this facility. Essentially all were recorded on magnetic tape, and reduced to appropriate formats off-line on a general-purpose computer.

A major portion of the patterns was measured and reduced in contour format, ranging in size from solid angles of 100° × 100° with 0.5° data point grid dimensions to smaller, more detailed plots with much smaller data-point intervals. These patterns, measured for both receiving and transmitting modes at array-beam positions distributed throughout the defined array coverage volume, included more than 50 of the 100° × 100° size. The recording method utilized a horizontal mechanical-raster motion of the array test pedestal moving at a rate of 2° per second. Recording of the three monopulse receiving functions of the array was done simultaneously through the three-channel coherent test-receiver systems driving a single magnetic tape unit.

Figure 3.A.2-3 AN/SPY-1 Array on Test Positioner.

Figures 3.A.2-4 and 3.A.2-5 are typical examples of the contour data showing the regions near a main beam, where Figure 3.A.2-4 is the monopulse reference sum pattern, and Figure 3.A.2-5 the corresponding monopulse elevation pattern. In these examples, the array beam position is at elevation broadside, but scanned 47.7° off broadside ($\alpha = 90°$, $\beta = 90°$) in azimuth. These sample patterns, spanning less than 100° in each direction, are plotted in equal increments of α and β coordinates, where α is the angle relative to a horizontal line on the array face, and β is measured relative to a vertical line on the array face. These patterns are plotted on a high-speed line printer, utilizing numerical values for levels 0 dB through −8 dB, then alphabet codes from A through F representing 5-dB increments down to −40 dB, the minimum level plotted. Companion patterns, not shown, were also reduced which plotted and identified the magnitude and location of only sidelobe peaks to the nearest 0.1 dB for use in detailed performance analysis. In general, the peak sidelobes of all receive channels were below the −23 dB specification requirements relative to the peak of the reference pattern, as evaluated over the radar operating frequency band and the defined array coverage volume. The rms sidelobe level for regions beyond the solid angle displayed in Figures 3.A.2-4 and 3.A.2-5 is below the specified value of −49 dB. The factor K_m^6 of the monopulse error channels averages 1.67.

Test procedures closely related to the recording of contour patterns provided a thorough evaluation of the array-coverage characteristics. Again, the array test pedestal was operated in a raster scan mode, but with the array beam automatically counter-steered such that the peak of the main beam was effectively maintained pointed at the boresight tower. The result is a detailed contour map of array main-beam gain *versus* array beam position, an example of which is shown in Figure 3.A.2-6. In the case illustrated, gain values relative to beam broadside are printed out at 1° intervals in both angular dimensions with 0.5-dB contour amplitude intervals. The total angular range displayed is 125° in β, 100° in α, representing an evaluation of some 11,000 different beam positions steered in the 1° increments. The diagonal boundaries at the four corners are essentially the edges of array real space. It is noted that the gain characteristic closely follows the expected cos θ function. The ellipses overlaid on the contour plot show lines of constant θ; the triangular overlay represents the defined angular coverage sector.

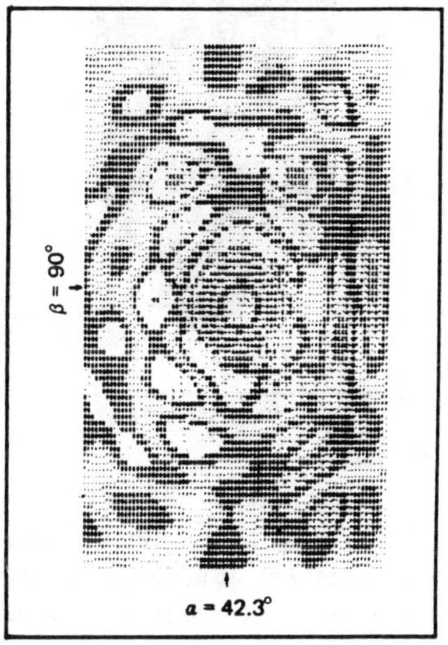

Figure 3.A.2-4 Contour Pattern, Monopulse Sum.

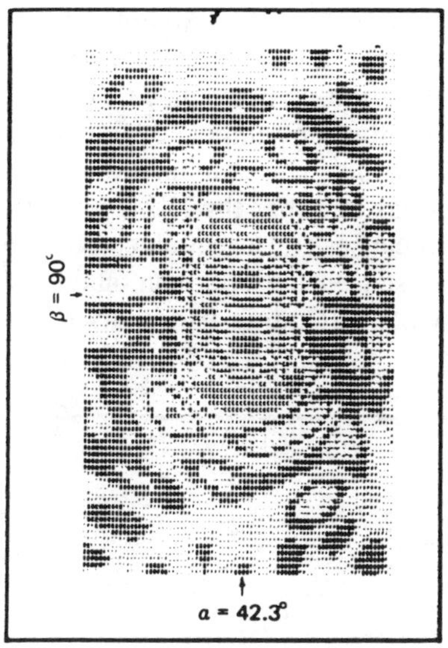

Figure 3.A.2-5 Contour Pattern, Monopulse Elevation Difference.

A major group of patterns, totaling more than 300 distributed over the array coverage volume, evaluated angle-tracking performance. In this case, detailed patterns were recorded over small angular sectors about one beamwidth in extent in both α and β coordinates. The three-channel test receivers were set up to directly output the amplitude ratio of the monopulse difference to reference patterns along with the associated phase angles, nominally $\pm 90°$. Recorded data also included the commanded beam positions and the actual array positions indicated

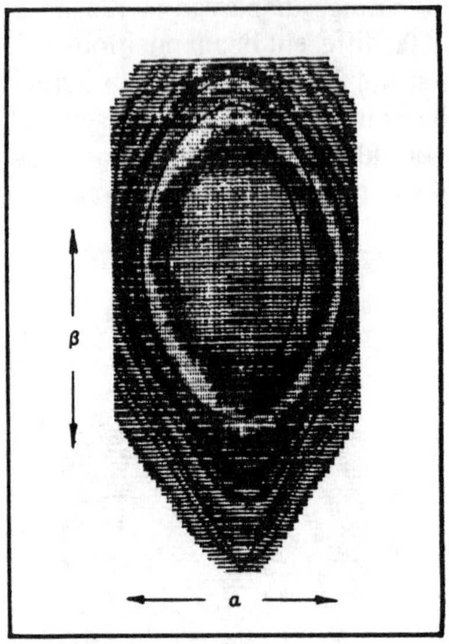

Figure 3.A.2-6 Array Coverage.

by the test pedestal digital encoders. Data reduction of these patterns included a least-squares linear fit to the total set of normalized error patterns measured for each of the two α and β angular coordinates. The output of this data reduction included biases and random errors in the zero crossing (nulls) of the normalized monopulse difference patterns, along with the total error of indicated-target angular position within the 3-dB beamwidth points of the reference beam. The latter includes all the variances of measured data points about the best-fit, mean-normalized difference pattern function. An example of the computer reduced data (in terms proportional to array cosine space) is shown in Figure 3.A.2-7. This figure illustrates the distribution of normalized difference pattern data in the β coordinate with the mean value superimposed, showing a small bias and variance of the monopulse null position. The standard deviation of null position is approximately 0.02 beamwidths in either the α or β directions. The standard deviation of total error from all causes for targets located anywhere within the -3-dB main-array beamwidth, based on the data summarized in Figure 3.A.2-7, is approximately 0.04 beamwidth excluding fixed-bias terms which are removed in radar system application.

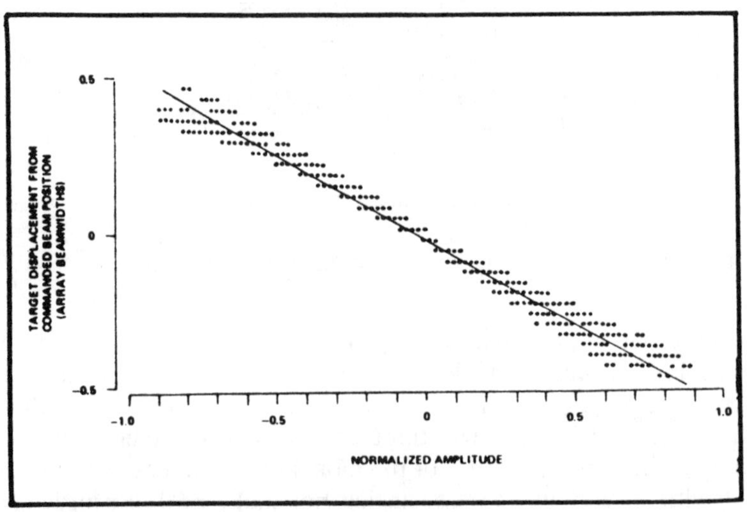

Figure 3.A.2-7 Composite of Measured Monopulse Patterns, β Coordinate.

Other tests, not described in this paper, included transmit-mode and receive-mode gains, transmitter-to-receiver isolation, radar waveform bandwidth characteristics, and polarization properties. In general, it was demonstrated that all characteristics satisfied the needs of the AN/SPY-1 radar.

References

[1] Taylor, T.T., "Design of Circular Apertures for Narrow Beamwidth and Low Sidelobes," *IEEE Trans. Antennas and Propagation,* Vol. AP-8, January 1968, pp. 17–22.

[2] Bayloss, E.T., "Design of Monopulse Antenna Difference Patterns with Low Sidelobes," *BSTJ,* May–June 1968, p. 623.

[3] Campbell, J.J, and B.V. Popovich, "A Broad Band, Wide-Angle Scan Matching Technique for Large Environmentally Restricted Phased Arrays," *IEEE Trans. Antennas and Propagation,* Vol. AP-12, July 1972, pp. 421–427.

[4] Landry, N.R., "Microwave Phase Shifters," RCA Reprint RE-15-3-5, 1969.

[5] Goodrich, H.C., and R.C. Tomsic, "Flux Monitoring Boosts Accuracy of Phased Array Radar Systems," *Electronics,* November 23, 1970, pp. 77–80.

[6] Barton, D.K., *Radar System Analysis,* Artech House, 1976.

Chapter 4
HYBRID AND MONOLITHIC MICROWAVE INTEGRATED CIRCUITRY (MMIC)

by D. G. Laighton

4.1 HYBRID TECHNOLOGY

A good departure point for discussing solid-state phased array radar hardware is the US Air Force PAVE PAWS system (built by Raytheon). The reason for this is that PAVE PAWS is the world's only active-aperture, solid-state, phased array radar. We are careful to say, ''active-aperture, solid-state, phased array radar,'' and when so described PAVE PAWS is the only system of its kind in the world. There are other solid-state and other phased array radars, however.

There are two of the original PAVE PAWS systems, vintage *circa* 1976. Two more have been completed in 1987 and 1988 for a total of four of these PAVE PAWS sites, built as originally designed. The two more recent systems are in Georgia and Texas, with a similar solid-state upgrade to the older BMEWS system having been built in Thule, Greenland. When we speak of PAVE PAWS and its BMEWS antecedents, we are talking about a two-faced, active-aperture, solid-state, phased array radar system. Each system, as shown in Figure 4.1, consists of two array faces, both having 1796 active elements. Almost double that number of antenna elements are used, but many of these are dummy elements employed for antenna pattern control and not driven by solid-state transceivers. The overall aperture, at UHF, is about 100 ft in diameter, so this is a physically large radar system.

The key element in the system is the solid-state transceiver module. Figure 4.2 is a photograph of the transceiver module used in the PAVE PAWS array, consisting of two nested boxes, with the bottom box containing the transmitter circuitry, and the upper box containing the receiver and control components. The transmitter consists of four 100-W output transistors powered by two drivers, which in turn are powered by a predriver. The module has circulators for output VSWR protection, and a filter (not shown) under the box. The bottom of the transmitter butts up to a cold plate through which we cycle liquid coolant. The top box has a four-bit phase shifter, a two-stage low-noise amplifier (LNA), and a transmit-receive (T/R) switch hidden by the corner of the cover. The transceiver module also has a logic driver assembly, which decodes row-column logic commands. The beam-steering computer is not wired individually from the beam-steering computer to every module, but rather on a row-column basis in the array. This is done so that each module can be addressed individually by knowing its intersecting point between row and column in the wiring scheme without individual module control wires.

The reason for showing PAVE PAWS is that it represents the largest existing base of experience for solid-state transceiver modules used in a phased array radar [1]. Currently, for the first two PAVE PAWS systems (which have been running for 10 years as of this writing), Raytheon has built 8000 modules, including spares.

LAIGHTON
CHAPTER 4

Figure 4.1 PAVE PAWS Radar (Courtesy of Raytheon Co.).

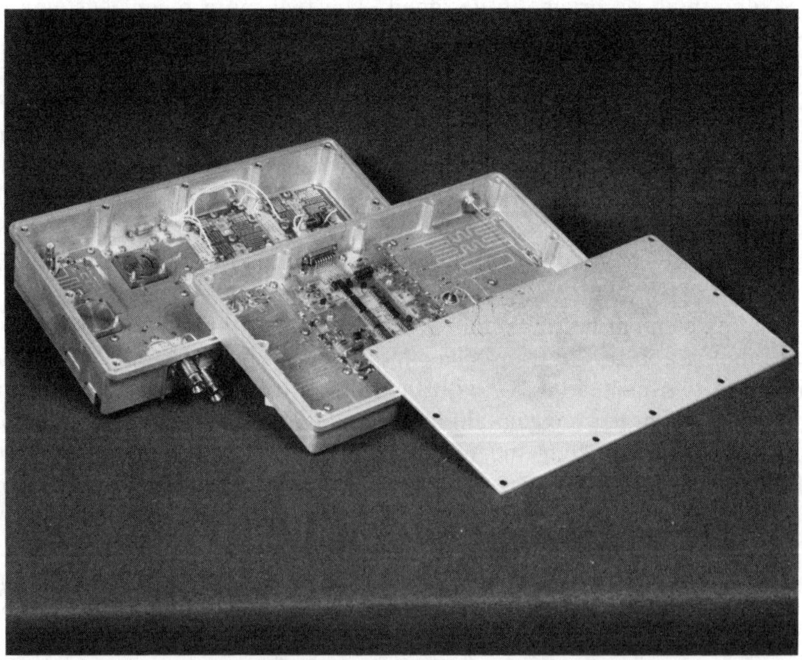

Figure 4.2 PAVE PAWS Transceiver Module (Courtesy of Raytheon Co.).

The original mean-time-between-failure (MTBF) projection for the modules in the array was approximately 100,000 hours, and the systems typically experienced about 150,000 to 200,000 hours in actual operation. What this means for the big PAVE PAWS system shown in Figure 4.1 is that the radar can be on the air 24 hours per day, 365 days per year. PAVE PAWS can run for about a year and a half to two years before replacement of any modules is required, which is unprecedented in a radar of its size and power level. The modules themselves are extremely reliable. By using the PAVE PAWS system as a reference point for solid-state phased arrays, we may discuss GaAs MMIC-based modules for future systems.

Figure 4.3 is a block diagram of a typical solid-state radar transceiver module. Typically, input is fed through an array manifold, and passed through a phase shifter, which is common to both the transmitting and receiving paths. A T/R switch may select between the transmitting and receiving paths. When in the transmitting mode, the signal goes through a power amplifier and then out through a circulator to the antenna. In this manner, energy from the several thousand antenna elements in the array is radiated and combined in space. The received signal comes in, is circulated into the LNA, then goes through the T/R switch, which has been switched into the receiving direction at that point, and finally out through the common phase shifter.

There are some aspects to this module that we must point out here. One is that there is a possible severe mismatch as a function of scan angle in a phased array radar. In an active-aperture phased array radar, as it scans off boresight, VSWRs as high as 5.0:1 may be seen. Such a mismatch would be very harmful to the power amplifier, if it were allowed to be seen at the output of the high-power transistor stages. Likewise, when low noise figure is achieved in a receiver, input match is generally traded off, so we will have severe interaction between the antenna element and the LNA. In this case, we would use a *circulator,* which is a nonreciprocal ferrite device that provides isolation when terminated in 50 Ω. Therefore, we will have T/R switches, not so much to switch from transmitting to receiving mode, but to terminate the third port of the circulator to obtain isolation as well as antenna duplexing.

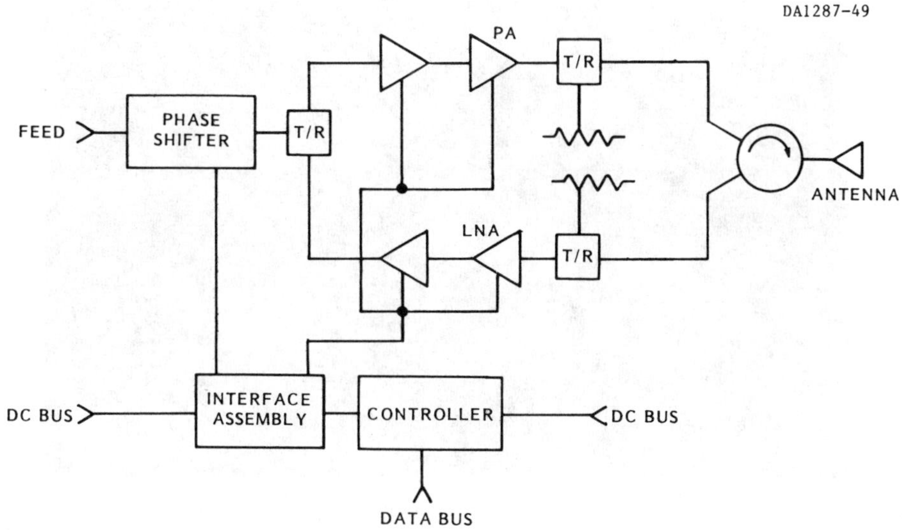

Figure 4.3 Typical Radar Transceiver Module.

In terms of the basic configuration of the transceiver module, the key in active-aperture phased array design is to generate the power as close to the antenna element as possible. Other types of phased array systems may also be designed. Present systems, for instance, are generally retrodirective using only phase shifters at each radiating element. Beam steering is achieved electronically by incrementally adjusting the phase shifters in the array, but then the phase shifter loss must be accepted in both transmitting and receiving modes. For most systems applications so configured, the penalty for using a so-called ''passive-aperture phased array'' of that type is about 6 to 10 dB in terms of two-way loss. By putting a transceiver module at the antenna element, we may be able to transmit the power very efficiently. There will be no loss after the power amplifier, except for the T/R switch and circulator. The noise figure may also be established, at the antenna element, with only the circulator and T/R switch loss. Using the active aperture concept, a full radar transceiver is being put at the array face to improve transmitter efficiency, establish noise figure, and achieve intertialess scanning.

Figure 4.4 is a photograph of a Grumman Aircraft concept for a space-based radar. The Grumman concept was to fold up to a compact drum-like package in the space shuttle, and then to unfold like a huge umbrella in space. There is a boom or a mast that can be seen in the figure, and a huge wire wheel with a membrane antenna. This membrane is typically 100 m in diameter, which gives the reader an idea of how long the boom is. Imbedded in this membrane are about 130,000 low-power (<1 W) transceiver modules. This was one early concept for a space-based phased array radar. Figure 2.27 (Chapter 2) shows a near-term technology concept in the form of a corporate-fed with 25,000 moderate power (<10 W) modules. Whether either concept is practical depends on the development of lightweight, small, and inexpensive solid-state transceiver modules.

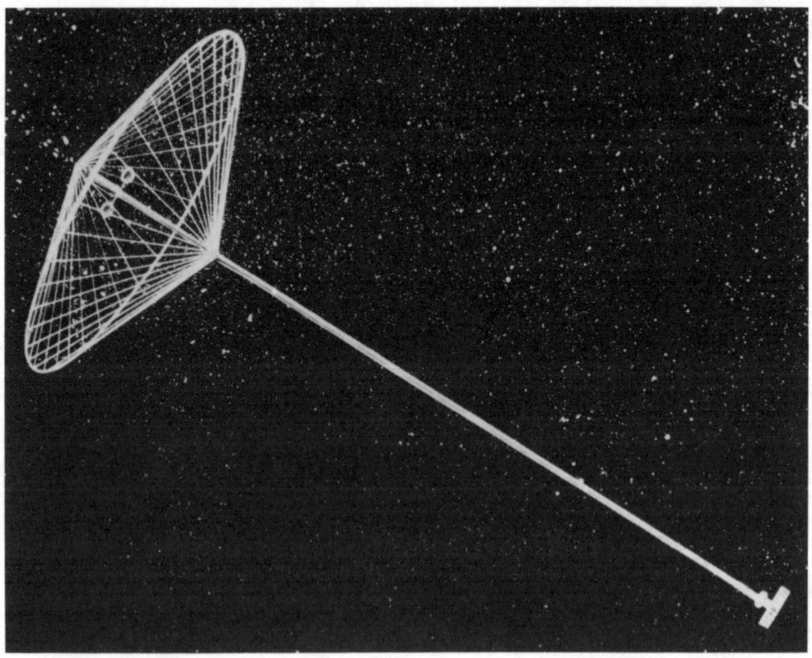

Figure 4.4 Grumman Space-Based Radar.

Figure 4.5 shows a hybrid L-band module. It is fundamentally the same technology as used for PAVE PAWS. The hybrid module uses a *distributed* technology, where there are transmission-line components and soldered discrete active devices. In this case, the transceiver module is built much like the block diagram of Figure 4.3. There is a four-bit phase shifter on microstrip, a two-stage power amplifier using FETs, and a two-stage LNA. There is also an L-band circulator and an aluminum block, representing the logic circuitry that drives the module. The module is about 3-$\frac{1}{2}$ inches on a side. It weighs about five ounces and, optimistically, will probably cost something like $2500 per unit in production. These numbers are fundamentally inconsistent with building a space-based radar, for example. First, 130,000 of them, at $\frac{1}{2}$ pound apiece, probably represent more weight in modules than the shuttle can even carry into orbit. These modules are relatively inefficient, and at quantities of 130,000 costing $2500 apiece, they are financially burdensome, even for a space-based radar system. What we may conclude is that a technology is needed which is about an order of magnitude smaller, lighter, and cheaper than anything similar that has been designed. The module sizes are not shown here, but the PAVE PAWS module is about 14-inches long by 8-inches wide by 3-inches deep, and it weighs about 14 pounds. At L-band, we may get some scale for frequency, but the module is still 3-$\frac{1}{2}$ inches by 3-$\frac{1}{2}$ inches by about $\frac{1}{2}$ inch, and weighs about $\frac{1}{2}$ pound. At the present time, using MMIC technology, it is possible to take all of these transceiver module functions and put them on two chips that are roughly $\frac{3}{10}$ of an inch on each side.

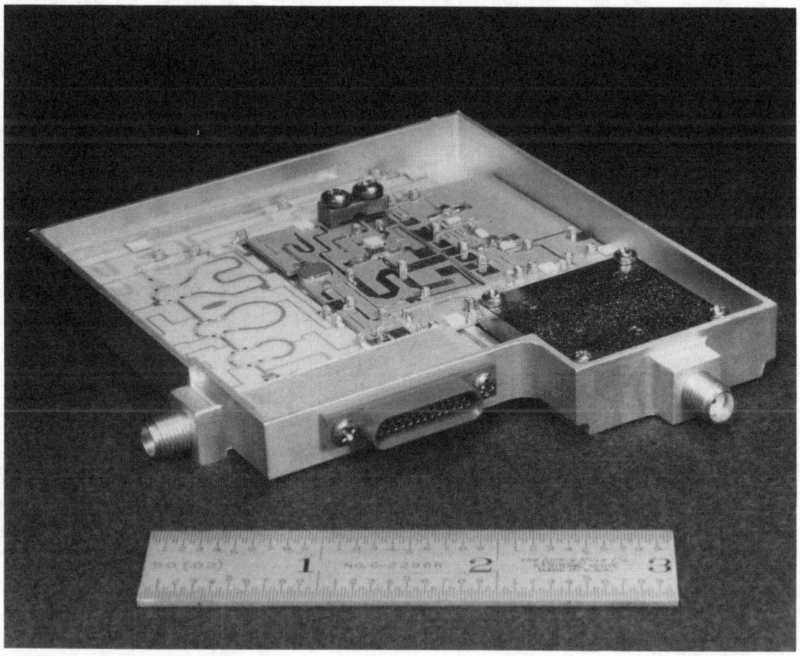

Figure 4.5 Hybrid L-Band Module (Courtesy of Raytheon Co.).

4.2 MMIC TECHNOLOGY

Figure 4.6 is a photograph of an early MMIC technology, which showed promise, but was later abandoned. This is a silicon-on-sapphire monolithic microwave integrated circuit. One chip is a four-bit phase shifter, while the other chip has the three-stage receiving LNA and a three-stage power amplifier (PA) integrated on one chip. So, all of the functions, except for the digital control and circulator, of that 3-$\frac{1}{2}$-inch square box can be done in a space of about $\frac{3}{10}$ by $\frac{6}{10}$ of an inch.

In the basic field-effect transistor (FET) we take a semi-insulating [2], substrate, grow or deposit a conductive channel region on it in some manner, and then put three terminals on it such that we may vary the field on the gate (center terminal) to vary the amount of current that flows through the device. Basically, we can look at the FET as an electrically conductive resistor, which can be controlled by varying the voltage on the gate of the device. We may turn off the gate voltage, in which case it will conduct current from source to drain, up to the limit of what the device can physically handle in terms of its physical geometry and doping level. The other extreme is the field which can be developed in the channel region to the point where current is restricted and the device is completely shut off. So, the two limits of operation are *saturated* drain current with the gate voltage off, and *pinch-off,* where the gate voltage is at a level high enough to shut off completely the electrically conductive path.

Figure 4.6 Silicon-on-Sapphire MMIC Transceiver Module (Courtesy of Raytheon Co.).

The general schematic symbol for the FET is shown in Figure 4.7, along with a greatly simplified model for the device having resistive and capactive effects associated with the gate, source, and drain. A key parameter in the definition of FET characteristics is G_M or transconductance.

In defining the performance of a FET device, the gate length determines the cut-off frequency, to a first approximation. When we talk about a one-micron (1 μm) gate device, we generally mean a device which has 1 μm in this short dimension, and this device in gallium arsenide (GaAs) is generally an X-band transistor. In terms of gate width, we define the ability to handle current, or the *figure of merit* for the device, which is the power per unit periphery. When we talk about a one-millimeter (1 mm) device, we mean 1 mm of gate width, where the gate length can be 1 μm, for instance.

Thus, if we consider the basic FET device in a little more detail, its power capability is determined by the *drain current* and the *breakdown voltage* between drain and gate. Drain current, since channel depth is limited by the ability to pinch-off a device, is increased by increasing gate width or periphery. The voltage breakdown will be determined by the weakest point in the device's gate width or periphery.

Gate periphery is usually increased by making wide "fingers." To illustrate we may begin by laying out the planar and vertical geometry of a typical X-band power device. (Raytheon's gates are laid out in interdigitated fashion.) Drains are interspersed and then source connections are made. In terms of an X-band device, something on the order of 150 μm is about the maximum width of a finger. For instance, if a device is needed that puts out 2 W, and $\frac{1}{2}$ W/mm is the basic figure of merit for power per unit periphery then 4 mm of periphery are required. These 4 mm are provided in 150 μm chunks at X-band. Therefore, 26 or 27 gate fingers would have to be used to add up to 4 mm, which leads to interconnection problems, and brings up the problem of how close together can we put these gate fingers before heating effects from individual fingers interact. Another factor which enters into the design of the device is that grounding of one of the terminals is essential. A grounded source is typically needed on power devices, so we must make a ground connection on the sources of the device.

Next, we would have to interconnect all of the sources with a minimum of inductance between them. This particular device uses what is called *air-bridge technology* to interconnect sources, and overlays the source contact from source finger to source finger by "jumping" over the gates and the drains. This is done in the device processing [3] by defining a patch of *photoresist* over the gate-drain region, depositing gold over that photoresist, plating the gold to make it physically strong, and then removing the photoresist from underneath. The final device, shown in Figure 3.8, uses air-bridge technology, where the sources are connected to the ground plane through *via holes,* which are put in from the back of the wafer as part of the fabrication sequence. The sources are grounded at the end of the cell, then they jump over and hit the source pads in the center of the device.

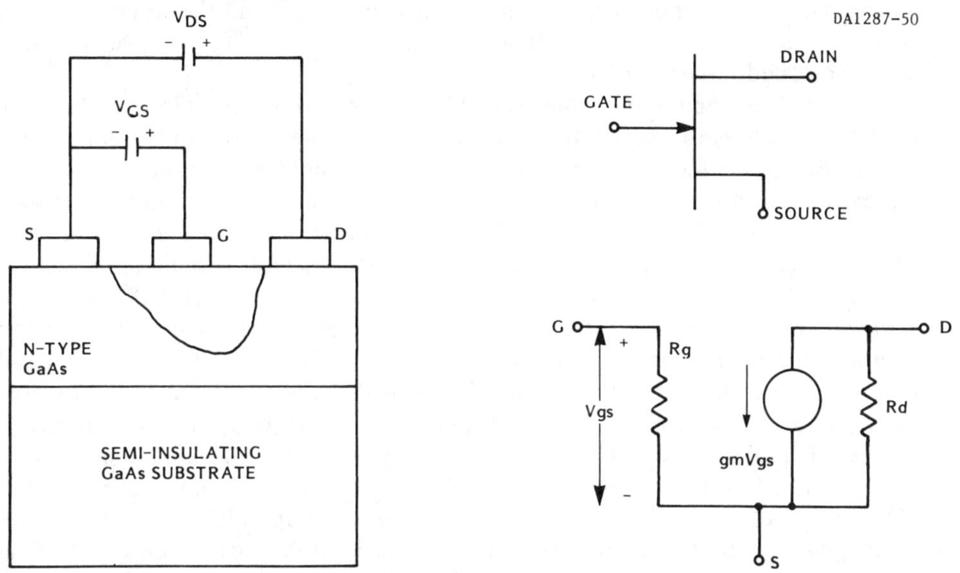

Figure 4.7 Basic Field-Effect Transistor.

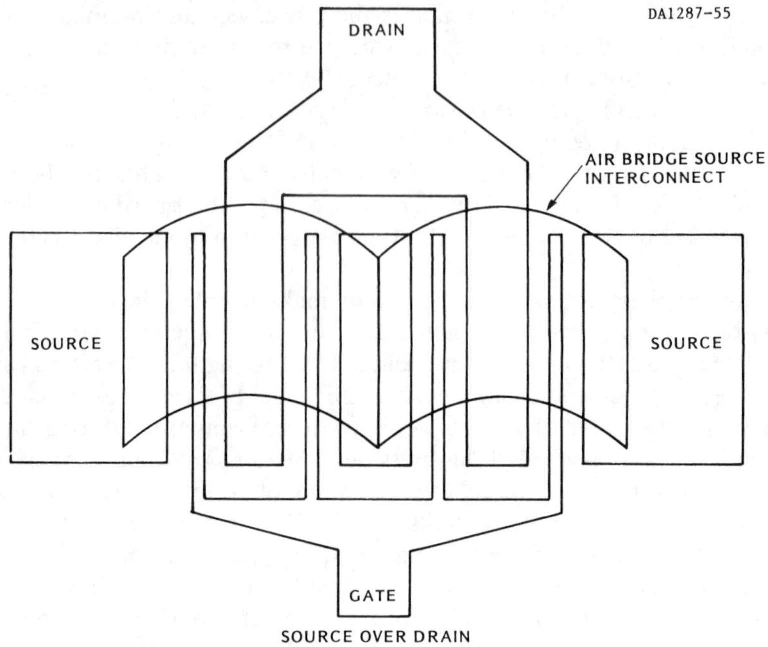

Figure 4.8 Geometry of FET Using Air Bridges Sources.

Now that we have explained the FET concept, let us consider how the FET is used in an actual amplifier design [4]. Most of the FETs used today are in the *class-A* mode. If we look at the drain characteristics of a device, as shown in Figure 3.9, we can work between two limits: the saturated drain current and the pinch-off of the device. In between the two limits, we can vary the gate source voltage to give current and voltage swings as shown in the curve. Class-A operation is characterized by a condition where we can bias the device somewhere in the middle, then swing the gate voltage to the voltage and current "knees" of the device characteristics. In other words, we may swing to the knee of the saturated drain current, and to the knee of the breakdown characteristic.

Class-B typically requires that we bias the device further downward so that during half of the input cycle we are beyond pinch-off and the device will not conduct. The other half of the cycle will run up the load line curve to give operation in the so-called *linear region*.

Class C is defined by being biased beyond pinch-off to the point where we have to work toward overcoming the pinch-off voltage to go into the conducting region. So, if we look into the basic classes of amplifier operation, class-A is typically used in GaAs MMIC design. The theoretical efficiency limit of class-A amplifiers is 50%. S-band FET amplifiers typically exhibit anywhere from 35 to 40% drain efficiency, and 25% drain efficiency is typical in X-band amplifiers.

Bob Pucell (Raytheon Research Division, Consulting Scientist) has written some good papers on this subject [5]. Typically higher device gain is required to operate in class-B than is necessary in class-A. Therefore, almost all X-band devices to date have operated in class-A. These devices do not have the available stage gain to be reduced by 5 or 6 dB in the interest of better output efficiency. Increased breakdown capabilities are necessary because the gate source is biased beyond cut-off. This adds numerically to the drain voltage in the reverse-bias part of the operation. Class-C has a theoretical efficiency of up to 100%, but this occurs when power output is zero! Therefore, it is a relatively trivial case. Practical efficiencies can be found in class-C amplifiers on the order of 50 to 60%, provided that the aforementioned breakdown conditions are not exceeded.

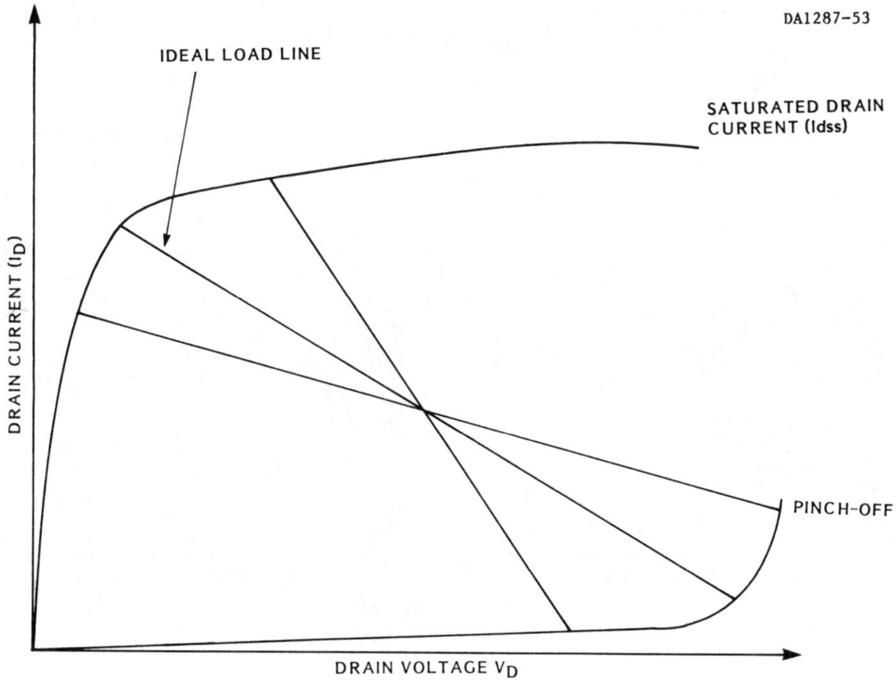

Figure 4.9 FET dc Characteristics.

Class-D is a switching amplifier. Switching amplifiers can also be almost 100% efficient. The key here is to design a device to have a cut-off frequency that is much higher than the frequency of operation, and to treat the harmonics efficiently. At the present time, the industry does not make any devices that have enough excess frequency capability to allow for true class-D operation, although classes C or B can be overdriven to make them appear to be somewhat like class-D. In terms of class of operation, virtually all MMICs to date use class-A or class-AB, and typically run on the order of 15 to 25% in efficiency. There is a possibility of operating classes B or C in the lower frequency MMICs that are used for L- and S-band radar PAs.

The dc FET characteristics were shown previously in Figure 4.9 (see [6]). Basic output power is the amount of voltage swing, times the amount of current swing, divided by eight. Input power at dc is a "fixed" or Q point, in old-time terminology, as defined by the bias current and the bias voltage. Obviously, drain efficiency is the quotient of the two numbers. Present limitations on power are defined by the knee or saturated drain current (ID_{ss}) current and the breakdown voltage (V_{bd}) of the power device. In some instances a good power device has shown breakdown as high as 25 V, but this has not been done repeatedly. Approximately 12–16 V is the average result. The most efficient load line, and the highest amount of power that the device has shown, are given where the load line intersects with both the current and breakdown voltage knee points.

One of the things done at Raytheon, for example, to improve designs is to consider a large body of load-pull data. We look at the device and what it does in terms of power output for constant power input and constant bias conditions. When a mismatch is applied to the device, we typically find a maximum power point and lesser power points that tend to follow somewhat concentric contours about the maximum power point (see Figure 4.10). Alternatively, we look at an idealized version of the collective load characteristics. Drain current and drain voltage were previously shown in Figure 4.9, along with several possible load lines. The maximum power point occurs when we match the load line that fits

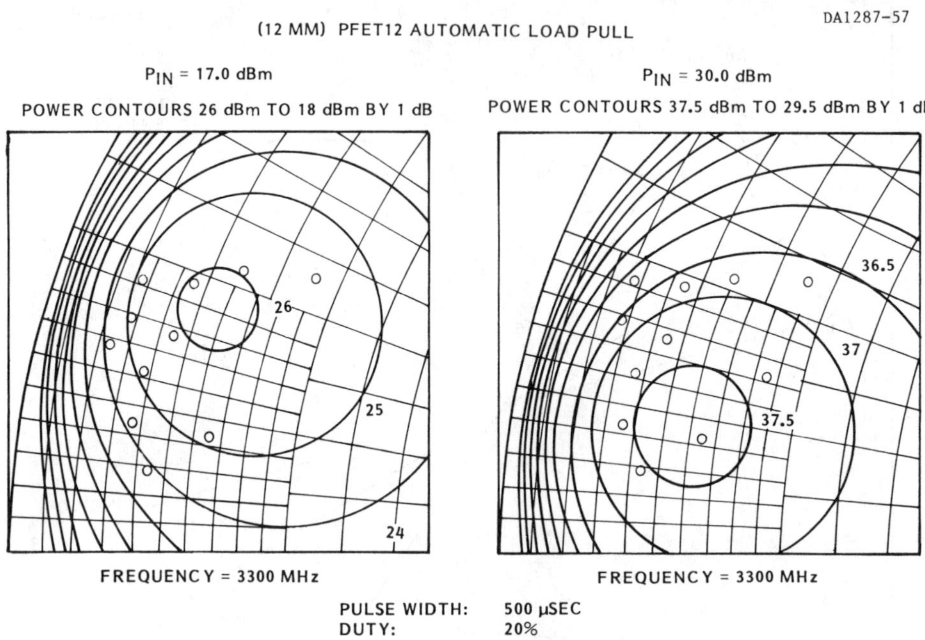

Figure 4.10 Pair of Load-Line Contours (12-mm Gate Width).

between the two knees. A mismatch in either direction is represented by the other two load lines of Figure 4.9. It is either in the direction to limit voltage swing, or in the direction to limit current swing. If the mismatch were 2:1, due to a load line having a higher impedance by a factor of two, then the current swing would be half as much for the same voltage swing. Consequently, the output power would be half as much (-3 dB) as would the power efficiency. The same drop in output power and efficiency occurs if the mismatch were 2:1 due to a lower impedance load. Figure 4.10 gives a pair of load-pull curves showing the effect of mismatch on an S-band FET power device for two input power levels. Often, the device is not nominally chosen to operate at the maximum output power point. This is because of the sharp fall-off in output when the mismatch goes to the left of the maximum output point. To avoid this sharp fall-off, the device is nominally set to operate to the right of the maximum. Figure 4.9 showed how varying the load line through mutual coupling in a phased array can limit either current or voltage swing, thereby degrading both power output and efficiency.

The active device is but one part of MMIC design. The other part of the MMIC design is how to make the passive circuitry that matches these devices. There are two major technologies in use. The most familiar are *distributed matching* techniques. In a distributed circuit, advantage is taken of the dielectric constant. Dielectric characteristics of GaAs are used to define transmission-line passive components. Figure 4.11(a) shows distributed components of this type, including a T/R switch with two embedded FETs. There is a quarter-wavelength at X-band between the FET and the junction point. If the FET is turned on, and there is a short circuit on one side, located a quarter-wavelength away, it will look like an open circuit. The appearance is that the line with the short-circuited FET is disconnected and all circuit energy flows through the other path. The quarter-wave line at X-band represents about $\frac{1}{10}$ of an inch in actual length. At L-band, the comparable quarter-wavelength would be almost $\frac{9}{10}$ of an inch in length, given the dielectric constant of GaAs. To overcome this, at the lower frequencies, we could synthesize these quarter-wave transmission lines with inductors and capacitors.

Thin-film metal oxide capacitors and thin-film spiral inductors are made to synthesize the quarter-wave distributed lines shown in Figure 4.11(a). Figure 4.11(b) shows an exact equivalent circuit of the split-T network, having two quarter-wave equivalent lines. The lumped element quarter-wavelength is synthesized with a shunt capacitance, a series inductance, and a shunt capacitance. Thus, at L-band, the size is reduced by at least an order of magnitude. This allows a large amount of circuitry to be put on a GaAs MMIC chip at L-band.

DA1287–56

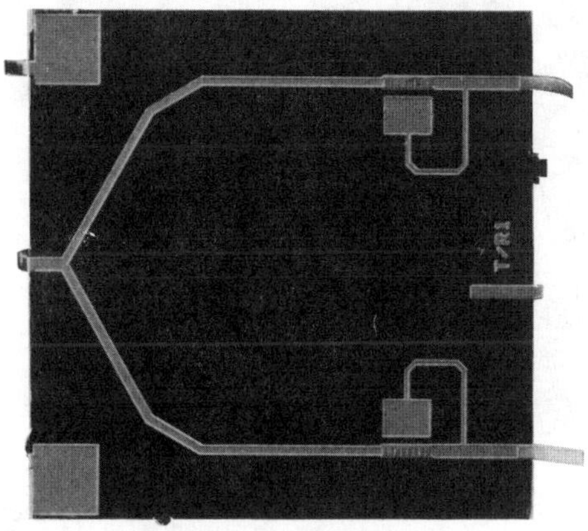

DISTRIBUTED CIRCUIT

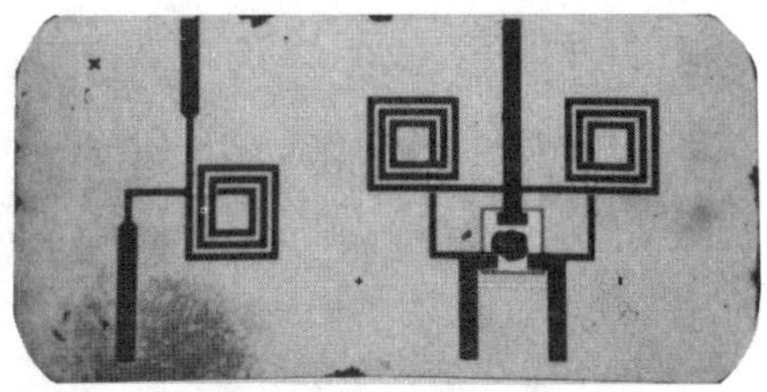

LUMPED INDUCTORS, CAPACITORS AND RESISTORS

A) DISTRIBUTED MICROSTRIP
QUARTER-WAVE LENGTHS OF LINES
AND STUBS: X-BAND 0.11 IN, L-BAND 0.86 IN.

B) LUMPED ELEMENT
INDUCTOR AND CAPACITOR SIZES:
0.02 IN x 0.02 IN.

Figure 4.11 Distributed and Lumped Circuit Elements: (a) Lumped Element (Inductor and Capacitor Sizes, 0.02″ × 0.02″); (b) Distributed Microstrip (Quarter-Wavelengths of Lines and Stubs, X-Band 0.11″, L-Band 0.86″).

4.2.1 L-Band and S-Band MMICs

Our original L-band transceiver module, shown in Figure 4.5, was $3\text{-}\frac{1}{2}$ inches on a side. The chips shown in Figure 4.12 represent about $\frac{3}{10}$ of an inch in one dimension and about $\frac{2}{10}$ of an inch in the other. There is a five-bit L-band phase shifter, an L-band $\frac{1}{2}$-W driver, a 1-W PA, and an L-band three-stage LNA.

To show some of Raytheon's earlier work at S-band, Figure 4.13 is a five-bit phase shifter that was made in the 1980s. The curve shows the phase shifter's bit states to be switched in. The rms phase shift error for this case was approximately 5°.

The 22.5° and 45° bits were designed by using loaded-line techniques, where FETs acted as active switches to change the loading on the line, which was brought about by an inductor. In the other two bits, 90° and 180°, switched-line phase shifter bits were synthesized. These used FETs as switches, and a path would be selected through either a high-pass (C-L-C) or low-pass (L-C-L) filter path.

Yield is inversely proportional to periphery. The bigger the device is made, the less chance there is of making it successfully. We should not try to put everything in one chip. Several years ago, Raytheon tried to build a 10-W, two-stage amplifier on a single GaAs chip. After about a dozen wafers with very poor yield, we re-evaluated the functionality that could be put on a single chip. Since

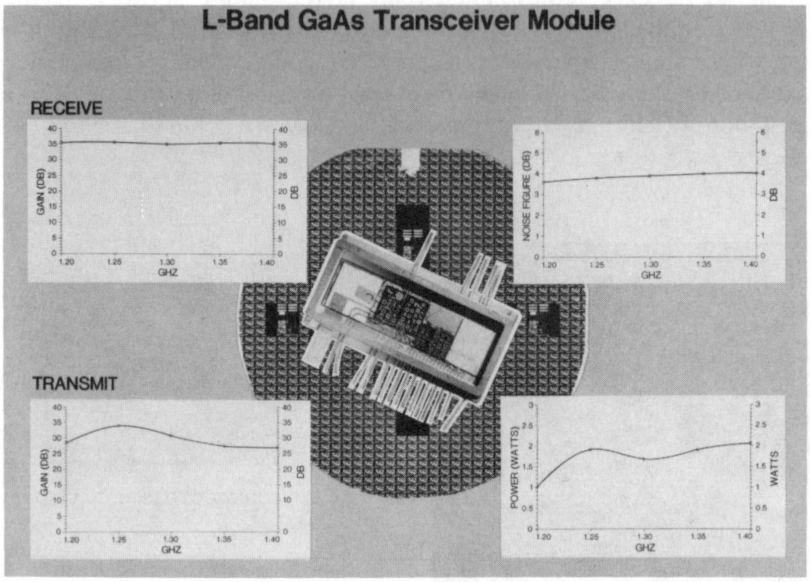

Figure 4.12 **Four-Chip L-Band GaAs Transceiver Module (Courtesy of Raytheon Co.).**

that time, chips have been broken into manageable chunks. Figure 4.14 is the latest design of a 10-W S-band amplifier. It has been broken up into two chips, each having a 12-mm output and a 6-mm driver. These two chips are then mounted on the same base carrier. A tremendous advantage in yield is gained by breaking the 10-W chip into two half-size chips. The effect on yield is probably close to a four-to-one increase.

There is a definite need for packaging different from that which had previously been done with PAVE PAWS. Multilayer ceramic packages have been built on *thermkon* bases (thermkon being a sintered metal consisting of copper and tungsten), which match the thermal expansion coefficients of GaAs. We could then put together a package, such as shown in Figure 4.15, which is an entire 10-W transceiver module in a package that is about $\frac{4}{10} \times 1$ inch. A five-bit phase shifter is included, which has a built-in variable attenuator. There is also a transmitter driver, which puts out $\frac{1}{2}$ W, a four-stage LNA, which has about a 3-dB noise figure, and two PA chips (previously shown in Figure 4.14). The complete module package shown in Figure 4.15 includes HEXFETs used to turn off the drain supply during receiving. (The HEXFETs turn off the drain supply in the transmitter amplifier chips while in the receiving mode.)

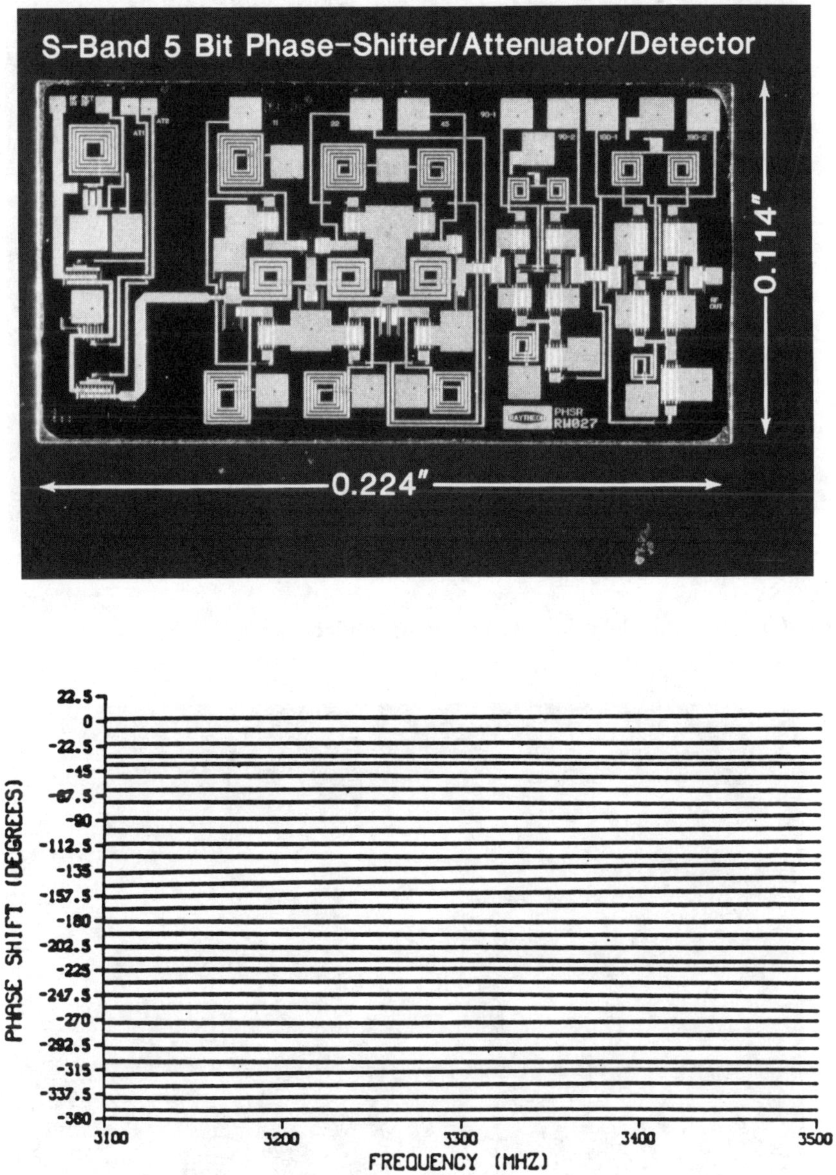

Figure 4.13 Four-Bit MMIC S-Band Phase Shifter: (a) Chip; (b) Phase Shifter Bit States.

LAIGHTON
CHAPTER 4

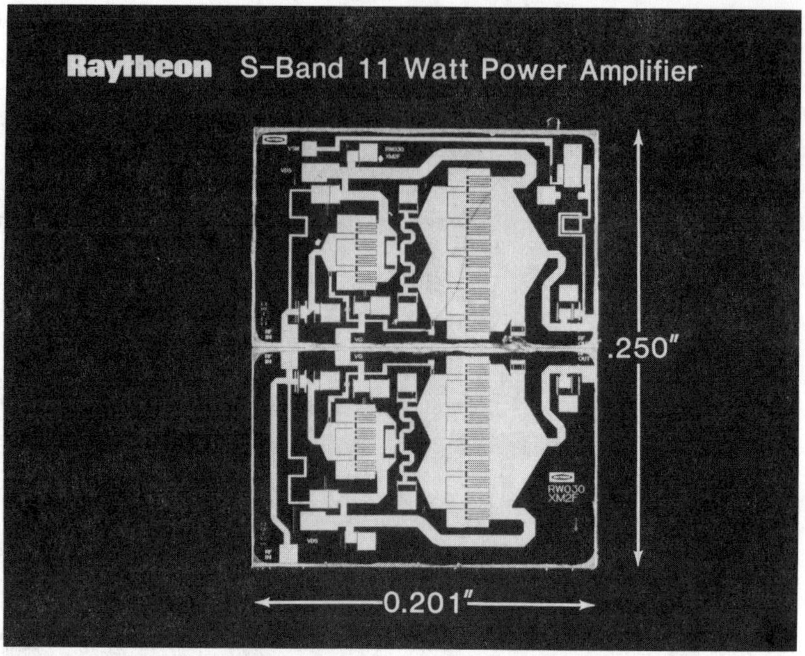

Figure 4.14 **S-Band 11-W MMIC Power Amplifier.**

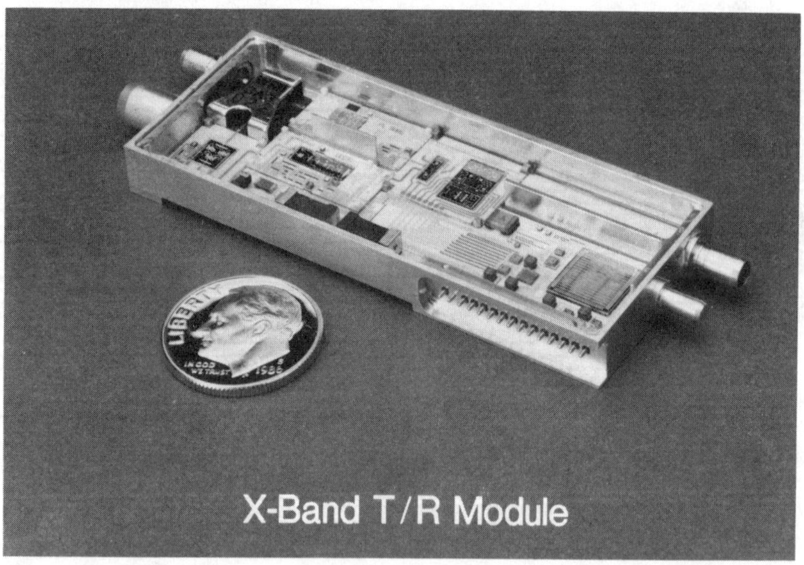

Figure 4.15 **S-Band 10-W Transceiver Module in 0.4″ × 1.0″ Package, with Four RF Chips, Two HEXFETs on the Right.**

4.2.2 X-Band MMICs

Airborne fire-control radars, space-based and ground-based imaging radars, and certain satellite communications systems can all use X-band solid-state transceiver modules in active apertures. The earliest of this class of module utilized bipolar power amplifiers at L-, S-, or C-band with varactor-diode frequency multipliers of the proper multiplication factor. This resulted in an automatic penalty in terms of module size, weight, and efficiency. Since that development, GaAs

FETs have allowed the use of direct amplification at X-band, hence greatly improving the practicality of transceiver modules at this frequency range.

The largest production run of modules built to date using discrete GaAs FETs and hybrid assembly techniques is the 2000 modules built by Texas Instruments (TI) on the Solid-State Phased Array (SSPA) program for AFWAL. This represents the largest data base on complete transceiver modules at X-band, and will provide valuable data on the use of these modules in airborne active aperture radar systems when the modules are inserted into the antenna subsystem. In addition to electrical performance and reliability data, the SSPA modules also provide the first meaningful data on assembly costs. Although the costs were initially high, toward the end of the SSPA program, they reflected a meaningful "learning curve" and started to appear more encouraging.

Ultimately, most observers in the industry believe that significant use of GaAs MMICs must be implemented to achieve costs in the range of several hundred dollars per module. Raytheon and TI are working jointly on a number of major radar transceiver module programs to introduce GaAs MMIC circuit functions into X-band transceiver modules. A photograph of a module developed jointly by Raytheon and TI, under RADC sponsorship, is shown in Figure 4.16. This module uses GaAs MMIC chips for all of the microwave functions as well as the phase shifting, low-noise amplification, power amplification, and power amplifier driver functions. The module also contains significant digital control capability, including internal solution of a common module beam steering algorithm, and each module calculates its own phase shift command, based on a common azimuth-elevation steering pattern that is fed to all modules. See Chapter 2 (Section 2.6.6.12). This module reflects the present technology in modules for airborne and imaging radar systems at X-band, and probably can be produced in quantity for significantly less of a unit cost than comparable hybrid modules.

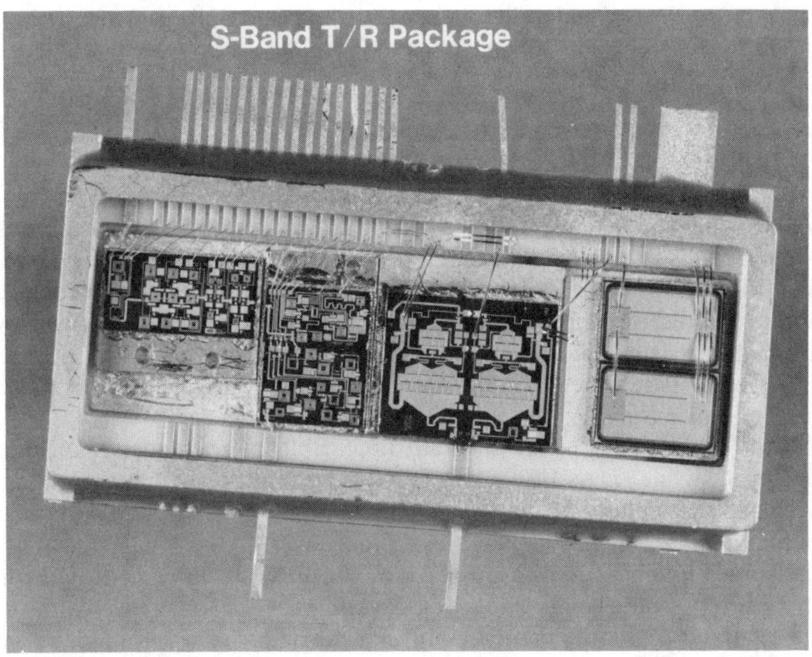

Figure 4.16 X-Band T/R Module (Courtesy of Raytheon Co.).

4.2.3 Yield *versus* Cost

At present, a large portion of the projected cost of MMIC module is in the relatively low yields of the GaAs MMIC chips. These yields are typically in the 5–10% range when we consider processing line yields, breakage, poor starting material, dc probe yields, and assembly yields into RF good circuits. These yields will improve significantly with time and production volume to the point where overall yields in the 30–40% range will become common.

At the point where MMIC chip yields result in low chip cost, the cost of a transceiver module will be dominated by the cost of other components, assembly, test, and quality assurance. The other components include circulators, housings, digital control, and power conditioning components. These components are already produced in large volume, and thus will follow a shallower learning curve with production quantities than will the MMIC chips. Assembly, test, and quality control can be most effectively reduced by the extensive use of automation [7].

At present, the industry has only begun to utilize batch assembly processes with robotic component insertion and transfer to successive work stations. At an even more primitive state is the automatic microwave testing capability to support the throughput of an automated assembly line. Virtually all present quality control inspection is done by human inspectors working with microscopes. The use of pattern-recognition equipment to assess microwave circuit quality in a meaningful way is yet to be demonstrated.

Present MMIC functional chips, such as phase shifters or amplifiers, cost several hundred to several thousand dollars apiece, and these chips dominate the cost of MMIC modules. With improved technology and increasing quantities, these costs will reduce dramatically. Every company which is now involved in producing GaAs MMIC chips can show projections that result in production costs for functional chips in the range of $10 to $50. The path to achieving such costs is quite clearly that yields must improve and wafer processing costs will reduce with increasing quantity. At present, however, if the MMIC chips were cost-free, a typical transceiver would still cost several thousand dollars. Thus, it is in the non-MMIC areas to which industry must look for achieving meaningful reduction in production costs.

4.3 SOLID-STATE DEVICE CAPABILITIES

A key question often asked is, "What power level is rational to consider across the frequency spectrum of interest to radar designers?" An answer to this question (this writer's opinion, at least) is shown in Figure 4.17. Since it is quite common to parallel several solid-state devices to achieve desired power levels, the information is presented on the basis of a single power device. There are currently two generic types of solid-state devices with practical application: the silicon (Si) bipolar transistor and the GaAs FET.

Silicon bipolar devices are generally applicable in the frequency region from HF to S-band, while GaAs FETs may be used at frequencies into the millimeter-wave region. Particularly at the frequencies through L-band, silicon bipolar devices are available, having capabilities in the range of 100 W and greater. The reader should note that the assumption of the bipolar curve is that the pulse conditions are typically solid-state radar driven, involving long pulsewidths (>100 μs) and relatively high duty factors (>10%). Transistors of several hundred watts in capability have been reported for specific applications such as identification friend or foe (IFF), but these generally involve short pulsewidths and low duty factors. The limitation on bipolar devices is generally the peak junction temperature that can be maintained in a system, with its resultant effect on reliability.

GaAs FET devices, conversely, are primarily limited by the capability to make a device with acceptable yield, and hence acceptable cost. The GaAs FET is generally designed with gate-to-gate spacing such that a reasonable junction temperature can be maintained; therefore, the power is limited by the ability to place enough gates in parallel to achieve a desired power level. Total chip size, consistent with semiconductor yield, is therefore the major limitation on device power. At present, chip sizes on the order of $\frac{1}{4} \times \frac{1}{4}$ inch are reasonable for MMIC PAs, and result in the types of yields mentioned above. The curve shown in Figure 4.17 represents a perception of the total output power per MMIC power amplifier device at the present time. As processing technology improves and larger wafers come into widespread use, the power per device will inevitably rise.

Power combining of individual devices of either type can be used to achieve any practical power level; however, combining *losses* and device *gain* are what bound the word "practical." A rule of thumb for power combining is to use 0.3-dB loss per binary combining step as a first estimate. This loss includes both the *resistive loss* of the combiner step and a *similarity loss* for the vector addition of two amplifier outputs. Another good rule of thumb is that an overall paired-amplifier stage must have greater than 6.0 dB of gain to achieve an efficient one-driving-four architecture. This overall gain must be inclusive of splitting and combining losses. Of course, gain decreases with increasing frequency in solid-state devices.

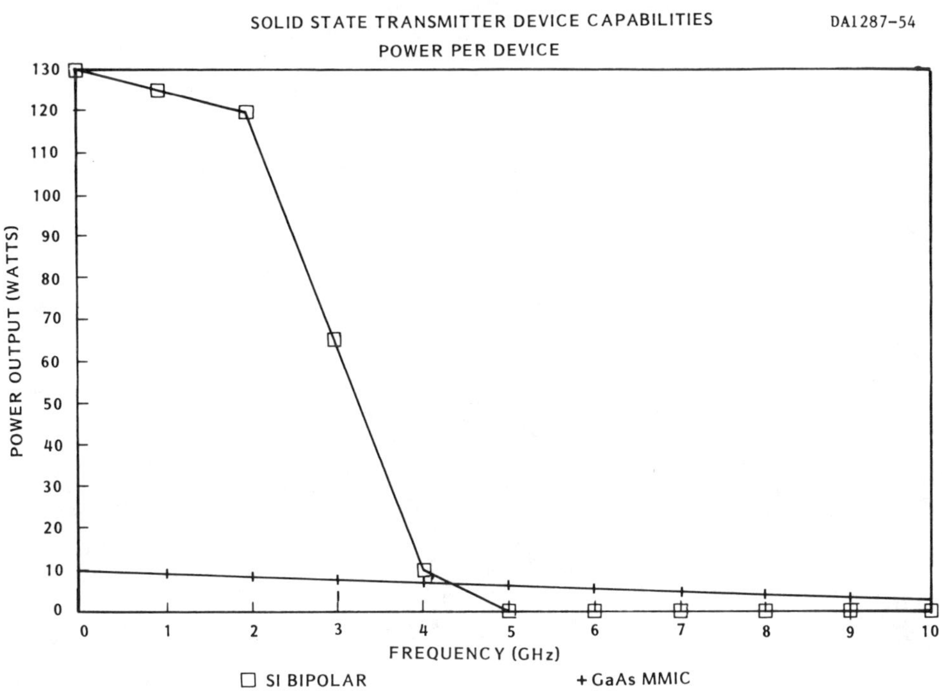

Figure 4.17 Solid-State Transmitter Device Capabilities.

LAIGHTON
CHAPTER 4

REFERENCES

1. Ostroff, E.O., *et al., Solid-State Radar Transmitters,* Artech House, Norwood, MA, 1985.
2. Soares, R., *et al., GaAs MESFET Circuit Design,* Artech House, Norwood, MA, 1988.
3. Hoffmann, R.K., *Handbook of Microwave Integrated Circuits,* Artech House, Norwood, MA, 1987.
4. Frey, J., and K. Bhasin (eds.), *Microwave Integrated Circuits,* 2nd ed., Artech House, Norwood, MA, 1985.
5. Pucel, R.A., *Monolithic Microwave Integrated Circuits,* IEEE Press, New York, 1985.
6. Ferry, D.K. (ed.), *Gallium Arsenide Technology,* Howard W. Sams, Indianapolis, 1985.
7. Rubin, O., *The Design of Automatic Control Systems,* Artech House, Norwood, MA, 1986.

Chapter 5
MONOPULSE PRINCIPLES AND TECHNIQUES

by Samuel M. Sherman

This chapter is a brief overview of *monopulse,* a technique of wide practical importance as well as theoretical interest. In order to cover so broad a subject in a limited space, emphasis is placed on fundamental principles and basic designs, with many details necessarily omitted. A more comprehensive treatment of the subject, including theory and mathematics, practical considerations, more detailed design descriptions, additional topics, and extensive references, can be found in the book, *Monopulse Principles and Techniques,* by Samuel M. Sherman (Artech House, 1984).

5.1 INTRODUCTION

Monopulse, also known as simultaneous lobe comparison (or simultaneous lobing), is a technique for determining the angular location of a radar target or a radiating source. A more specific definition will be given later. Monopulse is used in certain types of radars, and also has non-radar uses. As compared with other techniques for determining angular location, monopulse is capable of greater accuracy and also has other advantages.

5.1.1 General Principles

Radars are usually classified as search radars or tracking radars. A search (or surveillance) radar searches a designated volume of space in order to detect targets that may be present. A tracking radar automatically keeps its beam axis pointed toward a selected target. Search and track are sometimes combined as selectable or time-shared modes in a single multifunction radar. Monopulse is most commonly found in tracking radars (or in the tracking mode of multifunction radars), but can also be used for special search applications. It can be used in a single angle coordinate or in both coordinates.

Complete three-dimensional tracking of a target requires range as well as two angles. Monopulse is not involved in range tracking, but does require an input from the range tracker in order to track a target efficiently in angle.

The monopulse processor output indicates how far, and in which direction, the target is off the beam axis. This output can be used open-loop, in combination with the known direction of the beam axis, to determine target angle, or the output can be fed back in a closed loop to control the direction of the beam so as to keep the axis pointed as closely as possible toward the target. Open-loop operation

---

Editor's Note: Most of the content of this chapter is extracted from *Monopulse Principles and Techniques,* by Samuel M. Sherman, published by Artech House (Norwood, MA, 1984), with some revisions and rearrangement to make the chapter self-contained.

requires a calibration function, which is essentially linear for small angles, but may depart significantly from linearity for larger angles. To improve accuracy, open-loop and closed-loop operation are employed simultaneously in some radars, the open-loop output being used as a fine correction to the closed-loop angle indication.

Radars, in general, periodically transmit radio-frequency pulses or other waveforms and receive echoes produced by objects (targets) in the beam. If the radar includes monopulse, the monopulse operates only on reception. Hence the monopulse can also be used in a passive mode, in which the radar does not transmit, but only "listens".

Monopulse does not necessarily operate on "one pulse", as the name appears to imply. The name was originally coined because monopulse has the inherent *capability* to produce a two-coordinate angle measurement from each pulse, and in some radars this capability is indeed utilized. For practical reasons, however, most monopulse radars use integration or averaging of multiple pulses to obtain each angle measurement. As a matter of fact, monopulse does not require the use of pulses at all. It can operate with other waveforms, including CW, modulated CW, and noise-like waveforms. For brevity, the word "pulse" will be used in this chapter to describe the waveform, but it will be understood to include other waveforms unless a specific distinction is being made.

Applications of monopulse are not limited to radar, or even to electromagnetic radiation; they include passive direction-finding, communications, missile guidance, radio astronomy, and sonar. This chapter, however, is concerned with radar applications, in which monopulse has reached a particularly high state of development.

5.1.2 Review of Sequential Lobing

To explain how monopulse works, we begin with another angle-tracking technique, known as *sequential lobing* or *sequential lobe comparison,* which was an earlier development (but is still in use). It has two subcategories: *lobe switching* and *conical scan*. In lobe switching, illustrated in Figure 5.1-1 for one angle coordinate, let us say elevation, the beam does not point directly at the target, but alternates on successive pulses between two directions, indicated by the numbers 1 and 2 in the diagram, which are offset on opposite sides of the axis of symmetry by an angle on the order of one-half beamwidth. The switching of the beam direction is usually done electronically. If the target is on the crossover axis, illustrated by the solid line, equal echo amplitudes are received in beams 1 and 2. If the target is above the crossover axis, as indicated by the dashed line, the echo from beam 1 is greater than from beam 2. The voltage difference $v_1 - v_2$ in this case is positive, which means that to center the crossover axis on the target the antenna must point higher. The voltage difference is used in a servo loop to position the antenna automatically.

Lobe switching in both angular coordinates, which permits complete angle tracking, requires four successive beam positions, as shown in cross section in Figure 5.1-2(a). The echo amplitudes in the upper pair of beams are compared with those in the lower pair to obtain angle information in one coordinate, and a corresponding comparison between the right pair and the left pair provides angle information in the other coordinate.

Instead of changing the beam direction electronically, as in lobe switching, the same effect can be obtained by mechanical rotation or nutation of the antenna feed around the crossover axis. In that case, instead of moving the beam in four discrete steps, it is preferable to move it continuously in a circular path centered around the crossover axis, resulting in conical scan, which, as mentioned above, is the other subcategory of sequential lobing. Although the beam motion in conical scan is continuous, the target is illuminated only when it is reached by each

SHERMAN
CHAPTER 5

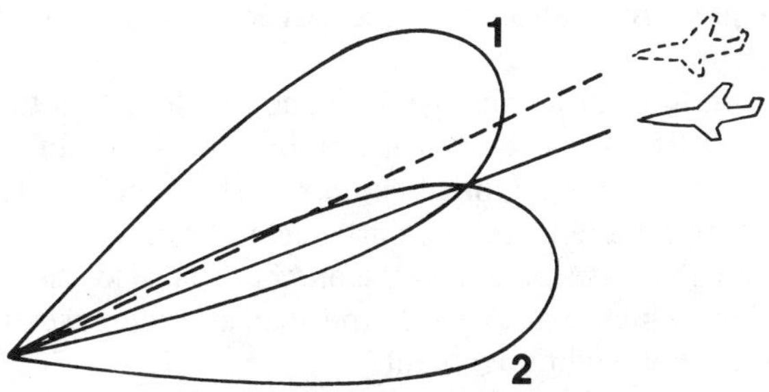

Figure 5.1-1 Sequential Lobing in Elevation.

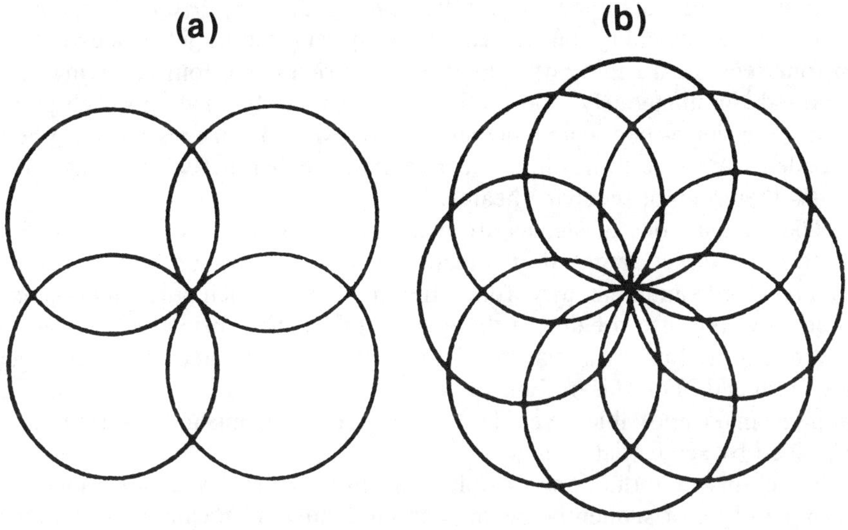

Figure 5.1-2 Two Types of Sequential Lobing:
(a) Lobe Switching in Traverse and Elevation;
(b) Conical Scan.

transmitted pulse, as illustrated in Figure 5.1-2(b). When the target is anywhere but on the crossover axis, the echo strength is modulated at the scanning frequency, being greatest when the beam axis is closest to the target direction. Demodulation of the received signal yields the modulation envelope, which, when normalized and compared with a reference sinusoid at the scanning frequency in a pair of phase-sensitive detectors, yields two outputs indicating the two orthogonal components of deviation of the crossover axis from the target. These outputs are used as inputs to the tracking servos for the respective angular coordinates.

Both lobe switching and conical scan have given successful service since World War II and are still used. However, they have certain limitations. A principal source of error in sequential lobing—both lobe switching and conical scan—is the fluctuation that occurs in the echo strength of most targets. This fluctuation is often rapid enough to cause significant variation in echo strength from pulse to pulse. A sequential-lobing radar cannot distinguish between pulse-to-pulse variations in echo amplitude due to target fluctuations and those due to target displacement from the antenna crossover axis; therefore the fluctuations cause tracking errors.

Another disadvantage of sequential lobing is the limitation on the data rate imposed by the necessity of obtaining at least four successive echo pulses for each two-coordinate angle estimate. This can be a serious limitation in tracking targets that have large angular acceleration.

In the case of conical scan there is also the disadvantage of mechanical rotation, which causes wear and tear and increases the difficulty of maintaining accurate boresight alignment.

5.1.3 Evolution of Monopulse

The effort to overcome these problems led to the invention and development of monopulse. The fundamental distinction between monopulse and sequential lobing is this: In monopulse, multiple receiving beams are formed *simultaneously* rather than sequentially. The name *simultaneous lobing* (or *simultaneous lobe comparison*) expresses this distinction better than the synonymous name *monopulse,* but the latter name is most commonly used, at least in radar applications.

One category of monopulse, called *amplitude-comparison* monopulse, is similar to lobe switching, but instead of comparing the target echoes obtained from four sequential beam positions (up-down, right-left), four receiving beams are formed simultaneously, and the comparisons can be made on each pulse if desired. (For single-coordinate monopulse, only two simultaneous receiving beams are needed.) A single transmitting beam suffices, being just wide enough to encompass the group of receiving beams.

The comparisons of signals are made in the form of voltage *ratios*. Since pulse-to-pulse target amplitude fluctuations have no effect on voltage ratios of signals received simultaneously from different lobes on each pulse, monopulse is theoretically free of error due to that cause. In practice, because of design and cost constraints, this error may not be completely eliminated, but it is at least greatly reduced. For this reason, and also because of other refinements that monopulse makes possible, overall angle accuracy is significantly better than can be obtained by sequential lobing.

Since angle information is available from every received pulse instead of requiring a cycle of sequential beam positions, monopulse can provide a higher data rate than sequential lobing.

The advantages of monopulse over the other angle-tracking techniques are obtained at the price of somewhat greater equipment complexity and cost. For some applications, the simpler systems are quite adequate and monopulse is not needed.

Most tracking radars in use today are of the monopulse type. Throughout the world, hundreds of monopulse radars have been built and installed on land, at sea, on aircraft, and in space. Among the large variety of functions that they perform are fire control, missile launching and guidance, strategic military applications, space applications, intelligence, and instrumentation for monitoring of various tests and exercises.

5.1.4 An Illustrative Monopulse Radar

Before entering into the more detailed aspects of monopulse, we will show a simplified example of one form of monopulse in order to illustrate the basic principles. This particular type of monopulse belongs in the general class known as *amplitude-comparison.* Another class, *phase-comparison,* will be examined later, as will also other configurations of amplitude-comparison monopulse.

The antenna of this illustrative radar consists of a mechanically steered paraboloidal reflector fed by a cluster of four feed horns in the focal plane, symmetrically offset about the axis, as shown in Figure 5.1-3. In the side view, part (a), two of the horns are visible. When viewed axially from the center of the reflector, the horns appear as in part (b) of the figure. (The horn apertures are customarily represented diagrammatically as squares, but in practice they can be either squares or rectangles, depending on polarization requirements.) Echo signals received from a target by the four beams are all in the same phase, but are unequal in amplitude unless the target is on the common crossover axis, which is also the axis of the parabolic reflector (the "dish", as it is often called). From the ratios of the amplitudes, the two angular components of the target direction can be determined.

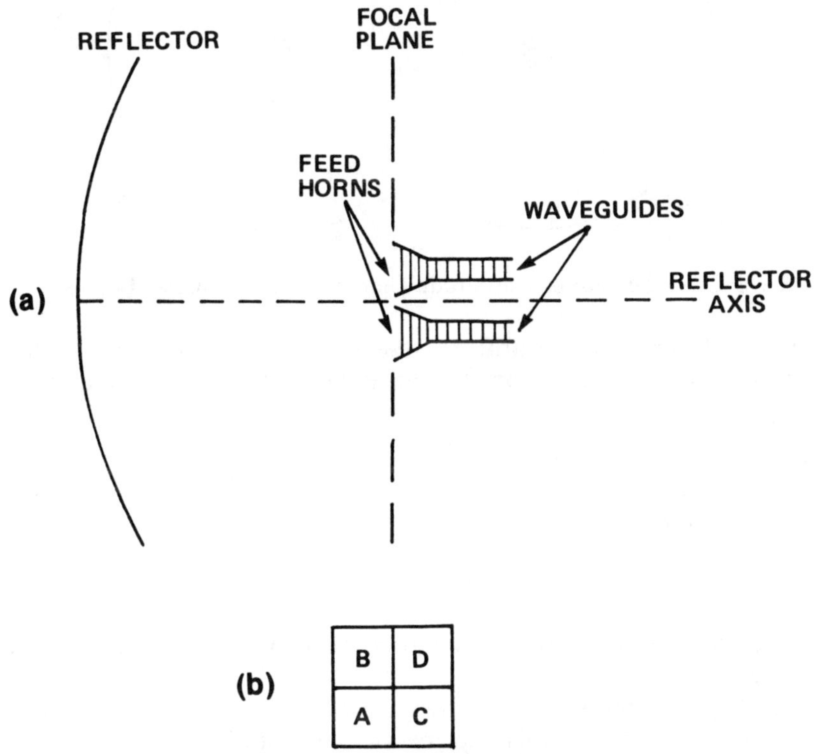

Figure 5.1-3 Reflector and Feed Horns:
(a) Side View;
(b) Axial View of Feed Horns from Reflector.

The four beams are shown in cross section in Figure 5.1-4. In theory, the outputs of the four horns could be connected to four identical receivers, the outputs of which could then be compared (that is, their ratios could be taken). However, four receivers, even if supposedly identical, will vary unequally as a function of time, signal level, radio frequency, and environmental conditions, causing errors in the measurements.

What is usually done, therefore, is to form the sum, the elevation difference, and the *traverse* difference. Traverse angle differs from azimuth angle: Traverse is measured in the slant plane containing the line of sight to the target, while azimuth is measured in the horizontal plane (or in the deck plane if the radar is on a ship or aircraft). The mathematical relation between traverse and azimuth will be presented later. Elevation and traverse are the "natural" angular coordinates for the monopulse antenna. The sum and the two differences of the horn outputs are formed at RF, ahead of the receivers, by RF combining devices known as hybrid junctions. Being electrically passive and mechanically compact and rigid,

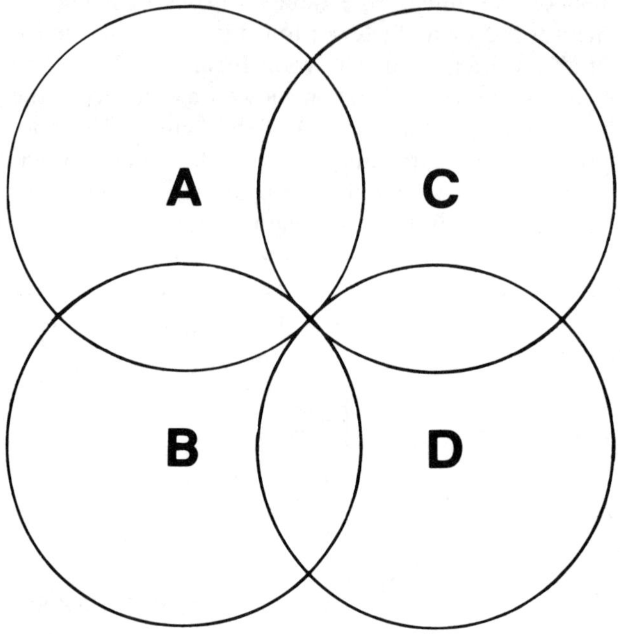

Figure 5.1-4 Cross Section of Amplitude-Comparison Monopulse Beams.

these devices have much less drift than the active circuits in receivers; therefore the null axis (the "boresight" axis) is more stable. The sum and the differences are

$$\text{Sum:} \quad s = \tfrac{1}{2}(A + B + C + D), \tag{5.1-1}$$

$$\text{Traverse difference:} \quad d_{tr} = \tfrac{1}{2}[(C + D) - (A + B)], \tag{5.1-2}$$

$$\text{Elevation difference:} \quad d_{el} = \tfrac{1}{2}[(A + C) - (B + D)]. \tag{5.1-3}$$

The factor $\frac{1}{2}$ in these equations is due to equality of total input power and total output power, assuming no losses.

The shapes of the sum and difference patterns (voltage *versus* angle) are shown in Figure 5.1-5 for either traverse or elevation. Because the sum pattern is a single beam centered on the monopulse axis, it has maximum gain in that direction, both in transmission and reception; this characteristic is desired for acquisition (detection) of targets and for tracking of targets on or near the axis. The sum signal is also used for ranging and to obtain the ratio of the difference to the sum, as will be explained. The beamwidth of a monopulse antenna is conventionally defined as the one-way half-power beamwidth of the sum pattern, unless otherwise specified.

The difference pattern has the desirable characteristic of being zero when the target is on axis and approximately proportional to angular displacement from the axis (with appropriate sign), for small angles. This makes it suitable as an input to a servo or for open-loop measurements. For reasons that will be explained, however, it must be used in the form of the difference-to-sum ratio, d/s, rather than d alone. The ratio d/s is plotted in Figure 5.1-5.

Also shown in the figure are the voltage patterns v_1 and v_2 of the pairs of squinted beams from which the sum and each difference are formed. Usually there is no access point in the RF hardware at which the v_1 and v_2 patterns can be measured, but these patterns can be deduced from the measured sum and difference patterns by inversion of equations (5.1-1) and either (5.1-2) or (5.1-3), keeping in mind that v_1 means either $C + D$ or $A + C$, and v_2 means either $A + B$ or

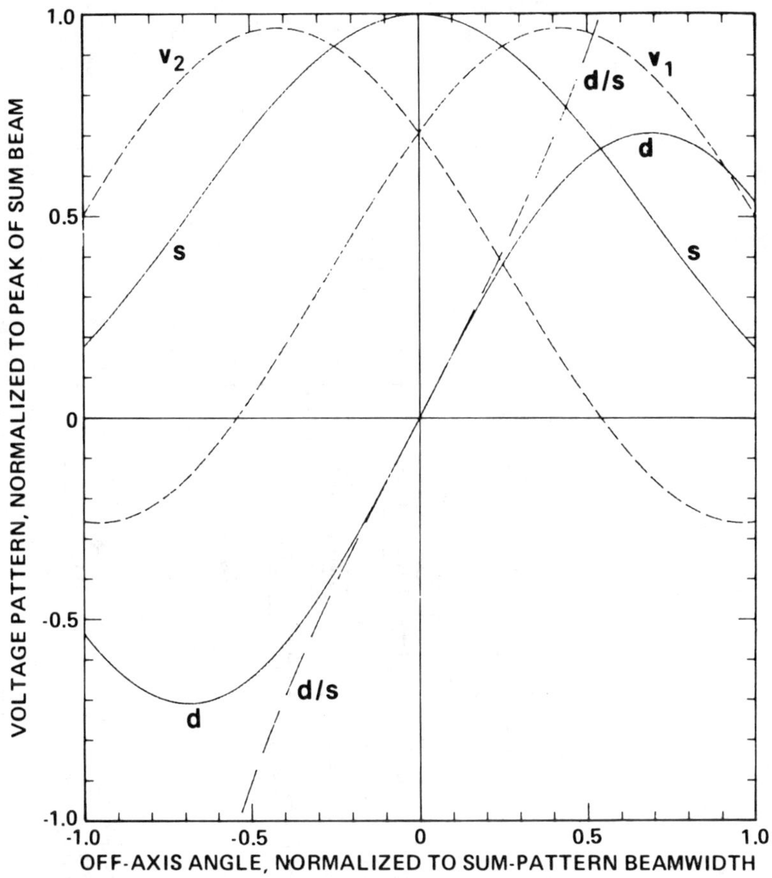

Figure 5.1-5 Amplitude-Comparison Monopulse Patterns in either Coordinate: Sum(s), Difference (d), Squinted Beams (v_1 and v_2), Normalized Difference (d/s).

$B + D$, depending on which angular coordinate is being considered.

Figure 5.1-6 is a simplified functional diagram of the illustrative monopulse radar. The duplexer permits two-way operation, so that in the microwave combining network the same channel that forms the sum pattern for reception is also used in the reverse direction for transmission. Thus the transmitting pattern is the same as the sum pattern in Figure 5.1-5. In reception the sum signal is used for detection, ranging, and displays, in addition to the monopulse function. The receivers convert the RF voltages to IF, then amplify and filter the IF voltages.

The monopulse processor can take any one of several forms. A common feature of all monopulse processors is that they respond to voltage *ratios* or phase *differences* or both, rather than absolute values. The inputs to each processor are the sum signal and the difference signal for one coordinate. The desired output is the ratio of the difference voltage amplitude to the sum voltage amplitude, multiplied by the cosine of the relative phase angle between the two voltages, denoted by δ_{tr} in traverse and δ_{el} in elevation.

Since the illustrative system is a mechanically steered tracking radar, the monopulse processor outputs (after coordinate transformation) become inputs to the servo-amplifier channels, and the servos turn the antenna until a null is reached. Coordinate transformation is needed because the mechanical rotation coordinates of the antenna mount are usually azimuth and elevation, while the monopulse coordinates are traverse and elevation relative to the antenna axis.

While the servos attempt to keep the tracking axis on the target, they may lag because of high angular rates or accelerations, or may deviate because of wind gusts or other disturbances. The resulting error can be largely removed by an

SHERMAN
CHAPTER 5

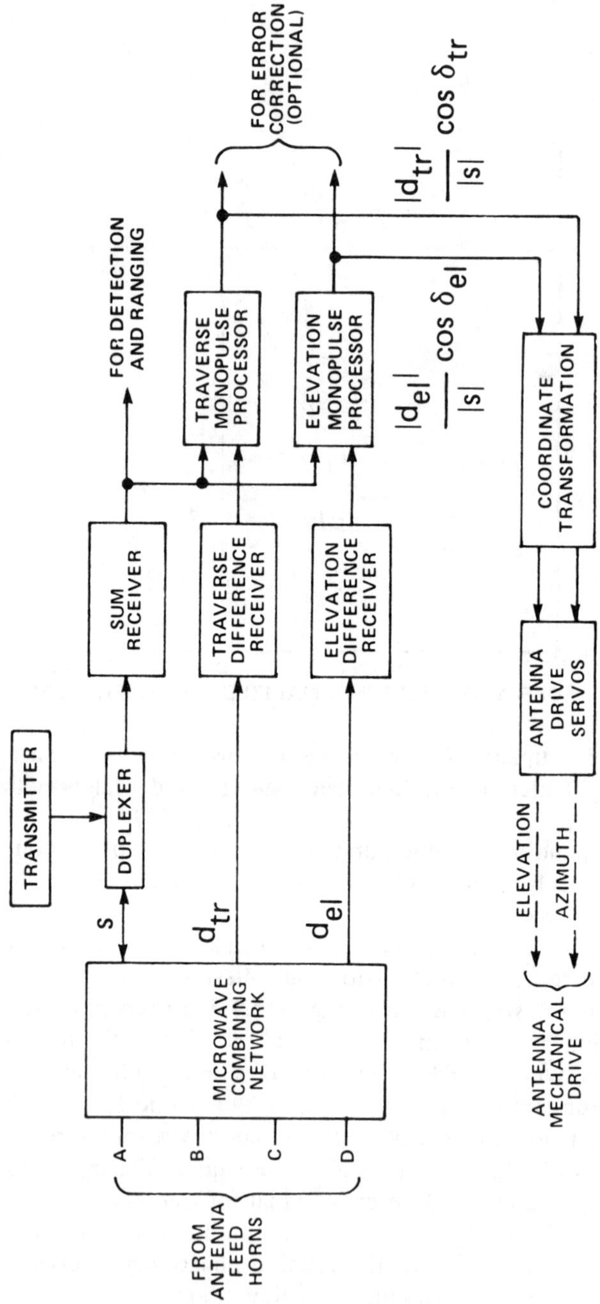

Figure 5.1-6 Functional Diagram of One Form of Monopulse Radar.

optional feature called the *error-correction* signal, indicated as a pair of auxiliary outputs at the right side of Figure 5.1-6. The residual monopulse output which the servo is unable to null out is measured and converted to the corresponding off-axis angle by a known calibration function. This angle, added to the pedestal shaft rotation angle, gives a fine correction to the value of the target angle.

A monopulse radar must be boresighted when installed and at prescribed intervals thereafter. Boresighting consists essentially of pointing the antenna so as to obtain a null on a test target at a known direction, then adjusting the angle readouts to the correct values in that direction.

5.1.5 Summary of Monopulse Advantages and Disadvantages

Some of the advantages of monopulse over lobe switching and conical scan have already been mentioned. Those advantages plus others are summarized below:

(a) Errors due to target amplitude fluctuations are eliminated or greatly reduced.

(b) The higher data rate permits the use of a wider bandwidth when needed to reduce lag error or loss of track on an accelerating target.

(c) The single-pulse capability, although not generally utilized, is highly important in some surveillance radars and in multifunction radars with electronically steered antennas, which perform interleaved tracks of several targets by switching the beam direction rapidly from one to another. Each dwell on each target may be so short as to allow time for only a few pulses or even a single pulse.

(d) Since the sum beam, both in transmission and reception, is pointed directly at the target rather than to one side or the other, there is no reduction in signal-to-noise ratio.

(e) Monopulse is free of the mechanical vibration and the wear and tear of conical scan.

(f) Since monopulse transmission is unmodulated, as contrasted to the modulation inherent in the other techniques, it is less susceptible to electronic intelligence and countermeasures.

(g) In certain applications, information about the nature of a target can be derived from analysis of pulse-by-pulse amplitude fluctuations of the echo. This is difficult or impossible in sequential lobing because of the modulation imposed by the radar itself. In a monopulse radar, the received sum signal can be used for this purpose because it is unmodulated.

The principal disadvantage of monopulse is that in order to realize its inherently superior capability, it needs more equipment and more painstaking design, adjustment, and calibration than does sequential lobing. In particular, monopulse requires a more complicated feed system and RF circuitry. Furthermore, whereas sequential lobing needs only one receiver channel, two-coordinate monopulse requires three well-matched channels, although the number can be reduced to two if some sacrifice of data rate is acceptable. In addition, other components, such as the pedestal and the servos, must meet more demanding specifications in order to support the accuracy available from monopulse. As a result, monopulse tends to be more expensive than other angle-tracking techniques.

5.2 TERMINOLOGY AND DEFINITIONS

Certain terms commonly used in monopulse need to be defined or explained, because, even if they are familiar in general usage, they may have specialized meanings when applied to monopulse, and these meanings may differ from the literal interpretations. Three important terms, including the word "monopulse" itself, are discussed below. Others will be explained as they occur in the remainder of the chapter.

5.2.1 Definition of Monopulse

As we explained in a previous section, the term *monopulse* does not necessarily imply "one pulse". That name was coined because monopulse does have the capability to extract two-coordinate angle information from one pulse, but monopulse is not necessarily used in that manner, and in fact it is not limited to pulsed waveforms. As mentioned above, a more descriptive name is simultaneous lobing or simultaneous lobe comparison, since the *simultaneity,* rather than the "one pulse", characterizes monopulse.

The following definition, taken from IEEE Standard 686-1982, *Radar Definitions,* published by the Institute of Electrical and Electronics Engineers, is presented here with minor changes for tutorial purposes:

> A technique in which information concerning the angular location of a source or target is obtained by comparison of signals received simultaneously in two or more antenna patterns, as distinguished from techniques such as lobe switching or conical scanning, in which angle information is derived from a series of sequential antenna patterns. The simultaneity of the patterns makes it possible to obtain a two-coordinate angle estimate from a single pulse, although multiple pulses are usually employed to improve the accuracy of the estimate, provide Doppler resolution, or simplify the processing. The monopulse principle can be used in continuous-wave as well as pulsed radar, and also in non-radar applications.

5.2.2 Sum and Difference Patterns

In the type of system described earlier, sum and difference voltages are obtained by addition and subtraction of the outputs of two feed horns (or two pairs of feed horns). The sum and difference patterns have even symmetry and odd symmetry respectively about the axis, as illustrated in Figure 5.1-5. The important characteristic of these patterns is their shape, not the method by which they are obtained. There are other ways of generating patterns having these same forms of symmetry and similar shapes, not necessarily involving addition and subtraction. Such patterns are commonly referred to as "sum" and "difference" even when these names are not literally correct, because they have the same shapes and serve the same functions as true sum and difference patterns. Alternative terms are "reference" and "error" patterns (or voltages). "Reference" is preferable to "sum" for generality, particularly when addition of voltages is not involved. The word "error", however, is subject to possible ambiguity, because the presence of an "error" voltage does not necessarily imply an error in measurement, and *vice versa.* Whichever terms are used, they should be clearly defined to avoid the possibility of misunderstanding.

5.2.3 Traverse and Azimuth

Traverse angle was introduced in Section 5.1.4 with the statement that it is measured in the slant plane containing the target, while azimuth is measured in the horizontal plane. (On a movable platform such as a ship, the deck plane rather than the horizontal plane is taken as the reference.) The relationship will now be made more explicit.

Figure 5.2-1 illustrates a common type of antenna mount for a tracking radar; it is known as an azimuth-elevation (''az-el'') mount, because the axes are designed for rotation in those two coordinates. The monopulse output, however, responds to displacement of the target from the antenna boresight axis in traverse and elevation, not azimuth and elevation. Let A and α' be the azimuth and traverse displacements respectively of the target from the boresight axis, and let E be the elevation angle of the target measured upward from the horizontal plane.* These angles are illustrated in Figure 5.2-2, from which the following relation can be derived:

$$\sin A = \sin \alpha' \sec E, \tag{5.2-1}$$

which for small A is approximated by

$$A \simeq \alpha' \sec E. \tag{5.2-2}$$

Equation (5.2-2) is the basis of the ''secant correction'' in mechanically steered tracking radars. The monopulse traverse output is approximately proportional to α', but the servo requires an input approximately proportional to A. The conversion is accomplished by generating a multiplying factor proportional to the secant of the elevation angle. The approximation suffices at elevation angles up to about 85°. The secant correction can be done by a secant-wound potentiometer or by computer.

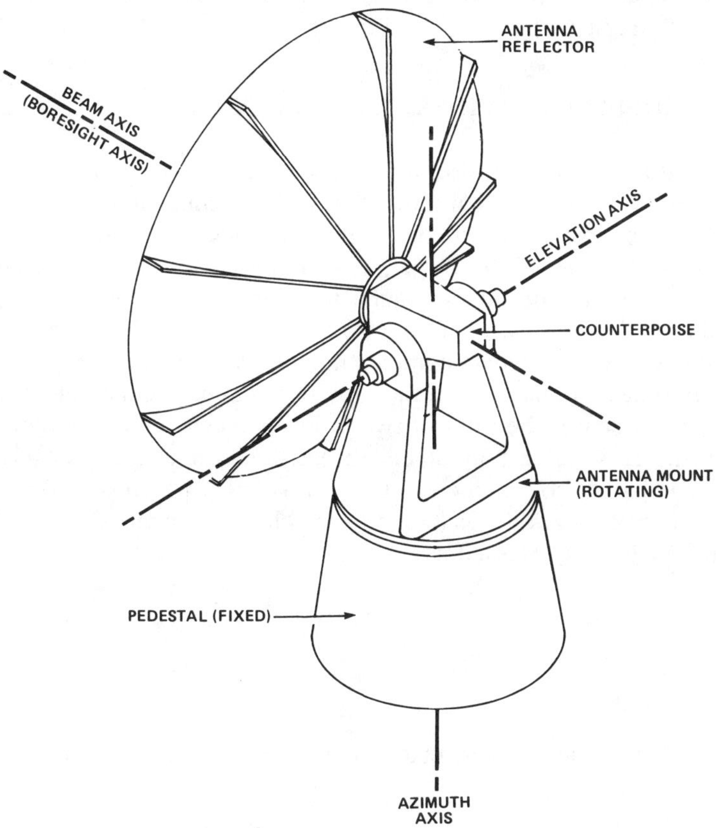

Figure 5.2-1 Axes of Mechanically Steerable Antenna.

*The ''elevation'' angle to which the monopulse output responds is not exactly the same as E except when the traverse displacement is zero, but the difference is usually very small.

SHERMAN
CHAPTER 5

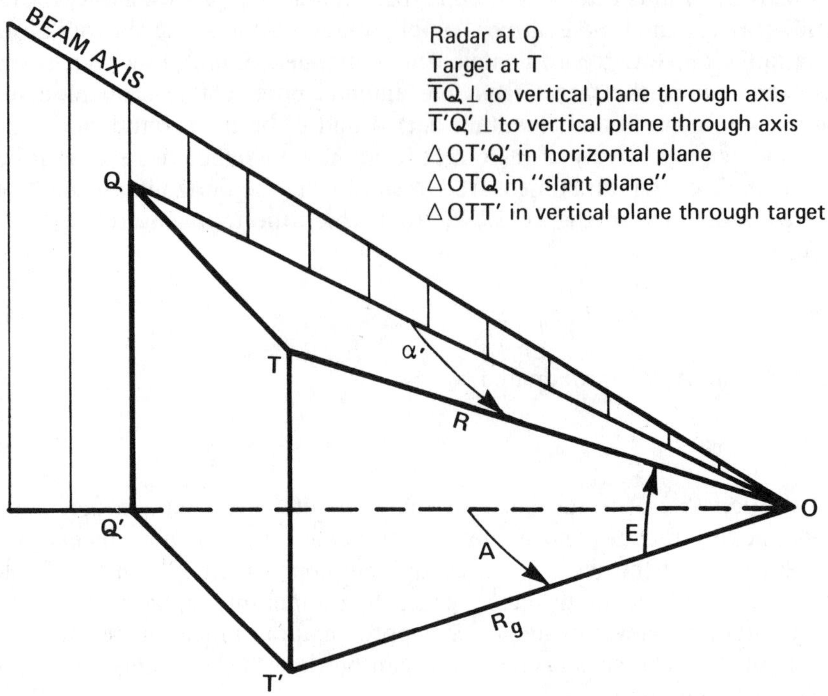

Figure 5.2-2 **Geometric Relations among Angles: Azimuth (*A*), Elevation (*E*), and Traverse (α').**

5.3 THE MONOPULSE OUTPUT AS A COMPLEX RATIO

As stated above, monopulse processing involves voltage amplitude ratios, phase differences, or both, rather than values of individual amplitudes or phases. If the angle indication depended on individual values, it would vary with the nature, size, and range of the target, but voltage amplitude ratios and phase differences depend only on target angular location.

Any alternating voltage, current, or electromagnetic field component can be represented mathematically by a complex number, which can be expressed in the form of magnitude and phase, or in the form of a real part and an imaginary part.

In a general sense, the ratio of two voltages means their *complex ratio*. In monopulse, the two voltages involved, for each coordinate, are usually the difference voltage in that coordinate and the sum voltage, which is common to both coordinates. The phasor voltages (complex envelopes) of a monopulse difference, d, and sum, s, can be expressed as

$$d = |d| \exp(j\delta_d), \tag{5.3-1}$$

$$s = |s| \exp(j\delta_s), \tag{5.3-2}$$

where δ_d and δ_s are phase angles relative to any arbitrary common reference. The complex ratio is

$$\frac{d}{s} = \frac{|d|}{|s|} \exp[j(\delta_d - \delta_s)] = \frac{|d|}{|s|} \exp(j\delta), \tag{5.3-3}$$

where δ is the relative phase:

$$\delta = \delta_d - \delta_s. \tag{5.3-4}$$

Thus the complex ratio has a magnitude equal to the ratio of the magnitudes of the two voltages and a phase equal to their relative phase.

In the type of monopulse configuration illustrated in Section 5.1.4 (amplitude-comparison monopulse), the relative phase δ of d and s from a single point target, under idealized conditions, is either 0° or 180°, depending on the sense of the angular displacement. Under these idealized conditions the ratio d/s is pure real. In practice, however, d and s can have *any* relative phase, because of noise, interference, unresolved targets, and equipment imperfections. Therefore d/s must be regarded as a complex quantity.

Using a well-known mathematical equivalence, we can expand equation (5.3-3) as follows:

$$\frac{d}{s} = \frac{|d|}{|s|} e^{j\delta} = \frac{|d|}{|s|} \cos \delta + j \frac{|d|}{|s|} \sin \delta. \tag{5.3-5}$$

Thus the real and imaginary parts of d/s are

$$\mathrm{Re}(d/s) = \frac{|d|}{|s|} \cos \delta \tag{5.3-6}$$

and

$$\mathrm{Im}(d/s) = \frac{|d|}{|s|} \sin \delta. \tag{5.3-7}$$

Referring to Section 5.1-4 and Figure 5.1-6, we see that the output of the monopulse processor in the radar model described there is only the real part (in-phase component) of the complex ratio d/s. Circuits that perform this operation will be described in a later section. The imaginary part can also be extracted if desired, but normally this is not done, because the target ideally contributes only to the real part, while the imaginary part is due to noise and other disturbances. For targets on or near the axis, the cosine factor prevents the quadrature component of noise (half of the total noise power) from contributing to the error in angular measurement, and thus reduces the variance of the error by a factor of two. The cosine factor also provides sense (up or down, left or right).

Under idealized conditions, when the relative phase of d and s is either 0° or 180°, the real part represented by equation (5.3-6) is the same as d/s itself; as a function of angle, it has the form shown by the d/s curve in Figure 5.1-5.

In some types of monopulse radars, the relative phase of d and s under idealized conditions is ±90° (quadrature phase) instead of 0° or 180°. In that case the ratio d/s is pure imaginary. The quadrature-phase relation can be converted to the in-phase relation by phase-shifting either s or d by 90°. Alternatively, the processing can easily be modified by replacing the real part of the ratio (cosine factor) by the imaginary part (sine factor) to produce the output shown in equation (5.3-7) instead of (5.3-6).

In the equations and discussion above, and throughout this chapter, voltage d and phase δ can refer to either angular coordinate. When it is desired to distinguish one coordinate from the other, subscripts *tr* for traverse and *el* for elevation are used, as in Figure 5.1-6.

5.4 COMPONENTS USED IN MONOPULSE

Certain components are either unique to monopulse, or they are designed or used in a special way for monopulse. A few important components of these types are discussed here. Monopulse processors, because of their importance and the multiplicity of forms they can take, are treated separately in Section 5.6.

5.4.1 Antenna Mounts

Antenna mounts of mechanically steered monopulse radars are basically the same as those of other mechanically steered radars, but the requirements are often more severe. Because monopulse is capable of higher angular accuracy than other methods of radar angle measurement and tracking, the mounts must be designed to more demanding specifications in order not to limit the attainable accuracy.

Various types of mounts have been used for monopulse. A common type is the elevation-over-azimuth two-axis mount illustrated in simplified form in Figure 5.2-1. Because of the geometric relations expressed by equations (5.2-1) and (5.2-2), azimuth rates and accelerations can be quite large when the target is at high elevation, even when the traverse rates and accelerations are moderate. The mounts must have sufficiently powerful drives to follow the target motion.

Figure 5.4-1 is a photograph of the antenna and mount of the AN/FPS-16 monopulse tracking radar. It is similar to the sketch in Figure 5.2-1, but the actual equipment includes additional components such as drive motors, antenna feed assembly, and certain microwave devices.

Figure 5.4-1 AN/FPS-16 Antenna, Side View (RCA).

5.4.2 Antennas

A single antenna can be shared by one or more transmitters and one or more receivers. Although monopulse action takes place on reception, the radar must also transmit in order to illuminate the target and produce echo signals, unless the monopulse is being used in a passive mode to track a radiating source or a target that is being illuminated by a different source.

Three main categories of antennas used in monopulse radar (as well as in other radars) are lenses, reflectors, and arrays, each of which can be subdivided into various types. The principal distinction between monopulse antennas and similar antennas used for other purposes is the nature of their feeds, which will be described separately in another subsection.

One type of lens, made of a dielectric material, is similar to an optical lens, as illustrated in Figure 5.4-2(a). Another type is the metal-plate lens, composed of plates parallel to the *E*-vector, spaced so that the phase velocity in the waveguides formed by adjacent pairs of plates is greater than in air. That type of lens is therefore concave, as illustrated in Figure 5.4-2(b). In both parts of Figure 5.4-2 a single feed horn is shown in order to illustrate the action of the lens. When used for monopulse, the lens would have multiple feed horns in the focal plane, displaced from the axis.

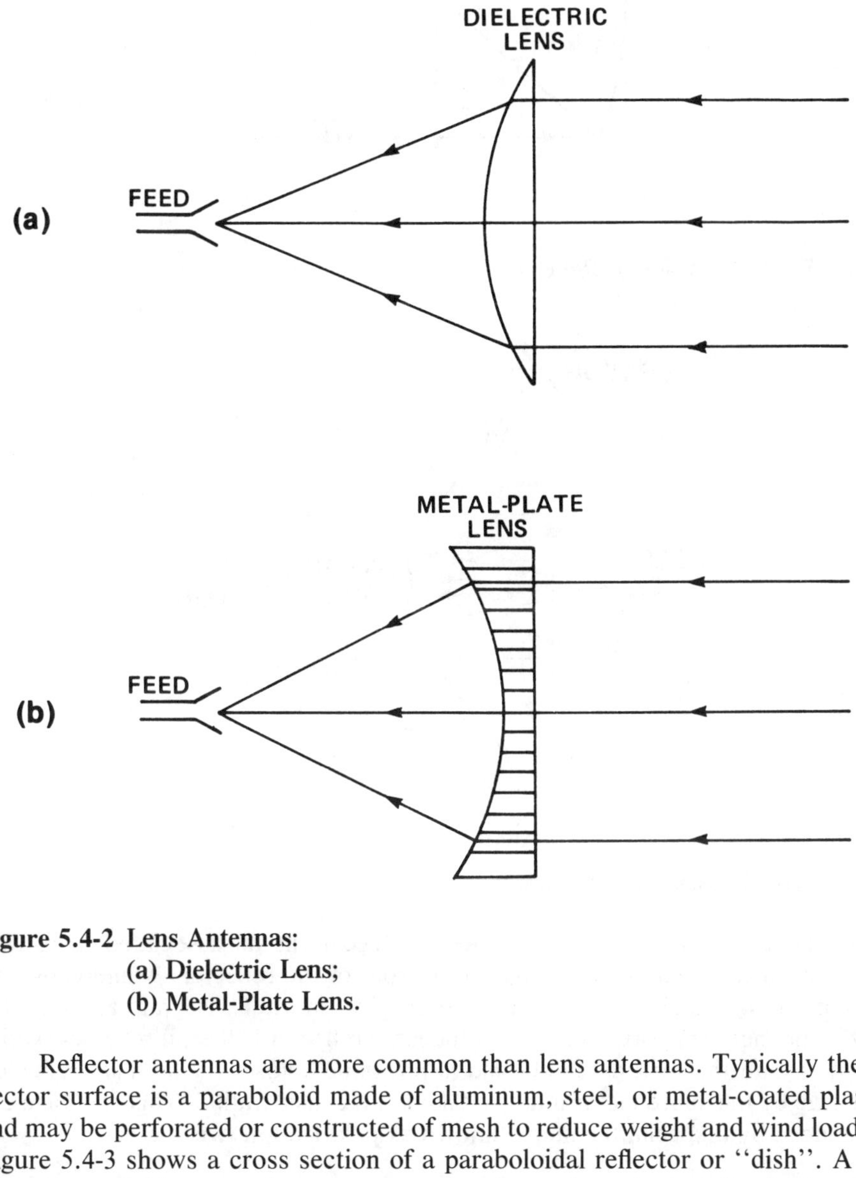

Figure 5.4-2 Lens Antennas:
(a) Dielectric Lens;
(b) Metal-Plate Lens.

Reflector antennas are more common than lens antennas. Typically the reflector surface is a paraboloid made of aluminum, steel, or metal-coated plastic, and may be perforated or constructed of mesh to reduce weight and wind loading. Figure 5.4-3 shows a cross section of a paraboloidal reflector or "dish". A disadvantage of this type of antenna, compared to a lens, is that the feed and its waveguides and support structure are in front, creating blockage which causes some loss of power, increased sidelobes, and an error in angle measurement. These effects are minimized by making the feed assembly as light and compact as possible.

Another type of antenna used in some monopulse systems is the Cassegrain or double-reflector antenna, which is the microwave analog of the optical technique used in some telescopes. As illustrated in Figure 5.4-4, the main reflector is a paraboloid and the subreflector is a hyperboloid. The advantages over a single-reflector antenna are a more compact antenna structure and reduction of length of transmission lines to the feed. These features are especially useful in large

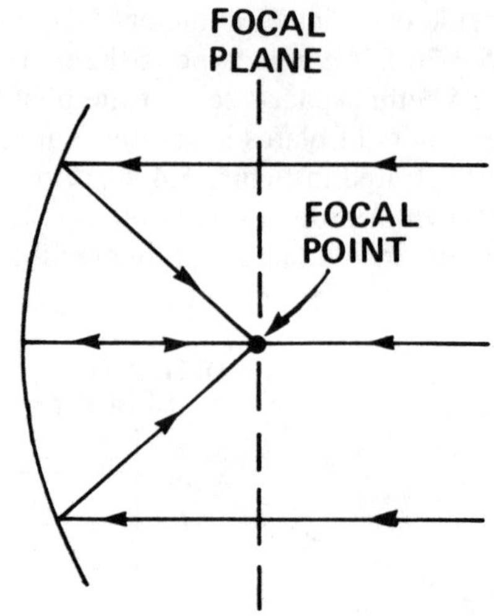

Figure 5.4-3 Paraboloidal Reflector.

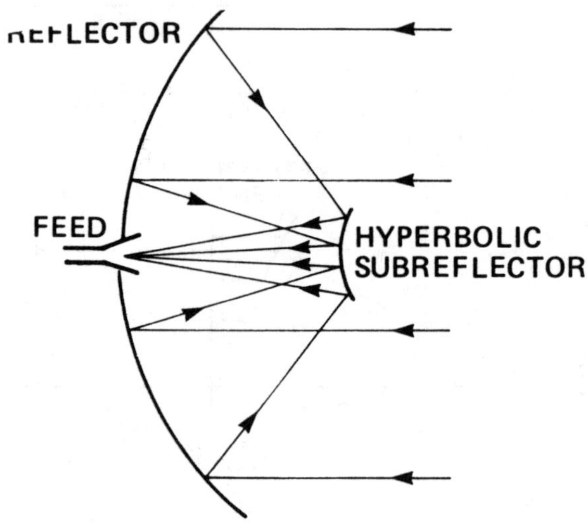

Figure 5.4-4 Cassegrain Antenna.

monopulse antennas, where the weight of the feed horns and associated microwave and mechanical components, if placed in front of the reflector, would cause deflection due to angular acceleration and gravity. In the Cassegrain antenna, the only component supported in front of the main reflector is the subreflector, which is lighter than the feed assembly and is positioned closer to the main reflector. Blockage by the subreflector must be taken into consideration. This blockage can be reduced by a special technique known as polarization twisting.

5.4.3 Feeds

A *feed* is that part of an antenna that distributes power from the transmitter to the lens, reflector, or array elements in a desired manner or collects the power captured by the lens, reflector, or array elements for transfer to one or more receivers. Although the feed is technically part of the antenna, it is discussed separately because of the importance of feed design in monopulse.

A widely used type of monopulse feed, mentioned earlier, consists of a cluster of four feed horns symmetrically offset from the focus, as illustrated in Figure 5.1-3. The horns are usually connected by short waveguides to a device called a *comparator,* which forms the sum and two differences. The sum and the differences then go to receivers. The horns, the comparator, and the connecting waveguides are combined in a compact, rigid assembly.

The square shape of the horns shown in Figure 5.1-3 does not necessarily portray the actual shape, but is a customary diagrammatic representation. The horn apertures may be square for dual polarization or rectangular for a single linear polarization. In the latter case, the polarization vector is parallel to the short side of the rectangles.

Although the early versions of four-horn feeds clearly demonstrated the advantages of monopulse over sequential lobing, it was soon recognized that optimum performance had not been reached. Important optimization objectives are: (a) to maximize the sum-pattern gain without exceeding specified sidelobe levels, and (b) to maximize the slope of the difference pattern at the null point, which determines the sensitivity, without exceeding specified sidelobe levels. Theory has shown that to achieve both of these objectives, the total feed aperture for the difference pattern should be about twice as wide as the total feed aperture for the sum pattern, in each coordinate. Total feed aperture refers to the dimensions of the entire cluster of horns, not to the dimensions of each horn. Since, in the four-horn feed, the same total feed aperture is used both for the sum and the difference, it was not possible in the early designs to achieve both objectives; the best that could be done was some optimum compromise.

However, subsequent design improvements have made it possible to approach the optimum patterns more closely. A technique of particular importance is the generation of multiple propagation modes in the feed horns and adjustment of their relative amplitudes and phases to shape the desired illumination functions across the face of the feed aperture. A degree of independence between sum and difference illuminations is obtained because the even and odd modes contribute only to the sum and difference illuminations respectively. Although the total physical aperture is the same for sum and difference, the *effective* aperture for the sum pattern is reduced by shaping the illumination function so that most of the power is concentrated near the center. This technique has been applied successfully to improve the performance of four-horn feeds and has also been used in other feed configurations.

Another approach to simultaneous optimization is the use of different physical apertures for the sum and difference feeds. One such design is the five-horn feed illustrated in Figure 5.4-5. The center horn, labeled R in the diagram, is used for transmission and reception of a reference signal, which serves the same functions as the sum signal in four-horn monopulse. The traverse difference signal is derived from horns A and B and the elevation difference signal is derived from horns C and D. This design has shown excellent performance but is not optimum, because the difference illumination function, which should be continuous over the entire feed dimension, has a gap in the center in the area of the reference horn.

Progress is continually being made in development of more sophisticated feed designs, and performance approaching the theoretical optimum is now achievable.

5.4.4 Devices for Forming Sums and Differences

Sums and differences of RF voltages are obtained by the use of certain passive microwave devices known as hybrid junctions (or "hybrids" for short), of which there are several different types. Hybrids are also used for other radar (as well as non-radar) functions, but here we are concerned in particular with monopulse applications. Hybrids used in monopulse have four ports—two for inputs and two for outputs. If two voltages v_1 and v_2 are applied in the proper manner at the input ports, outputs proportional to their sum and difference will

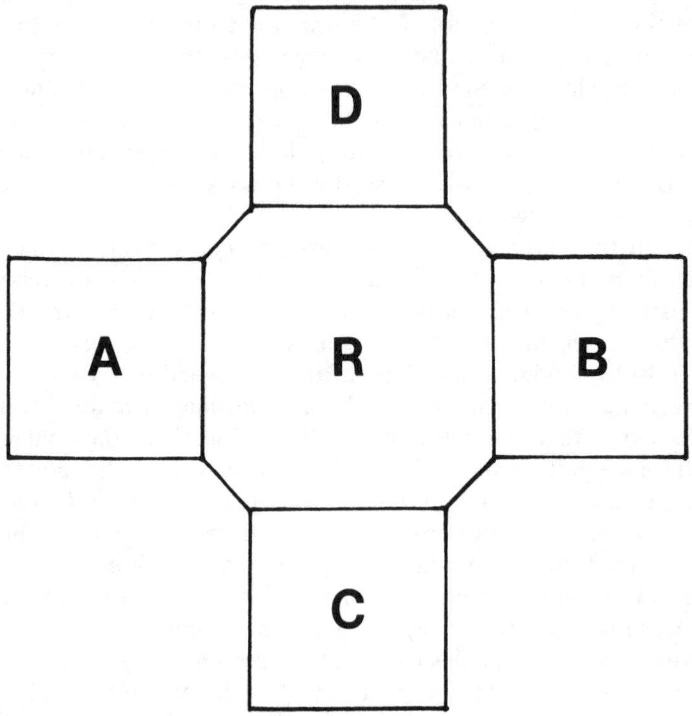

Figure 5.4-5 Five-Horn Monopulse Feed.

appear at the output ports. Three important types of these devices are the *Magic-T,* the *hybrid ring junction,* and the *3-dB directional coupler.*

The Magic-T is illustrated in Figure 5.4-6. An input at port 1 produces outputs of equal amplitude and phase at ports 3 and 4. An input at port 2 produces outputs of equal amplitude but opposite phase at ports 3 and 4, because, although the path length is the same, the electric vector is opposite in direction. Therefore, simultaneous input voltages at 1 and 2 will produce their sum at 3 and their difference at 4, or *vice versa,* depending on the polarity convention of the inputs. Input ports and output ports can be interchanged.

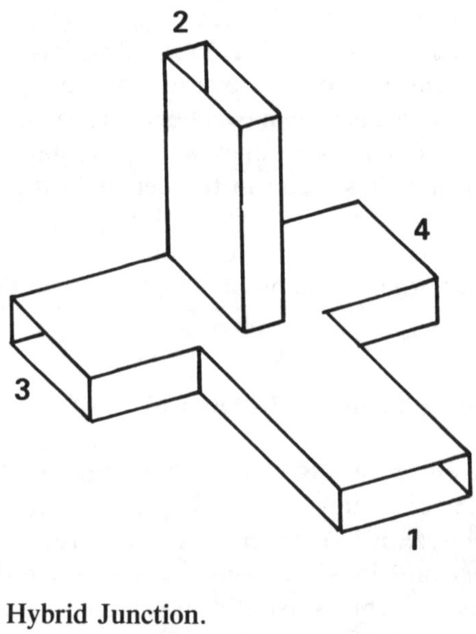

Figure 5.4-6 Magic-T Hybrid Junction.

To be precise, the outputs are not equal to the sum and difference, but are *proportional* to the sum and difference:

$$s = k(v_1 + v_2), \tag{5.4-1}$$

$$d = k(v_1 - v_2). \tag{5.4-2}$$

The proportionality constant k turns out to be $1/\sqrt{2}$, as can be verified by squaring both sides of the two equations and adding to obtain*

$$s^2 + d^2 = v_1^2 + v_2^2. \tag{5.4-3}$$

This is as expected, because the device is passive and nearly lossless, and thus total output power must equal total input power. Since the proportionality constant is the same for the sum and difference, it does not affect their ratio.

The hybrid ring junction (also known as a "rat-race") is illustrated in Figure 5.4-7. It is essentially a waveguide shaped in the form of a ring. If ports 1 and 2 are used as inputs, outputs proportional to the sum and difference appear at ports 3 and 4, with the same proportionality constant as for the Magic-T. Input and output ports can be interchanged. The principle of operation is that each input splits into two components traveling in opposite directions in the waveguide. The

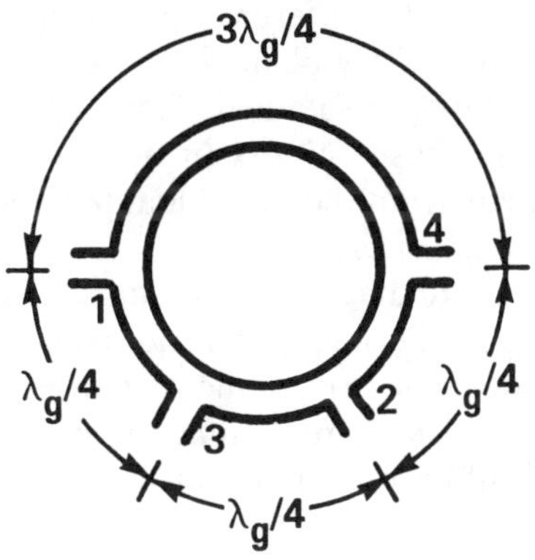

Figure 5.4-7 Hybrid Ring Junction ("rat-race").

lengths of the sections of the waveguide are such that the two inputs combine in phase at one of the output ports (producing their sum) and in opposite phase at the other output port (producing their difference). Because the lengths of the sections are certain fractions of the wavelength in the guide at a specified frequency, this type of device is frequency-sensitive and has limited bandwidth.

The 3-dB directional coupler is illustrated in Figure 5.4-8. It consists of two contiguous, parallel waveguides with a common wall. Holes or slots are made in the common wall, causing a portion of the power to pass from one waveguide to the other. In the particular version shown in the figure there are two holes spaced a quarter-wavelength apart. A portion of the power flowing through each waveguide passes through each hole and propagates in both directions in the other waveguide. Ports 1 and 2 can be used for inputs, and 3 and 4 for outputs, or *vice versa*. The 3-dB directional coupler differs in one important characteristic from the Magic-T and the hybrid ring junction. In order to obtain the sum and difference of two voltages v_1 and v_2 by means of a 3-dB directional coupler, either v_1 or v_2 must be phase-shifted by 90° before it is applied as an input; the sum and the difference (with the same proportionality factor) will then appear at the output

*In this simplified explanation, it is assumed that v_1 and v_2 are in phase and therefore s and d are in phase, in which case it can be assumed that all four are real. It can easily be verified that the proportionality constant is $1/\sqrt{2}$ even when the phases are not equal.

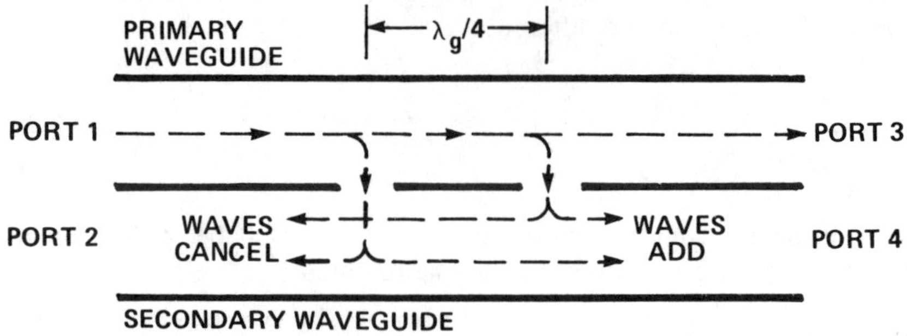

Figure 5.4-8 Two-Hole Directional Coupler.

ports, but will differ from each other in phase by 90°, while in the other two devices, the sum and difference emerge in phase with each other. Because of this peculiarity, the 3-dB directional coupler is also known as a *quadrature hybrid* or *90° hybrid*.

5.4.5 Comparators

Any one of the hybrid junctions described in the preceding subsection can be used to obtain the sum and the difference of two input voltages. However, a typical monopulse feed, as already explained, consists of four horns, from which a sum and two differences must be obtained. This requires multiple hybrids. The whole assembly of hybrids, phase shifters (if needed), and waveguide sections needed to convert the signals from the individual feed horns into a sum and two differences is called the *comparator*.

Figure 5.4-9 is a schematic diagram of an illustrative comparator of a monopulse tracking radar with a four-horn feed. The horns, labeled A to D, are arranged as shown at the bottom (the same arrangement as in Figure 5.1-3). Four hybrids are needed. In this illustration they are assumed to be Magic-T hybrids. The first hybrid gives the sum and difference of A and B, and the second gives the sum and difference of C and D. The third hybrid gives the sum and difference of A − B and C − D, and the fourth gives the sum and difference of A + B and C + D. The final outputs are the sum, the traverse difference, and the elevation difference. An additional output is available from the third hybrid. This output is (A − B) − (C − D), which can also be written (A + D) − (B + C). It is called the second difference or the diagonal difference. Except possibly for certain special applications, this output has not been found useful. In order not to disturb the impedance matching of the entire comparator, this output must be terminated in a matched dummy load, as shown in the diagram.

5.4.6 Receivers

For two-coordinate angle measurement or tracking, three separate receiver channels are normally needed for the sum and the two differences, although it is possible to reduce the number of receivers to two by time-sharing traverse and elevation at the expense of a reduction in data rate or of some other performance feature. These receivers perform the same basic functions as in any radar—namely, conversion from RF to IF, filtering to match the waveform spectrum, and amplification. However, not only must the individual monopulse receivers perform correctly over a wide dynamic range of input signal levels and under varying conditions of temperature, frequency, and passage of time, but the receivers must also remain closely matched to each other in gain and phase under those conditions.

SHERMAN
CHAPTER 5

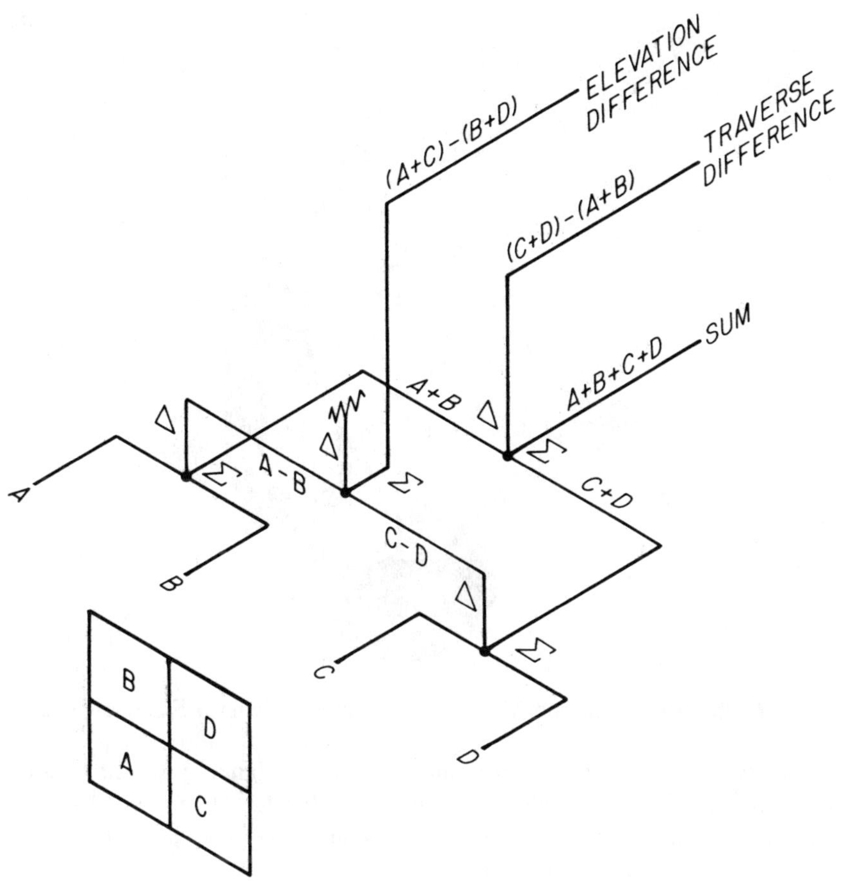

Figure 5.4-9 Comparator for Four-Horn Monopulse Feed Using Magic-T Hybrids (Factors of $1/\sqrt{2}$ and 1/2 Omitted).

Some monopulse radars use pilot pulses, injected at the front end of the receivers, to maintain amplitude and phase matching. Disparities among the receivers are automatically detected and corrected by adjustable attenuators and phase shifters, or in the monopulse processing if done by computer.

5.5 AMPLITUDE-COMPARISON AND PHASE-COMPARISON MONOPULSE

All the descriptions given so far pertain to a class known as *amplitude-comparison* monopulse. There is also another class known as *phase-comparison* monopulse. The distinctions between the two classes are explained below.

5.5.1 Physical Construction

In a reflector or lens antenna, the physical distinction is obvious: In amplitude-comparison there is a single main reflector or lens with a cluster of feed horns. In phase-comparison there is a cluster of four reflectors or lenses side by side in a square or cloverleaf arrangement, joined together at their edges to form a rigid assembly, each reflector or lens having a single feed horn.

Figure 5.5-1 shows a model of an early phase-comparison antenna. A corresponding method used in some array antennas is to divide the array into four quadrants, collect the contributions of the elements in each quadrant, and process the resulting signals in the same manner as if they were obtained from the four quadrants of an antenna such as the one pictured in the figure.

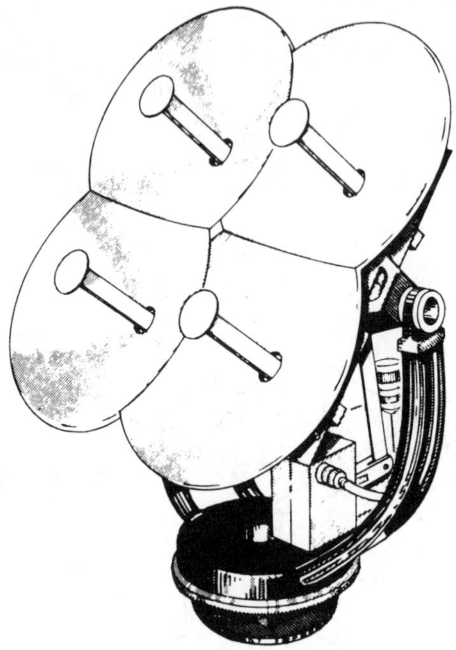

Figure 5.5-1 Phase-Comparison Monopulse Antenna (General Electric Co.).

Other types of monopulse array antennas are designed differently, and may belong to either class. In general, it is not possible to tell, merely by outward physical appearance, to which class an array antenna belongs. Array antennas are discussed further in Section 5.9.

5.5.2 Antenna Patterns

Amplitude-comparison and phase-comparison antenna patterns are distinctive. Figure 5.5-2 illustrates a pair of beams of each class, for one angular coordinate. In amplitude-comparison, the beams have a common phase center, but are squinted symmetrically away from the axis. A target produces voltages of the same phase but different amplitudes in the two beams; the amplitudes are equal only when the target is on the axis. In phase-comparison the beams are parallel and identical, except for a lateral displacement of their phase centers on opposite sides of the axis. A target produces voltages of the same amplitude but different phases in the two beams, because the pathlengths from the target to the two phase centers are different; the phases are equal only when the target is on axis.

5.5.3 Relative Phase of Sum and Difference

The terms amplitude-comparison and phase-comparison should not be interpreted literally. It has already been explained that in amplitude-comparison monopulse there is usually no direct comparison of amplitudes of signals from the individual feed horns; instead the comparison is done indirectly by means of the sum and difference. Similarly, in phase-comparison there is usually no direct comparison of phases, but the desired information is obtained by processing the sum and difference.

In Figure 5.5-3, phasors v_1 and v_2 represent the voltages from two squinted beams (or two pairs of squinted beams) in amplitude-comparison monopulse. Since they are in phase with each other, their sum and their difference are also in phase

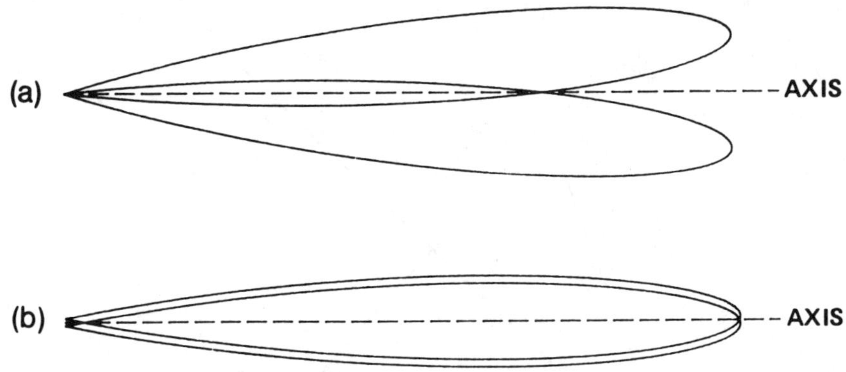

Figure 5.5-2 Monopulse Beams, Shown in One Coordinate:
(a) Four-Horn Amplitude-Comparison;
(b) Phase-Comparison.

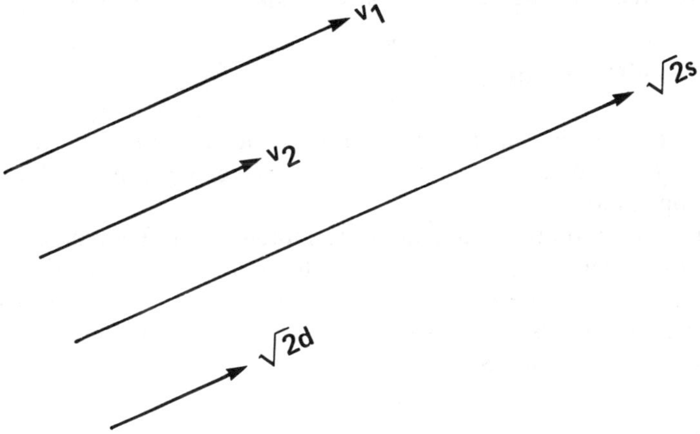

Figure 5.5-3 Phase Relationships in Amplitude-Comparison Monopulse.

with each other, as the phasor diagram shows. (For brevity, "in phase" is interpreted to include 180° relative phase.) The reason for the factor $\sqrt{2}$ appearing before s and d has been explained in Section 5.4.4. The in-phase relationship is the idealized or nominal condition, which in practice is disturbed to some degree by noise, unresolved targets, system imperfections, and other factors.

In Figure 5.5-4, phasors v_1 and v_2 represent the voltages from two beams (or two pairs of beams) in phase-comparison monopulse. They have the same amplitude but different phases, assuming that the target is off-axis. From the geometrical construction it is seen that the s and d phasors are perpendicular to each other, which means that the sum and difference signals differ in phase by 90°. As in the case of amplitude-comparison, this phase relationship represents the nominal condition, which is subject to various disturbing factors.

In Figures 5.5-3 and 5.5-4 the phase shift of the voltages in passing through the comparator is ignored, since it is the same for the sum and the difference and does not affect their relative phase.

Whether the nominal phase of the sum and difference is 0° or 90° is of no importance as far as operation of the radar is concerned. By means of phase shifters the relative phase can be transformed from the one value to the other, or an equivalent transformation can be readily accomplished in the monopulse processor.

SHERMAN
CHAPTER 5

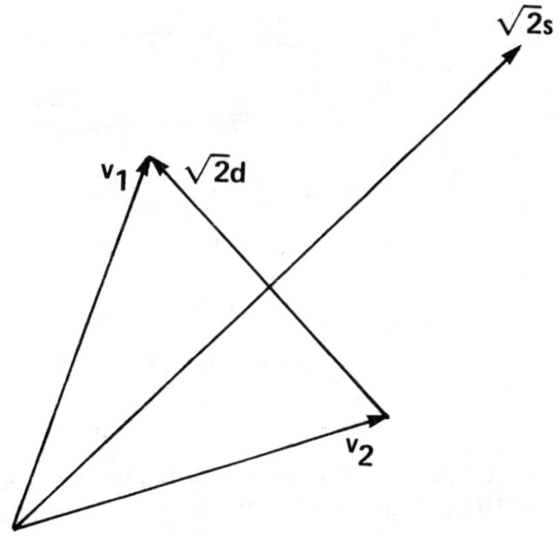

Figure 5.5-4 Phase Relationships in Phase-Comparison Monopulse.

5.5.4 Additional Comments

Generally speaking, neither class of monopulse can be said to be superior to the other. The choice is a matter of design convenience or suitability for each particular application.

A source of potential confusion in terminology can be avoided by keeping in mind that the amplitude-comparison or phase-comparison classification refers to the antenna, not to the monopulse processing. The processing may involve amplitude measurements, phase measurements, or both, regardless of the class to which the antenna belongs.

5.6 MONOPULSE PROCESSORS

The monopulse processor is that portion of a monopulse radar that operates on the voltages derived from the simultaneous antenna patterns to produce the monopulse outputs. In most cases the voltages on which the processor operates are the sum and the two differences, although in some amplitude-comparison radars the voltages are those from the squinted beams.

As explained in an earlier section, the desired output, for amplitude-comparison monopulse, is usually the real part of the complex ratio d/s, expressed by equation (5.3-6), and for phase-comparison monopulse the desired output is the imaginary part, expressed by equation (5.3-7). To simplify the discussion that follows, it will be assumed that the antenna is of the amplitude-comparison class; only a simple modification such as a 90° phase shift or computation of the sine instead of the cosine is needed for phase-comparison monopulse.

For closed-loop tracking the processor output becomes the input to a servo after coordinate conversion if needed, as shown in Figure 5.1-6. For open-loop measurements, the processor output must be converted to angle by a calibration function. As can be seen from the d/s curve in Figure 5.1-5, this function is very nearly linear for small angles from the axis, but usually departs significantly from linearity at angles one-quarter beamwidth or more from the axis.

There are many types of monopulse processors, of which three examples are described below. Several others are discussed in the author's earlier work [1].

5.6.1 Processor Using Amplitudes and Phases of *s* and *d*

One obvious method of implementing equation (5.3-6), illustrated in Figure 5.6-1 for one angle channel, is to measure the amplitudes of the sum and difference voltages at the outputs of linear receivers, measure their phases relative to a common reference (or alternatively, measure the phase of the difference relative to the sum), then compute the monopulse output by equation (5.3-6). It will be recalled that $|s|$ and $|d|$ are the amplitudes of the sum and difference, δ_s and δ_d are the phases of s and d relative to a common reference, and δ, which is equal to $\delta_d - \delta_s$, is the phase of d relative to s.

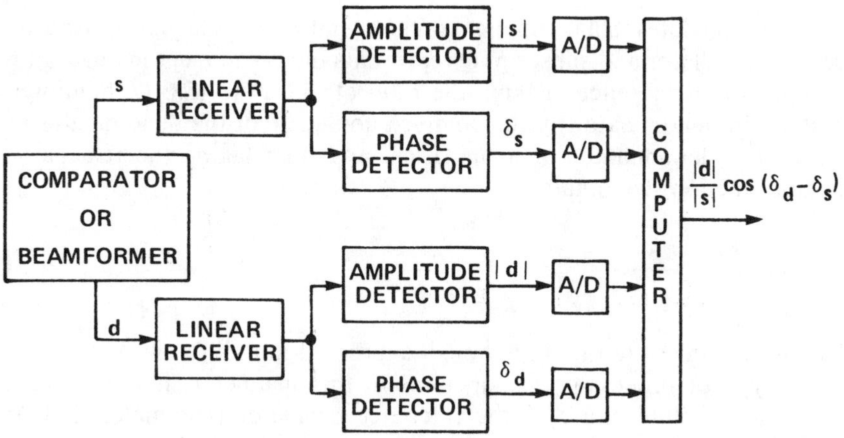

Figure 5.6-1 Processor Using Phases and Linear Amplitudes of *s* and *d*.

A variation of this design uses logarithmic rather than linear amplifier-detectors in order to increase the dynamic range. This processor is illustrated in Figure 5.6-2.

In either version of the design, as well as in most other processors, the computational load can be reduced for some applications by averaging the sum and difference amplitudes and their phase difference over several pulses before doing the final computation.

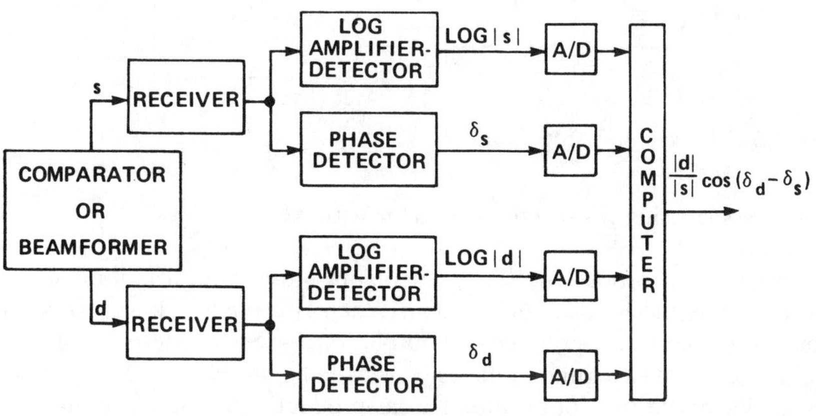

Figure 5.6-2 Processor Using Phases and Logarithmic Amplitudes of *s* and *d*.

5.6.2 Processor Using *I* and *Q*

Instead of expressing the phasor sum and difference voltages in terms of their amplitudes and phases, it is equally valid to express them in terms of their real and imaginary parts:

$$s = \text{Re}(s) + j\,\text{Im}(s) = s_I + js_Q \tag{5.6-1}$$

and

$$d = \text{Re}(d) + j\,\text{Im}(d) = d_I + jd_Q. \tag{5.6-2}$$

Subscript *I* indicates that component of the signal which is in phase with a common reference oscillator and subscript *Q* represents the component in quadrature (90°) phase with the reference. Taking the ratio of (5.6-2) to (5.6-1), multiplying numerator and denominator by the complex conjugate of the denominator in order to clear the denominator of imaginaries, and then taking the real part of the resulting fraction, we obtain

$$\text{Re}(d/s) = \frac{d_I s_I + d_Q s_Q}{s_I^2 + s_Q^2}, \tag{5.6-3}$$

which is equivalent to (5.3-6) although expressed in a different form.

A type of processor that operates in this manner is illustrated in Figure 5.6-3. The absolute phase of the reference oscillator is immaterial. If it were changed, all four of the *I* and *Q* components would change, but the real and imaginary parts of d/s would remain unchanged.

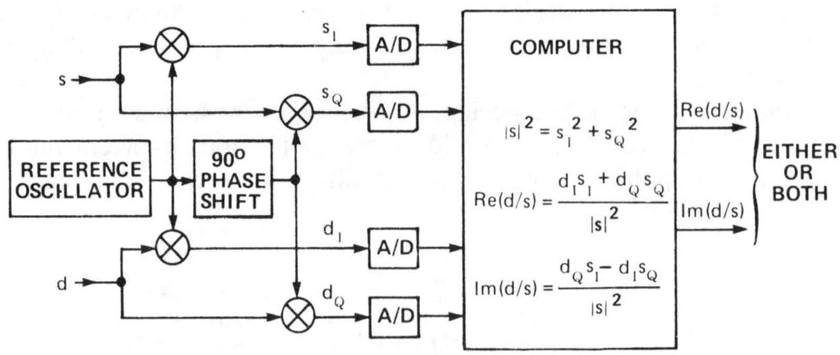

Figure 5.6-3 Processor Using *I* and *Q*.

5.6.3 Processor Using Dot-Product Detector with AGC

This type of processor, shown in Figure 5.6-4, uses a multiplying device which performs the following operation: When two input voltages e_1 and e_2 are applied to the device, it produces an output $|e_1|\,|e_2|\cos\delta$, where δ is the relative phase between the two input voltages. It is called a *dot-product detector* because of the analogy with the dot-product (scalar-product) operation in vector algebra. Other names for the same device are product detector, synchronous detector, and phase-sensitive detector. If the sum and difference voltages were used as the inputs, the output would be $|s|\,|d|\cos\delta$. In order to convert this to the desired output expressed by equation (5.3-6), it is necessary to divide by $|s|^2$. This is done by dividing s and d by $|s|$ before those voltages are applied to the dot-product detector. The division is accomplished by *automatic gain control* (AGC) derived

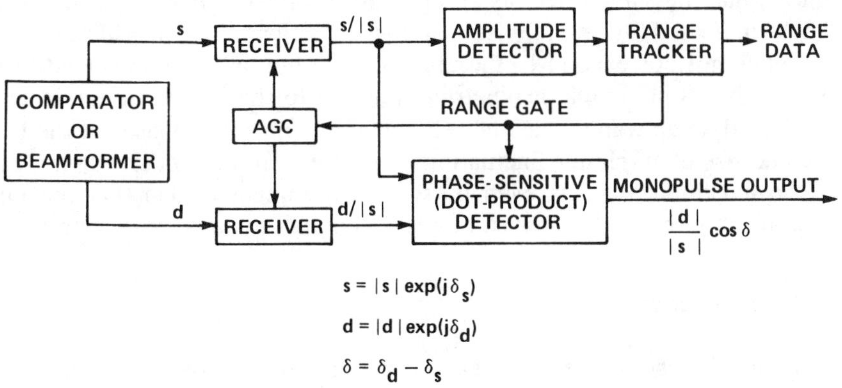

Figure 5.6-4 Processor Using Dot-Product Detector with AGC.

from the sum channel and controlling both the sum and difference channels. The AGC keeps the sum voltage at a constant voltage which we can consider unity (since the scale factor does not matter), but phase is preserved. Thus the sum voltage after AGC is $s/|s|$ and the difference voltage after AGC is $d/|s|$.

In some systems an approximate rather than exact dot-product detector is used. Its advantage is the simplicity of the circuit. The disadvantage is that, although the approximation performs identically to the exact version under ideal conditions ($\delta = 0°$ or 180°, and off-axis angle not too large), it behaves differently under other conditions.

Even if an exact dot-product detector is used, a disadvantage of AGC is that it has a time lag, which causes an error under some conditions. Nevertheless, the dot-product processor with AGC is used successfully in many monopulse tracking radars for applications where it is not necessary to obtain angle information on every pulse or during short dwells.

5.6.4 Choice of Processor

In the author's earlier work [1] the performance, advantages, and disadvantages of each type of processor mentioned above, and several others not mentioned here, are analyzed in some detail. Included are designs intended to reduce the number of receiver channels from three to two or even to one.

Generally speaking, the more closely the performance approaches the exact desired output expressed by equation (5.3-6), under a variety of conditions, the more complicated and expensive the design tends to be. Therefore the choice is usually the least complicated and least expensive method that will meet the needs of the intended mission.

5.7 RESPONSE TO UNRESOLVED TARGETS

Targets are said to be unresolved (with respect to angle measurement) when they are so close to each other that the radar cannot do accurate angle measurement or tracking of each individual target, but produces an angle indication based on the composite sum and difference signals. As a rule of thumb, targets become unresolved when they are approximately within a beamwidth* of each other in angle and within the same range resolution cell. They may be closely spaced, separate objects—for example, two or more aircraft flying in close formation—or multiple scattering points on a single target. A jammer within the same beam

*The term "beamwidth" means the width of the sum-pattern beam unless otherwise defined.

as a target has the same effect as an unresolved target. Under those conditions, the angle indicated by monopulse—or by any other angle-measurement technique—does not correspond to the angle of any one of the unresolved targets, and in fact bears no simple geometrical relation to the individual target angles. The indicated angle wanders as the relative amplitudes and phases of the targets change because of amplitude fluctuations, relative motion, or other causes. Sometimes the indicated angle is even outside of the angular span that the targets subtend at the radar. The error can be large enough to cause loss of track.

5.7.1 Phasor Relations

The mechanism of the error can be understood by considering the case of two targets close enough to be unresolved. Figure 5.7-1 is a phasor diagram for one angular coordinate; the contributions of the first target to the combined sum and difference signals are denoted by subscript a and those of the second target by subscript b. Although the two targets are in the same range resolution cell, in general their ranges are not exactly equal, and there will be some phase difference between s_a and s_b. (Even if the ranges are equal, there may be a phase difference because of different backscatter characteristics.) The total sum signal is the resultant s. For illustration, suppose the first target is on the positive side of the beam axis, so that d_a is in phase with s_a, and that the second target is on the opposite side, so that d_b is in phase opposition to s_b. The total difference signal d is the resultant of d_a and d_b. The assumption that the two targets are on opposite sides is not a necessary condition for the conclusion that follows, but it facilitates the diagrammatic demonstration.

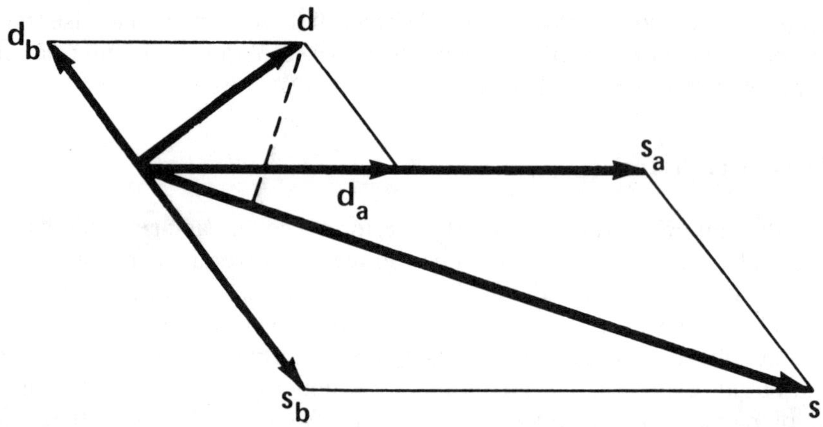

Figure 5.7-1 Sum and Difference Phasors for Two Unresolved Targets.

It is clear from Figure 5.7-1 that, in general, d has a component in phase quadrature with s, shown by the dashed line, as well as a distorted in-phase component. In other words, the ratio d/s is complex. It is easy to see also that if s_a and s_b are 180° out of phase and nearly equal in magnitude, the resultant s is small and the ratio d/s can become very large. In general, the real part of the ratio d/s, which is what the monopulse processor is designed to extract as the basis for angle measurement, is not equal to d/s for either target alone.

5.7.2 Two-Target Plots

The mathematical analysis, which is not difficult, is omitted here, but the results are plotted in Figure 5.7-2. The horizontal coordinate is the relative phase angle ϕ between the individual sum-signal contributions of the two targets. The

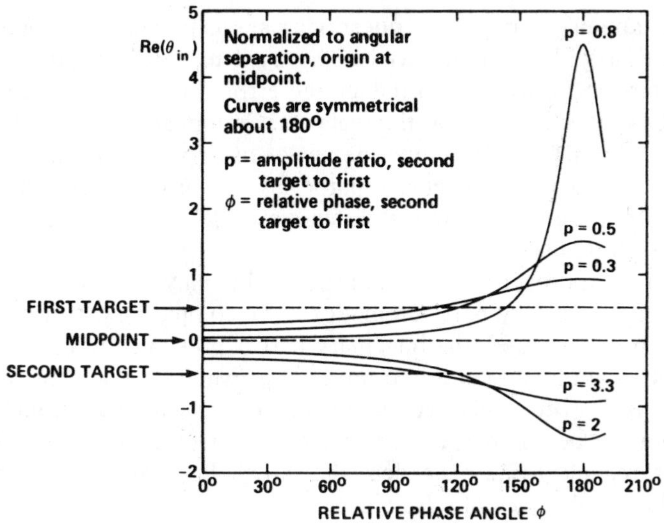

Figure 5.7-2 Real Part of Complex Angle *versus* Relative Phase for Two Unresolved Targets.

vertical coordinate, $\mathrm{Re}(\theta_{in})$, is the angle indicated by the monopulse output; it is the real part of the complex ratio d/s, converted to angle on the assumption that the relation between monopulse output and angle is linear. The vertical scale is normalized so that the angular separation between the two targets is unity, and the midpoint is zero. Hence the actual indicated angle, measured from the angular midpoint, is the number read from the vertical scale multiplied by the angular separation of the targets. Parameter p is the amplitude ratio of the second target to the first.

When the phase difference between the two targets is small, the indicated angle lies between the angles of the two targets. At larger values of the phase difference the indicated angle lies outside of the span of the two targets, reaching its greatest deviation when the phase difference is 180°. The more nearly equal the targets are in amplitude (that is, the more closely p approaches 1), the larger the deviation when $\phi = 180°$.

5.7.3 Glint

The unresolved targets need not be independent objects; they may be different scattering points on a single target. For example, the wingtips of some aircraft act as strong scattering points. The result, often observed in practice, is that when such a target is being tracked, the tracking direction wanders as the relative phase of the wingtip contributions varies because of slight changes in orientation of the target. This wander, known as *glint,* is a well known source of angular error. The indicated direction even lies outside the physical span of the target for a significant fraction of the time. Of course an actual target such as an aircraft generally has more than two scattering points. The deterministic analysis for more than two becomes unwieldy, but the general behavior is qualitatively similar to what has been described. If the number of unresolved targets is large, a probabilistic analysis is more appropriate.

Glint error will be discussed further in Section 5.10.

5.7.4 Use of Imaginary Part of Monopulse Ratio

As explained above, the presence of one or more undesired targets in the same resolution cell with a desired target not only causes the real part of the monopulse ratio d/s (from which angle information is derived) to be in error, but also creates an imaginary part. A question arises as to whether the imaginary part in combination with the real part can be used to determine the angles of the individual targets, or at least to reduce the error. The imaginary part can easily be measured in the same way as the real part. Certain approaches which in theory may make it possible to determine the angles of two unresolved targets by measuring both the real and imaginary parts have been proposed and investigated analytically, but there are practical difficulties, and so far no such technique has been put to actual use.

A simpler approach, also investigated only analytically or by simulation, is to monitor the imaginary part automatically by comparing it with a threshold. If the threshold is exceeded, the imaginary part is presumed to be due to contamination of the data by an undesired target in proximity to the desired target. The pieces of data in which this occurs are then discarded, the rationale being that missing data are better than contaminated data. However, since noise can also cause the threshold to be exceeded, the threshold must be set high enough to achieve a low probability of false alarms.

5.8 MULTIPATH

Multipath is the propagation of a wave from one point to another by more than one path. When it occurs in radar, it usually consists of a direct path and one or more indirect paths by reflection from the surface of the earth or sea. If the reflecting surface is smooth, the multipath is *specular* (mirror-like); if it is irregular, the multipath is *diffuse*. When the mean contour of the surface is smooth but there are small-scale irregularities, both specular and diffuse components are present. Both types of multipath, when they occur, produce errors in angle measurement (mainly elevation) in monopulse and other angle-measurement techniques.

Specular multipath causes erratic elevation tracking—sometimes even complete loss of track—at low target elevations within a beamwidth or so of the horizon, and because of antenna-pattern sidelobes, specular multipath can cause significant elevation error even at target elevations a few beamwidths above the horizon.

Diffuse multipath effects are somewhat noise-like and can be analyzed only by a rather complicated geometric and statistical approach. Although the resulting error is sometimes a significant component of the total angle error, it is not usually serious enough to disrupt tracking.

The discussion that follows pertains to specular multipath. Diffuse multipath has been treated at length in other sources.

5.8.1 Multipath as a Special Case of Unresolved Targets

The mechanism of specular multipath can be illustrated by a simplified model, represented by Figure 5.8-1. The reflected ray reaching the radar is equivalent to a ray from the image, which is below the surface at a location symmetrical to that of the target. When the target direction is a half-beamwidth or less above the horizon, the target and the image are unresolved because they are within a beamwidth of each other, and this is the condition that causes serious errors. In the figure, the vertical scale is exaggerated for clarity; typical beamwidths range from a fraction of a degree to a few degrees. A flat, horizontal reflecting surface is assumed in this illustration. Earth's curvature is not difficult to take into account, and this ought to be done in an actual analysis. However, the flat-earth assumption simplifies the geometry for the purpose of illustrating the principles and the qualitative effects, and even gives a good quantitative approximation at short ranges.

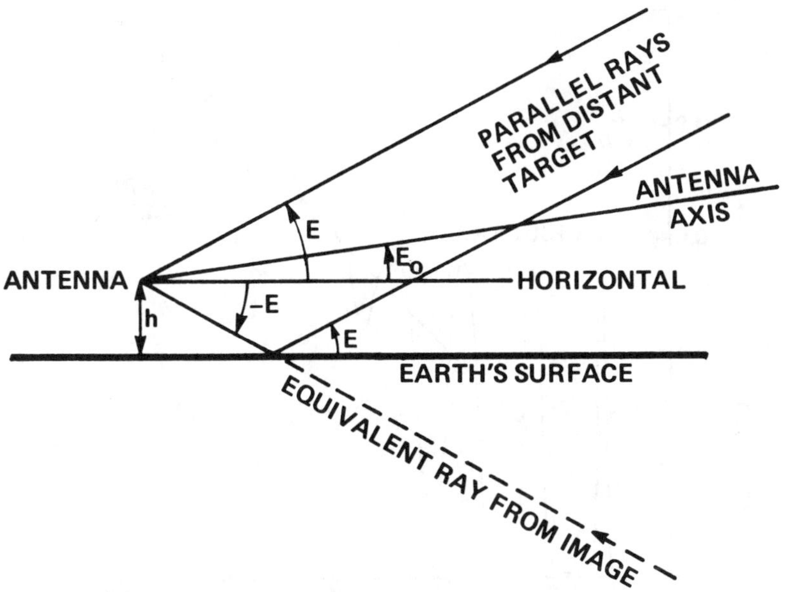

Figure 5.8-1 Geometry of Target and Image, Flat-Earth Approximation.

Specular multipath is a special case of the general two-target problem discussed in Section 5.7. In the general two-target case, the targets can be independent. In multipath, however, the target and its image are not independent: They are related in elevation angle, amplitude, and phase. The relationships depend on the geometry of the radar, target, and surface, on the reflection characteristics of the surface, and on the antenna patterns.

5.8.2 Effect on Tracking

Figure 5.8-2 illustrates the effect of specular multipath on closed-loop tracking. A flat earth was assumed and an analytical model of typical monopulse sum and difference patterns was used in the calculations. Plotted horizontally is the true elevation angle of the target. Plotted vertically is the monopulse null direction, which is the same as the beam-pointing direction (the tracking direction) if the target elevation varies slowly enough to allow the servo loop to reach equilibrium at each point. The upper dashed 45° line is the direction in which the beam would point if the image were absent and the lower dashed line would be the corresponding tracking direction if the image could be present without the target.

In the upper half of Figure 5.8-2 (above zero elevation) the radar is tracking the target, with perturbations caused by the image. The peaks occur when the target and image signals in the sum channel are in phase opposition. The minima occur when they are in phase. The variation in relative phase is due mainly to path-length difference as a function of elevation angle, and to a lesser extent to variation of the phase of the reflection coefficient as a function of grazing angle.

The lower part of the figure shows the locus of other null directions. The solid portion of each loop is stable and is due to tracking of the image, with perturbations by the target, rather than *vice versa*. The dotted portion of each curve is unstable; that is, if the antenna were pointed in that direction, a slight deflection would cause the beam to move away from that direction toward one of the stable directions.

Consider a radar with a mechanically steered antenna attempting to track a target which is approaching at a constant low altitude. The target elevation angle

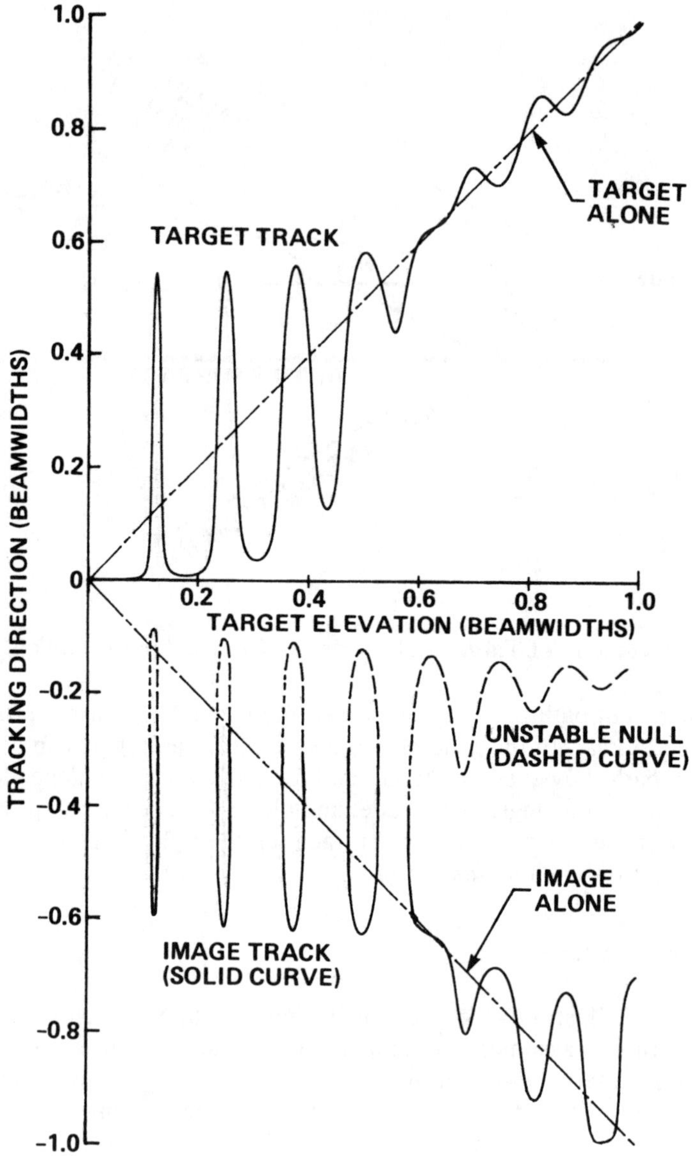

Figure 5.8-2 Closed-Loop Tracking in Multipath.

is near zero at long range and increases as the target approaches (moving from left to right in Figure 5.8-2). The oscillations in beam-pointing direction sometimes become quite violent, and the inertia of the antenna can carry it from the upper stable direction to the lower one. This phenomenon, known as "nose-diving", has often been observed in practice. The track may then transfer back to the upper null or may be lost altogether.

5.8.3 Multipath Remedies

A number of remedial techniques have been developed or proposed to alleviate the low-angle tracking problem. These techniques differ in degree of sophistication, in the types of applications for which they are suited, and in effectiveness. No technique yields the same elevation accuracy in multipath that is attainable without multipath, but the errors can be reduced and tracking can be extended to a lower angle than would be possible otherwise.

Some of the techniques do not attack the multipath problem directly, but mitigate its effects by selection of the radar design or operating parameters, such as beamwidth, sidelobe level, or servo bandwidth, to the extent permitted by other system considerations. Reduction of elevation beamwidth permits tracking to a lower angle, but requires an increase in antenna size or in frequency. One solution that has been implemented in an experimental radar is the use of two frequencies sharing the same antenna. High-angle tracking is done at a frequency which is optimum for that purpose and low-angle tracking at a second frequency in a higher band, producing a narrower beamwidth. Low sidelobes are helpful when the target elevation is high enough to put the image in the close-in sidelobes, where the elevation tracking error is not disruptive but still significant, and low sidelobes also reduce the error from diffuse multipath and clutter. However, low sidelobes are generally obtained at the price of some reduction of gain and widening of the beam. Smoothing of the elevation fluctuations by reducing the loop bandwidth is beneficial if the fluctuations are rapid.

A technique that prevents violent oscillations in elevation tracking of low-angle targets is to disable the elevation tracking loop and keep the beam at a fixed elevation angle, while maintaining track in traverse and range. A beam elevation angle of about two-thirds or three-quarters of a beamwidth is typical. The radar is switched to this ''*low-E*'' mode when the target enters the troublesome multipath region, and is switched back to normal three-coordinate tracking when the target is at higher elevation angles. There are several forms of the low-E mode, differing in some details but basically similar. In the simplest form, no elevation information is available, but the traverse and range tracking provides at least partial information about target location, the rationale being that it is better to forgo elevation information, which is erroneous anyway, than to risk complete loss of track. If the target is approaching the radar at constant altitude, the expectation is that it will eventually reach an elevation angle that permits easy transition to full track. In a somewhat more refined form of the low-E mode, the open-loop monopulse elevation output is measured and used for crude estimation of target elevation. In any case, this output is monitored to determine when the elevation has increased sufficiently to permit switching to closed-loop elevation tracking. Similarly, observation of the closed-loop behavior determines when to switch from normal track to the low-E mode.

Another technique that gives some reduction of multipath error is closed-loop tracking with the patterns modified so that the direction of the elevation difference-pattern null (the tracking axis) is offset downward from the direction of the sum-pattern peak by a fraction of a beamwidth. This technique, known as *offset-null tracking,* puts the target and the image on the underside of the sum pattern, where the steep fall-off gives increased rejection of the image contribution relative to the target contribution.

The *symmetrical-ratio* technique uses a pair of special elevation patterns instead of conventional sum and difference elevation patterns. These patterns are designed so that their ratio is an *even* (symmetrical) function of angle, approximately parabolic in shape, with a null on axis, in contrast to the usual monopulse ratio, which is an odd function of angle. The elevation angle of the axis of symmetry is held fixed in a horizontal direction,* and open-loop elevation measurements are made while normal tracking is maintained in range and traverse. A simple analysis shows that under ideal conditions, the open-loop elevation output is the same as it would be if the image were absent, and by use of a calibration function this output can be converted to elevation angle of the target.

A more sophisticated technique is closed-loop tracking with a *double-null tracker*. The difference pattern has not only a primary null, but also a second null at a lower elevation angle. The control loop steers the primary null toward the target direction and at the same time offsets the second null downward from the first by an angle computed by the laws of reflection from a level surface, so that the second null is directed at the image.

A technique that makes use of the imaginary part of the monopulse ratio in addition to the real part that is normally used is known as the *complex indicated*

*More precisely, the axis is pointed in a direction midway between the target and image directions.

angle technique. It is essentially a form of the low-E mode described above, but with the added feature of a calibration function by which target elevation can be determined from the measurements of the real and imaginary monopulse outputs. By using both parts, ambiguities are reduced and a more accurate elevation estimate can be obtained.

Space does not permit detailed discussion of these techniques and others that have been used or attempted. Each technique has certain limitations and disadvantages as well as useful features. The choice of a particular technique depends on the nature of the radar, its functions, and its required accuracy, as well as on practical considerations, including cost.

5.9 MONOPULSE IN ARRAY ANTENNAS

Up to this point, examples and illustrations have been drawn from monopulse systems using reflector or lens antennas. In such a system the sum and difference patterns are determined by the design of the antenna, including the feed and associated microwave components, and the patterns are steered by mechanical rotation of the antenna.

Here we will discuss implementation of monopulse principles in *array antennas*.

5.9.1 Principle of Operation

An array antenna is composed of radiating (and receiving) *elements* arranged in the form of an array. The elements, of which there are usually a large number, may be dipoles, horns, open ends of waveguides, or slots in waveguides. In reception, the sum and the two difference patterns are formed by combining the contributions of the individual elements in certain ways, and the shapes of the patterns are determined by the respective amplitude illumination functions—that is, the relative weights assigned to the individual element contributions in the process of combining them. Separate illumination functions are used for the sum and the two difference patterns. In transmission, the transmitter output power is distributed among the elements, and the pattern (which need not be the same as the receiving sum pattern) is determined by the illumination function or weighting used in transmission. The patterns are steered by imposing an appropriate linear phase progression (phase taper) from element to element along the aperture in the desired direction. Usually the same phase taper is used for the transmitting pattern and the receiving patterns, so all the patterns are steered simultaneously in the same direction. The principal advantage of array antennas is their ability to change beam direction almost instantaneously, without inertia. The principal disadvantages are complexity and cost.

5.9.2 Methods of Steering

Steering requires the ability to change the phases of the individual elements, and an antenna with this capability is commonly called a *phased array*. For rapidity and flexibility, the phase control is usually done electronically. There are three principal methods: (a) *phase steering* by electronically controlled changes in the setting of a variable phase shifter in the feed line of each element or subarray, (b) *time-delay steering* by changes in the settings of variable time-delay devices (needed in wideband systems), and (c) *frequency steering* in series-fed arrays where the incremental path length between successive elements causes a frequency-dependent phase increment.

5.9.3 Space Feeds

There are two categories of feeds for array antennas. A *space feed* (or *optical feed*) is used in certain kinds of arrays, as well as in all reflector and lens antennas. It radiates power to—or receives power from—the reflector, lens, or array through open space. When used in an array antenna, a space feed operates in the same manner as in a reflector or lens antenna, as illustrated in Figures 5.4-2, 5.4-3, and 5.4-4, but the array has the additional capability of rapid beam steering.

A lens type of array (or *transfer array*), illustrated in Figure 5.9-1(a), has elements on the opposite parallel faces, coupled by electronically controlled phase shifters. A reflector type of array, illustrated in Figure 5.9-1(b), has elements on one face; each element is connected through a phase shifter to a short-circuit termination, which reflects the incoming wave back in the opposite direction. In both cases the phase shifters convert the curved phase front into a plane wave front in transmission, and *vice versa* in reception, in addition to performing the steering.

The feed for a space-fed array can be any of the types of monopulse feeds described in earlier sections of this chapter, such as a four-horn feed.

5.9.4 Constrained Feeds

The other category of feeds for arrays consists of various kinds of *constrained feeds,* which distribute power to, or collect power from, the elements of the array through waveguides or transmission lines. The desired amplitude and phase distribution can be achieved in various ways, usually by means of power dividers or combiners, transmission line lengths, and phase shifters. An advantage of constrained feeds is the high degree of control that they permit in the design of illumination functions to obtain desired sum and difference patterns. In addition, they have no spillover and no blockage.

A relatively simple method of obtaining monopulse operation is to divide the array into symmetrical quadrants. The outputs of the elements in each quadrant are summed to produce four signals, which are then combined to form a sum and two differences. This arrangement is equivalent to that of the four-reflector phase-comparison system described in Section 5.5.1 and illustrated in Figure 5.5-1, but it has the additional capability of rapid steering. However, such an arrangement does not permit optimization of both the sum pattern and the difference patterns.

A monopulse feed arrangement that permits essentially independent shaping of the sum and difference patterns so that each can be optimized is illustrated in simplified form in Figure 5.9-2. Part (a) of the figure shows one column of the array. Each symmetrical pair of elements is connected to a hybrid to obtain the sum and difference of the pair. The sums of all the pairs are added with an appropriate set of weights in the column sum combiner to produce the column sum, and the differences of the pairs are added with a different set of weights in the column difference combiner to produce the column difference.

Part (b) of the figure shows how the sum and difference signals of the entire array are obtained. The column sums from all the columns are combined in symmetrical pairs, each pair being connected to a hybrid that produces their sum and difference. The sums and differences of all the column sum pairs go to the array sum combiner and the traverse difference combiner respectively, where they are added with independent sets of amplitude weights to obtain the array sum voltage s and the traverse difference voltage d_α. Likewise, the column differences from all the columns are combined in symmetrical pairs, each pair being connected to a hybrid that produces their sum and difference. The difference is terminated in a dummy load. The sums of all the column difference pairs are

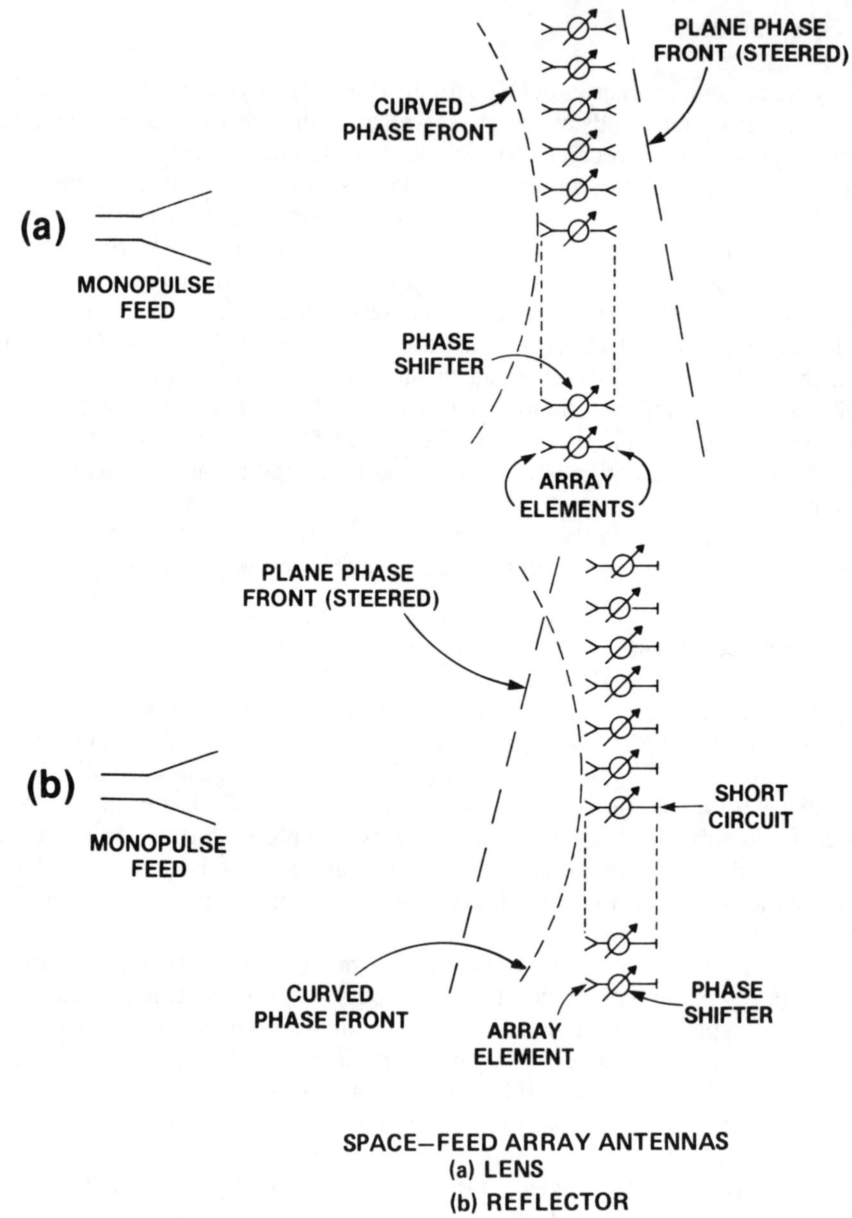

Figure 5.9-1 Space-Feed Array Antennas:
(a) Lens;
(b) Reflector.

added with appropriate weights in the elevation combiner to obtain the elevation difference voltage d_β for the entire array.

In addition to the two examples described above, there are various other ways of obtaining monopulse patterns in constrained-feed arrays. Although it is possible to design arrays for amplitude-comparison monopulse, most array antennas, including the examples above, are of the phase-comparison class, as can be verified by tracing the phase relationships between the sum signal and each difference signal. The sum and each difference are found to be in phase quadrature.

5.10 SOURCES OF ERROR

As was pointed out in Section 5.1, error due to target amplitude fluctuation, which affects sequential-lobing techniques for angle tracking, is eliminated or at

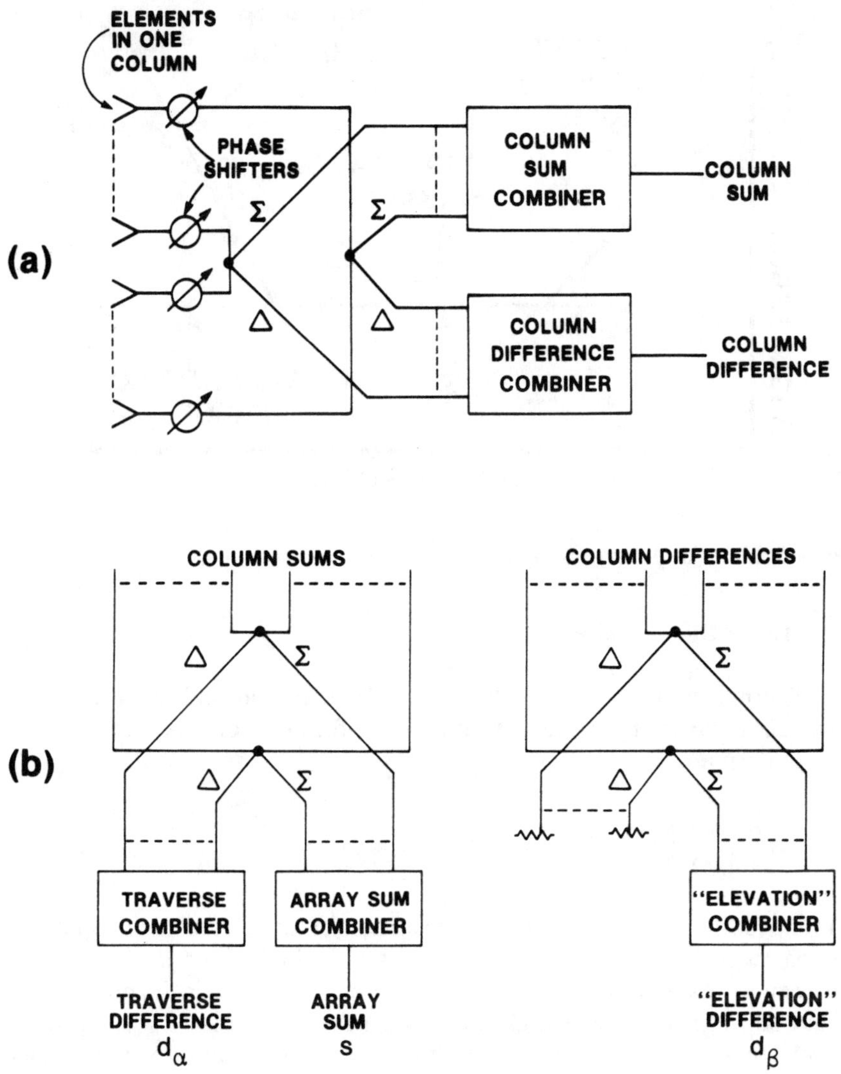

Figure 5.9-2 Array with Nearly Independent Sum and Difference Illuminations:
(a) Formation of Column Sum and Difference;
(b) Formation of Array Sum and Differences.

least greatly reduced by the use of monopulse. There are, however, several other sources of error that affect all angle-tracking techniques, including monopulse, although the errors are generally smaller in monopulse than in the other techniques. Four principal types of errors are discussed below and illustrated in Figure 5.10-1.

5.10.1 Glint

As explained in Section 5.7, typical radar targets have multiple scattering points, which cause the tracking direction to wander as the relative amplitudes and phases change as a result of target motion. This error, known as glint, behaves in a quasi-random manner and is usually quantified in terms of an rms value, which is proportional to the linear extent of the target in the coordinate of interest. Glint error expressed in terms of angle, for a given target, is therefore inversely proportional to target range, as shown by the line labeled $1/R$ in the log-log plot of Figure 5.10-1.

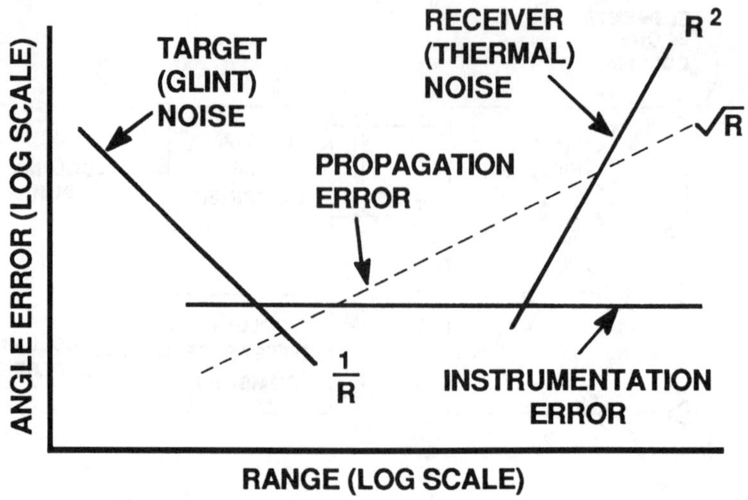

Figure 5.10-1 **Angle Errors** ***versus*** **Range.**

5.10.2 Error Due to Noise

Thermal noise is always present in the sum and difference channels, and causes an error in the angle indication. The formula for this error is given here without derivation:

$$\sigma_\theta = \frac{\theta_{bw}}{k_m \sqrt{2nS/N}} [1 + (d/s)^2]^{1/2} \tag{5.10-1}$$

In this equation, σ_θ is the standard deviation of angle error due to noise, θ_{bw} is the beamwidth (of the sum pattern), k_m is the slope of the d/s monopulse ratio *versus* angle, expressed in volts per volt per beamwidth, n is the number of pulses integrated to form a single angle estimate, S/N is the ratio of signal power in the sum channel to noise power in the difference channel, and d/s is the monopulse ratio at the target angle. (This ratio is zero if the target is exactly on axis.) Since the signal power is inversely proportional to the fourth power of range, the error is proportional to the square of range, as indicated by the line labeled R^2 in Figure 5.10-1.

5.10.3 Propagation Error

Under some atmospheric conditions, an error due to inhomogeneities in the propagation path becomes significant. This error, approximately proportional to the square-root of range, is represented by the dashed line labeled $\sqrt{R}$ in Figure 5.10-1.

5.10.4 Instrumentation Error

There are various mechanical and electrical errors which are independent of range. They include biases, servo noise, misalignments, and various other types of error. They are lumped together here under the generic name of "instrumentation error", indicated by the horizontal line in Figure 5.10-1, which forms a "floor" to the combined system error. In a well designed and maintained monopulse radar, this floor can be 1/200 of a beamwidth or lower.

REFERENCE

SHERMAN
CHAPTER 5

1. Sherman, S.M., *Monopulse Principles and Techniques,* Artech House, Norwood, MA, 1984.

Chapter 6
LOW-SIDELOBE PHASED ARRAY AND REFLECTOR ANTENNAS

by H.E. Schrank

6.1 INTRODUCTION

Low-sidelobe antennas have become an increasingly important component of high-performance electronic systems, particularly those operating in heavy clutter or jamming environments. This chapter discusses the critical developments that have made low-sidelobe antennas a reality despite the fact that in the 1960s, most antenna designers thought that −35 to −40 dB sidelobes were the lowest practical, whereas today −50 dB or lower sidelobes are not uncommon. This breakthrough in sidelobe suppression technology not only directly reduced interference (including clutter, electromagnetic interference (EMI), and electronic countermeasures (ECM)), but also makes other techniques (such as sidelobe cancelation, adaptive nulling, and decoys) more effective and economically practical.

This chapter reviews the historical background of modern low-sidelobe antenna technology; presents some tutorial information, including definitions of terminology; outlines the principles necessary for achieving low sidelobes; then discusses how these principles can be applied to future radar trends, including wider bandwidths, multifunctional agility, and active aperture arrays.

The first ultra-low-sidelobe breakthrough occurred at Westinghouse in the late 1960s, in the development of the slotted waveguide array for the Airborne Warning and Control System (AWACS). Other early low-sidelobe designs that followed were also arrays using various waveguide techniques to achieve the precision amplitude and phase control necessary for low sidelobes. After the feasibility of low sidelobes was demonstrated with array antennas, designers of reflector antennas began to apply the same basic principles to achieve some very successful low-sidelobe reflector antennas, primarily for communications systems, but also for radar applications. Some of these recent successful designs will be discussed here.

6.2 HISTORICAL BACKGROUND

The distributions necessary to achieve sidelobes below −40 dB have been known for many years. These distributions include such names as Hamming, Dolph-Chebyschev, Taylor, Bickmore-Spellmire, and Blackman. However, several factors stood in the way of achieving the desired theoretical sidelobe levels. In the "good old days," the usual antenna sidelobe design approach was to choose an aperture distribution having theoretical sidelobes 5 to 10 dB below the desired level, and then to hope that when the antenna was built the actual sidelobes would come close to the desired level. With this "safety margin" design technique, sidelobes on the order of −35 dB were achieved, but −40 dB or lower were seemingly unachievable. As recently as the mid-1960s, some authors wrote of the impracticality of achieving very low sidelobes. Here are two typical quotes:

> *Practically it is common to design . . . antennas for sidelobe levels of 20 to 30 dB . . . a 20 dB level is considered good, and 30 dB is excellent. A level of 40 dB is considered very difficult to achieve.* (Blake [1], *Antennas,* 1966)

> *A word of caution should be given concerning the ability to achieve in practice low sidelobes with extremely tapered illuminations . . . there is a practical limit beyond which it becomes increasingly difficult to achieve low sidelobes . . . the economic limit . . . seems to be of the order of 35 to 40 dB.* (Skolnik [2], *Introduction to Radar Systems,* 1962)

What was not fully appreciated in the past was the *accuracy* of amplitude and phase distributions necessary to achieve the theoretical low-sidelobe patterns predicted by a particular distribution. Also, the effect of *mutual coupling* between array elements was not thoroughly understood, and techniques for affordable *precision* machining had not yet been developed. In addition, the development of pattern test ranges suitable for measuring low sidelobes was in its infancy.

Nowadays, it is commonplace to achieve ultra-low (below −40 dB) sidelobes, not by luck, but by proper design and fabrication techniques. How and when did this seemingly impossible breakthrough in sidelobe suppression take place?

6.3 LOW-SIDELOBE ARRAY ANTENNAS

6.3.1 The Breakthrough in Sidelobe Suppression

In the mid-1960s, the various technologies matured in the Overland Radar Technology program, which was the forerunner of the AWACS radar. Ultra-low sidelobes were crucial to reduce clutter. A computer-aided design program was devised at Westinghouse by Phil Hacker and Dick McComas for slotted waveguide arrays in which mutual coupling was taken into account.

Network analyzers and range equipment were being developed that allowed precision tests, and computer-aided machining had become a reality. The combination of these developments resulted in the achievement of patterns that were within 1 or 2 dB of theoretical, down to levels below −50 dB. This approach successfully demonstrated the feasibility of achieving ultra-low-sidelobe arrays without need for empirical adjustments or design iterations. It also constituted a real breakthrough in technology, and provided the sidelobe suppression necessary for the AWACS mission.

Figure 6.1 shows Phil Hacker with an engineering prototype model of an ultra-low-sidelobe slotted waveguide array. It was built using eight precision S-band waveguides fed at one end by a tee-junction manifold and terminated by short circuits at the other end. This constitutes a resonant design which produced good patterns, but had very narrow bandwidth. Later designs used a traveling-wave approach to increase the bandwidth.

The resulting AWACS array is shown in Figure 6.2. It has roughly 4000 radiating slots and can be electronically scanned in elevation, while being mechanically scanned in azimuth. This antenna is the first ultra-low-sidelobe array in operational use. Figure 6.3 shows the famous E-3A aircraft with the rotating radome or "rotodome" containing the antenna.

One of the key components of the AWACS antenna is the ferrite rotary-field (Fox) beam-steering phase shifter for scanning in elevation. This high-power phase shifter achieved rms phase accuracies of less than 1° and is essentially frequency independent. The phase shifter was developed jointly by Microwave Applications Group (MAG) and Westinghouse, and is manufactured by both companies.

The AWACS sidelobe achievements constituted a notable advance in antenna technology. When the antenna community thought otherwise, Hacker and McComas [3,4] (and many other contributors) showed that it could indeed be done. However, because of the classification of the program, the success story could not be publicized for years, and it took the antenna community a long time to realize what had been achieved. I think that it is time to take the "bushel" off this bright and significant event in the history of antenna technology.

Figure 6.1 **P.S. Hacker with First Successful Ultra-Low Sidelobe Array.**

Figure 6.2 **AWACS Antenna.**

SCHRANK
CHAPTER 6

Figure 6.3 The E-3A (AWACS) Aircraft in Flight.

6.3.2 Review of Principles and Terminology

To achieve sidelobes lower than those produced by uniformly illuminated apertures, it is well known that a tapered amplitude distribution is necessary. This results in a wider beamwidth and a correspondingly lower directive gain. The wider half-power beamwidth is measured by a *beam-broadening factor* and the reduced gain is expressed as a *taper loss* (or its equivalent "aperture efficiency"), both relative to the uniformly illuminated apertures.

The basic principle, that *lower sidelobes are obtained at the cost of reduced antenna gain and widened beamwidths,* is well known. Various amplitude distributions, such as Chebyschev, Taylor, and others are attempts to minimize (or optimize) the cost of obtaining low sidelobes. Along with the low-sidelobe amplitude distribution, it is also necessary to have a flat (planar) phase distribution. If these two distributions are exactly achieved in an antenna, the theoretical low-sidelobe pattern will inevitably be obtained because the far-field pattern is the Fourier transform of the complex distribution. However, if amplitude or phase errors exist, then the resulting pattern will be degraded, which means that the sidelobes will be higher. This is what invariably happened in the "good old days," before the causes and effects of errors were well understood.

Now, let us review how antenna sidelobe levels are designated and try to define what we mean by low and ultra-low sidelobes. Sidelobe levels can be expressed in several ways, as illustrated in Figure 6.4, relative to the peak of the main lobe or relative to isotropic ("absolute"). Usually, a single number, such as "30-dB sidelobes," means that the peak of the highest sidelobe is 30 dB below the main lobe. *Average sidelobe* levels are often more important and meaningful. As an example, if 10% of the radiated power is in the sidelobes, the average sidelobe level is −10 dBi; if 1%, −20 dBi, and so forth. Average sidelobe levels are usually expressed in dBi (relative to isotropic), while peak sidelobe levels are usually expressed in dB relative to the main lobe peak.

Having reviewed these basics, let us attempt to define what we mean by low and ultra-low sidelobe levels. These are unofficial definitions because IEEE Antennas and Propagation Society has not yet defined such terms. I suggest

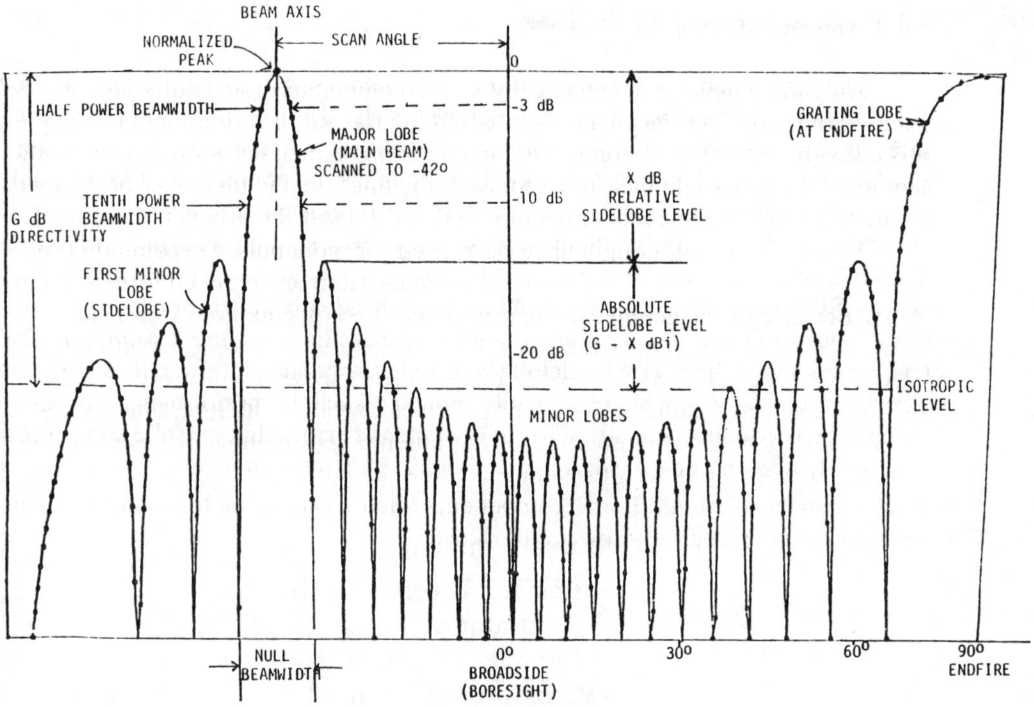

Figure 6.4 Typical Phased Array Pattern Terminology.

defining low sidelobes as 30 to 40 dB below main-lobe peak and ultra-low sidelobes as lower than 40 dB. In terms of absolute (average) levels, low sidelobes could be defined as between −5 and −20 dBi, with ultra-low levels as being less than −20 dBi (Figure 6.5). These exact values are negotiable, but some such definitions can be standardized.

dB		RELATIVE LEVELS	AVERAGE LEVELS
0			
-10			
-20	ORDINARY	(-13 TO -30 DB)	0 TO -5 DBI
-30			
	LOW	(-30 TO -40 DB)	-5 TO -20 DBI
-40			
	ULTRALOW	(BELOW -40 DB)	< -20 DBI
-50			
-60			

Figure 6.5 Sidelobe Definitions (Unofficial).

6.3.3 How to Achieve Low Sidelobes

The key to achieving low and ultra-low sidelobes is to recognize and control the sources of errors that degrade sidelobe levels. These error sources fall into three main categories: *mutual coupling, systematic* (or correlated) *errors,* and *random errors* (see Figure 6.6). Mutual coupling can be compensated in the computer-aided design program used to determine design parameters for the array elements, such as slots, or dipoles. Systematic errors, such as array flatness or manifold coupling must be controlled by tolerances and careful fabrication processes. Systematic errors generally affect *near-in* sidelobe levels or cause periodic error lobes to occur. Random error sources generally affect the *far-out* sidelobe levels, and hence the average sidelobe levels of an antenna pattern. Often, random errors are the factors which ultimately limit the sidelobe performance of an antenna. Figure 6.7 shows a partial list of the types of errors that must be considered in low-sidelobe antenna designs.

- Mutual Coupling
- Systematic Errors
  - Nonflatness
  - Cumulative Coupling
  - Cross Polarization
- Random Errors
  - Phase Shifter
  - Slot Machining
  - Coupling

Figure 6.6 Illumination Error Sources.

Phase Errors

1. Phase Shifters
   - Insertion Phase
   - Differential Phase
   - Quantization Phase
   - Temperature
   - Driver Characteristics
2. Element Placement Error
3. Mismatch
4. Beamformer
5. Radome

Amplitude Errors

1. Phase Shifter - Insertion Loss $f(\Delta\phi)$
2. Beamformers and Couplers
3. Polarization

Failed Elements

Figure 6.7 Types of Errors.

To give the reader a feel for the quantitative effects of random errors, consider the combination of random amplitude and random phase errors (see Figure 6.8, after Ruze [5]). It appears from this that a −10 dBi average sidelobe level requires a 10° rms phase-only tolerance or a 1.5 dB rms amplitude-only tolerance. Combinations of phase and amplitude may be obtained from the circular curves (e.g., 7° rms phase with 1 dB rms amplitude).

However, these two random errors are not the only error sources, and the others must also be considered and budgeted. For example, if digital phase shifters are used, there is also a bit quantization error, which is pseudorandom in nature. For a four-bit phaser, this error alone is 6.5° rms. A list of random error sources that applies to most phased arrays is shown in Figure 6.9. Since the various random errors are uncorrelated, their variances add, as shown in the equation at the bottom of the figure. For a −20 dBi average sidelobe level, all variances must add to a total of less than 0.01. This leads to stringent tolerances on components, if ultra-low sidelobes are to be realized.

A general principle of how various error sources affect the pattern sidelobes is illustrated in Figure 6.10. The principle is this: widespread errors (such as array nonflatness) cause large sidelobe degradation over small angular regions, while localized errors cause smaller degradations over wider angular regions.

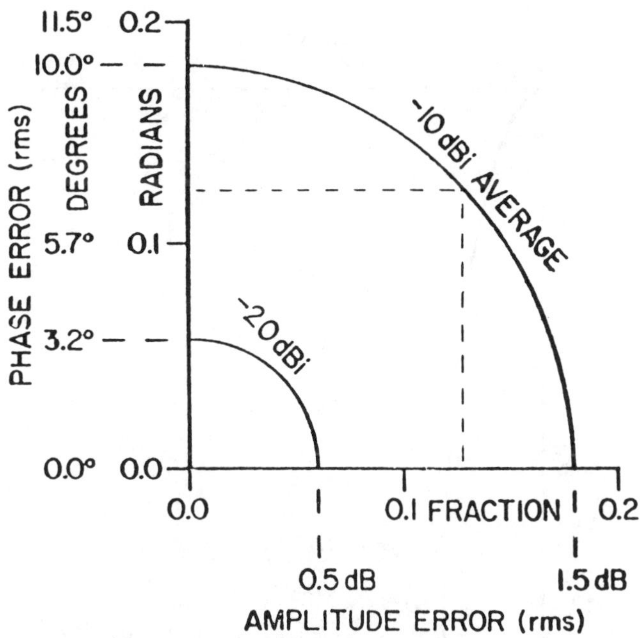

$$\overline{SL_o}\,(\text{dBi}) = 10\,\text{LOG}_{10}\,\pi\,(\overline{\phi^2} + \overline{\Delta^2})$$

Figure 6.8 **Combination of Random Errors (from Ruze [5]).**

SCHRANK
CHAPTER 6

Insertion Phase	(σ_{IP})
Differential Phase	(σ_{DP})
Phase Quantization	(σ_Q)
Element Location	(σ_D)
Insertion Loss Variation	(σ_{IL})
Coupler Amplitude Variation	(σ_{CA})
Coupler Phase Variation	(σ_{CP})
VSWR Variation	(σ_Γ)
Curvature	
Temperature	
Polarization	

Note: Since independent variances add, the total variance is

$$\sigma_T^2 = \sigma_{IP}^2 + \sigma_{DP}^2 + \sigma_Q^2\ {}^{c}\ \sigma_D^2 + \sigma_{IL}^2 + \sigma_{CA}^2 + \sigma_{CP}^2 + \sigma_\Gamma^2$$

For an average sidelobe level 20 dB below element gain,

$$\sigma_T^2 < .01$$

Figure 6.9 Summary of Errors.

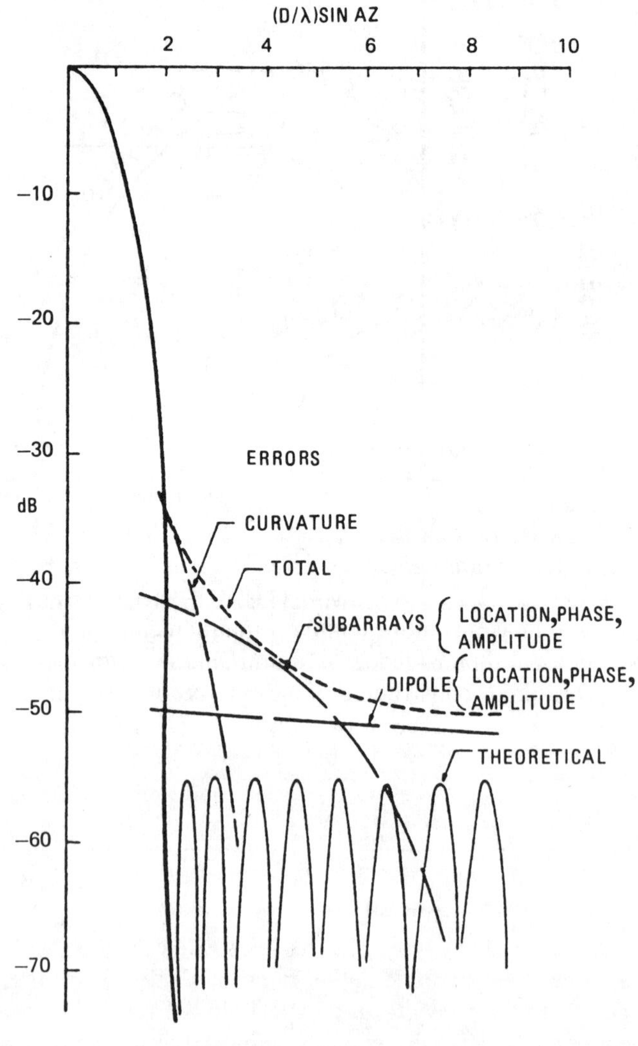

Figure 6.10 Effects of Various Errors.

6.3.4 Slotted Waveguide Arrays

It is no accident that slotted waveguide arrays were the first to achieve ultra-low sidelobes. Every point in the distribution is individually determined by slot location and shape, and these can be readily milled with exactitude. The transmission line between slots is a very short, simple, and precise length of waveguide. Perhaps even more importantly, since all the slots are fed serially, design errors occur gradually over the length of the array. Such gradual errors cause error sidelobes that fall under or near the main beam, where they are unimportant or masked by contour deflections.

As with conventional arrays, a variety of slots is suitable for low sidelobes. The slots can be broadwall, sidewall, series or shunt, and they can be spaced to be resonant or nonresonant for wider bandwidth. Slots are either displaced or tilted to achieve coupling. Tilting requires a means of suppressing cross-polarized radiation. Edge slots are attractive in that they allow the use of a controlled trough depth to suppress such radiation. Since the slots are generally between one-half and one free-space wavelength apart, which is nearly half a guide wavelength, they normally require phase reversals between neighbors. Periodic errors introduced by the reversals must be suppressed.

One of the gradual "errors" might be considered to be the slot-to-slot phase accumulation at an off-center frequency. Although large, it causes no sidelobes, only a beam shift. For applications that can tolerate or even employ such a shift, it is not a detriment. For others, such as those with wide instantaneous bandwidth, the steering, or "squint" as it is known, requires a different type of array.

6.3.5 Other Low-Sidelobe Array Designs

The now well understood and fairly advanced technology of low-sidelobe antenna design has been applied to many other antennas since the pioneer AWACS design. Recent applications of low-sidelobe technology to other Westinghouse systems include:

- The F-16 radar antenna, which is a gimballed fixed-beam array, using broadwall rather than narrow-wall slots (illustrated in Figure 6.11);
- The Electronically Agile Radar (EAR) array, which uses dielectric filled circular waveguide radiators with a different (dual-mode) ferrite phase shifter and has two-axis scan capability (illustrated in Figure 6.12);
- The TPS-70 slotted waveguide array, replacing the TPS-43 reflector antenna for 3D radar surveillance (Figure 6.13).

All of the above designs used a form of waveguide technology and achieved moderate bandwidth performance.

Other recent developments in low-sidelobe arrays include the Selenia RAT-31S, the Marconi-Martello, and the Thomson-CSF TRS-2215 antennas described in Chapter 2. These are planar arrays of dipole or helix elements, rather than waveguide radiators.

6.3.6 A Broadband Dipole Array

To demonstrate that low-sidelobe technology is not restricted to waveguide arrays, an experimental ultra-low-sidelobe array of broadband dipoles was designed and built at Westinghouse. This 512-element fixed-beam array (no phase shifters) is shown in Figure 6.14. It achieved ultra-low-sidelobe performance, as illustrated by the typical measured pattern shown in Figure 6.15, over almost an octave band (800 to 1400 MHz).

To achieve this wideband performance, a corporate air-dielectric ("suspended") stripline feed manifold was necessary. In a corporate manifold, the path lengths to each radiator are made equal by means of a branching tree network of

SCHRANK
CHAPTER 6

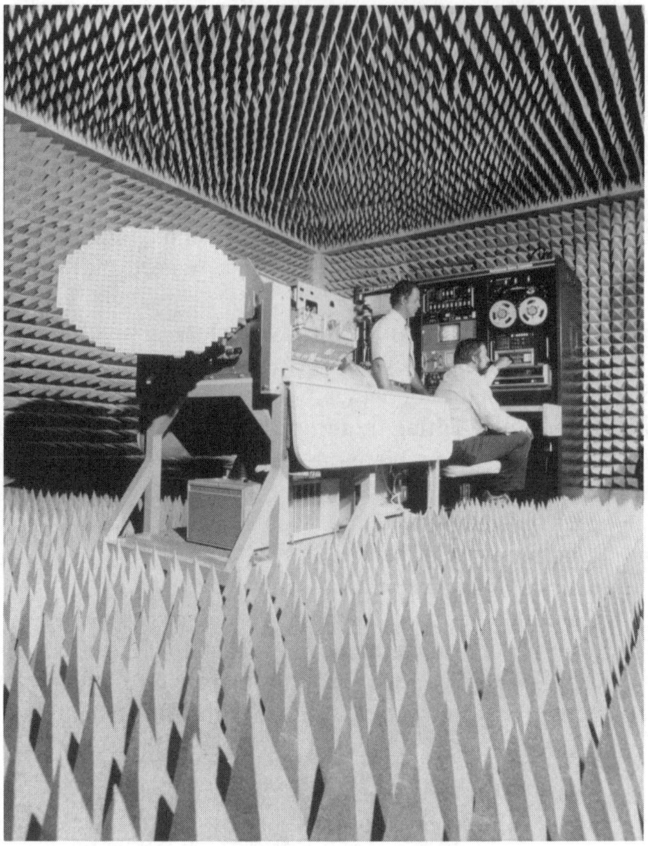

Figure 6.11 Slotted Waveguide Array Antenna for the F-16 Radar.

Figure 6.12 Two-Axis Scanned Array for the Electronically Agile Radar (EAR).

Figure 6.13 **TPS-70 Slotted Waveguide Array Antenna.**

Figure 6.14 **Broadband Ultra-Low-Sidelobe Dipole Array (Fixed Beam).**

SCHRANK
CHAPTER 6

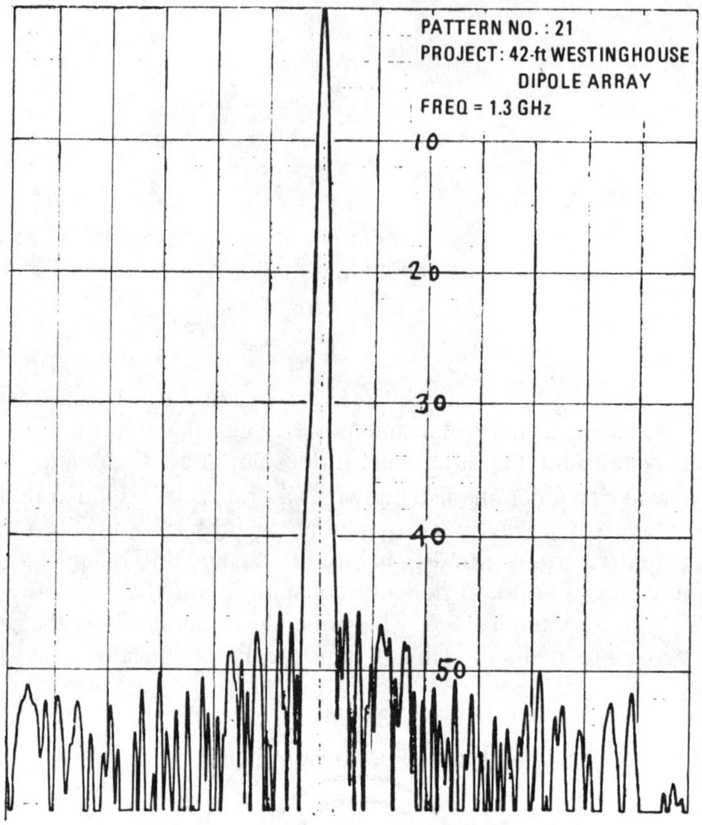

Figure 6.15 Typical Pattern of Ultra-Low-Sidelobe Dipole Array.

lines and couplers, resulting in a very wideband capability. The suspended stripline was necessary to achieve the phase and amplitude precision required for ultra-low sidelobes. This precision cannot be obtained by using printed circuit-board techniques because of dielectric constant variations inherent in large boards. For example, a 10-ft length of line on a circuit board made of dielectric constant 2.50 ± 0.01 has a 70° phase error spread at S-band.

An entirely new situation exists in this case as compared to the slotted waveguide, since the path to adjacent elements can pass through entirely different couplers and lines, although path lengths are designed to be equal. Errors are no longer gradual, but are less systematic, so that large wide-angle sidelobes result from modest tolerances. Considering the many individual components in the path to a radiator, exceptional precision is required in each. Couplers must be directional (four-port) rather than reactive types, and power division must be exceptionally flat across the band.

6.4 LOW-SIDELOBE REFLECTOR ANTENNAS

Although the first ultra-low sidelobe antennas were phased arrays, a number of recent reflector designs have achieved noteworthy success in sidelobe suppression, using various techniques by workers in the US, Europe, and Japan. The reasons that the successful breakthroughs in low-sidelobe technology occurred on phased array designs rather than on reflectors is not surprising, and is discussed above [10,11]. For either type of antenna, the same basic principle holds: provide the necessary aperture distributions (amplitude and phase) and you shall get the desired low-sidelobe pattern. Success or failure in either case depends on how well various error sources can be controlled in the design, fabrication, and assembly of the antenna.

6.4.1 Basic Reflector Limitations

SCHRANK
CHAPTER 6

Figure 6.16 illustrates a basic axisymmetric reflector (paraboloidal dish) antenna with a point-source feed at the focus. The feed illuminates the reflector with its primary pattern to produce an amplitude (and phase) distribution in the aperture plane, which then Fourier transforms into a secondary pattern in the far field. The primary illumination that misses the reflector produces spillover lobes at angles corresponding to the *F/D* ratio geometry, and diffraction effects at the reflector edge produce wide-angle and rearward lobes. The feed, which is typically a simple waveguide horn at microwave frequencies, produces a monotonically shaped amplitude distribution as shown solid, whereas a low-sidelobe distribution generally requires an inflected shape as shown dashed. The distribution shape and the edge taper determine the achievable near-in sidelobe levels. However, further degradation of achievable near-in sidelobe suppression is caused by blockage of the antenna aperture by the feed and its support structure.

Another factor which contributes to the basic limitations of sidelobe control is the reflector surface accuracy. This affects both the phase and amplitude distributions in the aperture plane, and can degrade wide-angle lobes as well as the near-in sidelobes [12].

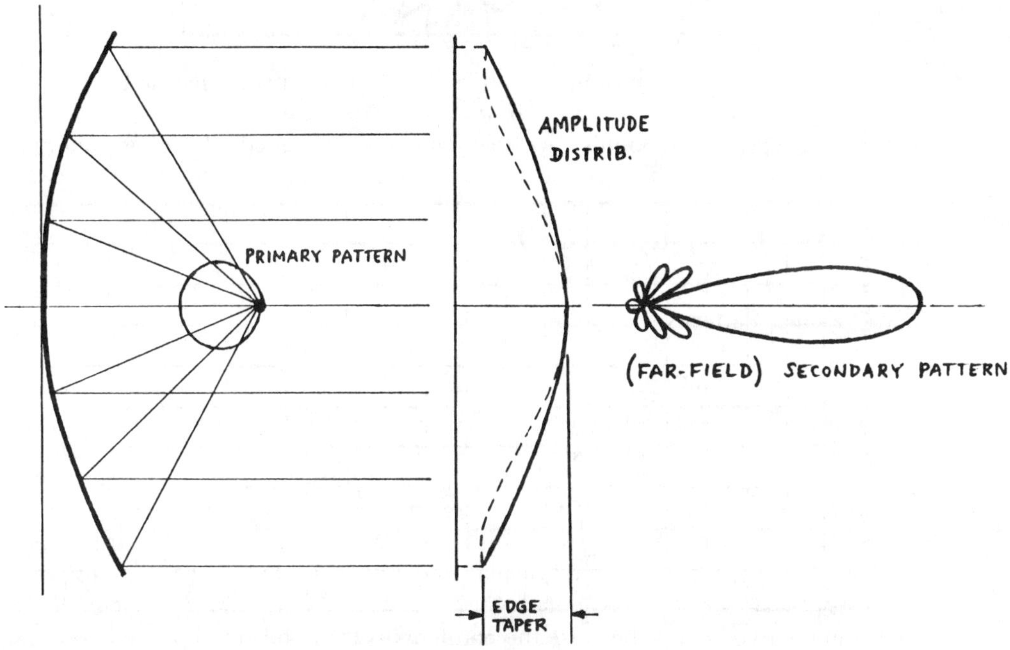

Figure 6.16 Basic Axisymmetric Reflector Showing Typical Amplitude Distribution from a Simple Feed Compared with Desired Low-Sidelobe Distribution (Dashed).

Figure 6.17 (from an old classic reference [13]) illustrates a typical pattern produced by early focal-fed axisymmetric paraboloids. Near-in sidelobe levels of −20 to −25 dB were typical, with −30 dB or lower difficult to achieve. Front-to-back lobe ratios were generally about 40 to 50 dB, depending on the antenna size (gain) and the edge illumination (taper).

In summary, we see that the generic axisymmetric single reflector antennas with simple feeds have four basic limitations in terms of sidelobe suppression:

1. Feed pattern shaping limitations;
2. Aperture blockage effects;
3. Spillover and edge diffraction effects;
4. Surface accuracy effects.

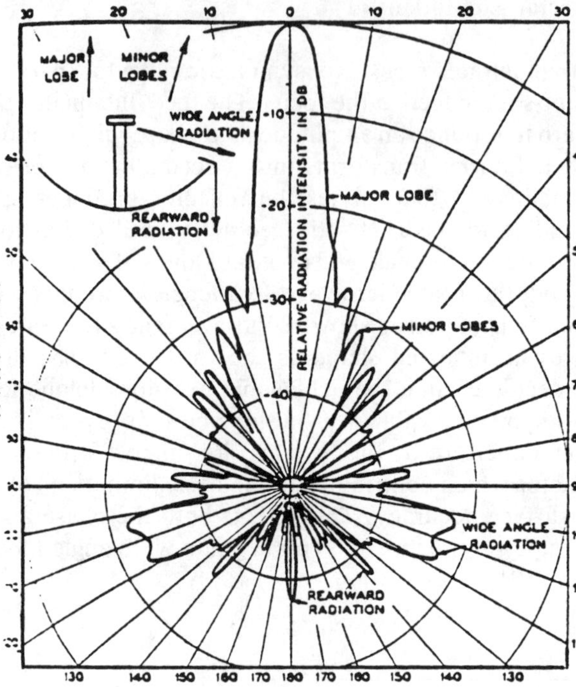

POLAR PLOT

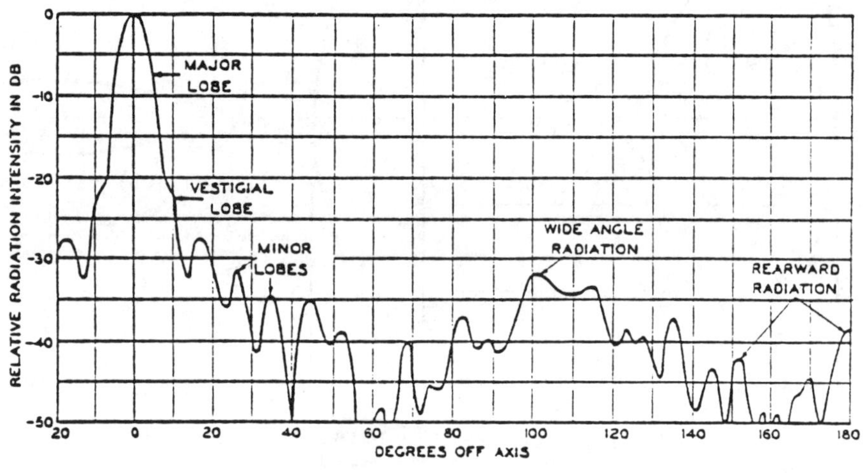

RECTANGULAR PLOT

Figure 6.17 Typical Pattern for Focal-Fed Axisymmetric Paraboloid (from Cutler, King, and Kock [13]).

Dual-reflector systems, such as the axisymmetric Cassegrain and Gregorian geometries, essentially suffer from the same limitations, except that spillover and edge diffraction can occur at both the primary and the subreflector edges. However, some improvements can be realized; for example, aperture blockage can be avoided by application of polarization twist schemes, and low-sidelobe amplitude illuminations can be more precisely designed by special shaping of the two reflectors. These techniques will be discussed later.

6.4.2 Early Sidelobe Suppression Efforts

A number of reflector edge treatments have been used to reduce spillover and diffraction backlobes. As shown in Figure 6.18, lining the rim of the dish with microwave absorber provides some minor improvement, and "castellating" the rim [14] can improve the front-to-back ratio significantly. Similar backlobe suppression along with spillover reduction has been achieved for many years by the use of absorber-lined shrouds or "tunnels." These are cylindrical tubes of metal (or dielectric) lined on the inside with microwave absorbing material to catch the spillover radiation and reduce diffraction effects. Although this technique is not new, some relatively recent work by Dybdal and King [15] at 92 GHz shows impressive results (see Figure 6.19) for a tunnel with length of twice the aperture diameter. Better than 60-dB front-to-back ratios can be achieved.

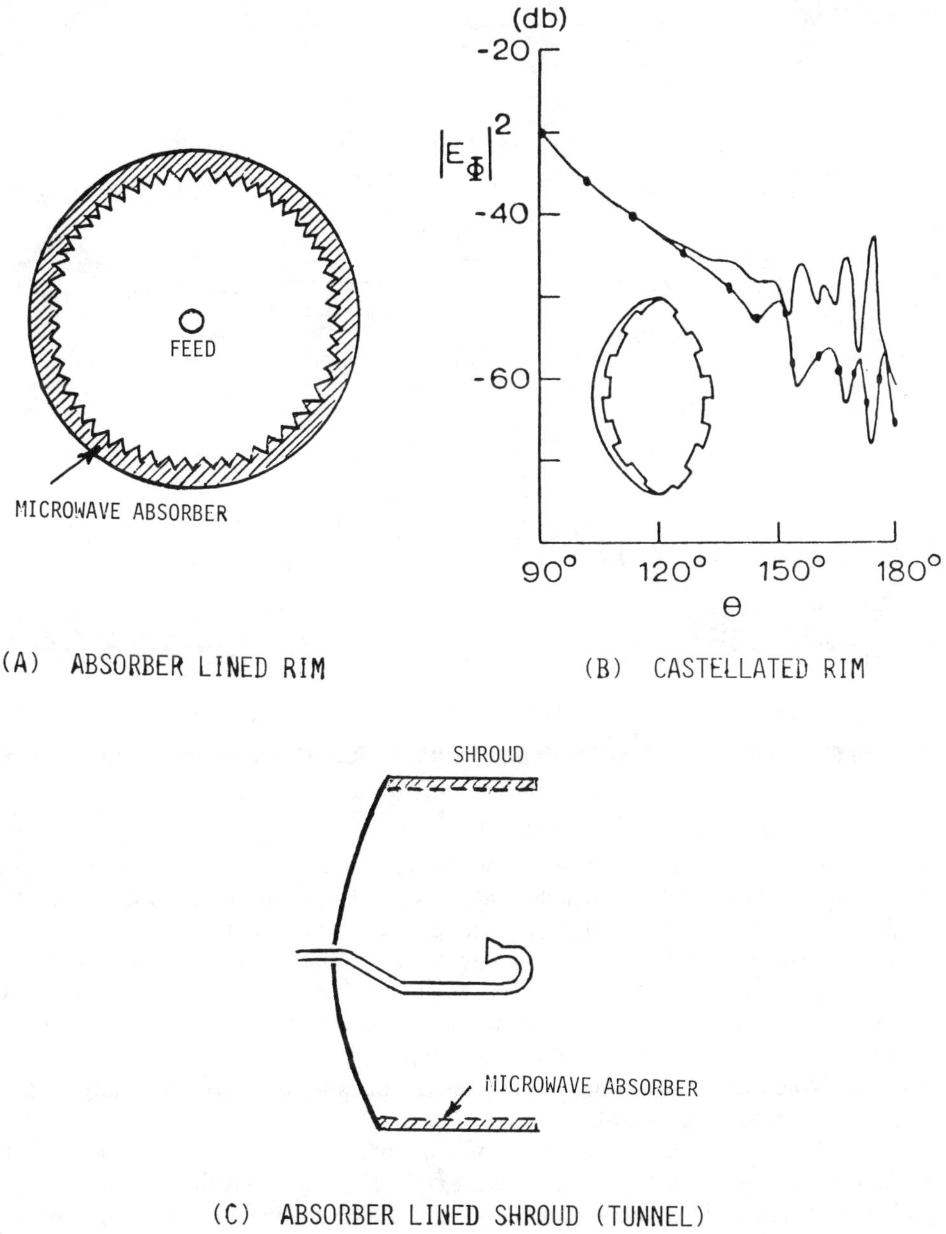

Figure 6.18 Reflector Edge Treatments to Reduce Spillover and Diffraction Lobes.

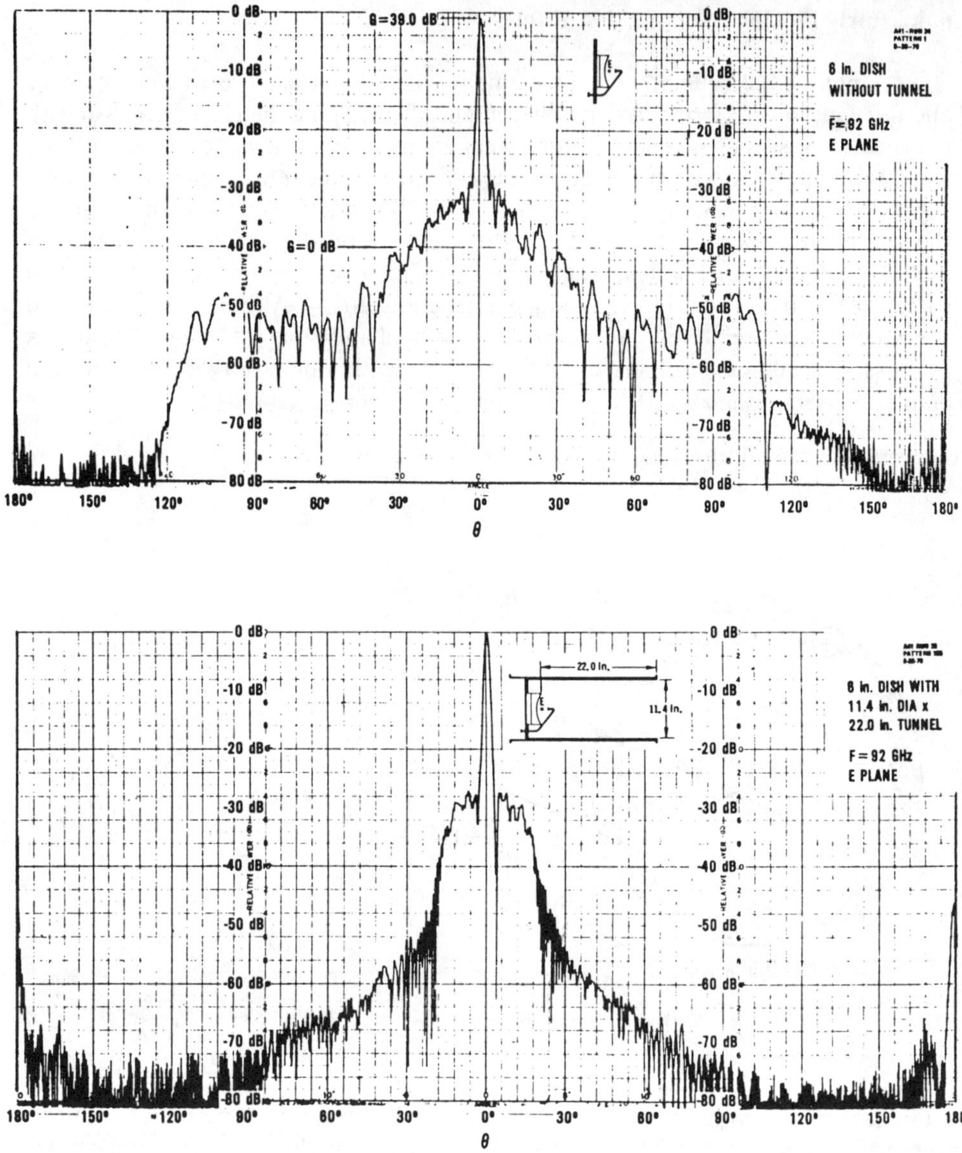

Figure 6.19 Patterns of Dish, with and without Long Absorber-Lined Tunnel (from Dybdal and King [15]).

Another successful technique for reducing wide-angle and rear lobes is the well known horn parabola, which is an offset-fed portion of a paraboloid fed by a flared waveguide horn with walls extending to the parabola, except at the exit aperture (see Figure 6.20). Rectangular horn parabolas are commonly used for commercial microwave links, allowing back-to-back antennas with well over 70 dB isolation [16]. A characteristic diffraction lobe at 90° was successfully reduced by means of saw-toothed (multiple-edge) blinders at the sides of the horn aperture [17,18]. Note that the resulting sidelobe envelope is superior to a comparable dish with an absorber-lined shroud.

Although the near-in sidelobes of horn parabolas are not very low (about −20 dB), the wide-angle sidelobes are extremely low, such that a conical horn parabola antenna designed at Bell Telephone Laboratories had one of the lowest noise temperatures ever achieved. According to Blake [1], this antenna (oriented toward the zenith) achieved a noise temperature of 4.5 K (at 5.65 GHz), of which only 2 K was caused by ground radiation entering through minor lobes. The

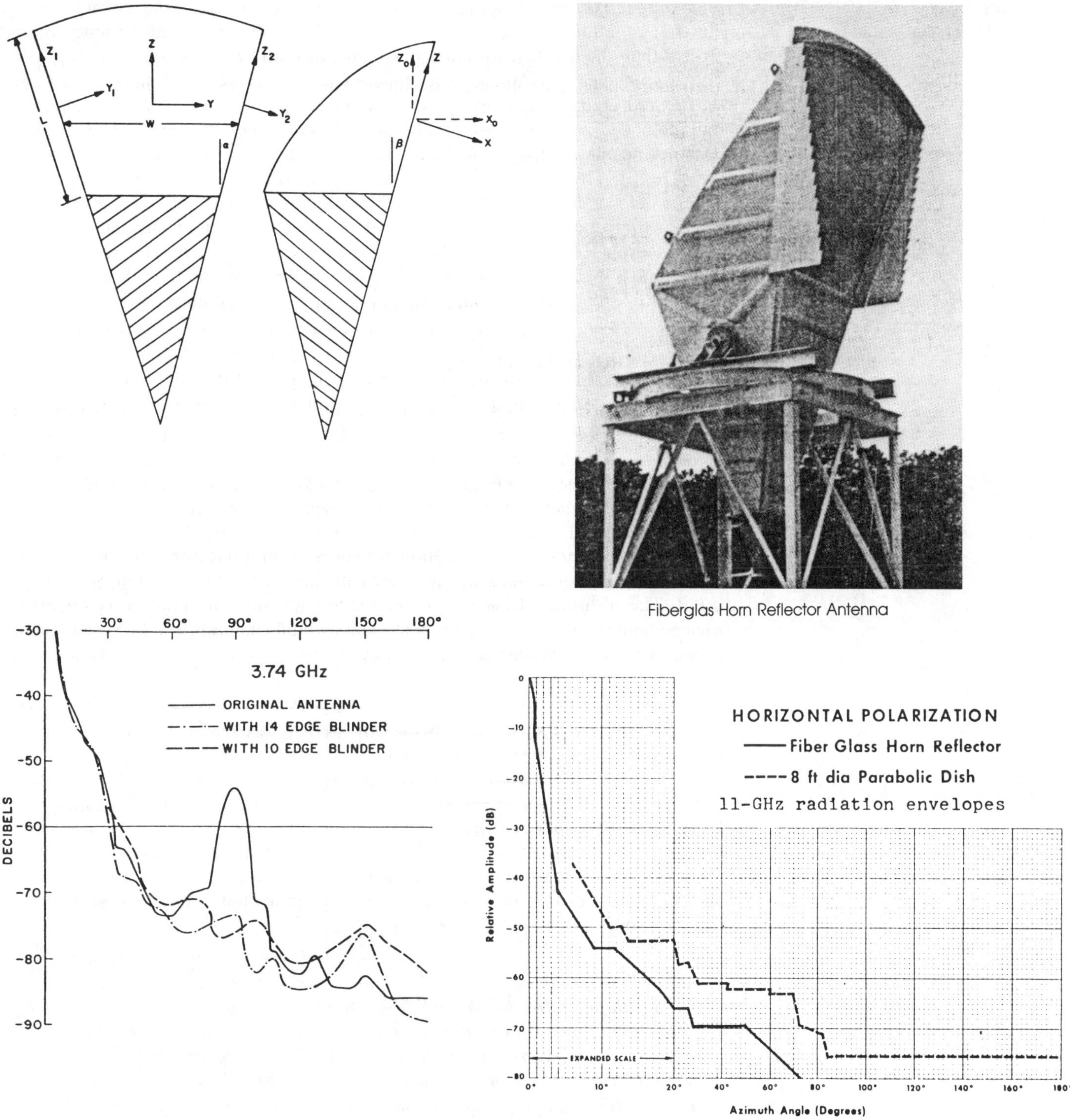

Figure 6.20 Horn Parabola Showing Wide-Angle Sidelobe Suppression (from Thomas [17] and Siller [18]).

remaining 2.5 K was the zenith "sky" noise (tropospheric and cosmic). This illustrates the fact that for some systems, wide-angle suppression is much more critical than the levels of the first few near-in sidelobes.

Another early effort to reduce sidelobes in one preferred plane (cut) is by the method of aperture shaping. An example of this method is seen in comparing the sidelobe levels of a uniformly illuminated square aperture, where the sidelobes in the principal planes are −13 dB, but in the diagonal planes the highest lobes are −26 dB. Thus, a diamond-shaped aperture (the square turned 45°) provides lower sidelobes in one preferred plane (usually azimuth) at the expense of higher sidelobes in other planes. Various other shapes, including cosine-squared have been used by Koch [19] to achieve single-plane sidelobe levels below −30 dB, but for most system applications this approach is not satisfactory. Most systems require low sidelobes in all angular space, not just in one plane.

6.4.3 More Recent Approaches

The polarization-twist technique has been successfully used [20] to avoid blockage effects on near-in sidelobes of Cassegrain and similar dual-reflector antennas. Figure 6.21 from Dahlsjo [21] shows a monopulse antenna design that achieved low sidelobes in both the main and cross-polarization patterns. However, this technique is limited to single linear polarization antennas, and thus cannot be used for systems requiring polarization diversity or circular polarization.

Another way to avoid aperture blockage is by means of offset feeding, as illustrated for single- and dual-reflector antennas in Figure 6.22. Offset feeding, together with the use of more advanced feeds, has proven to be a very effective way of achieving low and ultra-low sidelobe reflector antennas. An excellent tutorial review of this technology has been published by Rudge and Adatia [22] in the 1978 IEEE *Proceedings*.

While offset feeding complicates the analysis and the construction of such antennas, its potential for achieving high performance reflectors is significant, particularly using dual reflectors. Not only are the sidelobe limitations of aperture blockage avoided, but larger, more sophisticated (multimode, corrugated, and array) feeds can be used to achieve the desired low-sidelobe aperture distributions. In addition, the synthesis technique of Galindo-Israel *et al.* [23] can be applied to achieve not only low sidelobes, but also low cross-polarization and high aperture efficiency. Better than −32 dB first sidelobes and aperture efficiencies approaching 70% [24] are reported by Rudge for both predicted and measured results. Quoting Rudge:

> *These results indicate that a good efficiency sidelobe compromise is realizable with offset systems. Computer-aided design methods . . . combined with accurate vector-radiation models for the primary feed, are now developed to the point where fine optimization . . . is both practical and realistic. . . . performance in the near-in region can be predicted to good accuracy for low sidelobe application. . . . the far-field sidelobes decay rapidly with angle from the boresight and the wide-angle radiation can be largely attributed to primary feed spillover . . . this can be dealt with by a combination of primary feed design and hooding techniques . . .* [Here "hooding" means use of absorber-lined shields.]

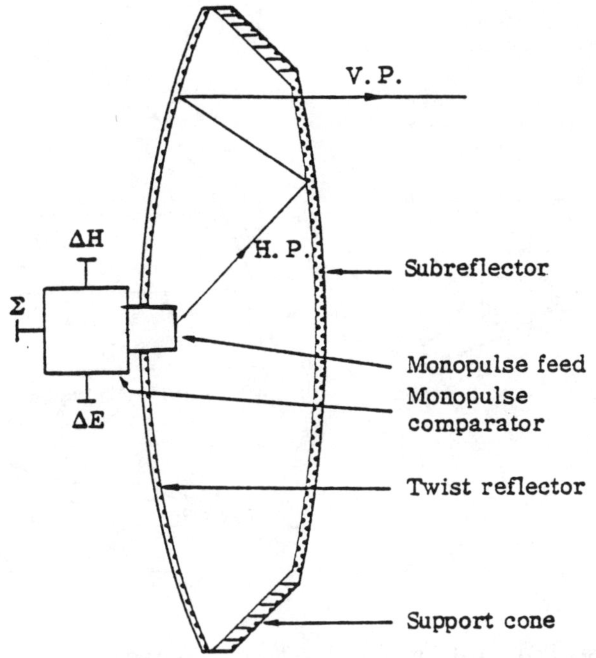

CASSEGRAIN ANTENNA

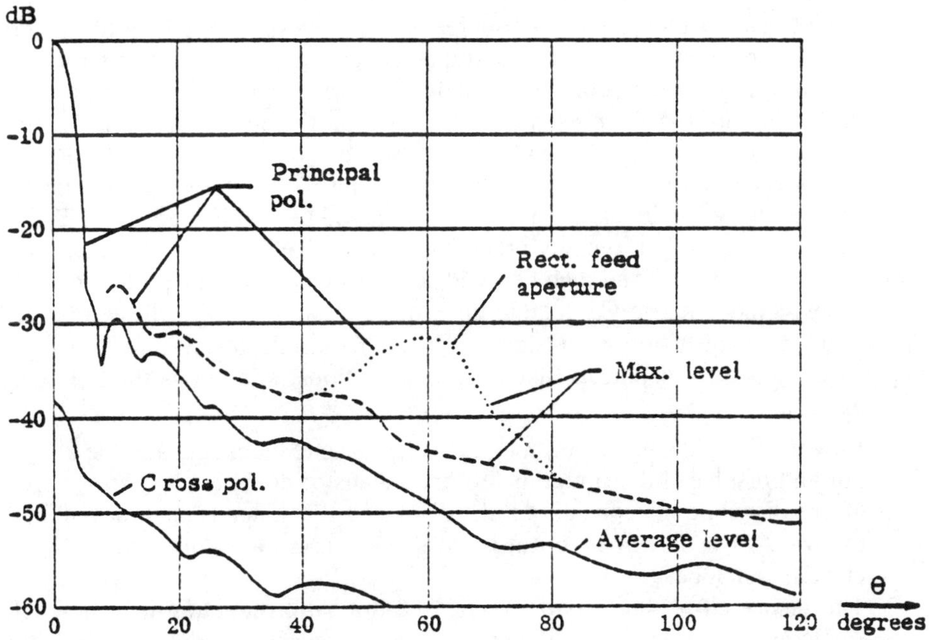

MEASURED SUM PATTERNS

Figure 6.21 Polarization-Twist Low-Sidelobe Cassegrain Antenna (from Dahlsjo [21]).

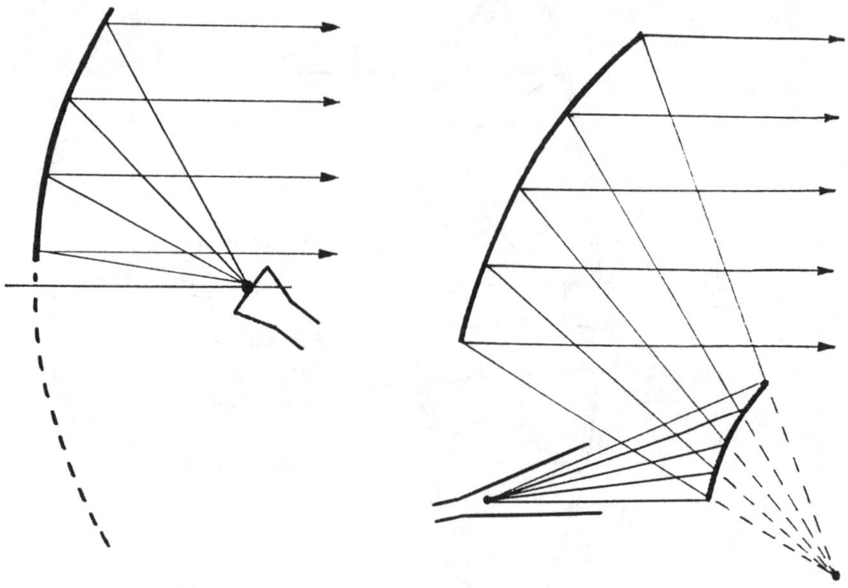

Figure 6.22 Offset-Fed Single and Dual Reflectors to Avoid Aperture Blockage.

6.4.4 A Possible Alternative to Offset Feeding

An interesting alternative to offset feeding is suggested by A.C. Ludwig [25]. For circular apertures with circular central blockage, Ludwig has derived modified Taylor distributions (called doughnut and displaced distributions) that provide low wide-angle sidelobes in the presence of blockage, at the cost of allowing the first near-in sidelobe to be higher. Figure 6.23 shows the computed patterns for three cases of blockage radius (i.e., r_b = 0.05, 0.10, and 0.20). Note that beyond the first sidelobe, very low levels are achieved compared to the blocked 40-dB Taylor case. Ludwig argues that in many practical systems, having low wide-angle sidelobes is most important, and high first and second sidelobes can be tolerated. For such cases, axisymmetric Cassegrain antennas can be used with these modified displaced Taylor distributions designed into the feed-reflector system. Ludwig's paper also considers displaced Blackman distributions, and draws the following conclusions:

1. Low wide-angle sidelobes can be achieved despite blockage, and the doughnut and displaced distributions are one means of doing it.
2. Of the distributions considered, the displaced Blackman is best for reducing the wide-angle sidelobes, but it is worse in terms of aperture efficiency and close-in sidelobes.
3. The displaced $p = 1$ Taylor provides a dramatic reduction in wide-angle sidelobes, good aperture efficiency, and good close-in sidelobes, with the exception of the first sidelobe.

6.4.5 Some Successful Radar Antenna Designs

In the low-sidelobe reflector designs described briefly here, the first by Fante [26] is a basic analytic study with experimental verification, and the second by Scudder and Yorinks [27] is a developmental design for a surveillance radar application. The remaining examples are all for satellite communication applications, where a strong motivation for low-sidelobe designs is the recent decision to decrease the spacing of geosynchronous satellites from 4° to 2°.

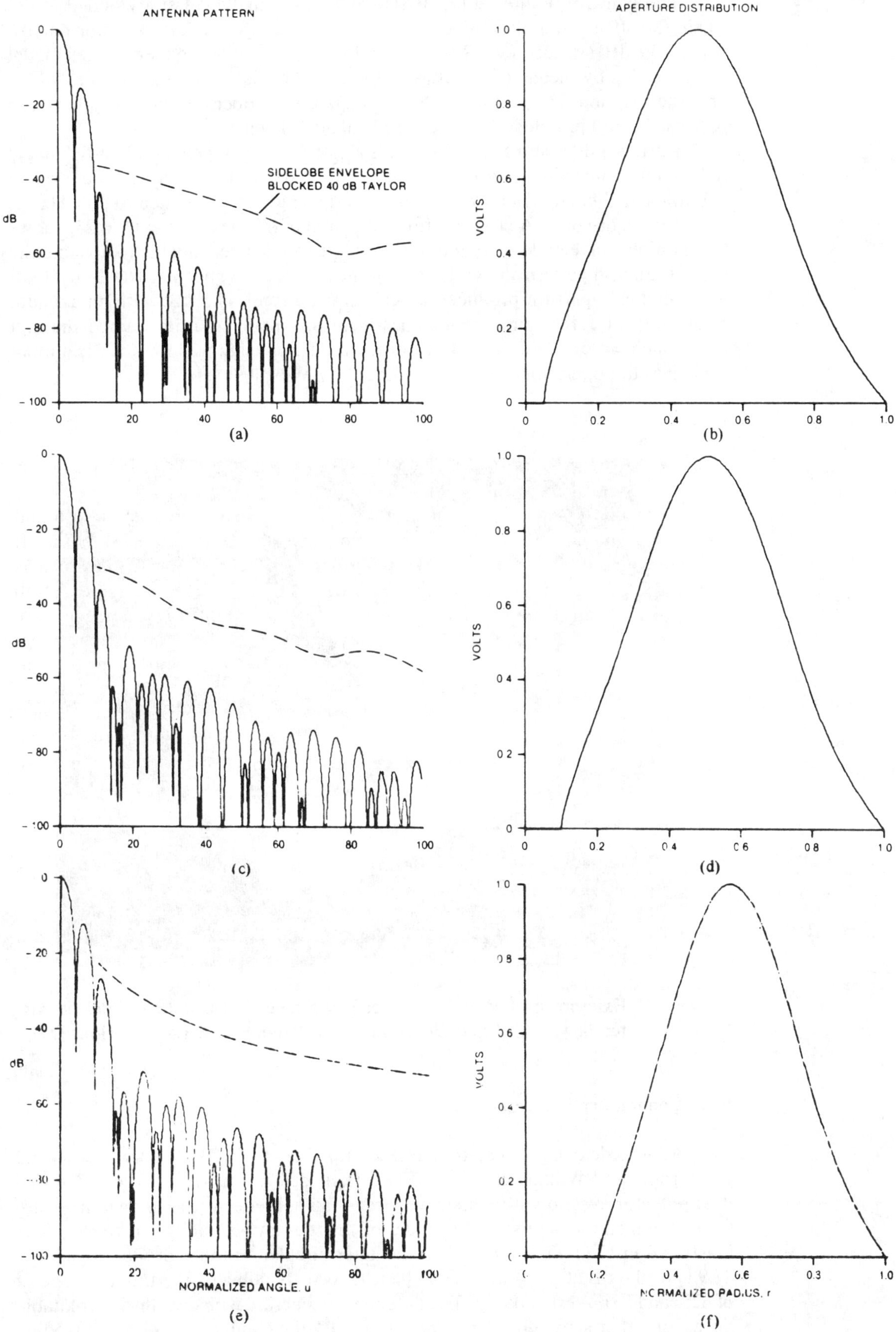

Figure 6.23 Computed Patterns for Blocked Circular Apertures, Using Displaced $p = 1$ Taylor Distribution: (a),(b) $r = .05$; (c),(d) $r = .1$; (e),(f) $r = .2$ (from Ludwig [25]).

The work by Fante [26] at RADC is a basic analytical study aimed at designing an offset parabolic cylinder and feed to obtain a secondary radiation pattern with −50 dB sidelobes at S-band (3.1 to 3.6 GHz). The experimental model achieved this by means of a multiprobe-fed rectangular waveguide feed in which only the TE_{10} and TE_{30} modes were excited in their proper proportions. Measured patterns came very close to the desired sidelobe level.

R.M. Scudder and L.H. Yorinks [27] at RCA developed an offset-fed single reflector antenna with multiple stacked elevation beams for a tactical radar system operating at S-band. Their design was tested using a model scaled to 36 GHz, as shown in Figure 6.24. A cross-sectional diagram of this design, Figure 6.25, shows the multiple waveguide feed and the use of absorber-lined shields to achieve the typical azimuth pattern shown for one of the six elevation beams. The 24-by-14.4-inch elliptical aperture produced a peak directivity of 45.5 dB, with an azimuth beamwidth of 1.4°. Azimuth sidelobes were less than −50 dB beyond the first two, which were −38 and −44 dB, or lower. This design was for a tactical radar similar to the operational AN/TPS-43 built by Westinghouse.

Figure 6.24 Experimental Model of Offset-Fed Single Reflector Low Sidelobe Antenna with Multiple Elevation Beams (from Scudder and Yorinks [27]).

6.4.6 Communication Antenna Designs

A low-sidelobe earth station antenna for the 4 to 6 GHz band was designed by Burdine and Wilkinson [28] at GTE. Their design incorporates an offset-fed dual reflector system with a concave Gregorian subreflector, as shown in Figure 6.26. A full-size demonstration model of this design was built and tested. It is illustrated in Figure 6.27. Its 7.6-m diameter circular aperture produces gains of 47.9 and 50.8 dB at the 4 and 6 GHz bands, corresponding to aperture efficiencies of 62% and 55%, respectively. The antenna can operate with dual linear, rotatable linear, or either sense circular polarizations. By the combination of offset feeding to eliminate aperture blockage, reflector shaping with good surface accuracies,

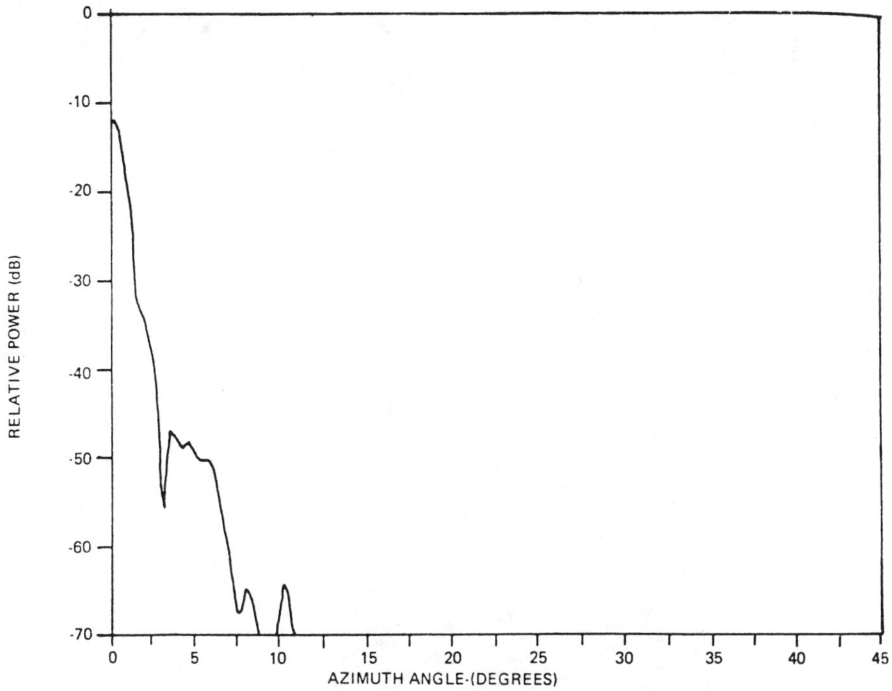

Azimuth Pattern of Baseline Antenna

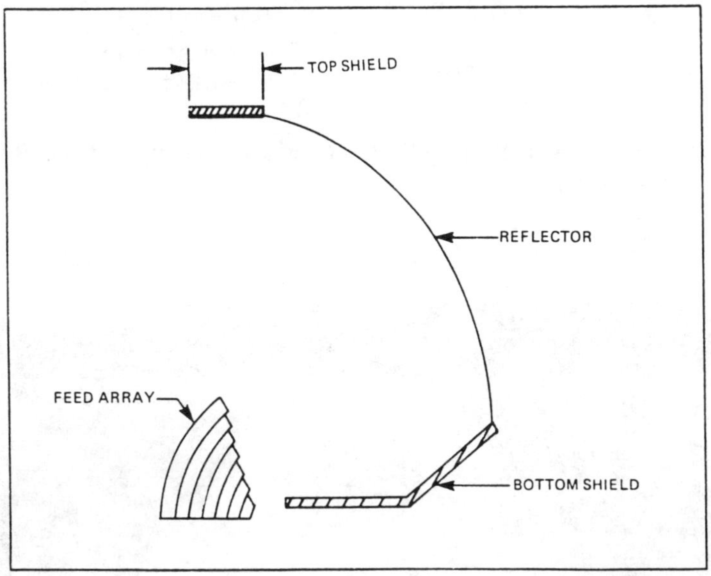

Side View of Reflector showing Location of Top and Bottom Absorber Panels

Figure 6.25 Reflector Cross Section and Typical Measured Azimuth Pattern.

and the use of strategically located absorber-lined shields, the azimuth and elevation pattern sidelobes have been successfully suppressed well below the G = 32–25 log θ maximum sidelobe envelope recommended by FCC, INTELSAT, and CCIR regulations at the time (1979) of this demonstration. (Recently this requirement has been tightened to 29–25 log θ, which the GTE design also meets by a good margin.) A typical azimuth pattern measured at 6 GHz is shown in Figure 6.28. Cross-polarization is better than −34 dB in both bands.

SCHRANK
CHAPTER 6

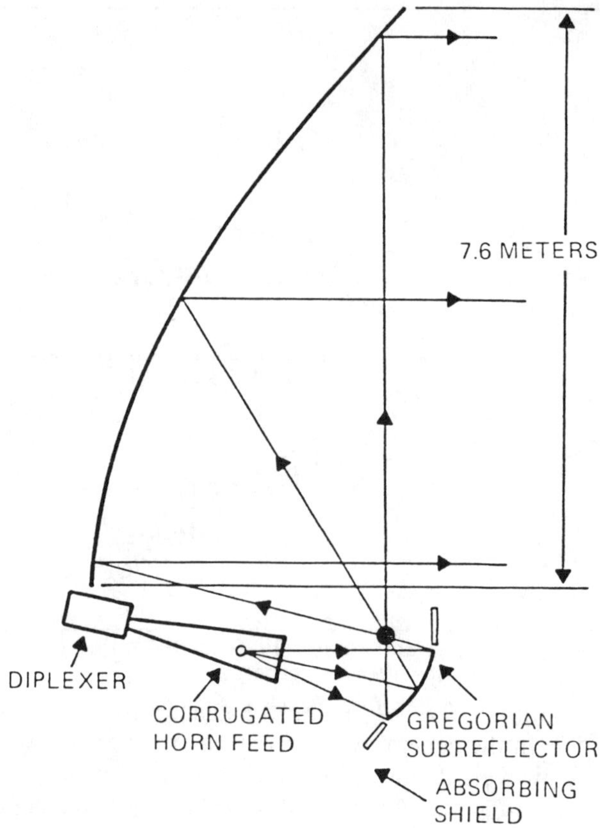

Figure 6.26 Low-Sidelobe Dual-Reflector Gregorian Antenna (from Burdine and Wilkinson [28]).

Figure 6.27 Offset-Fed Dual-Reflector Antenna with Concave Gregorian Subreflector (from Burdine and Wilkinson [28]).

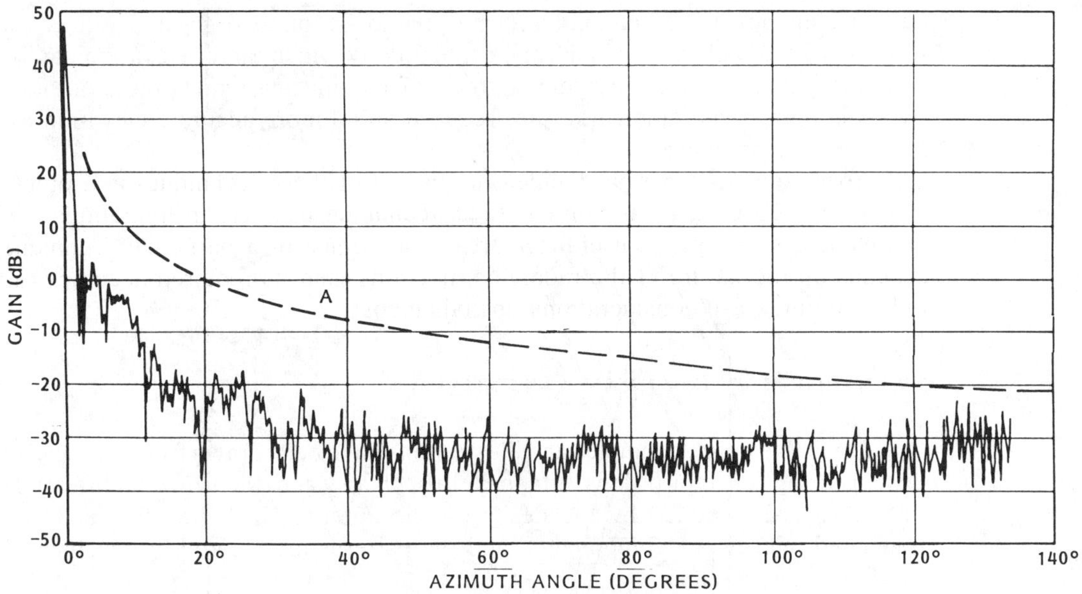

MEASURED PEAK ENVELOPE PATTERN OF AN ASYMMETRIC GREGORIAN ANTENNA (D = 7.6 M): f = 6175 MHz; D/λ = 100 (AZIMUTH PLANE)

Figure 6.28 Typical Measured Azimuth Pattern of Asymmetric Gregorian Antenna (from Burdine and Wilkinson [28]). (*Note*: Curve A Is 1979 FCC-INTELSAT-CCIR Maximum Recommended Sidelobe Level, G = 32–25 log θ [CCIR *Rec.* 465].)

In Japan some successful designs have also been achieved. M. Mizusawa, S. Urasaki, and H. Tanaka of Mitsubishi have demonstrated [29] an offset-fed dual reflector Cassegrain antenna for earth-terminal application. Their design is shown in Figure 6.29 along with a table summarizing some results measured at 21.6 GHz for a 1-m diameter circular aperture. Their design incorporated a corrugated conical feed horn, a shaped convex subreflector, and a shaped concave main reflector to achieve an aperture distribution with rotational symmetry about the antenna axis and good cross-polarization cancelation. Typical measured patterns in Figure 6.30 show excellent sidelobe suppression for both linear polarizations in the transverse (azimuth) plane.

Another successful Japanese design is reported by Y. Mizuguch [30] *et al.* of Kokusai Denshin Denwa. This is an offset-fed dual-reflector Gregorian antenna having a circular aperture 66 wavelengths in diameter. By using a concave ellipsoidal subreflector, this design achieves a more favorable geometry compared with an equivalent Cassegrain design. The subreflector can be larger and the main reflector can be smaller for the same projected aperture diameter. Also, the bandwidth is improved. Optimum placement of the feed and reflectors is used to achieve low cross-polarization. This results in cancelation of the cross-polarization components otherwise generated by the asymmetric illumination of the offset reflector.

A comparison is made by Mizuguch *et al.* [30] between using a conventional conical horn and a dielectric loaded horn. The dielectric loaded horn produced better axial symmetry of the main beam, lower cross-polarization lobes, and better sidelobe characteristics. The sidelobes with this design are also shown to be better than those of an equivalent Cassegrain antenna by about 3 to 5 dB. The authors attribute this to the reduced spillover around the larger Gregorian subreflector. The use of absorber shielding to improve sidelobes in the longitudinal (elevation) plane is also presented.

A number of British designs of dual reflector antennas with fairly low sidelobes are described by Rudge and Adatia [22] in their excellent review paper,

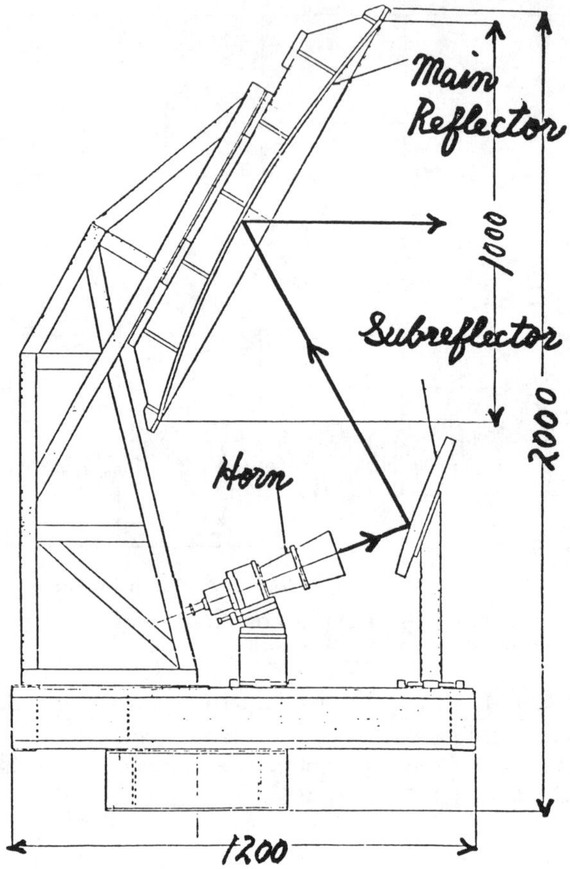

MEASURED PERFORMANCE

Frequency	21.6 GHz	First Sidelobe Level	-27 dB
Gain	45.5 dB	Cross polarization component (peak)	-31 dB
Aperture efficiency	70%		

Figure 6.29 Low-Sidelobe Dual-Reflector Cassegrain Antenna; Dimension Shown Is in mm (from Muzusawa *et al.* [29]).

among which is an interesting open Cassegrain antenna based on a Bell Laboratory design [31] and built as a 6-m diameter aperture earth station antenna at the University of Birmingham for satellite communication experiments with the UK Post Office. The near-in sidelobes of this design are not impressive (about -25 dB), but wide-angle levels are well below -40 dB. A more recent British paper by R.I. Henderson [32] of Marconi provides an exceptionally well written discussion of the design of an offset Gregorian antenna for transportable earth stations operating in the 11 to 14 GHz bands. This excellent paper reviews the new FCC and CCIR requirements for sidelobe envelope suppression ($G < 29$–$25 \log \theta$ for $1° \leqslant \theta \leqslant 48°$, and $G < -13$ dBI for $48° \leqslant \theta \leqslant 180°$) and develops the design rationale for meeting these levels. The Gregorian dual-reflector geometry illustrated in Figure 6.31, is shown to be superior to the Cassegrainian, and design details are clearly presented. The advantages of an elliptical rather than circular aperture outline are presented in terms of not only the mechanical packaging considerations for transport, but also in terms of the rate of decay of sidelobes in the preferred plane of the geostationary orbit (i.e., horizontal plane). The resulting design has a main reflector 5.6-m wide by 2.8-m high, giving it a gain equivalent to that of a 4-m circular aperture (nominally 52 dB), with enhanced

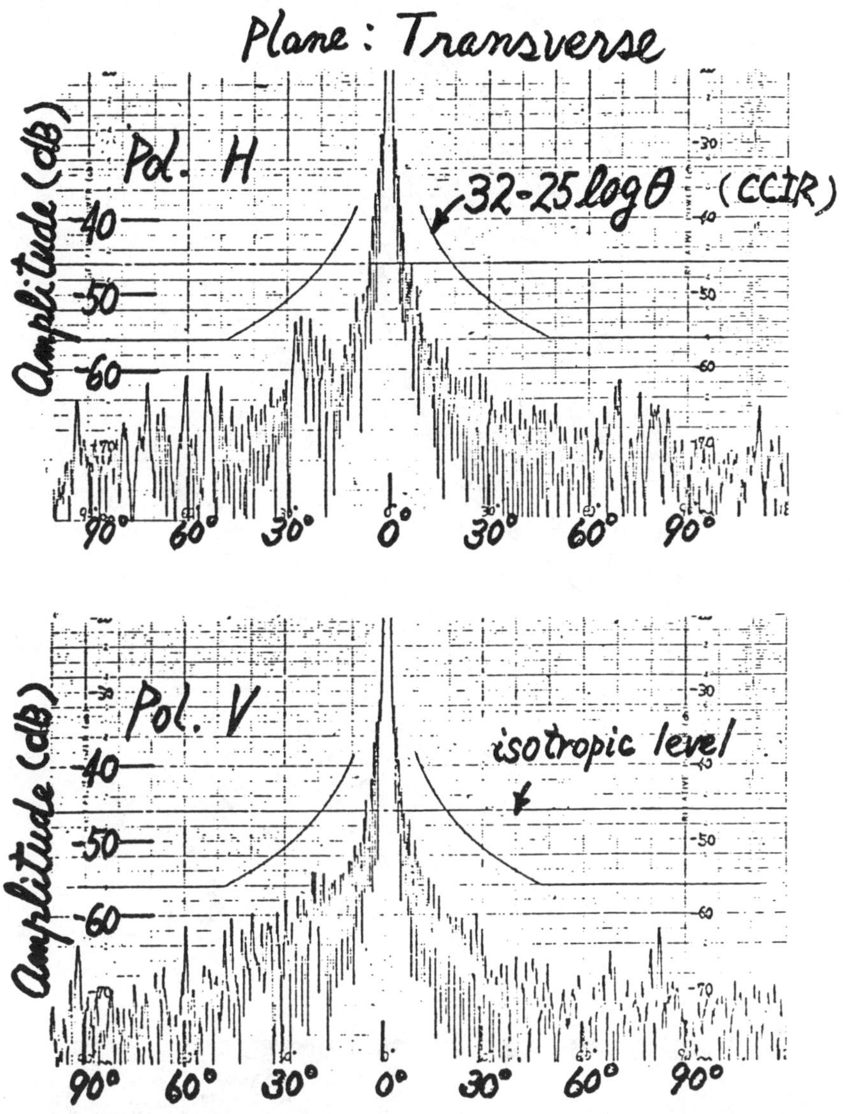

Figure 6.30 Measured Azimuth Patterns of Low-Sidelobe Cassegrain Antenna (from Mizusawa *et al.* [29]).

sidelobe performance in the preferred horizontal plane. Subreflector spillover is virtually eliminated by the use of an absorbing "guard ring" between the feed and the subreflector. A picture of the full-scale antenna is shown in Figure 6.32 and measured radiation patterns are shown in Figure 6.33.

Up to now we have described only ground-based antenna designs. A low-sidelobe satellite antenna design for the joint Franco-German direct broadcast satellites (project TV-SAT/TDF-1) is described in a recent paper by Fasold and Lieke [33] of the MBB Space Division. This antenna operates in the 11 to 12 GHz band and must provide an elliptical coverage of France and Germany from the $-19°$ West synchronous orbit position. Its circularly polarized (LHC) pattern must comply with the 1977 WARC requirements for transmitting pattern shape, sidelobe levels, and polarization purity (as well as pointing accuracy and EIRP). In addition, it must fit under the Ariane rocket shroud. The resultant transmitting antenna design is an offset-fed single reflector illuminated by an elliptical cross-section corrugated feed horn. The reflector has an elliptical outline 2.7-m wide

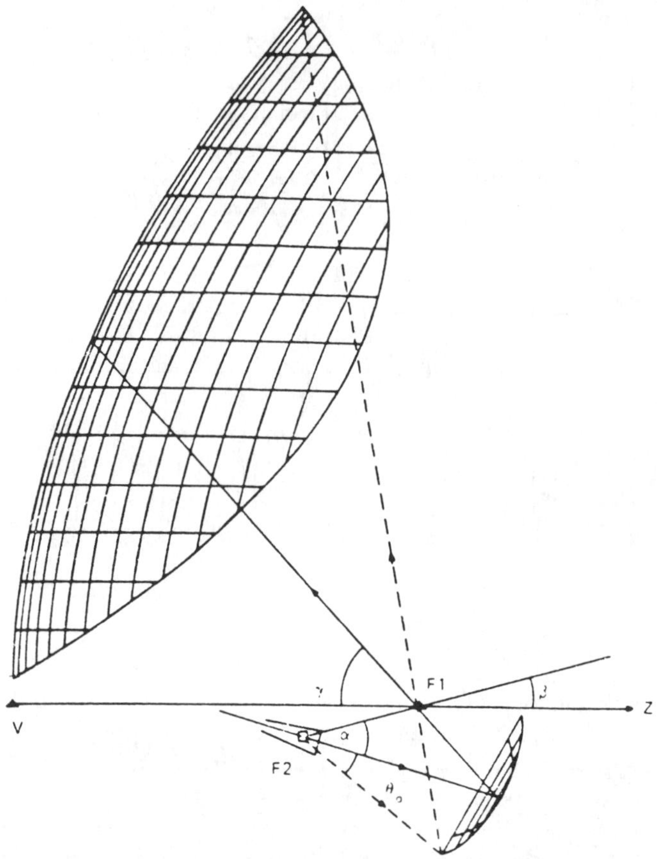

Figure 6.31 Offset Gregorian Antenna, Side View (from Henderson [32]).

Figure 6.32 Elliptical Offset Gregorian Antenna for a Transportable Earth Station.

SCHRANK
CHAPTER 6

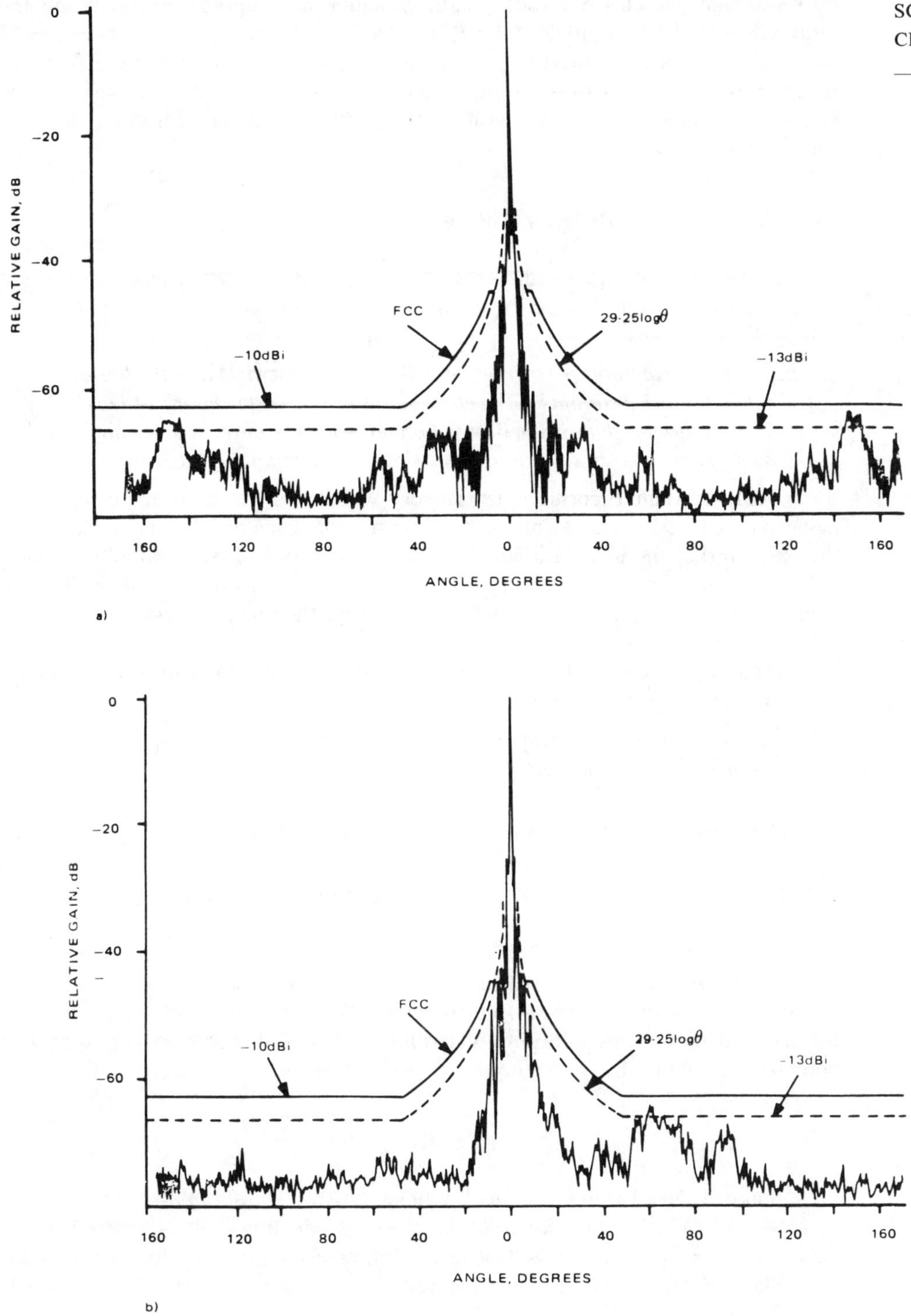

Figure 6.33 Measured Radiation Patterns at 14 GHz for the First Model; Top—Azimuth Cut, Bottom—Elevation Cut.

by 1.4-m high, with a 1.5-m focal length. Sidelobes are suppressed well below the required −31 dB first sidelobe level, and cross-polarization is well below −33 dB. The antenna also provides a beacon tracking mode at 17 GHz. Effects of reflector surface distortions are discussed. A separate similar receiving antenna is also mentioned. Both antennas have their reflectors hinged for deployment in orbit.

6.4.7 Conclusions for Reflector Antennas

In the above design examples we have seen significant advancements in the design of low-sidelobe reflector antennas. As mentioned by Rudge and Foster [24]:

> *The ultimate sidelobe performance which can be achieved with these [offset reflector] antennas has yet to be determined; in theory at least there is no fundamental reason why offset reflector antennas cannot equal or even exceed the best available sidelobe performance of arrays.*

While the above prediction may very well come true in terms of principal plane pattern performance, my array design colleagues at Westinghouse are of the opinion that the intercardinal sidelobe performance of (substantially) rectangular aperture arrays will be very difficult to duplicate with circular or elliptical aperture reflectors. Perhaps this will just challenge the reflector designers to prove the array designers wrong.

It can be concluded that low sidelobe reflector antennas can be designed by using the following design principles:

1. Blockage elimination (unless Ludwig's approach is used);
2. Good surface accuracies;
3. Special feeds;
4. Some absorber shielding.

6.5 FUTURE CHALLENGE OF LOW-SIDELOBE TECHNOLOGY

The future challenge of this dynamic technology area is to design and to build active aperture arrays with low sidelobes for advanced airborne, satellite, and ground or ship systems. Also, combining low sidelobes with adaptive, conformal, and digital beam-forming capabilities is a challenge that will keep our antenna community busy for the next couple of decades or more.

REFERENCES

[1] Blake, L.V., *Antennas,* Artech House, Norwood, MA, 1984.

[2] Skolnik, M., *Introduction to Radar Systems,* McGraw-Hill, New York, 1962.

[3] McComas, H.R. and P.S. Hacker, "Optimized Distribution for Symmetrical Planar Arrays," presented at the Seventeenth Annual Symposium on USAF Antenna R&D, Urbana, IL, November 1967.

[4] Hacker, P.S., "Low Sidelobe Electronic Scanning Antenna for AWACS Radar," *Proc. 19th Tri-Service Radar Symp.* Vol. II, July 1973.

[5] Ruze, J., "Pattern Degradation of Space Fed Phased Arrays," M.I.T. Lincoln Laboratory Report SBR-1, December 1974.

[6] Evans, G.E. and S.G. Winchell, "A Wideband, Ultralow Sidelobe Array," *Proc. Antenna Applications Symp.,* Urbana, IL, September 1979.

[7] Evans, G.E. and W.F. Hoover, Jr., "S-Band, Polarization Diverse, Low-Sidelobe Array," *Proc. Antenna Applications Symp.,* Urbana, IL, September 1980.

[8] Kiss, M.J. and B.A. Sichelstiel, "High Performance Electronically Scanned Array for Airborne Multimode Radar," *Proc. 24th Annual Tri-Service Radar Symp.,* June 1978.
[9] Hacker, P.S. and H.E. Schrank, "Range Distance Requirements for Measuring Low and Ultralow Sidelobe Antenna Patterns," *IEEE Trans. Antennas and Propagation,* Vol. AP-30, No. 5, September 1982.
[10] Schrank, H.E., "Low Sidelobe Phased Array Antennas," *IEEE Antennas and Propagation Society Newsletter,* April 1983, pp. 5–9.
[11] Evans, G.E. and H.E. Schrank, "Low Sidelobe Radar Antennas," *Microwave Journal,* July 1983, pp. 109–117.
[12] Ruze, J., "Antenna Tolerance Theory—A Review," *Proc. IEEE,* Vol. 54, 1966, pp. 633–640.
[13] Cutler, C.C., A.P. King, and W.E. Kock, "Microwave Antenna Measurements," *Proc. IRE,* Vol. 35, December 1947, pp. 1462–1471.
[14] James, G.L. and V. Kerdesmelides, "Selective Reduction in Back Radiation for Paraboloidal Reflector Antennas," *IEEE Trans. Antennas and Propagation,* Vol. AP-21, November 1973, pp. 886–887.
[15] Dybdal, R.B. and H.E. King, "Performance of Reflector Antennas with Absorber-Lined Tunnels," *IEEE Symp. Antennas and Propagation,* Seattle, June 1979.
[16] Crawford, A.B., D.C. Hogg, and L.E. Hunt, "A Horn-Reflector Antenna for Space Communication," *BSTJ,* Vol. 40, 1961, pp. 1045–1116.
[17] Thomas, D.T., "Design of Multiple-Edge Blinders for Large Horn Reflector Antennas," *IEEE Trans. Antennas and Propagation,* Vol. AP-21, March 1973, pp. 153–158.
[18] Siller, C.A., Jr., "A Fiberglass Pyramidal Horn Reflector Antenna for Point-to-Point Microwave Communications," *IEEE AP-S Symposium Digest,* 1974, pp. 239–240.
[19] Koch, G.F., "Flachenstrahler mit Kleinen Nebenmaxima," *FTZ,* October 1954, pp. 498–509.
[20] Hannan, P.W., "Microwave Antennas Derived from the Cassegrain Telescope," *IEEE Trans. Antennas and Propagation,* Vol. AP-9, March 1961, pp. 140–153.
[21] Dahlsjo, O., "A Low Sidelobe Cassegrain Antenna," *IEE Conf. Proc. "Radar—Present and Future,"* London, October 1973.
[22] Rudge, A.W. and N.A. Adatia, "Offset Parabolic Reflector Antennas—A Review," *Proc. IEEE,* Vol. 66, December 1978, pp. 1592–1618.
[23] Galinda-Israel, V., R. Mittra, and A. Cha, "Aperture Amplitude and Phase Control of Offset Dual Reflectors," *IEEE Trans. Antennas and Propagation* Vol. AP-27, March 1979, pp. 154–164.
[24] Rudge, A.W. and P.R. Foster, "Low Sidelobe Radar Antennas," *Military Microwaves Conf. Proc.,* London, October 1978.
[25] Ludwig, A.C., "Low Sidelobe Aperture Distribution for Blocked and Unblocked Circular Apertures," *IEEE Trans. Antennas and Propagation,* Vol. AP-30, September 1982, pp. 933–946.
[26] Fante, R.L., "Design of Parabolic Cylinder Reflector Systems With Low Sidelobes," RADC-TR-77-204, June 1977.
[27] Scudder, R.M. and L.H. Yorinks, "Low Sidelobe Reflector Antenna for Tactical Radars," *Antenna Applications Symp.,* Urbana, IL, September 1979.
[28] Burdine, B.H. and E.J. Wilkinson, "A Low Sidelobe Earth Station Antenna for the 4- to 6-GHz Band," *Microwave Journal,* Vol. 23, No. 11, November 1980, pp. 53–58.
[29] Mizusawa, M., S. Urasaki, and H. Tanaka, "An Offset Shaped Reflector Cassegrain Antenna," *IEEE AP-S Symp. Digest,* Palo Alto, CA, June 1977.

SCHRANK
CHAPTER 6

[30] Mizuguch, Y., M. Akagawa, and H. Yokoi, "Offset Dual Reflector Antenna," *IEEE AP-S Symp. Digest,* Amherst, MA, October 1976.

[31] Cook, J.S., E.M. Elam, and H. Zucker, "The Open Cassegrain Antenna: Part 1: Electromagnetic Design and Analysis," *BSTJ,* Vol. 44, 1965.

[32] Henderson, R.I., "Elliptical Offset Gregorian Antenna for a Transportable Earth Station," *GEC Journal of Research,* Vol. 2, No. 3, 1984.

[33] Fasold, D. and M. Lieke, "A Circularly Polarized Offset Reflector Antenna for Direct Broadcasting Satellites," *13th European Microwaves Conf.*, Nurnburg, 1983.

[34] Schrank, H.E., "Low Sidelobe Reflector Antennas," *IEEE Antennas and Propagation Society Newsletter,* April 1985, pp. 5–16.

Chapter 7
FUNDAMENTALS OF MMW RADAR SYSTEMS

by E.K. Reedy and J.C. Wiltse

7.1 INTRODUCTION

Some of the operational fundamentals of MMW radar systems are presented in this chapter. In some ways, the chapter will serve as an overview of MMW radar, in general. Topics covered range from history of MMW radar to applications of MMW radar and considerations of specific component characteristics and limitations. Particular emphasis is placed on the propagation environment and its effect on MMW radar applications and performance because of the extreme importance of this area in understanding both the advantages of MMW radars and their limitations.

Those readers seeking more details on the topics covered here are referred to *Principles and Applications of Millimeter-Wave Radar,* N.C. Currie and C.E. Brown, Editors, Artech House, 1987, from which this chapter was extracted [1].

7.2 HISTORY OF MMW RADAR

The first reported activity in MMW technology began in the 1890s [2]. From then until the 1930s, researchers developed various components and sources, and measured the bulk absorption and spectroscopic properties of numerous materials. The principal source used to generate MMW energy during these early days was the spark-gap generator, and radiometers or thermopiles were used to detect MMW energy. In 1934 the first resonant spectral frequency measured in the microwave or MMW regions (the ammonia inversion at 27.3 GHz) was observed by Cleeton and Williams using an early type of magnetron, rather than a spark-gap generator. In 1936 they operated such a tube as high as 47 GHz (6.4-mm wavelength).

The late 1930s saw the development of several new radiofrequency sources based on various vacuum tube configurations. These included the klystron (invented during 1937 by the Varian brothers and reported by them in 1939) and the cavity magnetron (invented in 1939 by Boot and Randall). These sources were immediately used in the radars of World War II, and early applications occurred at microwave frequencies rather than MMW frequencies. Late in the war the microwave traveling wave tube also was developed. The packaged silicon crystal detector-mixer was developed in the early 1940s (primarily at Bell Laboratories), and by the latter part of 1944 the production rate for these devices at Western Electric was in excess of 50,000 units monthly.

Editor's Note: The material for this chapter was extracted from *Principles and Applications of Millimeter-Wave Radar,* N.C. Currie and C.E. Brown, Editors, Artech House, 1987 [1]. Sections 7.1 and 7.3 through 7.7 were written by E.K. Reedy. Section 7.2 was written by J.C. Wiltse. Certain material in Reedy's sections was in turn excerpted from Reedy and Ewell (1981[8]).

The typical airborne radar in use by the Americans and British in the latter part of World War II operated in X-band (near 10 GHz). However, as the war progressed, frequencies were pushed above 20 GHz to K-band and beyond. The impetus to move to higher frequencies was prompted in part by the desire to obtain better angular resolution from apertures limited in size. Late in the war a new radar design operating near 24 GHz was produced. It was quickly found that this naive choice of frequency was very unfortunate because of the increased atmospheric attenuation (nearly an order of magnitude greater than that at X-band), resulting from the fact that a broad absorption band due to water vapor is centered near 22.3 GHz. Because of a lack of knowledge of atmospheric absorption effects by the radar designers, these radars were rendered far less effective than originally anticipated.

Prior to World War II only one discrete microwave or MMW spectral transition had been measured. During the war, Beringer, using a crystal harmonic generator driven by a centimeter-wave klystron, obtained enough second-harmonic signal to measure the 5-mm wavelength (60 GHz) absorption of oxygen, but did not resolve its fine structure. Thus, the period through World War II closed out with the stage set for the extension of coherent radar techniques into the millimeter frequency region.

After the war the new sources were heavily used to develop the discipline of microwave and MMW spectroscopy. For the first time, molecular spectra could be studied with the use of coherent radiation. The resolving power was remarkable and permitted precise determination of molecule sizes and shapes. In addition, there was interest in MMW applications to radar, radio communication, and radiometric sensing. Gordy's group at Duke University was one of the first to take advantage of this capability. He obtained from Raytheon a few of the first klystrons developed for frequencies from 30 to 50 GHz and, starting in 1947 and continuing for many years, he and a succession of graduate students made many high-resolution spectral measurements.

Other very successful programs in MMW technology were undertaken at the University of Illinois, Lincoln Laboratory, the Johns Hopkins University, and Bell Laboratories, where an interest had developed even before World War II as an outgrowth of early waveguide work. While the members of the Duke group were developing their spectroscopy methods, a Columbia University group was experimenting with the filtered harmonic energy obtained directly from high-powered magnetrons. As early as 1949 they announced the detection of harmonic energy in the 1.5–3.0 mm wavelength region utilizing a Golay cell. Later, however, they discontinued the magnetron measurements and instead used the klystron harmonic-multiplier approach.

At the University of Texas, Straiton and Tolbert were leading a very active group which became well known for its research and measurements on atmospheric propagation and the development of radars and radiometers. Straiton's group produced a set of curves of atmospheric attenuation (for horizontal propagation) which have since become classics. The curves, derived from theoretical analyses of absorption of O_2 and H_2O, as originally performed by Van Vleck, were supplemented with experimental data obtained at various frequencies by several groups, including the University of Texas and Bell Laboratories.

During the 1950s and early 1960s many other institutions were involved in MMW research. For instance, at the Georgia Institute of Technology, in Atlanta, there was considerable emphasis on antennas (particularly the geodesic lens) and on radar designs and, by 1959, a 70-GHz surveillance radar was developed to the point of being given U.S. Army nomenclature (AN/MPS-29) [3,4]. Research on high-power sources continued at the Columbia University Radiation Laboratory; by 1954, the magnetron was extended upward to 115 GHz with a 3.3-kW output power, but its lifetime and duty cycle, and hence (average) power output, were very low. In 1960, the US Army sponsored an extensive development program on electron tube sources, including switching, duplexing, and ferrite devices [5].

Throughout the 1970s, the MMW radar field continued to expand in level of effort, diversity, and technical progress. Although the advent of solid-state sources was a major boost to MMW system applications in the 1970s, recent progress in new vacuum tube sources such as extended interaction oscillators (EIOs), extended interaction klystron amplifiers (EIKAs), and gyrotrons have provided a similar boost to this technology field. By 1978 the US Army's STARTLE (Surveillance and Target Acquisition Radar for Tank Location and Engagement) radars (developmental) were using all-solid-state transmitters consisting of a 94-GHz phase-locked Gunn oscillator chain whose frequency-controlled output was used to injection lock the IMPATT transmitter; this technique provided 4-W peak power in a pulse mode. Various modulation formats were employed: short-pulse spread-spectrum or coherent MTI (moving target indication); or pulse compression with frequency agility. Figure 7.1 further illustrates this type of progress.

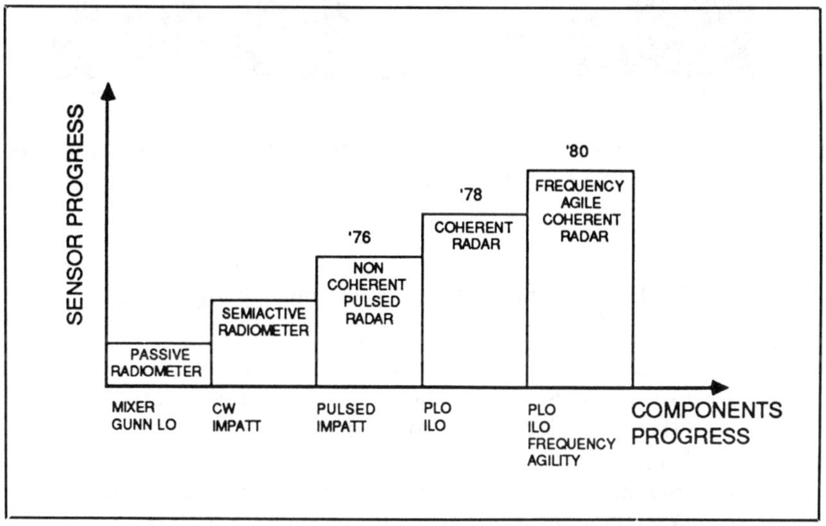

Figure 7.1 MMW Sensor Technology Development Reflects Progress of Solid-State Components (Courtesy of J. Kuno)

Representative of the most advanced state-of-art MMW radars is the High Power Coherent Radar operating at 95 GHz (HIPCOR-95) developed by the Georgia Institute of Technology for the US Army Missile Command for research into advanced target detection and missile guidance techniques. The HIPCOR-95 radar (see Figure 7.2) employs two selectable output stages to support a broad range of research objectives. A high-power, moderate-bandwidth transmitter section uses an extended interaction amplifier as its final output stage to achieve 2 kW of peak output power. A moderate-power, high-bandwidth mode utilizes a current technology traveling wave tube as its final output stage to achieve a transmitted bandwidth in excess of 2 GHz [7]. This radar is described in greater detail in Section 7.6.2.

Table 7.1 summarizes the early MMW technological milestones and Table 7.2 is included to provide examples of early MMW radars.

REEDY & WILTSE
CHAPTER 7

Figure 7.2 High-Power Coherent 95 GHz (HIPCOR-95) Radar (Courtesy of the Georgia Institute of Technology)

Table 7.1 Significant MMW Technology Milestones

Date	*Event*
1936	First resonant spectral measurement (ammonia inversion at 27.3 GHz), by Cleeton and Williams
1937	Invention of the klystron oscillator by Russell and Sigurd Varian
1939	Invention of the cavity magnetron by Boot and Randall
1942	Development of the crystal mixer at Bell Laboratories
1946	First MMW magnetron developed
1947	First K_a-band radars (AN/SPS-7 surface radar developed by the US Navy, and AN/TPQ-6 cloud height measuring radar developed by Bendix for the US Air Force)
1960	Development of MMW klystrons and magnetrons by US Army Signal Corps Laboratories
1962	Gunn effect discovered in GaAs by J. B. Gunn
1964	IMPATT oscillator developed separately at Bell Laboratories and in the Soviet Union
1978	Commercial availability of the extended interaction klystron oscillator (EIKO or EIO)

Table 7.2 Examples of Early MMW Radars
(Adapted from S.L. Johnston [6], © 1977 Microwave Journal)

Year	*Identification*	*Application*	*Band*	*Sponsor*	*Source*
1947	AN/SPS-7	Surface Search	K_a	Navy	—
1947	AN/TPQ-6	Cloud Height	K_a	Air Force	Bendix
1948	AN/APS-36	Search	K_a	Navy	Philco
1950	AN/SPN-10	Aircraft Landing	K_a	Navy	—
1952	Instrumentation Radar	Sea Clutter Measurements	K_a	Navy	Georgia Tech
1954	AN/APQ-58	Terrain Clearance	K_a	Air Force	—
1955	Unknown	Air Surveillance	K_a	United Kingdom	Decca
1956	AN/APQ-56	Mapping	K_a	Air Force	Westinghouse
1957	AN/TPQ-11	Cloud Height	K_a	Air Force	Bendix
1958	AN/BPS-8	Surface Search	K_a	Navy	Fairchild
1959	AN/MPS-29	Gnd Surveillance	W	Army	Georgia Tech
1959	Lacrosse	Missile Tracker	K_a	Army	Cal/Martin
1961	AN/APQ-97	Target Location	K_a	Army	Westinghouse
1962	AN/APQ-89	Terrain Clearance	K_a	Navy	NAFI
1964	AN/APN-161	Mapping	K_a	Air Force	Sperry
1965	MRL-1	Weather	K_a	USSR	—
1968	Lunar Radar	Moon Measurement	K_a	Air Force	MIT
1968	AN/APQ-137	Ground Targets	K_a	—	Emerson
1968	AN/APQ-144	Multimode	K_a	Air Force	GE
1968	River Radar	Ship Navigation	K_a	Fed. Rep. of Germany	—
1969	—	Meteorological Research	K_a	Rep. of South Africa	—
1970	AN/SPN-42	Aircraft Landing	K_a	Navy	Motorola
1973	AN/APQ-122	Dual-Band Multimode	K_a	Air Force	TI
1975	Instrumentation Radar	Land Clutter Measurements	W	Army	Georgia Tech
1976	TRAK X	Dual-Band Low Altitude Tracking	K_a	Navy	NRL
1976	Flycatcher	Dual-Band Air Defense	K_a	The Netherlands	HSA
1977	—	Automobile Braking	K_a	Dept. of Transportation	—

7.3 OVERVIEW

7.3.1 Key System Operational Considerations [8]

Full understanding of the operational characteristics of MMW radar requires an appreciation for the location of the MMW band within the electromagnetic spectrum and the resulting effects on the propagation of MMW energy. Figure 7.3 illustrates the electromagnetic spectrum and identifies the region of conventional radar operation, along with standard radar operating band designations. Radars certainly can be designed and developed at any frequency at which electromagnetic energy can be generated and controlled; however, the primary operational region for radars which must penetrate the atmosphere extends from below 300 MHz to perhaps 300 GHz or from 3×10^8 Hz to 3×10^{11} Hz, a three-order-of-magnitude frequency range. This region of the spectrum has been further subdivided into standard radar bands and given standardized letter designations which are also shown in Figure 7.3.

Differences of opinion exist among radar researchers about exactly what region constitutes millimeter waves. One obvious definition would be the region between 1 mm (300 GHz) and 10 mm (30 GHz). However, the IEEE Standard [9] defines millimeter waves as the region between 40 GHz and 300 GHz, excluding the operationally important and very active region around 35 GHz. This book will consider operation near 35 GHz in discussions of general MMW radar system performance and characteristics.

Other band designations have been proposed for regions of the MMW spectrum. For example, Kulpa and Brown [10] used the term *near millimeter* to refer to the region between 100 GHz and 1000 GHz. In some references, *submillimeter* refers to the region between 300 GHz and 3000 GHz. Other overlapping and sometimes confusing band designations have appeared in the literature. For example, see Figure 7.4 from Bhartia and Bahl [11], where US and United Kingdom frequency band designations are summarized. The reader can readily see that there is definitely room for both confusion and debate over the correct terminology to use. Extreme care must always be exercised to ensure that the reader understands which terminology is being utilized by a specific author. For consistency, the standard band designations shown in Figure 7.3 will be used.

Figure 7.5 provides a representation of the clear-air atmospheric attenuation between 3 cm and 0.3 μm wavelengths. Note that the atmospheric attenuation generally monotonically increases from 3 cm to 0.3 mm, after which it flattens and then decreases in the infrared region. Between 10 GHz and 500 GHz, where the attenuation strongly increases as a function of frequency, there is a series of peaks or regions of higher attenuation, and valleys or regions of lower attenuation, compared to that in the immediate vicinity. The peak frequencies are often referred to as "walls" and the valleys as "windows."

Most applications of radar involve achieving as long a detection range as possible; consequently, radar research and development have been concentrated in the areas of lower attenuation at the MMW window frequencies of 35 GHz, 95 GHz, 140 GHz, and 220 GHz. Very little radar development has occurred above 220 GHz. Water vapor and oxygen molecule electromagnetic interaction resonances provide regions of abnormally large absorption which, in turn, cause the relative maxima or walls occurring at approximately 60 GHz, 120 GHz, and 180 GHz. Radars have also been developed at these frequencies to take advantage of the covertness provided by the high atmospheric absorption, and to operate in exoatmospheric space where atmospheric absorption does not exist.

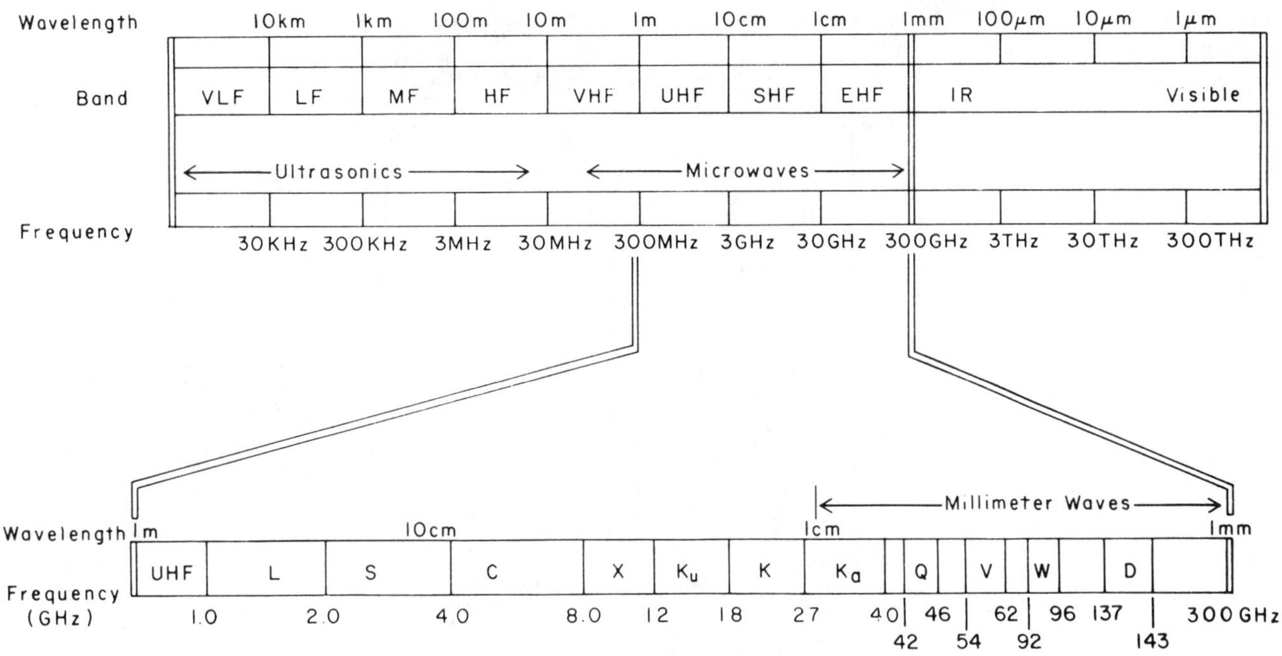

Figure 7.3 Electromagnetic Spectrum with Radar Band Designations.

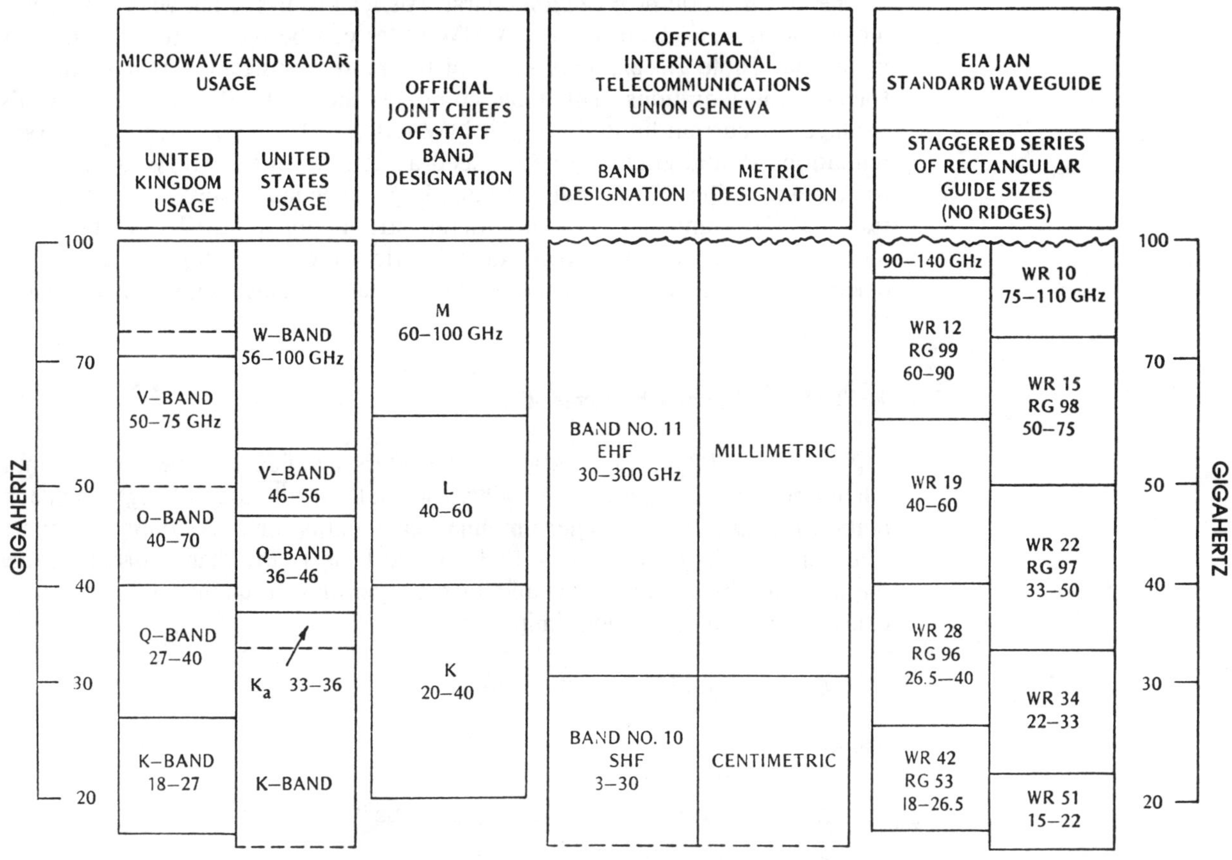

Figure 7.4 Frequency Band Designations [11].

REEDY & WILTSE
CHAPTER 7

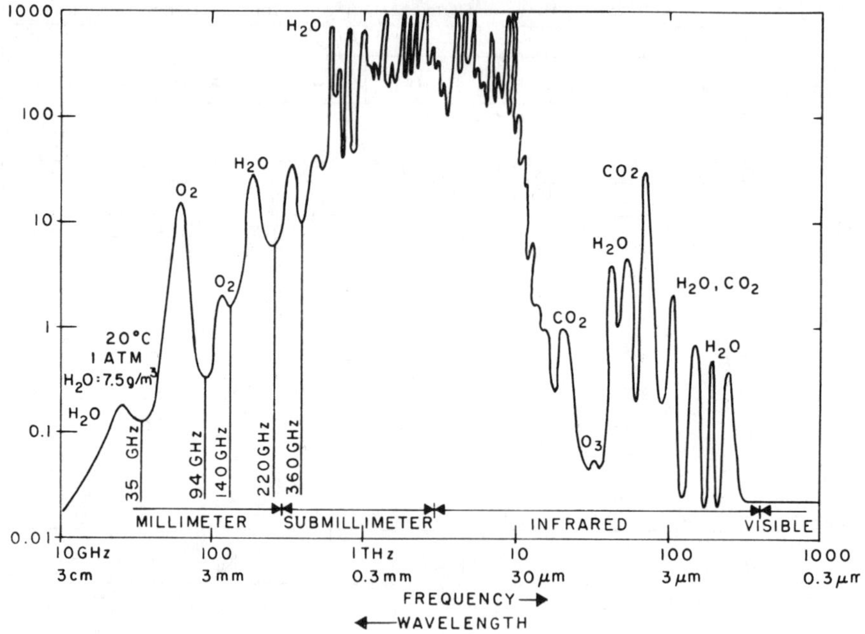

Figure 7.5 Clear Weather Atmospheric Attenuation (from Preissner [17]).

Many of the key operational features of MMW radar are directly dictated by the atmospheric propagation characteristics discussed above. Obviously, not all the operational features of a MMW radar can be related to its frequency of operation. Table 7.3 presents some of the more important MMW radar system tradeoff considerations. This table compares the relative advantages and disadvantages or current limitations of MMW radars. The importance of the first two limitations should gradually be reduced as activity in the MMW area continues to increase and leads to production and large-scale manufacture and standardization of MMW components, along with the introduction of MMW circuit integration and increased use of microstrip and dielectric waveguide components. Additional discussion of other MMW radar operational characteristics is given in the following section.

7.3.2 MMW Radar Characteristics

As Table 7.3 indicates, MMW radar offers significant operational advantages when compared to microwave radar, especially in the area of high angular resolution resulting from smaller antenna beamwidths for a fixed antenna aperture size. Antenna half power or 3 dB beamwidth, θ, in the plane corresponding to the antenna dimension, is related to the operating frequency by the following expression for a diffraction limited antenna [12]:

$$\theta = k\lambda/l \text{ (radians)} \tag{7.1}$$

where

k = constant ($4/\pi$ for -25 dB sidelobes),

λ = wavelength,

l = aperture dimension.

Solving (7.1) for several aperture dimensions and frequencies yields Table 7.4 which provides antenna beamwidth (in degrees) as a function of radar operating

Table 7.3 Millimeter-Wave Radar System Trade-off Considerations

Advantages	*Limitations*
Physically Small Equipment	Component Cost High
Low Atmospheric Loss[a]	Component Reliability and Availability Low
High Resolution Angular Doppler Imaging Quality Classification	Short Range (10–20 km) Weather Propagation[b]
Small Beamwidths High Accuracy Reduced ECM Vulnerability Low Multipath and Clutter High Antenna Gain	
Large Bandwidth High Range Resolution Spread Spectrum ECCM Doppler Processing	

[a]Compared to IR and Visual Wavelengths
[b]Compared to Microwave Frequencies

Table 7.4 Antenna Beamwidth *versus* Frequency

Diffraction Limited Antenna

Beamwidth $= \Theta = k\lambda/l$ (Radians)
$k =$ Constant ($4/\pi$ for -25 dB Sidelobes)
$\lambda =$ Wavelength
$l =$ Aperture Dimension

Antenna Beamwidth (in Degrees) as a Function of Radar Frequency

Aperture Size (m)	*Frequency* (GHz) 10	35	95	140	220
0.01	—	62.5	23.1	15.6	9.9
0.1	21.9	6.3	2.3	1.6	1.0
1.0	2.2	0.63	0.23	0.16	0.1

frequency. Equation (7.1) and Table 7.4 also illustrate that θ is inversely proportional to frequency ($\theta \propto 1/f$). Therefore, for a fixed antenna aperture, a radar operating at 9.5 GHz (X-band) would radiate an antenna beam 10 times larger than a radar operating at 95 GHz. In other words,

$$\theta_{95} = \frac{\theta_{9.5}}{10} \tag{7.2}$$

The availability of significantly smaller antenna beamwidths results in several of the most important operational advantages attributable to MMW radar. Among the more important of these are those given as (1) through (6) below.

(1) ***High Antenna Gain with Small Aperture***

Skolnik [12] gives the gain, G, of an antenna relative to an isotropic radiator as

$$G = 4\pi A_e/\lambda^2 \tag{7.3}$$

where A_e = effective area of antenna aperture. Thus, the gain of an antenna increases in proportion to the frequency of operation squared, again for a fixed aperture. As shall be shown later, antenna gain is one of the most important radar parameters in determining target detection performance.

(2) ***High Angular Tracking Accuracy***

A radar's angular tracking accuracy is directly related to its antenna beamwidth, θ, as shown below for a thermal noise limited case:

$$\sigma_t = (k_t\theta)/[(S/N)n]^{1/2} \tag{7.4}$$

where

σ_t = *root-mean-square* (rms) angle tracking error,
k_t = constant depending on type of tracking,
(S/N) = signal-to-noise ratio at receiver output,
n = number of pulses integrated.

Thus, all other things equal, a MMW radar can be expected to achieve a reduction in tracking errors when compared to a lower frequency radar, as a direct result of smaller antenna beamwidth.

(3) ***Reduced Electronic Countermeasures (ECM) Vulnerability***

Smaller radar antenna beamwidths provide less opportunity for a jammer to inject energy into the radar's main beam and thus reduce the radar's susceptibility to jamming. Also, higher antenna gains reduce vulnerability to jamming through the antenna sidelobes. The basic cross-range resolution of a radar, d_x, is defined in terms of the antenna half power beamwidth and distance or range to the target cell as follows:

$$d_x = \theta_x R \tag{7.5}$$

where

d_x = cross-range resolution,
θ_x = half-power beamwidth,
R = range to target cell.

The higher angular resolution available with MMW radars allows spatial isolation of a target in a volume clutter background such as might be present in a chaff cloud or rain background. This isolation provides enhanced target-to-clutter-plus-noise ratios and improved detection performance.

(4) *Reduction in Multipath and Ground Clutter at Low Elevation Angles*

A MMW radar with small beamwidth will typically have less ground intercept than a lower frequency radar with larger beamwidths. Since ground intercept is reduced, multipath propagation conditions and ground clutter effects are correspondingly reduced.

(5) *Improved Multiple Target Discrimination and Target Identification*

Again, a radar's ability to separate multiple, closely-spaced targets and to provide information for target identification is closely coupled to its resolution. Thus, MMW radar has inherent advantages in these areas.

(6) *Mapping Quality Resolution Possible*

Figure 7.6 illustrates the level of target resolution attainable from an airborne, real aperture 95 GHz radar. This figure shows a radar image of a US Navy ship taken with a noncoherent 95 GHz radar at a range of 122–152 m. The radar beamwidth was approximately 10 mr.

In addition to the high angular resolution of a MMW radar, another important characteristic is the radar's ability to measure and resolve target motion through the doppler effect. Any target motion in the radar's beam will cause a shift in the received signal frequency in accordance with the following relationship:

$$f_d = 2V_r/\lambda \tag{7.6}$$

where

f_d = shift in transmitted frequency (doppler frequency),

V_r = radial target velocity,

λ = wavelength of transmitted signal.

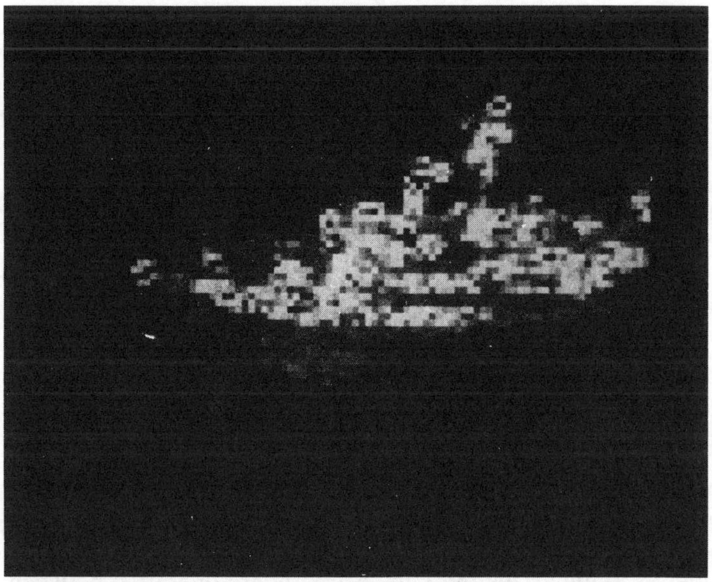

Figure 7.6 Radar Image of a Ship Taken with an Airborne, Noncoherent 95 GHz Radar.

Table 7.5 presents the doppler frequency shift (given in hertz per meter per second of target radial velocity) for several radar operating frequencies. For a target having a 30 m/s radial velocity with respect to the radar, a radar operating at 10 GHz (X-band) would experience a 2.0-kHz frequency shift (i.e., the received signal frequency would be 10,000,002 kHz for a target radially approaching the radar). A 95-GHz radar observing the same target would experience a 19-kHz frequency shift. There is a rule of thumb that 1 mi/hr produces approximately 300-Hz doppler at 100 GHz; this is easy to scale in speed or carrier frequency (e.g., a 500-mi/hr aircraft target would produce a doppler of about 150 kHz). Also near 100 GHz, radial speeds between a fraction of 1 mi/hr and 50 mi/hr produce frequencies that fall directly into the human hearing range. Thus, a homodyne radar system operating near 100 GHz and receiving echo signals from personnel, machinery, and vehicles has the full range of "noises," and sounds like the real world (observed remotely). Such large doppler shifts from relatively slow moving targets provide the capability for increased target detection and perhaps recognition of such target features as skin vibration, and second and higher order velocity signatures. These features aid in the automatic classification of targets based on radar signatures.

Table 7.5 MMW Radar Characteristics: Doppler Frequency Properties

Doppler Frequency Shift:

$$f_d = 2\, V_r/\lambda$$

V_r = Radial Target Velocity
λ = Wavelength of Transmitted Signal

Radar Frequency (GHz)	*Doppler Shift* (Hz/m/s)
10	66.7
35	233.3
95	633.3
140	933.3
220	1466.7

Even with the advantages given in Table 7.3 and discussed above notwithstanding, a MMW radar should not be considered a panacea for all radar surveillance and tracking problems. For terrestrial applications, MMW radar is basically a short-range, high-resolution sensor which may be severely degraded in some adverse weather environment. However, for many applications, it represents an excellent compromise between the performance characteristics of microwave and lower frequency radar on the one hand, and optical and infrared sensors operating at higher frequencies, on the other. Table 7.6, which is adapted from Bhartia and Bahl [11], compares the relative performance of MMW radar with its microwave and optical counterparts for several important radar operating characteristics. MMW radar performance usually falls between the extremes of the microwave and optical sensors.

7.3.3 Typical Applications

Table 7.7 summarizes the known US radars operating between 70 and 140 GHz prior to approximately 1975 [13]. Considerable activity at several different organizations is evident. Applications of these radars can be generally subdivided

Table 7.6 Radar System Performance Comparison
(Adapted from Bhartia and Bahl [11] © 1984 John Wiley and Sons)

Radar Characteristics	*Microwave*	*Millimeter Waves*	*Optical*
Tracking Accuracy	Fair	Fair	Good
Classification or Identification	Poor	Fair	Good
Covertness	Poor	Fair	Good
Volume Search	Good	Fair	Poor
Adverse Weather Performance	Good	Fair	Poor
Performance in Smoke, Dust	Good	Good	Poor

Table 7.7 Early MMW Radar
Known US Radars Operating between 70 and 140 GHz (prior to 1975)

Identification	*Application*	*Frequency*	*Sponsor and Source*
AN/APQ-62	Side Looking Mapping Radar	70 GHz	—
JR-9	Search Mapping Radar	70 GHz	WPAFB and TRG and Raytheon
AN/BPS-8	Search Radar	70 GHz	—
AN/MPS-29	Search and Surveillance Radar	70 GHz	USAECOM and Georgia Institute of Technology
Experimental	Aircraft Obstacle Avoidance Aircraft Instrument Landing	70 GHz	USAECOM and Nordon Division, United Aircraft
Experimental	Obstacle Avoidance Sea Clutter Measurement	95 GHz	NADC, Warminster, PA
Experimental	Space Object identification	95 GHz	Aerospace Corp., El Segundo, CA
Experimental	Arctic Terrain Avoidance	95 GHz	Navy and Applied Physics Lab., Silver Springs, MD
Experimental	Airborne Applications Instrument Landing Short Range Weapon Delivery Sensor Cueing	95 GHz	WPAFB and Goodyear Aerospace Corp., Litchfield Park, AZ
Experimental	Low Altitude Aircraft Tracking Target Acquisition Basic Millimeter Wave Radar Studies	70 GHz	Ballistic Research Labs., Aberdeen Proving Grounds, MD
Experimental	Noise Modulated Radar for Clutter Suppression and FM/CW for High Range Resolution	94 GHz	Ballistic Research Labs., Aberdeen Proving Grounds, MD
Experimental	Bistatic CW Radar for Cross Section Measurements	140 GHz	Ballistic Research Labs., Aberdeen Proving Grounds, MD
Experimental Rapid Scan	Ranging, Target Acquisition Command Fusing, Fire-control, and Navigation	70 GHz	Harry Diamond Laboratories and Georgia Institute of Technology
Experimental	Instrumentation for Basic Millimeter Radar Studies; Backscatter Studies, *et cetera*	95 GHz	Georgia Institute of Technology
Experimental	Instrumentation for Basic Millimeter Radar Studies; Backscatter Studies, *et cetera*	70 GHz	Georgia Institute of Technology
Experimental	Monopulse Tracking Investigations	70 GHz	Norden Division, United Aircraft

into the following three areas:

- Surveillance and target acquisition
- Instrumentation and measurement
- Fire control and tracking.

In addition to continuing development of MMW radars to satisfy requirements in these areas, emphasis in the last ten or so years has shifted to include development of radars to satisfy requirements in the additional area of guidance and seekers. Table 7.8 identifies four generic areas for application of MMW radar. Some specific examples of applications in these areas will be presented later in this chapter. Other examples are covered in Part IV of Currie and Brown [1].

Table 7.8 Generic Applications for MMW Radar

- Surveillance and Target Acquisition
- Instrumentation and Measurements
- Guidance and Seekers
- Fire Control and Tracking

7.4 FUNDAMENTAL CONSIDERATIONS

7.4.1 Functional Elements

The functional elements of a MMW radar are basically the same as those of the microwave radar, namely the transmitter, receiver, signal processor, and antenna. However, major differences in the operational and performance parameters of these functional elements and the passive components which connect these elements result in significant differences in the system level performance of a microwave and MMW radar.

Block Diagram

A basic block diagram for a typical MMW radar (or, for that matter, any other radar) is shown in Figure 7.7. In its simplest form, a basic radar consists of an antenna, which acts as a transducer between the radar and the space into which the radar is radiating energy, and the power source or transmitter and receiving or signal processing subsystems. Obviously, a functional radar must include several other components, not all of which are illustrated in Figure 7.7. Some of the most important include the modulator for forming the basic pulse shaping; a synchronizer to provide system timing; a duplexer to switch the antenna between the transmitter and the receiver; some form of receiver protection device; a display or indicator acting as the interface between the radar and the user of the information and data provided by the radar, antenna servo control (and tracking circuit or processor if required); various power supplies, conditioners, and converters. In addition, these elements must be linked together, usually by conventional means of connection (wires, printed circuits, *et cetera*), and some form of waveguide, microstrip, or MMW integrated circuit techniques.

Significant progress continues to be made in developing components and the basic technology to support radar programs in the MMW region. However, the performance characteristics of most of the critical millimeter components tend to decrease with increasing frequency. A wide range of components is currently available from commercial vendors up to, and including, the 95 GHz window. Above 95 GHz—at 140 GHz and, most especially 220 GHz—generally available components consist primarily of laboratory devices lacking the basic structural and dielectric integrity required for military and commercial use in a harsh field environment, although progress is currently underway to improve off-the-shelf

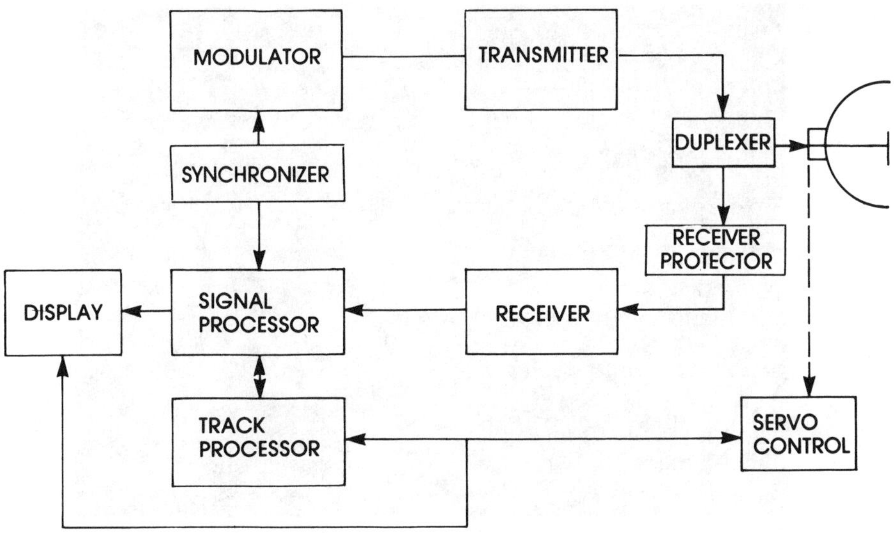

Figure 7.7 Radar Block Diagram.

and catalog devices at these frequencies. The cost of components remains high, however.

Transmitter

One of the primary performance-determining parameters in any radar design and development program is the amount of power available from the radar transmitter and the type of source available (i.e., amplifier, coherent or noncoherent oscillator, *et cetera*). This is particularly true for MMW systems, since they are invariably power limited. The transmitter normally consists of a high-power oscillator, perhaps a magnetron, or a lower power oscillator coupled to a high-power amplifier such as a *traveling wave tube* (TWT) or the newer *extended interaction klystron amplifier* (EIKA). In a pulsed radar, the transmitter generates a high-power MMW frequency during the time period that the modulator is producing a drive or control pulse—usually on the order of a few nanoseconds to hundreds of microseconds. In effect, the dc modulator pulse has been converted to a pulse of millimeter wavelength energy which is then passed through the duplexer to the antenna where it is shaped and radiated.

Sources of MMW transmitters can be either solid-state or thermionic devices. The solid-state sources consist primarily of the IMPATT and Gunn devices. Thermionic sources include *magnetrons, traveling wave tubes* (TWT), *klystrons, extended interaction klystron oscillators* (EIKO) and *amplifiers* (EIKA), *backward wave oscillators* (BWO), and *gyrotrons*. Shown in Figure 7.8 are several MMW source devices. Summarized in Table 7.9 are typical power levels for five devices that are extensively used as power sources in MMW radar—the magnetron, EIKO/EIKA, TWT, gyrotron, and solid-state devices (Gunn and IMPATT). All of these devices have improved power outputs or operate at higher frequencies and, in some cases, have longer lifetimes than previous versions of the device. The increasing availability of both TWTs and EIKAs supports the design and development of fully coherent *master oscillator power amplifier* (MOPA) transmitters at 95 GHz with sizable output powers to complement previously available coherent solid-state transmitters with much lower power.

Additional, more detailed consideration of both solid-state and thermionic, or high-power transmitters are presented in Chapters 8 and 9 of Currie and Brown [1].

REEDY & WILTSE
CHAPTER 7

Figure 7.8 Photograph of Several Commonly Used High-Power MMW Transmitter Sources. A 5-W Gunn Oscillator Is Located in the Center, while Surrounding it from Left to Right Is an EIO with a Samarium-Cobalt Magnet, an EIO with an Alnico Magnet, a 500-W, 70-GHz Magnetron, a 6-kW, 95-GHz Magnetron, and a 1-kW, 95-GHz Magnetron.

Table 7.9 MMW Radar Component Considerations

Typical Transmitter Source Power Levels (Peak)				
	Frequency (GHz)			
Power Source	35	95	140	220
Magnetron	50–100 kW	1–6 kW	—	—
EIK/EIKA	2–3 kW	1–2 kW	200 W	60 W
TWT	3 kW	1–2 kW	—	—
(Water Cooled)	(30 kW)	(6 kW)		
Gyrotron[a]	340 kW	500 kW	180 kW	150 kW[b]
Solid State (Gunn, IMPATT)	28 W	22 W	5 W	1 W

[a]Higher power levels have appeared in the literature.
[b]Estimate

Receiver

The receiver in a MMW radar converts the received signal, collected by the antenna, from a frequency corresponding to the approximate transmitted frequency (depending on doppler effects) to a lower or intermediate frequency (IF) where it can then be more conveniently filtered, amplified, and processed. While direct detection of millimeter wavelength signals is a possibility, the lack of sensitivity of such an approach effectively argues against its use in most applications. The translation of the incoming signal to an intermediate frequency and the subsequent processing of the lower frequency signal is termed a heterodyne receiver. A low-noise mixer-amplifier combination is normally used to accomplish the down-conversion and first-level amplification. This device usually represents the key element in the receiver chain in determining the radar's detection performance, dynamic range, sensitivity, and noise properties.

Figure 7.9 shows the current status of MMW receiver technology. Considerable progress has been made in recent years in developing improved performance for mixer-amplifier combinations. Generally the double sideband noise figure increases rather rapidly with frequency, ranging from perhaps 3 dB at 35 GHz to 5 dB at 95 GHz, and 6.5 dB at 140 GHz. Noise figures of approximately 13 dB have been reported at 220 GHz, although work at this frequency has been somewhat limited. More detailed treatment of MMW radar receivers and their properties are presented in Chapters 10 and 13 of Currie and Brown [1].

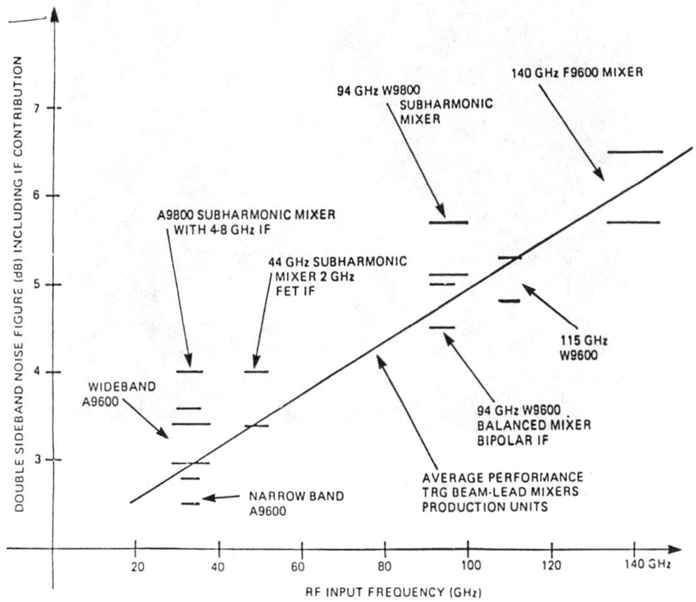

Figure 7.9 Current Status of MMW Receiver Technology.

Antenna

Most of the antenna technology and techniques currently in use at MMW involve direct extension of lower frequency approaches. In many applications, a microwave antenna design whose performance characteristics have been well established through design, theoretical prediction, and measurements at the lower frequencies will simply be "scaled" up in frequency to satisfy requirements at MMW frequencies. For this reason, research directed toward developing MMW antenna techniques and associated technology has been limited.

Extensive use has been made at MMW of the classical reflector antennas, such as the front-fed paraboloidal dish shown in Figure 7.10 and the Cassegrain-fed reflector shown in Figure 7.11. A Cassegrain-feed arrangement avoids the waveguide losses associated with the more conventional front-fed reflector which can become excessive at MMW frequencies.

Horn and lens antennas are perhaps more popular at MMW than at microwave frequencies. These antennas avoid the aperture blockage and its associated effect on sidelobes common to the reflector antennas. Lens antennas, in particular, have seen limited application at microwave frequencies, due to size and weight limitations. However, the size and weight of a MMW lens antenna are much less than those of its microwave counterpart. That advantage, when coupled with advantages of lens antennas over reflector antennas in angle scanning, spillover loss, and fabrication tolerances, makes the lens antenna much more capable of meeting requirements at MMW than lower frequencies. A zoned dielectric lens antenna for use at 95 GHz is shown in Figure 7.12.

REEDY & WILTSE
CHAPTER 7

Figure 7.10 Front-Fed Paraboloidal Dish MMW (35 GHz) Antenna.

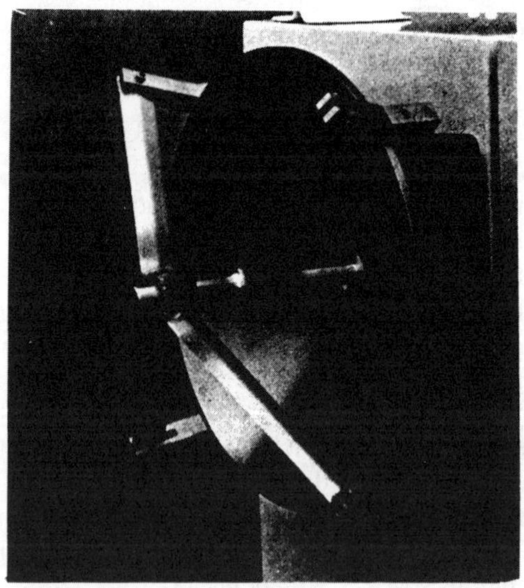

Figure 7.11 Cassegrain 95-GHz Antenna.

Slotted waveguide antennas, which are composed of arrays of parallel waveguide runs having either broad-wall or narrow-wall slots as the radiating elements, have several characteristics which have resulted in their utilization at MMW. These include the capability for electronic scanning (by changing the transmitted frequency), ease of fabrication, well known and predictable radiation performance, and the elimination of a complex feed structure and phase shifters such as that required for a phase-scanned array antenna.

Figure 7.12 Zoned Dielectric Lens.

Other antenna techniques, such as the microstrip antenna, dielectric rod, and leaky-wave antenna combine small size, electronic scanning capability, compatibility with integrated circuit construction techniques, and other important characteristics in such a manner as to make them desirable for MMW applications. Table 7.10 summarizes the antenna techniques which have seen the most application at MMW. More complete discussions of MMW antenna technology are presented in Chapters 11 and 17 of Currie and Brown [1] and in Bhartia and Bahl [11].

Table 7.10 MMW Antenna Techniques

- Reflector Antennas
  - Front Fed
  - Dual Reflector (Cassegrain)
  - Offset-Fed
  - Shaped Reflector
- Lens Antennas
  - Zoned Dielectric
  - Luneburg
- Horn Antenna
  - Flared
  - Multimode
  - Corrugated
  - Lens-Corrected
- Dielectric Rod Antennas
- Slotted Waveguide Antennas
  - Broad Wall
  - Narrow Wall
- Leaky Waveguide Antennas
  - Fast Wave Structure
  - Slow Wave Structure
- Microstrip Antennas

Passive Components

As previously indicated, a MMW radar requires an array of components which are normally passive to transport, control, attenuate, filter, isolate, *et cetera,* the MMW energy. These passive components include such devices as waveguide, connectors, couplers, transitions, attenuators, hybrids, filters (preselectors), circulators, couplers, isolators, terminations, matching networks, and duplexers. Many of these devices are simply microwave designs of a particular component scaled up in frequency (and usually down in size). However, frequency scaling poses problems, particularly at the higher MMW frequencies of 140 GHz, 220 GHz and above, where manufacturing tolerances and increased losses can severely affect performance. In these situations, optical and quasioptical techniques and structures (dielectric waveguide, slabs, gratings, lenses, *et cetera*) have provided practical solutions. Also, hybrid approaches, involving a combination of both waveguide and optical technologies, have been successfully developed which combine the advantages of the two techniques and circumvent many of their limitations.

At frequencies above 35 GHz, however, considerable work is still required to develop low cost, wide bandwidth, and high isolation components that are lightweight, low power, and complement the small size, lightweight advantages of millimeter waves.

7.4.2 Environment

For radars operating at millimeter wavelengths, the environment has a significant effect on the overall radar system performance, directly influencing such performance parameters as target detection range, track accuracy, target discrimination, *et cetera*. Environment, as used in the discussion to follow in this section, refers to both the electromagnetic reflection and attenuation properties of the propagation path between the radar and the target (propagation effects), and the nature of the electromagnetic energy reflected back toward the radar by objects in the vicinity of the target (usually termed clutter if these returns interfere with detection of the desired target). Much of the discussion presented in this section has been extracted from Reedy and Ewell [8].

Propagation Effects

No attempt will be made in this section to review and present data that completely describe MMW propagation effects. Only that information required to make this chapter reasonably self-contained or to introduce the reader to basic concepts is included here.

Attenuation and Reflectivity

Figure 7.13 illustrates, in block diagram form, the major factors affecting the signal received at any radar. Of particular interest are those attenuation and reflection or scattering influences present in the propagation path which, for terrestrial radar, is normally the atmosphere. The primary attenuation-producing factors for MMW radar are the molecular absorption of water vapor and oxygen in a relatively clear atmosphere, absorption of condensed or suspended water in the form of droplets in rain, fog, clouds, *et cetera,* and scattering from those water droplets. Backscatter of energy from the suspended water droplets in fog and clouds and from rain can severely limit radar performance, since the target return signal will be embedded in and corrupted by this noiselike reflected energy. Suspended particulate matter, such as dust particles and smoke, may also influence MMW propagation.

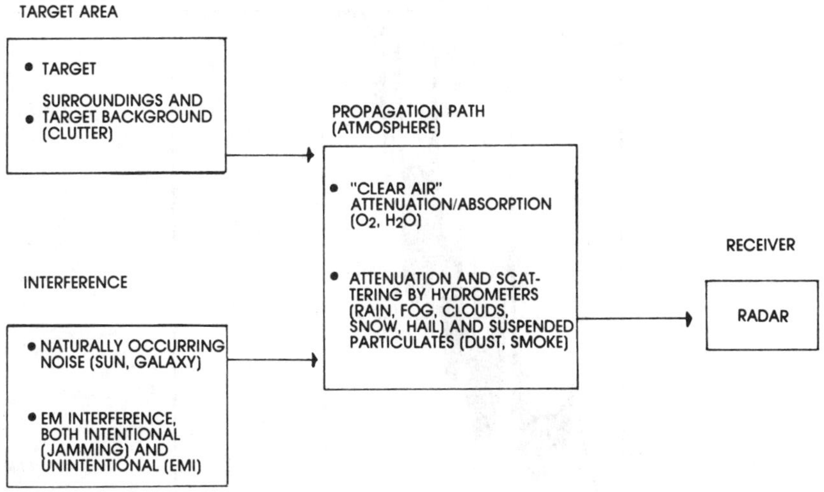

Figure 7.13 Major Factors Affecting Received Radar Signal.

Attenuation by Atmospheric Gases, Rain, and Fog

Although the transmission of electromagnetic energy through the atmosphere is a very complex phenomenon, certain aspects of this phenomenon have been extensively studied and documented [10, 14, 15]. However, much of the basic propagation data necessary to predict the performance of MMW sensors accurately either does not exist or the uncertainty associated with existing data is large enough to make accurate, analytical prediction of the sensor performance extremely difficult.

Figure 7.14, after Strom [16], illustrates this uncertainty by shading the window region two-way attenuation levels. Note that at 220 GHz the clear-air attenuation varies between approximately 1.5 dB/km and, perhaps, 11 dB/km, depending on such parameters as atmospheric pressure, temperature, and water density or humidity. Large regions of uncertainty are present at all of the window frequencies. To reduce this level of uncertainty and to predict radar performance more accurately, Kulpa and Brown [10] point out that a rather detailed specification of the conditions under which the sensor is expected to operate is required. For example, the geographic location of expected operation can be important since geographic as well as seasonal variations in humidity and temperature can affect clear air attenuation values by as much as 10 to 15 dB/km.

Nevertheless, it is still worthwhile to document propagation effects on millimeter waves as completely as possible so that equipment performance can be estimated as accurately as possible using available data. Attenuation produced by basically clear air atmospheric gases at 20°C, one atmosphere pressure, and 7.5 g/m^3 water vapor is shown in Figure 7.15 along with attenuation produced by certain hydrometers (rain and fog) [17]. Attenuation produced by rain increases as a function of frequency with maxima occurring in the 100–200 GHz region, after which it remains relatively constant. Ice clouds generally produce negligible attenuation at these frequencies [17]. We refer the reader to later chapters of [1] or to Kulpa and Brown [10] for a more complete discussion of MMW propagation in both clear air and in the presence of naturally occurring hydrometers such as rain, snow, and hail.

The variation in currently available propagation data bases at millimeter waves cannot be overemphasized. For example, Table 7.11 (taken from Kulpa and Brown [10]) summarizes the variation in total one-way attenuation at 220 GHz in a 100-m fog and 4-mm/hr rain using the best available data. Variations of 10–15 dB in attenuation are present in the data, illustrating the difficulty in analyzing performance of millimeter equipment and the necessity for specifying very accurately the operational environment in which the radar is expected to perform.

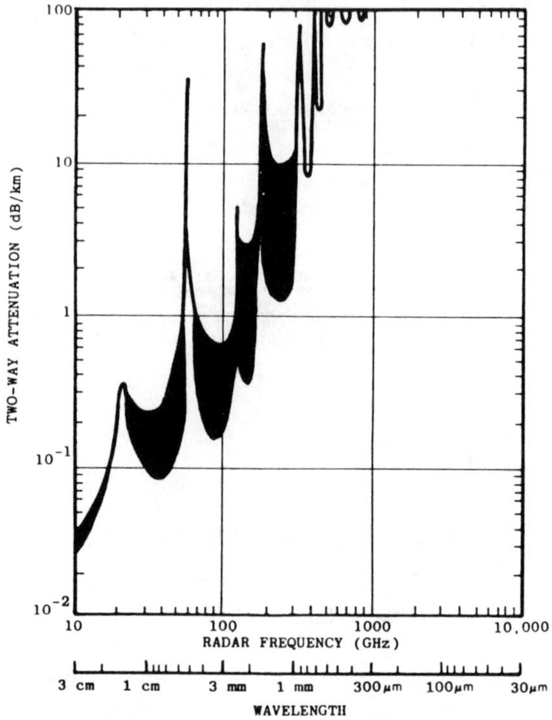

Figure 7.14 Clear Weather Attenuation Variation for Horizontal, Sea-Level Path (from Strom [16]).

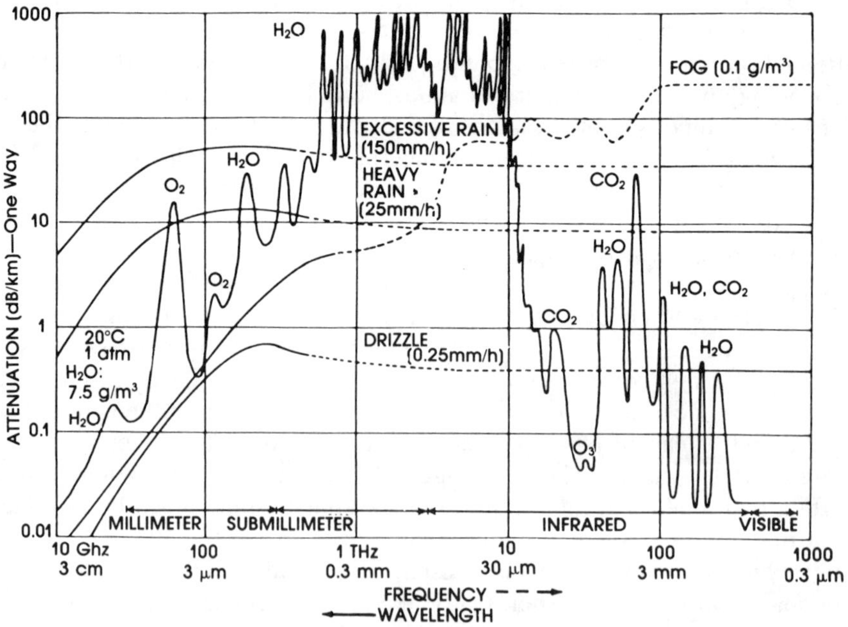

Figure 7.15 MMW and Near-MMW Attenuation by Atmospheric Gases, Rain, and Fog (from Preissner [17]).

Table 7.11 Variation in Attenuation at 220 GHz*

Condition	*One-Way Attenuation* (dB/km)
Clear Air	1.6 – 11.2
100-m Fog	0.4 – 4.7
TOTAL	2.0 – 15.9
Clear Air	1.6 – 11.2
4-mm/hr Rain	1.0 – 7.0
TOTAL	2.6 – 18.2

*From Kulpa and Brown [10].

Rain Backscatter

Only a very limited number of theoretical and experimental investigations have attempted to quantify the radar backscattering properties of rain at millimeter wavelengths (Richard and Kammerer [18], Wilcox and Graziano [19], Currie *et al.* [20], Strom [16]). The basic source of experimental data relating rain radar backscatter characteristics to rain rates at the low millimeter wavelengths and X-band is a series of measurements conducted at McCoy AFB, Florida, in August and September 1973 by the US Army Ballistic Research Laboratories, Aberdeen, Maryland. The resulting data from these measurements are recorded and documented in Currie *et al.* [20] and Richard and Kammerer [18]. A summary of the data from these experiments is shown in Figure 7.16, where the measured average backscatter coefficient of rain is given as a function of rain rate for the operating frequencies at 9.375, 35, 70, and 95 GHz. Also, shown in Figure 7.16 are data for 45 GHz taken at a different time and location [21]. One interesting property of rain reflectivity is apparent from Figure 7.16. In the vicinity of 45 GHz, a significant change in the raindrop, electromagnetic wave interaction mechanism occurs, as evidenced by the change in the slope of the rain reflectivity *versus* frequency data and the decrease in MMW backscatter coefficient as the frequency increases from 45 GHz to 95 GHz. Data shown in Figure 7.16 with the exception of the 45 GHz data, were taken from Richard and Kammerer [18]. Their method of data reduction resulted in a larger and, thus, somewhat more conservative value for the backscatter coefficient, especially at the lower rain rates. (See Chapter 5 of Currie and Brown [1] for a discussion of this experiment.)

Dust, Smoke, and Other Obscurants

Even though one of the primary advantages of MMW sensors over their electro-optical counterparts is enhanced propagation characteristics through dust, smoke, and other battlefield and naturally occurring obscurants, very limited definitive data exist to specify attenuation and backscatter when propagation takes place under these conditions. Allen and Simonson [22] and Morgan *et al.* [23] presented some of the first comparative data for attenuation by smoke and dust. More recently, spurred on by the increased interest in, and usage of, the millimeter region, more investigators have examined the effects of smoke and dust on millimeter propagation [23,24].

REEDY & WILTSE
CHAPTER 7

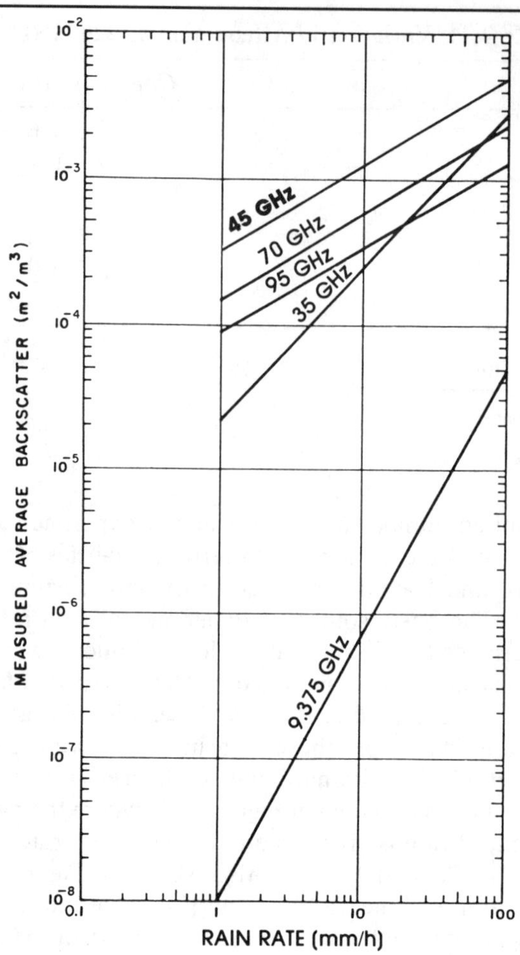

Figure 7.16 Rain Reflectivity Coefficient *versus* Rain Rate, Vertical Polarization (from Richard and Kammerer [18]).

All of the above references confirm that suspended dust and smoke in the atmosphere in amounts typical of a battlefield or naturally occurring environment produce a practically imperceptible amount of attenuation at frequencies of 140 GHz and below. Knox [25] found that the only obscurant to affect transmissivity significantly during exercises conducted at Smoke Week II was dust and debris produced by exploding charges of C4 material that had been buried under several centimeters of dirt. Maximum attenuation measured during these trials was approximately 0.4 dB/km at 140 GHz.

Petito and Harris [26] observed that only during periods when exploding artillery rounds hurled debris (principally soil and metallic fragments) into the air was there significant attenuation and backscatter from artillery bursts when observed with a 95-GHz radar.

Turbulence Effects

Refractive index inhomogeneities in the propagation path of a MMW signal causes certain propagation phase shifts to result, which in turn result in propagation scintillation effects, angle of arrival fluctuations, and depolarization. Only scintillation and angle of arrival fluctuations are thought to be significant at millimeter wavelengths, however. Theoretical and experimental investigations of these effects at 94 and 140 GHz have shown that the scintillation effects are not likely to affect MMW systems except at the extreme limits of performance, but that atmospheric turbulence effects may result in angle of arrival variations of perhaps 0.35 mr, which is approximately the same level of accuracy required of many systems [27].

Multipath Effects

When an RF signal is incident on the surface of the earth, a portion of the incident electromagnetic energy is forward scattered; at the target, the energy that reaches the target by the direct path from the transmitter to the target and by reflection from the surface of the earth combines vectorally and can add either in or out of phase. This same phenomenon occurs on the return path from the target to the radar, where it affects tracking accuracy as well as signal strength. Effects produced by these reflected signals, collectively termed "multipath," can produce fluctuations in signal strength from a target, or in the case of a tracking system, can introduce considerable tracking errors. The magnitude of these multipath-related effects is related to the amount of energy incident on the surface, the reflection coefficient of the surface, the amount of energy that reaches the target by the direct path, and the relative phase of the direct and indirect components.

While the analysis of a general multipath situation may be quite complex, considerable insight into the process may be gained by a simplified analysis. In such a simplified analysis, the surface is considered to be a randomly rough surface having known dielectric properties. The voltage reflection coefficient for a smooth surface of the same dielectric material can be calculated, and this modified by a factor to compensate for roughness of the surface in order to obtain an approximate description of its forward scattering properties. The forward scattered reflection coefficient for a smooth uniform dielectric surface, ρ_0, can be calculated directly from the Fresnel equations.

The reflection coefficients are then modified by the specular scattering factor, ρ_s, where

$$\rho_s^2 = \exp[-(4\pi\sigma_h)\sin\gamma/\lambda^2] \tag{7.7}$$

where σ_h is the rms deviation of the surface height, λ the wavelength and γ the grazing angle. Note that a surface is considered rough when ρ_s^2 is less than 0.5 or $\rho_s \leq 0.7$; this is approximately the Rayleigh roughness criterion.

The reflection coefficient for specular reflection is given by

$$\rho = \rho_0\rho_s. \tag{7.8}$$

Equation (7.8) gives the specular reflection coefficient for rough surfaces; there will remain a diffuse component, not as significant in target fading but important in determining tracking error.

In general, use of MMW radar may offer the advantage of a reduction in multipath effects, primarily because for a given aperture size, the antenna beamwidth decreases with increasing frequency, and since narrow beamwidths can be used to illuminate the target without directing as much energy toward the reflecting earth surface. Also, since roughness is dependent on σ_h/λ and since λ is small for MMW radars, a given surface appears rougher as the frequency increases, thus decreasing the specular scattering factor and decreasing ρ.

Clutter Characteristics

When a radar system illuminates the earth's surface (land or sea), a portion of the energy is scattered forward, giving rise to multipath effects described above;

in addition, a portion of the energy is reflected back toward the radar system. These unwanted signals are usually referred to as "clutter," and can seriously affect overall system performance for those situations where an appreciable amount of energy is backscattered toward the radar. Historically, such clutter has been described in terms of its radar cross section per unit area (σ°), a dimensionless quantity that, when multiplied by the resolution cell size of the radar system, gives the clutter radar cross section. Particularly when high resolution radar systems are utilized, clutter no longer appears to be homogeneous with a uniform spatial distribution, and the concept of σ° should be used with some care. However, it is still a quite useful quantity, and one almost universally utilized to report the magnitude of clutter backscatter.

Another problem in characterizing clutter returns in a radar system is that the return is a complex function of a number of parameters, including frequency, polarization, depression angle, clutter type, and system resolution. Analysis and reporting of clutter data are further complicated by the fact that clutter returns are not constant, but fluctuate with time; thus, in addition to the average value, some measure of both the amplitude fluctuations and the temporal behavior (or spectral width) of the clutter must also be provided if a useful description of clutter is to be obtained.

The properties of land clutter can vary widely. This variability is partially due to the fact that land clutter is present in a large number of forms, such as trees, fields, desert, snow, and cultured areas; even within such broad categories, different clutter regions can produce substantial differences in reflectivity. In addition, local environmental conditions such as wind velocity, wind direction, and moisture content can affect clutter returns.

Figure 7.17 presents a collection of MMW surface backscatter data taken from several sources and compiled by Dyer and Hayes [28]. In this figure, the surface backscatter coefficient, σ°, is presented as a function of wavelength covering the frequency band from 10 GHz to approximately 100 GHz. Data for several different types of terrain backscatter are presented. Also, varying grazing angles are represented. The terrain backscatter data of Figure 7.17 illustrate a relatively weak dependence of σ° on the radar frequency. In contrast, however, a relatively wide spread in the terrain backscatter or reflectivity coefficient values occurs for fixed frequency and terrain type. This last characteristic substantiates the general observation that the properties of land clutter tend to vary widely, even within the same type of surface vegetation and for a constant frequency. Also, note in Figure 7.17 the tendency for the terrain backscatter coefficient to increase slowly as a function of frequency in the MMW region. This is especially true for a relatively smooth surface such as a concrete or asphalt runway or road surface.

A characteristic of surface backscatter which is extremely important in determining the performance of a particular radar over a range of operational scenarios is how the surface reflectivity varies as a function of grazing angle or depression angle. The terms grazing angle, depression angle, or, sometimes, incidence angle are normally used to refer to the angle of incidence or intercept of the radar's illumination pattern and the surface of reflection, where grazing angle is measured relative to the horizontal surface plane, and vertical incidence is measured with respect to the normal to the surface. Figure 7.18 illustrates the variation of σ° as a function of grazing angle for several types of surface vegetation and a fixed radar frequency of 35 GHz. A rather strong dependence of σ° on grazing angle is evident in this figure. This figure also illustrates one of the limitations of the current MMW surface reflectivity data bases—a lack of representative data for low ($<10^\circ$) grazing angles.

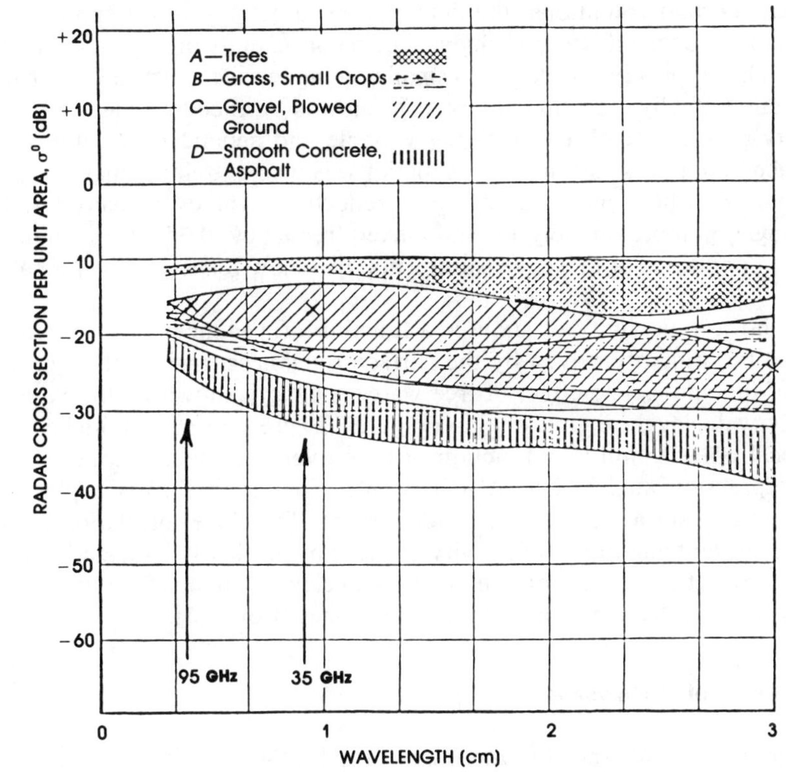

Figure 7.17 Wavelength Dependence of Terrain Backscatter (from Dyer and Hayes [28]).

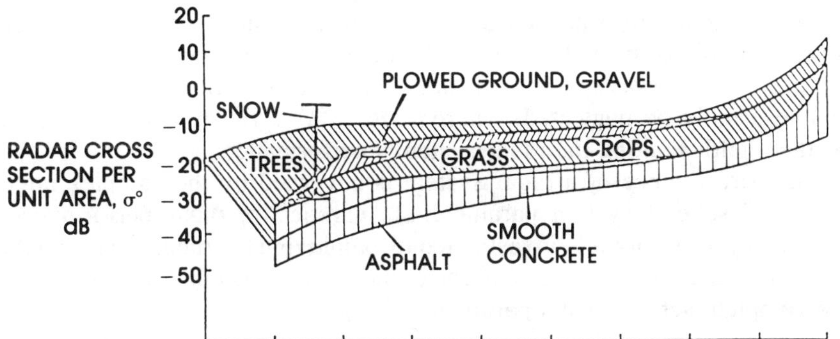

Figure 7.18 Radar Cross Section *versus* Grazing Angle (35 GHz) (from Dyer and Hayes [28]).

Under certain conditions, the detection of targets against a snow covered background has been difficult to achieve with a consistently high probability using a MMW radar (typically a seeker on a small homing missile). Figure 7.18 provides one indication of why detection of targets in snow is a difficult problem. The radar reflectivity of snow at MMW is highly variable and dependent on atmospheric and surface conditions and, in some situations, is also extremely high, as shown in Figure 7.18, where the range of snow reflectivity values measured at fixed grazing angle of approximately 17° and a fixed frequency of 35 GHz is indicated. Under the conditions of these measurements, the snow reflectivity varied almost 30 dB.

While the reflectivity of snow is a function of a number of parameters, it appears to be primarily driven by the surface roughness, and by the free water content of the snow. The effect of free water is clearly shown by the variations in reflectivity during the course of a day; Figure 7.19 presents σ° for both horizontal and vertical polarization as a function of time over a 24-hour diurnal cycle for several depression angles [29]. Also presented in Figure 7.19 is a plot of air temperature and surface snow free water content. The close correlation between free water content and radar reflectivity of the snow is clearly indicated in Figure 7.19. The properties of clutter at millimeter wavelength are considered in considerably more detail later in Chapter 5 of Currie and Brown [1].

7.4.3 Measures of Performance

Quantitative and tractable measures of a MMW radar's performance are required to provide a basis for comparison and evaluation of relative performance of several radars, as well as a basis on which the optimization of a particular radar design can be based. Different measures of performance are used for different radar applications (search, track, *et cetera*) and operational scenarios (benign or jamming environment, *et cetera*). However, several performance measures, such as maximum detection range, probability of detection, and probability of false alarm are almost invariably used in specifying a radar's performance.

MMW radar performance, like its microwave counterpart, is governed by, and can be predicted from, the basic radar range, beacon, and jamming equations. In the case of microwave radars, the atmospheric attenuation term in these equations can usually be neglected, whereas, for MMW radar, it may be the most important factor limiting the radar system's performance.

The performance of a radar system is governed by the complex interrelationships of several system parameters. Analytically, radar performance and tradeoff analysis is normally investigated by employing the radar range equation—a mathematical expression that can relate the radar maximum range performance to the complete set of system parameters:

$$R_m^4 = (P_t G^2 \lambda^2 \sigma 10^{-0.2\alpha R})/[(4\pi)^3\,(kT_0B)F_n(S/N)_1 L_s] \qquad (7.9)$$

where R_m is the maximum radar range corresponding to minimum single-pulse signal-to-noise ratio, $(S/N)_1$, P_t is the peak radar transmitted power, λ is the radar wavelength, B is the receiver bandwidth $\approx 1/\tau$ (for matched receiver), τ is the pulsewidth (integration time for a CW radar), G is the antenna gain, σ is the target radar cross section, k is the Boltzmann's constant (1.38×10^{23} joule/K), T_0 is the

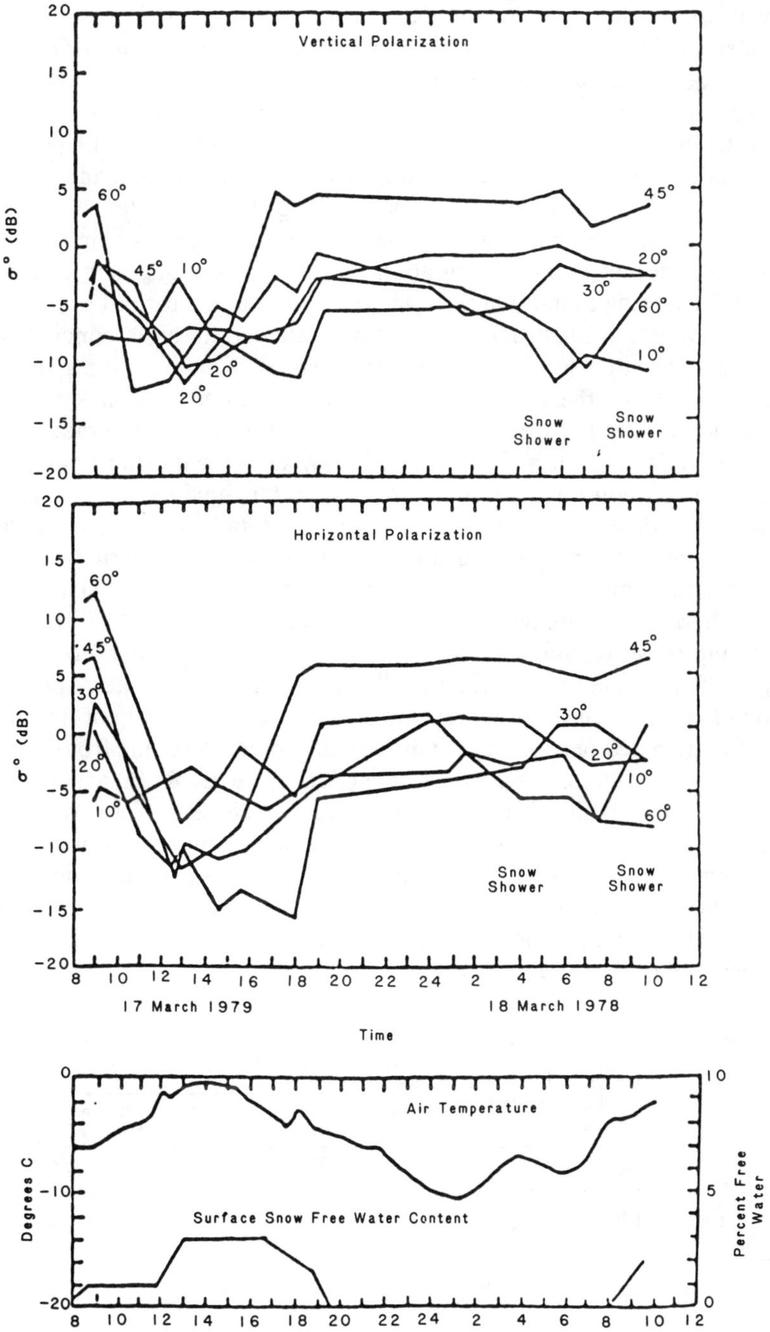

Figure 7.19 MMW Snow Reflectivity *versus* Time over a 24-Hour Cycle (from Hayes, Currie, and Scheer [29], © 1980 IEEE).

standard reference temperature (290 K), F_n is the receiver noise figure, α is the one-way atmospheric attenuation coefficient (dB/km), $(S/N)_1$ is the minimum equivalent receiver output single-pulse signal-to-noise ratio required for the radar's function (detection or tracking), and L_s is the system losses.

This form of the radar range equation incorporating a single-pulse signal-to-noise ratio does not measure the overall effectiveness of the radar since the effects of integration are not included. Integration can significantly enhance the performance of a radar for both detection and tracking. However, the single-pulse S/N represents a quantity that is widely used in the literature to compare and predict the performance of radars. Integration or signal processing gain can usually be included by adding a multiplicative factor in the numerator of (7.9).

It is sometimes instructive to demonstrate certain basic concepts and limitations through the use of a system performance calculation for a simple radar. As an example of the use of the radar range equation for a simple, first-cut calculation of radar performance, consider the set of postulated radar parameters shown in Table 7.12 for a short range, MMW target acquisition radar. To avoid an iterative solution for determining maximum range (since the propagation loss is range dependent), for convenience a constant total loss factor L_T, independent of range is defined which includes both system and atmospheric losses. For this calculation, assume $L_T = 6$ dB. Also, assume no integration or signal processing gain; furthermore, a circular antenna aperture having a gain of 37 dB is assumed.

Using these parameters, the maximum expected radar range is plotted in Figure 7.20 as a function of target RCS. Indications of the general range of cross section of several classes of targets are also included at the top of this figure. The band of anticipated performance indicates the spread resulting from system compromises, atmospheric loss, range dependence, target RCS, and possible field degradation that might be reasonably expected. These performance curves are for clear weather and represent, for the parameters chosen, an upper boundary on performance, since such factors as clutter have not been included. The range performance achieved in this example is strongly influenced by the assumed constraints on the antenna aperture size; choice of a larger aperture would result in a corresponding increase in range.

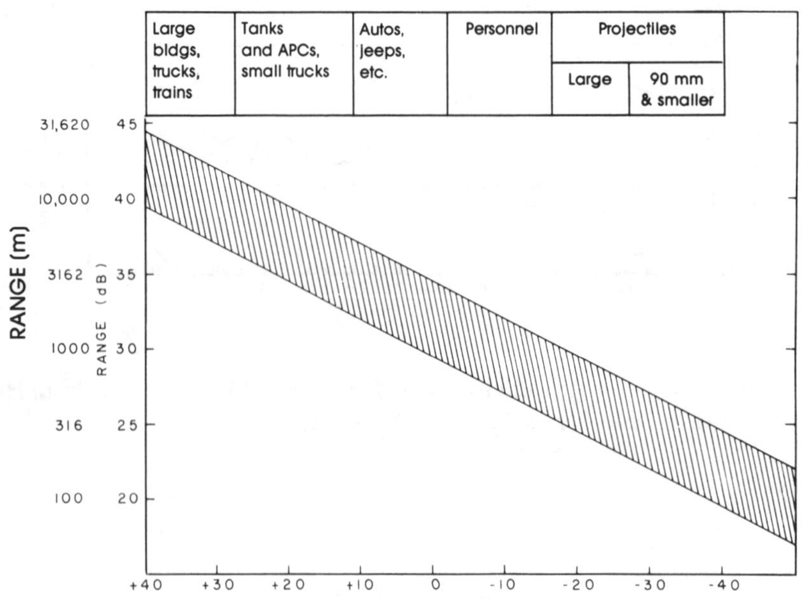

Figure 7.20 Theoretical Maximum Target Detection Capability of a Postulated MMW Radar.

Table 7.12 Postulated Radar Parameters for Radar Range Calculation

Parameter	*Value Assumed*
Wavelength λ	3 mm
Transmitter Power P_t	4 kW
Antenna Gain G	37 dB
Pulse Length τ	50 ns
Bandwidth B	20 MHz
Noise Figure F_n	10 dB
Losses L_T	6 dB
Signal-to-Noise $(S/N)_1$	13 dB

7.5 ELECTRONIC WARFARE

Increasing utilization of MMW sensors, such as radar, is resulting in the development of specific actions, equipment, and techniques to counter these sensors or limit their effectiveness. Because of the unique presentation of the example radar's performance, Figure 7.20 provides a good example of the effectiveness of one particular approach—reduction, suppression, or modification of the target's radar signature. Consider the performance of this radar against a tank with an RCS of 20 dBsm. The maximum detection range is predicted to be approximately 10 km from Figure 7.20. If, through the use of radar cross-section reduction techniques (such as radar camouflage using mesh, pads, blankets, and other radar-absorbing materials, shaping of the primary scattering surfaces of the tank, and other cross-section reduction techniques), the cross section of the tank were reduced by 20 dB from +20 dBsm to 0 dBsm, then the detection range of the radar would drop from 10 km to approximately 3 km—a 70% decrease in the range performance of this particular radar. Obviously, such a performance decrease would significantly decrease the operational effectiveness of this radar, perhaps to the point of rendering it operationally ineffective.

Table 7.13 summarizes most of the major countermeasure threats to MMW radar. In addition to the passive threats, such as the one discussed above, two other categories of countermeasures are presented. Active techniques include both intentional (enemy produced) and unintentional (usually naturally occurring) *electronic countermeasure* (ECM) or jamming. *Electronic support measures* (ESM) include such techniques as wide bandwidth reconnaissance surveillance-warning receivers and direction-finding systems.

However, MMW radar has inherent advantages to counter such threats. Several characteristics of MMW radar which help reduce the susceptibility of these radars to electronic countermeasures threats are given in Table 7.14. Narrow mainbeams (high spatial resolution) and potentially low antenna sidelobes with well designed MMW radar antennas are probably the most important of the items in Table 7.14. Low sidelobes and narrow beamwidths make intercept and jamming extremely difficult; high spatial resolution reduces the effectiveness of chaff and the deleterious effects of volume clutter such as rain. Covert operation, with low propagation overshoot, can be achieved at some frequencies (near or on one of the high atmospheric attenuation bands, perhaps at 60 GHz as an example) reducing the vulnerability of these systems to standoff intercept and limiting the use of long range surveillance or jamming. More conventional and proven electronic counter-countermeasures (ECCM) such as spread-spectrum and frequency agility, are potentially very effective at MMW because of the availability of large operational bandwidths thereby making MMW systems even more resistant to active noise and other bandwidth-limited countermeasures.

Table 7.13 Major MMW Radar Countermeasure Threats

MMW RADAR THREATS		
ECM (*Active*)	*ECM* (*Passive*)	*ESM*
Intentional Noise Deception Unintentional Mutual Interference EMI	Chaff RAM/Signature Modification Foliage/Natural Cover Camouflage Screens False Targets (Confusion) Decoys (Target-Like) Clutter Rain, Snow, Hail Ground Sea Atmospheric Contaminants Fog Smoke Dust	Direction Finding Elint Receivers Defense Suppression ARM

Table 7.14 MMW Radar Advantages in a Countermeasures Environment

- Narrow Beamwidths with Small Apertures
- Low Sidelobes
- Small, Lightweight Equipment (High Mobility)
- Large Bandwidths (Spread Spectrum; Frequency Agility)
- Covert Operation (Low Propagation "Overshoot")
- High Resolution

7.6 MMW RADAR APPLICATIONS

As an overview of the application of MMW technology to radar, and to summarize the discussion of MMW radar presented in this chapter, potential applications and earlier MMW radar developments will be reviewed in the four generic categories of application previously defined:

- Surveillance and Target Acquisition;
- Instrumentation and Measurement;
- Seekers and Terminal Missile Guidance;
- Fire Control and Tracking.

7.6.1 Surveillance and Target Acquisition

There are a number of applications where MMW radars operating at short-to-moderate ranges are quite attractive for surveillance and target acquisition. The advantages of millimeter waves for such applications include small size and weight coupled with high resolution in both azimuth and range, providing excellent resolution of the area under surveillance. Often, for such applications, a rapid scan over an extended angular sector is desirable; the unique capabilities of an electromechanically scanned geodesic lens have resulted in several prototype systems being fabricated for operation at millimeter wavelengths that incorporate such a scanning concept. The basic operation of the scanning geodesic lens is well known and has been covered elsewhere [8]; characteristics of two 70 GHz radar systems, which employ such antennas, are summarized in Table 7.15. These radars are typical of early MMW radar development efforts.

The resolution achievable with a short transmitted pulse and narrow azimuthal beamwidth permits a considerable amount of detail to be displayed to the operators. In addition, the rapid scan rates achievable (up to 70 scans/second) permit a virtually flicker-free display to be realized without the necessity for digital or analog storage techniques. Geodesic antennas may also rapidly stop the scan for ''searchlighting'' on a given target in order to investigate the doppler signatures that might be associated with specific targets of interest within the surveillance area. The fact that the system rapidly scans an extended area also permits the utilization of track-while-scan techniques, thus permitting accurate tracking of a number of targets simultaneously while maintaining area surveillance. Figure 7.21 is a photograph of the antenna and transmitter-receiver portion of one such surveillance radar, the AN/MPS-29 [30]. The degree of terrain mapping detail achievable with this radar permitted navigation of vehicles using only the radar generated information without use of optical information.

Table 7.15 Characteristics of Two 70-GHz Surveillance Radars

	AN/MPS-29	Rapid Scan Radar
Frequency	70 GHz	70 GHz
Pulse Width	50 ns	20-45 ns
PRF	10 kHz	5-25 kHz
Azimuth Beamwidth	0.2°	0.55°
Elevation Beamwidth	0.3° (shaped)	3.5°
Peak Power	15 kW	500 W
Scan Sector	30°	45°
Scan Rate (scans/second)	Up to 40	Up to 70

Figure 7.21 Antenna and Transmitter-Receiver of a 70 GHz Rapid Scan System Used for Battlefield Surveillance (from Long and Allen [30]).

7.6.2 Instrumentation and Measurement

Accurate and meaningful MMW radar performance prediction and analyses require an extensive data base of readily available data on signature and radar cross section (see Chapter 8) for targets and clutter of interest. This requirement, in turn, implies the need for instrumentation and measurement radars designed especially to perform calibrated reflectivity measurements over a wide variety of propagation conditions, target types, and clutter backgrounds. If this data set is to be complete, these radars must cover the complete frequency spectrum of interest. Instrumentation radars must maintain extremely high standards for overall system measurement accuracy (amplitude, time, phase, frequency, and polarization). Internal and external calibration must be included. As might be expected, highly accurate instrumentation-quality radars are relatively complicated and expensive to design, build, and operate.

Only limited propagation and radar reflectivity data are available for the potential radar operating band around 140 GHz. One of the reasons for this fact is that there are virtually no existing instrumentation radars capable of acquiring data at this frequency, characterizing target signatures and clutter reflectivity. To overcome this limitation, Georgia Tech has recently completed the development of a 140 GHz dual polarized instrumentation radar, shown in Figure 7.22. The radar incorporates a pulsed EIKO transmitter, dual channel superheterodyne receiver, and separate dual polarized transmit and receive lens antenna to improve isolation. The major radar operational characteristics are given in Table 7.16 and

Figure 7.22 140 GHz Dual-Polarized Instrumentation Radar.

an overall system block diagram appears in Figure 7.23. The radar is intended to provide calibrated reflectivity measurements for radar cross section, parallel- and cross-polarization, relative phase, and polarization scattering matrix investigations. Potential applications for the radar and resulting data include MMW radar phenomenology studies, radar target reflectivity signatures, clutter and false target characterization, environmental effects, and reflectivity model development and validation.

Another unique instrumentation radar has recently been developed at Georgia Tech. This is the HIPCOR-95, for *high-power, coherent radar operating at 95 GHz* of Figure 7.2. This radar, which was described briefly in Section 7.2, incorporates the heretofore difficult to achieve characteristics of combined high power with coherent operation in an integrated, MMW radar operating at 95 GHz. Furthermore, the radar is unique in that it combines both an EIKA and a TWT amplifier in one master oscillator, power amplifier coherent chain. The radar is also dual polarized, with two operational modes: high power (1–2 kW peak) and moderate power (100 W peak). Coherent frequency agility over 2 GHz bandwidth is available along with two pulsewidths of 50 and 10 ns.

Table 7.16 Georgia Tech Portable Instrumentation Grade 140-GHz Radar

(a) Pulsed EIO Transmitter	
Antenna	Dual Polarized Horn Lens
Gain	44 dB
Beamwidth	1.0°
Polarization	H, V, RHC, or LHC (Fixed or Pulse-to-Pulse Agile)
Peak Power	200 W
Pulse Width	50, 100, 400 ns (Selectable)
Pulse Repetition Frequency	10 kHz (Maximum)
RF Losses	3 dB
(b) Dual-Channel Superheterodyne Receiver	
Antenna	Dual Polarized Horn lens
Gain	44 dB
Beamwidth	1.0°
Polarization	H and V, or RHC and LHC
System Polarization Isolation	>20 dB
Noise Figure	7 dB
IF Bandwidth	50 MHz
Detection	
Amplitude	Logarithmic
Relative Phase	Quadrature
Dynamic Range	70 dB
Output Signals	
Video, Logarithmic Amplitude	Each Receiver Channel
Video, Relative Phase	Between Receiver Channels (I and Q)
Sample/Hold Boxcar	Each Log Video
Sample/Hold Boxcar	Each I and Q Video

REEDY & WILTSE
CHAPTER 7

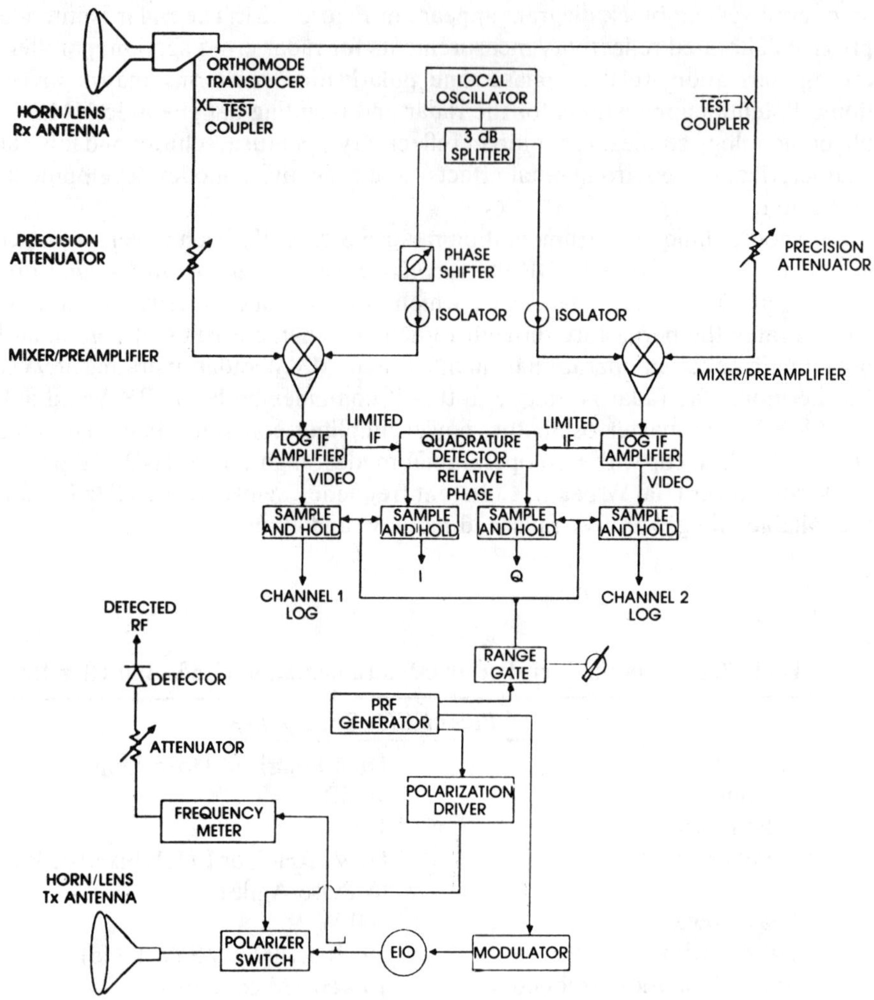

Figure 7.23 140 GHz Instrumentation Radar Block Diagram.

The salient features of the HIPCOR-95 radar are given in Table 7.17 and a complete block diagram in Figure 7.24. The radar incorporates a Varian VKB 2449 series EIKA (see Figure 7.25) and a Hughes 982H TWT (see Figure 7.26). HIPCOR-95 will be used by the US Army Missile Command, Redstone Arsenal, Alabama, to evaluate a 95-GHz MOPA transmitter using the most recent EIKA and TWT devices, to evaluate high-range resolution performance parameters for a high-power MMW radar, to establish baseline MTI performance parameters for a high-power MMW radar, and to provide a high technology test bed for evaluating advanced target detection, discrimination, and classification techniques, and obtaining target-clutter reflectivity data.

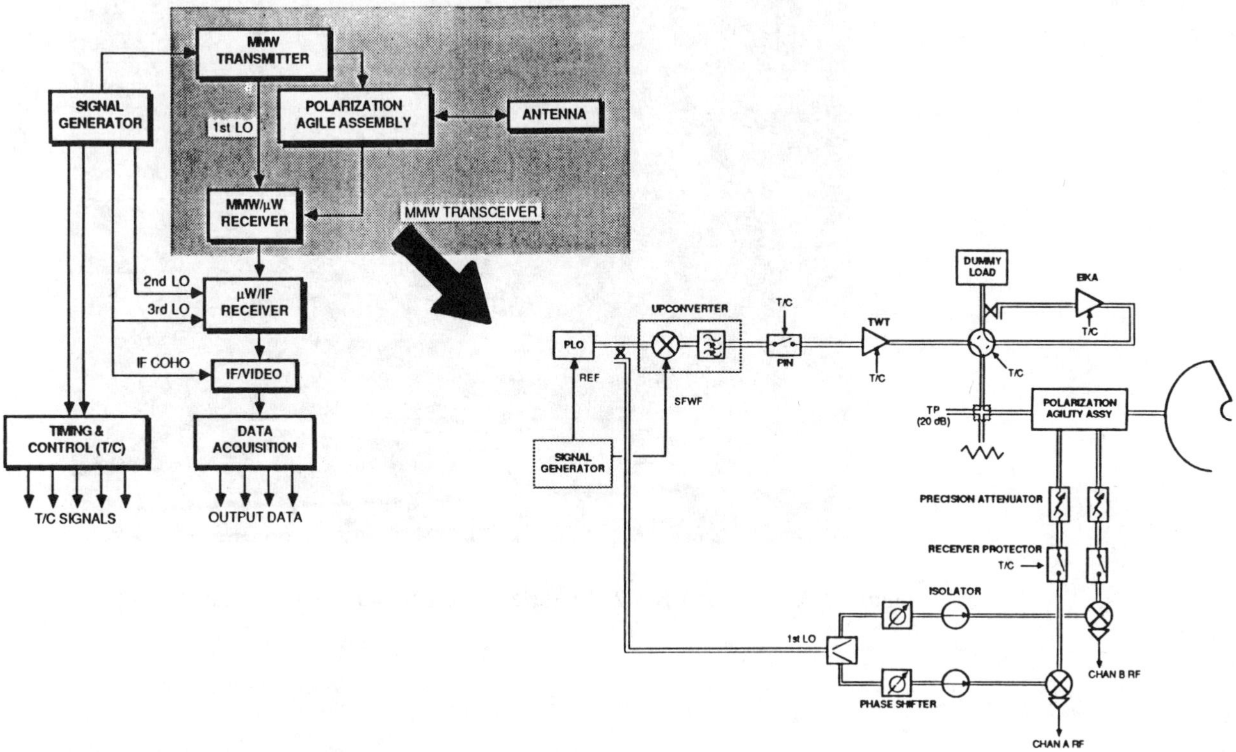

Figure 7.24 HIPCOR-95 MMW Transceiver Block Diagram.

Table 7.17 HIPCOR-95 Salient Features

Coherent	I and Q Outputs
Polarization Agile (Transmit)	H/V, RHC/LHC
Dual Polarized (Receive)	Co-Polarization Cross-Polarization
Operational Modes:	
High Power	1 kW (2 kW Goal) 1 W (Average)
Moderate Power	100 W 0.1 W (Average)
Coherently Frequency Agile	2 GHz BW at Moderate Power Level
Pulse Width	50, 10 ns
High Range Resolution	0.5 ft (Off-Line Processing) at Moderate Power Level
Multiple Range Samples	1 per Channel (6 Total)

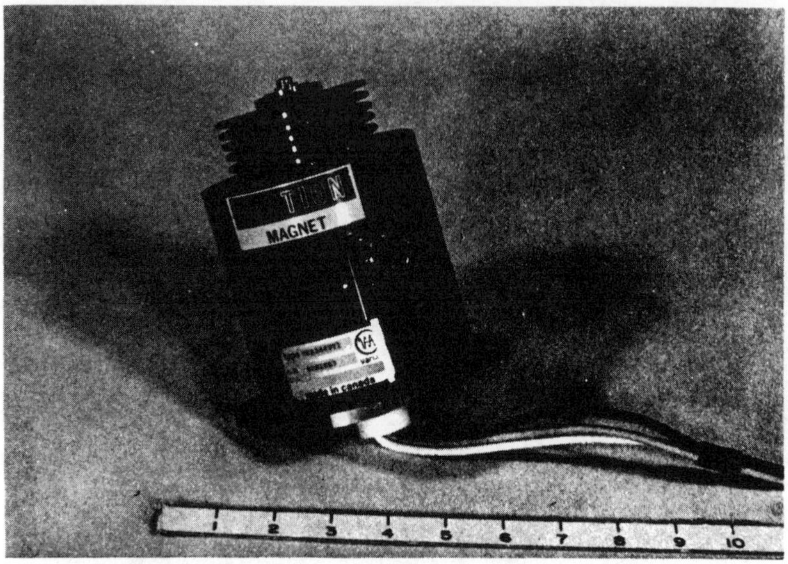

Figure 7.25 Varian VKB 2449 Series EIKA (Courtesy of Varian Corp.).

Figure 7.26 Hughes 928H TWT Amplifier (Courtesy of the Georgia Institute of Technology).

7.6.3 Guidance and Seekers

The size and weight advantages inherent in MMW sensors make them ideal for applications involving missile seekers and terminal guidance. In fact, this appears to be the most active research area currently incorporating MMW sensors (both passive and active). Passive, fully active, semiactive, beamrider, and dual-mode sensors and guidance techniques are all being investigated and evaluated for MMW applications.

Shown in Figure 7.27 is a 94 GHz missile seeker for air-to-ground applications incorporating either an active monostatic pulsed radar mode or a passive or bistatic CW radar mode. The antenna is a 25.4-cm diameter Cassegrain with a conical scan feed arrangement. The radar is all-solid-state, utilizing an IMPATT transmitter (Bernues *et al.* [31]). Dual-mode operation allows the seeker to circumvent active radar "aim point wander" and target-induced "glint" track errors in the terminal engagement phase by switching to the passive mode.

Figure 7.28 is a conceptual drawing of a MMW beamrider guidance concept, a technique receiving considerable attention for application at millimeter wavelengths. In the most simplified terms, a MMW beamrider guidance system can be defined as a system for guiding missiles to a target that incorporates a narrow pencil beam directed at the target along which a missile flies. Receiving and guidance equipment in the missile determines the missile's position with respect to the beam center and develops commands to control the missile's aerodynamic control surfaces in such a manner as to keep the missile trajectory in the middle of the beam.

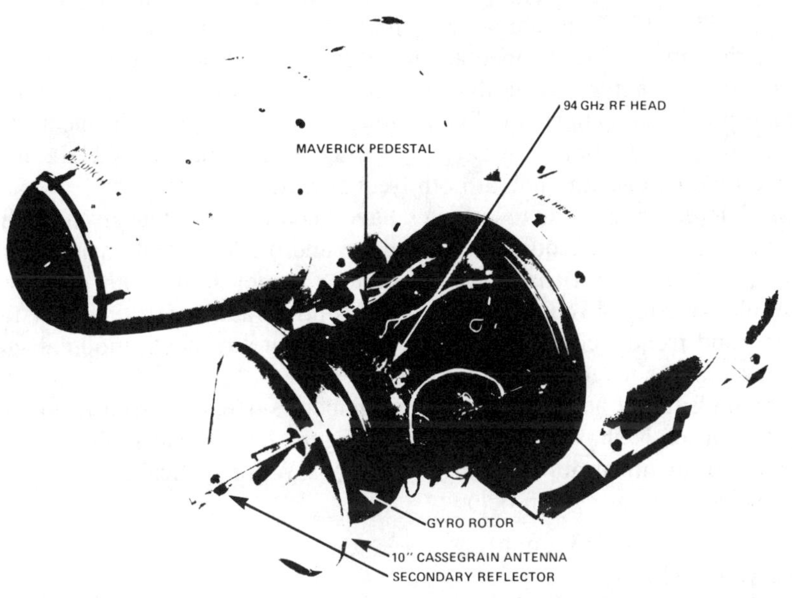

Figure 7.27 **A 95-GHz Missile Seeker (from Bernues, King, and Kaswen [31], © 1979 *Microwave Systems News*).**

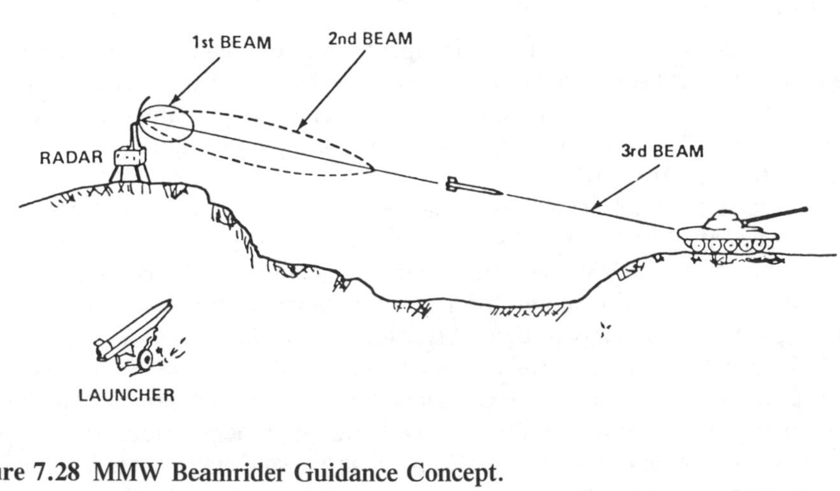

Figure 7.28 **MMW Beamrider Guidance Concept.**

7.6.4 Fire Control and Tracking

Short-range, closed-loop, fire control applications, such as that illustrated in Figure 7.29, where extremely high spatial accuracy, small size and weight, and high mobility are the prime system requirements, are well suited to MMW radar [32]. Such systems might serve as point defense antiaircraft weapons, ship defense systems, or for close-in battlefield air defense applications. Radars supporting these applications should provide some inherent target acquisition capability and, at least, first-order target classification capability.

Multipath propagation and ground main beam clutter are classical problems associated with low-angle tracking for air defense fire control systems—in particular, gunfire control systems. In this environment, MMW radar can provide significant, potential advantages due to its narrow beamwidths and high resolution properties. The US Army has investigated and, in various research and development programs, has developed advanced fire control radar systems to test the feasibility of millimeter-wave radar to direct and control antiaircraft gun systems with high precision against low-flying aircraft and helicopters. Typical of these radar systems is a dual-band tracker, operating at Ku band and 94 GHz, utilizing a common monopulse antenna at both frequencies.

Millimeter-wave radar techniques have been investigated for such diverse applications as a target acquisition (detection and tracking) and engagement system (limited fire control) having capabilities commensurate with current main battle tank fire-control requirements, a fire-control radar operating near the absorption band frequencies for ship defense, and an advanced, modular gunfire-control system for use on fast patrol boats.

The tank target acquisition and engagement system referred to in the preceding paragraph is shown conceptually in Figure 7.30 (Backus [33]). This radar is intended to demonstrate the feasibility of MMW for tactical systems. Preliminary specifications are given below:

Frequency: 94 GHz (3.2 mm)
Bandwidth: 11 mr
Average Power: 0.1–0.5 W
Antenna Aperture: 35.6 cm
Field of View (Scan Sector): 15° × 7.5° in wide mode; 5° × 2.5° in narrow tracking mode
Frame Time: > 2 s
Target Detection: 3000 m with 100-m visibility
Target Tracking: 0.5 mr accuracy at 2 km with 100-m visibility

The radar is intended to complement, and be integrated with, the infrared tank thermal sight such that both systems share the same display. Spread-spectrum waveforms and both coherent MTI and area MTI techniques are being investigated for detection of moving targets. The radar incorporates a solid-state IMPATT transmitter power source.

One of the simplest and most creative applications of MMW radar to a closed-loop fire control problem is discussed in Strom [16]. In this reference, a design is developed for a MMW radar fire control system for the Army infantryman and his rifle. The capability and performance parameters of this radar were examined and were found to compare favorably with the human eye.

The millimeter-wave radar application examples discussed in this section are certainly not all-inclusive or complete, but were selected primarily to illustrate the variety of operational problems involving target acquisition, tracking, guidance, and instrumentation where MMW radar has some inherent advantages, and shows promise for providing improved performance over either higher- or lower-frequency radars.

Figure 7.29 An Artist's Concept of a Short-Range Antiaircraft Weapons System Incorporating a MMW Radar-Directed, Rapid-Fire Cannon Mounted on an Armored Personnel Carrier.

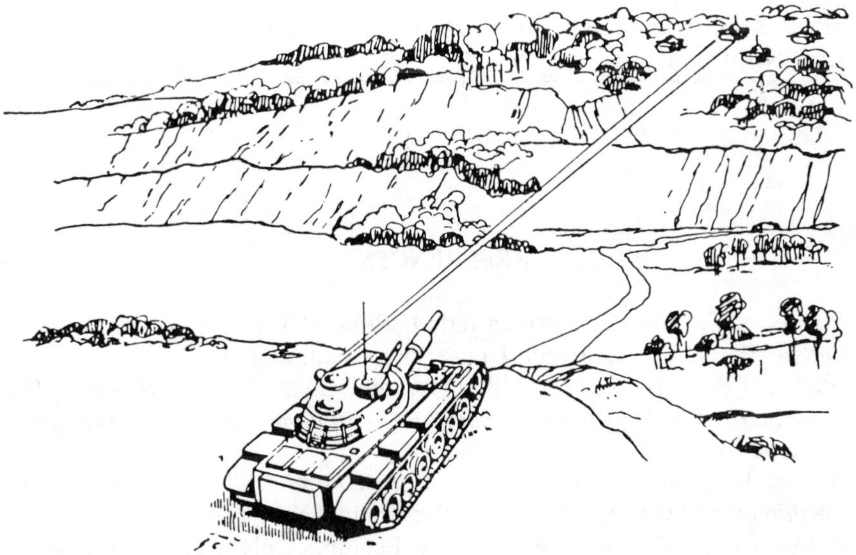

Figure 7.30 Surveillance and Target Acquisition Radar for Tank Location and Engagement (STARTLE) (from Backus [33], © 1979 *Journal of Electronic Defense*).

7.7 FUTURE DIRECTIONS

As the discussion in the last section has indicated, extensive utilization of the MMW frequency spectrum for radar applications is currently taking place. The future is certainly bright also. Where is the future for MMW radars? What directions will radar applications take? Table 7.18 lists some of the expected directions for MMW radar applications and research. Fully coherent, high-power radars are currently being developed, as evidenced by HIPCOR-95, but these are still early in their research and development cycle. Dual-frequency approaches, especially combining MMW radar and EO/IR sensors for missile guidance are currently being investigated. The increased emphasis on exoatmospheric (space) based defense systems, under the Strategic Defense Initiative (SDI) program, bodes well for MMW sensors because propagation limitations are virtually eliminated outside the earth's atmosphere.

The last item listed in Table 7.18 may well be the most important, however, in determining the extent of future MMW radar development and application. Fabrication costs must be significantly reduced for MMW radar to realize its full potential.

Table 7.18 MMW Radar Future Directions

- Fully Coherent, High Power
- Dual Frequency Detection and Tracking Sensors (Dual Mode)
  - Millimeter/Microwave
  - Millimeter/EO (IR)
- Higher Power
- Packaging Improvements and Component Development
  - Integrated Components
  - Smaller Size, Lower Weight
  - Solid State
- Electronic Scanning
- IFF, Target Recognition, and Classification
- ECM/ECCM/ESM
- Space Applications (SDI)
- Cost Reduction

REFERENCES

1. Currie, N.C., and C.E. Brown (eds.), *Principles and Applications of Millimeter-Wave Radar,* Artech House, Norwood, MA, 1987.
2. Wiltse, J.C., "History of Millimeter and Submillimeter Waves," *IEEE Transactions on Microwave Theory and Techniques,* Vol. MTT-32, September 1984, pp. 1118–1127.
3. Button, K.J., and J.C. Wiltse, eds., *Infrared and Millimeter Waves,* Vol. 4, "*Millimeter Systems,*" Academic Press, New York, 1981.
4. Wiltse, J.C., "Millimeter-Wave Radar Features Unique Characteristics and Designs," *Microwave Systems News,* Vol. 14, May 1984, pp. 58–76.
5. Hersch, H.J., E.J. Kaiser, F.E. Kavanaugh, and I. Reingold, "Electron Tubes and Devices in the 4.3-mm Frequency Range," *IRE Transactions,* Vol. MIL-4, No. 4, October 1960, pp. 481–492.

6. Johnston, S.L., "Millimeter Radar," *Microwave Journal,* Vol. 20, November 1977, pp. 16–28.
7. Lane, T.L., *et al.,* "Coherent 95 GHz High Power Radar," *SPIE Technical Symposium East Conference,* No. 544, April 1985.
8. Reedy, E.K. and G.W. Ewell, "Millimeter Radar," Chapter 2 of *Infrared and Millimeter Waves,* Vol. 4, *Millimeter Systems,* Kenneth J. Button and James C. Wiltse, eds., Academic Press, New York, 1981.
9. IEEE Standard 521–1976, November 30, 1976.
10. Kulpa, S.M. and E.A. Brown, "Near-Millimeter Wave Technology Base Study, Volume I, Propagation and Target/Background Characteristics," Special Report Number HDL-SR-79-8 on Contract DA 1L161102AH44, Harry Diamond Laboratories, Adelphi, MD, November 1979.
11. Bhartia, P. and I. Bahl, *Millimeter Wave Engineering and Application,* John Wiley and Sons, New York, 1984.
12. Skolnik, M.I., *Introduction to Radar Systems,* McGraw-Hill, New York, 1962.
13. Dyer, F.B., *et al.,* "Review of Millimeter Wave Radar Development at Georgia Tech," Technical Report 77-01, Georgia Institute of Technology, Atlanta, Georgia, 1977.
14. Downs, A.R., "A Review of Atmospheric Transmission Information in the Optical and Microwave Spectral Regions," Memorandum Report No. 2710, USA Ballistic Research Laboratories, December 1976.
15. McCartney, E.J., "Scattering: The Interaction of Light and Matter," Report No. AB-1272-0057, Sperry Rand Corporation, February 1966.
16. Strom, L.D., "Applications for Millimeter Radars," Report 108, System Planning Corporation, 1973.
17. Preissner, J., "The Influence of the Atmosphere on Passive Radiometric Measurements," AGARD Conference Reprint No. 245, *Millimeter and Submillimeter Wave Propagation and Circuits,* 1978.
18. Richard, V.W., and J.E. Kammerer, "Rain Backscatter Measurements and Theory at Millimeter Wavelengths," Report No. 1838, USA Ballistic Research Laboratories, October 1975.
19. Wilcox, F.P., and R.S. Graziano, "Millimeter Wave Weather Performance," *Millimeter Waves Techniques Conference,* Vol. 1, Naval Electronics Laboratory Center, San Diego, CA, March 1974.
20. Currie, N.C., F.B. Dyer, and R.D. Hayes, "Analysis of Radar Return at Frequencies of 9.375, 35, 70, and 95 GHz," Technical Report No. 2 on Contract DAAA 25-73-C-0256, Georgia Institute of Technology, February 1975.
21. Nespar, J.D., "Dual Frequency, Dual Polarization Backscatter Measurement of Rain," *Proceedings of the Tenth DARPA/Tri-Service Millimeter Wave Conference,* Harry Diamond Laboratory, Adelphi, MD, 1986.
22. Allen, G., and B. Simonson, "Attenuation of Infrared Laser Radiation by HC, FS, WP, and Fog Oil Smokes," Technical Report EATR 4405, U.S. Army Edgewood Arsenal, May 1978.
23. Morgan, R.L., J.D. Stattler, and G.A. Tanton, "Results of MIRADCOM Workshop on Millimeter and Submillimeter Atmospheric Propagation Applied to Radar and Missile Systems," *Proceedings of the Eighth DARPA/Tri-Service Millimeter Wave Conference,* Eglin AFB, FL, April 1979.
24. Martin, E.E., "Radar Reflectivity of Dust Cloud Lofted by Misers Bluff II High Explosive Test," Final Technical Report, GIT Report A-2104, Georgia Institute of Technology, March 1979.
25. Knox, J.E., "Effects of Smoke Obscurants on Millimeter Waves," *Proceedings of the Eighth DARPA/Tri-Service Millimeter Wave Conference,* Eglin AFB, FL, April 1979.

26. Petito, F.C., and R.L. Harris, "Millimeter Wave Radar Transmission through High Explosive Artillery Barrages," *Proceedings of the Eighth DARPA/Tri-Service Millimeter-Wave Conference,* Eglin AFB, FL, April 1979.
27. McMillian, R.W., "Atmospheric Turbulence Effects on Millimeter Wave Propagation," *IEEE EASCON-79 Conference Record,* Vol. 1, 1979.
28. Dyer, F.B., and R.D. Hayes, "Computer Models for Fire Control Radar Systems," Technical Report No. 1 on Contract DAAA 73-C-0256, Georgia Institute of Technology, Atlanta, Georgia, 1973.
29. Hayes, R.D., N.C. Currie, and J.A. Scheer, "Reflectivity and Emissivity of Snow and Ground at MM Waves," *IEEE International Radar Conference Record,* Washington, DC, April 1980.
30. Long, M.W., and G.E. Allen, "Combat Surveillance Radar," Final Report on Contract DA36-039-SC-74870, Georgia Institute of Technology, 1960.
31. Bernues, F.J., R.S. King, and M. Kaswen, "Solid State Oscillators, Key to Millimeter Radar," *Microwave Systems News,* Vol. 9, No. 5, May 1979.
32. Reedy, E.K., and J.L. Eaves, "Millimeter Wave Instrumentation Radar Technology—Considerations and Applications," *Military Electronics Exposition '79,* Anaheim, CA, 1979.
33. Backus, P.H., "Electro Optics," *J. Electronic Defense,* Vol. 11, No. 3, March 1979.

Chapter 8
RADAR CROSS SECTION

by Eugene F. Knott

8.1 OVERVIEW

President Jimmy Carter announced in August 1980 that the US government had initiated a Department of Defense program known as the "stealth bomber" or the "stealth aircraft." There is no doubt that a great deal of research had been accomplished well before the announcement, but this public admission of the existence of a rather large, technically advanced program sparked the curiosity of scientists, engineers, politicians, and the general public. The stealth aircraft was alleged to be invisible to radar, and technical and nontechnical people alike soon learned that radar stealth was characterized by what was called a low "radar cross section."

The technical community, in the meantime, had discovered the need to become acquainted with a new technology. Scientists and engineers of all disciplines were forced to deal with new and unfamiliar concepts, although they might have been proficient in their own fields. Government and private industry dispatched managers and engineers to attend short courses tailored specifically for this demand for "instant education." The civilian sector was equally interested, hoping for a space age countermeasure to police radars.

Notwithstanding the various courses that have been assembled to disseminate technical information on radar technology in general, and on stealth technology in particular, this chapter represents one abbreviated version. The information is presented at a relatively high level (i.e., not in great detail) in three parts that attempt to answer three questions about radar cross section:

What is it?
Can you predict it?
Can you reduce it?

The last of the three, it might be noted, is but one aspect of the much broader concept of stealth.

Our response to the first question is an introduction to the concept of *radar cross section* (RCS). RCS is a measure of the power "scattered" by an obstacle, and therefore how large an echo signal that a radar will likely sense. An expression can be derived for the RCS in terms of the strengths of the incident and scattered electric fields, and the range from the radar to the obstacle. However, because

---

Editor's Note: Much of the information presented in this chapter appeared in three chapters of E. Knott, J. Shaeffer, and M. Tuley, *Radar Cross Section: Its Prediction, Measurement and Reduction* (Artech House, 1985). Because the original lecture was based on three chapters of *Radar Cross Section*, the task was to boil them down to manageable size for this book. Further condensation of the three chapters was necessary to make room for new material. Thus, while this chapter has been created specifically for this book, and some new material developed, readers familiar with *Radar Cross Section* will encounter many words, phrases and passages they have already seen.

the formal definition gives little insight into the characteristics of typical radar targets, examples of the echo properties of a collection of targets are shown. These include simple shapes, such as spheroids, flat plates, and the like, and more complicated objects, such as ships, airplanes, and insects. A single chart (Figure 8.23) is presented near the end of that section, summarizing the general echo strengths of several classes of target.

The answer to the second question is a resounding "Yes!" and several RCS prediction methods are described. The ability to estimate the echo properties of military targets is vital in the design of modern radar and weapon systems, and literally dozens of computer codes have been developed over the last 20 years to meet this demand. RCS prediction methods are split in the discussion into two categories: *exact* and *approximate*.

The exact methods attack the solution of either the differential or the integral equations describing the electromagnetic fields in all space, given the boundary conditions the fields must satisfy on the surface of an obstacle. Solutions of the differential form (the wave equation) are difficult to obtain and are restricted to very simple geometric structures; therefore, these solutions have little practical application to actual targets. The solution of the integral equation (called the *method of moments*) reduces the integral equations to a collection of homogeneous linear equations, and is restricted to relatively small targets, usually a few wavelengths long at best. However, even though the exact methods can seldom be applied to realistic targets, their solutions for simple structures often impart a great deal of insight into the scattering (echoing) process.

The approximate methods are all based on optical concepts, in which the wavelength is assumed to be much shorter than any target scattering feature. This makes the methods relatively easy to formulate and the methods themselves give reasonably reliable results. Most of the computer codes developed for realistic targets are based on these "high frequency" approximations. Each of the approximate methods has its own particular claim to fame—and none is without its shortcomings. Also, as we shall see, none of them is capable of routinely predicting the contributions of surface traveling waves for arbitrary structures.

As to the third question, the reduction of the echo of an object, be it an aircraft or the family sedan, is a study in compromises. Most engineering problems can be reduced to a collection of options, and we need only choose the option that comes nearest to meeting the requirements within the constraints specified. Certainly, the RCS can be reduced—but at a price. Although the price usually entails the sacrifice of otherwise desirable features, such as speed, range, or strength, the price always can be translated into dollars. Some of the cost must be invested in the acquisition of unusual materials and the development of new fabrication techniques, and a significant investment must be made in engineering research and development before a single part is fabricated.

Of the four known ways to reduce target echo, only two—shaping and radar absorbing materials—have been used with any reasonable effectiveness in the past. The objective of shaping is to deflect the reflected radar wave in some other direction except back toward the radar, and, as the name implies, absorbing materials soak up radar energy, thereby reducing the amount available to bounce back to the radar. Both methods have been tested in civilian and military applications, and perhaps not surprisingly, the civilian and military exploiters of radar stealth both agree that the cost is high. Cost, however, should always be judged by how badly the product is needed. While there appears to be no market for a \$300, radar-absorbent, stone shield for personal automobiles, a billion-dollar stealth bomber seems to be essential for national defense.

8.2 WHAT IS IT?

8.2.1 Formal Definition of RCS

The notion of assigning an area to the radar echo of an obstacle can probably be traced to antenna engineers. An essential quantity of interest in radio communications is the signal-to-noise ratio (S/N), for if the signal is not significantly stronger than the receiver noise level, it may be impossible to extract any intelligence from the signal, nor indeed even to detect the signal itself. Because noise is the result of a random process, the only meaningful characterization of it is power, in watts or milliwatts, for example. It is natural, therefore, to characterize radio signals in terms of power as well, so that important gauges of system performance, such as signal-to-noise ratio, can be measured or estimated.

A fundamental characteristic of a radio wave, however, is its power *density*, measured in watts per square meter, for example. This is a measure of the rate of energy transfer through an imaginary surface perpendicular to the direction of propagation of the wave. We extract energy from the wave by exposing an antenna to it, and power is channeled to a receiver. Obviously, if the wave can be characterized by a power density, and if we seek to measure or to estimate the power collected by the antenna and delivered to the receiver, we must multiply the power density of the incident wave by the capture area of the antenna. Thus, radio and antenna engineers long ago adopted the notion of characterizing the antenna in terms of an area.

This "capture" area may or may not be strongly related to the physical area of the antenna. Indeed, early antennas (and many modern antennas as well) were merely thin wires stretched between masts or poles, with physical areas that were virtually nil. Yet, they were reasonably effective collectors of radio energy, and engineers were forced to assign them "effective areas" that were more descriptive of their power collection capabilities than their physical areas. When engineers began bouncing short radio waves off obstacles and observing echoes, it was natural that they attempted to describe the echo strength in terms of an area. It was even called "echo area," instead of "radar cross section," at first.

If radio engineers could estimate the power collected by an antenna of some effective area exposed to a wave of known power density, they could certainly estimate the echo area of a target from the known (or measured) power density of the reflected wave. Engineers did so by assuming the target to be an isotropic scatterer. Assume, for example, that an obstacle (target) is exposed to an incident wave whose electric and magnetic field strengths are E_i and H_i. The power density of the wave (W_i) is [1]:

$$W_i = \frac{1}{2} E_i H_i = \frac{1}{2} Y_0 |E_i|^2 \tag{8.1}$$

where $Y_0 = 0.00165$ mho is the admittance of free space. While it is assumed that the radar is well removed from the target, so that the incident wavefronts are sensibly spherical, the field intensities E_i and H_i are those measured at the target, not at the source.

If we now assign the reflecting obstacle some effective capture area σ, then the power "collected" by the obstacle is

$$P = \sigma W_i = \frac{1}{2} \sigma Y_0 |E_i|^2 \tag{8.2}$$

where σ, of course, will be the radar cross section of the target. If we assume that all of this "captured" power is radiated uniformly in all directions, the scattered power density at some large distance R from the target is

$$W_s = P/(4\pi R^2) = \sigma Y_0|E_i|^2/(8\pi R^2) \tag{8.3}$$

Our obvious interest is in the power density of the scattered wave as it returns to the radar, consequently R will be the range from the radar to the target.

Like the incident power density, the scattered power density can be written in terms of an electric field strength E_s:

$$W_s = \frac{1}{2} Y_0 |E_s|^2 \tag{8.4}$$

We can equate (8.3) and (8.4) to determine σ,

$$\sigma = 4\pi R^2|E_s|^2/|E_i|^2 \tag{8.5}$$

Note that the RCS so derived depends on the distance R, implying that different values may be observed at different distances. By convention, the RCS is defined for an incident *plane* wave, whereupon R is forced to approach infinity:

$$\sigma = \lim_{R\to\infty} 4\pi R^2|E_s|^2/|E_i|^2 \tag{8.6}$$

Thus, formally, at least, the radar cross section is independent of the distance R. An equally valid definition results when the incident and scattered electric field strengths are replaced by the incident and scattered magnetic field strengths.

The scattered field strength falls off as $1/R$ in the three-dimensional world, but there is considerable motivation to examine two-dimensional scattering, for which the cylindrical wave is the counterpart of the spherical wave. While it is doubtful that cylindrical waves actually exist in nature, the solutions of two-dimensional problems are extremely useful because of the simplicity afforded by the elimination of one spatial variable. In passing from three dimensions to two, however, the point source becomes a line source, and the echo strength becomes a width instead of an area. The formal definition in this case is [2]:

$$\sigma = \lim_{\rho\to\infty} 2\pi\rho\, |E_s|^2/|E_i|^2 \tag{8.7}$$

8.2.2 Bistatic Scattering

Whether gauged in the three- or two-dimensional world, the radar cross section (or width) describes the intensity of the scattered wave back at the source. Waves are scattered in all directions, of course, hence the intensity in the direction to the source (the *backscattering* direction) represents but one point of a more extensive *bistatic* pattern that includes all scattering directions. The scattering in the *forward* direction is often as important as that in the backward direction, because it is the forward scattered wave which combines with the incident wave to form the shadow behind the obstacle. The forward scattering cross section (or forward scattering width for two-dimensional bodies) is proportional to the square of the projected area (or the projected width for two-dimensional bodies) of the body. Thus, except for good reflectors like corners or flat plates viewed at normal incidence, the forward scattering is typically orders of magnitude larger than the backscattering. This is illustrated in Figures 8.1 through 8.5.

Figure 8.1 is the bistatic scattering pattern of an infinite, circular, metallic cylinder that is 10λ in circumference, which is a two-dimensional structure. The

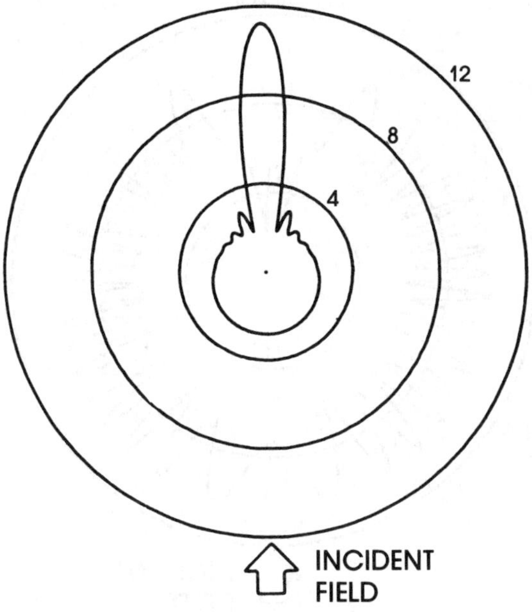

Figure 8.1 Bistatic scattering of an infinite cylinder 10λ in circumference.

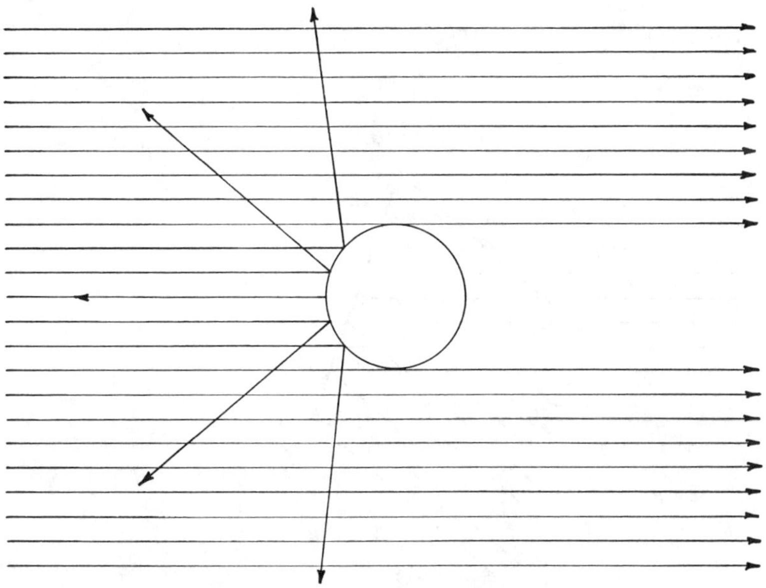

Figure 8.2 The cylinder blocks incident rays from the shadow zone.

pattern was generated from the exact solution of the wave equation for an incident plane wave with its electric field aligned along the cylinder axis. The direction of incidence is from the bottom of the figure, as shown, and except for the strong lobe in the forward direction, the scattered field strength is nearly uniform in all directions. Although not demonstrated here, the forward lobe becomes narrower and more intense as the cylinder becomes larger, and the region of uniform scattering intensity consequently extends closer to the forward direction.

The large forward scattering lobe is required to cancel the incident field and thereby create the necessary shadow behind the cylinder. Figure 8.2 is intended to show this. The incident field propagates from left to right in the illustration, and a few rays are shown reflected from the illuminated front side of the cylinder.

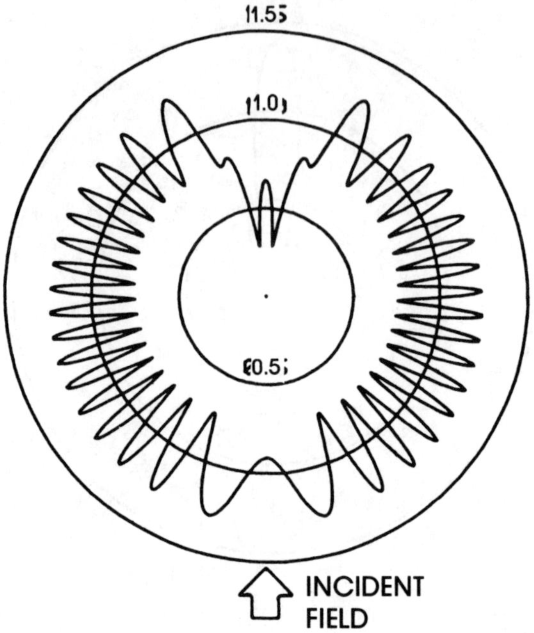

Figure 8.3 **Total field in the vicinity of the cylinder along a circular path 10λ from the cylinder axis.**

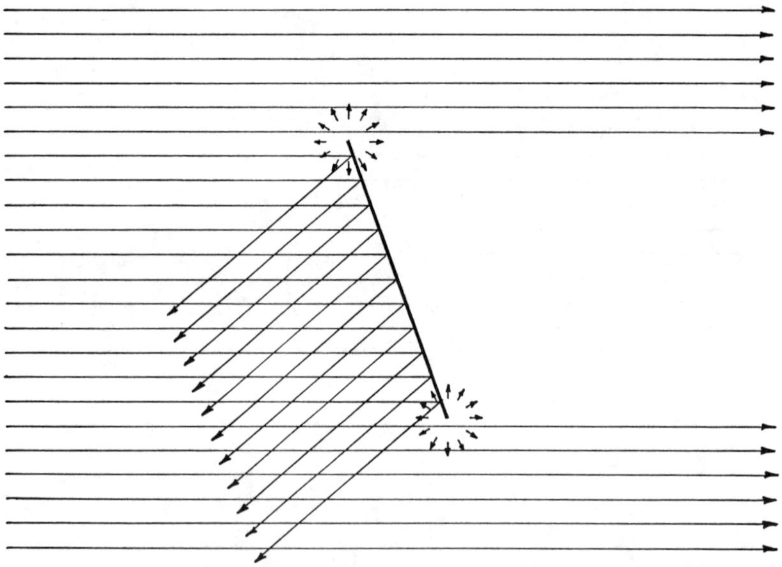

Figure 8.4 **A flat plate reflects most of the incident energy in the specular direction.**

These rays are all equally intense for very large cylinders, as predicted by the theory of *geometric optics* (GO) discussed in Section 8.3 below. The diagram shows no rays propagating behind the cylinder, suggesting that a deep shadow exists there.

Figure 8.3 is a plot of the total field (incident field plus scattered field) probed around the cylinder in a circular path 10λ from the axis. This is obviously an interference pattern resulting from the alternate cancellation and reinforcement of the incident and scattered waves as the relative phase between the two change along the circular path. The incident field strength has a value of unity, and it can be seen that the nulls and peaks of the pattern reach values of about 0.75 and 1.25, respectively. The total field in the forward direction does not drop to zero;

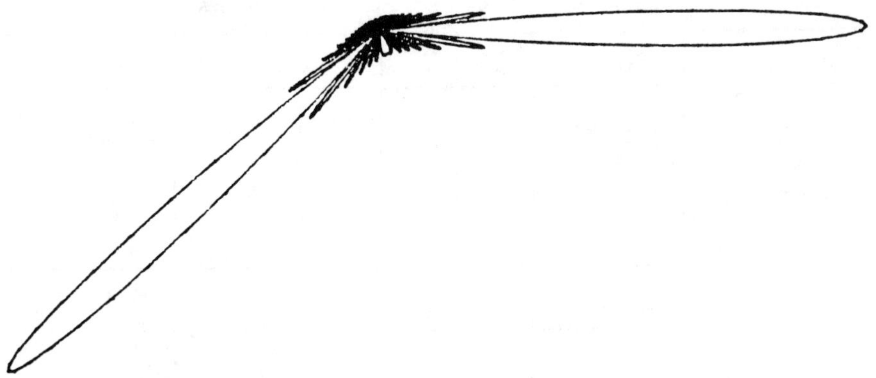

Figure 8.5 The specular and forward scattering lobes of a flat plate are equally intense. Physical optics was used to compute this pattern.

hence, the shadow at this distance from the cylinder is not a deep one. The finite value there is due to diffraction of energy from the rounded sides of the cylinder.

The bistatic fields scattered by a flat plate are quite different. Figure 8.4 suggests that most of the rays striking an oblique plate are reflected in the same specular direction instead of uniformly in all directions, although the edges of the plate may have omnidirectional scattering properties. In addition to a strong forward lobe, therefore, one also expects a strong lobe in the specular direction. As shown in Figure 8.5, this indeed occurs. The theory of physical optics was used to generate this pattern, and note that the specular lobe is as intense as the forward lobe. The peak amplitudes of both are proportional to the square of the projected area of the plate, and the sidelobes can be attributed to diffraction from the edges of the plate.

8.2.3 RCS Characteristics of Simple Bodies

These two examples suggest that the size and shape of the target, and its orientation with respect to the incident wave, are important parameters influencing the scattered fields, and therefore the RCS. Equally important is the material composition of the target, but that discussion is undertaken in Section 8.4.

There are few, if any, simple targets in the practical world of radar target encounters. Neither the armored vehicle under attack by a radar guided missile, nor the automobile in the beam of a police radar can be called simple, at least as regards the mathematical description of their surface profiles. Nevertheless, objects as complex as these often can be described as collections of relatively simple shapes, such as spheroids, plates and cylinders, that approximate the overall target profile. Moreover, if we are interested in measuring the RCS of complicated objects, the measurements are often calibrated by comparing the echo levels of the test object with those of a calibration target. Because the echo strength of the calibration target must be known with a high degree of certainty, the calibration target is almost always a simple one. Therefore, the examination of the properties of simple objects is easily justified.

Table 8.1 is a summary of the characteristics of a few simple objects, arranged in the order of decreasing echo strength. Corner reflectors head the list as the strongest scatterers because of multiple reflections between their faces. The faces of both the dihedral and trihedral corner reflectors are assumed to be angled 90° apart; consequently, the wave leaving the corner is directed back toward the radar after the final bounce. The magnitude of the echo is proportional to the square of the effective aperture presented to the incident wave. This effective area is somewhat less than the projected area of the corner because not all the energy

KNOTT
CHAPTER 8

Table 8.1 Hierarchy of Scattering Shapes

Geometry	Type	Freq. Dep.	Size Dep.	Formula	Remarks
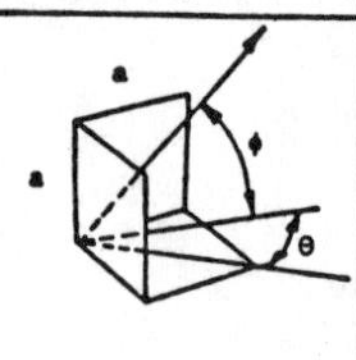	Square trihedral corner retro-reflector	F^2	L^4	Maximum $\sigma = \frac{12\pi a^4}{\lambda^2}$	Strongest return; high RCS due to triple reflection
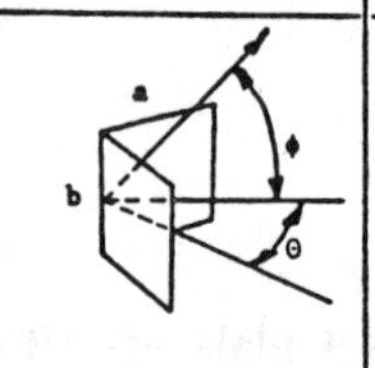	Right dihedral corner reflector	F^2	L^4	Maximum $\sigma = \frac{8\pi a^2 b^2}{\lambda^2}$	Second strongest; high RCS due to double reflection, tapers off gradually from the maximum with changing θ and sharply with changing ϕ.
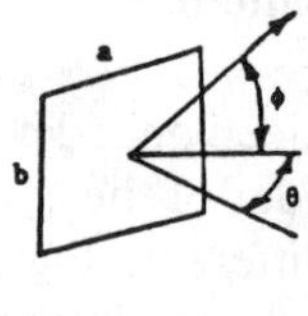	Flat plate	F^2	L^4	Maximum $\sigma = \frac{4\pi a^2 b^2}{\lambda^2}$ Normal Incidence	Third strongest; High RCS due to direct reflection, drops off sharply as incidence changes from normal.
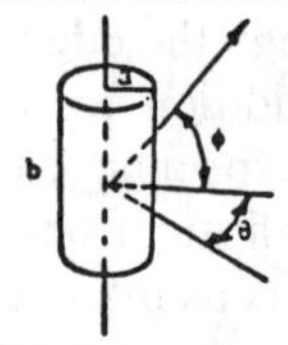	Cylinder	F^1	L^3	Maximum $\sigma = \frac{2\pi a b^2}{\lambda}$ Normal Incidence	Prevalent cause of strong, broad RCS over varying aspect (θ), drops off sharply as azimuth (ϕ) changes from normal. Can combine with flat plate to form dihedral corner reflector.
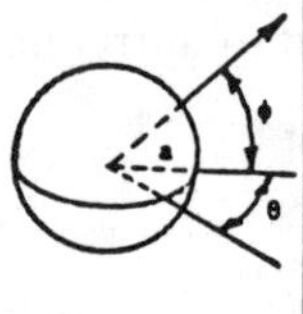	Sphere	F^0	L^2	Maximum $\sigma = \pi a^2$ Normal Incidence	Prevalent cause of strong, broad RCS peaks other than those due to large openings in target body. Energy defocused in two directions.
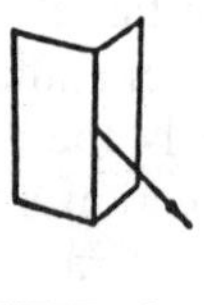	Straight edge normal incidence	F^0	L^2	$f(\theta,\phi)L^2$ θ - aspect θ_{int} - interior dihedral angle between faces meeting at edge	Limiting case of 2-dimensional curved plate mechanism as radius shrinks to 0. Prevalent cause of strong, narrow RCS peaks from supersonic aircraft.
	Curved edge normal incidence	F^{-1}	L^1	$f(\theta,\theta_{int})\, a\lambda/2$ $a \geq \lambda$	Limiting case of 3-dimensional curved plate mechanism as principal radius shrinks to 0. The function f is the same as in mechanism 3.
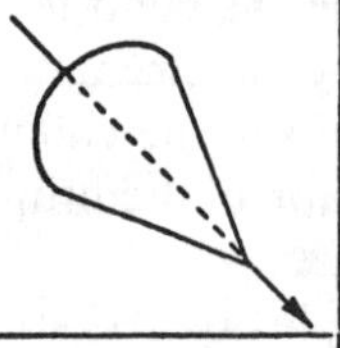	Apex	F^{-2}	L^0	$\lambda^2 g(\alpha,\beta,\theta,\phi)$ α,β - interior angles of tip θ,ϕ - aspect angles	Limiting case of previous mechanism as a shrinks to 0. For $\alpha=\beta$, the tip is that of a cone. For $\alpha=0$, the tip is the corner of a thin sheet, or fin.
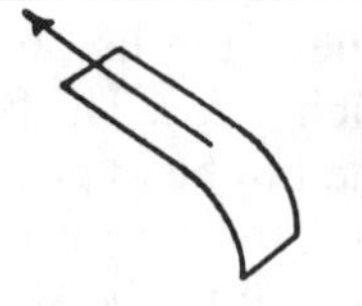	Discontinuity of curvature along a straight line, normal incidence	F^{-2}	L^0	$\frac{\lambda^2}{64\pi^3}\left(\frac{L}{a}\right)^2\left\{1+\left(\frac{dy}{dx}\right)^2\right\}^{-3/2}$ $a \geq \lambda$ $(1/a)$ - jump in reciprocal of dy/dx - slope of surface w.r.t. incident ray	Strongest of an infinite sequence of discontinuities. Very weak mechanism which together with 6 shares dominance of nose-on RCS of cone sphere.

Table 8.1 Hierarchy of Scattering Shapes (continued)

Geometry	Type	Freq. Dep.	Size Dep.	Formula	Remarks
	Discontinuity of curvature of a curved edge	F^{-3}	L^{-1}	$f(\theta,\phi)\,\frac{\lambda^3 b}{a^2}\left(1+\left(\frac{dy}{dx}\right)^2\right)^{-3/2}$ $f(\theta,\phi)$ - function of aspect b - radius of edge $>\lambda$	Important mechanism for traveling wave backscatter where RCS of discontinuity is augmented by gain of traveling wave structure. Dependences are based on dimensional considerations.
	Discontinuity of curvature along an edge	F^{-4}	L^{-2}	$g(\theta,\phi)\lambda\left(\frac{1}{a}\right)^2\left(1+\left(\frac{dy}{dx}\right)^2\right)^{-3/2}$ $g(\theta,\phi)$ - function of aspect	Important mechanism for traveing wave backscatter where RCS of discontinuity is augmented by gain of traveling wave structure. Dependences are based on dimensional considerations.

reflected by one face of the corner is captured by another. However, the computation of the effective area is a simple geometric exercise, one reason the corner reflector is often used as a calibration target. The echo amplitude can easily be reduced if the faces of the corner are tilted slightly away from each other, as we shall presently see. The reduction increases with the electrical size of the corner (in wavelengths) for a given tilt, the prime reason for imposing tight fabrication tolerances on electrically large corners used as RCS calibration standards.

The strength of the echo from a flat plate presented perpendicular to the incident wave (also called the *line of sight,* or LOS), is also proportional to the square of the area. In this case the effective area and the physical area are the same, but the formula in the rightmost column holds only for normal incidence. When the plate is presented obliquely to the incident wave, the echo is primarily due to diffraction from the edges and the corners of the plate, which are altogether different scattering mechanisms that appear further down the list.

The plate is followed by the cone frustum oriented with one slanted surface perpendicular to the line of sight. In the approximate formula in the rightmost column, s is the length of the slanted side of the frustum and a is the mean radius of curvature at the specular line running down the side of the frustum. For the special case in which the cone angle becomes zero, the frustum becomes a cylinder, whence s is the length of the cylinder. If the cylinder is a circular one, a is its radius; if the cylinder is an elliptical one, a is the radius of curvature at the specular line. As noted in the leftmost column, the surface of a cone frustum is singly curved.

The surface of a spheroid is doubly curved, and the formula given for its RCS in the rightmost column is the geometric optics approximation. In this expression, a_1 and a_2 are the radii of curvature of the spheroid at the specular point where the surface normal is parallel to the line of sight, and the approximation is within a few percent of the exact value when both radii exceed two wavelengths.

The echo mechanism for the flat plate, and the frustum and the spheroid with their surface normals parallel to the line of sight, is a surface phenomenon. The dependence of the echo on wavelength varies as λ^{-n}, where n is the number of infinite radii of curvature at the specular point. At normal incidence, the specular "point" for a flat plate is the entire surface, for a frustum it is a line, and only for the spheroid is it truly a point. An interesting parallel may be noted when there is no surface presented perpendicular to the line of sight, as in the next three cases.

The first of the three cases is that of a straight edge perpendicular to the line of sight. A straight edge has no curvature along its length (infinite radius of curvature), and its radius of curvature in a plane perpendicular to the edge is zero. The RCS formula in the rightmost column is an approximate value for edge-on incidence in the plane of the surface terminated by the edge and shows no dependence on wavelength. Thus, the straight edge aligned perpendicular to the line of sight and the doubly curved surface of the spheroid both exhibit the same dependence on frequency.

If the edge is a curved one, the RCS is proportional to the radius of curvature, a, in the plane of the edge; the other radius of curvature is zero, of course. It should be noted that the radius of curvature is measured at the point along the edge where the plane containing the line of sight is perpendicular to the plane containing the edge. The apex or vertex appearing at the bottom of the list is a degenerate case of an edge curved so sharply that both radii of curvature are zero. Note that the RCS in this case has no dependence whatever on any dimension of the scattering feature; indeed, a vertex is a point characterized by no physical dimension at all.

It is clear from the table that if we are able to control the surface geometry of our target, and the orientation of surface features with respect to the radar line of sight, we can bring about a significant reduction in the echo strength. However, the latter implies that we can forecast the direction to the radar, which in some scenarios may include virtually all directions. In that event, the control of surface orientation—a form of shaping—offers scant echo reduction possibility. Also, while the table may suggest that the least detectable targets are those having only edges and vertices, such targets have no volume, hence no payload.

Some of these characteristics are illustrated more graphically in the figures that follow. Because of its radial symmetry, the sphere is the simplest possible three-dimensional object. The RCS of the perfectly conducting sphere can be expressed as the sum of an infinite series of spherical Bessel functions, and the results are charted in Figure 8.6 as a function of ka, the electrical circumference of the sphere. The ordinate is the RCS normalized with respect to the projected area of the sphere, which is also the geometric optics result listed in Table 8.1.

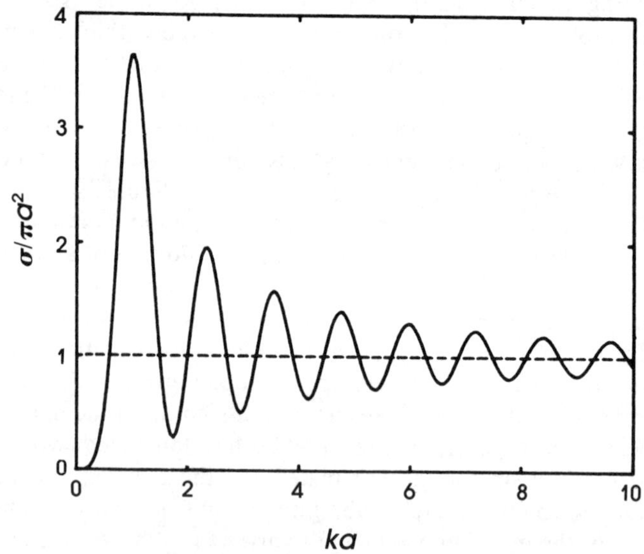

Figure 8.6 RCS of a perfectly conducting sphere (exact formulation).

Note that the RCS rises to a strong, resonant value near $ka = 1$, and then traces out a damped oscillation as the electrical size of the sphere increases. The oscillations converge on the optics value rapidly enough that by the time the sphere is 10λ in circumference ($ka = 10$), the optics approximation is within 0.5 dB of the exact value. The oscillations can be attributed to creeping waves that skirt the rear of the sphere, shedding energy at every point. Part of this energy is launched back toward the radar at the shadow boundaries, and the creeping wave contribution goes in and out of phase with the specular contribution from the front of the sphere. It is this alternate cancelation and reinforcement that creates the undulations in the RCS with increasing size.

The path length traversed by the creeping waves obviously increases with increasing size, and the creeping waves also become weaker. This is why the oscillations converge to the specular value for large spheres. While the creeping wave is a striking natural phenomenon worthy of study, it rarely plays an important role in the scattering by complex military targets such aircraft and ground vehicles. On the other hand, it may be important for simpler weapons, such as re-entry vehicles.

The RCS of a prolate spheroid expectedly behaves in much the same way as that of a sphere, although the creeping wave may not actually propagate all the way around the back of the body. Figure 8.7 is a comparison of measured and predicted echoes from a 2:1 prolate spheroid at end-on incidence as a function of the smaller circumference of the body. Note that this is a logarithmic plot and that the RCS is normalized with respect to the square of the wavelength instead of the projected area of the spheroid. As a result, the undulating pattern rises with increasing *ka*. Senior presented measured and predicted results for a 10:1 prolate spheroid [5] exhibiting much the same kind of pattern.

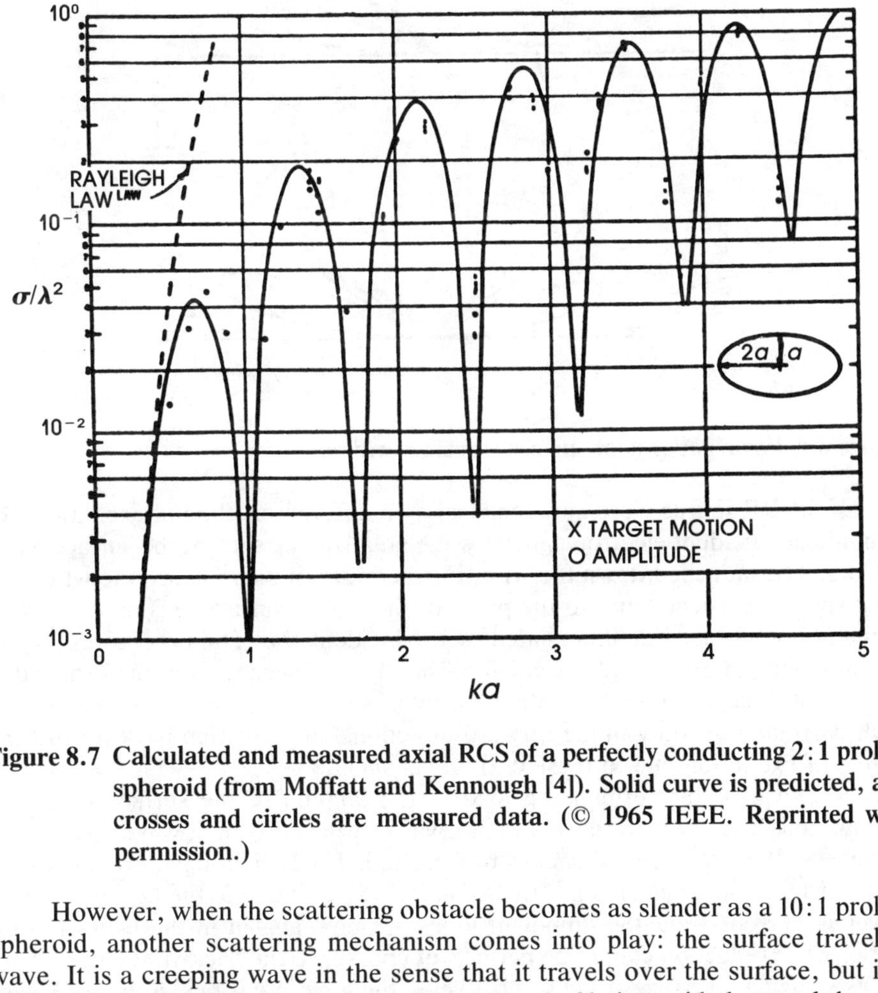

Figure 8.7 Calculated and measured axial RCS of a perfectly conducting 2:1 prolate spheroid (from Moffatt and Kennough [4]). Solid curve is predicted, and crosses and circles are measured data. (© 1965 IEEE. Reprinted with permission.)

However, when the scattering obstacle becomes as slender as a 10:1 prolate spheroid, another scattering mechanism comes into play: the surface traveling wave. It is a creeping wave in the sense that it travels over the surface, but it is reflected by a discontinuity in the surface instead of being guided around the rear of a smooth body. The reflected surface wave travels back toward the front of the body, and depending on the length and curvature of the surface, can give rise to *surface traveling wave* contributions. The surface traveling wave is launched down the sides of a long smooth metallic structure, only if there is a longitudinal component of the incident electric field along the surface.

Indeed, the surface traveling waves launched along straight or gently curved surfaces at relatively small grazing angles are not unlike those excited on long, end-fed wire antennas, as suggested in Figure 8.8. The antenna is stretched between masts not too far above the ground, with one end terminated to ground by a suitable load to prevent reflections, and the other end excited by an RF source. If it were possible to isolate the antenna from the effects of the ground, the current wave launched down the wire by the source would radiate a pattern like that shown in the upper diagram. In the absence of ground effects, this would be a pattern of revolution.

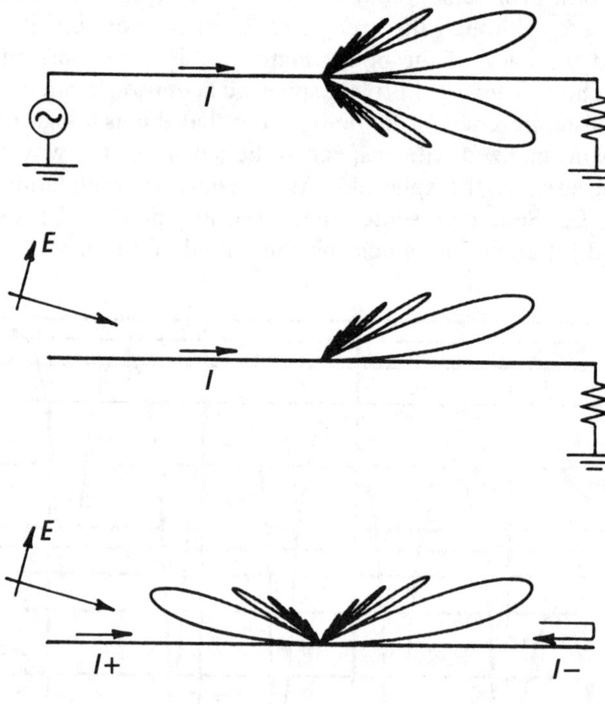

Figure 8.8 Polar patterns of surface traveling waves.

If the RF source were now replaced by a different method of excitation, for example, an incident electromagnetic wave polarized as shown, the surface wave could still be launched down the wire. Furthermore, if we replaced the wire with a flat surface perpendicular to the plane of the figure, surface currents would be excited only on that side illuminated by the incident wave, and the lower part of the radiation pattern would essentially disappear, as suggested in the center diagram. Note that because the antenna (now a surface) remains terminated by a load, currents flow only in the forward direction. The radiation back toward the source of excitation due to these currents is small.

Removal of the terminating load at the far end of the surface creates a discontinuity there which reflects the forward surface traveling wave, giving rise to currents that travel in the backward direction. The radiation pattern due to the backward traveling currents has the same form as that due to the benign forward traveling currents, but the dominant lobes are now angled in the backward direction. The reflection cannot be perfect, of course, so the backward currents are not as strong as the forward currents. Thus, the lobe intensities in the backward direction are not as strong as those in the forward direction. This is indicated in the diagram by somewhat reduced lobe intensities. Nevertheless, the backward traveling wave contribution gives rise to significant echo levels, as we shall see.

Peters used an end-fire antenna analysis to successfully predict the effects of the surface traveling wave launched on an electrically long metallic ogive (a body of revolution formed by rotating a circular arc about its chord) [6]. Figure

8.9, taken from Peters' classic paper, clearly illustrates the surface traveling wave lobes, which appear on the left side of the diagram. Note that his analysis yields quite good results for the first two or three peaks, evidence that the mechanism is understood and, at least for simple structures like this, that reasonably accurate predictions can be made. As an aside, the reader is invited to examine the ordinate scale at the left side of the chart. The vertical scale ranges from 3.5 dB per division at the top of the chart to 5.5 dB per division near the bottom, suggesting that the equipment available for RCS measurements thirty years ago was not nearly as good as it is today. This does not detract from Peters' work, however; sound engineering is timeless.

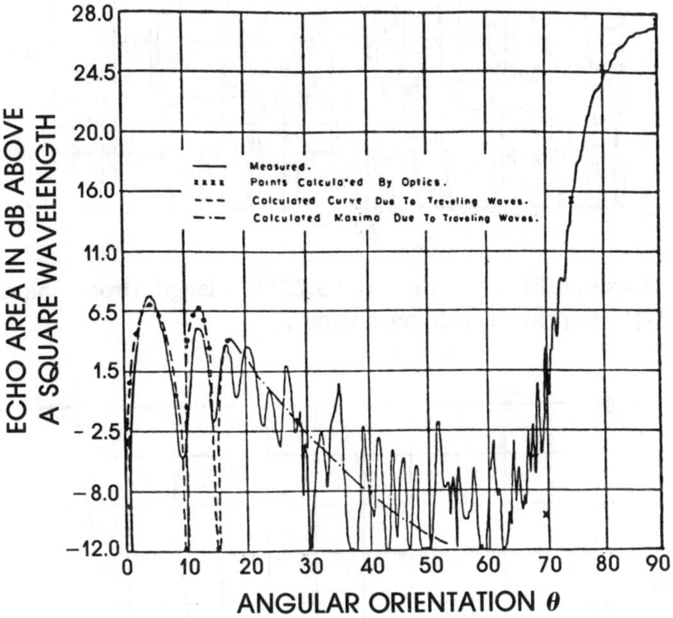

Figure 8.9 Comparison of measured and predicted RCS of a 39λ, 15° metallic ogive (from Peters [6]). (© 1958 IEEE. Reprinted with permission.)

The surface traveling wave contributions to the RCS can be nearly as strong as those observed for broadside incidence, as illustrated by the measurements of thin wires by Chang and Liepa [7]. Figure 8.10 is the RCS pattern of a wire 5.221λ long, which is only one of a sequence of several dozen patterns they recorded. End-on incidence is at the extreme left and right sides of the pattern, broadside is at the center, and the dashed line across the pattern represents a level of one square wavelength. The central peak in the center of the pattern is the broadside echo and it attains a level of about 6.5 dB above a square wavelength (about $4.7\lambda^2$). Aside from the broadside lobe, there are no other lobes more intense than those due to the surface traveling wave contributions, located about 22° either side of end-on incidence. Their intensities are 3 dB above a square wavelength ($2\lambda^2$), illustrating that the surface traveling wave contribution can be significant. The progression of decreasing lobes inward from the two prominent lobes are also due to surface wave effects, but their intensities are much weaker than that of the first lobe.

Figure 8.11 shows how the broadside RCS varies with the electrical length of the wire. The echo does not rise smoothly, but increases in a series of steps, with sharp local peaks at odd multiples of a half wavelength. The response is quite flat over distinct plateaus between the first few resonances so that, over a length change of nearly one wavelength between resonances, the broadside RCS is sensibly constant. Although not shown on this diagram, the resonances become

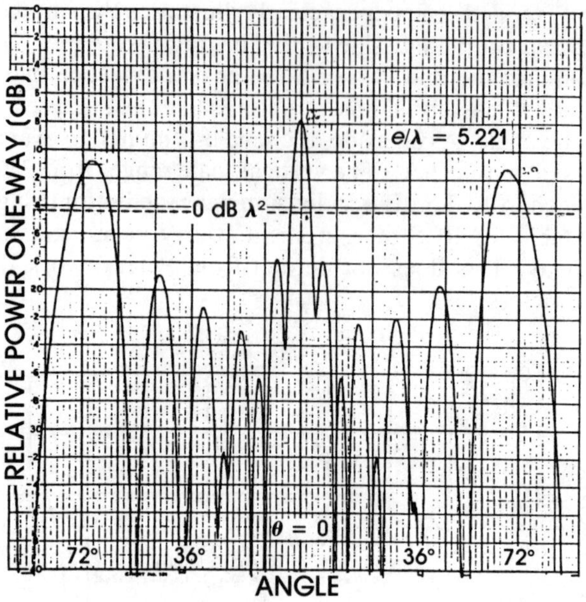

Figure 8.10 **Measured RCS of a thin wire 5.221λ in length (from Chang and Liepa [7]). (Reprinted with permission.)**

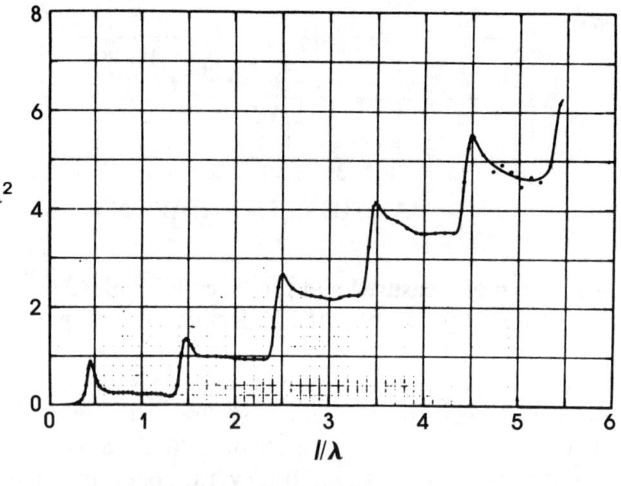

Figure 8.11 **Measured broadside RCS of thin wires as a function of wire length (from Chang and Liepa [7]). (Reprinted with permission.)**

less pronounced as the wire becomes thicker, and the RCS rises essentially monotonically as the square of the length.

Chang and Liepa traced the locations of the various lobes in their measured patterns, plotting them on the chart of Figure 8.12. The uppermost trace is the location of the first (dominant) surface traveling wave lobe noted above for the 5.221λ wire, and we have added a dashed line indicating a position predicted by the formula [8, 9]:

$$\theta = 49.35\sqrt{\lambda/l} \text{ (degrees)} \tag{8.8}$$

The formula assumes a uniform surface traveling wave on a body of length l, assumed to be at least several wavelengths long, but it can be seen that the error in the prediction of the location of the lobe is less than 3% for wires as short as one wavelength.

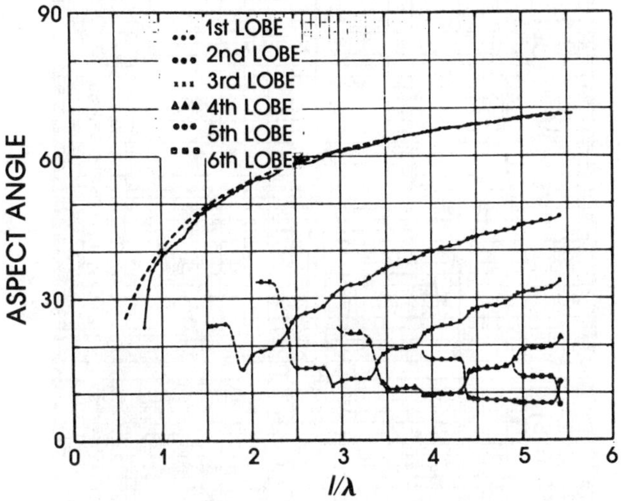

Figure 8.12 Measured lobe locations for a thin wire (from Chang and Liepa [7]). (Reprinted with permission.)

The surface traveling wave should also be observed in the RCS patterns of right circular cones, such as those of Figures 8.13 and 8.14, from Knott and Senior [10]. These patterns were recorded by rotating the cone about a vertical axis perpendicular to the cone axis, with horizontal and vertical incident polarizations. The cone half-angle was 15° and the base was 12.575λ in circumference, and these particular patterns constituted only one of several dozen sets measured at different frequencies. The upper pattern in each figure is of the bare cone, and the lower pattern is of the cone with a pad of absorbing material cemented to its base.

The length of the slanted surface of the cone was about 7.74λ; (8.8) therefore suggests that the surface traveling wave lobe should appear at an angle of about 17.7° away from the surface grazing angle. Unfortunately, this is barely 3° from nose-on incidence (at the center of the diagram), and the surface traveling wave contribution in this region is masked by diffraction from the base of the cone. Although the pad of absorber attached to the base of the cone was not exposed to the incident wave over the central part of the pattern, it reduced the amplitudes of the first few lobes off nose-on incidence. This is evidence that part of the echo contribution in the nose-on region is due to multiple diffraction across the base of the cone.

At nose-on incidence the entire base of the cone contributes to the RCS, which is the reason that the echo rises to a peak there. As the cone swings to either side, the dominant echo sources are the "flash" points where the plane of incidence intersects the base, but these contributions are weaker than that of the entire base as seen from nose-on. As the cone swings beyond the backward half-angle, one flash point disappears and the RCS reaches its lowest levels in the pattern. As the cone swings closer to normal incidence to the slanted side, the tip now becomes a dominant scatterer.

In the intermediate region more or less midway between nose-on incidence and incidence normal to the slanted side, the echo for vertical polarization is off the bottom of the chart, suggesting that the contributions from the flash points are much smaller than for horizontal polarization. It must be remembered, however, that some of the scattering features in this region (for horizontal polarization, Figure 8.13) are higher order sidelobes of the surface traveling wave mechanism, which do not exist for vertical polarization.

The cross-base diffraction mechanism can be illustrated by radar image measurements, one of which is shown in Figure 8.15, a pattern generated at the

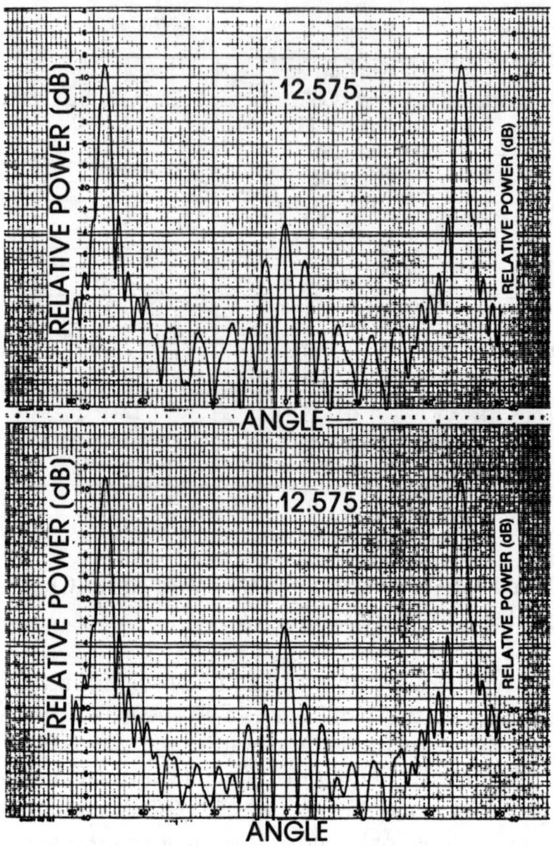

Figure 8.13 **RCS of a 15° half-angle metallic cone for the incident electric field in the plane of incidence. The upper pattern is for the bare cone and the lower for the cone with an absorber pad cemented to its base (from Knott and Senior [10]). (Reprinted with permission.)**

US Navy Pacific Missile Test Center at Point Mugu, CA. Images like these are generated by collecting scattering data with wideband, coherent, swept- or stepped-frequency instrumentation radar as the target is slowly rotated. The data must be processed off-line by using a double Fourier transformation; the range variation is generated by processing the recorded data over a window of frequencies, and the cross-range variation by processing those results over an aspect angle window. Thus, it requires some target rotation to produce plots like that of Figure 8.15 for a wedge.

Note that although the tip of the wedge is not a dominant scatterer, its echo can be measured. As stated earlier, the base is a strong scatterer. However, in addition to these obvious echo mechanisms, others appear to lie *behind* the wedge. These are due to diffraction across the shadowed base of the wedge, and because the cross-base path increases the total length of the path traversed by the energy, the cross-base diffraction contributions appear later in time. The radar has no way of diagnosing the origin of these contributions, except to place them in range according to their time of arrival. Note that doubly diffracted contributions can be separated from singly diffracted contributions. Thus, complex scattering mechanisms may be found, even on simple structures.

The three-dimensional display of Figure 8.16 is another way to plot the scattering behavior of objects, which is, in this instance, that of a rectangular flat plate. The amplitude in a given scattering direction is indicated by the height of the pattern, and the base axes represent the angles to the radar as measured in orthogonal planes perpendicular to the plane of the plate. The large hump in the center of the diagram is the broadside echo, and, as we have seen, its amplitude

KNOTT
CHAPTER 8

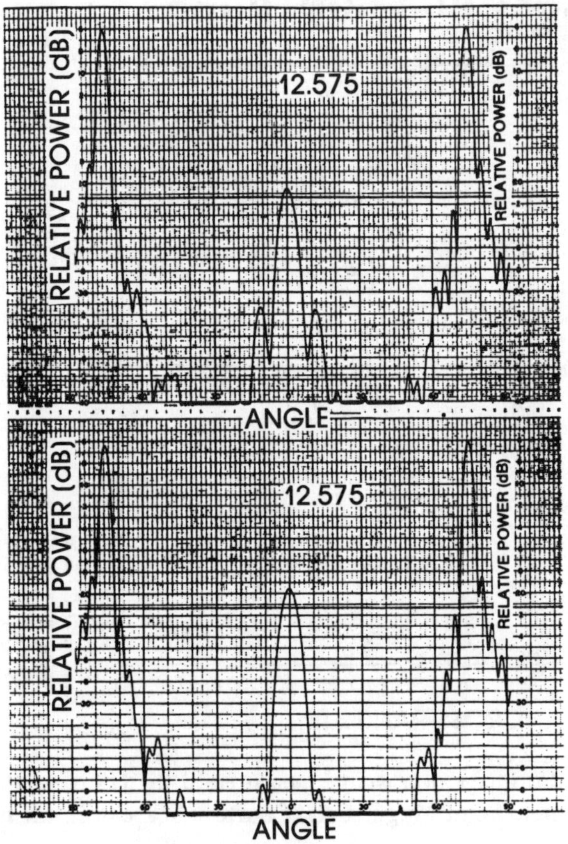

Figure 8.14 RCS of a 15° half-angle metallic cone for the incident magnetic field in the plane of incidence. The upper pattern is for the bare cone and the lower for the cone with an absorber pad cemented to its base (from Knott and Senior [10]). (Reprinted with permission.)

is proportional to the square of the area of the plate and inversely proportional to the square of the wavelength.

Descending away from the central broadside peak at right angles to each other are a pair of undulating ridges that result when the plate is moved away from broadside incidence, but in such a way that a pair of edges remains perpendicular to the line of sight. The edges oblique to the line of sight are essentially invisible, and the undulations along the ridges are caused by the alternate cancellation and reinforcement of the returns from the two edges perpendicular to the line of sight as the relative phase between the two echoes changes. The peak amplitudes along these ridges is proportional to the square of the length of the two interfering edges, reduced by a trigonometric factor that involves the projection of the direction of incidence onto the plane of the plate.

The interference structure persists into the intermediate regions away from the broadside and principal plane returns, but the amplitudes in those directions are significantly lower than in the principal directions. In these regions there are no edges perpendicular to the line of sight, and the echo is due to the four vertices of the plate. The undulations are caused by changes in the relative phase angles between the returns from the four corners of the plate. The peak values of these echoes do not depend on the size of the plate, and are instead proportional to the square of the wavelength.

Thus, the flat plate exhibits three different kinds of scattering behavior, depending on the angle of incidence. At normal incidence, all elements of the surface are illuminated in phase, and the entire surface "lights up." As the plate is tilted away from the line of sight, but with a pair of edges held perpendicular

KNOTT
CHAPTER 8

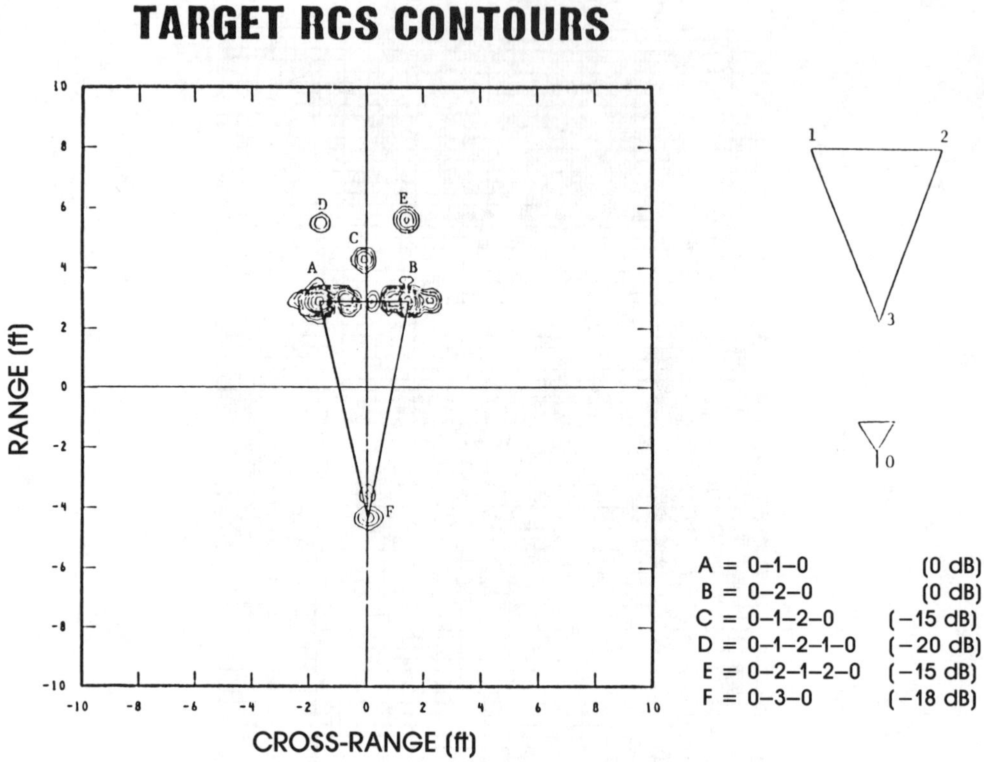

Figure 8.15 Radar image of a wedge measured at the US Navy Pacific Missile Test Center, Point Mugu, CA. (Courtesy of Dean L. Mensa.)

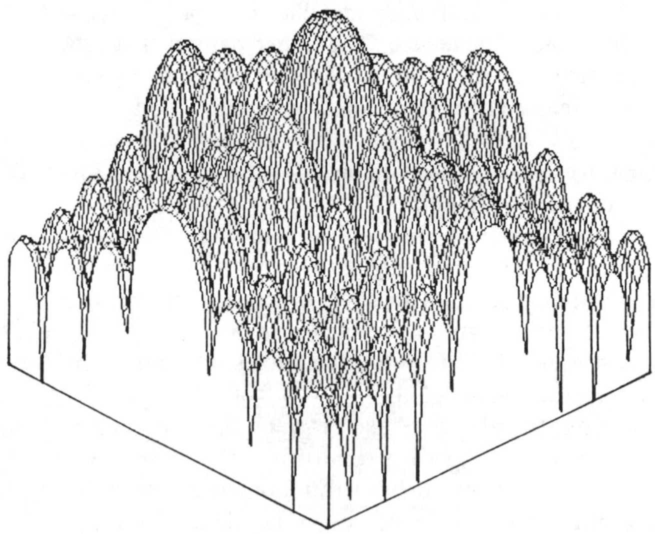

Figure 8.16 RCS of a flat plate. The central peak is the broadside echo.

to the line of sight, the bright surface degenerates to a pair of bright lines at the edges of the plate. When the plate is tilted at right angles to the first angle of motion, the two bright lines degenerate to a collection of four bright points, one at each vertex of the plate. In much the same way, a complex target oriented with no surface perpendicular to the line of sight can be visualized as a collection of bright points.

Corner reflectors are made by assembling two or more flat plates at right angles to each other, but the dominant echo mechanism is now a double or triple

KNOTT
CHAPTER 8

internal bounce. Figure 8.17 illustrates the double bounce mechanism; it requires only a simple geometric exercise to convince oneself that as long as the re-entrant side of the corner faces the general direction of the incident wave, a ray impinging on one of the faces suffers two internal bounces, and then heads back in the direction from whence it came. This is why corner reflectors have large echo areas.

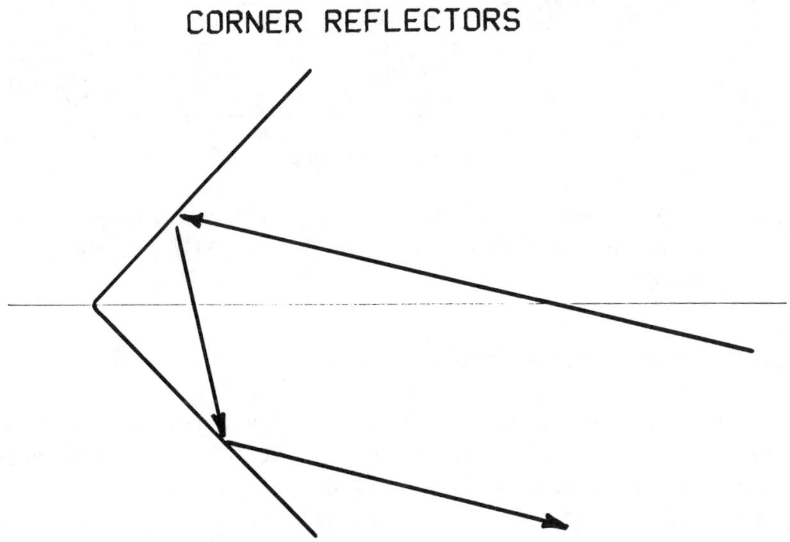

Figure 8.17 Double-bounce mechanism from corner reflector.

Figure 8.18 is the RCS pattern of a dihedral corner reflector with square faces about 5.6λ along a side. The broad central part of the pattern is due to the double-bounce mechanism, and the small ripples seen in the central part of the pattern are due to the sidelobes of a single-bounce mechanism like that of the flat plate pattern of Figure 8.16. The "Batman ears" of this pattern are the single-bounce specular echoes from the individual plates comprising the corner reflector. When the angle between the faces is expanded to 100°, thereby destroying the orientation responsible for the double bounce, we obtain the pattern shown in Figure 8.19. The double-bounce mechanism has been spoiled, and over some parts of the pattern the RCS has been reduced by 15 dB or more. Some targets, notably naval surface ships, have literally hundreds of unintentional corner reflectors distributed over them.

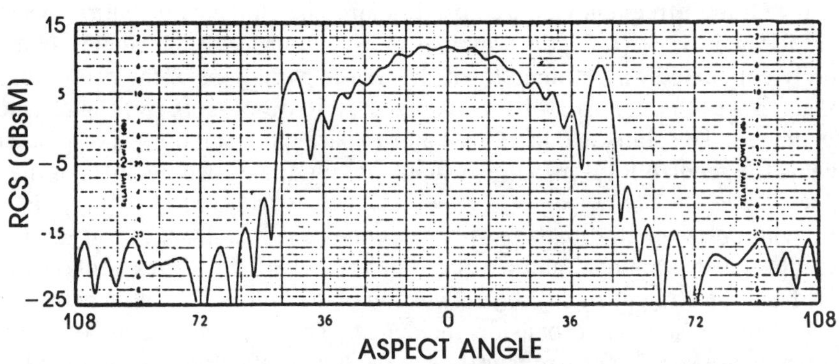

Figure 8.18 RCS pattern of a 90° dihedral corner with square faces 17.9 cm along a side measured at 9.4 GHz (from Knott [11]). (Reprinted with permission.)

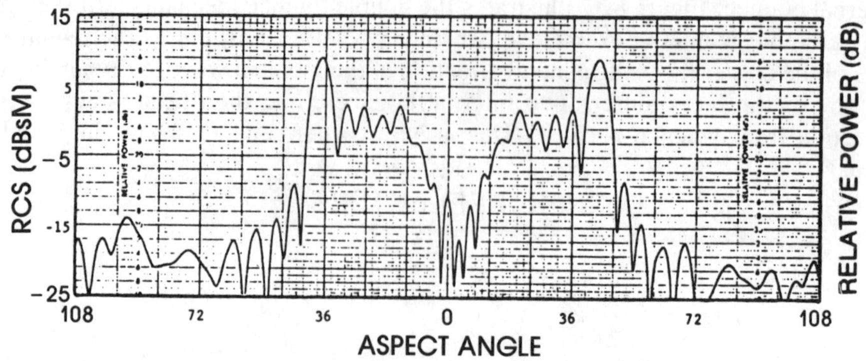

Figure 8.19 **RCS pattern of a 100° dihedral corner with square faces 17.9 cm along a side measured at 9.4 GHz (from Knott [11]). (Reprinted with permission.)**

8.2.4 RCS Characteristics of Complex Targets

The RCS of a complex body is obviously much more complicated than that of the simple objects discussed above. The complication is mainly because there are literally hundreds, and sometimes thousands, of scatterers distributed all over the target; some of them are large, like the doors and fenders of an automobile, and many of them are smaller, such as headlamps, taillights, rearview mirrors, and seams. Even for slight changes in aspect angle, the relative phase angles between the members of this multiplicity of scattering centers may change by several multiples of 360°; hence, the sum of the echoes of individual contributors varies rapidly. The RCS variations for large targets are so rapid, in fact, that it makes no sense at all to attempt to measure them. Instead, the mean RCS and its standard deviation over some aspect angle range are more useful characterizations.

This is the case for the ship RCS patterns shown in Figure 8.20. Even at the lower frequency (2.8 GHz), the ship was over 1500λ long, and for targets of this size the peaks in the lobe structure are separated by barely a 0.05° change in aspect angle. Moreover, the undulating sea surface causes a temporal variation in the echo signal that cannot be removed; consequently, only the statistical variations of the echo strength are meaningful.

The traces in the polar plots of Figure 8.20 are the 20, 50, and 80 percentile levels of the calibrated signals recorded as the ship steamed in a circle in the beam of a multifrequency shore-based radar. The statistics of the recorded data were formed over two-degree windows in aspect angle, and adjacent points were merely connected with straight lines. The actual variations of the echo signals are not given, but the signals probably changed by 20 dB or more over small fractions of a degree. The statistical data, by contrast, are quite smooth. The 50 percentile level is the median RCS, which is usually within one or two decibels of the mean value.

The port and starboard broadside echoes are the strongest on the pattern, attaining levels of more than a square mile (64.1 dBsm). These echoes usually arise from the straight sided, central portion of the hull as presented perpendicular to the line of sight at broadside incidence. The echo strengths at the higher frequency are somewhat stronger than at the lower, and the echoes in the forward port and aft starboard quarters are stronger than in the other two quarters. Both patterns seem to be biased (shifted) about one degree to port, which may be due to an error in the measurement of the ship's orientation.

The significant scatterers on a ship depend on the distance from the radar to the ship. For all except broadside incidence, the hull is *not* the dominant

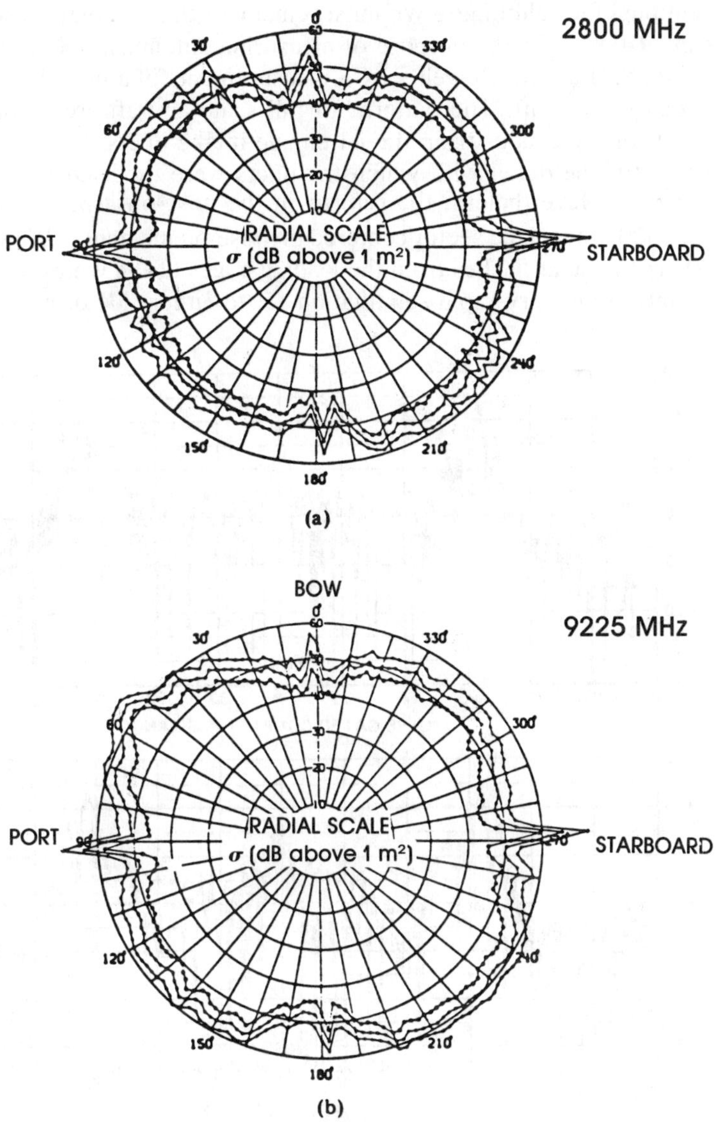

Figure 8.20 **RCS pattern of a large naval auxiliary ship. Upper chart measured at 2.8 GHz and lower chart measured at 9.225 GHz, both for horizontal incident polarization (from Skolnik [12]). (© 1980 McGraw-Hill Book Co. Reprinted with permission.)**

scatterer, even when the hull is not occluded by the horizon. The superstructure and masts are large scatterers, and every ship has an assortment of equipment and fixtures topside exposed to the radar. These include railings, lockers, vents, stairways, catwalks, and dozens of antennas. Although the echo from any one fixture may be negligible compared to the ensemble, the sheer number of items significantly increases the backscattered signal.

Although airplanes are smaller than ships, they also have rapidly varying RCS patterns. There is, unfortunately, a rather small selection of patterns available for display in the open literature, and we cite here the measurements of a $\frac{1}{15}$ scale model of a commercial jetliner published in 1970 [13]. The model was measured at 10 GHz; hence, the pattern corresponds to a full scale frequency of about 667 MHz. The actual purpose of the measurements was to demonstrate that RCS patterns collected on a compact range could be as accurate as those recorded on a conventional far-field range. We must remember that the major use of compact ranges in 1970 was for the purpose of measuring antennas, not RCS patterns.

Figure 8.21 shows Howell's RCS measurements of a model of a Boeing 737 commercial jet aircraft. The patterns of full scale aircraft are dominated in the nose-on region by echoes from the jet engine intake ducts, which are re-entrant structures, but the detail with which the ducts were modeled in the test object is not known. Nevertheless, the returns in the nose-on aspects are the largest of any in a 120° yaw plane sector centered on nose-on incidence. We would expect to see sharp specular flashes from the leading edges of the wings, but the pattern showed only some very rapid scintillation 15° to either side of nose-on.

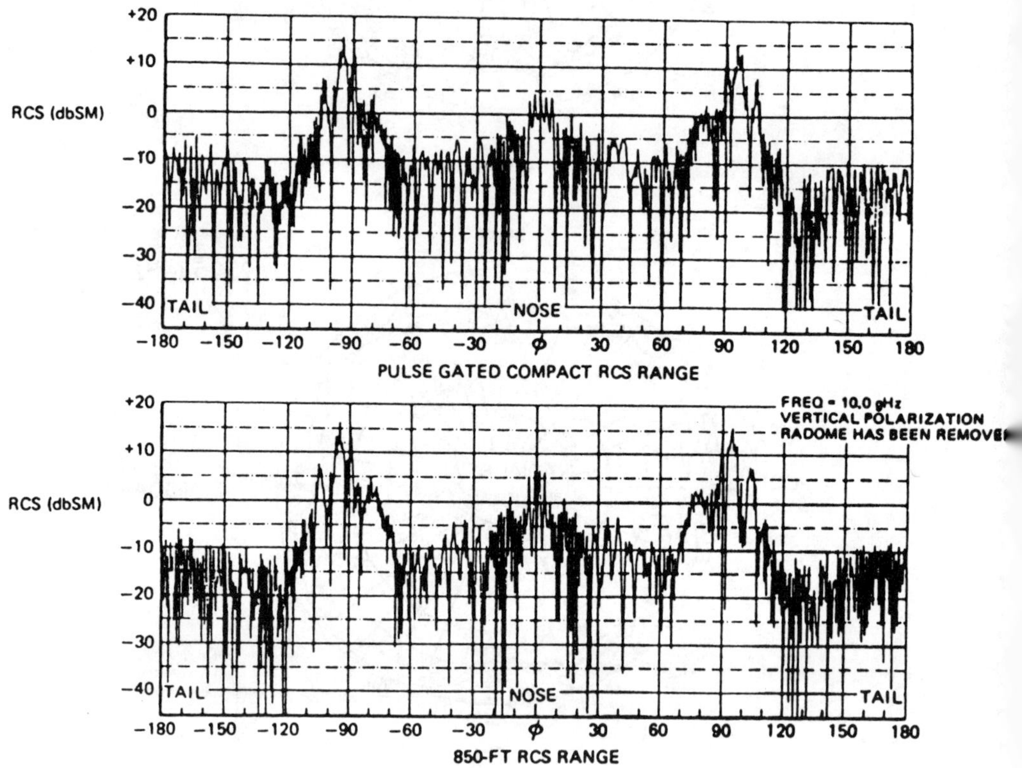

Figure 8.21 **RCS pattern of $\frac{1}{15}$ scale model Boeing 737 commercial jetliner measured at 10 GHz (from Howell [13]). (© 1970 IEEE. Reprinted with permission.)**

The side of the fuselage is a very narrow flash at precisely 90° on either side of the pattern, preceded by a very broad lobe centered at 80°. The latter is probably due to the curved forward part of the fuselage. A pair of broad, intense lobes lie approximately 95° and 105° from nose-on, and may be from the engine cowlings. The mean RCS attains its lowest level in the region from 120° to 150°, and then rises slightly to its tail-on value. The latter is some 10 dB lower than the nose-on value, suggesting that the scattering from the engine exhaust ducts is not as strong as it is from the intake ducts. The pattern of a full scale aircraft measured at this frequency (10 GHz) would have signal scintillation rates 15 times as much as shown in Figure 8.21, of course, and depending on the frequency dependence of the particular scattering mechanisms involved, the amplitudes in a given aspect angle region could be higher, lower, or about the same as those measured at the scaled frequency.

Although insects are several orders of magnitude smaller than ships or airplanes, we regard them as complex objects because their scattering surfaces defy mathematical description, and incident radar waves may penetrate their dielectric bodies. There is nothing to prevent our measuring their echoes, however, and a list of measured values appears in Table 8.2 [14]. The original results were reported in cm^2, but we have converted the reported values to dBsm for convenience.

Table 8.2 Insect RCS at 9.4 GHz*

	Length (mm)	*Diameter* (mm)	σ_L (dBsm)	σ_T (dBsm)
Blue-winged locust	20	4	−30	−40
Army worm moth	14	4	−39	−49
Alfalfa caterpillar butterfly	14	1.5	−42	−57
Honey bee worker	13	6	−40	−45
California harvester ant	13	6	−54	−57
Range crane fly	13	1	−45	−57
Green bottle fly	9	3	−46	−50
Twelve-spotted cucumber beetle	8	4	−49	−53
Convergent lady beetle	5	3	−57	−60
Spider (unidentified)	5	3.5	−50	−52

*From Hajovsky *et al.* [14].

Ten insect species and an unidentified arachnid were measured at a radar frequency of 9.4 GHz at both broadside and nose-on incidence. These are the columns labeled σ_L and σ_T, respectively, where the subscripts L and T stand for longitudinal and transverse. The insects in the upper third of the chart were much longer than they were thick; consequently, their broadside (longitudinal) echoes were an order of magnitude stronger than their nose-on (transverse) echoes. The insects in the lower part of the chart were two or three times as long as they were wide, and we see a corresponding reduction in the variation of the echo from the nose-on to the broadside value. The "roundest" of the test specimens was the spider at the bottom of the list, which registered a nose-on to broadside echo variation of only 2 dB.

The insects were small targets that had to be placed very close to the instrumentation antenna in order to yield measurable signals, but the authors do not state the actual distance used. It was probably of the order of a few antenna diameters at most. A problem faced by the experimenters was how to keep the animals from moving during a measurement. The easiest solution—to dispatch them and measure the echo from the dead carcasses—was unacceptable, because the dielectric constant of the carcasses would change as the bodies lost their moisture. The experimenters solved the problem by drugging their test targets, which kept the creatures both immobile and alive, and their dielectric constants truly constant.

Blacksmith and Mack faced essentially the same problem in their measurements of ducks and chickens at 400 MHz [15], but chose to train their test targets instead of drugging them. Accustomed to conventional RCS testing, in which an inert test object is placed on a foam plastic support column, and then rotated to obtain the RCS pattern, Blacksmith and Mack quickly learned that chickens and ducks are far from inert. The animals refused to remain stationary for even a few degrees of a complete 360° rotation, and the experimenters had to be content with a quick measurement at one aspect angle per session. This was repeated enough times to form the statistical summary shown in Figure 8.22, which suggests that the RCS of an adult duck is about −12 dBsm. The authors presumably concluded their testing by dispatching and consuming their test targets.

We conclude this section with the summary chart of Figure 8.23. It attempts to show, with no great accuracy, how the RCS of various classes of targets varies

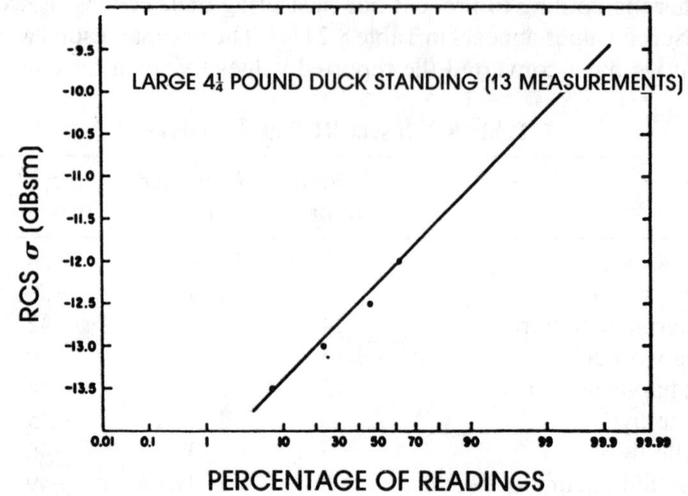

Figure 8.22 **RCS of adult ducks measured by Blacksmith and Mack [15]. (© 1965 IEEE. Reprinted with permission.)**

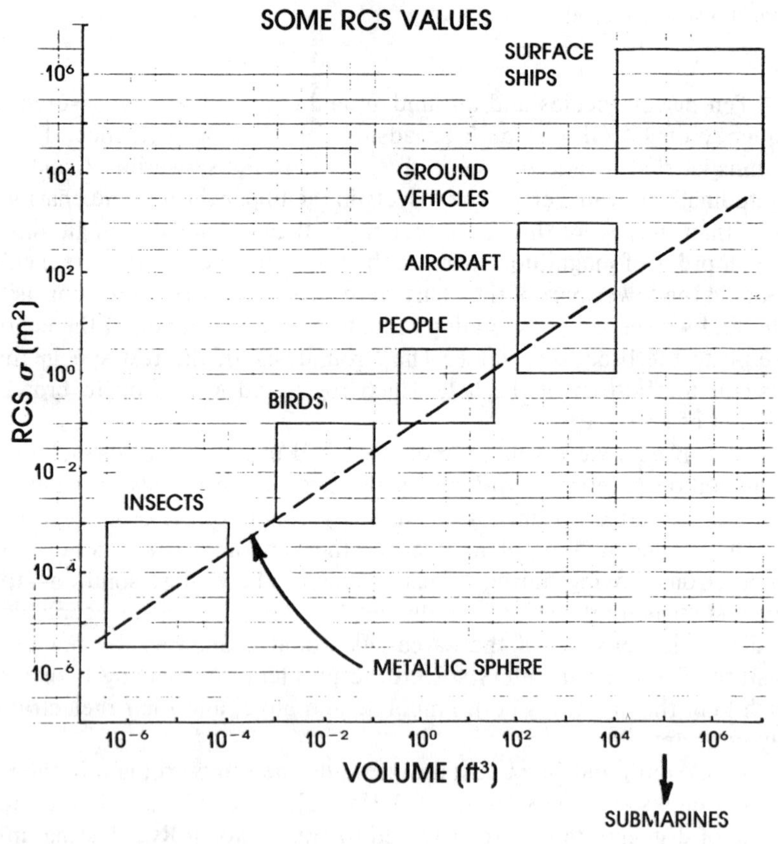

Figure 8.23 **RCS of common objects.**

with size. The information comes from a variety of unreferenced sources that include measurements and predictions at many sundry wavelengths. The boundaries sketched for each class of target are crude at best, and were chosen only to focus attention on specific kinds of target. The interested reader will find in the open literature many examples that fall within the boundaries sketched, but he will also find some that do not.

The abscissa on this log-log chart is the volume of the test object in cubic feet and the ordinate is the RCS as expressed in dB relative to a square meter.

The smallest objects on the chart are insects, and the largest are surface ships. The dashed line represents the echo of a metallic sphere for comparison, and the chart suggests that the RCS of familiar objects is more or less proportional to the volume of the object. There is one exception: patrolling submarines are perfectly invisible to radar. This leads one to a perfectly logical way to reduce the echo of surface ships.

8.3 CAN YOU PREDICT IT?

The question of whether or not we can predict the RCS of an object should probably be preceded by the query, "Why try to predict it at all?" One obvious answer is that, like the mountain to be climbed, the problem is there to be solved. It is a challenge to be surveyed, appraised, skirted, approached, and finally attacked and wrestled to the ground. Questions of practicality aside, scholars will be with us always, and describing the behavior of electromagnetic waves on, near, and far from obstacles will spark their curiosity, and challenge their intellectual skills, for a long time to come.

A more practical reason is that it is usually cheaper to make a few calculations than to measure a test target, especially in an era when digital computers are big and fast. The capability to estimate the echo properties of a given target is especially important if there is no physical object to measure, but merely a set of drawings or sketches on a drafting board. As in the design of any major system, we must resort to engineering calculations to assess the effects of the many parameters that may influence the final configuration of the system concept. The RCS of a weapon system is rapidly becoming a critical factor in the vulnerability and survivability of the system, and the ability to predict or estimate this property is essential in the system development process.

Of no lesser importance is the capability to estimate the RCS of a target being a useful echo reduction and diagnostic tool. Sometimes it can be used to identify scattering centers that otherwise may have been thought to be insignificant. In other cases, an estimate of the scattering properties of actual test targets makes it easier to plan for testing. Finally, RCS predictions can provide the confidence we often need that a measurement has not been contaminated by measurement or calibration errors.

Whatever the motivation, RCS predictions are based on (8.6), or one of its forms, which reduces the problem to that of computing the scattered field E_s at a distance R, given an incident field strength E_i on the target. Depending on the resources available, the accuracy desired, and the complexity of the target, we may or may not be able to choose the method used. For example, if the RCS of a small calibration sphere is needed with an accuracy of 0.1 dB, we are forced to compute the echo strength by using the exact solution. On the other hand, if the target is complicated, such as the ship discussed above, we must use one or more of the available high-frequency approximations. In the sections that follow, we briefly discuss both the exact and approximate methods of computing RCS.

8.3.1 Exact Methods

Based on a very careful examination of the experiments of a few pioneers, and molded by his insight, James Clerk Maxwell decided that electromagnetic waves in source-free space behave according to the description:

$$\nabla \times \mathbf{E} + \mu \frac{\partial \mathbf{H}}{\partial t} = \mathbf{0} \tag{8.9}$$

$$\nabla \times \mathbf{H} - \epsilon \frac{\partial \mathbf{E}}{\partial t} = \mathbf{0} \tag{8.10}$$

$$\nabla \cdot \mathbf{E} = \mathbf{0} \tag{8.11}$$

$$\nabla \cdot \mathbf{H} = \mathbf{0} \tag{8.12}$$

These are Maxwell's famous equations, in which, physical chemist Oscar Sandus once declared, "everything is soluble." Without manipulating them any further, we immediately see that, by virtue of the vector curl operation in (8.9) and (8.10), the electric and magnetic fields are at right angles to each other. Moreover, those two equations declare that a time-changing magnetic field gives rise to an electric field, and that a time-changing electric field gives rise to a magnetic field. Equations (8.11) and (8.12) state that both fields are divergence free. These are the characteristics of propagating electromagnetic waves, covering the spectrum from "dc to daylight." Indeed, if we assume a harmonic time dependence, such as $\exp[j\omega t]$, with the actual field quantities given by the real part of that time dependence, one cannot discern negative from positive frequencies. This being the case, we may expand our spectrum to cover all frequencies from minus daylight to plus daylight.

Assuming the harmonic time dependence $\exp[-j\omega t]$, we may perform the time derivative implied in either (8.9) or (8.10) and substitute it in the other equation, making use of one or the other of (8.11) or (8.12). Then, suppressing the assumed time dependence because it is common to all field quantities, we find the relationships:

$$\nabla^2 \mathbf{E} + k^2 \mathbf{E} = \mathbf{0} \tag{8.13}$$

$$\nabla^2 \mathbf{H} + k^2 \mathbf{H} = \mathbf{0} \tag{8.14}$$

where $k = 2\pi/\lambda$ is the free-space wavenumber. Both expressions have the same form (the wave equation, a second-order differential equation), and either may be used in an exact solution.

We must find a coordinate system in which a coordinate surface (coordinate = a constant) coincides with the surface of the obstacle of interest. Given a specification of the material properties of the obstacle, we then force the fields to satisfy prescribed boundary conditions at the surface of the obstacle. Metallic bodies are usually assumed to be perfectly conducting, in which case the tangential component of the electric field vanishes on the surface, and the difference between the magnetic intensities just inside and just outside the surface must be twice the induced surface currents. For bodies made of nonconducting materials, the tangential electric and magnetic fields must be continuous as one crosses the surface from just inside the body to just outside, or *vice versa*.

Whatever the boundary conditions in force, however, there are only a few coordinate systems that allow a solution. These include familiar systems, such as rectangular, cylindrical (including circular, elliptic, and parabolic), conical, spherical, and a few other coordinate systems. We can readily see that few targets of practical interest fall into any of these categories. Indeed, we can find no solution of the wave equation for the fields scattered by the truncated right circular cylinder, a particularly simple finite structure.

Moreover, even if our target of interest is simple enough that a coordinate system can be found to describe its surface profile, the functions involved in the solution are often cumbersome, and even downright difficult to compute, especially if the body is electrically large. The solution for the infinite circular cylinder, for example, is expressed in terms of angular trigonometric functions and radial Bessel functions, which pose no particular problem for modest arguments. The infinite series representation eventually must be truncated to a polynomial, but

Bessel functions still must be generated for as many orders as remain significant in the series. One usually resorts to recursive relationships to generate a Bessel function of a given order from the known values of two previous orders, but this eventually generates a number larger than the computer can handle (decimal overflow). While there are ways around these kinds of difficulties, some of them rather ingenious, one soon becomes well aware of the numerical limitations of even the simplest of the exact solutions of the wave equation.

An alternative formulation is the Stratton-Chu integral equations for the fields in source-free regions,

$$\mathbf{E} = \oint_s \{-jkZ_0\phi\,(\hat{\mathbf{n}} \times \mathbf{H}) + (\hat{\mathbf{n}} \times \mathbf{E}) \times \nabla\phi + (\hat{\mathbf{n}} \cdot \mathbf{E})\nabla\phi\}\,\mathrm{d}S \tag{8.15}$$

$$\mathbf{H} = \oint_s \{jkY_0\phi\,(\hat{\mathbf{n}} \times \mathbf{E}) + (\hat{\mathbf{n}} \times \mathbf{H}) \times \nabla\phi + (\hat{\mathbf{n}} \cdot \mathbf{H})\nabla\phi\}\,\mathrm{d}S \tag{8.16}$$

where $\phi = \mathrm{e}^{-jkr}/4\pi r$ is known as the free-space Green's function, r is the distance between the surface patch of integration and the observation point, **E** and **H** are the total fields at the surface patch dS of integration, $\hat{\mathbf{n}}$ is the unit surface normal there, and Z_0 and Y_0 are impedance and admittance of free space, respectively. The equations are obviously coupled, and the solution is by no means trivial, even for simple bodies. The tangential surface field components in these equations are interpreted as electric and magnetic currents induced by an incident wave, and the normal components are interpreted as induced electric and magnetic charges. The requirement that charge be conserved makes it possible to express all the terms in terms of induced currents. That, however, does not yet constitute a solution.

If the induced current distribution were known at all points on the surface of the obstacle, the currents could be integrated, numerically if not analytically, to calculate the scattered fields. The currents are *not* known, but we may do something very clever: we may divide the surface into a collection of small patches of discrete size and assume that the patches are small enough that the current on each is sensibly constant, or at least can be described by a very simple function. In the simplest case, we can let this function be the same for all surface patches, but we assign it a weighting function (i.e., a coefficient) that remains to be determined. The surface integrals can then be written as the sum of a finite number of terms.

We now force the observation point down to a particular patch on the surface. For any one patch, therefore, the **E** or **H** field on the left-hand side of (8.15) or (8.16) becomes that due to radiation from the currents on all the other patches, plus the incident field strength and a "self" current. The self current, which is unknown, will appear on the right-hand side of the equation. When the observation point is stepped to each of n patches created on the surface, a system of $2n$ linear homogeneous equations in $2n$ unknowns is generated (unknown electric and magnetic currents on each of the n patches). The coefficients of the unknown terms involve only the geometry of the path between the two patches, the orientation of their surface normals with respect to that path, and the weighting function assigned to the current distribution there. This being a linear system of equations, the unknown currents can be found, and the fields at any point in space can be determined by inserting them in the equivalent of (8.15) and (8.16) and summing them.

The procedure just described is the essence of what has become known as the method of moments. In some cases, simplifications can be made in the procedure. We may impose an impedance boundary condition, for example, which explicitly couples the electric and magnetic fields by some electrical property of the material comprising the surface of the body. While this ploy decouples the two equations, thereby halving the number of unknowns, the electric and magnetic fields may not necessarily satisfy the assumed relationship over the entire surface of the body. A special case is the perfectly conducting body, which can support no magnetic currents, again decoupling the equations, but with less influence on the confidence in the solution than offered by the impedance boundary condition.

The method of moments is not without some difficulties, on the other hand. For one, the self term may have troublesome singularities that require special treatment. Probably the most annoying difficulty is the demand for computer resources for bodies of even modest size. This is because the surface patches must be relatively small—of the order of 0.1 to 0.2λ along a side at most. If patches are chosen larger than this, the currents on the patch may no longer be as constant as assumed.

If there are m unknowns to be found, one must compute the m^2 elements of a matrix $m \times m$ in size. Once the matrix elements have been "filled," the matrix must be inverted, for which there are several procedures available. But on the average, the time required to fill and invert (or otherwise solve) the matrix rises approximately as the cube of the number of surface patches chosen for solution. The amount of time needed to solve the problem for a given body may be more than we are willing to invest, as illustrated in Figure 8.24.

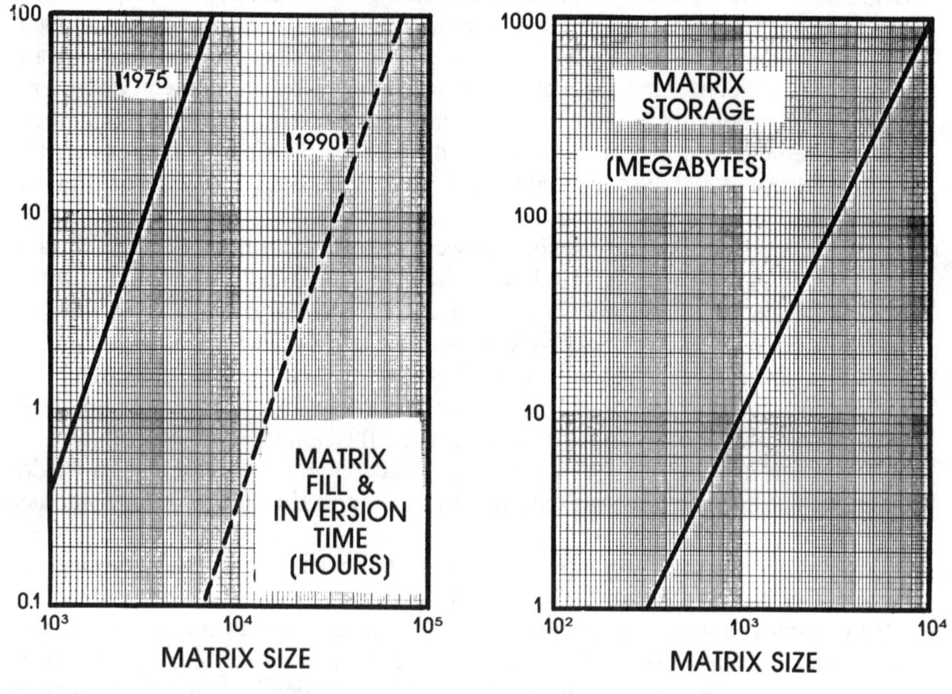

Figure 8.24 Time and storage requirements for matrix inversion.

This figure shows an estimate of the time and memory requirements needed for finding the currents on a body as a function of the matrix size. (The matrix size here is taken to be the number of surface patches used to sample the induced surface currents.) In 1975, it required about 0.4 hour (25 minutes) to fill and invert a 1000 $\times$ 1000 element matrix, and at the rate of 10 bytes per complex element (five bytes each for the real and imaginary parts), 10 megabytes were needed to contain the matrix in memory.

Also shown on the diagram is a forecast of computer performance some 15 years later (the year 1990), under the assumption that advances in digital processing technology will decrease the matrix fill and inversion time by an order of magnitude. Thus, by 1990, we might expect to be able to fill and invert a 10,000 $\times$ 10,000 element matrix in about 40 minutes which, according to the chart at the right, will require one gigabyte (a billion bytes) of memory for the matrix alone. We tacitly assume that access to any one of those one billion bytes of information will be as fast in 1990 as it was in 1975.

Consider now the target shown in Figure 8.25, the versatile Northrop F-5E fighter. The three short lines sketched on the figure are the wavelengths, drawn approximately to scale, of the radiation from radars operating at 0.4, 1.2, and 10 GHz, respectively. Note that the very short wavelength at the highest of the three frequencies appears only as a dot. At the longest wavelength (for 400 MHz), the airplane is about 20λ long, and at the shortest it is about 500λ. At the top of the figure is an estimate of the surface area of the F-5E, listed as 800 ft^2. (A more accurate estimate can be obtained from Northrop engineers, but the precise value is not critical in this analysis.)

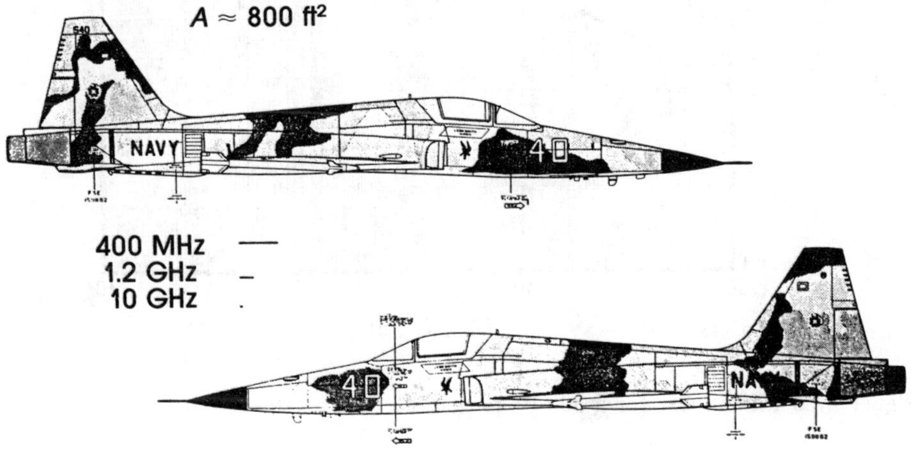

Figure 8.25 Northrop F-5E fighter.

Experience has shown that the obstacle surface should be sampled at rates not less than about five times per wavelength, or about $\frac{1}{25}$ of a square wavelength per patch. Thus, depending on the frequency of interest, the surface of a target like the F-5E may require as many as ten million surface patches, or perhaps even more, for adequate sampling of the surface fields. This is illustrated in Figure 8.26.

Let us now assume that 10 hours is about as much computer time as anyone in 1990 is willing to invest in the routine solution of scattering problems for targets like this. The leftmost chart of Figure 8.24 suggests that this much time could be expended on the filling and inversion of a matrix approximately 31,800 × 31,800 elements in size (total of one billion elements, or 10 billion bytes of memory at 10 bytes per element). However, if this represents the limit of our resources—or the limit of our willingness to *invest* resources—then predictions of the scattering from the F-5E must be restricted to frequencies of about 1.3 GHz or less. This covers but a fraction of the spectrum of practical interest, as there are dozens of tactical radars operating at much higher frequencies than this. Thus, although the method of moments offers the promise of very good estimates of the RCS of tactical targets, those estimates will probably be limited to frequencies below about 1 GHz for a long time to come.

It would be foolish to dismiss the method of moments as a useless tool for this one reason, however. Indeed, the method is an extremely powerful tool in the hands of skilled analysts seeking more insight into scattering phenomenology. It can be used to conduct inexpensive "numerical experiments" for fictitious two-dimensional bodies of almost any surface profile. While there are no two-dimensional bodies in nature, two-dimensional results often display the same characteristics that three-dimensional bodies do, and a great deal can be learned from

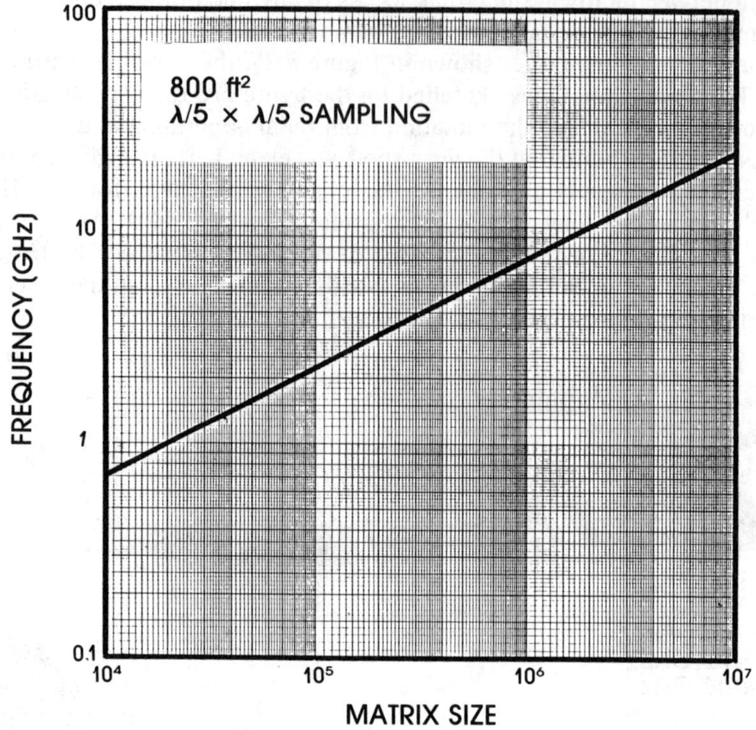

Figure 8.26 Matrix size for a 800-ft^2 target.

carefully designed experiments, whether conducted for real bodies or fictitious simulations thereof. Indeed, the method of moments yields no analytical formulas at all, but simply a set of numbers for a given set of "test" conditions. In order to pry further into the physics of the problem, the experimenter must decide which parameters to change for the next run, just as he does for physical experiments in the real world. The method of moments is an experimental tool that produces numbers, not formulas, and it should be used as such.

8.3.2 High Frequency Approximations

Given the limitations in the exact formulations for the electromagnetic scattering by finite bodies, it is natural to turn to approximations that significantly simplify RCS computations. Most of the approximate methods yield quite acceptable results with little sacrifice in accuracy, and this is their attraction. Indeed, the exact formulation for a perfectly conducting sphere requires the computation and summation of a sequence of complicated spheroidal Bessel functions, the number of which increases with the size of the sphere. Every student, incidentally, should be assigned this computation for a sphere 10λ in circumference using nothing more than a desk calculator: he or she would soon learn to appreciate the utility of the geometric optics formula. That approximation requires but two numbers: π and the radius of the sphere; the approximation becomes more accurate as the sphere becomes larger. The approximation neglects the contributions of creeping waves, but in the RCS world, where estimates within 3 or 5 dB of the truth are often acceptable, there are far more uncertainties to contend with than the creeping wave contribution. A powerful defense of the approximate methods can easily be presented.

In the pages that follow, we shall briefly describe the essence of six high frequency methods, from the theory of geometric optics to the diffraction coefficients of Mitzner. The sequence is in the order of increasing complexity, and the last four are relatively recent in their development (within the last 30 years). The six are given in Table 8.3.

Table 8.3 High Frequency Methods

Acronymn	*Method*	*Developer*
GO	Geometric optics	unknown
PO	Physical optics	unknown
GTD	Geometrical theory of diffraction	J.B. Keller
PTD	Physical theory of diffraction	P.Ia. Ufimtsev
MEC	Method of equivalent currents	A. Michaeli
ILDC	Incremental length diffraction coefficient	K. Mitzner

We hasten to point out that in each case a great deal of work has been performed by scholars and researchers whose names do not appear in the list. While we would wish to recognize all who have contributed to the understanding and characterization of scattering phenomenology, we cannot possibly do so in the short space allocated for this summary. With due apologies to any who may think that their work should be recognized, we simply have not the room here.

The high frequency theories all assume that the characteristic dimensions of the body (such as length, width, or radii of curvature) are large compared to the wavelength, except as noted. As such, the methods are all called "high frequency," "large body," or "optics region" techniques. These terms all imply that the target or scattering feature of interest is large compared with the incident wavelength, and not necessarily that the frequencies are in the HF region, that the body is physically large, or that optical wavelengths are involved. The only meaningful yardstick here is the size of the body in wavelengths, and for the high frequency theories, this generally means "at least a few wavelengths." It includes most targets of tactical interest.

Geometric Optics (GO)

The theory of *geometric optics* is probably the simplest of the high frequency methods. The theory is based on three simple concepts. First, rays are reflected mirror-fashion from a surface, which is to say, a reflected ray lies in the plane containing the incident ray and the surface normal, and subtends the same angle with respect to the surface normal as does the incident ray. Second, rays that pass through surfaces from one medium into another are "bent" (refracted), a consequence of the boundary conditions the fields must obey from one side of the surface to the other. Third, energy flows in a direction perpendicular to surfaces of constant phase [9].

Rays from sources at a finite distance diverge, or spread away from one another, when reflected by flat or convex surfaces. The ray intensity is modified by two factors. One is the classical Fresnel reflection coefficient accounting for the difference in material properties at the surface, and the other is the spreading of energy away from the surface. The amount of spreading is related to the curvature of the body surface, the curvature of the incident wavefronts, the angles by which the principal planes are rotated out of the plane of incidence, and the angle of arrival of the incident ray. The mathematical relationships have all been worked out for us, however, and we need not repeat them here (see [9,16]). Rays may also converge to a line or a point (called a *caustic*), as in the case of focusing lenses or reflectors, but once past that line or point, they diverge again. As we shall see, there are only one or two cases for which the caustic actually lies at the point of interest in RCS computations.

In applying the theory, one may invoke the conservation of energy along a tube of rays, which yields an estimate of the power density at the "output" end of the tube in terms of the power density at the "input" end. This ratio is expressed in terms of the radii of curvature of the wavefront at the output end of the tube and the distance of the source from the surface. Those curvatures may be expressed in terms of the curvature of the surface at the point of reflection and the angle of the incident ray from the normal. Although the expression can become complicated, a considerable simplification is brought about by allowing the observation point to recede to infinity, as required by (8.6). When this is done, the RCS will be found to be the rather simple formula,

$$\sigma = \pi a_1 a_2 \tag{8.17}$$

where a_1 and a_2 are the principal radii of the body surface at the specular point.

This expression is valid for perfectly conducting surfaces only, for which the reflection coefficient is -1. The formula for dielectric objects is not quite as simple, because rays can penetrate the body and be reflected by surfaces at the rear, and several internal reflections may have to be be taken into account. Nevertheless, the result is striking because of its simplicity and because it holds for scattering in bistatic directions as well as back to the source. Note that there is no dependence on the frequency of the wave, as was pointed out earlier for doubly curved surfaces. The application of the formula is quite straightforward: simply find the specular point on the body where the angle of reflection equals the angle of incidence, find the principal radii of curvature there, and insert the values in (8.17).

The formula is quite good for smooth objects whose radii of curvature at the specular points are not too small (at least a wavelength or so). The geometric optics estimate for the RCS of a sphere, for example, is within 0.5 dB of the exact value for a sphere only 3.2λ in diameter, not a very large one. Thus, (8.17) is an attractive formula by virtue of its simplicity and versatility. However, there are extremely important cases for which it utterly fails: when one or both radii of curvature are infinite, which we have seen is the case for flat plates and cone frusta. Our point of observation in these cases—the far field in the specular direction—happens to be a caustic, and we obtain infinite results for the RCS, which are obviously erroneous. Despite the attractive simplicity of geometric optics, it fails us when the body curvature is too small or too large.

Thus, although geometric optics is a very attractive and simple ray-tracing scheme, it suffers several drawbacks:

- the body radii of curvature must be large;
- the theory fails when the specular point is near an edge;
- it yields infinite results for flat and singly curved surfaces.

Physical Optics (PO)

There is, however, another method which is quite suited for bodies with one or more infinite radii of curvature: *physical optics*. Indeed, physical optics is easier to apply to flat and singly curved surfaces than it is to doubly curved surfaces.

The theory is based on the so-called *tangent plane approximation,* which simplifies the Stratton-Chu integral equations (8.15) and (8.16). A tangent plane is brought to bear against the body at the surface patch of integration dS, and the fields on the actual surface are then assigned precisely the value they would have had had the tangent plane itself been the surface of interest. Those fields are none other than the geometric optics fields for an infinite interface, and although the surface is usually taken to be perfectly conducting, this is not essential. For metallic surfaces, the tangential electric field is sensibly zero, and the tangential magnetic field is twice the tangential component of the incident magnetic field. If the body is a conductor, the fields over all surfaces shaded from the incident field are taken to be precisely zero. The integrals, therefore, extend only over the illuminated portions of the surface.

A second simplification is the far field approximation, in which it is assumed that the distance from the body to the point of observation is much larger than any body dimension. Thus, the distance from the surface patch of integration to the observation point can be expressed in terms of a constant plus a small component along the direction to the observation point. The gradient of the Green's function may then be replaced by the Green's function itself, times a constant, for a considerable simplification of the integral. Only one of the two equations, (8.15) or (8.16), is needed, for if the electric field is calculated at the far field point, the magnetic field is automatically known, and *vice versa*. The physical optics approximation of (8.15), for example, is

$$\mathbf{E}_s = \frac{jkZ_0 e^{-jkR_0}}{2\pi R_0} \int_s \hat{\mathbf{s}} \times [\hat{\mathbf{s}} \times (\hat{\mathbf{n}} \times \mathbf{H}_i)] e^{-jk\mathbf{r}\cdot\hat{\mathbf{s}}}\, dS \tag{8.18}$$

where R_0 is the distance from the object to the far field observation point, Z_0 is the impedance of the medium (usually free space), $\hat{\mathbf{s}}$ is a unit vector aligned along the scattering direction, $\mathbf{H}_i$ is the incident magnetic field strength, and $\mathbf{r}$ is the position vector of the patch of integration dS. The problem is now a matter of carrying out the integration.

The integral can be evaluated analytically only for a handful of very simple structures, notably flat plates and circular cylinders, and it should always be remembered that even when analytic evaluation is possible, the integral itself is an approximation. The result for a flat plate is

$$\sqrt{\sigma} = \frac{jkLW}{\sqrt{\pi}} \hat{\mathbf{n}} \cdot \hat{\mathbf{e}}_r \times \hat{\mathbf{h}}_i e^{-jk\mathbf{r}_0\cdot(\hat{\mathbf{i}}-\hat{\mathbf{s}})} \cdot \frac{\sin[\frac{1}{2}k\mathbf{L}\cdot(\hat{\mathbf{i}}-\hat{\mathbf{s}})]}{\frac{1}{2}k\mathbf{L}\cdot(\hat{\mathbf{i}}-\hat{\mathbf{s}})} \cdot \frac{\sin[\frac{1}{2}k\mathbf{W}\cdot(\hat{\mathbf{i}}-\hat{\mathbf{s}})]}{\frac{1}{2}k\mathbf{W}\cdot(\hat{\mathbf{i}}-\hat{\mathbf{s}})} \tag{8.19}$$

where $\mathbf{W}$ and $\mathbf{L}$ are the length vectors of the edges of the plate, $\hat{\mathbf{n}}$ is the surface normal of the illuminated side of the plate, $\hat{\mathbf{h}}_i$ and $\hat{\mathbf{e}}_r$ are the magnetic and electric polarizations, respectively, of the transmitter and receiver, $\hat{\mathbf{i}}$ and $\hat{\mathbf{s}}$ are unit vectors along the directions of incidence and scattering, respectively, and $\mathbf{r}_0$ is the position vector of some convenient origin on or near the plate. This bistatic result is reasonably accurate in the forward and specular directions, but can become seriously in error in other directions.

An example is the comparison in Figure 8.27 made by Ross [17]. The particular plate measured was five wavelengths along a side, and was measured for both vertical and horizontal incident polarizations. The solid traces are the measured patterns and the dashed traces are the physical optics predictions. For broadside incidence (at the left of the patterns), the agreement between measurement and prediction is excellent, but the agreement deteriorates as the scattering direction swings further away from the specular direction. Near edge-on incidence, the difference between theory and experiment exceeds 10 dB, and the theory becomes totally unreliable at precisely edge-on incidence. Note also that the physical optics predictions are independent of polarization (the dashed lines are identical in both patterns), whereas the measured patterns show a significant dependence on polarization, especially in the edge-on region.

The theory of physical optics also generates fictitious returns that represent the contributions of shadow boundaries, as may be seen from a comparison of

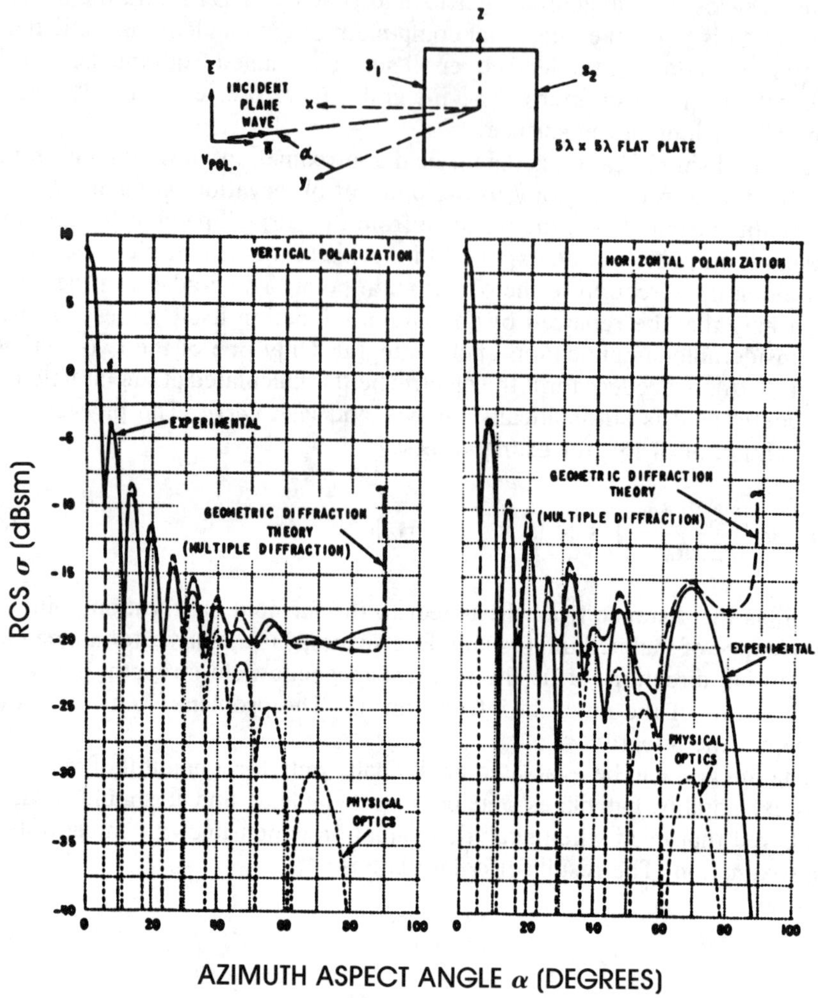

Figure 8.27 RCS of rectangular plates (from Ross [17]). (© 1966 IEEE. Reprinted with permission.)

Figures 8.28 and 8.29. Both figures trace the scattering width of a perfectly conducting, infinite circular cylinder as a function of the electrical circumference, ka. The solid trace in Figure 8.28 represents a stationary phase evaluation of the integral, while the dashed trace represents an analytic evaluation in which no approximations were made. (The latter result is a linear combination of Bessel and Struve functions of the first kind of order 1.) The small undulations in this trace are due to shadow boundary contributions adding constructively and destructively with the specular return from the front side of the cylinder. Indeed, the periodicity of the oscillations is that of two sets of scattering centers separated by precisely the radius of the cylinder—the separation along the line of sight between the specular reflection from the front of the cylinder and the shadow boundaries at its sides. Both traces are independent of the incident polarization.

Compare this with the traces in Figure 8.29, which were computed from the exact series representation of the solution of the wave equation for the cylinder. The series representation consists of a sequence of Bessel functions of the first and second kinds extending to infinite orders, and the results shown here were truncated when the next term added to the series was more than six orders of magnitude smaller than the sum itself. When the incident electric field is parallel to the cylinder axis (solid trace), the curve rises monotonically with increasing size. When the incident magnetic polarization is parallel to the cylinder axis, the

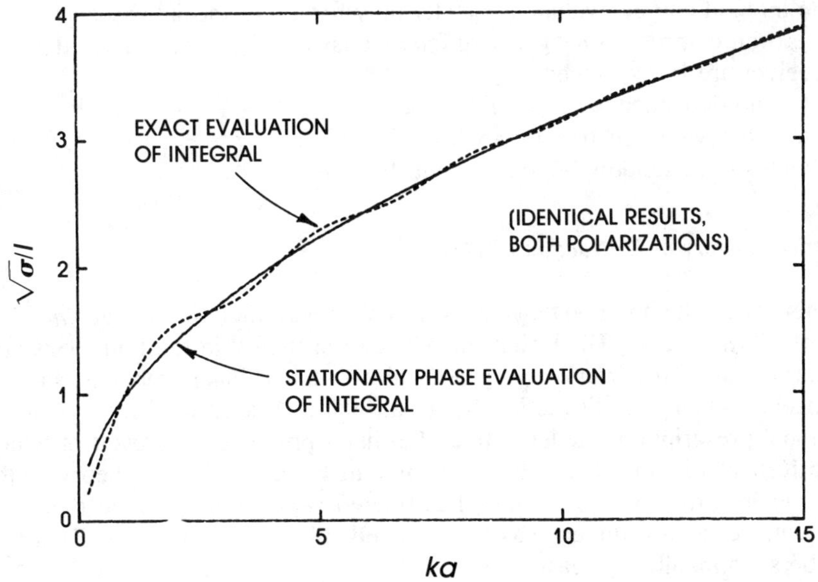

Figure 8.28 **Physical optics result for an infinite conducting right circular cylinder.**

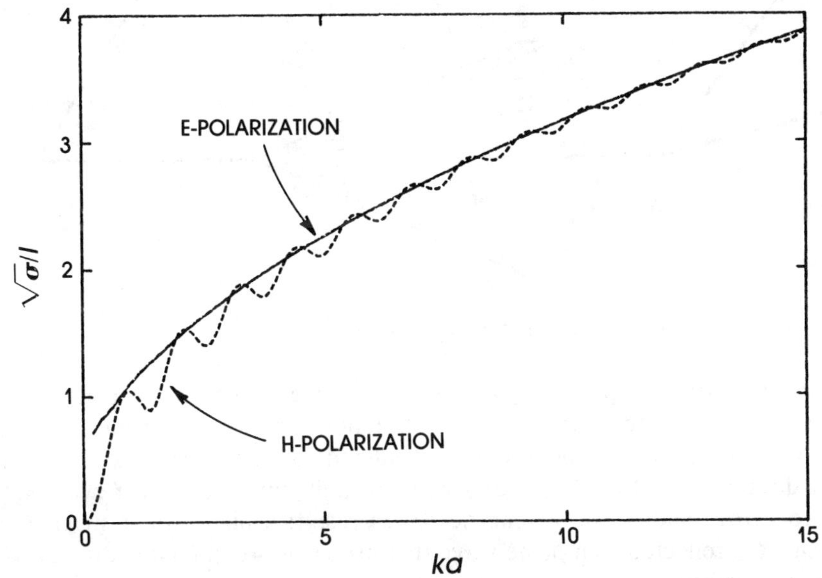

Figure 8.29 **Exact result for an infinite conducting circular cylinder.**

response (dashed trace) shows a series of undulations that become weaker as the cylinder becomes larger. The undulations are due to a creeping wave that skirts the rear of the cylinder in identically the same way the creeping wave skirts the rear of a sphere.

Thus, the exact solution shows a dependence on polarization not matched by the solution yielded by physical optics. Moreover, although the physical optics prediction has undulations like the exact solution for H-polarization, they are erroneous. Because the undulations in the physical optics prediction are due to a nonexistent mechanism, we may be better off accepting the stationary phase approximation instead of the more rigorous one that generated the Struve functions. We conclude that physical optics, while certainly better than geometric optics for some bodies, has serious defects:

- the integral can be evaluated exactly only for a few special cases;
- it is nonreciprocal (we get a different answer when the transmitter and receiver are interchanged);
- it has no dependence on polarization;
- it fails for wide nonspecular angles;
- it yields false shadow boundary contributions.

Geometrical Theory of Diffraction (GTD)

There is another method that can overcome these difficulties: the *geometrical theory of diffraction* (GTD). Keller introduced the theory in the mid-1950s [18] as the three-dimensional version of canonical two-dimensional problems, such as Sommerfeld's result for diffraction by a half-plane [19]. In devising his three-dimensional prescription, Keller applied Fermat's principle, the result of which was the Keller cone of diffracted rays shown in Figure 8.30. The theory states that an incident ray excites a cone of diffracted rays, all subtending the same angle β with respect to the edge as that subtended by the incident ray. That this should be so is manifestly reasonable.

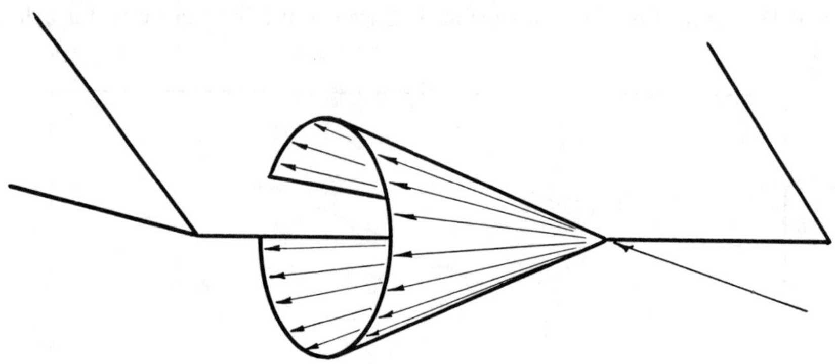

Figure 8.30 The Keller cone of diffracted rays.

Consider, for example, the physical optics reflection by a flat plate. Most of the energy is scattered in the specular direction, and the larger the plate becomes, the more narrowly the energy is confined to a small solid angle in the specular direction. As the plate grows without limit, that solid angle eventually shrinks to zero, yielding the geometric optics result: a plane wave striking an infinite sheet is reflected as a plane wave that propagates in a unique direction—the specular direction.

Let us now shrink the transverse dimension so that the sheet becomes a strip. As the width becomes smaller, energy will be spread out in the transverse plane, but not in the longitudinal plane, and in the limit, the strip becomes a wire. Because of the infinite (or at least very large) longitudinal dimension of the surface, energy is still confined to very narrow directions centered on the specular direction. But in the absence of any breadth to the surface, the "specular" direction is any that makes the same angle with respect to the wire as did the incident wave. And because the transverse dimension of the surface has become very small, the scattering pattern in the transverse plane becomes omnidirectional. For a wire, therefore, we reason that the scattered energy is confined to a thin cone of directions.

Consider, then, the half-plane, a semi-infinite surface. The scattered field can be thought to arise from surface currents that behave essentially like those on an infinite sheet, plus filamentary currents concentrated near the edge. The filamentary edge currents will generate a cone of diffracted energy like that of the wire, and the surface currents will produce a sharp narrow beam in the specular direction from the surface. There will also be strong scattering in the forward direction required to produce the shadow in the shaded region behind the sheet.

These properties are embedded in the following formula for the diffracted field due to an edge, which may be the edge of a wedge as well as the special case of a half-plane:

$$\mathbf{E}_s = -\frac{\Gamma\, e^{-jkr}}{\sin^3 \beta}\{(\hat{t}\cdot \mathbf{E}_i)(X - Y)\hat{\mathbf{s}} \times (\hat{\mathbf{s}} \times \hat{\mathbf{t}}) + Z_0(\hat{t}\cdot \mathbf{H}_i)(X + Y)\hat{\mathbf{s}} \times \hat{\mathbf{t}}\} \quad (8.20)$$

where the diffraction coefficients are

$$X = \frac{\frac{1}{n}\sin\frac{\pi}{n}}{\cos\frac{\pi}{n} - \cos\frac{\psi_s - \psi_i}{n}} \quad (8.21)$$

$$Y = \frac{\frac{1}{n}\sin\frac{\pi}{n}}{\cos\frac{\pi}{n} - \cos\frac{\psi_s + \psi_i}{n}} \quad (8.22)$$

In these expressions, Γ is a divergence factor accounting for the spreading of energy away from the point of diffraction, r is the distance from that point to the point of observation, β is the Keller cone half-angle (and the angle of arrival of the incident ray), $\hat{t}$ is a unit vector aligned along the edge, and Z_0 is the impedance of the medium, usually free space.

Keller did not write the diffraction coefficients X and Y explicitly in this form, and the notation is instead due to Senior and Uslenghi [20]. The angles ψ_i and ψ_s are the projections of the directions of incidence and scattering, respectively, onto a plane perpendicular to the edge (see Figure 8.31), and n is the exterior angle α of the wedge normalized with respect to π, (i.e., $n = \alpha/\pi$). Note that both the electric and magnetic field of the incident wave appear in (8.20), imparting a polarization dependence not present in the physical optics expressions cited earlier for a flat plate. Because of the explicit treatment of the edge, the Keller GTD prescription is inherently more accurate than the physical optics formulas for non-specular observation directions.

Despite these improvements, however, GTD suffers some limitations. For one, the direction of observation must lie along a generator of the Keller cone erected at an edge. As in the theory of geometric optics, no energy is conveyed to the point of observation unless that point is pierced by a ray. And the obverse is nearly as bad: We obtain an infinite result (a caustic) when the observation point is pierced by an infinity of rays. This occurs more frequently than we wish; in computing the backscattering from a ring discontinuity, for example, we find that an infinity of rays passes through any observation point placed along the axis of the ring. The junction of a nose cone with a cylindrical body is an example of a ring discontinuity.

Furthermore, the diffraction coefficients given in (8.21) and (8.22) fail in the transition regions where the denominators become very small. This occurs in the diffraction coefficient X when the direction of observation is close to the shadow boundary, and in the diffraction coefficient Y when it is close to the reflection boundary. These two cases are illustrated in Figure 8.32 for the case of the edge of a half plane illuminated at an angle of 60°. Note that Y (dashed curve) becomes singular along the reflection boundary at 120°, while X (solid curve) becomes

KNOTT
CHAPTER 8

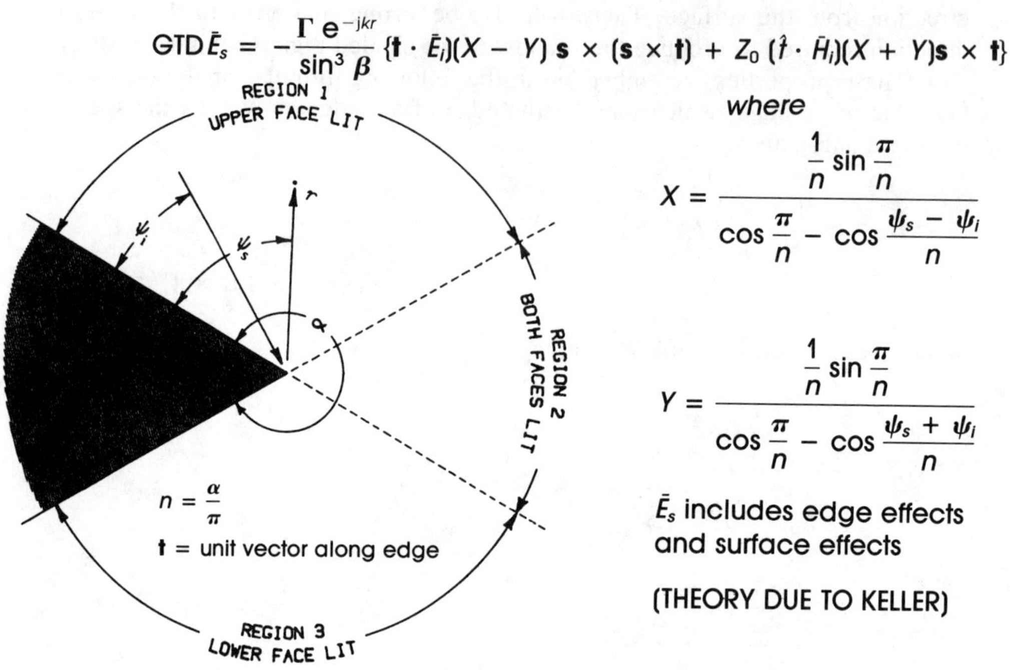

Figure 8.31 Projections of the incident and scattering directions onto a plane perpendicular to the diffracting edge.

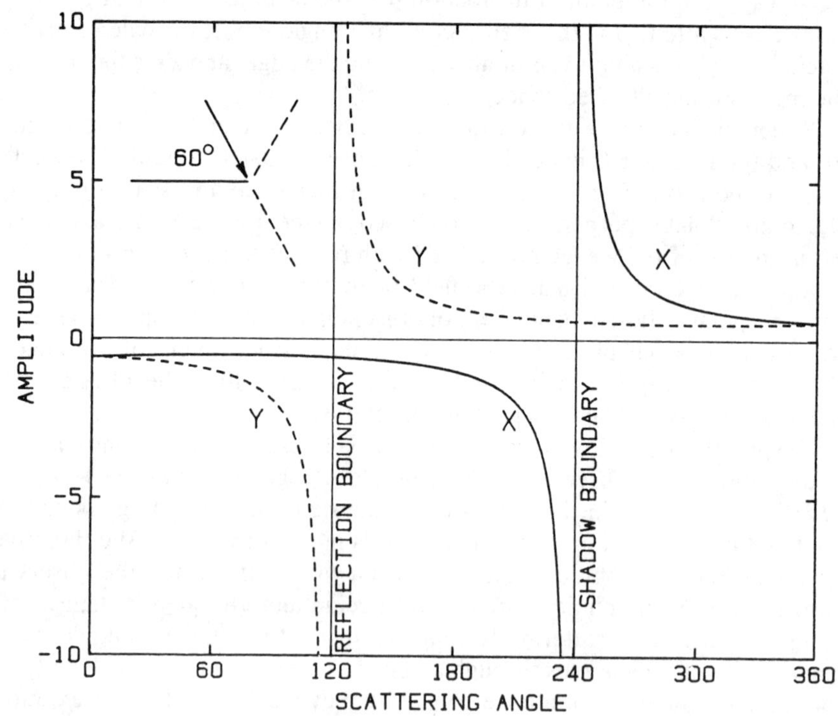

Figure 8.32 Example of Keller's diffraction coefficients for a half-plane.

singular along the shadow boundary at 240°. Thus, although GTD offers the promise of more accurate predictions away from the specular direction from the surface of a finite plate, for example, it fails completely along the specular direction itself. Although the annoyances of the caustics and singularities of the geometrical theory of diffraction can be overcome, in some cases, by clever manipulations, they will certainly appear in explicit problems and must be dealt with.

Despite these shortcomings, however; GTD remains a very powerful tool in the analyst's kit, much to Keller's credit and original insight. The theory is simple in its formulation, being a ray tracing scheme that exhibits the proper polarization dependence. Its faults are:

- the point of observation must lie on a Keller cone;
- an infinity of rays can come together to form a caustic;
- the diffraction coefficients are singular along the shadow boundary and reflection boundary.

Physical Theory of Diffraction (*PTD*)

There is a prescription that eliminates the third of these undesirable features. Introduced by P.Ia. Ufimtsev at about the same time that Keller presented his GTD, it is called the physical theory of diffraction (PTD) [21].

Like Keller, Ufimtsev relied on the exact solution of a canonical two-dimensional problem (the wedge), and applied the results to special structures, such as the half-plane and the strip [21]. He synthesized the solution of the strip problem by overlapping a pair of half-planes so that their edges coincided with the edges of the strip. We must ignore, of course, the contributions from those portions of the half-planes extending beyond the boundaries of the strip.

The physical theory of diffraction is based on the canonical solution for the perfectly conducting wedge, of which the half-plane is a special case. Ufimtsev postulated that the currents induced on the wedge surfaces by an incident wave could be separated into "uniform" and "nonuniform" components, the first being none other than the classical physical optics currents discussed above. The nonuniform component, by contrast, represents currents concentrated near the edge, which Ufimtsev called the "fringe" currents and which therefore radiated what he called "fringe" waves. He did not perform an analytical integration of those currents to obtain the scattered field, but instead started with the solution of the canonical wedge problem, which was available at the time.

From that solution for the total field (incident field plus scattered field), he first subtracted the incident field. This left only the scattered field which, he reasoned, can arise from only two sources: the uniform physical optics currents and the nonuniform fringe currents. His objective being to isolate the contributions from the fringe currents alone, he then calculated the physical optics contribution and subtracted it from the solution. What remained was the contribution from the edge alone (via the fringe currents), and which can be expressed in terms of Keller's diffraction coefficients.

One of Ufimtsev's tasks was to decide how to handle the semi-infinite dimension of the upper and lower surfaces of the wedge. The physical optics integral demands that the surface currents be integrated over the illuminated wedge surfaces, hence one limit of integration was the edge itself, while the other extended to infinity. He did what any of us might do: he ignored the contribution from the infinitely remote dimension. The result, therefore, is a physical optics contribution that appears to come from *the edge alone*. When expressed in the form of diffraction coefficients like Keller's, the physical optics contributions of the uniform currents are:

$$X_1 = -\tfrac{1}{2}\tan[(\psi_s - \psi_i)/2] \tag{8.23}$$

$$Y_1 = -\tfrac{1}{2}\tan[(\psi_s + \psi_i)/2] \tag{8.24}$$

$$X_2 = \tfrac{1}{2}\tan[(\psi_s - \psi_i)/2] \tag{8.25}$$

$$Y_2 = -\tfrac{1}{2}\tan[\alpha - (\psi_s + \psi_i)/2] \tag{8.26}$$

where the angles ψ_i and ψ_s are those defined in Figure 8.31, α is the exterior angle of the wedge, and the subscripts 1 and 2 identify these contributions with either the upper or lower face of the wedge, respectively.

Ufimtsev defines two diffraction coefficients, f and g, depending whether the wave incident on the edge is polarized with its electric or magnetic field along the edge, respectively. In terms of the physical optics coefficients above, and the canonical solution for the infinite wedge, Ufimtsev writes the diffracted field as

$$E_Z = E_{0Z}\, f\, e^{-j(kr+\pi/4)}/(2\pi kr)^{1/2} \tag{8.27}$$

$$H_Z = H_{0Z}\, g\, e^{-j(kr+\pi/4)}/(2\pi kr)^{1/2} \tag{8.28}$$

$$f = \begin{cases} (X - Y) - (X_1 - Y_1), & 0 \le \psi_i \le \alpha - \pi \\ (X - Y) - (X_1 - Y_1) - (X_2 - Y_2), & \alpha - \pi \le \psi_i \le \pi \\ (X - Y) - (X_2 - Y_2), & \pi \le \psi_i \le \alpha \end{cases} \tag{8.29}$$

$$g = \begin{cases} (X + Y) - (X_1 + Y_1), & 0 \le \psi_i \le \alpha - \pi \\ (X + Y) - (X_1 + Y_1) - (X_2 + Y_2), & \alpha - \pi \le \psi_i \le \pi \\ (X + Y) - (X_2 + Y_2), & \pi \le \psi_i \le \alpha \end{cases} \tag{8.30}$$

where X and Y are none other than the diffraction coefficients appearing in (8.21) and (8.22). Ufimtsev did not use this notation in his papers; we have written the diffraction coefficients in this form to make it easier to compare them with Keller's coefficients.

Note that Ufimtsev's f and g are Keller's $(X - Y)$ and $(X + Y)$ with certain combinations of physical optics terms subtracted, depending on whether the upper or lower surface of the wedge, or both, is illuminated. Happily enough, the unsubscripted diffraction coefficients (X, Y) and the subscripted physical optics terms (X_1 or X_2, Y_1 or Y_2) become singular in the same directions, but because of sign differences, they cancel each other. Therefore, Ufimtsev's diffraction coefficients remain well behaved along the shadow and reflection boundaries where Keller's become infinite. This is illustrated in Figures 8.33 and 8.34.

The case chosen for illustration is the half-plane used in Figure 8.32, illuminated at an angle of 60°; in these figures, we indicate the physical optics terms with primes instead of subscripts. In Figure 8.33, Keller's X becomes infinite at the shadow boundary while Ufimtsev's $(X - X')$ passes through zero and is well behaved. Similarly in Figure 8.34, Keller's Y becomes infinite at the reflection boundary while Ufimtsev's $(Y - Y')$ passes through zero and is well behaved. Ufimtsev's coefficients are obviously more attractive than Keller's.

Because the physical optics terms have been deliberately eliminated in Ufimtsev's diffraction coefficients, such contributions must be explicitly accounted for later in a practical computation, such as for a three-dimensional edged structure. Paradoxically, the very terms Ufimtsev subtracts from the canonical result must be added back for practical computations. By contrast, the physical optics contributions are implicitly embedded in Keller's formulation and need not be accounted for separately. On the other hand, if the scattering direction approaches one of the transition regions, Keller's GTD yields no useful result while Ufimtsev's PTD does, even if physical optics terms must be handled separately.

There is yet another paradox: if the physical optics contribution for a single edge is singular along the shadow and reflection boundaries, as expressed in (8.23) through (8.26), why does not (8.19), the physical optics formula for a flat plate, suffer the same singularities? Indeed, does not that expression include the contributions from four edges, not just one?

It certainly does, but for a given scattering direction, whether it be along a reflection boundary or a shadow boundary, the contribution from any edge element (e.g., along the near edge of the plate) is accompanied by a contribution from an

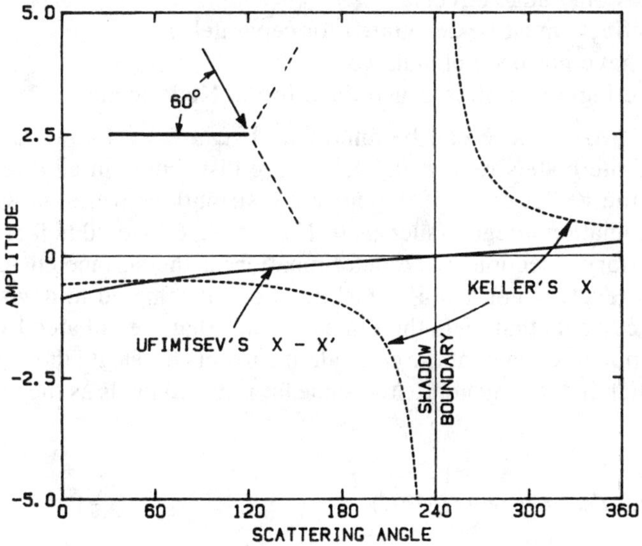

Figure 8.33 Comparison of Keller's X coefficient with Ufimtsev's $(X - X')$.

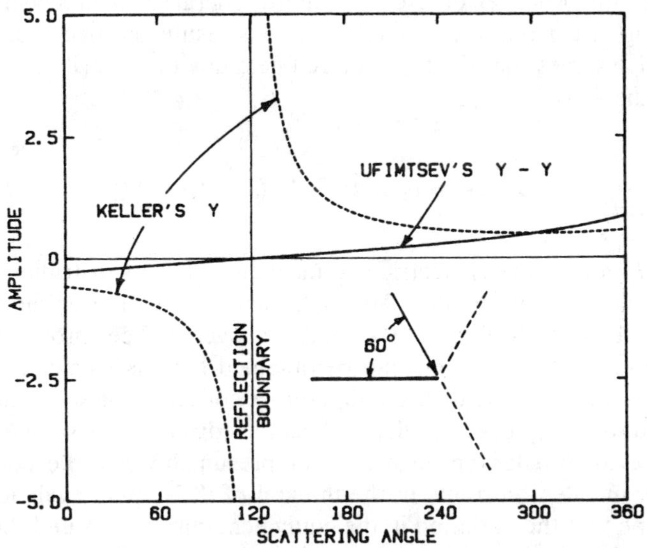

Figure 8.34 Comparison of Keller's Y coefficient with Ufimtsev's $(Y - Y')$.

element somewhere along an opposite edge. The singularities of those contributions will oppose each other, and a finite result will be obtained. This happens because the plate is flat. Suppose, on the other hand, that we can deform a rectangular or polygonal plate in such a way that its edges remain straight, but with none sharing a plane with any other edge. The singularities in the physical optics terms of the individual edges would no longer cancel one another, and the theory would fail us for those directions.

Whether or not physical optics fails us in such a case, Ufimtsev's physical theory of diffraction has several attractions:

- it has the simplicity of Keller's GTD;
- it exhibits the proper polarization dependence;
- it has no singularities in the transition regions.

To exploit those features, one must be willing to accept the following undesirable features:

- it includes only edge effects;
- surface effects must be accounted for separately;
- caustics have not been eliminated;
- the scattering direction is constrained to the Keller cone.

The last two of these can be annoying, because we know that the fields scattered by finite bodies remain finite, and are distributed in all directions, not just those on the Keller cone. If the current distributions were known, an evaluation of the radiation integral (such as (8.15) and (8.16)) would in fact yield finite results in arbitrary directions. The determination of the surface currents, as we have seen, however, is not a simple task, and one is tempted to devise fictitious "equivalent" currents that yield the same far fields that are radiated by the actual currents. It is not essential that the equivalent currents duplicate the real currents, but it *is* essential that they generate the same far radiated fields as the real currents do.

Method of Equivalent Currents (MEC)

The use of equivalent currents is by no means new: Ryan and Peters devised equivalent currents to overcome the backscattering caustic on the axis of a ring discontinuity [22], and Knott and Senior extended that concept to bistatic cases [23]. More recently, Michaeli devised a far more general set of equivalent currents, and with enough rigor that they include previous results as special cases [24].

Michaeli assumes that the far scattered field due to a distribution of surface currents has the form:

$$\mathbf{E}_d = \frac{\mathrm{j}k\,\mathrm{e}^{-\mathrm{j}kR_0}}{4\pi R_0}\int [Z_0 I_e\,\hat{\mathbf{s}} \times (\hat{\mathbf{s}} \times \hat{\mathbf{t}}) + I_m\,(\hat{\mathbf{s}} \times \hat{\mathbf{t}})]\,\mathrm{e}^{-\mathrm{j}k\mathbf{r}\cdot\hat{\mathbf{s}}}\,\mathrm{d}S \tag{8.31}$$

where I_e and I_m are fictitious electric and magnetic currents distributed over two flat surfaces that meet at an edge. Michaeli focuses attention on an elementary length along the edge as if it were a small transverse slice through an infinite wedge. If the equivalent currents can be obtained for this elementary slice, he reasoned, they may then be distributed along a collection of such slices placed side-by-side to form any edge profile, such as the edge at the base of a cone. The far field scattered by such an edge contour presumably may be computed by inserting the equivalent currents in the integral of (8.31), and evaluating the integral. We hope that the variation in the equivalent currents around the profile is simple enough that the integral may be evaluated analytically, but even if that is not the case, we can always exploit the ubiquitous computer to perform a numerical evaluation.

Michaeli thus considers the case of the infinite wedge, and integrates the induced currents along a pair of narrow surface strips (one each on the upper and lower wedge surfaces) perpendicular to the edge. The two limits of integration are the end of the strip at the edge, and the other end at infinity. Like Ufimtsev, Michaeli ignores the contribution from the infinitely remote end, which he calls an asymptotic evaluation. Upon comparing the result with the assumed form of the integral in (8.31), he finds the equivalent currents to be

$$I_e = -\frac{\mathrm{j}2\,(\hat{\mathbf{t}}\cdot\mathbf{E}_i)\,D_e}{kZ_0\sin^2\beta_i} - \frac{\mathrm{j}2\,(\hat{\mathbf{t}}\cdot\mathbf{H}_i)\,D_{em}}{k\sin\beta_i} \tag{8.32}$$

$$I_m = \frac{\mathrm{j}2\,(\hat{\mathbf{t}}\cdot\mathbf{H}_i)\,D_m}{kY_0\sin\beta_i\sin\beta_s} \tag{8.33}$$

where D_e, D_m, and D_{em} may be called diffraction coefficients, and β_i and β_s are the angles of incidence and scattering, respectively, subtended at the edge. Michaeli did not write the equivalent currents using the (D_e, D_m, D_{em}) notation; the notation is introduced here only for convenience in discussing the results.

Although we do not display the diffraction coefficients here (see [9], p. 138, for example), they contain the projections of the angles of incidence and scattering onto a plane perpendicular to the edge, ψ_i and ψ_s, and the angles β_i and β_s. Actual currents are induced only by an excitation of the edge, whereas the currents in (8.32) and (8.33) depend on the scattering direction as well. Therefore, the equivalent currents are indeed fictitious. However, we do not require that they be real, only that they yield the correct far field result, which they do—most of the time, at least. Furthermore, because the method of equivalent currents is not a ray-tracing procedure, the caustic elimination problem (one of the original motivations for devising equivalent currents) has disappeared.

Note that the electric equivalent current has two terms, one generated by the component of the incident electric field along the edge and the other by the component of incident magnetic field along the edge. This fact was not discovered in previous explorations of equivalent currents, but was demonstrated by Mitzner some 10 years earlier by using a different approach. The magnetic equivalent current conversely is excited only by the incident magnetic field.

The equivalent currents allow us to compute the fields scattered by an arbitrary edge in any direction $\hat{s}$, as suggested in (8.31). This represents a considerable improvement over GTD and PTD, for which the scattering direction is constrained to a generator of the Keller cone. Perhaps expectedly, the diffraction coefficients D_e, D_m, and D_{em} are more complicated than those of Keller and Ufimtsev because arbitrary directions are now involved. Moreover, they sometimes behave strangely, suddenly becoming imaginary as certain boundaries are crossed. The significance of this behavior remains to be explored.

However, despite their elegance, Michaeli's diffraction coefficients, like Keller's, are singular in the transition regions near the shadow and the reflection boundaries, hence that particular problem has not been solved by the method of equivalent currents. In later papers, Michaeli shows how to eliminate some of the singularities, but with only one exception, the elimination is none other than the removal—*à la* Ufimtsev—of the physical optics surface contributions. Nevertheless, Michaeli's method of equivalent currents brings about two significant improvements:

- the elimination of caustics;
- unlimited scattering directions.

The theory still leaves us, however, with the singularities in the transition regions.

Incremental Length Diffraction Coefficient (ILDC)

Although Michaeli shows how to eliminate those singularities by subtracting the physical optics contributions, Mitzner did precisely the same thing some ten years earlier [25]. We therefore cite Mitzner's work on his *incremental length diffraction coefficients* (ILDC) as an example of the removal of the singularities. Mitzner's ILDC happens to have the same relationship to Michaeli's diffraction coefficients as Ufimtsev's have to Keller's, and the comparison is rather striking.

In his report, Mitzner focuses immediately on the incremental form of the diffracted field, with the understanding that the scattering from any edge contour can be obtained by integrating the ILDC over the illuminated portion of the contour. He expresses the far diffracted field in terms of a dyadic diffraction coefficient which, when multiplied by a unit polarization vector aligned along the incident electric field, yields the scattered electric field, a vector quantity. The dyadic diffraction coefficient is expressed in terms of its components along two pairs of unit vectors, one pair perpendicular and parallel to the plane of incidence, and the other pair perpendicular and parallel to the plane of scattering. Thus, Mitzner's diffraction coefficients depend, as indeed they must, on the direction of incidence and the direction of observation. We shall label them $D_{\|}$, $D_{\perp}$, and $D_{\|\perp}$ to distinguish them from Michaeli's diffraction coefficients.

The comparison of the diffraction coefficients of Michaeli and Mitzner requires considerable manipulation because of differences in the notation and the geometry set up by the two, and the reader may find it helpful first to examine Knott [26]. Mitzner performs essentially the same surface field integration performed by Michaeli, confining attention to a pair of strips, one each on the upper and lower surfaces of the wedge, both perpendicular to the edge. Following Ufimtsev's example, Mitzner ignores the contribution from the (infinitely remote) far end of the strip, and specifically subtracts the physical optics surface current contributions. The reader will find that although we refer to the three diffraction coefficients $D_{\|}$, $D_{\perp}$, and $D_{\|\perp}$, these coefficients are not explicitly displayed anywhere in Mitzner's report.

Although not explicitly written out here, Mitzner's ILDC are related to Michaeli's diffraction coefficients by the following:

$$D_{\|} = D_e - D'_{\|} \tag{8.34}$$

$$D_{\perp} = D_m - D'_{\perp} \tag{8.35}$$

$$D_{\|\perp} = D_{em} \sin \beta_i - D'_{\|\perp} \tag{8.36}$$

where the unprimed quantities on the left are Mitzner's coefficients, the unprimed quantities on the right are Michaeli's coefficients, and the primed quantities on the right are a set of physical optics diffraction coefficients. (All of these quantities are written out explicitly in Chapter 5 of [9].) Thus, *Mitzner's results relate to Michaeli's in identically the way Ufimtsev's results relate to Keller's*. The difference between the two pairs of theories is that the coefficients of Keller and Ufimtsev hold only for scattering directions on the Keller cone, while those of Mitzner and Michaeli are valid for arbitrary directions. This is a rather remarkable parallel.

We have seen that Michaeli's coefficients, like Keller's, become singular in the transition regions. The primed physical optics terms on the right of (8.34) through (8.36) also become singular, but with the opposite sign. Therefore, the singularities of the primed terms cancel those of the unprimed terms, and Mitzner's coefficients are well behaved in the transition regions while Michaeli's are not. Thus, just as in the application of the physical theory of diffraction, the results describe the diffraction from an edge, and *only* an edge; surface scattering (from the surfaces that meet to form the edge) must be handled by some other means, such as geometric optics or physical optics.

Mitzner's ILDC seems to overcome most of the limitations of earlier (and later) attempts to characterize diffraction from an edge. The scattering direction need not be confined to the Keller cone, the diffraction coefficients are well behaved in the transition regions, and there are no caustics to contend with. This is not to say that further refinements are unnecessary, however. How does one handle, for example, the corners of a simple structure like a flat plate? Although plate edges can be constructed of hundreds of tiny edge elements pieced together end-to-end, and Mitzner's incremental result integrated along all of the illuminated edge, the corners represent abrupt changes in the edge geometry. It must be remembered that all of the results cited thus far—from Keller and Ufimtsev to Mitzner and Michaeli—are based on the canonical wedge result, *which has no corners, only an edge*. Thus, the fields diffracted by the corners of any edged structure cannot be accounted for in these theories. Moreover, the surface traveling wave is not automatically included in these prescriptions, although it can be treated by means of a special application of one of them.

Finally, singularities appear in the diffraction coefficients of all four theories, and special efforts are made in two of them to exploit the cancellation effect available by the application of the simple theory of physical optics. The singularities can be traced to the fact that all four theories rely on the results of a single canonical problem: diffraction by an infinite wedge. When we excite an infinite surface with the infinite power of a plane wave, we must expect infinite fields *somewhere*. The canonical problem of the wedge is useful because the solution is available, but the extension of a tractable two-dimensional solution to the three-dimensional world comes not without a price.

8.4 CAN YOU REDUCE IT?

8.4.1 The Four Methods of RCS Reduction

There are only four ways to reduce the radar echo of an object:

- shaping;
- radar absorbing materials;
- passive cancelation;
- active cancelation.

Each involves a different philosophic approach and exploits different aspects of the encounter between the radar and the object. To exploit shaping, we must be assured that the radar threat will appear within a definable and limited cone of directions. Radar absorbing materials are typically used when shaping is not a useful or implementable option. Deep reductions in radar echo are possible by the use of passive cancelation, but only over narrow aspect angle sectors and for narrow bandwidths. Active cancelation depends on the installation of energy sources on the object and requires careful control of the reradiated energy. Of the four, only shape control and radar absorbing materials have been of any tactical use. We briefly consider each of these options below.

The objective of shaping is to orient target surfaces so that they reflect an incident wave in directions other than back to the radar. This is more difficult than one might suspect, for several good reasons. First, it implies that one has a reasonably good idea of where the radar is likely to appear. While this may be the case in some tactical encounters, it is definitely not the case for all. Second, shaping is effective only if one is prepared to sacrifice higher echo signal levels over some aspect angles to gain lower echo signals over critical threat sectors. This is because a reorientation of the surfaces of an object for favorable reductions over the specified sector is inevitably accompanied by an increase in the RCS over other sectors. Third, it is seldom possible to change the surface profile of a tactical vehicle without impairing the performance of the vehicle in its intended mission.

Almost every major system placed on either the military or civilian market is the result of a management decision to manufacture a product and to sell it (or field it) at a profit. Because of the large investment involved, that decision is reached only after a detailed market study shows that a profit is possible and likely. In the case of military products, profit is not measured in dollars, but in terms of some other standard, such as an expansion of the national defense capability or the improvement of one's bargaining position in strategic international negotiations. No matter how profit is measured, however, the product must be defined, a concept or configuration evolved, the concept verified by extensive testing, a prototype developed and engineered, and production eventually undertaken. Each stage of this development requires a major investment of resources, and unforeseen defects in the product at any stage of development are costly to remedy.

Therefore, whether it be a toaster, an automobile, or a space vehicle, it is difficult to change the final configuration of the product once it is in production. Indeed, changes of any kind are painfully costly; the replacement of faulty fuel cells in a single spacecraft may not be much different than the recall of 200,000 automobiles for the replacement of a defective $10 accelerator cable. The costs could well be comparable, but in either case the cost might have been avoided had the prototypes been more fully tested. Consider, then, the cost of adding a radar deflecting plate here, or rounding a corner there, simply to reduce the echo of the product over some aspect angle sector. The chances of this happening are small, primarily because of the cost of the change itself.

While some changes might be worth the cost, systems are designed to meet specific requirements. Unless RCS specifications are included in the mission requirements of the product, shaping is not a useful RCS reduction measure. Thus, shaping is not a practical alternative for systems already in production or in service in the field. This method of RCS control is best implemented in concept definition phase of system design.

The same is true of radar absorbing materials, although many people would believe otherwise. While commercial radar absorbers of several different varieties have been available for decades, they are not necessarily suitable for after-the-thought application. Many of them are not rugged enough for the environment they would be used in, some are too heavy to be practical, most are difficult to apply to doubly curved surfaces, and all are difficult to maintain in the field. Nevertheless, like shaping, if the RCS requirement is established in advance of product development, radar absorbers can be a very useful and practical way to reduce radar echoes.

As the name implies, radar absorbers depend on the absorption of incident energy, thereby reducing the amount reflected back to the radar. Some absorbers depend on the ohmic loss of energy in carbon fibers or particles embedded in a lossless binder, while others depend on a kind of magnetic hysteresis. Although carbon is classed as a semiconductor, its conductivity is too high to be useful in the bulk form in an absorber. The effective conductivity is reduced by dispersing carbon as particles or fibers on the surfaces of a nonconducting matrix. The particles must touch one another so as to form long conducting paths, and if the paths are broken up, for example, by physical flexure, the material loses its effectiveness.

The lossy element in magnetic materials is usually some form of iron, the cheapest of which are carbonyl iron and iron ferrites. Other compounds and alloys have also been used, the more common of which have been nickel and zinc. The desirable magnetic properties of these materials fade away when the frequency becomes too high, and their performance virtually disappears when the ambient temperature rises above the Curie point. Carbonyl iron oxidizes before it reaches the Curie point, and is therefore temperature limited. Ferrites, on the other hand, can withstand higher temperatures because they are oxides to begin with. All the magnetic materials are heavy, which forces one to use them carefully, especially on airplanes. They are attractive for suppressing the surface traveling wave, however, because relatively thin layers may be used.

Passive cancelation exploits the fact that the amplitude and phase of the scattering from a pair of echo sources can be adjusted so that the individual contributions are nearly the same in amplitude, but opposite in phase, thereby canceling each other. The method is restricted to relatively small objects because there are simply too many opportunities for the returns from multiple scattering centers on complicated bodies to add up in phase, resulting in just as many reinforcements as there are cancelations. Thus, passive cancelation, if effective at all, works best for bodies that are not electrically large.

Passive cancelation may be implemented by machining slots in the surface of the candidate structure and installing loads (cavities) of the required characteristics within the body and connected to the slots. An alternative method is to attach loaded loops or dipoles to the body. It is usually possible to choose the structure and the required load to provide a null in the scattered field for a given aspect angle, and if the loads are purely reactive, the reduction achieved for one direction of incidence is usually accompanied by an enhancement in some other direction. Moreover, because of the necessity to achieve a 180° phase angle between a natural echo contribution and one deliberately introduced to cancel it, the canceling effect disappears as the frequency of the incident wave moves off the design frequency. Thus, passive cancelation has not proven to be a useful technique.

Active cancelation implies the radiation of a signal of the proper amplitude, frequency and waveform so as to achieve the same result as passive cancelation. As suggested by its name, the method requires an energy source, but tailoring the source to provide the desired signal characteristics is a very difficult task. A general application of the concept would require a "smart" system that knows the RCS pattern of the body over a range of frequencies. The system would have to be adaptive enough to radiate a signal of the required characteristics in the required directions. It is doubtful that any truly adaptive systems have been constructed that can accomplish this. The cost of designing and building active cancelation systems with the required characteristics is probably high enough that more conventional methods will remain attractive, despite their shortcomings.

8.4.2 Shaping as a Means of RCS Reduction

The effective use of shaping to reduce the echo characteristics of a target depends on several factors, many of which involve the mission of the vehicle. It is generally accepted that a change in the shape or surface profile of a vehicle will reduce the effectiveness of the vehicle, and in many cases, the change is not acceptable. The intakes of jet engines, for example, are large echo sources in the nose-on sector, and the echo can be reduced if the intake aperture is shielded from the incident wave by a conical screen installed over the aperture. The tip of the cone should face forward, and the mesh size of the screen should be less than one-tenth of the wavelength of the incident wave. Intake air can pass through the screen, but incident radar waves are deflected away from the radar. Therefore, the intake screen would appear to be an acceptable treatment for a jet engine duct.

This is not the case, however. As usually happens when we fail to consult experts from other disciplines, a solution that seems apparent to us is unacceptable when other features of the design are considered. The aerodynamicist will quickly point out that the intake has been designed to meet specific performance requirements, and although air will indeed pass through the screen, the screen will block enough air flow to significantly impair engine performance. The degradation will impair the mission capability of the aircraft to the degree that the screen is an unacceptable echo reduction method. This simple example holds for a surprisingly wide variety of cases.

Thus, shaping or configuration changes are not always easy to exploit, whether the target of interest is a real one or only a design concept. And in those cases when it can be safely applied to a vehicle, it is useful only if one can identify specific threat sectors. Because a change in the shape of an object has the effect of shifting echoes from one angular location to another, shaping is useful only if one is prepared to accept higher echo levels in one sector in return for reduced levels in another. If the possibility of the appearance of a threat is equally likely for all directions, then shaping is not a useful echo reduction method and should not be considered.

This shifting of echoes from one aspect to another can be illustrated by comparing the returns of three simple targets already mentioned earlier; the flat plate, the right circular cylinder and the sphere. Let us assume that the cylinder and sphere are just large enough to enclose a square flat plate, as suggested in Figure 8.35. The radar cross sections of the three objects are

$$\sigma = \begin{cases} \dfrac{l^2}{\pi}\left[kw\cos\theta\,\dfrac{\sin(kw\sin\theta)}{kw\sin\theta}\right]^2 & \text{(flat plate)} \quad (8.37) \\ kal^2 & \text{(cylinder)} \quad (8.38) \\ \pi a^2 & \text{(sphere)} \quad (8.39) \end{cases}$$

where l is the vertical dimension of the plate or cylinder, $w = l$ is the width of the plate, a is the radius of the cylinder or sphere, and θ is the aspect angle shown in Figure 8.35. To just enclose the square plate with the cylinder or sphere, their radii must be $a = l/2$ and $a = l/\sqrt{2}$, respectively. Note that the echo strengths of the cylinder and sphere are independent of the angle θ because both objects are rotationally symmetric in the plane of the measurement.

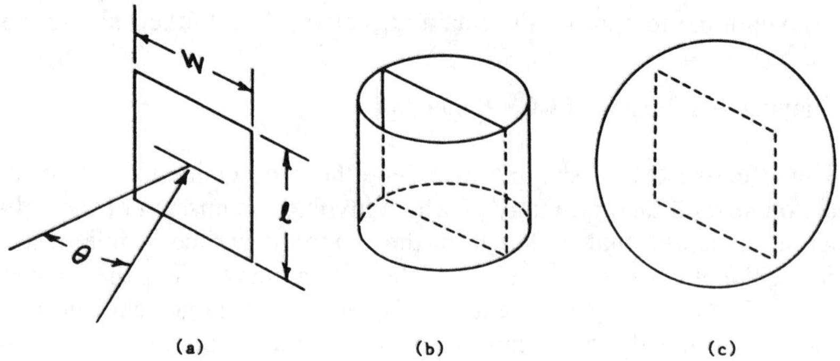

Figure 8.35 Radar cross section of a flat plate (a) can be reduced by replacing it with, or enclosing it by, a cylinder (b) or a sphere (c).

The plotted results shown in Figure 8.36 are normalized with respect to the square of the plate length. The RCS pattern of the plate attains its maximum value at broadside incidence and the sidelobes of the pattern trail off to either side of this peak. The amplitude falls off by more than 30 dB as the aspect angle swings out to 15° on either side of the broadside return, making it clear that a substantial reduction could be achieved if the plate were never presented to the radar at angles within 15° of the broadside echo. The cylinder and sphere returns, by contrast, are constant in this plane, with their echoes being 20 and 30 dB below that of the square plate, respectively. Substantial echo reductions are clearly possible for certain target orientations.

The reductions, however, are accompanied by enhancements. Enclosing the plate with a cylinder, for example, does indeed reduce the main lobe and the first two sidelobes, but the RCS of the cylinder is larger for all aspects outside this range. Although the pattern in Figure 8.36 does not reveal the return from the sphere because of the limited aspect angle window plotted, it will exceed that of the plate by several dB in the edge-on sector. Judging whether these enhancements are worth the reduction achieved in the broadside region depends on other factors,

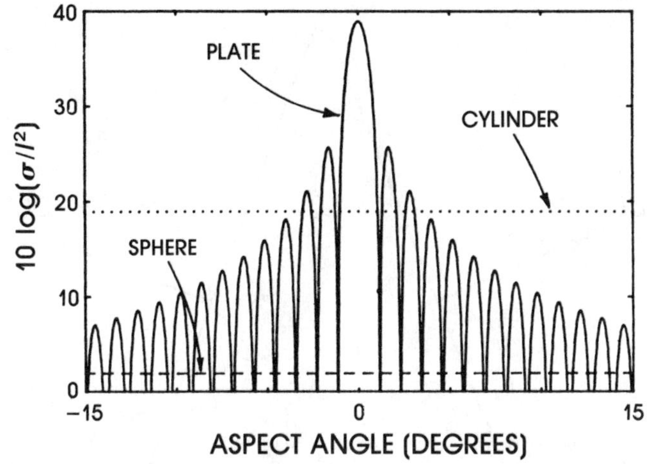

Figure 8.36 RCS patterns of a flat plate, a cylinder, and a sphere.

of course. Those factors must be evaluated individually for specific targets, because the evaluation criteria vary from one system to the next.

The fact that shaping enhances the echo over some aspect angles while reducing it over others is sometimes a dilemma: Is it better to have a lower RCS over wide angular sectors or a higher RCS over narrow angular sectors? The answer depends on the vulnerability of the particular system of interest, and probably cannot be given without a detailed assessment of the threat. It is altogether possible in some cases that a few intense spikes may be tolerable in the RCS pattern if they are narrow and lie far enough from sensitive directions.

Aside from the dichotomy of whether narrow spikes may be tolerated over given angular sectors, we must decide whether the indicated shape is practical. An example is the threat to ships at sea, which may be from other ships or from aircraft. Given that ships are always at sea level, that aircraft are limited to finite altitudes, and that the ocean is vast, the threat is likely to appear within a degree or two of the horizon, but at almost any bearing (aspect angle). Because vertical surfaces of ships form dihedral corner reflectors with the mean sea surface, the tilting of all vertical surfaces inward a few degrees would spoil the corner effect and bring about an echo reduction for threats near the horizon. Rotating those tilted profiles about a vertical axis would presumably generate a surface that would be effective for threats appearing at any azimuth angle. The resulting shape, however, is a collection of upward pointing cones, the simplest of which is shown in Figure 8.37. While this shape might be the optimum profile of a ship from the radar detection standpoint, we may be assured that the US Navy will never build one like it.

The primary threat sector for fast moving aircraft is not the entire azimuth sector, and in many cases the US Air Force identifies instead a forward cone of angles. The reason is presumably that threats appearing to the side and rear of the aircraft have a much lower probability of intercepting the aircraft than those appearing to the front. Restricting attention to one kind of echo—the leading and trailing edges of the wings—we may select any one of the shaping concepts shown in Figures 8.38 through 8.42.

Figure 8.38 shows what might be a conventional planform for a subsonic airplane. Note that the leading edges are swept back slightly and that the trailing edges are perpendicular to the longitudinal axis of the airplane. Assuming that the radius of curvature of the leading edge is not less than a few wavelengths at the lowest anticipated threat frequency, we can expect the leading edges to be dominant scatterers in directions perpendicular to the edge. If those directions

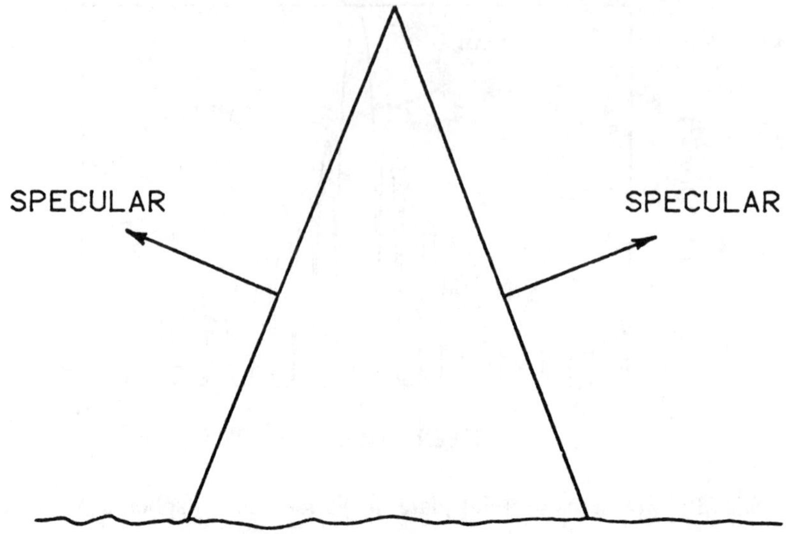

Figure 8.37 Optimum, if impractical, ship profile.

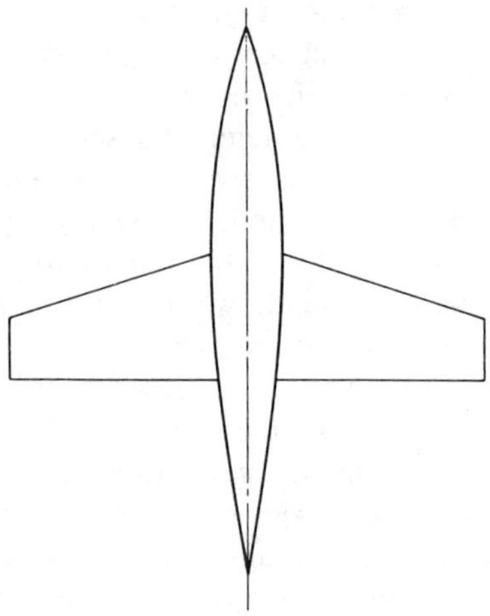

Figure 8.38 Wing planform for generic subsonic airframe.

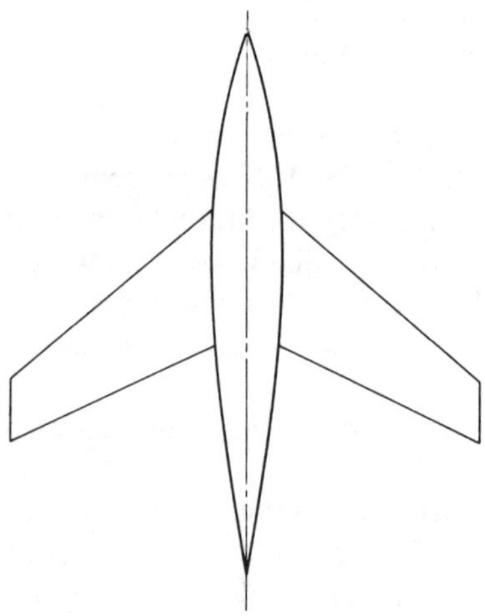

Figure 8.39 Planform with greater sweep angle.

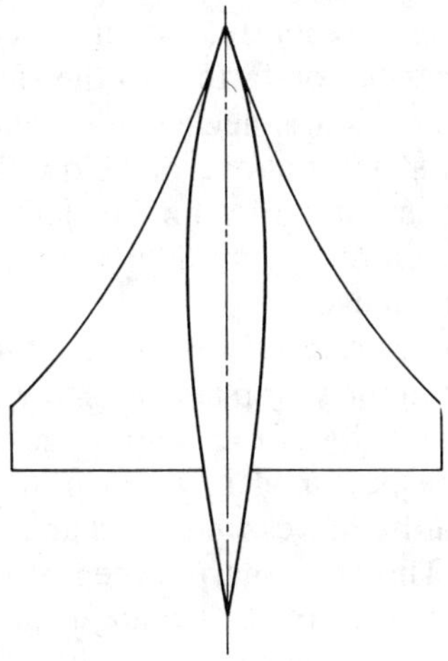

Figure 8.40 Curved leading edge at a greater sweep angle.

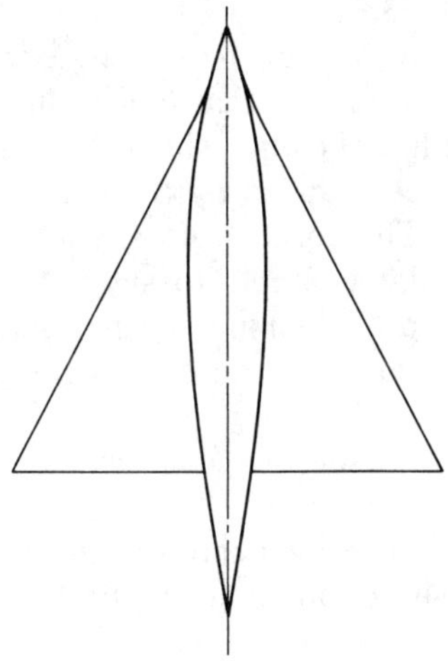

Figure 8.41 Delta wing.

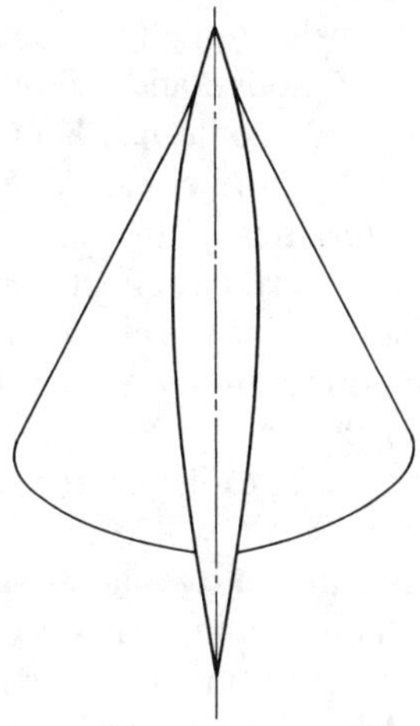

Figure 8.42 Modified delta wing with rounded corners.

fall in the forward threat sector over which reduced echoes are desired, the leading edges may be swept back more sharply, as suggested in Figure 8.39, thereby shifting the leading edge contribution farther to the side.

The trailing edge contribution is negligible for incidence in the yaw plane, and for incidence at angles slightly above or below the yaw plane, the trailing edge can be a strong scatterer, primarily for incident electric polarizations perpendicular to the edge. Therefore, Figure 8.39 shows the trailing edge swept to the rear as well as the leading edge.

The intensity of the echo from the leading edge may be reduced by giving the leading edge a curvature in the yaw plane, as shown in Figure 8.40. However, while this may reduce the intensity of the echo, it spreads the leading edge echo out over a broader range of aspect angles, in much the same way as the cylinder echo of Figure 8.36, which, although less intense than that of the flat plate, extends over a wider aspect angle. The very sharp sweep of the leading edge shown in Figure 8.41 increases the intensity of the leading edge echo because the edge is longer, but the edge contribution appears well out of the forward sector. Note that because the trailing edges in Figures 8.40 and 8.41 are perpendicular to the axis of the airplane, trailing edge contributions appear very close to nose-on incidence for aspects slightly above and below the yaw plane, as they will for the generic planform in Figure 8.38.

We could reduce the intensity of the trailing edge contribution by curving the trailing edge, for the same reason that the leading edge of Figure 8.40 was curved. The result might be like the edge shown in Figure 8.42. However, while the trailing edge echo might be reduced by the planform curvature, it would be spread out in azimuth angle. The corners of the leading edges of the wings have been rounded in this particular example to suggest a reduction in the surface traveling wave contribution from the long, straight leading edges. Each attempt to reduce the echo is accompanied by an enhancement in one way or another, and the designer will find himself testing several configurations in an attempt to arrive at an optimum planform using the analytical and experimental tools at his disposal.

The use of shaping to reduce the echoes from the fuselage of an airplane or missile requires a similar assumption about the probable direction of the appearance of a threat. If threats appearing at or near the broadside aspects are important, *chines* may be considered, as shown in Figure 8.43. The upper diagram represents a transverse section through the fuselage, and in the plane of the diagram, echoes will be strong in all directions, from the nadir up through the yaw plane to the zenith. If the vulnerable direction is near the yaw plane, chines may be added to the fuselage as shown in the center diagram. The configuration shown will reduce the echo for angles as much as 40° above and below the yaw plane because the specular reflection from the original profile has been eliminated by the surfaces that have been drawn out to meet at the chine. While there will be a reflection from the chine, it has the characteristics of an edge, which, as we have seen, are less intense than those of specular surface reflections.

Nevertheless, the lower surface of the profile in the center diagram offers the possibility of specular reflections in downward directions from the nadir to as much as 30° on either side of the nadir. One way to reduce that wide angular coverage in the downward direction is to flatten the bottom of the fuselage, as in the lower diagram of Figure 8.43. This will reduce the echo over most of the downward half-circle, but will enhance it over a narrow range of aspects centered on the nadir. We accept the enhancement in a narrow cone of downward directions as the price paid for a reduction over a much wider range of angles.

The flat bottom suggested by the lower diagram of Figure 8.43 has been included as one of a collection of shaping options illustrated in Figure 8.44. The sharp nose presents a very small echo in the forward direction, and the long, straight leading edges of the airfoil are swept sharply to the rear, pushing the edge

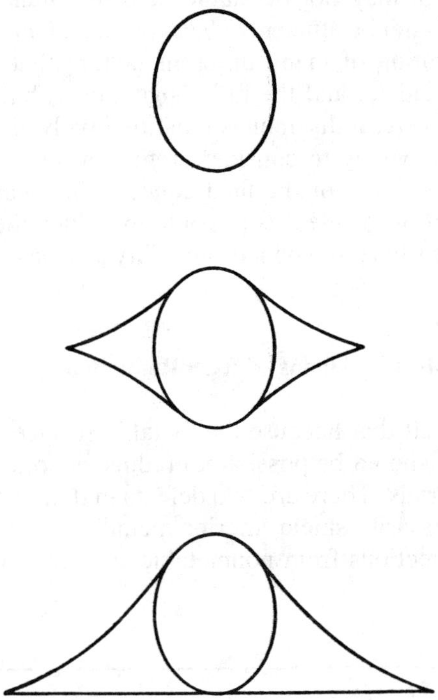

Figure 8.43 Fuselage shape treatments.

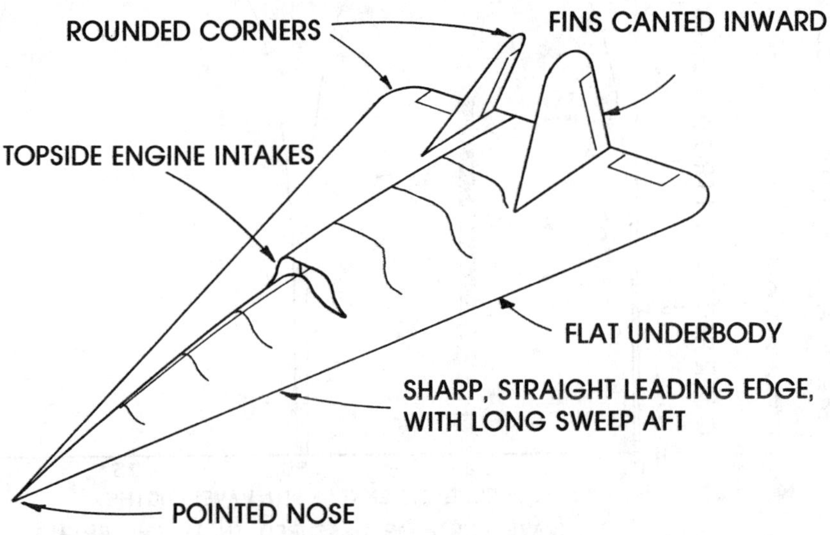

Figure 8.44 Some shaping options for an air-breathing vehicle.

reflections well out of the forward sector. The corners of the airfoils are rounded to minimize discontinuities that may give rise to traveling wave echoes, and the rear fins are canted inward to minimize their broadside returns. In the configuration shown, the engine intake is shown on the top side of the fuselage, minimizing the return from the duct and the lip of the duct for observation directions below the yaw plane. The overall configuration of the vehicle would be quite different, of course, had the primary threat been from above the yaw plane.

Being presented for illustrative purposes only, the shape of the vehicle in Figure 8.44 is by no means optimum. We have suggested a few ideas for the control of the echo properties only, ignoring important aerodynamic considerations in the process. Depending on the relative locations of the centers of pressure and gravity, the design may not be stable, and depending on the speed of the vehicle, the shape may not be efficient. The evolution of an optimum configuration requires the consideration of many different factors that influence the mission effectiveness of the vehicle, and the radar signature is but one of those factors. Experts representing several disciplines must be involved in the design process, and each one must be willing to consider compromises in performance in order to assure the overall success of the final concept. We leave this topic with the "triple-zero" observation: while it is possible to reduce the radar echo to nearly zero, a vehicle with nearly zero payload capability also has a utility value of nearly zero.

8.4.3 Radar Absorbents as a Means of RCS Reduction

It is commonly felt that because the metallic surfaces of man-made objects are good reflectors, it should be possible to reduce echo of an object by building it of nonmetallic materials. There are two defects in that observation: nonmetallic materials do not necessarily shield interior metallic structures from an incident wave, nor are the reflections from nonmetallic surfaces small. Figures 8.45 and 8.46 illustrate this.

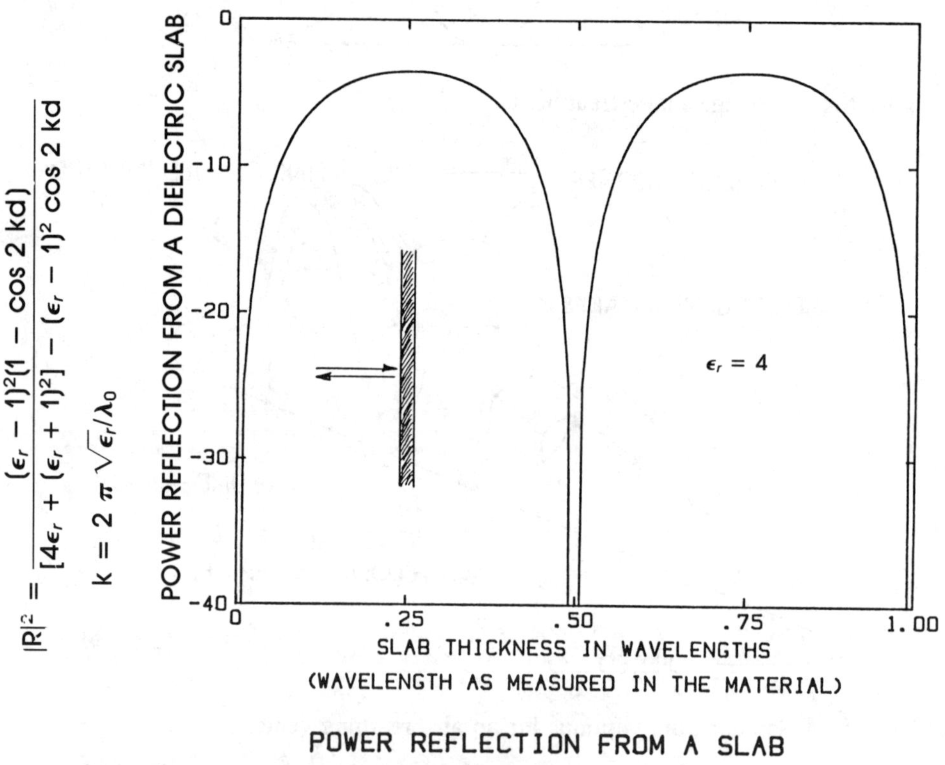

Figure 8.45 Power reflection from a dielectric slab.

Figure 8.45 shows the amplitude of the reflection from a dielectric slab as a function of the electrical thickness of the slab, as measured in terms of the wavelength in the slab material. The total reflection is the phasor sum of two reflections, one from the front face of the slab and one from the rear face. The net reflection rises to peak values when the slab thickness is an odd multiple of a quarter wavelength, and becomes very small when the thickness is an integral multiple of a half wavelength. Note that the maximum strength of the reflection is only 4.4 dB below that of a perfectly reflecting surface for this particular case (dielectric

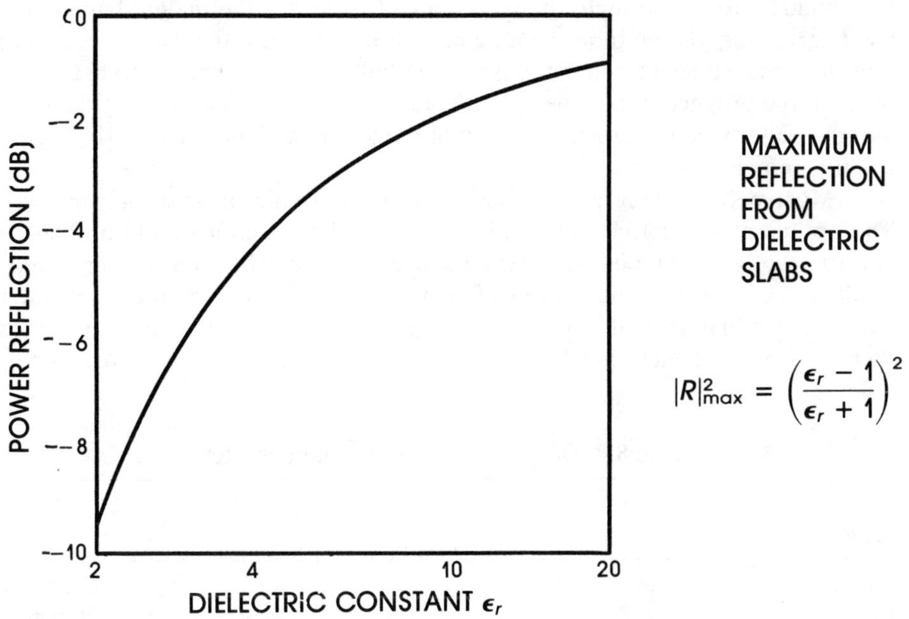

Figure 8.46 Maximum reflection from dielectric slabs.

constant = 4). This represents a power reflection coefficient of 36%, which is not much of a reduction.

Figure 8.46 is a plot of the maximum reflection coefficient (for slab thicknesses that are an odd multiple of a quarter wavelength) as a function of the dielectric constant. Note that the power reflection can be as much as 10 dB below that of a perfect reflector, but only for dielectric constants of the order of two or less. It is barely 3 dB down for a dielectric constant of six, which is probably typical of the electrical properties of composite materials that might be chosen for load-bearing structures. Thus, nonmetallic materials do not necessarily have low radar echo properties.

Nonmetallic materials with these properties are not effective for two reasons: their dielectric constants are significantly higher than that of free space and there is no loss mechanism present to soak up the electromagnetic waves. To be useful as radar absorbents, their dielectric constants—or more generally, their indexes of refraction—must have imaginary components. This is because the index of refraction, which is a measure of the propagation constant of the material, must have an imaginary component. It is the imaginary component of the index of refraction that describes the absorption of energy.

The imaginary part of the index of refraction is provided by all four components of electric and magnetic susceptibilities, because

$$k = \omega\sqrt{\mu\epsilon} = k_0\sqrt{\mu_r\epsilon_r} = \omega\sqrt{\mu_0\epsilon_0}\,\sqrt{\mu_r\epsilon_r} \qquad (8.40)$$

where ϵ_0 and μ_0 are the permittivity and permeability, respectively, of free space, and ϵ_r and μ_r are the relative permittivity and permeability of the material. The imaginary components of the latter represent a kind of *molecular friction* exhibited by the molecules of the material as they attempt to keep up with the polar reversals of the impressed electric or magnetic field. The electrical conductivity of the material is a dominant "friction factor" in the loss of a material.

The consumption of energy in radar absorbents due to conduction losses is quite analogous to the consumption of energy in an ohmic resistor: the currents induced in the material by the incident wave are in phase with the electric field, and power is extracted from the wave and delivered to the material. However, if the conductivity is too high, the wave cannot penetrate the material and deposit its energy. Nor, on the other hand, can it be too low, lest the wave enter and be reflected from some internal boundary essentially undiminished in strength. Thus, the material properties must be carefully tailored to first induce the incident wave into the material itself with a minimum of reflection, and then to absorb whatever is left as efficiently as possible.

Most metals are far too conductive to be of any value as absorbing materials. The optimum conductivity of a unit volume of the material should probably be approximately the admittance of free space (0.00265 mho), but as suggested in Table 8.4, even the conductivities of semiconductors like carbon and silicon are much higher than this. If one of these metals is to be used as the basis for loss, therefore, some means must be provided for "thinning" out the loss per unit volume.

Table 8.4 DC Conductivity of Common Metals

Metal	*Conductivity* (mhos/meter)
Silver	62,000,000
Copper	58,000,000
Gold	41,000,000
Steel	1,200,000
Nichrome	1,000,000
Graphite	71,000
Silicon	1,200

One way to spread out the loss is to combine the lossy element in a lossless matrix in such a way as to maintain conducting chains in the material. While the effective dielectric constant of mixtures is not easy to compute given the electrical properties of the components of the mixture, William Cuming has used the following guideline:

$$\log K'_{eff} = \sum V_i \log K'_i \qquad (8.41)$$

where K'_{eff} is the effective dielectric constant of the mixture, K'_i is the dielectric constant of the ith component, and V_i is the volume fraction of that component in the resultant mixture. The relationship in (8.41) is obviously logarithmic, and a plot of it for a two-component mixture of polystyrene and titanium oxide is shown in Figure 8.47 for illustration. The relationship, which applies to complex dielectric constants as well as real ones, suggests that the conductive properties of the mixture may be controlled via the relative proportions of the components that are combined to produce the final product.

In general, therefore, the lossy components of effective radar absorbing materials must be distributed throughout a lossless matrix in order to reduce the effective conductivity of the bulk material. Common lossy materials are carbon in any of its many forms, carbonyl iron and iron ferrites, and alloys of magnetic materials such as iron and nickel with other metals. The lossless binders or matrices used in these mixtures are typically urethane foams, natural and synthetic rubbers, animal hair, and other organic materials (plastics). Ferrites, on the other hand, may be sintered into hard tiles.

Probably the simplest form of radar absorber is the Salisbury screen, patented by W.W. Salisbury in 1952 [27]. The device described by Salisbury is a graphite-impregnated sheet of canvas stretched over a wooden frame in front of a metal backing plate. As suggested in the upper diagram of Figure 8.48, the distance between the lossy layer and the metal plate should be $\lambda/4$. In his patent, Salisbury notes that the lossy layer may also be fastened to one side of a sheet

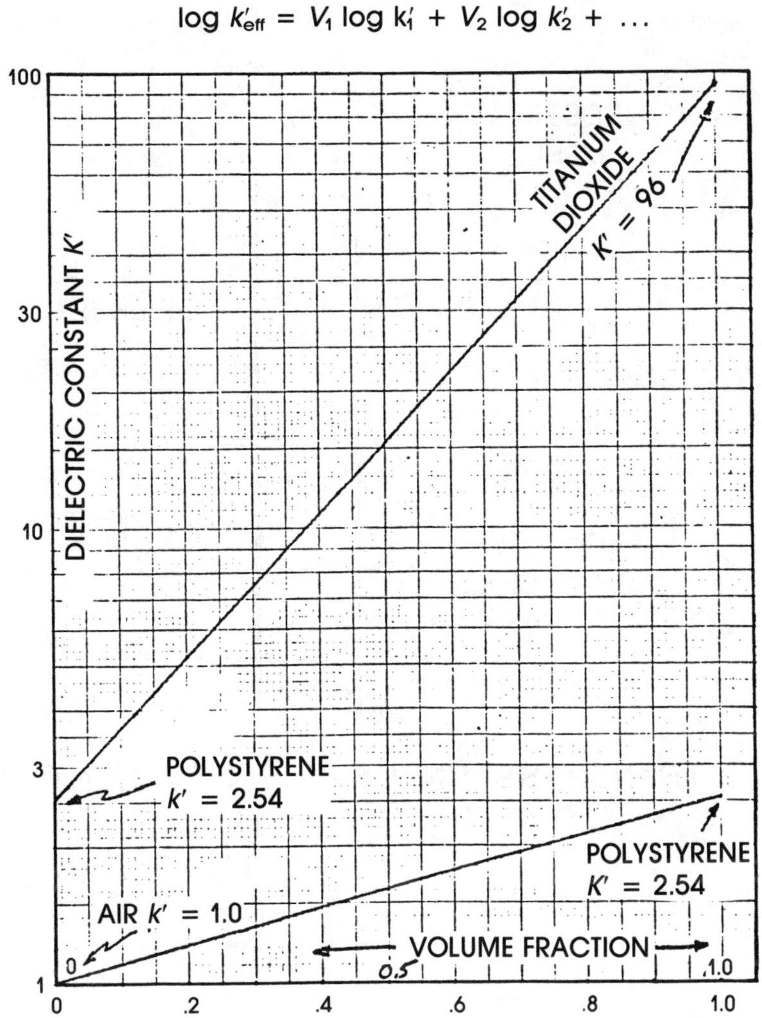

Figure 8.47 **A mixture law used by William Cuming.**

of a plywood sandwich in which one of the plies is a metal sheet. This produces a very sturdy structure, but the dielectric constant and thickness of the plywood layer in front of the metal sheet must be chosen for an electrical thickness of $\lambda/4$ in the wood. Salisbury filed for the patent in 1943, and the nine-year delay he experienced in being awarded the patent was no doubt directly related to the tactical value of his invention at the time.

E.B. McMillan patented a similar device in 1959, and he even included performance data for oblique angles of incidence [28]. Like Salisbury's screen, McMillan's absorber had a lossy layer at the front side of the material and a metallic screen at the back, but the spacing between the two was provided by a flexible layer of foam rubber. A flexible product known was MX-410 was based on this patent and marketed commercially for several years. Designed for operation at X-band frequencies, the material was barely 0.3-in thick. McMillan's patent data shows performances as good as −30 dB, and interestingly enough, that performance was achieved for an incidence angle about 30° off the normal. This may be because the test sample was slightly thicker than optimum for the frequencies used in the tests.

The operation of the Salisbury screen is best explained in terms of the transmision line equivalent shown in the lower diagram of Figure 8.48. The lossy screen is represented as a resistance attached to the transmission line a quarter wavelength in front of a short circuit representing the metal backing plate. We

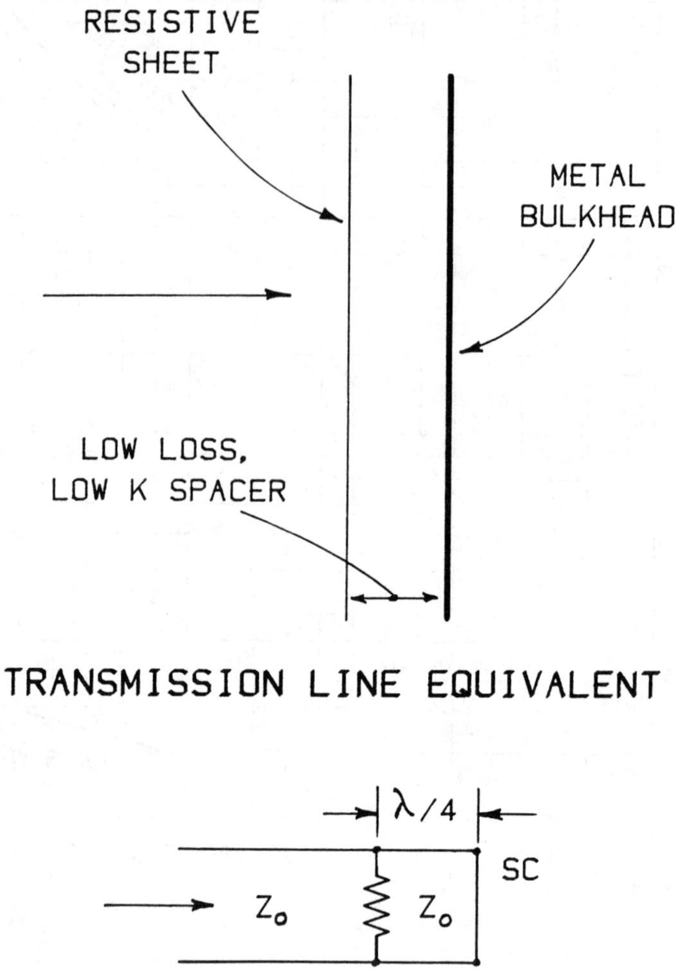

Figure 8.48 The Salisbury screen.

know from transmission line theory that the short circuit is transformed to an open circuit a quarter wavelength toward the generator, therefore the effective load is the resistance paralleled by an open circuit. If the sheet has a resistance of 377 Ω/square, the transmission line is terminated by a matched load (its characteristic impedance), and there is no reflection from the load.

This represents an ideal case, of course, and if the spacing between the resistive sheet and the metal backing plate is not maintained at an odd multiple of $\lambda/4$, or if the resistivity of the lossy sheet is not identically 377 Ω/square, there will be a measurable reflection from the Salisbury screen. This is illustrated in Figure 8.49 for a spacing of 0.5 in, which is a quarter-wavelength for a radar operating frequency of 5.9 GHz. The reflection coefficient for a 377-Ω sheet is very nearly zero at this frequency (off the bottom of the logarithmic chart), and rises to values of about −20 and −11 dB for sheet resistivities of 300 and 200 Ω, respectively. The performances of all three resistive sheets deteriorate as the frequency moves away from the design frequency of 5.9 GHz. Thus, the Salisbury screen has a limited performance bandwidth and, like most radar absorbers, the performance itself is sensitive to the properties of the materials used to construct the absorber.

The Dällenbach layer is another simple radar absorbing material. It is a single, homogeneous slab made of materials that may have magnetic and dielectric

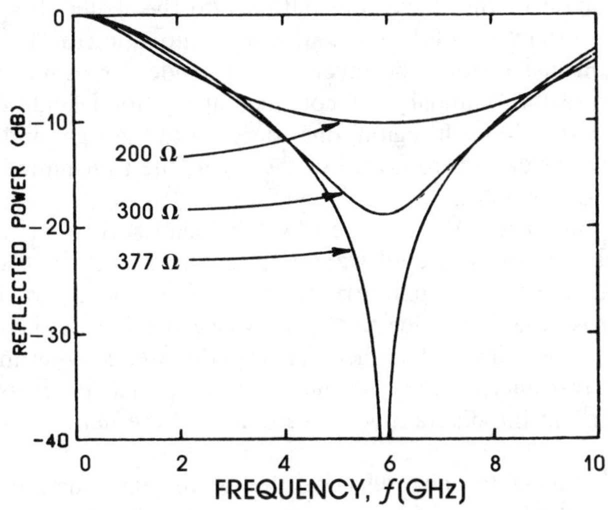

Figure 8.49 Performance of a Salisbury screen tuned for 5.9 GHz.

losses. The layer is typically backed by a metallic surface, and the losses of the material attenuate waves that travel forward in the material as well as those that are reflected and travel back toward the front. The design of Dällenbach layers involves a careful "tuning" of the material thickness and properties so as to achieve a cancelation between the reflections from the front face and the metallic backing layer. This is illustrated schematically by the amplitudes of the fields inside and outside the layer shown in Figure 8.50.

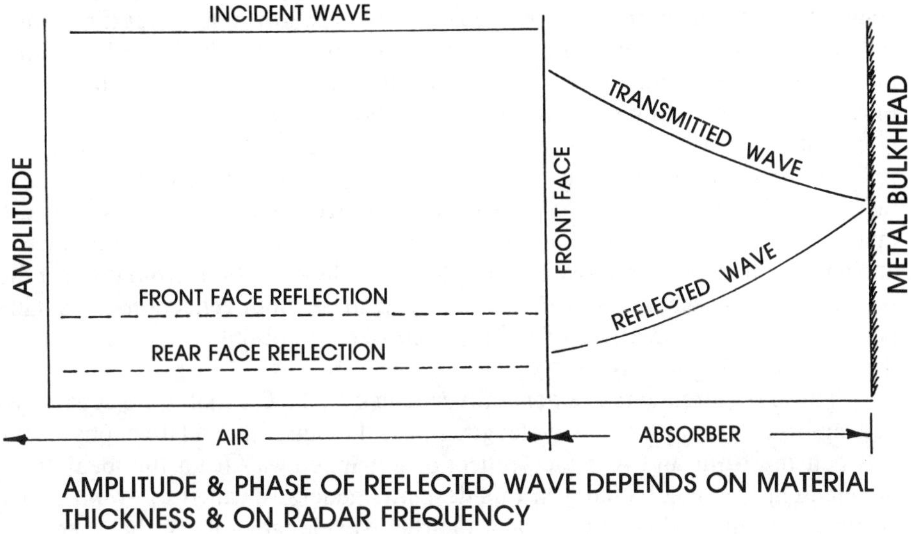

Figure 8.50 The Dällenbach layer.

The left side of the diagram represents the space (air) just outside the Dällenbach layer, while the right side represents the space occupied by the layer itself. The amplitude of an incident plane wave is indicated at the top left side of the diagram, and because air is essentially lossless, the strength of the incident field is constant to the left of the front face of the layer. Part of the incident wave is transmitted into the layer and part is reflected; the amplitude of the field reflected at the front face of the material is indicated near the bottom of the diagram on the left side.

The amplitude of the wave transmitted into the material is shown in the diagram to be less than that of the incident wave, accounting for the abrupt change in intensity from just outside the layer to just inside. The *total* fields must be continuous across the boundary, of course, but the total fields on either side consist of two waves in each region, one traveling to the right and one traveling to the left. Once inside the front face of the layer, the transmitted wave decays in amplitude due to material losses.

The transmitted wave is reflected by the metal surface at the rear of the absorbing layer, and the reflected wave is attenuated as it travels back toward the front face of the layer. Upon arriving at the front face, part of the wave is transmitted across the front face to the air space outside the layer and part is reflected back toward the rear surface. There is theoretically an infinity of such reflections and transmissions across the front face, each progressively weaker than the last, and in the diagram we have indicated the net amplitudes of all of them.

The amplitude of the net reflection due to the rear surface of the metal-backed layer is indicated as the constant level near the bottom of the diagram on the left side. In the air just outside the absorbing layer, therefore, the total field can be represented as the sum of the incident field, a field due to a reflection from the front face of the layer, and a reflection due to the reflection from the metal backing surface. The reflected waves travel to the left in the diagram and the incident wave travels to the right, therefore the total field will exhibit the characteristics of a standing wave pattern. It is, of course, the net reflected wave that interests us.

Because the net reflection from the Dällenbach layer is due to two mechanisms, the design of the layer must consider them both. While it is possible to choose materials that minimize the front face reflection, such materials are not generally useful absorbers. For example, a low-loss, low-density plastic foam can reduce the front face reflection to levels of the order of −40 dB or less, but the material will not attenuate the wave transmitted into the layer, and the rear face reflection will be strong. Conversely, a high-loss material can be chosen that attentuates the transmitted wave to insignificant levels by the time the wave is reflected and emerges back at the front face. However, the mismatch at the front face of such a material will give rise to strong front face reflections. Thus, the material properties required to reduce the reflections from the front face are altogether different from those needed to reduce the reflections from the rear face. An actual design should strive, therefore, to make both reflections as small as possible, with the thickness of the layer chosen, if possible, in such a way that the two are 180° out of phase.

These requirements are generally satisfied by layers that are electrically an odd multiple of a quarter-wavelength thick. Because the relative phase angle between the front and rear face reflections swings away from the ideal 180° as the incident frequency changes, this design is generally called a *resonant layer*. While layers any odd multiple of a quarter wavelength satisfy the cancelation requirement, we invariably choose the first multiple simply to make the material as thin as possible.

By way of illustration, Figure 8.51 depicts the performance of two Dällenbach layers, one of which has both magnetic and dielectric losses, and the other with dielectric losses only. While one material is four times as thick as the other, both are resonant at about 3.75 GHz. We emphasize that these are hypothetical materials with frequency-invariant properties chosen for illustration only. Materials can be found that have electromagnetic properties like those listed, but only over limited bandwidths.

The pure dielectric material whose performance is indicated by the upper trace is 0.2-in thick, has a relative dielectric constant of 16 and an electric loss angle of 20°. It is very nearly a quarter-wavelength thick at 3.75 GHz (as measured

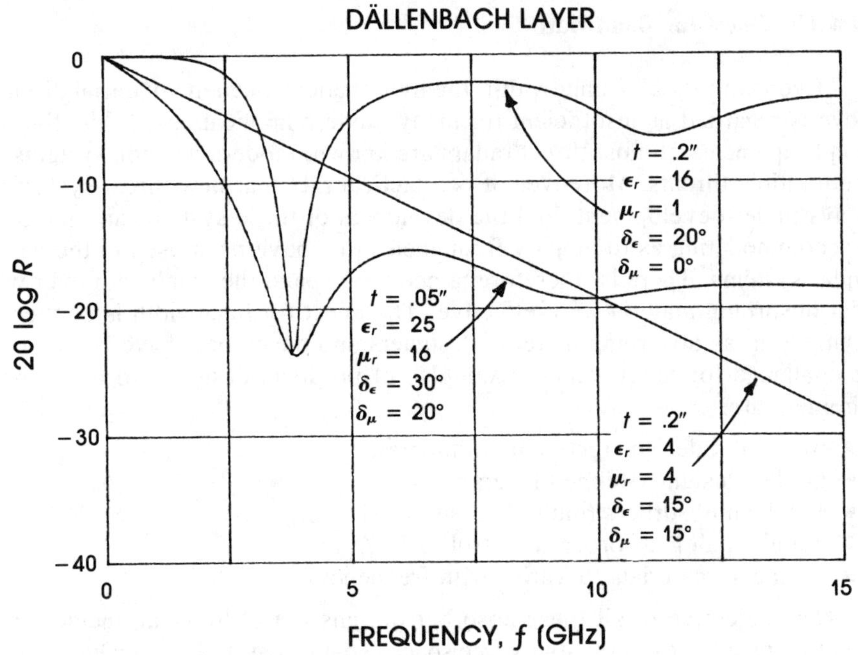

Figure 8.51 Comparison of a pair of hypothetical Dällenbach layers.

in the material), and at this frequency it has a -20 dB bandwidth of perhaps 5%. Note that it has a second "resonance" corresponding to a thickness of $3\lambda/4$ (near 12.5 GHz), but that the performance at this frequency is barely -6 dB. This is because the attenuation of the wave transmitted into the material and reflected back toward the front face is excessive, and the rear face reflection is too weak to be effective in canceling the front face reflection.

The material whose performance is illustrated by the lower trace may be considered a magnetic material even though it has dielectric as well as magnetic losses. It has a higher dielectric constant and more dielectric loss ($\epsilon_r = 25$ and a loss angle of 30°), and its relative permeability is 16 with a magnetic loss angle of 20°. Because the material's index of refraction is several times larger than that of the pure dielectric material, the magnetic material is only one-fourth the thickness of the pure dielectric material. Note that resonant performance is very nearly the same as that of the pure dielectric (about -23 dB), but it has about five times the bandwidth. Its second "resonance," if indeed it can be called that, occurs near 9 GHz, but at a performance level close to -19 dB, in comparison to the -6 dB level noted for the pure dielectric material.

More important is the fact that the performance of the magnetic material never rises above the -15 dB level for frequencies above about 3 GHz. Thus, the magnetic material is superior, being thinner and having better performance over wider frequency ranges than the pure dielectric material does. This is primarily because the intrinsic impedance of this material is closer to that of free space (0.8 *versus* 0.25), with the result that the front face reflection is lower. Indeed, the front face reflection of a hypothetical material having an intrinsic impedance of 1.0 is indicated by the diagonal trace running across the diagram. The front face reflection of such a material is precisely zero, and the reflectivity level shown in the diagram is entirely due to the metal backing behind the rear face. The performance of this fictitious material increases without limit as the frequency increases because the wave in the material traverses more wavelengths, and therefore suffers more attenuation, as the frequency increases. Scientists have yet to discover a substance with such broadband characteristics.

8.4.4 The Quest for Bandwidth

Even with its 25% bandwidth, the hypothetical magnetic material discussed above is regarded as insufficient for many modern applications. While the operation frequencies of most threat radars are known, modern weapon systems may face multiple threats. Moreover, it is a tactical fact that new threat systems are always under development, and the developers of those systems are not always so accommodating as to employ frequencies that have been used in the past for similar systems. There is, therefore, a need to expand the bandwidth over which radar absorbing materials are effective. The quest for bandwidth is neither new nor modern, as absorbing material designers and developers have been meeting the challenge for many years. Examples of products designed to meet the requirement are:

- pyramidal designs (geometrical tapers);
- graded designs (stepped layers);
- Jaumann absorbers (multiple resistive sheets);
- circuit analog absorbers (control of R, L, and C);
- magnetic materials (μ varies with frequency).

The objective of all these absorber designs is to "fool" an incident wave into thinking it "sees" essentially free space. That is, carefully designed absorbers present an impedance to the incident wave which is intentionally not much different from that of free space. Some of the absorbers rely on a variation of electrical properties while others rely on a geometrical variation. An example of the latter is the familiar pyramidal absorbers used in microwave anechoic chambers.

Pyramidal absorbers may be fabricated by the deposition of a lossy layer on the surfaces of cones or pyramids of lossless material, or by the infusion of lossy material throughout the bulk of a lossless matrix of pyramids. The latter are by far the more common. In either case, the tips of the pyramids should be pointed toward the source of the incident wave. The performance of pyramidal absorbers is generally better the longer the physical taper, and very long pyramids are required for acceptable performance at low frequencies; pyramids as long as 12 ft have been manufactured for applications in the UHF bands.

Although the pyramids used in current production have the familiar four-sided configuration, Tiley described six-sided pyramids in his 1949 patent [29]. Tanner was awarded a patent for flat and pyramidal absorbers in which the fibers of mat-like materials were coated with a rubbery carbon-laden coating [30]. Most of the pyramidal materials marketed today are made of carbon-loaded, open-cell plastic foams. The material must be painted with fire-retardant coatings to meet fire safety codes, which probably degrades the performance somewhat. If the absorber is handled and installed carefully, so as to preserve the sharpness of the pyramidal tips, extremely broadband performance (100 MHz to 100 GHz) can be achieved.

A physical geometry is usually easier to control than the electrical properties of an absorber, which is probably one reason for the good broadband performance of pyramidal absorbers. We can assemble a heuristic explanation of the performance of pyramidal absorbers by slicing a single pyramid along a plane perpendicular to the axis of the pyramid (parallel to the incident wave front). That slice will expose a square, transverse area of the pyramid which can be compared with the area of the base of the pyramid. If we weight the area of the slice according to the bulk admittance of the material, and the remaining area with the admittance of air (free space), the net admittance of the slice is the weighted sum:

$$Y = [A_m Y_m + (A_b - A_m) Y_0]/A_b \tag{8.42}$$

where A_m is the area of the slice through the pyramid, A_b is the base area of the pyramid, Y_m is the intrinsic admittance of the bulk pyramid material, and Y_0 is the admittance of free space.

The area A_m, and hence the admittance of the slice, increases with the square of the distance inward from the tip. Thus, the effective normalized admittance will decrease as the slice moves toward the base of the pyramid:

$$Y/Y_0 = 1 + (y_m - 1)(z/h)^2 \tag{8.43}$$

where y_m is the normalized admittance of the bulk material of which the pyramid is made, z is the distance inward from the tip, and h is the height of the pyramid. A plot of the normalized impedance of the slice (the reciprocal of (8.43)) is shown in Figure 8.52 as a function of position within the material. According to this heuristic picture, the impedance at the plane of the tips is the same as that of free space, and it decreases monotonically toward the base of the pyramid. The rate of decay depends on the bulk admittance (conductivity) of the material of which the pyramidal absorber is made, the higher the admittance, the faster the decay. The wave will be reflected by the discontinuity at the base of the pyramid, but if the pyramid is deep enough, the reflected wave will be very weak by the time it emerges at the front of the absorber.

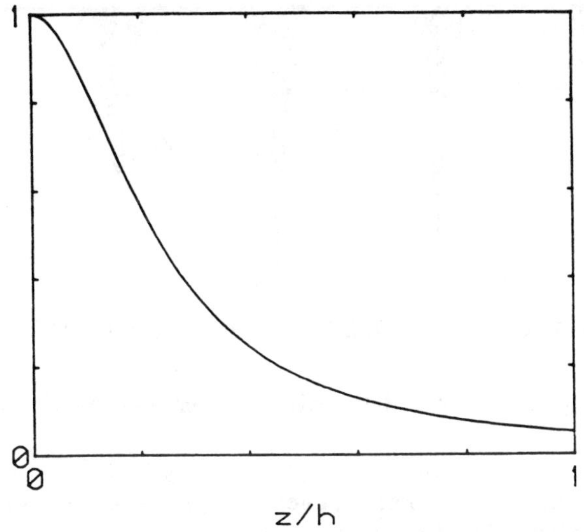

Figure 8.52 Normalized impedance of a pyramid as a function of distance from the tip.

If we are to "fool" the incident wave successfully into thinking that it "sees" free space as it propagates into the pyramid, the impedance change ought to be gentle. This in turn suggests that several factors are to be observed in the construction of the absorber:

- the bulk material of the pyramid should be as homogeneous as we could make it;
- the material should not be excessively conductive;
- the pyramids should be long and slender;
- the tips should be sharp and protected from damage.

The geometrical tapering concept can be applied to other structures as well, and dipped honeycomb is an example. Honeycomb is a common material used in the fabrication of lightweight structural panels and in multilayered absorbers. When used in structural panels, a honeycomb spacer between load-bearing skins serves to carry the shear stresses while the skins sustain the tensile and compressive loads; the honeycomb in these instances may be metallic. When used in the fabrication of absorbers, nonconducting honeycomb layers may be used as low loss, low dielectric constant spacers to separate thin lossy sheets. A tough, rigid Salisbury screen may be fabricated, for example, by replacing the flexible foam spacer of McMillan's resonant material by a layer of honeycomb material.

It is also possible to dip the honeycomb in a carbon-loaded bath, which, after drying, imparts some loss characteristics to an otherwise lossless material. It may then become just one more of the many materials available to the absorber designer. A measure of geometrical tapering can be provided in the process by successive dipping, with the submersion being progressively deeper with each dip. The thickness of the resulting coating will vary from one side of the honeycomb layer to the other, as suggested in Figure 8.53. The process probably ought to be considered only for relatively thick layers, because the taper in the electrical loss is less important for thin layers than for thick ones, and the process is costly. It is not known if special adhesives are required to bond dipped honeycomb structures to other layers in an absorber design.

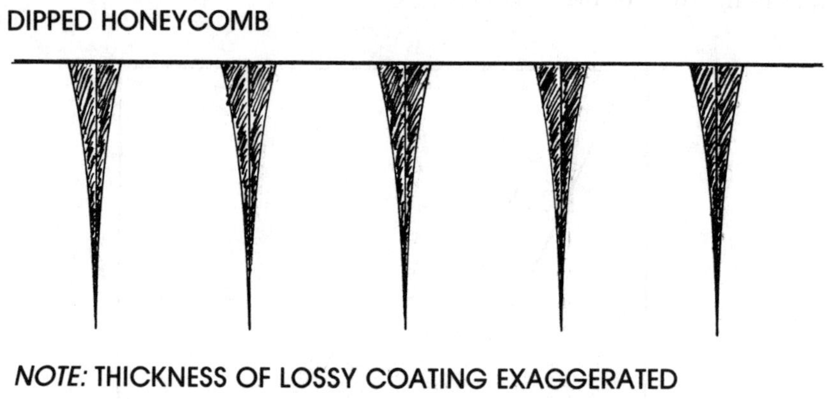

Figure 8.53 Dipped honeycomb can also be tapered geometrically.

While the change in intrinsic impedance from the front of an absorber panel to the rear should be as smooth and gentle as possible, this is not a requirement. Indeed, the design of any product, whether it be a pencil or an aircraft carrier, is a study in compromises, and cost is invariably one of the more important factors. The smooth, gentle variation in the effective impedance shown in Figure 8.52 results in high, broadband performance, but it is expensive and the material is very thick. The cost and the thickness both can be reduced by an order of magnitude by the use of the graded (stepped) design shown in Figure 8.54. Although one must expect a reduction in performance, the graded absorbers are reasonably effective for the price. The impedance characteristic is constant throughout any given layer, and the variation in impedance from the front of the absorber to the rear is achieved by bonding together several layers of differing characteristics. Despite the fact that the graded materials may have up to three or four layers, they are often flexible enough to be wrapped around curved surfaces. Typically made of flexible carbon-loaded urethane foams, they are widely used in many research projects, and may even be found in some permanent installations. Graded magnetic materials also have been manufactured, and have been used in research and in permanent installations as well.

The graded multilayer absorbers are nothing more than cascaded collections of Dällenbach layers, with the characteristics of each layer chosen to approximate an impedance taper from the front of the absorber to the back. It is not surprising,

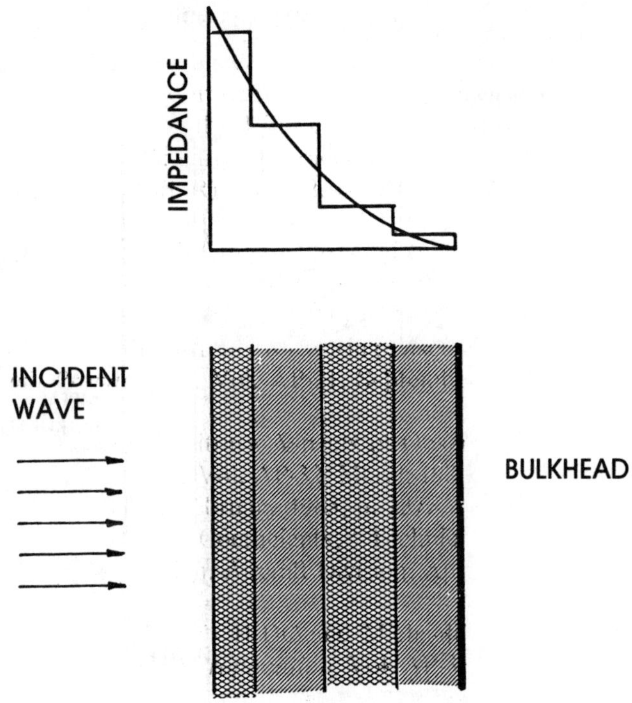

Figure 8.54 Graded absorbers approximate a smooth impedance change with series of fixed steps.

therefore, that multiple resistive sheets can be cascaded in much the same way, thereby improving the single-sheet Salisbury screen. This produces what is known as the Jaumann absorber depicted schematically in Figure 8.55, in which the sheets are separated by low-loss, low dielectric constant spacers. Optimum performance is obtained when the sheets are spaced $\lambda/4$ apart and when the sheet resistivity varies progressively from high values in the outer sheets to low values in the inner sheets. The transmission line equivalent of the absorber in the lower diagram makes it relatively easy to assess various design options for normal incidence.

Figure 8.56 shows how the bandwidth may be expanded by the use of multiple resistive sheets. The sheet resistivity for these multilayer absorbers varies quadratically (with the square of the sheet position) from the metal backing plate. The single-sheet design is the conventional Salisbury screen, of course, with a sheet resistivity of 377 Ω. The addition of successive sheets expands the bandwidth of the material, with four sheets turning in an impressive 300% bandwidth at the −20 dB level. As is always the case, one pays a price for this improvement in performance. First, because the structure is considerably more complicated than that of the conventional Salisbury screen, it is much more expensive to build. Second, because the layers are all spaced $\lambda/4$ apart, the absorber is thicker than the Salisbury screen. Nevertheless, the design concept is a useful solution of the quest for bandwidth. Note that the effects of bond lines and finite resistive sheet thicknesses have been ignored in computing the theoretical curves displayed in Figure 8.56; such effects should always be accounted for in the design of absorbers intended for production, of course.

Although the Jaumann absorber offers the theoretical promise of as much bandwidth as the designer is willing to pay for, the material eventually may become too thick to be practical, especially at low frequencies. A design concept that offers the possibility of reduced thickness is the circuit analog absorber, and the thickness reduction is achieved through a special design of the resistive layers. In contrast to the continuous coating that characterizes resistive sheets, circuit

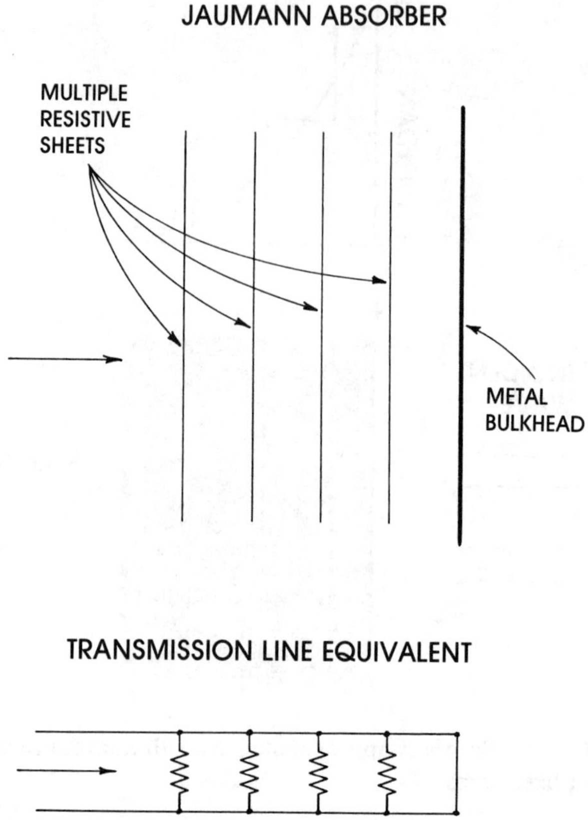

Figure 8.55 The Jaumann absorber is a collection of resistive sheets.

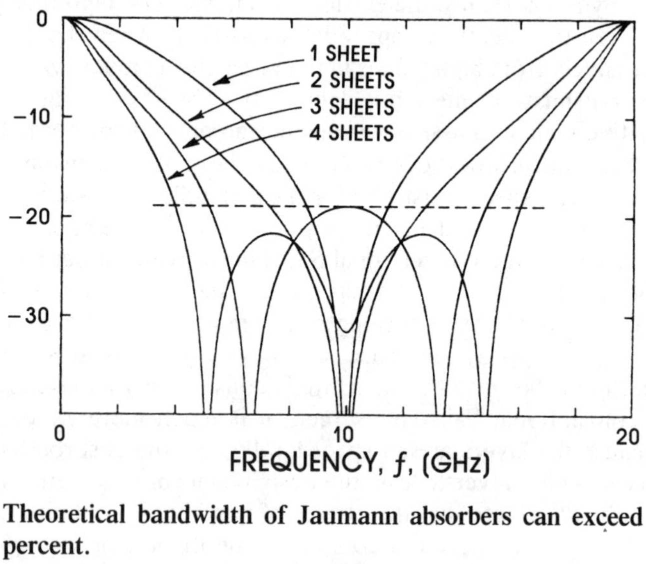

Figure 8.56 Theoretical bandwidth of Jaumann absorbers can exceed 200 or 300 percent.

analog sheets are made by the deposition of conductive geometric patterns on lossless films. The pattern is typically less than a wavelength in size, and is repeated in both directions across the sheet. Properly designed patterns impart reactive characteristics (i.e., inductance and capacitance) to the sheet in addition to pure resistance. In general terms, the thickness of the deposition controls the effective resistance of the pattern, the length (width) of the pattern controls the effective inductance, and the spacing between elements and the thickness of the film control the effective capacitance of the pattern. These characteristics allow one to "tune" the sheet in much the way one tunes an RLC circuit.

Like the Jaumann absorber design, several circuit analog sheets may be cascaded, as suggested in Figure 8.57, and the characteristics of the individual sheets may be varied so as to achieve the design objective. Again, a transmission line equivalent may be used to design the absorber, but in addition to the spacing between sheets and the resistance associated with a given sheet, one may also control the inductance and capacitance of a sheet. Devising a geometric pattern that yields the desired values of resistance, inductance and capacitance is a distinct and separate problem, on the other hand, but it has been accomplished with some success using Floquet mode expansions. One of the challenges in the design of the geometrical patterns is to obtain optimum performance for all incident polarizations and for wide variations in the angle of incidence.

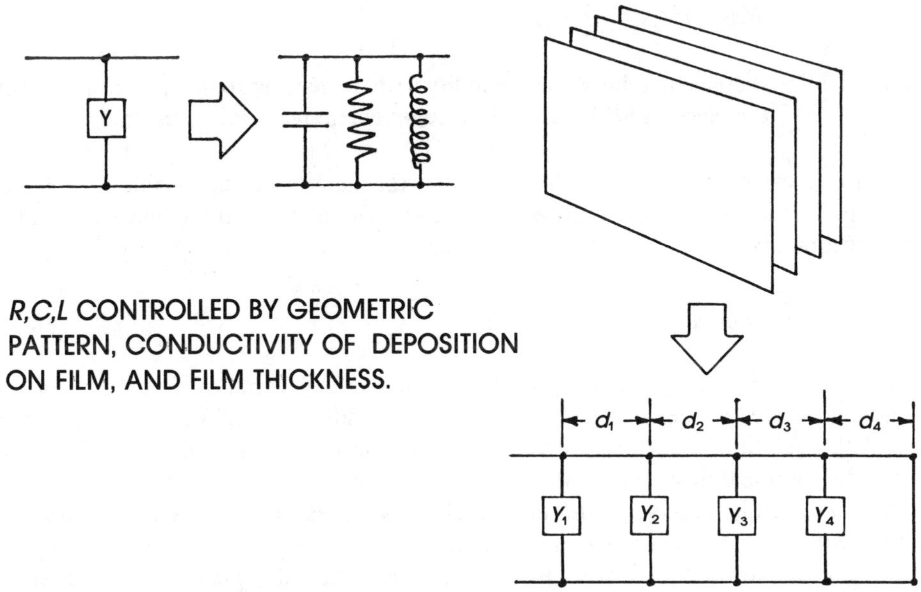

Figure 8.57 Circuit analog absorbers rely on thin sheets with geometric patterns deposited on them.

Pure dielectric radar absorbing materials work best when the material is installed where the electric field strength is a maximum, which is the reason the resistive sheet of the Salisbury screen ought to be spaced $\lambda/4$ away from a metallic backing plate. Absorbers that exploit the magnetic losses of compounds and alloys of iron and nickel, on the other hand, are to be placed where the magnetic field is a maximum. This is at the metallic backing surface itself, and at multiples of $\lambda/2$ off the surface. Consequently, magnetic materials tend to be much thinner than materials based on dielectric losses only.

While a homogeneous slab of magnetic absorber is another version of the Dällenbach layer, which, as we have seen, can be a resonant structure when the layer is electrically a quarter wavelength thick, the properties of some magnetic materials vary with frequency in such a way that the electrical thickness of the layer tends to be constant over a range in frequencies. This is illustrated in Figure 8.58. In this schematic diagram, the relative permeability rises with decreasing frequency, and often attains a peak at frequencies in the UHF band. Characteristics like these are desirable for two reasons: magnetic materials barely 1-cm thick provide reasonable echo reductions at frequencies for which non-magnetic absorbers would have to be ten times thicker or more, and the rising permeability

KNOTT
CHAPTER 8

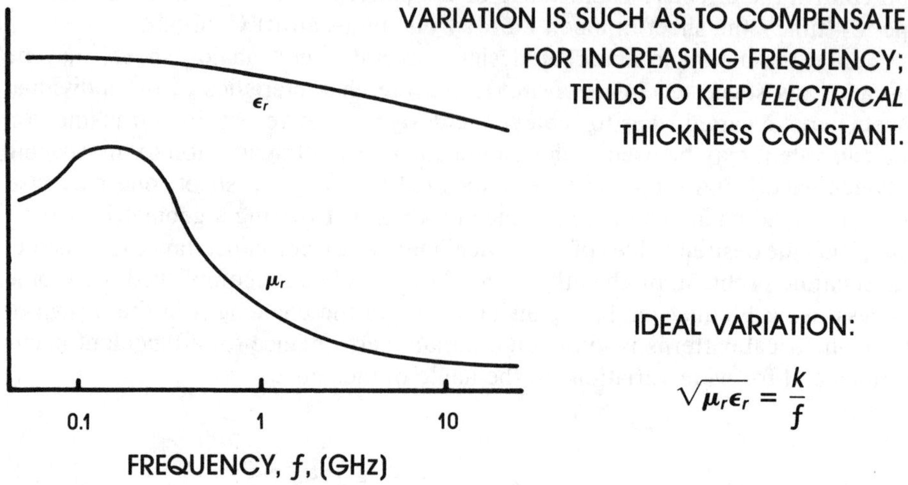

Figure 8.58 Increasing relative permeability with decreasing frequency tends to keep the electrical thickness of some magnetic materials constant.

with decreasing frequency extends the performance over a considerable bandwidth. The ideal variation that results in constant electrical thickness with changing frequency is

$$\sqrt{\mu_r\epsilon_r} = K/\omega \tag{8.44}$$

where K is a constant of proportionality and ω is the radian frequency of the incident wave. The left-hand side of (8.44) is the index of refraction of the material, and the relationship therefore includes the relative permittivity as well as the relative permeability. While no magnetic material has characteristics that follow the relationship precisely, the electrical thicknesses of some of them vary less than 25% over an octave frequency band.

An illustration of the low frequency capabilities of some magnetic materials may be found in the US patent of Grimes, *et al.* [31]. Among the information submitted by the inventors are reflectivity data for an iron ferrite. The power reflection from a sample only 6-mm thick is shown to be down more than 20 dB at 700 MHz, and at the −15 dB level, the material has a bandwidth of nearly 300%. This performance is remarkable for a material with thickness that is barely 1/50 of a free-space wavelength.

8.4.5 Civilian Applications of RCS Reduction

President Carter disclosed in August 1980 that the US had embarked on a program of stealth technology development, a fact known within only a few government agencies before the disclosure. The announcement aroused national interest in a technology that the existence of which was unknown to most, let alone its active development. Although the US population was well aware that radar systems were in widespread use in civilian life as well as in the military, it was not generally known that passive countermeasures were available to defeat radars. While we most certainly would not want to defeat navigation radars, or those used to control air traffic, police radars were another matter. Was there anything in this new technology that could be exploited?

The editors of *Car and Driver* magazine had been publishing articles on police radars for years, and probably more so than most, they also wondered what radar stealth had to offer the civilian sector. *Car and Driver* had kept its readers informed on several aspects of police radars in the past, ranging from accuracy and detection characteristics to "Fuzz Buster" detectors, and there was now a need to investigate radar stealth. Impelled by the new technology, Patrick Bedard published several articles on the subject.

Car buffs and engineering students can be particularly avid, and libraries were quickly and thoroughly researched for the latest developments on radar stealth. Unfortunately for many of us, articles on radar written by Bedard and others were soon razored out of Georgia Tech's stock of *Car and Driver,* and the account given below is based on but one of Bedard's articles [32].

Car and Driver had undertaken over the years to test several police radars, and had assembled a significant data base of radar performance and detection ranges. One particular list included the detection ranges of 14 different vehicles, but because that list appeared in one of the missing articles, we can cite only the three vehicles in Table 8.5. The detection range obviously increases with the physical size of the target, but it also depends on the "stealthiness" of the vehicle. On the basis of these tests and others, Bedard concluded that a Corvette is a "natural-born stealth car." The Corvette, it might be noted, has retractable headlights, smooth surface contours and stands very low to the ground. Until 1984, its radiator was tilted away from the vertical, a feature that some owners declared was responsible, at least in part, for the relatively low detectability of the vehicle.

Table 8.5 Detection Range of Three Vehicles

Allied moving van	9000 ft
Honda Civic	1200 ft
Corvette	600 ft

We have no idea how many entrepreneurs discovered the existence of radar absorbing materials, but at least three judged that there was a market for stealth kits based on absorption. *Car and Driver* listed the names of the kits ("The Barrier," "The Shadow," and "The Radar Eater"), and even purchased one for testing. The kits were essentially nothing more than stone shields lined with a commercial radar absorber, possibly one of the flexible, carbon-loaded urethanes mentioned earlier. If there is one feature that makes a product attractive, it is its universality, the fact that "one size fits all." The kit manufacturers exploited this appeal, and their products were mentioned with universality in mind.

As the reader may have discovered, however, one size usually fits none, and the kits did not smoothly conform to the profiles of the vehicles tested by *Car and Driver*. Bedard noted that they tended to wrinkle when installed, giving the car a "prune" look. We will not try to rate here the aesthetic value of the prune look, but the visual effect itself is a possible clue to the one of several reasons the kits may have turned in disappointing performances. First, if the absorbent layer used in the kit were one of those sold without a metallic backing, the air trapped under the wrinkles would detune the material. Second, commercial absorbing materials reduce specular reflections more effectively than any other kind of reflection. However, the forward profiles of some sleek automobiles present few specular surfaces to the incident police radar wave. The kits, therefore, would not likely be very effective on naturally stealthy cars. Third, the successful implementation of passive echo reduction is a battle against the inverse fourth-power dependence of the echo intensity on range. Because of the inverse fourth-power law, one must reduce the target echo by a factor of 16 in order to halve the detection range, all other factors being constant.

This implies two things: the absorber must cover all dominant echo sources on the vehicle, and it must have a performance of not less than 12 dB (10 log16). Apparently neither condition was satisfied in the *Car and Driver* tests, because Bedard noted that the detection range of a Mazda RX-7 was reduced by less than 100 ft. Assuming that the detection range of the untreated RX-7 was not less than that of the Corvette listed in Table 8.5, the kit either did not cover the dominant echo sources or if it did, its effective performance rating was a scant 3.2 dB (40 log 0.833). The former is the more likely reason. In any event, Bedard concluded that a 100-ft reduction in detection range at a price of $300 was "not a very good deal." We heartily concur.

As we have seen, however, absorption of the incident radar energy is not the only way to reduce the echo: shaping is another powerful tool in our bag of echo reduction tricks. In researching his article, Bedard found a Detroit engineer who was willing to discuss stealth, but who insisted that his identity be protected. The engineer, whom Bedard identified only as "*Deep Thwart,*" said that the detection range could be reduced by tilting the vehicle's radiator back by five degrees, an example of shaping discussed earlier. Significant amounts of energy apparently get past the grillwork of the car and illuminate the radiator, which behaves essentially like a flat plate. Tilting the plate (radiator) reflects the radar beam slightly upward instead of back to the radar.

However, Deep Thwart went on to explain a shaping device even more effective: a large metal shield installed in front of the car. Bedard does not describe the shield, but we presume it worked as suggested in Figure 8.59. It was probably tilted back a few degrees, thereby deflecting the radar beam away from the radar, and it was probably tall enough to hide the roof of the vehicle. According to Deep Thwart, the shield was effective: "The car pretty well disappeared," he said, "but it was hell to see where I was going." So, not surprisingly, the civilian world has discovered, like the military, that while stealth technology can be effectively developed and exploited, we must be prepared to pay the price. In the application cited by Bedard, it cost $300 to reduce the detection range by a scant 100 ft, and we must agree with him that it is "not a very good deal."

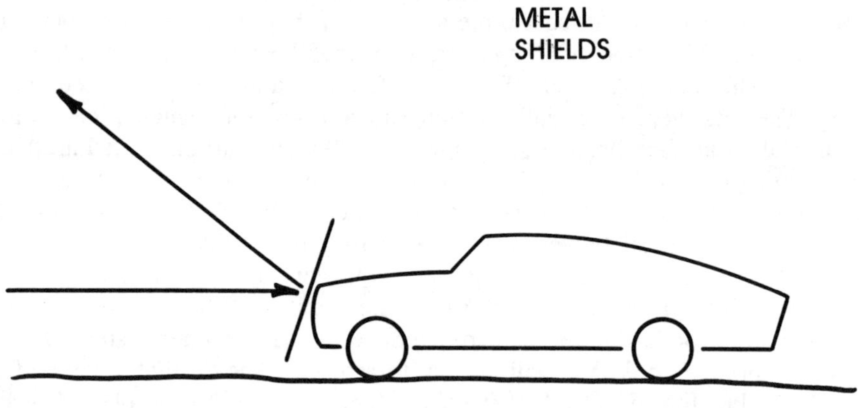

Figure 8.59 Metal shield for an automobile.

In all fairness, it must be pointed out that the $300 price must be weighed against the fine a speeder might be assessed in the event of his apprehension and conviction. If the fine is, say, $50 per violation, the $300 cost of a stealth kit is hard to justify (except possibly for elimination of the inconvenience caused by having to go to court and the possible increase in insurance rates resulting from the ticket). If the fine were to be $1000, the cost of the kit is much easier to justify. Finally, we emphasize that passive radar countermeasures are by no means the only tricks in the stealth bag. Radar jammers, while illegal in the civilian world, certainly have their place in military applications, and tactical countermeasures can be employed. Indeed, we know of at least two tactics guaranteed effective against police radars: observing posted speed limits and traveling on foot.

8.5 SUMMARY

This chapter has been an attempt to compress into a few pages a first-level response to three questions about radar cross section. Widespread interest in this general way of characterizing targets is attributable to the formal disclosure by the US government that it was committed to the development of stealth technology. The three questions were:

What is "radar cross section"?
Can you predict it?
Can you reduce it?

It was natural to organize the chapter, therefore, into three sections (plus introductory and concluding sections) that address these questions.

The reader was acquainted with the concept of RCS by means of measured and predicted examples of the characteristics of some simple and some complicated targets. Although an expression was derived for the RCS in terms of incident and scattered electric field strengths, the formal definition gives little insight into the characteristics of typical radar targets. Consequently the dependence of the echo strength on the size and shape of simple bodies, and on the frequency and direction of arrival of the incident wave, was summarized in a chart. Examples of the scattering by complex targets were presented in the form of measured results for a large surface ship, a model jetliner, insects, and domestic birds.

The second major section described the tools available for predicting the RCS of simple and complex bodies. The methods surveyed were not exhaustive, but they included exact and approximate formulations. The exact methods discussed attack the solution of the differential and integral forms of Maxwell's equations, and it was pointed out that neither kind of solution meets the need to predict the RCS of large complicated targets. Despite practical limitations in the exact methods, however, they are extremely useful in their respective areas of application, if only to give us an understanding of scattering mechanisms and a few test cases against which to gauge the results of approximate theories.

The approximate methods are all based on optical concepts, and two of them even have the word "optics" in their names. Of the six approximate methods discussed, four were described in terms of "diffraction coefficients," even though the developers of the theories may not have used such terms in their original developments. The first two diffraction theories were developed independently in the US and in the USSR at nearly the same time (in the mid-1950s). Respectively known as the geometrical and physical theories of diffraction, they allowed the analyst to predict the fields scattered by an edge anywhere on the Keller cone of diffraction. Somewhat later (in 1974 and 1984), Mitzner and Michaeli presented their theories, known respectively as the incremental length diffraction coefficient and the method of equivalent currents. Both theories eliminated the requirement that the point of observation be on the Keller cone, a significant improvement over previous diffraction theories. We concluded our discussion of the approximate methods of RCS computation by noting the remarkable parallel between the first and second pairs of edge diffraction theories.

In the third major section of the chapter we turned attention to the reduction of radar echoes. We emphasized that echo reduction is a study in compromises, and the price is usually high. Four methods of echo reduction were discussed, but only two of the four are of any practical importance at the moment. These are shaping and radar absorbing materials, and neither is an ideal solution. The shape of a target can rarely be changed significantly once the product is in production, and even during concept formulation, the shape or configuration of the vehicle can adversely affect its mission capabilities. The use of radar absorbing materials is only slightly less restrictive and, like shaping, is most effective when accommodated as a means of meeting the mission requirements during the formulation stage of the vehicle concept. We concluded our discussion of RCS reduction by summarizing the observations of a journalist on "stealth cars." We noted that civilian and military exploiters of radar stealth both agree that it is expensive.

REFERENCES

1. Stratton, J.A., *Electromagnetic Theory,* McGraw-Hill, New York, 1941, p. 281.
2. Bowman, J.J., T.B.A. Senior, and P.L.E. Uslenghi (eds.), *Electromagnetic and Acoustical Scattering by Simple Shapes,* North-Holland, Amsterdam, 1969, p. 7.
3. *Ibid.,* pp. 93 and 103.
4. Moffatt, D.L., and E.M. Kennough, "The Axial Echo of a Perfectly Conducting Prolate Spheroid," *IEEE Trans. Antennas and Propagation,* Vol. AP-13, May 1965, pp. 401–409.
5. Senior, T.B.A., "Axial Backscattering by a Prolate Spheroid," *IEEE Trans. Antennas and Propagation,* Vol. AP-15, July 1967, pp. 587–588.
6. Peters, L., Jr., "End-Fire Echo Area of Long, Thin Bodies," *IRE Trans. Antennas and Propagation,* Vol. AP-6, January 1958, pp. 133–139.
7. Chang, S., and V.V. Liepa, "Measured Backscattering Cross Section of Thin Wires," University of Michigan Radiation Laboratory Rep. No. 8077-4-T, May 1967.
8. Ruck, G.T., D.E. Barrick, W.D. Stuart, and C.K. Krichbaum, *Radar Cross Section Handbook,* Vol. 1, Plenum Press, New York, pp. 50–59.
9. Knott, E.F., J.F. Shaeffer, and M.T. Tuley, *Radar Cross Section: Its Prediction, Measurement and Reduction,* Artech House, Norwood, MA, 1985, p. 149.
10. Knott, E.F., and T.B.A. Senior, "CW Measurements of Right Circular Cones," University of Michigan Radiation Laboratory Rep. No. 011758-1-T, April 1973.
11. Knott, E.F., "RCSR Guidelines Handbook," Georgia Institute of Technology, Engineering Experiment Station, Rep. Contract No. N00039-73-C-0676, April 1976, AD A099 566.
12. Skolnik, M.I., *Introduction to Radar Systems,* McGraw-Hill, New York, 1980, p. 45.
13. Howell, N.A., "Design of Pulse Gated Compact Radar Cross Section Range," *IEEE G-AP International Program and Digest,* September 1970, pp. 187–195.
14. Hajovsky, R.G., A.P. Deam, and A.H. LaGrone, "Radar Reflections from Insects in the Lower Atmosphere," *IEEE Trans. Antennas and Propagation,* Vol. AP-14, March 1966, pp. 224–227.
15. Blacksmith, P., Jr., and R.B. Mack, "On Measuring the Radar Cross Sections of Ducks and Chickens," *Proc. IEEE,* Vol. 53, No. 8, August 1965, p. 1125.
16. Kouyoumjian, R.G., and P.H. Pathak, "A Uniform Theory of Diffraction for an Edge in a Perfectly Conducting Surface," *Proc. IEEE,* Vol. 62, No. 11, November 1974, p. 1459.
17. Ross, R.A., "Radar Cross Section of Rectangular Flat Plates as a Function of Aspect Angle," *IEEE Trans. Antennas and Propagation,* Vol. AP-14, May 1966, pp. 329–335.
18. Keller, J.B., "Diffraction by an Aperture," *J. Applied Physics,* Vol. 28, No. 4, April 1957, pp. 426–444.

19. Sommerfeld, A., "Mathematische Theorie der Diffraction," *Math. Ann.*, Vol. 47, 1896, pp. 317–374.
20. Senior, T.B.A., and P.L.E. Uslenghi, "High-Frequency Backscattering from a Finite Cone," *Radio Science*, Vol. 6, No. 3, March 1971, pp. 393–406.
21. Ufimtsev, P.Ia., "Approximate Computation of the Diffraction of Plane Electromagnetic Waves at Certain Metal Bodies: Part I. Diffraction Patterns at a Wedge and a Ribbon," *Zh. Tekhn. Fiz.* (USSR), Vol. 62, No. 8, 1957, pp. 1708–1718.
22. Ryan, C.E., Jr., and L. Peters, Jr., "Evaluation of Edge-Diffracted Fields Including Equivalent Currents for Caustic Regions," *IEEE Trans. Antennas and Propagation*, Vol. AP-17, March 1970, p. 275.
23. Knott, E.F., and T.B.A. Senior, "Equivalent Currents for a Ring Discontinuity," *IEEE Trans. Antennas and Propagation*, Vol. AP-21, September 1973, pp. 693–695.
24. Michaeli, A., "Equivalent Edge Currents for Arbitrary Aspects of Observation," *IEEE Trans. Antennas and Propagation*, Vol. AP-32, March 1984, pp. 252–258; *see also* correction in Vol. AP-33, February 1985, p. 227.
25. Mitzner, K.M., "Incremental Length Diffraction Coefficients," Technical Report No. AFAL-TR-73-296, Northrop Corporation, Aircraft Division, April 1974.
26. Knott, E.F., "The Relationship between Mitzner's ILDC and Michaeli's Equivalent Currents," *IEEE Trans. Antennas and Propagation*, Vol. AP-33, January 1985, pp. 112–114.
27. Salisbury, W.W., "Absorbent Body for Electromagnetic Waves," US Patent 2,599,944, awarded June 10, 1956.
28. McMillan, E.B., "Electromagnetic Wave Absorbing Dielectric Walls," US Patent 2,875,435, awarded February 24, 1959.
29. Tiley, J.W., "Radio Wave Absorption Device," US Patent 2,464,006, awarded March 8, 1949.
30. Tanner, H.A., "Fibrous Microwave Absorber," US Patent 2,977,591, awarded March 28, 1961.
31. Grimes, D.M., W.W. Raymond, R.J. Hach, and R.M. Walser, US Patent 3,938,152, awarded February 10, 1976.
32. Bedard, P., "Stealth Cars," *Car and Driver*, Vol. 26, No. 6, December 1980, pp. 140 *et seq.*

Chapter 9

THE DOPPLER WEATHER RADAR

by R.J. Doviak and D.S. Zrnić

9.1 INTRODUCTION

The capability of microwaves to penetrate cloud and rain has placed the weather radar in an unchallenged position for remotely surveying weather. Although visible and infrared cameras on satellites can detect and track storms, they cannot probe inside the storm's shield of clouds to reveal, as the radar does, the storm's internal structure and the hazardous phenomena that might be harbored therein. The doppler radar is the only remote sensing instrument that can detect tracers of wind and measure their radial velocities, both in the clear air and inside heavy rainfall regions veiled by clouds—clouds that disable lidars because optical radiation can be completely extinguished in several meters of propagation distance. This unique capability supports the doppler radar as an instrument of choice to survey the wind and water fields of storms and the environment in which they form.

In this chapter, radar principles for detection and ranging of discrete targets are summarized and extended to the more complex weather target, which is a conglomerate of discrete targets producing a continuous stream of echoes with rapidly fluctuating amplitude and phase. The origin of these fluctuations is shown, and the weather radar equation is developed. The reflectivity weighted doppler spectrum and its principal moments are related to the spatial distribution of wind and water. Finally, a few observations and analyses of weather radar data are presented. A comprehensive treatment of the techniques used in doppler weather radar and applications to observe both clear and stormy weather is given in the text by Doviak and Zrnić [33].

9.2 PRINCIPLES OF DOPPLER WEATHER RADAR

Figure 9.1 shows in block diagram the principal components of a simplified pulsed doppler weather radar. This is a schematic of a homodyne system in which there are no intermediate frequency circuits which are found in operational radars to improve performance. Nevertheless, this simplified homodyne radar illustrates the principle of a doppler radar.

The *stabilized local oscillator* (STALO) generates an extremely coherent *continuous wave* (CW) microwave signal, which is amplified by a pulse modulated klystron to produce a pulse of intense microwave power. This pulse is directed to the antenna by the transmit-receive (T/R) switch during the short 1 or 2 μs duration τ of the pulse. Pulsewidth τ is defined as the time between points having a power one-quarter of the peak (Taylor and Mattern [106]). Thereafter, the switch allows echoes entering the antenna to be directed to the mixer-detector, wherein the CW STALO signal is mixed (beat) with the doppler shifted echoes to produce

DOVIAK & ZRNIĆ
CHAPTER 9

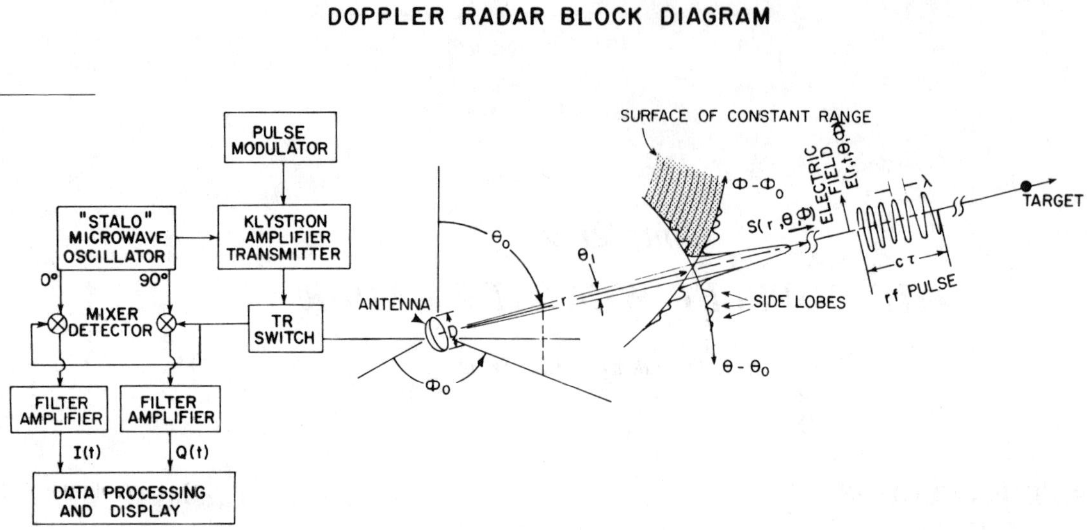

Figure 9.1 Simplified doppler radar block diagram (from Doviak and Zrnić [33]).

(as explained in Section 9.2.4) a sampled beat frequency in both the in-phase I and quadrature phase Q channels of the receiver. A train of microwave pulses is transmitted with spacings at the *pulse repetition time* (PRT) or pulse repetition interval (PRI), T_s, which corresponds to the sample spacing of the beat frequency signal. The pulse of microwave power density, $S(r,\theta,\phi)$ W/m^2, emanates from the antenna to illuminate hydrometeors, insects, and refractive index irregularities (targets) at range r as it propagates within a narrow beam, and a minute fraction of this radiation is scattered by these targets.

9.2.1 Beamwidth and Antenna Gain

The microwave pulse leaves the antenna in an essentially collimated beam of diameter D_a equal to that of the antenna-reflector. However, because of diffraction, the electromagnetic beam begins to spread at a range $r \approx D_a^2/\lambda$ into a conical one having a width θ_1, commonly specified as the angle within which the microwave radiation is at least one-half its peak intensity (3-dB width).

The radiation pattern $S(r,\theta,\phi)$ specifies the angular distribution of power density that emanates from the antenna. It is impossible to confine all the energy into a narrow conical beam, and some of it inevitably falls outside the main lobe and into the sidelobes (Figure 9.1).

The antenna reflector is usually a paraboloid of revolution and is illuminated by a source (horn) located at its focal point. The illumination is made to be nonuniform across the reflector to reduce sidelobe levels, and θ_1 of the circularly symmetric beam is then roughly given by

$$\theta_1 = 1.27\lambda/D_a \quad \text{(radians)} \tag{9.1}$$

The angular diameter of the first null circle is 3.27 λ/D_a (radians). The functional form of the illumination for which (9.1) holds is given by Doviak and Zrnić [33]. Equation (9.1) should be a good approximation for the 3-dB beamwidth of most weather radar antennas.

The microwave pulse is within a $c\tau$-thick spherical shell that expands (i.e., propagates) at a speed c. Thus, the pulsed power density $S_i(\theta,\phi)$ incident on targets must decrease inversely with r^2, even though peak power P_t' transmitted through any enclosing sphere is constant. Because transmitted power is not constant during the pulse, its power is defined as the instantaneous microwave power integrated over the duration of the pulse and divided by the pulsewidth τ. An instantaneous power is the power averaged over any one cycle of the transmitter frequency. Because of losses in the antenna and its protective radome, as well as in transmission lines, the pulse power P_t generated by the transmitter and delivered to the antenna port is larger than P'_t.

If P'_t were radiated equally in all directions, S would be equal to $P'_t/4\pi r^2$. However, the antenna focuses radiation into a narrow, angular region (i.e., the beam) where radiated power density $S = S_b$ is more intense than $P'_t/4\pi r^2$. The ratio:

$$\frac{S_b}{P'_t/4\pi r^2} \equiv g'_t \tag{9.2}$$

defines the maximum directional gain of the antenna. Measurement of P'_t *is difficult, and the antenna engineer therefore measures power* P_t delivered to the antenna's waveguide port and S_b at distances far (i.e., $r \geq 2D_a^2/\lambda$) from the antenna. In this case the computed gain g_t (without the prime) accounts for losses of energy associated with the antenna system (e.g., radome, waveguide). Then the power density S_i incident on a target at range r, and in the direction θ,ϕ (not necessarily within the beam), is given by

$$S_i(\theta,\phi) = \frac{P_t g_t f^2(\theta,\phi)}{4\pi r^2} \tag{9.3}$$

where $f^2(\theta,\phi)$ is the normalized power gain function (i.e., $f(\theta,\phi) = 1$ along the beam axis at θ_0,ϕ_0; Figure 9.1), and g_t is simply the antenna gain. S_i, as given by (9.3), is the power density incident on the hydrometeors that form the scattering media of cloudy air, if attenuation of radiation by the atmosphereic constituents (i.e., molecules and hydrometeors) can be ignored (see Section 9.4).

9.2.2 The Backscatter Cross Section of a Spherical Hydrometeor

The backscatter cross section σ_b of a hydrometeor (i.e., snow, hail, water drop, etc.) is an apparent area that intercepts a power $\sigma_b S_i$, which if radiated isotropically, produces at the radar antenna a power density:

$$S_r = \frac{S_i \sigma_b}{4\pi r^2} \tag{9.4}$$

equal to that scattered by the actual target. A hydrometeor is called a point target because its dimensions are much smaller than $c\tau/2$. The backscatter cross section σ_b for a spherical water drop of diameter D (large drops are oblate and their cross sections are treated in Sec. 8.5.3 of Doviak and Zrnić [33]) small compared with λ (i.e., $D \leq \lambda/16$) is

$$\sigma_b = \frac{\pi^5}{\lambda^4} |K_w|^2 D^6 \tag{9.5}$$

where $|K_w|^2$ is a parameter, related to the refractive index of the drop's water, that varies between 0.91 and 0.93 for wavelengths between 0.01 and 0.10 m, and is practically independent of temperature (Battan [9], p. 39). Ice spheres have $|K_w|^2$ values of about 0.18 (for ice density of 0.917 g cm^{-3}) independent of temperature as well as wavelength in the microwave region. Raindrop diameters can be as large as 8 mm. Therefore, only for wavelengths at about 10 cm and longer could the simple formula (9.5) be applied without limitation. When D exceeds

$\lambda/16$, σ_b has a complicated oscillatory dependence on D (see, for example, Battan [9], Fig. 4.2) and then interpretations of meterological parameters such as reflectivity factor (see Section 9.3.6) are open to error.

Battan [9] and Atlas [4] have reviewed the abundance of experimental and theoretical work that relates a particle's scatter cross section to its shape, size relative to wavelength when $D \geqslant \lambda/16$, temperature, and mixture of phases (e.g., water-coated ice spheres).

9.2.3 The Radar Equation

S_r, from a target at r,θ,ϕ, is intercepted by the antenna and the echo power P_r delivered to the antenna port is

$$P_r = S_r(r,\theta,\phi)A_c(\theta,\phi) \tag{9.6}$$

where A_c is the effective collection area of the antenna for radiation from direction θ,ϕ. Theoretically (Silver [101]; Sherman [99], p. 9–12),

$$A_c = \frac{g_r\lambda^2}{4\pi} f^2(\theta,\phi) \tag{9.7}$$

where g_r, the gain of the receiving antenna, is equal to g_t, if the transmitting antenna is used for echo reception and P_r is measured at the same location as P_t. For large-aperture antennas and targets along the beam axis, A_c is roughly the physical area of the antenna reflector. For example, the National Severe Storms Laboratory's (NSSL) doppler radar antenna has an $A_c = 42\ \mathrm{m}^2$, but its physical area is $65.7\ \mathrm{m}^2$. The significantly smaller A_c is caused principally by tapering the illumination of the reflector, which is equivalent to weighting the backscattered power density reflected from the parabolic surface to the horn located at the focus. Furthermore, radome and waveguide losses (omitted in our example) cause A_e to be smaller than the physical area.

Combining (9.3) to (9.7) gives the power of an echo from a point target having backscatter cross section σ_b:

$$P_r = \frac{P_t g^2\lambda^2\sigma_b f^4(\theta,\phi)}{(4\pi)^3 r^4 l^2} \tag{9.8}$$

where we have substituted $g_t = g_r = g$ and have introduced the one-way transmission loss l (defined according to the IEEE standard which is a reciprocal of the definition in Doviak and Zrnić [33]) to account for loss of wave energy due to absorption in and scatter from hydrometeors and air molecules along the path of wave propagation from the radar to the target (see Section 9.4). Equation (9.8) relates echo power P_r to the radar and target parameters.

As an example of the sensitivity that radar has for detection of some extremely small targets, consider that a 10 cm ($\lambda = 0.1$ m) weather radar receiver can detect echo power as weak as 10^{-14} W, its transmitted peak power is about 10^6 W, and its antenna gain is 4×10^4. Thus, using (9.8), the minimum detectable cross section at a range of 20 km is

$$\sigma_b(\min) = 2 \times 10^{-7}\ \mathrm{m}^2$$

That is, a single water drop with a diameter of 6.3 mm could be detected at 20 km. Thus, it would not take a very large number of drops to provide a detectable signal at longer range (e.g., 200 km)—and, at that, the transmitted power and receiver performance of weather radars are only moderate in relation to other advanced systems in use today (such as deep-space probes, for example).

9.2.4 Retrieving the Doppler Signal (the In-Phase and Quadrature Components)

Nonlinear (usually square-law) devices called *mixers* or *detectors* (Fig. 9.1) are used to retrieve a modulating signal and generate the in-phase and quadrature signals from which the target speed and direction (away from or toward the radar) are obtained. For the sake of simplicity, assume CW signals and a point target for which σ_b is a constant. A modulating signal (traces 3 and 4 of Figure 9.2) is produced when the STALO signal is summed with the usually weaker echo signal having a radian frequency ω equal to that transmitted plus a typically small doppler shift ($\omega_d \ll \omega$), if the target has motion relative to the radar. The modulating signal (at the doppler beat frequency $f_d = \omega_d/2\pi$) contains the doppler shift and phase information. If the device only translates the frequency from some high value to an intermediate one, the device is a *mixer-converter*. However, if the modulating signal is directly recovered, as in the homodyne receiver (Figure 9.1), where the STALO's frequency is equal to the transmitted frequency, the device can be properly called a *mixer-detector*. In most doppler weather radars, the doppler beat frequency is usually recovered in a two-step process: first, a conversion to an intermediate frequency, and then detection of the doppler signal. Although the mixer-detector recovers the doppler signal, there will also be high frequency components that need to be filtered to have an uncontaminated replica of the modulating signal.

The voltage V_m applied to the mixer-detector is the linear sum $V_m = V_l + V_e$ of two voltages. One is V_l, the STALO voltage given by $A_l \cos(\omega_l t + \psi_l)$ where ψ_l is the STALO's signal phase, and the other V_e is the echo voltage expressed as $A_e \cos[\omega(t - 2r/c) + \psi]$. The term $(t - 2r/c)$ shows the effect of the finite speed of wave propagation: any point on the transmitted wave will be returned to the receiver at a time $t = 2r/c$ later. The phase $\psi = \psi_t + \psi_s$ is the sum of the phase ψ_t of the transmitted signal and any phase shift ψ_s produced by the target upon scattering. The phase ψ_e of the echo voltage is defined as

$$\psi_e = -\frac{4\pi r}{\lambda} + \psi \tag{9.9}$$

in which $\lambda = 2\pi c/\omega = c/f$ is the wavelength of the transmitted signal. Figure 9.2 shows the signals in the homodyne Doppler receiver in which CW signals are assumed. We shall, for the sake of simplicity, assume that $\omega_l = \omega$ for the discussion in the remainder of this section. So that echoes can be received simultaneously while the klystron (Fig. 9.1) is transmitting, we need another but separate receiving antenna, isolated from the transmitting one and connected to the mixer-detectors. Then the T/R switch can be eliminated. For highly coherent doppler transmitters, and targets not changing their shape with time, ψ can be assumed to be a constant that is independent of time. If the target is moving with constant radial speed, v_r, ψ_e will change uniformly with time and the echo's radian frequency $\omega_e = \mathrm{d}\psi_e/\mathrm{d}t$ (equal to ω for stationary targets) is doppler shifted from ω by an amount

$$\omega_d = 2\pi f_d = \frac{\mathrm{d}\psi_e}{\mathrm{d}t} = -\frac{4\pi v_r}{\lambda}$$

Thus,

$$f_d = -\frac{2v_r}{\lambda} \tag{9.10}$$

shows the reverse relation between doppler velocity v_r and doppler shift f_d. That is, a positive radial velocity (outward moving target) generates a negative doppler shift, which decreases the frequency of the echo relative to that transmitted.

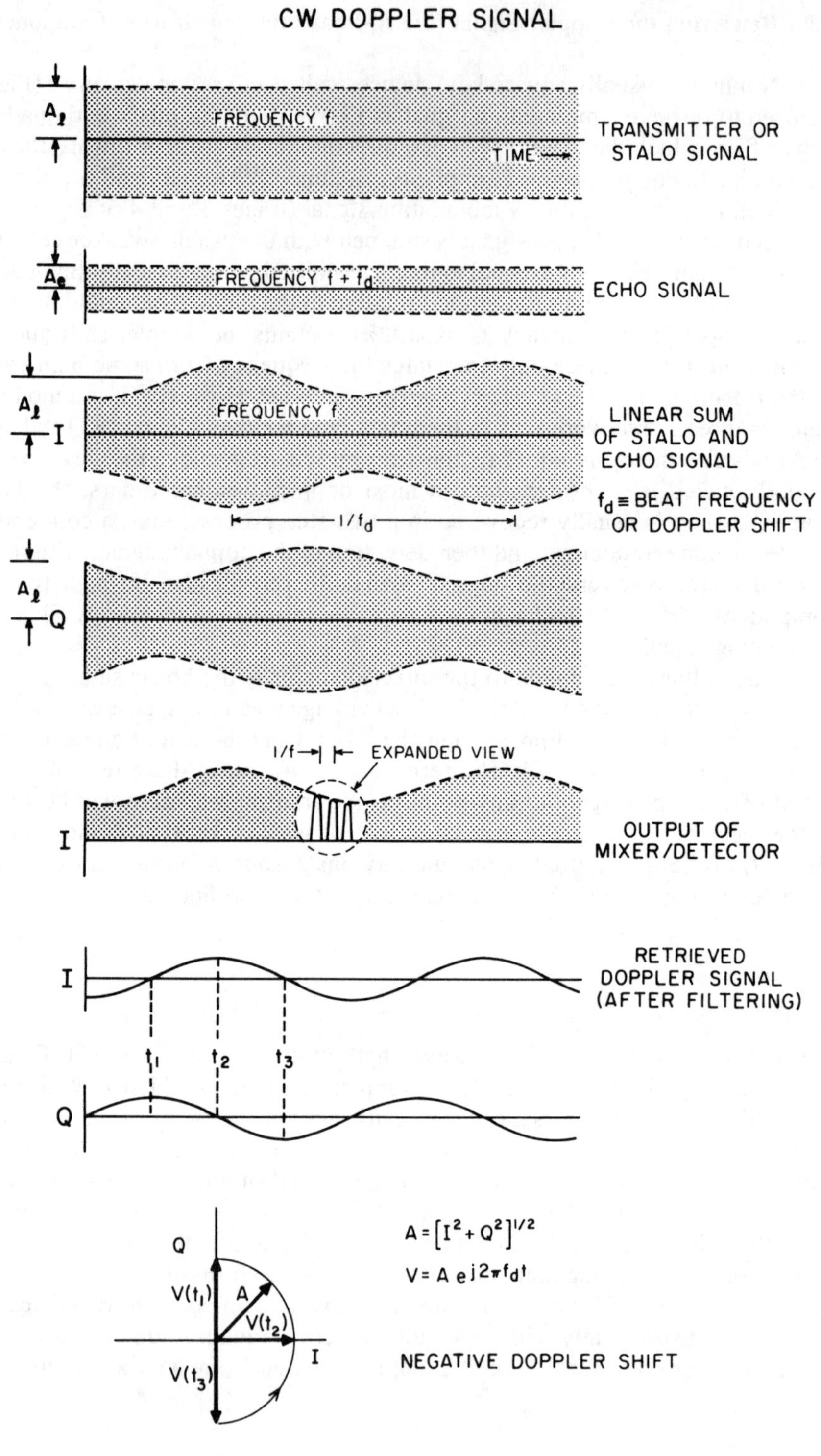

Figure 9.2 Signals at various points in a homodyne doppler receiver when a CW waveform illuminates a single-point target that moves at a uniform radial velocity.

We now write the sum $V_l + V_e$ as

$$V_m = A_l \cos(\omega t + \psi_l) + A_e \cos(\omega t + \psi_l + \psi_e - \psi_l)$$

and expand the second term to obtain

$$V_m = [A_l + A_e \cos(\psi_e - \psi_l)] \cos(\omega t + \psi_l) - [A_e \sin(\psi_e - \psi_l)] \sin(\omega t + \psi_l) \quad (9.11)$$

This can be conveniently transformed to

$$V_m = A_m \cos(\omega t + \phi) \quad (9.12)$$

where

$$A_m(t) = \sqrt{[A_l + A_e \cos(\psi_e - \psi_l)]^2 + A_e^2 \sin^2(\psi_e - \psi_l)}$$

$$\phi(t) = \tan^{-1}\left[\frac{A_e \sin(\psi_e - \psi_l)}{A_l + A_e \cos(\psi_e - \psi_l)}\right]$$

We have explicitly shown the time dependence of A_m and ϕ to emphasize that ψ_e is, in general, time dependent. In radar receivers the local oscillator signal level A_l is usually much larger than the echo signal amplitude A_e so that

$$A_m(t) \approx A_l\left[1 + \frac{A_e}{A_l}\cos(\psi_e - \psi_l)\right] \quad (9.13)$$

$$\phi(t) \approx \frac{A_e}{A_l}\sin(\psi_e - \psi_l) \quad (9.14)$$

Up to this point we really only have the two frequencies, ω, ω_e, because we have as yet not subjected the composite signal to the effects of the nonlinear device that will generate sum and difference frequencies as well as harmonics. The signal described by (9.12) would look like the third or fourth trace from the top of Figure 9.2, if ψ_e is linearly increasing with time (i.e., at a constant doppler shift ω_d).

Although we have both envelope modulation as well as phase modulation, we need only concern ourselves with $A_m(t)$ and not $\phi(t)$. The composite signal (9.12) applied to the mixer-detector circuit is then well approximated by

$$V_m(t) \approx A_l\left[1 + \frac{A_e}{A_l}\cos(\psi_e - \psi_l)\right]\cos(\omega t) \quad (9.15)$$

Now consider the mixer-detector to be a square-law device that has the following voltage-current relation:

$$i = aV_d + bV_d^2 \quad (9.16)$$

where a and b are constants that define the characteristic of the nonlinear device, which is the heart of the mixer-detector circuit (Figure 9.3).

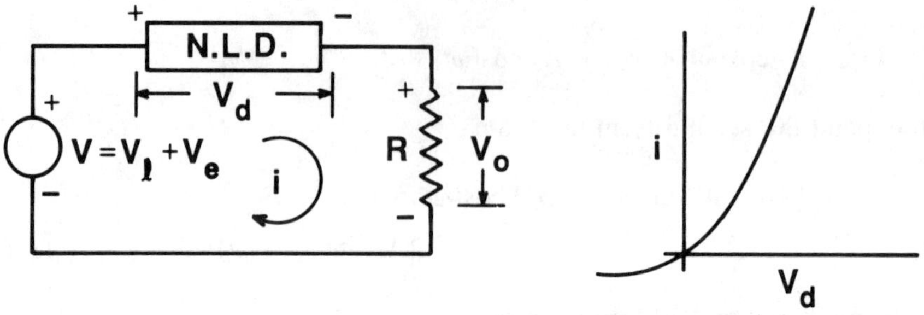

Figure 9.3 **A nonlinear device (N.L.D.) in a circuit in which two signals are summed to form $V(t)$. The nonlinear characteristic is sketched on the right and gives the current i for a given voltage V_d.**

If $b = 0$, the device is simply a resistor with resistance equal to $1/a$. The squaring term in (9.16) generates harmonics of ω and $\omega + \omega_d$ as well as frequencies that are multiples of the sum and differences of ω and $\omega + \omega_d$. The current passing through the resistor R then generates the observed output signal V_0. If $R \ll 1/a$, it can be shown that

$$i \approx aV + bV^2 \tag{9.17}$$

That is, the output voltage $V_o = iR$ must be small compared to the voltage V_d across the device in order to have the nonlinearity effective. Substituting (9.15) into (9.17), and solving for i, we obtain

$$\begin{aligned} i = {} & \frac{b}{2}[A_l^2 + 2A_lA_e\cos(\psi_e - \psi_l) + A_e^2\cos^2(\psi_e - \psi_l)] \\ & + a[A_l + A_e\cos(\psi_e - \psi_l)]\cos\omega t \\ & + \tfrac{1}{2}b[A_l^2 + 2A_lA_e\cos(\psi_e - \psi_l) + A_e^2\cos^2(\psi_e - \psi_l)]\cos 2\omega t \end{aligned} \tag{9.18}$$

The fifth trace in Figure 9.2 roughly depicts the signals of (9.18). (If the nonlinear device were linear for $V_d > 0$, but had $i = 0$ for $V_d < 0$, then Figure 9.2 would be an accurate depiction of the signals.) The last two terms in (9.18) are signals at the STALO frequency ω and its second harmonic. These signals can easily be filtered. The first term contains the detected modulation signal with its doppler information. Because $A_l \gg A_e$, the filtered signal (the first term in (9.18)) is

$$i_f \simeq \frac{b}{2}A_l^2 + bA_lA_e\cos(\psi_e - \psi_l) \tag{9.19}$$

The direct current component $b/2\ A_l^2$ can also be easily filtered to obtain the sixth trace of Figure 9.2. The phase ψ_l of the local oscillator is contained in (9.19), and if we have two mixer detectors and the phase of the STALO signal in one is changed by $\pi/2$ relative to the other (as in Figure 9.1), then we have from (9.19) the two phase quadrature components of the detected doppler signal (Figure 9.2; traces 6 and 7):

$$i_I = I = I_0\cos(-4\pi r/\lambda + \psi_t + \psi_s) \tag{9.20a}$$

$$i_Q = Q = Q_0\sin(-4\pi r/\lambda + \psi_t + \psi_s) \tag{9.20b}$$

Now ψ_t has been redefined so that it also contains the reference phase (i.e., the phase at time zero) of the STALO signal. If the filters and amplifiers in both I and Q channels have the same gain, then $I_0 = Q_0 = bA_lA_e$ and, furthermore, if both channels cause identical phase shift, then the phase or argument of I is the same as for Q, as already indicated by (9.20). The effects that different gain and phase shift (i.e., amplitude and phase imbalance), as well as time change in ψ_t, have on the performance of doppler receivers is discussed in some detail by Doviak and Zrnić ([33], Sec. 7.9).

Figure 9.2 depicts both the I and Q traces of the output signal for the case in which r is linearly changing with time. If these continuous signal components are sampled in time and the sampled values plotted as points on a polar (phasor) diagram in which I and Q values are respectively along the abscissa and ordinate axes, the points will be described by the tip of a vector (phasor) of amplitude $A = bA_eA_l$, which rotates, for increasing time, clockwise (Figure 9.2) to indicate a negative doppler shift or, as deduced from (9.10), a positive radial velocity. Notice that if only one of the I or Q were available, it would be impossible to determine the direction (toward or away from) of target motion, although target radial speed could be measured.

Now, if the transmitter is pulsed, the transmitted signal takes the form of the top trace sketched on Figure 9.4. However, the STALO voltage remains as a continuous wave. Point target echoes at the mixer-detector input are exact replicas of the transmitted pulse, but they are delayed by $\tau_s = 2r/c$. When these delayed echoes are linearly combined with the CW STALO signal, the resulting combination is an amplitude modulated sequence of pulses superimposed on the otherwise constant amplitude of the STALO signal (Figure 9.4, fourth trace from top). This is similar to the CW radar case, except that the beat frequency signal is now in sampled form. The output of the mixer-detector after filtering is depicted by the sixth trace on Figure 9.4. We thus see that, for the case of a pulsed radar, the doppler signal in the I and Q channels is a sampled version of the CW signals.

Figure 9.4 shows the I and Q waveforms and the resulting phasor diagram for echoes at three time positions. Whereas the echo phasor for CW radar describes a continuous path, the phasor for a pulsed radar is only observed at discrete instants separated by a time interval equal to the pulse repetition time T_s. Because the pulsed doppler radar is a sampling system, we have ambiquities in the simultaneous measurement of range and velocity (Section 9.2.5).

It is relatively easy to see from (9.20) that, for usual radar parameters (i.e., pulsewidth $\tau \approx 10^{-6}$ s) and meterological target velocities on the order of tens of meters per second, the change in the trigonometric functions is extremely small during the time τ of the echo pulse. Thus, echo phase change is measured during the longer time, $T_s \approx 10^{-3}$ s, from echo pulse to echo pulse, rather than during a pulse period. Because of this, the pulsed doppler radar samples phase as well as amplitude; samples are at $t = \tau_s + (m - 1)T_s$, where τ_s is the time delay between the mth transmitted pulse and its echo. τ_s is called "range time" because it is proportional to range (i.e., $\tau_s = 2r/c$). It is convenient to introduce another time scale, "sample time" in which time is incremented in discrete steps of length T_s after $t = \tau_s$ (see Figure 9.10 in Section 9.3). It is important to realize that echo phase and amplitude changes are examined in sample time space at the discrete instants mT_s for a target echo at range time τ_s.

Samples of $I(\tau_s,T_s)$ and $Q(\tau_s,T_s)$, taken from actual observations of stationary and moving targets, are shown in Figure 9.5. Traces 1, 2, and 3 in this figure are echo samples at three successive sampling intervals. Because $I(\tau_s,T_s)$ is near zero and decreasing, whereas $Q(\tau_s,T_s)$ has maximum amplitude and is also decreasing,

DOVIAK & ZRNIĆ
CHAPTER 9

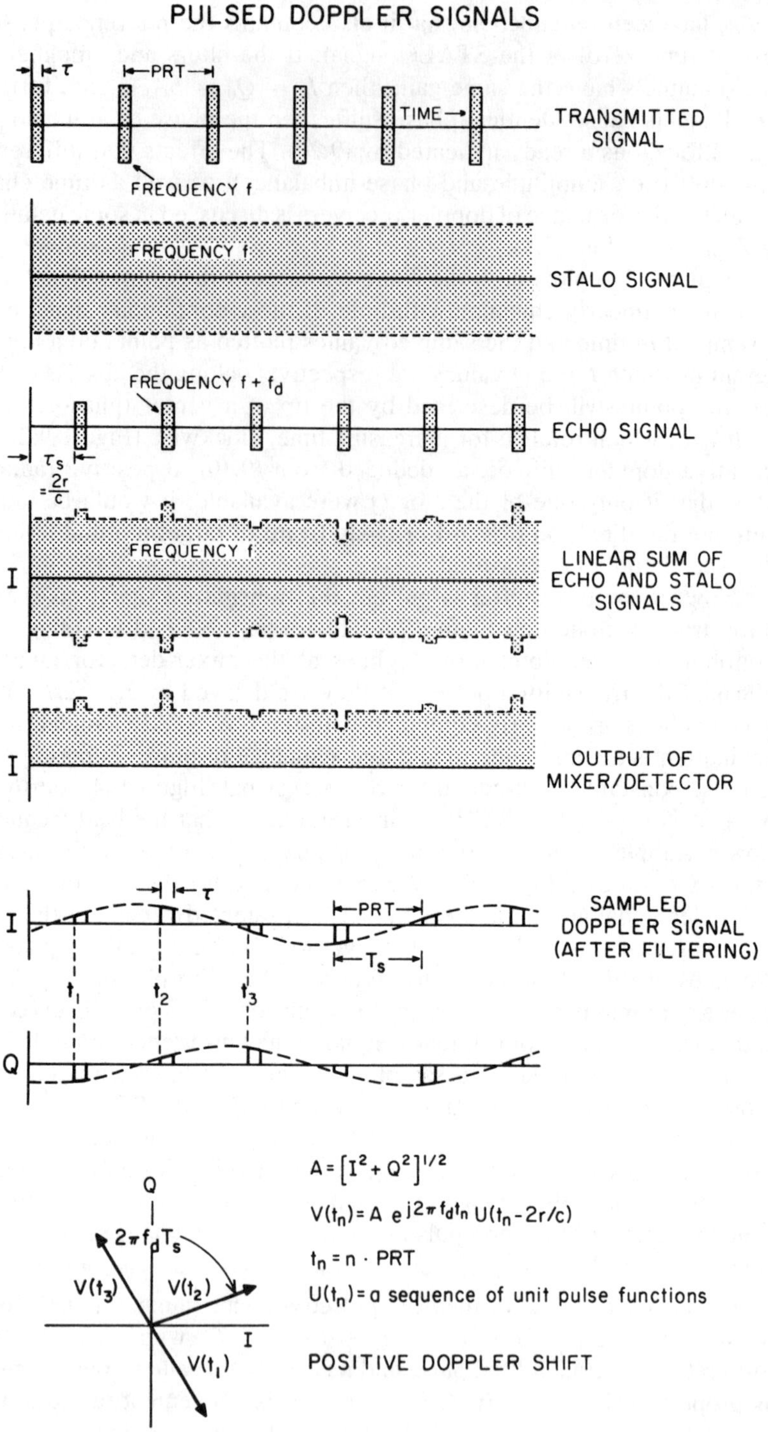

Figure 9.4 Signals at various points in a homodyne receiver when pulsed microwaves are transmitted and a single-point target is at range r and moving with a uniform radial velocity.

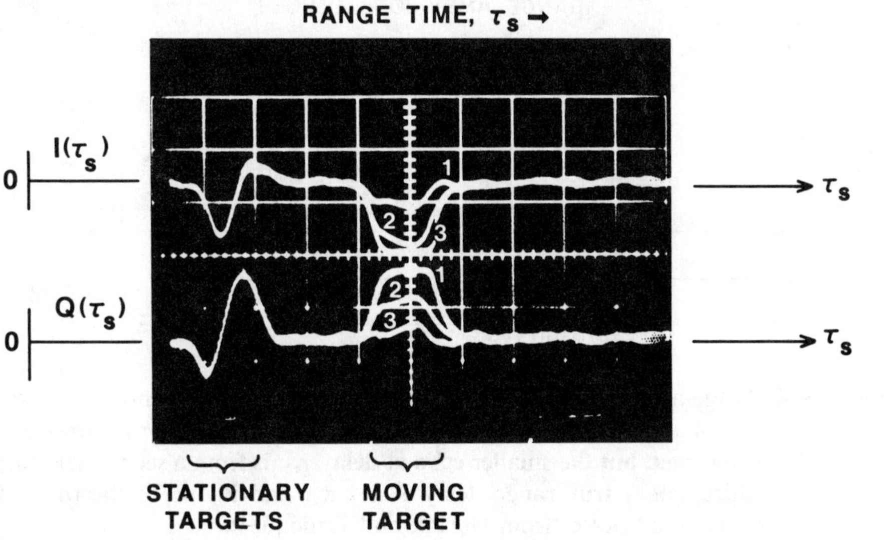

Figure 9.5 **Samples of in-phase $I(\tau_s)$ and quadrature phase $Q(\tau_s)$ components of echoes for moving and stationary targets. Echo traces *versus* τ_s are from three successive sampling intervals T_s, but have been superimposed to show the relative change of amplitude.**

the doppler shift of the moving target is positive, corresponding to a negative radial velocity or target motion toward the radar. The transmitted pulsewidth τ is 1.2 μs and because the moving target produces an echo pulse that nearly replicates the transmitted one, we can assume that there is only one moving target observed in the range interval (1500 m) displayed in Figure 9.5. On the other hand, the echoes associated with the stationary targets are not a replica of the transmitted pulse because more than one target lies in the range interval $c\tau/2$. Nevertheless, the I and Q samples do not change amplitude as a function of sample time mT_s (i.e., traces 1, 2, and 3 are identical for echoes from the stationary targets).

9.2.5 Ambiguities and Other Radar Measurement Limitations

Two kinds of ambiguities, range and velocity, plague measurement with pulsed doppler radars. When doppler radars are operated with uniform PRT (i.e., T_s) and targets have range r larger than $cT_s/2$, their echoes for the nth transmitted pulse are received after the $(n + 1)$th pulse is transmitted (Figure 9.6). Therefore, these echoes are received during the same interpulse time interval that targets at $r < cT_s/2$ return echoes from the $(n + 1)$th pulse. Thus, range r to the distant target has an apparent value $r' = r - (N_t - 1)r_a$ and is ambiguous with targets at $r = r'$. (N_t is the trip number and $N_t - 1$ designates the whole number of $cT_s/2$ intervals between the target and receiver; $r_a = cT_s/2$ is that range within which all targets must lie to have their ranges unambiguously measured.) We emphasize that r_a does not necessarily limit the range to which the pulsed doppler radar can achieve useful measurement. If the radar's STALO is phase coherent over many T_s intervals, the radar can accurately measure velocities of targets far beyond r_a, but interference occurs when echoes from targets in different trips are received simultaneously.

DOVIAK & ZRNIĆ
CHAPTER 9

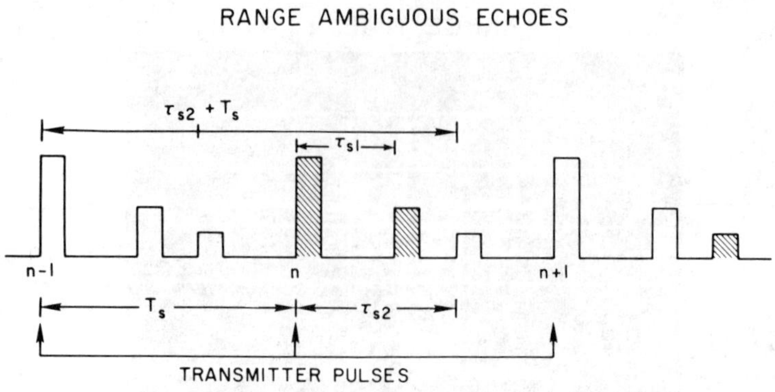

Figure 9.6 **Range-ambiguous echoes. The *n*th transmitted pulse and its echoes are cross-hatched. In this example, the larger echo at delay τ_{s1} is from a first-trip target, but the smaller echo at delay τ_{s2}, is from a second-trip target, which has a true range delay $T_s + \tau_{s2}$, and is due to the $(n - 1)$th transmitted pulse (from Doviak and Zrnić [33]).**

The second ambiguity relates to measurement of target velocity. As discussed in Section 9.2.4, the echo phase ψ_e is sampled at intervals T_s and its change over the interval T_s is a measure of the doppler frequency (9.10). Unfortunately, given a set of sampled phases (computed from I and Q samples), we cannot relate them to a unique doppler frequency. Figure 9.7 shows the same set of samples that could have resulted from any one of three signals having different doppler frequencies. Signals that fit the sample data set are called aliases, and $f_N = (2T_s)^{-1}$ is the Nyquist (or folding) frequency. All doppler frequencies between $\pm f_N$ are the principal aliases, and any frequency higher than f_N is ambiguous with one of those between $\pm f_N$. Thus, target velocities must lie within $\pm v_a(v_a = \lambda/4T_s)$ to avoid velocity ambiguity. However, transmission of pulses at two interlaced PRFs and signal processing can extend both the unambiguous velocity and range (Doviak and Zrnić [33]).

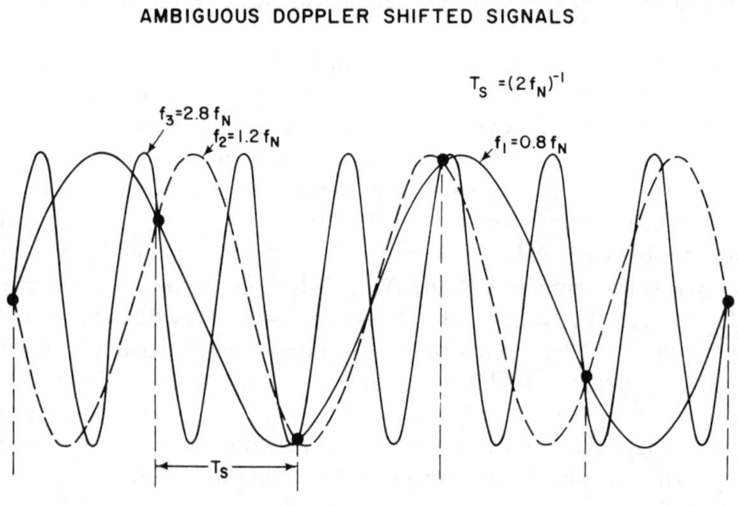

Figure 9.7 **Signals at three different doppler frequencies that yield, when sampled, the same set of data. These doppler frequencies are aliases of each other (from Doviak and Zrnić [33]).**

Receiver noise and the bandwidth of the filter and amplifiers (Figure 9.1) also limit receiver performance. In an ideal noiseless radar receiver, there is no lower limit in detecting the weakest echo. However, all receivers have a power detection limitation usually imposed by the random electrical signals associated with thermal agitation of electron motion in each receiver's front end (in Figure 9.1 it is the mixer). High-performance receivers have parametric amplifiers or other low-noise amplifiers before the mixer that add so little noise to the echo that microwave emissions from the earth's surface, from the atmosphere's constituents (e.g., water vapor, oxygen, *et cetera*) and from outer space can be the more significant contributors to the overall noise power.

Noise power N limits weather radar performance, and its value:

$$N = kT_{sy}B_n \tag{9.21}$$

is a characteristic constant of the radar, where k is Boltzmann's constant, (1.38×10^{-23} W s K^{-1}), B_n is the receiver's noise bandwidth, and T_{sy} is the effective system temperature at the antenna port. The product of B_n and the peak density (i.e., W/Hz) of the noise power spectrum gives a noise power equal to that observed. The system temperature includes contributions from environmental noise received by the antenna, thermal radiation due to losses in the antenna's transmission line and protective radome, as well as thermal noise within the receiver itself. For a more comprehensive discussion and methods to compute T_{sy}, the reader should refer to the text by Blake [12], pp. 2–29) or Doviak and Zrnić [33].

The echo pulse, as well as any signal, can be synthesized in terms of its Fourier spectral coefficients. The receiver amplifies only a band of these spectral or frequency components; that is, the receiver has a finite bandwidth. The receiver's filter (Figure 9.1) power gain $G^2(f)$ is usually a monotonically decreasing function of frequency, and its width, B_6, is best specified as the frequencies within which $G^2(f)$ is larger than one-fourth of its highest level—its 6-dB width (Taylor and Mattern [106]). Both positive and negative frequencies (positive and negative doppler shift) are passed by the filter, and B_6 is the width between the positive and negative frequencies at which $G^2(f)$ is one-fourth of its peak value. Noise bandwidth B_n is not so simply specified because it depends on $G^2(f)$ as well as the noise power spectrum, but in practice it is nearly equivalent to the 3-dB width of $G^2(f)$, if noise power is "white," (i.e., independent of frequency (Kraus [69], p. 265)).

The larger the B_6, the better is the fidelity of the echo pulse shape, but noise power increases in proportion to B_6. We also run the risk of accepting unwanted products of the mixer output. Usually, it is not important to detect the echo with nearly perfect fidelity, nor is it advisable to reduce noise excessively by decreasing B_6 because, as we show below, smaller filter bandwidth coarsens range resolution.

An important radar performance parameter is the ratio of peak echo power to noise power, the signal-to-noise power ratio (SNR), which can be shown to be (Doviak and Zrnić [31,33])

$$\mathrm{SNR} = \frac{(I_0^2 + Q_0^2)\,\mathrm{erf}^2(aB_6\tau/2)}{kT_{sy}B_n}, \tag{9.22}$$

for a Gaussian shaped $G(f)$, where I_0 and Q_0 are the prefilter echo amplitudes bA_eA_l (9.20), τ is the pulsewidth, $a = \pi/2\sqrt{ln2}$, and erf( ) is the error function (Abramowitz and Stegun [1]). It can easily be shown that SNR is maximum when $B_6 = 1.04/\tau$. If the receiver's $G(f)$ is not Gaussian but is perfectly matched to the spectrum of the detected echo pulse (assumed to replicate the rectangular transmitted pulse shape), the output of the receiver's filter would be a triangular pulse with a base having a duration of 2τ.

9.3 THE WEATHER ECHO

Section 9.2 delineated the radar signal properties of waves scattered from point targets. It has been assumed that no more than one target lies in the range interval $r_i + c\tau/2$, where r_i is the range to the ith target, and that the target dimensions are small compared to the smaller of $c\tau/2$ or $r_i\theta_1$. Targets satisfying these conditions are called *point targets*.

Each echo is then a replica of the transmitted pulse, but in passing through the receiver, these replicated pulses are distorted by the receiver (the mixers, filters, and amplifiers), which also add electronic noise that reduces the echo signal-to-noise ratio. These echoes (Figure 9.8 exhibits one), shaped by the receiver transfer function, are then displayed in video equipment to show their relative strengths and delay referenced to the time of the transmitted pulse. The echo can have either polarity or even be zero, but cannot be zero simultaneously in the I and Q channels of the receiver. Weather echoes from a very large number of hydrometeors are composites of echoes, each of which is similar to that sketched in Figure 9.8. Collectively these echoes are designated as *clutter*.

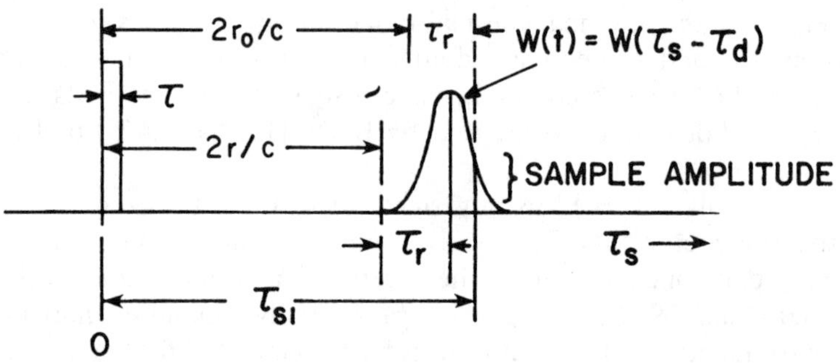

Figure 9.8 **The receiver's filtered output (I or Q) for an echo from a point target. The point target is located at r, and τ_r is the echo's delay through the radar. The range r_0 is where targets receive maximum weight from $W(r)$ when the sample gate is at τ_{s1} (adapted from Doviak and Zrnić [33]).**

Whereas Figure 9.8 shows the output signal for the case in which a single point target is present, Figure 9.9 shows the output signal for an actual radar receiver in which echoes from a multitude of targets (both airborne and ground) are received. The quasicontinuous distribution of echoes along the range axis is called clutter, to which both weather and ground targets contribute. For radars designed to detect point targets such as aircraft, clutter is a nuisance which often obscures observation. On the other hand, weather clutter contains the needed meterological information, and, for weather radars, ground clutter and echoes from aircraft, birds, *et cetera,* obscure meteorological observation.

Because of the large span of echo amplitudes, a logarithmic amplifier and video detector are often used in a separate channel of the radar receiver to measure echo intensity, and Figure 9.9 shows the logarithm of echo intensity A^2 (amplitude squared). In this way, both the strong echoes from nearby ground targets and weak echoes from atmospheric targets in clear air can be displayed simultaneously. The decibel range 10 log$[A^2]$ of echoes displayed in Figure 9.9 is about 90 dB; that is, the strongest echo is about 10^9 times more intense than the receiver noise power ($\approx 3 \times 10^{-15}$ W), which is actually larger than the power of atmospheric echoes at ranges beyond 60 km. Receiver noise level is an apparent echo power which, although fluctuating, has an average value independent of range time τ_s.

DOVIAK & ZRNIĆ
CHAPTER 9

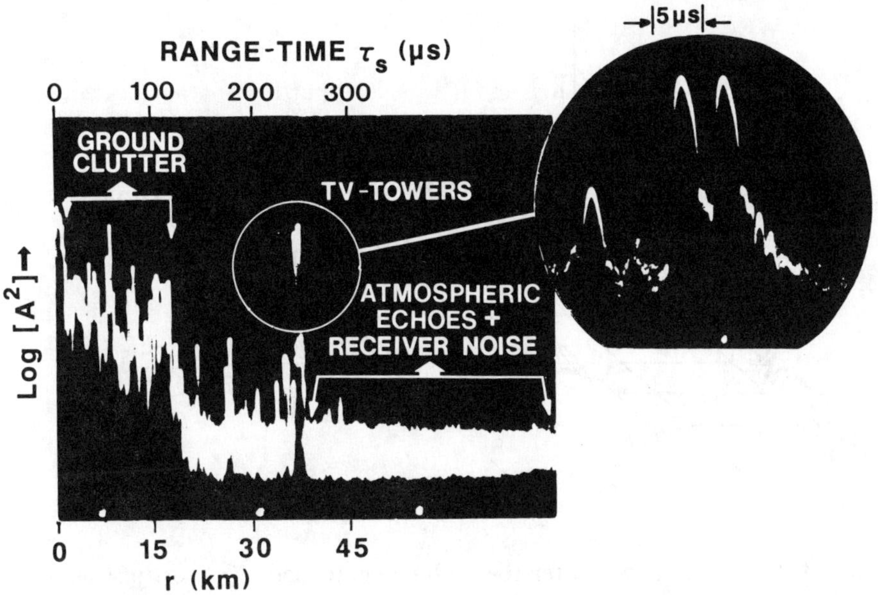

Figure 9.9 **Range-time delay τ_s and strength of echoes from discrete (e.g., TV towers at about 40 km) and distributed targets. Ground clutter is a conglomerate of indistinguishable echoes from targets on the ground. The vertical scale is proportional to the logarithm of the echo power. The transmitted pulsewidth τ is 1.2 μs. Inset shows an expanded view of the tower echoes (from Doviak and Zrnić [33]).**

Thus, the stretch of fluctuating amplitudes decreasing beyond about 40 km shows mainly echoes from airborne insects or atmospheric targets (e.g., refractive index irregularities) superimposed on receiver noise, whereas, for ranges inside 40 km, echoes from ground targets are added to the mixture. When atmospheric targets are spatially homogeneous, dense (e.g., many insects or irregularities in refractive index within the range extent $c\tau$ of the transmitted pulse), and moving relative to each other, the fluctuations of echo amplitude are exactly the same as receiver noise and so it is difficult to distinguish, as in Figure 9.9, noise from atmospheric echoes. However, doppler velocity measurements, not shown here, yield values at ranges beyond 40 km, equal to those expected from airborne targets moved only by wind.

Although the fluctuating signals of atmospheric echoes superimposed on the steady echo from ground targets (e.g., echoes from the TV towers in Figure 9.9) can have amplitudes larger than receiver noise (i.e., $\log A^2$ at ranges far beyond 60 km), their amplitudes appear to be less than noise because of the logarithmic scale. To illustrate, note that the receiver noise and atmospheric echoes are not evident on the expanded view of echoes from TV towers. Nevertheless, the amplitude of atmospheric echoes is not diminished in the presence of echoes from ground targets, and it becomes the purpose of the doppler processor to sort out the strong steady signals due to ground targets from those sometimes weaker echoes from the atmosphere.

Because we cannot resolve echoes from the individual weather targets (e.g., hydrometeors), we resort to sampling, at discrete range-time delays τ_s, the composite continuous echo signal that forms the phasor voltage $V(t) = I(t) + jQ(t)$. Figure 9.10 shows either the I or Q component of $V(t)$. Each voltage sample $V(\tau_s)$ is a weighted composite of discrete echoes from all the scatterers, with weights determined by the antenna's radiation pattern $f^2(\theta,\phi)$ and a range weighting function $W(r)$ of a form dependent on the filter's gain function $G(f)$ and the shape of

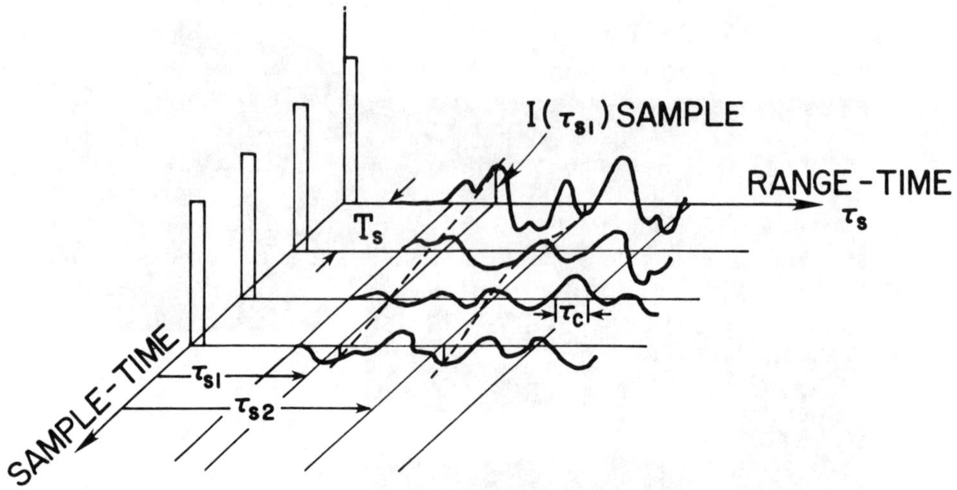

Figure 9.10 **Idealized traces for the inphase component $I(\tau_s)$, or, just as well, the quadrature phase, $Q(\tau_s)$ echoes from distributed targets. Each trace represents echoes from a single transmitted pulse. Dashed line indicates probable time dependence of echo samples taken at τ_{s1}, τ_{s2}; τ_c depicts signal correlation time along τ_s and is related to $B_6\tau$. Samples at fixed τ_s taken at T_s intervals are used to construct doppler spectra for scatterers located about the range $c\tau_s/2$ (from Doviak and Zrnić [33]).**

the transmitted pulse. The form of $W(r)$ is the mirror image of the filter's output when a replica of the transmitted pulse is applied to it (Zrnić and Doviak [122]). These weighting functions determine a resolution volume (Section 9.3.5) in space where targets contribute significantly to the echo sample at τ_{s1}.

Each scatterer in the resolution volume returns an echo signal like that sketched in Figure 9.8, which constructively or destructively (depending on the sign of the signal) interferes with signals from other scatterers to produce a composite phasor sample $V(\tau_{s1})$. The random size and location of scatterers cause the amplitude and phase of $V(\tau_{s1})$ to be random variables. The echo sample at another range time delay τ_{s2} is a composite signal from scatterers in a different resolution volume, and hence we expect $V(\tau_{s1})$ to differ from $V(\tau_{s2})$. Thus, $V(\tau_s)$ fluctuates as τ_s increases (Figure 9.10), even when the scatterer density is spatially uniform. The correlation between $V(\tau_s)$ samples taken at different τ_s is related to the radial dimensions of the resolution volume and the spacing $\delta\tau_s$, between samples (Doviak and Zrnić [33], p. 60).

In this section we determine the statistical properties of weather echo samples $V(\tau_s)$ and instantaneous power $P(\tau_s)$, relate the average of successive $P(\tau_s)$ samples to the backscatter cross section of targets within the resolution volume, and develop a form of the weather radar equation that accounts for range and angular weighting.

Figure 9.11 is an actual recording of weather echo samples of I or Q. In this figure range time, τ_s, is along the vertical and sample time, mT_s, is along the abscissa. The sample time dependence of echo amplitude at 16 range times τ_s (i.e., range gate locations) is shown in this figure, where zero amplitude for each trace is a horizontal line (for clarity, not shown in Figure 9.11) at the gate number location. The radar beam is vertical and the first gate location is at 600-m range (ground targets are the cause of the relatively constant signal level) and each successive trace is the composite echo from targets in resolution volumes of about 170-m depth spaced 600-m apart corresponding to a range-time separation of 4×10^{-6} s. In the first four or five range gate locations of this figure we see relatively high doppler shifts produced by some fast moving targets. The transient of high frequency doppler shift seen in gate number 12 at about mT_s = 350 to 500 ms is caused by echoes from lightning channels, which move with the velocity of the air, and hence produce a doppler shift different than that of precipitation which falls through the air. The mean squared value of $V(\tau_s)$, averaged over sample time, mT_s, in each of the traces is about the same even though echo amplitude might vary considerably from gate location to gate location. This is because the receiver in the radar which produced these traces had an automatic gain control circuit, not shown in Figure 9.1, which adjusts the output to have a nearly constant mean squared value at each range time.

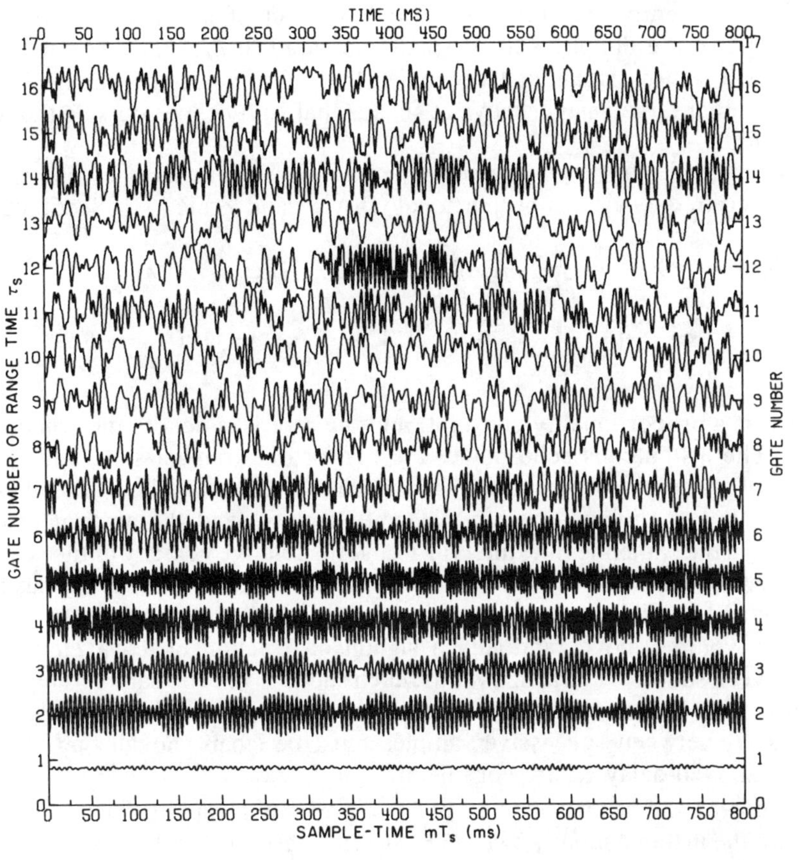

Figure 9.11 The I or Q samples of echoes at 16 range time locations $\tau_{s1} \ldots \tau_{s16}$ *versus* sample time mT_s. Sample-time interval T_s is 0.75 ms and the samples themselves have a duration less than 10^{-3} ms. Although the separation between sample points is several hundred times the sample duration, the points in this figure give a continuous trace because the resolution of the trace is coarse (i.e., ≈2 ms along mT_s).

9.3.1 The Echo Sample

As mentioned previously, the echo sample $V(\tau_s)$ is a composite of signals:

$$V(\tau_s) = \sum_i A_i W_i \, e^{-4\pi r_i/\lambda} \tag{9.23}$$

where $|A_i| = (I_i^2 + Q_i^2)^{\frac{1}{2}} = bA_l A_{ei}$ is the prefilter echo amplitude for the ith scatterer located at r_i, ϕ_i, and θ_i. The antenna weighting factor $f^2(\theta,\phi)$ is now assumed to be part of A_i, a complex number (i.e., it has amplitude and phase), which contains the phase shifts ψ_t and ψ_{si} (see (9.20)). W_i is the range weighting function for the ith scatterer, and its functional form is controlled by the filter and transmitted pulse shape. We shall not yet be specific about the functional form of W_i, which is to be discussed in Section 9.3.4. For now, it suffices to say that W_i is a complex number and has a range dependence such that it principally weights those targets residing near a range r determined by τ_s (i.e., $r = c\tau_s/2$). We also assume target velocity sufficiently small that W_i is independent of it. This is tantamount to assuming that the doppler shift is small compared to B_6; otherwise, fast moving targets will return signals at frequencies that fall outside the bandwidth of the filter. Weather targets never move at speeds that shift echoes outside the receiver bandwidth.

The sample, at range time τ_s, of echo power averaged over a cycle of the transmitted frequency f is proportional (Doviak and Zrnić [33], Section 2.1) to

$$\begin{aligned} P(\tau_s) &= \alpha\tfrac{1}{2}VV^* = \tfrac{1}{2}\alpha\sum_{i,k}^{N_s} A_i A_k^* W_i W_k^* \exp[-j4\pi(r_k - r_i)/\lambda] \\ &= \alpha\tfrac{1}{2}\sum_i |A_i|^2|W_i|^2 + \tfrac{1}{2}\alpha\sum_{i\neq k}^{N_s} A_i A_k^* W_i W_k^* \exp[-j4\pi(r_k - r_i)/\lambda] \end{aligned} \tag{9.24}$$

where α is a proportionality constant, and the asterisks denote the complex conjugate. The instanteous echo power $P(\tau_s)$ is for one transmission, and N_s is the number of scatterers. When scatters move relative to one another, the echo sample at range time τ_s differs for each transmitted pulse, and the amount of change depends on the sampling time interval T_s and the relative velocity of the scatterers. The equation defining $P(\tau_s)$ for successive transmission will have the same form, but the r_i and r_k will have changed owing to scatterers' motion. If scatterers within a sample volume move randomly a significant fraction of a wavelength (e.g., $\lambda/4$) between successive transmissions, each successive echo sample $V(\tau_s)$ (spaced T_s apart) will be uncorrelated. To measure the average radial speed of scatterers, the time T_s between successive samples must be small enough that sequential echoes, at fixed delay τ_s, are correlated.

Whereas the second sum in (9.24) represents a rapidly fluctuating contribution to the instantaneous power $P(\tau_s)$ as scatterers change their position in time, the first sum is relatively constant. Fluctuations are caused by target displacements, which alter the phase of each echo sample. Hydrometeor displacements on the order of a wavelength can cause large changes in the second sum, whereas displacements need to be on the order of a hundred meters or more (for typical weather radar operating parameters) before significant change occurs in the first term. Although the second sum can be significantly larger than the first (it has $N_s(N_s - 1)$ contributions, compared to N_s for the first sum) for some echo samples, its expected value is zero if its phase is uniformly distributed (in a probabilistic sense) across the 2π interval. Expected values are estimated by sample time averaging over a sufficiently long period such that a large number of uncorrelated samples are accumulated (see Doviak and Zrnic 1984, Chapter 6 for estimate accuracy). If, during the long period that samples are accumulated, scatterers displaced far from the region where they were receiving significant weight from

$f^2(\theta_i,\phi_i)W(r)$ are replaced by others having the same statistical properties, the sample time average of the first sum can then be a good estimate of the expected or mean power $\overline{P}(\tau_s)$. In practice, sample time averaging periods are much less than one second, and hence scatterers are displaced much less than a few tens of meters during the time of average power estimation. However, we stress that $\overline{P}(\tau_s)$ estimates can only be made because hydrometeors are moving relative to one another so that the average of the second sum in (9.24) is small compared to the first. An accurate estimate of $\overline{P}(\tau_s)$ is important because it relates to the estimates of liquid water in the resolution volume.

9.3.2 Signal Statistics

The I and Q components of the echo sample are random variables if the scatterer's positions are random. Let us consider two consecutive echo samples spaced T_s seconds apart. The first echo is given by (9.23), and the second can be written as

$$V(\tau_s,T_s) = \sum_i |A_iW_i| \cos\gamma_i - j\sum_i |A_iW_i| \sin\gamma_i \tag{9.25a}$$

where

$$\gamma_i = 4\pi r_i/\lambda + 4\pi v_i T_s/\lambda - \psi_i \tag{9.25b}$$

$v_i = \Delta r_i/T_s$ is the average radial velocity needed to move the ith scatterer by Δr_i, and ψ_i sums the phases ψ_{si},ψ_t, and possibly the phase of W_i. For example, W_i would produce an unwanted phase shift if the gain function $G(f)$ of the two filters in Figure 9.1 had different amplitudes. The range extent of the resolution volume is much larger than the wavelength (i.e., $c\tau/2 \gg \lambda$), and if there are many scatterers within it, we will naturally expect the first term of (9.25b) to have values uniformly distributed between $-\pi$ and π. Although the distribution of $4\pi r_i/\lambda$ is not uniform, its width usually spans many intervals of 2π so that multiple aliasing (of phases into the 2π interval) causes the distribution across 2π to be, for all practical purposes, uniform. Therefore, regardless of the v_i or ψ_i distributions, the phases γ_i are also uniformly distributed. This follows because the distribution of the sum of random variables is obtained after convolving (on a circle from $-\pi$ to π) the individual distributions. Because one of them is uniform, the distribution of the sum will always be uniform.

Now we are in a position to apply the central limit theorem to the real and imaginary parts of (9.25a). The theorem states that a sum of independent random variables tends to have a Gaussian distribution, if their number is large and none of the variables is dominant (i.e., no variable is much larger than the rest). Both conditions are typical for hydrometeors, and thus the $I(\tau_s,T_s)$ and $Q(\tau_s,T_s)$ have Gaussian distribution with zero mean. Figure 9.12a illustrates the distribution of I,Q values when sampled as in Figure 9.11.

Although the probability of the phase ψ_e of $V(\tau_s,T_s)$ is uniform in the interval $\pm\pi$, the change in phase from sample to sample is not necessarily uncorrelated because the phasor can follow the well ordered path sketched in Figure 9.12a. That is, the phase ψ_e can have well ordered changes in time, but its starting position has equal probability of being anywhere in the $\pm\pi$ interval. Although the mth I sample, I_m, is uncorrelated with Q_m, the random processes controlling the changes in the I and Q samples are correlated. The I and Q samples will follow a somewhat ordered path as shown in Figure 9.12a only if echo sample spacing T_s is sufficiently short that each sample is correlated with the previous one. If the path encircles the origin, there is a nonzero mean doppler shift. If stationary ground clutter is

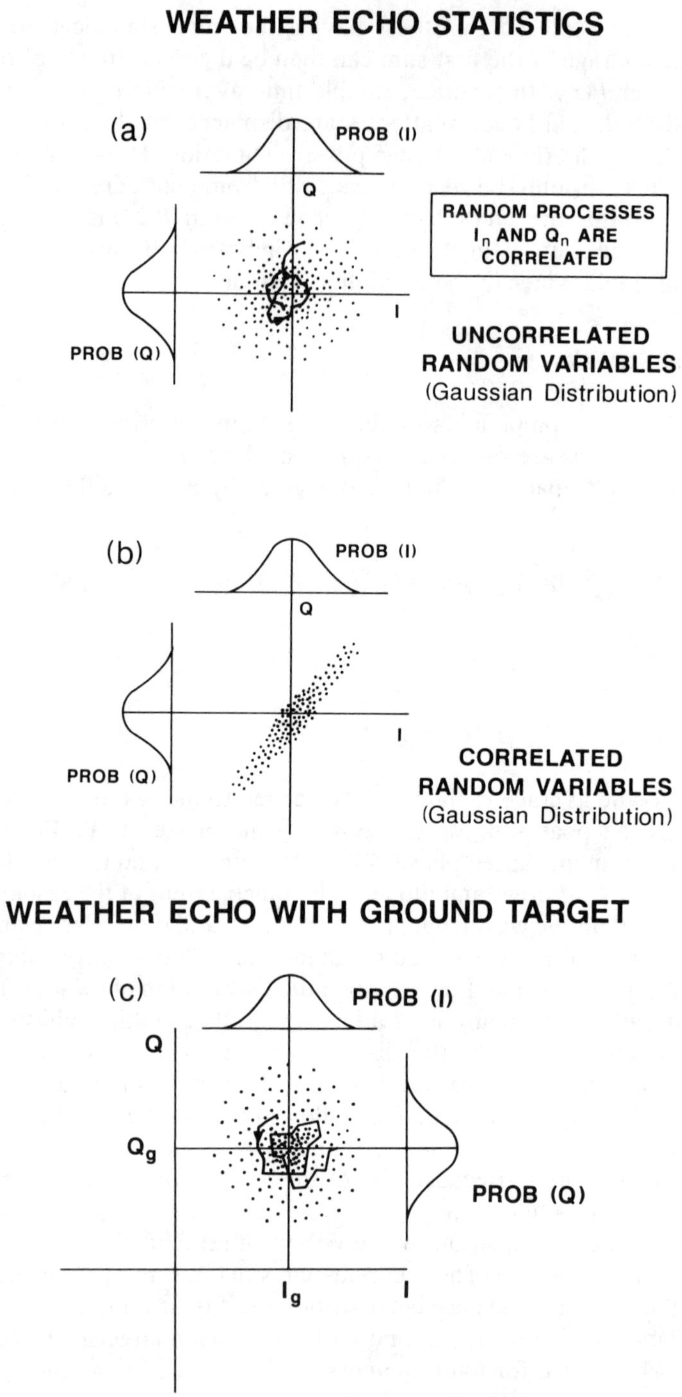

Figure 9.12 (a) The distribution of the amplitudes of I and Q for weather echo samples which are uncorrelated and have Gaussian distribution. (b) Same as in (a), but I and Q are correlated. (c) Distribution of I and Q for weather echo samples superimposed on ground clutter sample I_g, Q_g. Each increment of voltage along the path in (a) and (c) occurs over a PRT interval.

mixed with weather echoes, the distribution can be like that sketched in Figure 9.12c, where the path may encircle a point I_g, Q_g, which is the phasor of the ground target echo.

Because the sample I_m is uncorrelated with the sample Q_m (see Doviak and Zrnić [33], App. B), the distribution of I_m and Q_m is spherically symmetric, as shown in Figure 9.12a. If I_m was correlated with Q_m the distribution of I_m and Q_m can still be Gaussian, but the values of I_m and Q_m will scatter about some line and have a distribution like that sketched in Figure 9.12b.

It is worth noting that

$$I(\tau_s,T_s) = \sum_i |A_iW_i| \cos\gamma_i = |V(\tau_s,T_s)| \cos[\theta(\tau_s,T_s)] \qquad (9.26a)$$

$$Q(\tau_s,T_s) = -\sum_i |A_iW_i| \sin\gamma_i = |V(\tau_s,T_s)| \sin[\theta(\tau_s,T_s)] \qquad (9.26b)$$

The equalities in (9.26) follow because a sum of sinusoids can always be expressed in a sinusoid with a phase $\theta(\tau_s,T_s)$ and an amplitude factor (envelope) $|V(\tau_s,T_s)|$. However, this does not mean that I and Q have pure sinusoidal variation with time, but rather have changes like those shown in Figure 9.11.

Because the in-phase and quadrature components are independent random variables, the joint probability distribution of I and Q is the product of the individual probabilities:

$$p(I,Q) = (\tfrac{1}{2}\,\pi\sigma^2)\exp(-I^2/2\sigma^2 - Q^2/2\sigma^2) \qquad (9.27)$$

where σ^2 is the mean square value of I (equal to that for Q). Using well established procedures (Papoulis [86], p. 418), the probability distributions of $|V|,\theta$, and the power $P(\tau_s)$ can be obtained from (9.26) and (9.27). We can show that the phase θ is independent of the amplitude $|V|$ and is uniformly distributed, while the amplitude $|V| = (I^2 + Q^2)^{\frac{1}{2}}$ has a Rayleigh probability density:

$$p(|V|) = (|V|/2\sigma^2)\exp(-|V|^2/2\sigma^2) \qquad (9.28)$$

Because the power $P(\tau_s)$ is proportional to $I^2 + Q^2$, it follows from (9.27) that $P(\tau_s)$ is exponentially distributed with density:

$$p(P) = (\tfrac{1}{2}\,\sigma^2)\exp(-P/2\sigma^2) \qquad (9.29)$$

and a mean value $\overline{P}(\tau_s) = 2\sigma^2$. Constants of proportionality, and impedance factors that make the transition from (9.27) to (9.29) dimensionally correct, have been ignored. Radar receivers have gains and losses, and therefore the receiver needs to be calibrated with a known input power to relate accurately output $I^2 + Q^2$ to the echo power. The relationship between echo power and radar parameters is developed in Section 9.3.3.

In summary, although I and Q are independent random variables, the random processes controlling $I(\tau_s,mT_s)$ and $Q(\tau_s,mT_s)$ are not independent. This means that, in general, the expected value $E[I(\tau_s,mT_s)Q(\tau_s,kT_s) \neq 0$ for $k \neq m$ (see Doviak and Zrnić [33] for the proof). However, the correlation between two successive samples of the complex signal will be appreciably different from zero, only if the distribution of $4\pi v_iT_s/\lambda$ in (9.25b) is narrow compared to 2π, which is equivalent to saying that the distribution of the v_i is narrow compared to $\lambda/2T_s$ (i.e., the radar's Nyquist velocity interval).

9.3.3 The Weather Radar Equation

We now relate the mean echo power $\overline{P}(\tau_s)$ to the radar parameters and target cross section. The expected power contribution $E[P_i]$ from each scatterer to the mean echo power is, from (9.24),

$$E[P_i] = \alpha\tfrac{1}{2}|W_i|^2 E[|A_i|^2] \tag{9.30}$$

where $\alpha E[|A_i|^2]/2$ is the expected prefilter echo power for the ith scatterer, and hence can be directly expressed in terms of radar parameters and target cross section by using (9.8). We must use the expected value because hydrometeor cross sections are not constant (e.g., drops can have vibrational deformations and snowflakes wobble).

Now consider an elemental volume dV for which the expectation of $|A_i|^2$ (i.e., the average over ensembles of scatterer configurations) for drops within dV is allowed to depend on the location $\mathbf{r}$ of dV. Then the expected echo power $E[dP]$ from dV is

$$E[dP] = \tfrac{1}{2}\alpha \sum_i E[|A_i|^2]$$

which, by using (9.8), can be expressed as

$$E[dP] = I(\mathbf{r}_0,\mathbf{r})\, dV \int_0^\infty \sigma_b(D)N(D,\mathbf{r})dD \tag{9.31a}$$

where

$$I(\mathbf{r}_0,\mathbf{r}) = \frac{Cf^4(\theta - \theta_0,\phi - \phi_0)|W(r_0,r)|^2}{l^2(\mathbf{r})r^4} \tag{9.31b}$$

$$C = P_t g^2\lambda^2/(4\pi)^3 \tag{9.31c}$$

is an illumination function the value of which at $\mathbf{r}$ depends on the beam direction θ_0,ϕ_0, as well as the sampling gate delay τ_s that determines r_0, the range to a resolution volume (to be discussed in more detail in Sections 9.3.4 and 9.3.5). The term $N(D,\mathbf{r})\, dD\, dV$ is the expected number of hydrometeors within dV having diameters in the interval dD about D, and $\sigma_b(D)$ is the expected backscatter cross section for a drop of diameter D. $N(D,\mathbf{r})$, named the *drop size distribution,* is a probability density function which determines the expected volume density of hydrometeors having diameters between D and $D + dD$. Thus, $\sigma_b(D)N(D)\, dD$ is the backscatter cross section per unit volume associated with all the drops within dV of diameters between D and $D + dD$. The integral in (9.31) which depends upon $\mathbf{r}$, is defined as the reflectivity $\eta(\mathbf{r})$;

$$\eta(\mathbf{r}) = \int_0^\infty \sigma_b(D)N(D,\mathbf{r})\, dD \tag{9.32}$$

or the expected backscatter cross section per unit volume. Substituting (9.32) into (9.31) and integrating over all space, we obtain the following equation:

$$\overline{P}(\mathbf{r}_0) = \int_0^{r_2}\int_0^{\pi}\int_0^{2\pi} \eta(\mathbf{r})I(\mathbf{r}_0,\mathbf{r})\, dV \tag{9.33}$$

where

$$dV = r^2 dr \sin\theta\, d\theta\, d\phi$$

The upper limit for the r integration does not extend to infinity because targets beyond some range r_2 cannot return echoes soon enough to contribute to the echoes sampled at delay τ_s. Doviak and Zrnić ([33], Sec. 4.3.1) discuss how the range limit of integration for r is determined, but for now we need only recognize that the r^4 dependence in (9.31c) is not valid for $r < 2D_a^2/\lambda$, where echo power is somewhat independent of r. Thus, the integral (9.33) is an approximation valid for $r_0 > 2D_a^2/\lambda$.

In general, η and l are functions of **r**, but let us assume that over the volume where $f^4(\theta,\phi)|W(r)|^2$ has significant value the reflectivity and attenuation can be considered constant. Furthermore, we assume that the range r_0 to the volume is large compared to its range extent. Thus, we can approximate (9.33) by

$$\overline{P}(\mathbf{r}_0) \approx \frac{P_t g^2 \lambda^2 \eta}{(4\pi)^3 r_0^2 l^2} \int_0^{r_2} |W(r)|^2 \, dr \int_0^{2\pi} \int_0^{\pi} f^4(\theta,\phi) \sin\theta \, d\theta \, d\phi \tag{9.34}$$

If the antenna radiation pattern is circularly symmetric, and if it can be approximated by a Gaussian shape, we can show that

$$\int_0^{\pi} \int_0^{2\pi} f^4(\theta,\phi) \sin\theta \, d\theta \, d\phi = \pi\theta_1^2/8 \ln 2 \tag{9.35}$$

where θ_1 is the 3-dB width (in radians) of the one-way pattern.

9.3.4 The Range Weighting Function

The range-weighting integral in (9.34) causes a loss l_r in weather echo strength defined as

$$l_r^{-1} \equiv \frac{2}{c\tau} \int_0^{\infty} |W(r)|^2 \, dr \tag{9.36}$$

This loss is caused by the finite bandwidth of the receiver that filters some of the echo power. If the receiver's filter has Gaussian frequency response, l_r is (Doviak and Zrnić [33]):

$$\begin{aligned} l_r &= [\coth(\chi) - 1/\chi]^{-1} \\ \chi &= \pi B_6 \tau / 2\sqrt{\ln 2} \end{aligned} \tag{9.37}$$

This loss is a function of both transmitted pulsewidth τ and receiver bandwidth B_6. For radar receiver bandwidths matched to the pulsewidth (i.e., having B_6 equal to τ^{-1}), l_r is significant (i.e., $l_r \approx 1.7$ or 2.3 dB) and needs to be taken into account in the weather radar equation.

9.3.5 The Resolution Volume

The resolution volume V_6 is that volume over which the weighting function $f^4(\theta,\phi)|W(r)|^2$ has a value no less than one-quarter its peak value, or equivalently, that spatial region circumscribed by the 6-dB surface when the weighting function is expressed in decibel units. Thus, the resolution volume has a complicated axially symmetric shape with maximum angular width θ_1 and a range extent, along the beam axis, given by (Doviak and Zrnić [33], Sec. 4.3.2):

$$r_6 = \frac{c\tau/2}{\chi} \cosh^{-1}(2 + \cosh\chi) \tag{9.38}$$

when the filter's response $(G(f)$ is Gaussian. The receiver's bandwidth B_6 is "matched" to the rectangular transmitted pulse of width τ when $B_6\tau = 1$ and then r_6 is about 20% larger than $c\tau/2$, the resolution for $B_6 \rightarrow \infty$. From (9.38) we note that both B_6 and τ determine the range distribution of weights impressed on the cross sections of targets.

9.3.6 The Reflectivity Factors

Radar meteorologists need to relate reflectivity η to factors that have meteorological significance. If scattering particles are known or assumed to be spherical and have diameters that are small compared with wavelength (i.e., the Rayleigh approximation), we can substitute (9.5) into (9.32) to obtain

$$\eta = \frac{\pi^5}{\lambda^4} |K_w|^2 Z \tag{9.39a}$$

where

$$Z \equiv \int_0^\infty D^6 N(D)\, \mathrm{d}D \tag{9.39b}$$

is the reflectivity factor. Whenever the Rayleigh approximation does not apply, and this is usual for short-wavelength ($\lambda < 10$ cm) radars measuring thunderstorm precipitation, we write

$$Z_e \equiv \frac{\lambda^4}{\pi^5} |K_w|^{-2} \eta \tag{9.39c}$$

where Z_e is the equivalent reflectivity factor.

Values of Z or Z_e commonly encountered in weather observations span many orders of magnitude, so radar meteorologists use a logarithmic scale $10 \log_{10} Z$ (where Z is in units of $\mathrm{mm}^6/\mathrm{m}^3$). Precipitation produces Z values ranging from near 0 dBZ in drizzle (rainfall rates as small as 0.1 mm/hr) to values somewhat larger than 60 dBZ in regions of heavy rainfall (rates larger than about 100 mm/hr) and hail. Reflectivity factors do not relate radar echo power to meteorologically significant parameters such as rainfall rate or liquid water content because the drop size distribution $N(D)$ and the substance phase (i.e., $|K_w|^2$), need to be known. Although η or Z_e can be used in the radar equation, radar meteorologists have chosen Z_e.

9.3.7 A Useful Form of the Weather Radar Equation

Incorporating (9.35), (9.36), and (9.39c) into (9.34), we obtain the weather radar equation that gives average echo power in terms of $Z_e(\mathrm{m}^3)$:

$$\overline{P}(\mathbf{r}_0) = \frac{\pi^3 P_t g^2 \theta_1^2 c\tau |K_w|^2 Z_e(\mathbf{r}_0)}{2^{10}(\ln 2)\lambda^2 r_0^2 l^2 l_r} \tag{9.40a}$$

where all units are in the MKS system (i.e., meters, seconds, and watts); θ_1 is in radians; g is dimensionless.

However, it is common to express Z_e in units of $\mathrm{mm}^6\ \mathrm{m}^{-3}$, θ_1 in degrees, r in kilometers, λ in centimeters, τ in microseconds, P_t in watts, and $\overline{P}$ in milliwatts.

Thus, using units conventional to the radar meteorologist, the mean weather echo power for a resolution volume at r_0, θ_0, ϕ_0 is

$$\overline{P}(\text{mW}) = \frac{\pi^5 10^{-17} P_t(\text{W}) g^2 \tau(\mu\text{s}) \theta_1^2(\text{deg}) |K_w|^2 Z_e(\text{mm}^6/\text{m}^3)}{6.75 \times 2^{14} (\ln 2) r_0^2(\text{km}) \lambda^2(\text{cm}) l^2 l_r} \tag{9.40b}$$

9.3.8 Signal-to-Noise Ratio for Weather Echoes

The SNR for weather echoes is (9.40a) divided by (9.21). If the receiver's filter response $G(f)$ is Gaussian, SNR can be shown to depend on B_6 and τ as

$$\text{SNR} = \frac{C_0 \eta}{r^2 l^2} \left[\coth(aB_6\tau) - \frac{1}{aB_6\tau} \right] \frac{\tau^2}{B_6\tau} \tag{9.41}$$

where C_0 contains constants pertaining to the radar. If $B_6\tau$ is equal to one (a reasonable selection because this "matches" the receiver bandwidth to the transmitted pulse width producing a maximum SNR for point targets), SNR is proportional to the square of the transmitted pulsewidth. However, we can show that maximum SNR is obtained as $B_6 \to 0$, in contrast to point-target measurements, where we obtain a maximum SNR at $B_6\tau = 1$ (see (9.22)). Although SNR increases monotonically as B_6 decreases, range resolution r_6 worsens. Zrnić and Doviak [122] define an optimum B_6 as the value that maximizes SNR for a given resolution and conclude that a Gaussian shaped filter matched to a Gaussian shaped transmitted pulse produces the best resolution and the smallest noise bandwidth.

9.4 ATTENUATION

Were it not for electromagnetic energy absorption by water or ice drops, radars with shorter wavelength radiation would be much more utilized because of their superior angular resolution (9.1). However, short-wavelength (e.g., $\lambda \approx$ 3 cm) radiation suffers transmission loss that can be 100 times larger than that for $\lambda \geq 10$ cm.

9.4.1 Attenuation Due to Absorption and Scatter

Each drop absorbs an amount of power that can be expressed as

$$P_L = \sigma_a S_i$$

where σ_a is the absorption cross section that intercepts a power equal to that dissipated as heat in the drop. In contrast with (9.5), there is no simple formula for σ_a that can be used for different wavelengths and for different drop temperatures; hence, we must resort to numerical evaluations of complicated solutions or measurements.

In addition to power absorption, the incident wave suffers power loss from scatter. Analogous to the backscatter cross section is a total scatter cross section σ_{sc} that accounts for the total power scattered by a drop. Total scatter cross section is the area that, when multiplied by the incident power density, gives a power equal to that scattered by the target in all directions; σ_{sc} is proportional to the integral, over a sphere enclosing the target, of the scattered power density $S_r(\theta', \phi')$; for small spherical drops, $\sigma_{sc} = 2\sigma_b/3$.

Thus, the total power extracted from the incident wave is proportional to $\sigma_a + \sigma_{sc}$ defined as the attenuation cross section σ_t. Values of σ_t for four wavelengths are plotted on Figure 9.13. When drop diameters are small relative to the wavelength, the leading terms in the series solution formulated by Mie (*circa* 1908) for the scattering and absorption cross sections are

$$\sigma_{sc} = (2\pi^5 D^6 / 3\lambda^4)|K_w|^2 \tag{9.42a}$$

$$\sigma_a = (\pi^2 D^3/\lambda)\ \mathrm{Im}(-K_w), \tag{9.42b}$$

where Im is the imaginary part of $-K_w$. We can show that $\sigma_{sc} < \sigma_a$ when $D \lesssim \lambda/16$ (i.e., the drop diameter limit for Rayleigh scatter). Figure 9.13 shows that, for drops of the same diameter, shorter wavelengths suffer much more attenuation than longer ones, and the attenuation cross section significantly exceeds the Rayleigh cross section when normalized drop diameters ($\pi D/\lambda$) exceed a value of about 0.04. This value is significantly smaller than that at which the Rayleigh limit for scatter is reached. Although, at centimetric wavelengths, absorption loss dominates scatter loss for all rain rates, it has been shown that when wavelengths are less than one centimeter, scatter loss can dominate absorption at low rain rates (e.g., Doviak and Zrnić [33], p. 198).

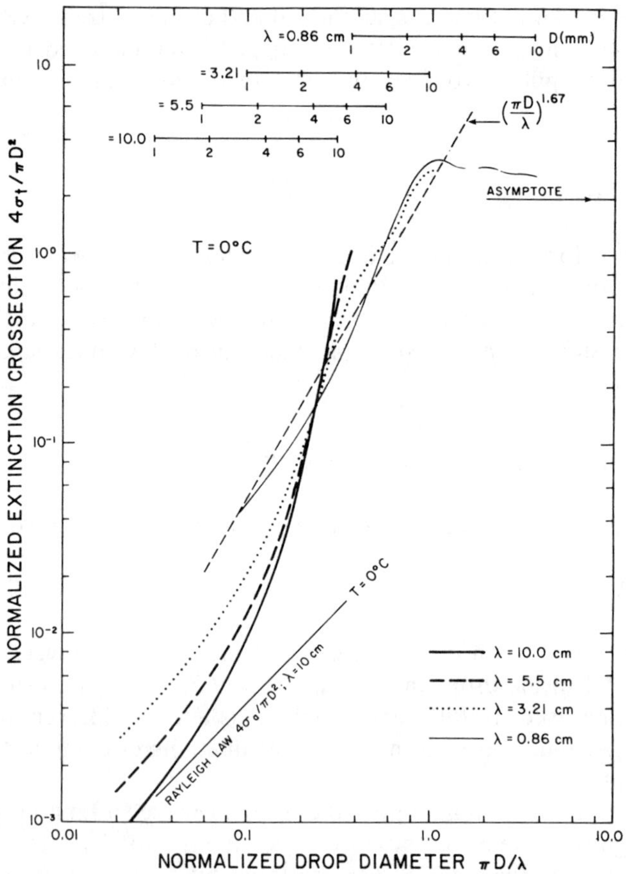

Figure 9.13 The normalized extinction cross section *versus* normalized drop diameter for spherical water drops at four wavelengths (data from Herman et al. [58]). The solid straight line is the asymptotic absorption cross section in the Rayleigh limit. The dashed straight line shows the power for normalized cross sections that fit attenuation data at $\lambda = 0.86$ cm (adapted from Doviak and Zrnić [33]).

If we assume that the presence of drops within an elemental volume $\Delta V(r)$ does not significantly alter the incident power density S_i within this volume (i.e., we can neglect scattering of the scattered field—the Born approximation), the power density change in a wave propagating a short distance Δr through the volume is

$$\Delta S_i = -\frac{\Delta r}{\Delta V} \sum_{n=1}^{N} (\sigma_{an} + \sigma_{scn}) S_i \tag{9.43}$$

where the summation extends over all N drops within ΔV, and σ_{an} and σ_{scn} are the absorption and scatter cross sections of the nth particle. The rate of power density loss is then

$$\frac{dS_i}{dr} = \lim_{\Delta r \to 0} \left(\frac{\Delta S_i}{\Delta r}\right) = -kS_i \tag{9.44a}$$

where

$$k = \lim_{\Delta r \to 0} \sum \frac{(\sigma_{an} + \sigma_{scn})}{\Delta V} \tag{9.44b}$$

$$= \int_0^\infty N(D)\sigma_t(D)\, dD \tag{9.44c}$$

is the *specific attenuation*, or (as denoted by some authors) the *attenuation coefficient*. In the limit $\Delta r \to 0$, S_i can be considered a constant within ΔV and thus placed outside the summation. Then, the power density at any range r is the integral solution of (9.44a):

$$S_i(r_2) = S_i(r_1) \exp\left(-\int_{r_1}^{r_2} k\, dr\right) = S_i(r_1) l^{-1} \tag{9.45}$$

where l is the transmission loss. The specific attenuation K of power density expressed in decibels per kilometer is defined as

$$K \equiv \frac{d}{dr_2}[10 \log l] = 4.34 \times 10^3 k \quad (\text{dB km}^{-1}) \tag{9.46}$$

9.4.2 Specific Attenuation in Rain and Cloud

Values of K for rain (i.e., K_r) have been tabulated by Bean and Dutton [10], pp. 192–296) and Figure 9.14 plots K_r for various rainfall rates and the corresponding reflectivity factors for three radar wavelengths.

When drop diameters are small compared with wavelength, attenuation becomes independent of drop size distribution and depends only on cloud liquid mass density. Data on cloud attenuation rate *versus* wavelength and temperature are given by Gunn and East [53]. Easy-to-use formulas are given by Doviak and Zrnić [33]. Although attenuation through ice clouds is usually negligibly small, it cannot be ignored for water clouds, where the specific attenuation is generally an order of magnitude larger. Clouds between the radar and a distant storm can produce measurable attenuation, even at $\lambda = 10$ cm (e.g., 1.8 dB for two-way propagation through 100 km of 1.0 g m^{-3} liquid water), and thus may cause underestimates of rain in the distant storm.

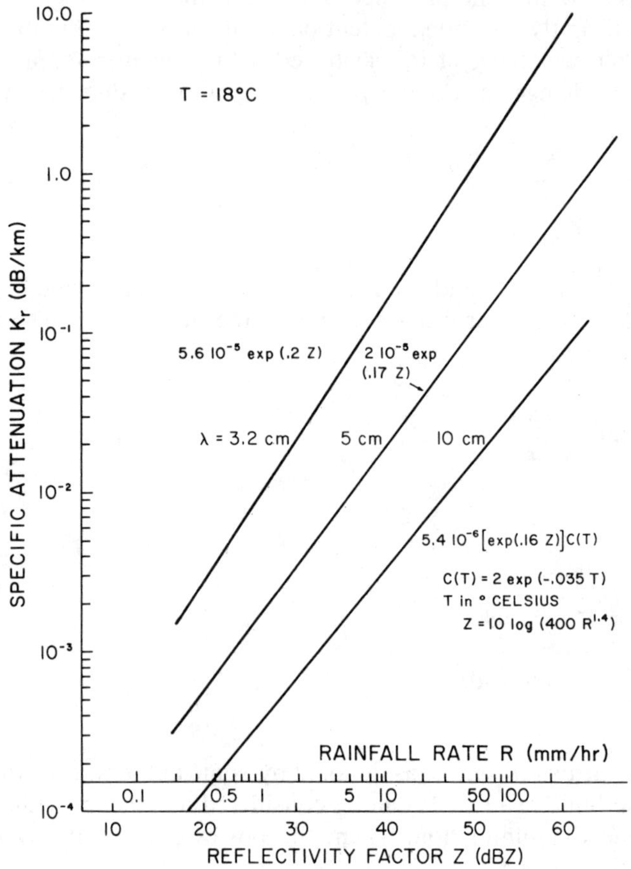

Figure 9.14 **Specific attenuation *versus* rainfall rate for propagation through rain showers (T = 18°C). The Laws and Parsons [72] drop size distribution is assumed. $C(T)$ is a temperature adjustment factor; $Z = 10 \log(400R^{1.4})$, and the units of Z are dBZ (from Doviak and Zrnić [33]).**

9.4.3 Attenuation in Air

Besides attenuation due to rain or cloud droplets, there is attenuation due to energy absorbed by the atmosphere's molecular constituents, mainly water vapor and oxygen. This gaseous specific attenuation k_g is not negligible when storms are far away ($r \gtrsim 60$ km) and beam elevation is low, even at $\lambda = 10$ cm, if accurate reflectivity measurements are required. Blake [12] gives expressions for attenuation due to absorption by water vapor and oxygen for wavelengths (3–10 cm) typically found in weather radars. Because molecular diameters are much smaller than λ, loss due to scatter is negligible. The two-way attenuation due to absorption in the atmosphere's gases at $\lambda = 10$ cm is given in Figure 9.15.

9.4.4 Rainfall Rate Estimation

Next to severe storm hazard warnings (see Section 9.6), rainfall rate (R) estimation is one of the principal applications of weather radar. Accurate rainfall rate estimation with radar is difficult because it does not give a direct measure of R, but rather of Z_e. So, assumptions need to be made, usually about the drop size distribution $N(D)$. There has been a considerable effort to establish whether some of the $N(D)$ parameters can be known for a given type of rain (i.e., stratiform, thunderstorm, *et cetera*). Measurements of $N(D)$ around the globe, under different

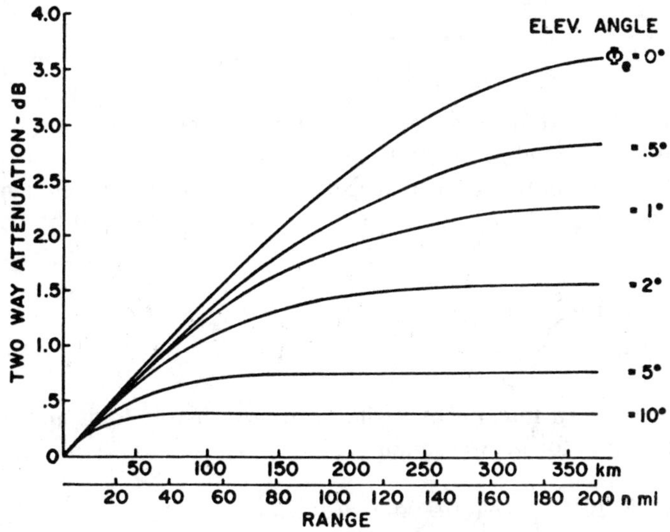

Figure 9.15 **Attenuation (two-way) for propagation at λ = 10 cm through a standard atmosphere (from Doviak and Zrnić [33]).**

climatic conditions, have been made, and Battan [9] lists no fewer than 69 R,Z relations derived from $N(D)$. There, it can be seen that for rain conditions which are supposedly the same (stratiform), the same Z can be associated with R = 33 mm hr^{-1} or R = 11 mm hr^{-1}, a possible 300% error, depending on which measured $N(D)$ is used. Because these R,Z relations were obtained from actual drop size measurement, they are accurate estimates of R and Z for each event. Thus, although both rains were classified as stratiform, they differed significantly in their $N(D)$ content.

Not only do variabilities in $N(D)$ cause errors in R estimated from R,Z_e relations, so, too, does inaccurate radar calibration. It is quite difficult to maintain radars to a calibration accuracy of better than 1 dB, so there can be a systematic bias in the radar measured reflectivity. Furthermore, there are many other factors that affect the accuracy of radar measurements of rain (Zawadski [115]). Some of these errors can be compensated by choosing an appropriate R,Z_e relation (see Doviak and Zrnić [33], Ch. 8).

Despite the superfluity of R,Z_e relations, many of them do not differ greatly at rainfall rates between 20 and 200 mm hr^{-1}. The R,Z_e relation

$$Z_e = 200\,R^{1.6}$$

has proved to be quite useful for stratiform rain. Although the use of this relation may produce large dispersion from actual rain rate measured by a gauge, the accuracy of *rainfall* measurements are greatly improved by averaging in space and time. For further discussions on the use of R,Z_e relations, see Doviak and Zrnić ([33], Ch. 8). We shall now briefly discuss an alternative technique, which although it has practical limitations, can give more accurate measurement of R than the R,Z_e methods.

Measurements made by communication engineers show that specific attenuation K_r and rainfall rate R appear to be uniquely related by a power law for a wide range of rainfall rates and rain types (e.g., stratiform, thunderstorm). Although K_r is dependent on drop size distribution $N(D)$, it has been observed that K_r and R are linearly related, for wavelengths around 1 cm, and most importantly nearly independent of $N(D)$. Atlas and Ulbrich [5] illustrated that the power law dependence of microwave attenuation on rainfall rate implies an effective power law dependence:

$$\sigma_t(D) = CD^n \tag{9.47}$$

of extinction (or attenuation) cross section on drop diameter. Thus, useful analytic formulas could be obtained for K_r by fitting a power law to the attenuation cross sections in Figure 9.13 for the range of drop diameters that contributes significantly to rain. The dependence of K_r (in dB km^{-1}) on $N(D)$ is obtained by combining (9.44b) with (9.46). Thus,

$$K_r = \frac{4.34 \times 10^3}{\Delta V} \sum_n \sigma_{tn}$$

can be expressed as an integral over the drop size distribution, and using the power law expression for σ_t, we obtain

$$K_r = 4.34 \times 10^3 C \int_0^\infty N(D) D^n \, dD \quad (\text{dB km}^{-1}) \tag{9.48}$$

Now, rainfall rate R is given by

$$R = \frac{\pi}{6} \int_0^\infty D^3 N(D) w_t(D) \, dD \quad (\text{m s}^{-1}) \tag{9.49}$$

where $w_t(D)$ is the terminal velocity of water drops (i.e., the speed of the drop relative to the air). In order for R to be expressed as an integral of a power law function of D we use a power law equation for w_t that fits observations. The formula:

$$w_t(D) \approx 386.6 D^{0.67} \quad (\text{m s}^{-1}) \tag{9.50}$$

fits observations quite well in the rain drop diameter range, 5×10^{-4} m $< D <$ 5×10^{-3} m, which contributes significantly to rainfall. Substituting $w_t(D)$ into the integral for R, we obtain

$$R = \frac{\pi}{6} (386.6) \int_0^\infty D^{3.67} N(D) \, dD \tag{9.51}$$

Comparison of this equation with (9.48) immediately shows that if the power exponent $n = 3.67$, then R and K_r are linearly related and, moreover, this relation is independent of drop size distribution. Data show that $n \approx 3.67$ for $\lambda = 0.86$ cm (see Doviak and Zrnić [33], Fig. 8.13). Any wavelength near 1 cm would similarly show that specific attenuation and rainfall rate relations were nearly independent of the drop size distribution. This conclusion is supported by data in Figure 9.16, which displays K_r *versus* R for four wavelengths. The circles represent K_r and R values numerically computed by Atlas and Ulbrich [5] using measured $N(D)$ (204 of those values obtained from three days of observation). The analytic formulas in the inset table of Figure 9.16 are the best fit to the data of the regression equation $\log K_r = \alpha + \beta \log R$. Note how closely packed the data are for $\lambda = 0.86$ cm despite the different drop spectra used, differences that result in R ranging from as little as 0.03 mm/hr to R as large as 60 mm/hr. However, at the longer λ of 3.22 cm the larger scatter is attributed to differences in $N(D)$. Thus, for wavelengths shorter than 3 cm, empirical formulas can be used to relate R to K_r, but at longer wavelengths, drop size spectra must be known to estimate rainfall rate with equally good precision. Moreover, specific attenuation would have to be measured more accurately at long wavelengths for equal precision in R because K_r was smaller.

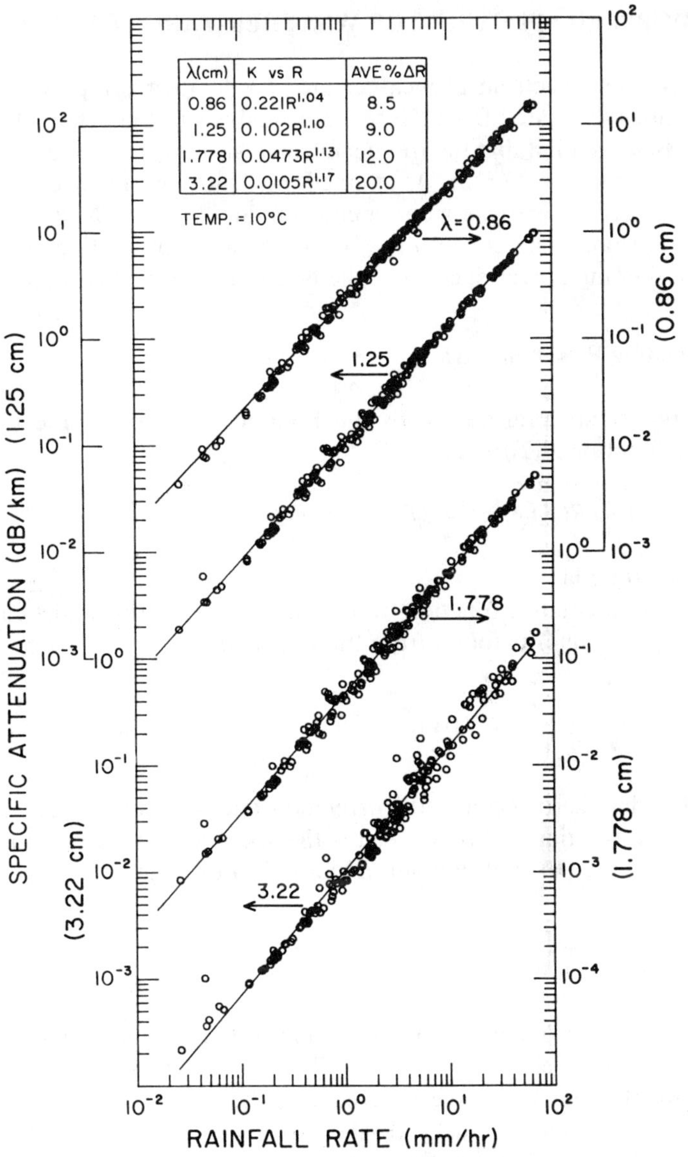

Figure 9.16 Scattergrams and regression lines of K_r *versus* R at indicated wavelengths (T = 10°C). The inset table gives the regression equations and average errors in R (see the text). (Courtesy of C.W. Ulbrich, Clemson University.)

Although specific attenuation measurement offers the potential of measuring rainfall rate with much greater accuracy than conventional radar methods using an empirical relation between Z_e and R, vertical winds (up- and downdrafts) can cause large errors. For example, it has been demonstrated that an updraft of 3 m s^{-1} can result in R errors of at least 50% (Ulbrich [107]). However, when the beam was close to the ground where vertical velocities were nearly zero, remarkably good agreement was obtained between rain dates inferred from K_r measurements and those directly measured with rain gauges beneath the beam, for both thunderstorm and frontal rain (Norbury and White [84]). Nevertheless, there are severe limitations on the range to which measurement can be made because of the large attenuations experienced by the short wavelengths required for the R, K_r technique to be accurate.

DOVIAK & ZRNIĆ
CHAPTER 9

9.5 THE DOPPLER SPECTRUM OF WEATHER TARGET VELOCITIES

The doppler spectrum of weather target velocities is a power spectrum of radial velocities estimated from a frequency analysis of the echo voltage samples $V(\tau_s,mT_s)$. Bear in mind that the frequency analysis is performed along the sample time axis for samples $V(\tau_s,mT_s)$, at fixed τ_s. Thus, we have discrete samples $V(\tau_s,mT_s)$, spaced T_s apart, of a continuous random process. Next, we shall make some general statements concerning spectral analysis of continuous random signals before relating power spectra to the fields of wind and reflectivity.

9.5.1 Estimating Power Spectra

The power spectral density is the Fourier transform of the signal's autocovariance function $R(T_l)$:

$$\overline{S}(f) = \int_{-\infty}^{\infty} R(T_l)\, e^{-j2\pi f T_l}\, dT_l \tag{9.52}$$

where T_l is a time lag.

The autocovariance function of a stationary signal (its statistical properties do not change in time) is found from the time average:

$$R(T_l) = \lim_{T\to\infty} \frac{1}{T} \int_{T/2}^{T/2} V^*(t)V(t + T_l)\, dt \tag{9.53}$$

Because $V(t)$ has zero mean, autocorrelation and autocovariance are identical. Note that conservation of power relates the spectral density per unit frequency $\overline{S}(f)$ to the density per unit doppler velocity $\overline{S}(v)$ through

$$\overline{S}(-v) = \frac{2}{\lambda}\, \overline{S}(f) \tag{9.54}$$

When the spectrum or the autocovariance is known, all pertinent information (e.g., mean doppler velocity) can be readily obtained. However, neither $\overline{S}(v)$ nor $R(T_l)$ is available; they must be estimated from the ensemble of samples $V(\tau_s,mT_s)$ spaced T_s apart. Because radars observe any resolution volume V_6 for a finite time, we are faced with estimating at a given range, $c\tau_s/2$, the power spectrum and its parameters from a finite number M of time samples $V(mT_s)$. (We have deleted the range time argument τ_s, and henceforth $V(mT_s)$ designates a sample sequence at the implicit delay τ_s, whereas T_l increments in steps of T_s.)

Samples $V(mT_s)$, often multiplied by a weighting function $d(mT_s)$, are Fourier transformed, and the magnitude squared of this transformation is a power spectrum estimate $\hat{S}(k)$ commonly known as the periodogram:

$$\hat{S}(k) = \frac{1}{M} \left| \sum_{m=0}^{M-1} d(mT_s)V(mT_s)\, e^{-j2\pi km/M} \right|^2 \tag{9.55}$$

where k and m are integers, and unit increments of k correspond to velocity increments of $2v_a/M = \lambda/(2MT_s)$. However, k need not be an integer, in which case $\hat{S}(k)$ is a continuous function of k with period M.

The finite number of time samples from which the periodogram is computed limits velocity resolution and creates an undesirable "window effect"; i.e., we may imagine that the time series extends to infinity, but is observed through a finite-length window MT_s. The magnitude squared of the data window transform is referred to as the spectral window and is significant because its convolution with the true spectrum equals the measured spectrum (Doviak and Zrnić [33], Sec. 5.15).

Figure 9.17 illustrates the spectrum of weather echo samples weighted with a uniform window and one with a von Hann (raised cosine) window, showing considerable difference in the spectral domain, especially in spectral skirts. The echo samples are identical in both cases. The dynamic range for spectral coefficients is about 30 dB for the data weighted uniformly, whereas there is more than a 45 dB of dynamic range for data weighted with the von Hann window, resulting in an estimated spectrum which better defines the true spectrum and its maximum velocity (60 m s^{-1}). Since the von Hann window has weights with a gradual transition between no data and data points, the spectral window has a less concentrated main lobe but significantly lower side lobes, enabling us to observe weak signals in the spectral skirts to more than 40 dB below the main peak (Doviak and Zrnić [33], Sec. 5.15). This is very significant when we are trying to estimate the peak winds of tornadoes or other severe weather (Zrnić [118]) within the resolution volume because power associated with high velocities in the spectral skirts is rather weak and will be masked by the strong spectral peaks that are seen through the window sidelobes, unless a suitable window is applied. The example in Figure 9.17 is from a tornadic circulation with translation; the broad spectrum results from high-speed circulatory motion within the resolution volume. The apparent lack of randomness of coefficients in the spectral skirts for the rectangularly weighted data is due to the larger correlation between coefficients. This correlation is attributed to the strong spectral powers that are seen through the nearly constant-level window sidelobes (Zrnić [118]).

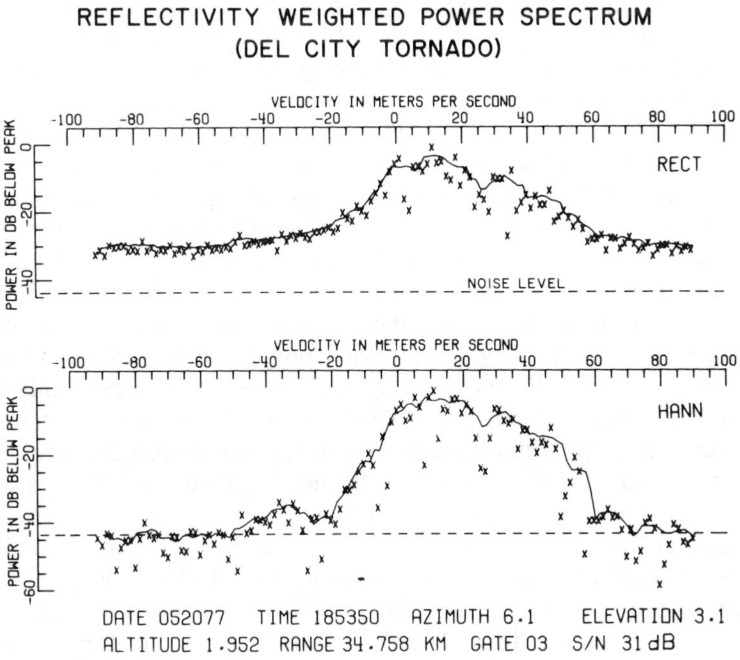

Figure 9.17 **Power spectra of weather echoes showing statistical fluctuations in spectral estimates (x). RECT (upper trace) signifies spectra of echo samples weighted uniformly, whereas HANN (lower trace) signifies samples weighted by a von Hann window function. Solid curves are five-point running averages of spectral estimates (from Doviak and Zrnić [33]).**

Besides the window effect, which is intimately tied to signal processing, there are several spectral artifacts due to the radar hardware, which are discussed by Doviak and Zrnić ([33], Sec. 7.9).

9.5.2 Doppler Spectra Related to Velocity and Reflectivity Fields

The doppler spectrum is a power-weighted distribution of the radial velocities of the scatterers that lie mostly within the resolution volume. The power weight depends not only on the reflectivity of the scatterers, but also on the weighting given to them by the antenna pattern, the transmitted pulse shape, and the receiver's response to it. Doviak and Zrnić [33] derive the relationship among the velocity and reflectivity fields, the resolution volume weighting functions, and the doppler spectrum. They show that the spectral density (i.e., power per unit velocity) $\overline{S}(v)$ of the doppler velocities is

$$\overline{S}(\mathbf{r}_0,v) = \iint_A \eta(\mathbf{r}) I(\mathbf{r}_0,\mathbf{r})\, |\text{grad } v(\mathbf{r})|^{-1}\, dA \tag{9.56}$$

where $\mathbf{r}_0$ and $\mathbf{r}$ are the radius vectors to the resolution volume and the elemental scattering volume. A is an area on a surface of constant doppler velocity $v(\mathbf{r})$, and $I(\mathbf{r}_0,\mathbf{r})$ is the illumination function (9.31b).

Equation (9.56) is fundamental and worthy of more discussion. The area A consists of all *isodop* surfaces (surfaces of constant doppler velocity); thus, it is a union of such surfaces. At each point $\mathbf{r}$ on the surface, the reflectivity is multiplied with the corresponding illumination function. The gradient term adjusts the isodops' contribution according to their density; hence, the closer the isodop surfaces, the smaller is the weight applied to the spectral components in the velocity interval between two isodops.

Because hydrometeors between the isodop surfaces $v, v + dv$ are moving relative to one another and being replaced with new hydrometeors, $S(\mathbf{r}_0,v)$ changes randomly, whereas (9.56) determines its average or expected value. If there are many hydrometeors of nearly the same cross sections within these isodop surfaces, $S(\mathbf{r}_0,v)$ has an exponential distribution because then the central-limit theorem applies to the sum of the voltages (i.e., the voltages I and Q have Gaussian distributions, see Section 9.3.2). Thus, the periodograms have spectral coefficients (e.g., denoted by x on Figure 9.17) and their power (on a linear scale) is exponentially distributed. Only by averaging a large number of these randomly varying periodograms (for weather echoes having stationary statistical properties) do we obtain, in the limit, a spectrum equal to $\overline{S}(\mathbf{r}_0,v)$ convolved with the window spectrum. The five-point running average (solid line on Figure 9.17) gives a better estimate of $\overline{S}(\mathbf{r}_0,v)$ at the expense of velocity resolution.

Doviak and Zrnić ([33], Sec. 5.2) have shown that, when shear and reflectivity fields are uniform, the power spectrum shape follows the shape of the illumination function (9.31b). Because the illumination function for weather radars can, in most cases, be approximated by a Gaussian function, the doppler spectrum will have nearly a Gaussian shape. It has been found that a large proportion of observed doppler spectra are Gaussian shaped and we may infer that the reflectivity and radial velocity shear are then uniform within the resolution volume V_6. However, another reason is related to the fact that several other spectral shaping mechanisms (e.g., window function, turbulence, *et cetera*) act independently, and thus the resultant spectral shape can be deduced by considering each mechanism separately, as we now demonstrate.

The observed spectral shape is associated with both velocity distributions as well as changes in echo amplitude or phase not caused by the relative motion of targets. Furthermore, target velocity is a linear sum of velocities associated with various kinds of motion (e.g., uniform wind, shear, turbulence, *et cetera*) and each contributes to the spread of the doppler spectrum (the uniform wind contributes spread because radial velocities vary across V_6). Shear produces a spread (proportional to its magnitude) of doppler velocities about the mean radial velocity of the uniform wind. Likewise, turbulence provides a spectrum of doppler velocities with a spread proportional to the intensity of turbulence. We now show that the observed power spectrum is obtained from the convolution of the respective spectra, whether they are associated with real or apparent velocity distributions.

The in-phase and quadrature phase signal components obtained from (9.23) at sample time $mT_s = 0$ are

$$V(\tau_s,0) = I_0 + \mathrm{j}Q_0 = \sum_i A_i(0)F_i(0)\,\mathrm{e}^{-\mathrm{j}\phi_i} \tag{9.57}$$

where $\phi_i \equiv 4\pi r_i/\lambda$ and $|F_i(0)|^2 = I(\mathbf{r}_0,\mathbf{r}_i)$ is the illumination (9.31b) impressed on the ith target. (Note that A_i does not incorporate the antenna weighting function, which is now part of $I(\mathbf{r}_0\mathbf{r}_i)$.) These components for subsequent echo samples at the same range time τ_s, but mT_s later, are

$$V(\tau_s,mT_s) = I_m + \mathrm{j}Q_m = \sum_i A_i(mT_s)F_i(mT_s)\,\mathrm{e}^{-\mathrm{j}\zeta_i} \tag{9.58}$$

where $\zeta_i = \phi_i + 4\pi v_i mT_s/\lambda$. It is important to keep in mind that the phases ϕ_i have a uniform distribution over the interval of width 2π, whereas the velocities v_i are usually concentrated around a mean velocity. Strictly, v_i is an average radial velocity that takes the ith scatterer to its new position in the time mT_s. Furthermore, we assume that the $A_i s$ are independent of the phases ϕ_i and velocities v_i (i.e., the scatterer's cross section and phase shift ψ_{si} due to scattering should not depend on its position or velocity), and the weight of the illumination function applied to any target does not change appreciably (i.e., $F_i(0) \approx F_i(mT_s)$) while the target moves during the time of spectral estimation (for an alternative and less restrictive condition, see Section 9.3.2).

A_i changes in time in a random way because the drop may oscillate or change its canting angle (i.e., the angle between the vertical and its axis of symmetry). Thus, the autocorrelation of echo samples is

$$\begin{aligned} R(mT_s) &= \mathrm{E}[V^*(\tau_s,0)V(\tau_s,mT_s)] \\ &= \sum_i\sum_k F_i^*F_k\,\mathrm{E}[A_i^*(0)A_k(mT_s)] \\ &\quad\times \mathrm{E}[\exp\{\mathrm{j}(\phi_i - \phi_k - 4\pi v_k mT_s/\lambda\}] \\ &= \sum_k C\sigma_{bk}(mT_s)|F_k|^2\,\mathrm{E}[\exp\{-\mathrm{j}4\pi v_k mT_s/\lambda\}] \end{aligned} \tag{9.59}$$

where

$$\sigma_{bk}(mT_s) \equiv \mathrm{E}[A_k^*(0)A_k(mT_s)]/C$$

and C is defined in (9.31c). When $m = 0$, we recognize that $\sigma_{bk}(0)$ is the expected backscatter cross section, or the time-averaged backscatter cross section, if the statistical properties are time independent. We can show, for Rayleigh scatter, horizontal propagation, and vertically polarized waves, that the average backscatter cross section of a vibrating drop is larger than that for a nonvibrating drop of equal volume. On the other hand, when waves are horizontally polarized, average backscatter cross section decreases when drops vibrate. The double summation reduces to a single one because the expectations with respect to the exponential argument are zero, except at $i = k$. As we can see, the autocorrelation is independent of the initial phases ϕ_i.

Examination of (9.59) reveals that there is a spectral broadening mechanism due to the change in hydrometeor shape or orientation with respect to the incoming electric field. This change is described by the autocorrelation $\sigma_{bk}(mT_s)$, which has a maximum at $m = 0$ and decreases as $m \rightarrow \infty$. Thus, $\sigma_{bk}(mT_s)$ produces a narrower $R(mT_s)$, and consequently a broader doppler spectrum than will be obtained without change in hydrometeor shape or orientation. The amount of spectral broadening due to this effect has not been quantified.

The spectral broadening mechanisms in (9.59) act through product terms. For example, the kth scatterer's velocity v_k is the sum of velocities due to mean wind, shear, and turbulence, which move the scatterer from one radial position to the next in the internal mT_s. However, because v_k appears in the exponent of (9.59), the velocities associated with mean wind, shear, and turbulence can each be placed into separate exponential functions that multiply one another, and then the spectrum (9.59) can be expressed as the convolution of the spectrum associated with each function. We can show that, under certain general conditions, the convolution of a number n of spectra approaches a Gaussian function as n increases, independent of the shape of each spectrum (Papoulis [86], p. 267). Because several mechanisms can contribute to the doppler spectrum, we see why the observed spectra often appear to be Gaussian in shape.

We now demonstrate that if hydrometeors are perfect tracers of homogeneous turbulent motion (i.e., they do not vibrate, wobble, or move relative to the air), the normalized power spectrum:

$$\overline{S}_n(\mathbf{r}_0,v) \equiv \frac{\overline{S}(\mathbf{r}_0,v)}{\int_{-\infty}^{\infty} \overline{S}(\mathbf{r}_0,v)\,\mathrm{d}v} \tag{9.60}$$

is equal to the probability of velocities of the turbulent flow. Note that the integral in the denominator is the total power and equals the volume integral (9.33).

If hydrometeors did not vibrate or wobble, then

$$\sigma_{bk}(mT_s) = \sigma_{bk}(0) \tag{9.61}$$

If a scatterer selected randomly has the same probability distribution of velocities $p(v)$ as any other scatterer (i.e., the turbulent flow is statistically homogeneous), we can express the expectation of the exponential as

$$\mathrm{E}[\mathrm{e}^{-\mathrm{j}4\pi v_k mT_s/\lambda}] = \int_{-\infty}^{\infty} p(v)\,\mathrm{e}^{-\mathrm{j}4\pi v mT_s/\lambda}\,\mathrm{d}v \tag{9.62}$$

Substituting (9.61) and (9.62) into (9.59), we obtain for the normalized correlation function:

$$\rho(mT_s) \equiv \frac{R(mT_s)}{R(0)} = \int_{-\infty}^{\infty} \mathrm{p}(v)\,\mathrm{e}^{-\mathrm{j}4\pi v mT_s/\lambda}\,\mathrm{d}v \tag{9.63}$$

where

$$R(0) \equiv \sum_k C\sigma_{bk}(0)\, I(\mathbf{r}_0,\mathbf{r}_k).$$

However, because the normalized correlation function is related to the normalized power spectrum as

$$\rho(mT_s) = \int_{-\infty}^{\infty} S_n(f)\,\mathrm{e}^{\mathrm{j}2\pi f mT_s}\,\mathrm{d}f \tag{9.64}$$

we can equate (9.63) to (9.64), and use (9.54) and the doppler shift relation, $f = -2v/\lambda$, to obtain

$$p(v) = \overline{S}_n(-v). \tag{9.65}$$

Thus, for homogeneous turbulence, the normalized power spectrum is equal to the velocity probability distribution. Although in this derivation we did not state that reflectivity is uniform, our assumption that the hydrometeor's size does not depend on its position or velocity implicitly assumes reflectivity to be uniform.

9.5.3 The Moments of the Doppler Spectrum

The first three moments of the doppler spectrum, (1) power or the area beneath the power spectrum, (2) mean velocity, and (3) velocity variance, are of principal interest to radar meteorologists. The methods used to estimate these moments and the accuracies of estimation are discussed in Doviak and Zrnić ([33], Ch. 6), whereas a detailed analysis of these estimators as well as maximum likelihood estimation is treated by Zrnić [126]. Here, we shall present some basic formulas that relate the normalized dopper spectrum (9.60) to the fields of reflectivity, velocity, and turbulence.

The doppler radar measured reflectivity is a spatial weighted averaged reflectivity $\overline{\eta}(\mathbf{r}_0)$,

$$\overline{\eta}(\mathbf{r}_0) \equiv \frac{\iiint \eta(\mathbf{r}) I(\mathbf{r}_0,\mathbf{r})\, dV}{\iiint I(\mathbf{r}_0,\mathbf{r})\, dV} \tag{9.66}$$

The integral in the numerator is the total power (i.e., (9.33)) whereas the value of the denominator can be estimated using (9.35) to (9.37). Now, the mean doppler velocity is defined as

$$\overline{v}(\mathbf{r}_0) \equiv \int_{-\infty}^{\infty} v \overline{S}_n(\mathbf{r}_0,v)\, dv \tag{9.67}$$

The spectral width $\sigma_v(\mathbf{r}_0)$ of the doppler spectrum is obtained from

$$\sigma_v^2(\mathbf{r}_0) = \int_{-\infty}^{\infty} [v - \overline{v}(\mathbf{r}_0)]^2 \overline{S}_n(\mathbf{r}_0,v)\, dv \tag{9.68}$$

which is the second moment of the doppler velocity spectrum about its mean velocity.

The relationship between the field of point velocities $v(\mathbf{r})$ and the power-weighted moment $\overline{v}(\mathbf{r}_0)$ is obtained by substituting (9.56) and (9.60) into (9.67):

$$\overline{v}(\mathbf{r}_0) = \frac{\iiint v(\mathbf{r})\eta(\mathbf{r}) I(\mathbf{r}_0,\mathbf{r})\, dV}{\iiint \eta(\mathbf{r}) I(\mathbf{r}_0,\mathbf{r})\, dV} \tag{9.69}$$

This shows that, unlike the $I(\mathbf{r}_0,\mathbf{r})$ weighted average reflectivity (9.66), $\overline{v}(\mathbf{r}_0)$ is the average of point velocities weighted by both reflectivity and the illumination function. Similarly, (9.68) reduces to

$$\sigma_v^2(\mathbf{r}_0) = \frac{\iiint v^2(\mathbf{r})\eta(\mathbf{r}) I(\mathbf{r}_0,\mathbf{r})\, dV}{\iiint \eta(\mathbf{r}) I(\mathbf{r}_0,\mathbf{r})\, dV} - \overline{v}^2(\mathbf{r}_0) \tag{9.70}$$

and it also corresponds to reflectivity and illumination weighted deviation of velocities from the mean doppler velocity. Most contributions to $\overline{v}(\mathbf{r}_0)$ and $\sigma_v^2(\mathbf{r}_0)$ usually come from the region wherein $I(\mathbf{r}_0,\mathbf{r})$ is large, namely the resolution volume V_6. Thus, $\overline{v}$ and σ_v^2 are considered to be values representative of the spatial averages over V_6.

The mean doppler velocity (9.69) depends on the distribution of scatterers' cross section within the resolution volume and its weighting functions. Thus, $\overline{v}(\mathbf{r}_0)$ cannot in general be equated to a spatial mean velocity. However, if reflectivity and illumination are symmetrical about the resolution volume center and if radial wind changes linearly across V_6, then $\overline{v}(\mathbf{r})$ is a very good approximation of the true radial component of wind at the volume center. Targets such as water drops having small mass quickly respond to horizontal wind forces and faithfully trace the wind. Stackpole [104] has shown that for energy spectrum of wind scales following a $-\frac{5}{3}$ law to at least 500 m, more than 90% of the rms wind fluctuations are acquired by the drops, if their diameters are less than 3 mm. When radar beams are at low elevation angles, hydrometeor terminal velocity w_t (i.e., the steady-state vertical velocity of the hydrometeor relative to the air) gives negligible error in the radial wind component. At high elevation angles, terminal velocity must be considered. (See Section 9.6.1(a)).

9.5.4 Spectral Width

The spectral width (i.e., the square root of the second spectral moment about the mean velocity) is a function of both the radar system parameters, such as beamwidth, bandwidth, and pulsewidth, and the meteorological parameters that describe the distribution of hydrometeors and their velocity within the resolution volume. Relative radial motion between targets broadens the spectrum. For example, turbulence produces random relative radial motion of drops. Wind shear can cause relative radial target motion, as can differences in the speed of fall for various sizes of drops. Drop oscillations or rotation of hydrometeors produce modulation of the echo signal which increases the spectral width. There is also a contribution to spectral width caused by the beam sweeping through space (i.e., the radar does not receive echoes from identically weighted targets on successive samples). This change in the location of the resolution volume V_6 from one transmitted pulse to another results in a decorrelation of echo samples and a consequent increase in the spectral width σ_v (Doviak and Zrnić [33], App. C). The echo samples become uncorrelated more quickly (independent of particle motion inside V_6) with faster antenna rotation. Thus, spectral width increases in proportion to the antenna angular velocity.

Because the cited spectral broadening mechanisms are independent of one another, the square of the velocity spectral width σ_v^2 can be considered as a sum of the variances contributed by each. That is,

$$\sigma_v^2 = \sigma_s^2 + \sigma_\alpha^2 + \sigma_d^2 + \sigma_t^2 + \sigma_h^2 \tag{9.71}$$

where σ_s^2 is due to shear, σ_α^2 to antenna motion, σ_d^2 to different terminal velocities for different sized drops, σ_t^2 to turbulence, and σ_h^2 to hydrometeor oscillation and rotation. Some authors add yet another term which explicitly accounts for spectral broadening due to finite beam width. Even when the wind is uniform, radial velocity will change across V_6. We shall consider this term to be part of σ_s because uniform wind produces angular shear of doppler (radial) velocities, and hence this contribution to spectral broadening is automatically taken into account by our definition of σ_s, to be given shortly. Because broadening caused by the mean relative motion between the radar and targets is small except in the case of an airborne radar, σ_α refers only to antenna rotation. The significance of the width σ_v for the weather radar design is discussed by Doviak and Zrnić ([33], Ch. 7).

The contribution σ_d^2 to the spectrum's second moment due to the different radial components of terminal velocities of the assorted-size drops is

$$\sigma_d^2 = (\sigma_{d0} \sin\theta_e)^2 \tag{9.72}$$

The width σ_{d0} is caused by the spread in terminal velocities of varous size drops falling relative to the air contained in V_6. Lhermitte [74] has shown that for rain $\sigma_{d0} \approx 1.0 \text{ m s}^{-1}$ and is nearly independent of the drop size distribution. The elevation angle $\theta_e = (\pi/2) - \theta_0$ is measured with respect to the beam center.

If the antenna pattern is Gaussian with a one-way half-power pattern width θ_1 and rotates at an angular velocity of α, the spectral width due solely to resolution volume displacement is (Doviak and Zrnić [33], App. C):

$$\sigma_\alpha^2 = (\alpha\lambda \cos \theta_e/2\pi\theta_1)^2 \ln 2 \tag{9.73}$$

Doviak and Zrnić [33] prove that the wind shear term can be separated into three contributions. If k_θ, k_ϕ, k_r are the shears in the directions θ, ϕ, r, then

$$\sigma_s^2 = \sigma_{s\theta}^2 + \sigma_{s\phi}^2 + \sigma_{sr}^2 = (r_0\sigma_\theta k_\theta)^2 + (r_0\sigma_\phi k_\phi)^2 + (\sigma_r k_r)^2 \tag{9.74}$$

where σ_θ^2 and σ_ϕ^2 are defined as the second central moments of the two-way antenna power pattern in the indicated directions, and σ_r^2 is the second central moment of the weighting function $|W(r)|^2$. A circularly symmetric Gaussian radiation pattern has

$$\sigma_\theta^2 = \sigma_\phi^2 = \theta_1^2/(16 \ln 2) \tag{9.75}$$

For a rectangular transmitted pulse and a Gaussian receiver frequency response under matched conditions (i.e., $B_6\tau = 1$),

$$\sigma_r^2 = (0.35c\tau/2)^2 \tag{9.76}$$

The width σ_t due to turbulence is somewhat more difficult to derive. For turbulence that is isotropic within V_6, Frisch and Clifford [43] have shown that σ_t^2 is related to the eddy dissipation rate ϵ. (A detailed discussion of the contribution of turbulence and shear to the spectra width is presented in Ch. 10 of Doviak and Zrnić's text [33], where examples of data fields are also given.)

9.6 OBSERVATION OF WEATHER WITH DOPPLER RADAR

The great utility of storm observation with a centimeter wavelength pulsed doppler radar derives from its capacity to map reflectivity $\bar{\eta}$ and mean radial velocity $\bar{v}$ inside the storm's shield of clouds. A three-dimensional picture of a single storm takes about two to five minutes of data collection time, not only because of antenna rotation limitations, but also because many echoes from each resolution volume must be processed to reduce the statistical uncertainty in the $\bar{\eta}$ and $\bar{v}$ estimates. Although storm structure can change significantly during this period with distortion in the radar's image of the true reflectivity and velocity fields, highly significant achievements have been made in depicting the structure and evolution of the thunderstorm.

Nonetheless, the meteorologically interesting variables are not $\bar{\eta}$ and $\bar{v}$, but parameters such as rainfall rate (on the ground) and wind. Pulsed doppler radar measures the reflectivity weighted average radial speed of hydrometeors, not air, in the resolution volume V_6, and under certain situations (e.g., with vertically directed beams) these speeds can differ significantly from the spatially averaged (over V_6) radial component of wind. Surface rainfall rate estimates are also not easily related to $\bar{\eta}$, and often radar reflectivity measurements are supplemented by surface rain gauges (Brandes [15]; Crawford [26]; Ahnert *et al.* [2]).

9.6.1 Dual-Doppler Radar

A single-doppler radar maps the wind component directed toward (or away from) the radar. A second doppler radar, spaced far from the first, provides a field of different radial velocities because the wind vector is projected on different radials. The two radial-velocity fields can be vectorially synthesized to retrieve the two-dimensional velocities in the plane containing the radials (Armijo [3]). The synthesis is usually accomplished on common grid points to which doppler velocity data are interpolated. However, radial velocities in each of the radar's resolution volumes surrounding a grid point are not measured simultaneously, but rather at times that can be separated by as much as the few minutes required for each radar to scan, for example, a storm. The respective resolution volumes are also usually quite different in size and orientation. Nevertheless, useful estimates of wind can be made on scales of air motion large compared with the biggest V_6 dimension in the region of interest, if the velocity field is quasi-stationary during the period required for data collection.

(a) *The Synthesis of Wind Fields*

Wind field determination from data of two doppler radars is greatly simplified, if the synthesis is performed in cylindrical coordinates with axis chosen to be the baseline connecting the two radars. That is, mean radial velocities (i.e., roughly the reflectivity weighted spatial average over V_6) at data points (i.e., centers of V_6) are interpolated to nearby grid points on tilted planes having the common axis (the COPLAN technique; Lhermitte [74]) depicted in Figure 9.18. Cartesian wind components, commonly used for meteorological interpretation, can be derived from these synthesized cylindrical components. Although we could solve directly for Cartesian wind components, this would necessitate solving an inhomogeneous hyperbolic partial differential equation to derive the vertical wind (Armijo [3]).

The mean doppler velocity must be corrected for the illumination function and reflectivity weighted mean (i.e., approximately the average over V_6) terminal velocity $\overline{w}_t$. Thus, the estimate of the radial component of air motion is

$$\overline{v}_{1,2} = \overline{v}'_{1,2} + \overline{w}_t \sin \theta_{e1,2} \tag{9.77}$$

where $\overline{v}'_{1,2}$ are the mean doppler target velocities measured by radars 1 and 2 at data points, $\overline{w}_t$ is positive, and θ_e is the elevation angle. To estimate $\overline{w}_t$ for each V_6, we could use the empirical expression (Atlas *et al.* [6]):

$$\overline{w}_t = 2.65\overline{Z}^{0.114} \left(\frac{\gamma_0}{\gamma}\right)^{0.4} \quad (\text{m s}^{-1}), \tag{9.78}$$

where the parenthetical term is that suggested by Foote and duToit [42] to account for height-dependent air density γ, and $\overline{Z}$ is the reflectivity factor averaged over V_6. This relation well represents experimental data over a large range of $\overline{Z}$ (i.e., $1 \leqslant \overline{Z} \leqslant 10^5$ mm^6 m^{-3}) for regions of liquid water, but large errors up to several meters per second in $\overline{w}_t$ estimates can be caused by erroneously relating regions of hail with a $\overline{w}_t$,$\overline{Z}$ relation appropriate for liquid water. Usually, there is little or no information to identify these regions uniquely, and errors in vertical wind, w_z, can result. However, it has been shown for typical arrangements of storms relative to the two-radar placement that these are significantly smaller than errors in $\overline{w}_t$ (Doviak *et al.* [36]).

DOVIAK & ZRNIĆ
CHAPTER 9

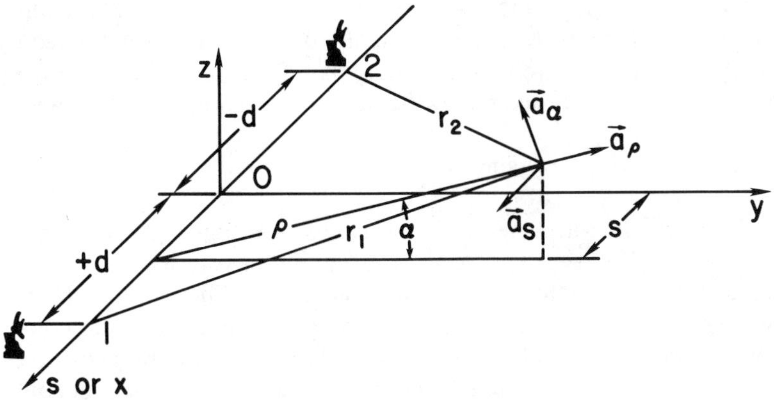

Figure 9.18 **Cylindrical coordinate system for dual-radar data analysis. The radars are located at the points 1 and 2; $\hat{a}_\rho$, $\hat{a}_s$, and $\hat{a}_\alpha$ are the unit normal defining directions of the three orthogonal velocity components (from Doviak and Zrnić [33]).**

The estimated radial velocities $\bar{v}_{1,2}$ of the air can be interpolated to uniformly spaced grid points in planes tilted at angle α to the horizontal surface containing the baseline. Interpolation filters the data and reduces variance. The cylindrical wind components in the ρ,s plane are related to $\bar{\bar{v}}_1,\bar{\bar{v}}_2$ as

$$w_\rho = \frac{(s + d)r_1\bar{\bar{v}}_1 - (s - d)r_2\bar{\bar{v}}_2}{2d\rho} \tag{9.79a}$$

and

$$w_s = \frac{r_2\bar{\bar{v}}_2 - r_1\bar{\bar{v}}_1}{2d} \tag{9.79b}$$

where $\bar{\bar{v}}_{1,2}$ are the interpolated doppler velocities of air. The double overbar signifies the two-step averaging process: (1) an average over V_6 and (2) an interpolation average of $\bar{v}_{1,2}$ typically over distances of 1–2 km. Although we have not placed bars over the wind components, the reader is to understand that these also are resolution and interpolation volume averaged values.

The wind component, w_α, normal to the plane, can be obtained by solving the continuity equation in cylindrical coordinates:

$$\frac{1}{\rho}\frac{\partial}{\partial\rho}(\rho\gamma w_\rho) + \frac{1}{\rho}\frac{\partial}{\partial\alpha}(\gamma w_\alpha) + \frac{\partial}{\partial s}(\gamma w_s) = 0 \tag{9.80a}$$

with the boundary condition being that $w_\alpha = 0$ at the ground. The air's mass density γ is given by

$$\gamma = \gamma_0 \exp[-(g_0 M\rho \sin\alpha)/(RT)] \tag{9.80b}$$

where g_0 is the gravitational constant (9.8 m s^{-2}), M the mean molecular weight of air (29 g mol^{-1}), R the universal gas constant (8.314 J mol^{-1} K^{-1}), and T the absolute temperature (K). Appropriate values of γ_0 and T can be obtained from surface instruments and upper-air soundings.

The accuracy with which wind vectors can be synthesized from the two radial velocity data fields depends principally on the angle β subtended by $\mathbf{r}_1$ and $\mathbf{r}_2$. Formulas for wind estimation accuracy as a function of ρ, s, and α are given by Doviak *et al.* [36]. For example, the sum of the variances of w_ρ and w_s is

$$\sigma_\rho^2 + \sigma_s^2 = (\sigma_1^2 + \sigma_2^2)/\sin^2 \beta \tag{9.81a}$$

where σ_1^2, σ_2^2 are the variances of $\bar{\bar{v}}_1$, $\bar{\bar{v}}_2$. Thus, for angles β less than about 30° (or more than 150°) error variance in w_ρ or w_s increases rapidly.

Large areas for vector wind estimation can only be obtained by increasing radar separation, which of course, worsens spatial resolution and consequently compromises need to be made (Davies-Jones [28]). Furthermore, because of earth's curvature, a larger separation reduces the area of common surveillance at low altitudes where most weather phenomena hazardous to human life reside. These considerations, and the need to observe the wind and reflectivity structure of storm cells near the surface as well as at their tops has led NSSL to select a radar separation of about 40 km. The location of the two NSSL Doppler radars is indicated by the asterisks on a plan position indicator (PPI) display (Figure 9.19) of reflectivity factors Z for storms that occurred in central Oklahoma on June 19, 1980. The PPI projects any radar data field from the conical surface on which it is acquired onto the plane surface of the indicator. The volume within which β lies between 30° and 150° is one that is swept out by rotating, about the baseline, a crescent-shaped area with vertices at the radar sites (see Figure 9.29 for an example of the area).

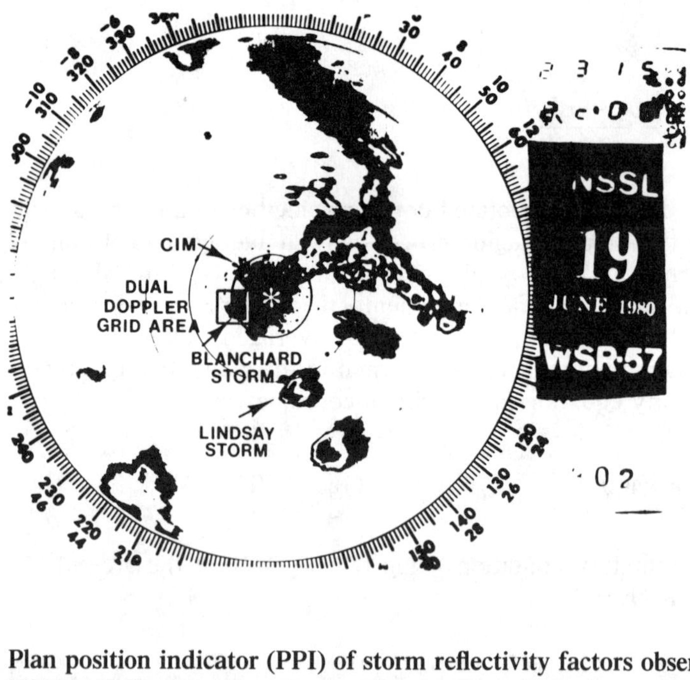

Figure 9.19 **Plan position indicator (PPI) of storm reflectivity factors observed with NSSL's WSR-57 radar. Gray scale shadings (light gray, dark, white, light gray,** ***et cetera*****) represent reflectivity levels differing by about 10 dBZ starting at 17 dBZ. Range marks are 40 km apart and elevation angle is 0°. The boxed area is where dual-doppler data from radars at CIM and NRO (collocated with the WSR-57) are synthesized to form wind vectors shown in Figure 9.20.**

The box on Figure 9.19, which encompasses the area of dual-doppler grid points, lies within the dual-doppler radar volume and encloses a storm cell that produced hail at least as large as 2.9 cm near the town of Blanchard, Oklahoma. The Blanchard storm cell is at the southwestern end of a line of cells that are displayed on Figure 9.19, but only the Blanchard storm (and one to the NE of the baseline) is in a suitable position for wind syntheses.

Of the three components of wind, the vertical component w_z is the most difficult to reliably estimate from dual-doppler radar observations. Neglecting errors in $\overline{w}_t$, the variance of w_α (roughly w_z for small α) is (Doviak *et al.* [36], *op. cit.*)

$$\sigma_\alpha^2 = 2\alpha\Delta\alpha\left(\frac{\rho}{2\Delta\rho}\right)\left[\frac{r_1^2 r_2^2}{4(\rho d)^2}C_1\right] \tag{9.81b}$$

where

$$C_1 = \frac{1}{\gamma^2(\alpha)\alpha}\int_0^\alpha \gamma^2(\sigma_1^2 + \sigma_2^2)\,d\alpha$$

where α is in radian units, and $\Delta\alpha$ and $\Delta\rho$ are grid point spacings. It has been our experience, from theory and observations of wind in the turbulent air of the fair weather boundary layer, that σ_1^2, σ_2^2 have a lower bound of about 1 m^2 s^{-2} (see Rabin and Zrnić [88] Fig. 6; see also Doviak and Berger [32], Table 2a). Thus, in thunderstorms where turbulence is stronger σ_1^2, σ_2^2 values may even be larger. Istok and Doviak [62], separating doppler velocity fields in a supercell thunderstorm into turbulent and ordered flows, show rms velocity associated with turbulence at scales smaller than about 3 km to be significantly larger than 3 m s^{-1} throughout much of the storm, and especially in regions of updrafts. Interpolation of data reduces velocity variance at the expense of filtering the ordered flow, which contains important small-scale features (e.g., tornadic circulations). Assuming $\sigma_1 \approx \sigma_2 \approx 1$ m s^{-1} and an upward integration of (9.80a), (i.e., from α = 0 to α), Doviak *et al.* ([36], *op. cit.*) demonstrate that σ_z^2, the variance of w_z, at an altitude of 14 km can be as large as 900 m^2 s^{-2}! However, this large variance can be reduced, if a boundary condition on w_z is known at the top of the storm and integration of (9.80a) is performed downward (as suggested by Bohne and Srivastava [13]. The ratio of σ_α for upward integration $\alpha_{\alpha u}$ to that for downward integration $\sigma_{\alpha d}$, obtained from (9.81b), is

$$\frac{\sigma_{\alpha u}}{\sigma_{\alpha d}} = \frac{\gamma(0)}{\gamma(\alpha)}$$

where $\gamma(0)$ is the air density at the storm bottom and $\gamma(\alpha)$ the density at the top. For storm tops near 14 km the ratio $\gamma(0)/\gamma(\alpha)$ is about five and hence downward integration should result in a $\sigma_{\alpha d}$ for vertical wind, at the earth's surface, of about 6–7 m s^{-1}. Distributions of estimated $w_z(0)$ at the ground (where w_z should be zero), computed from downward integrations of actual storm data assuming $w_z = 0$ at the storm top (i.e., where $Z \approx 0$), show for one storm a measured σ_z of about 9 m s^{-1} (Nelson and Brown [83]). Likewise, Vasiloff *et al.* [110] have, for another storm, observed a ±5–10 m s^{-1} error level. Thus, both observations are in close agreement with theory. The measured values, which are slightly larger than the 6–7 m s^{-1} theoretically estimated, suggest that σ_1, σ_2 are larger than 1 m s^{-1}, a reasonable deduction in view of the fact that this value is appropriate for fair weather. However, vertical wind errors can be reduced by imposing both a top and bottom boundary condition on w_z, if known, and using a variational adjustment technique to produce a wind field that satisfies both boundary conditions as well as the continuity equation (Ziegler [117]). Several techniques used to compute the vertical velocity are compared by Ray *et al.* [90].

(b) *Observation of a Thunderstorm*

Figure 9.20 shows the wind field derived from dual-doppler radar observations of an intense thunderstorm located in the boxed area of Figure 9.19. Horizontal and vertical grid spacings of 1 km are used and radial velocity data are interpolated to the grid points using a Cressman weighting function (Cressman [27]) with an interpolation radius of 1.5 km. The horizontal wind field at 1-km altitude *above ground level* (AGL) shows a flow from the east that appears to be relatively unperturbed. This wind field is the flow relative to cell motion, the vector velocity of which is indicated on (b). Although cell motion, computed by tracking cell reflectivity maxima, is to the ENE, the storm propagated SE as new cells developed on the storm's southeast edge (i.e., right rear flank of the storm) and old cells dissipated to the NE. Earlier, the Lindsay storm (Figure 9.19) was in the location of the Blanchard storm, but it, too, propagated to the SE.

The storm's velocity field, averaged over the field displayed at each altitude, approximates well the environmental wind at all corresponding levels as has been noted by Zrnić *et al.* [125] for severe storms. At 4-km AGL, the wind field is light and variable but curvature of flow is noted in the southwestern region of the cell. This region of large vertical vorticity is also seen at 7-km AGL about $X = -34$, $Y = -14$. Although there was no apparent closed circulation, one would appear if the flow were referenced to a faster cell velocity. This strong vorticity region defines a mesocyclone because it satisfies criteria stipulated by Donaldson [29] (see Section 9.6.2(b) for more discussion). Mesocyclones are precursors of tornadoes, but no tornado formed in this severe storm. At the 10-km altitude (d), the wind field appears to be diverted around an obstacle at $X = -36$, $Y = -10$. This obstacle is the strong updraft (Figure 9.20f,g), which brings low altitude air of different horizontal momentum to the upper regions of the storm. Near the top of the storm (e), the flow is strongly divergent because the updraft experiences a strong deceleration due to the negative buoyancy of the air as it penetrates the extremely stable air of the stratosphere. As can be seen from the cross sections in Figure 10.20f,g the storm reaches altitudes as high as 17 km and vertical wind at the 12-km altitude is as strong as 60 m s^{-1}! A horizontal rotor is evident on cross section CD at 12-km AGL and air from below 2-km altitudes flows from the SE into the storm's updraft as seen in the cross section GH.

(c) *Observations of Downbursts and Microbursts*

A downburst is defined as a "strong downdraft which induces an outburst of damaging winds on or near the ground" (Fujita [44, 45]). A microburst is a downburst when the distance between diverging centers of fastest outflow is less than 4 km. Downbursts are mostly associated with intense severe storms but they also have been observed, especially in the high plains of the US, below innocuous looking (low reflectivity) virga (Eilts and Doviak [40]; McCarthy and Wilson [78]). Observations made in regions where subcloud moisture contents are higher show many more microbursts having reflectivity factors larger than 40 dBZ. There is no apparent correlation between reflectivity factor in the divergent flow and the maximum velocity difference across it (McCarthy *et al.* [79]; Rinehart and Isaminger [95]). Evaporative cooling in dry air below cloud base reduces the reflectivity factor and therefore it might be better to correlate maximum reflectivity anywhere in the cell with surface outflow intensity.

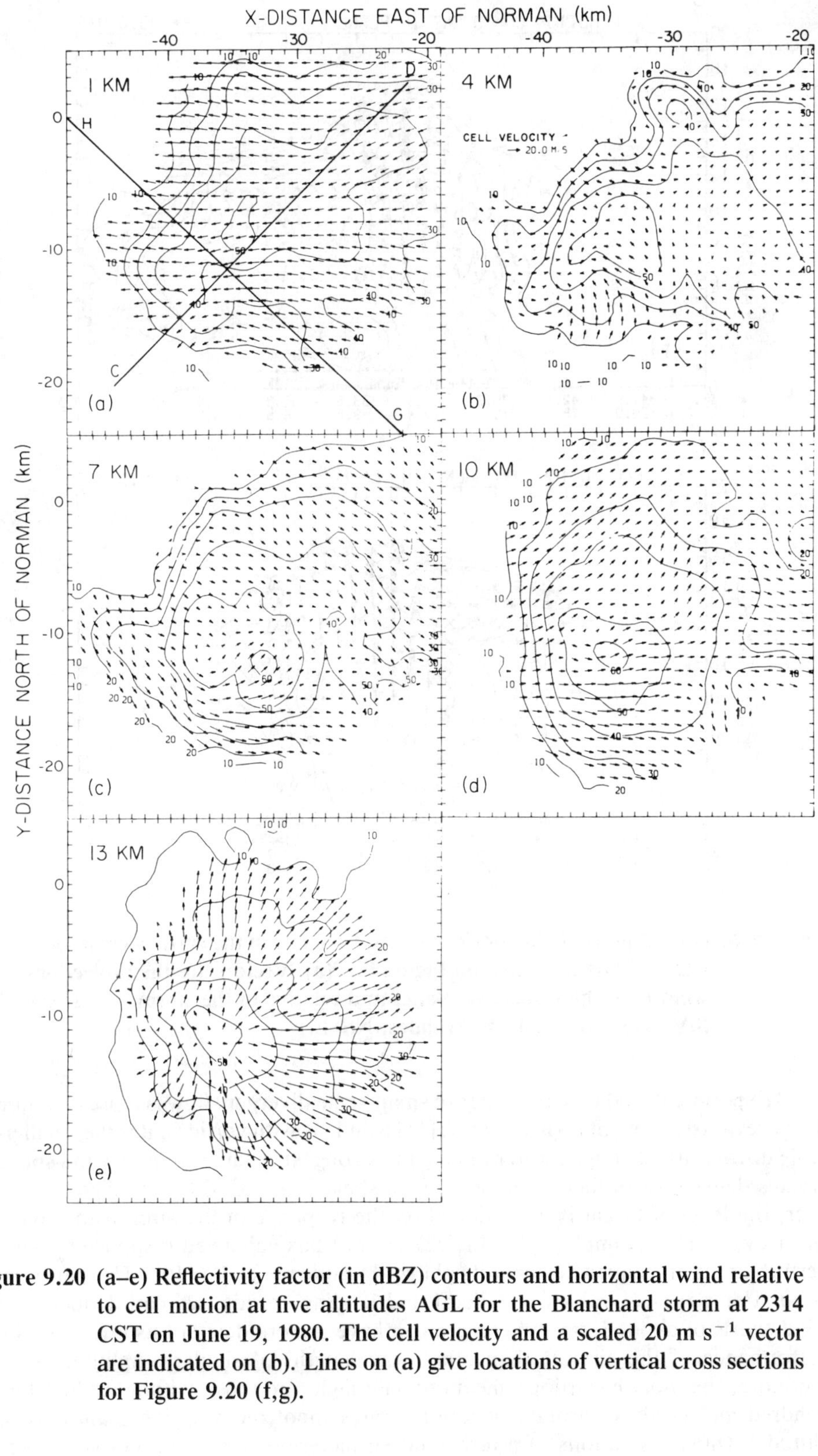

Figure 9.20 (a–e) Reflectivity factor (in dBZ) contours and horizontal wind relative to cell motion at five altitudes AGL for the Blanchard storm at 2314 CST on June 19, 1980. The cell velocity and a scaled 20 m s^{-1} vector are indicated on (b). Lines on (a) give locations of vertical cross sections for Figure 9.20 (f,g).

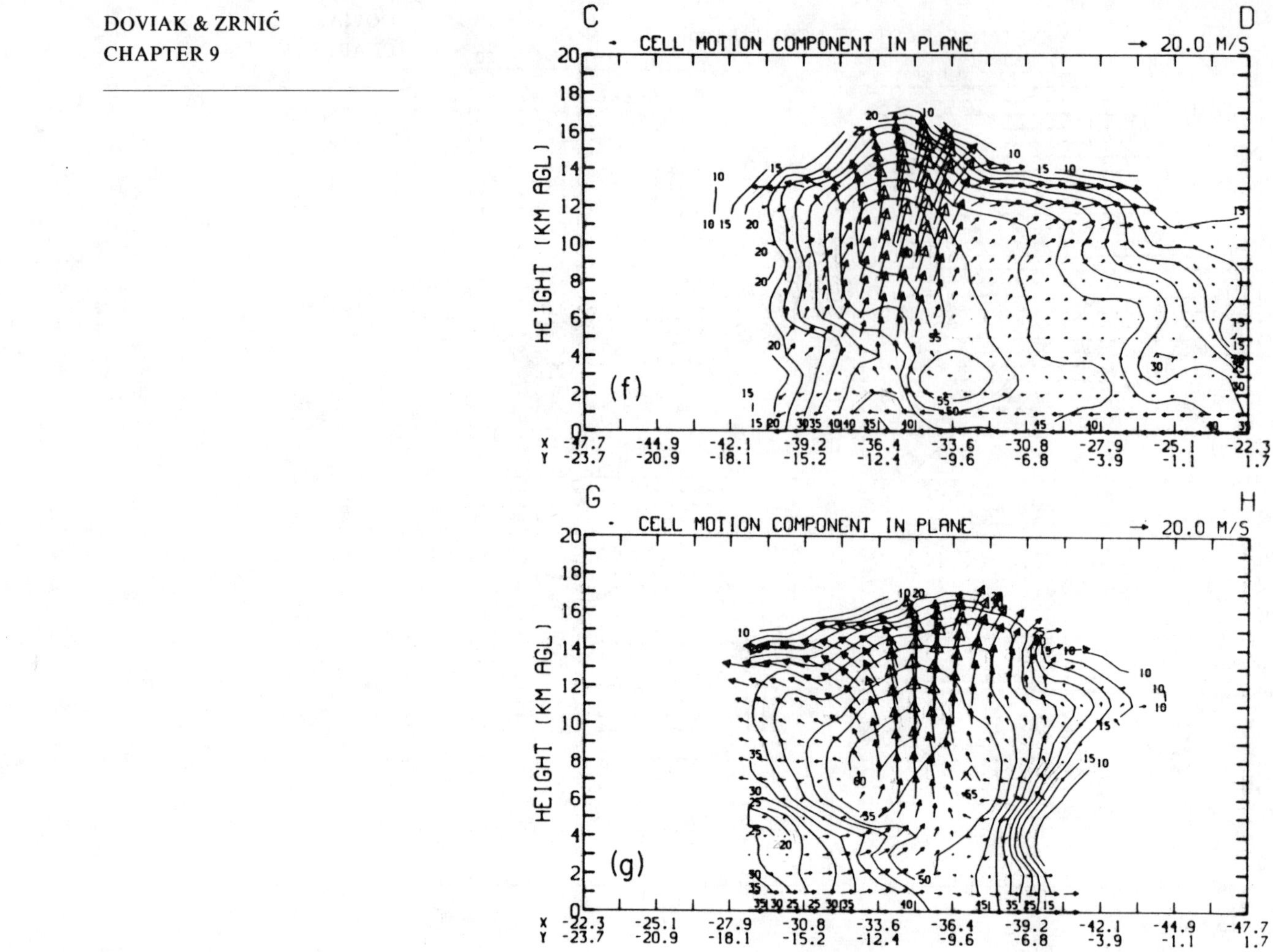

Figure 9.20 **(f,g) Reflectivity factor (in dBZ) and velocity in the vertical cross sections CD and GH indicated on Figure 9.20(a). Vectors are the projections of wind onto the respective planes and relative to the storm. (Courtesy of S.V. Vasiloff and E.A. Brandes, NSSL.)**

Experiments indicate that intense small-scale divergent flow is more common than previously thought (Wilson, *et al.* [111]) and, as with vortices, it is the smaller-scale downdrafts that appear to produce the strongest shear. Although wind speed creates damage to surface structures, wind shear is a hazard to safe flight. However, the level of threat is conditioned by the response of the aircraft to a wind shear event. For example, a Boeing 727 aircraft has enhanced response to wind perturbations at frequencies near its phugoid frequency of 3×10^{-1} Hz (period $\approx$ 30 s). Using typical values of takeoff and landing speeds, we can deduce that wind swirls of about 3-km wavelength can have a more deleterious effect on the performance of the aircraft than other wavelengths of similar amplitude. Furthermore, the most hazardous shears to safe flight are those within the first few hundred meters above ground, where aircraft cannot recover significant loss of altitude. Thus, resolutions of a few hundred meters or less are required to recognize these small-scale diverging flows, and only radars with high angular resolution ($\leq 1°$) and observations close to the radar will accurately measure the shear produced by microbursts.

Compounding the difficulty in making accurate measurements of hazardous wind shear is the short lifetime of the phenomena. For example, lifetimes of microbursts range from 5 to 15 minutes, but the period of severe wind shear lasts only from 2 to 4 minutes with an average velocity difference of 25 m s^{-1} across the divergent flow (McCarthy and Serafin [77]). Although these small scale hazards might be difficult to detect, they are usually embedded in larger-scale phenomena (e.g., mesocyclones, gust fronts), which have longer lifetimes and hence are more easily detected. Thus, detection of phenomena that might cause microbursts and other small scale hazards could be equally or more important than detection of the hazard itself (Doviak and Lee [34]).

Microburst winds derived from simultaneous observations with a 10-cm radar (the Federal Aviation Administration's transportable doppler weather radar operated by MIT's Lincoln Laboratory; Evans [41]) and a 5-cm radar (from the University of North Dakota) are shown in Figure 9.21. These radars were located approximately 15 km apart near Memphis, Tennessee, and the wind field displayed in Figure 9.21 is about 5 minutes after the time of maximum divergence. The undulating highly turbulent front of the outflow is clearly depicted.

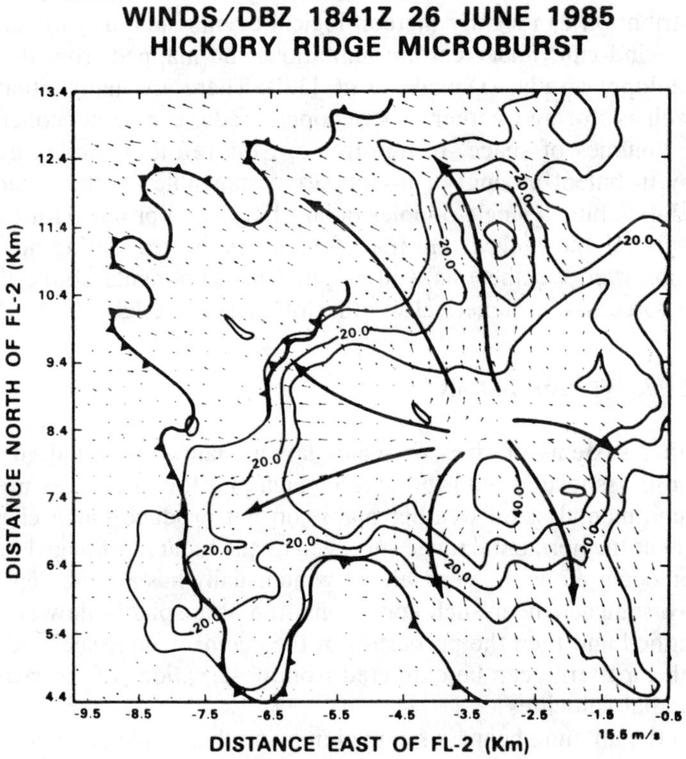

Figure 9.21 **Wind field, reflectivity factor (in dBZ) field, selected stream lines, and the outflow front of a microburst. One grid space length equals a wind vector of 15.5 m s^{-1}. Winds are relative to ground. Height above ground is 50 m. (Figure and analysis are the courtesy of M. Wolfson, MIT Lincoln Laboratory.)**

The use of doppler radar to forewarn the occurrence of a microburst along flight paths is a subject of numerous projects. Results of recent research reveal that there are a number of possible precursors before the formation of divergence at the surface. These include a rapidly descending reflectivity core, organized convergence near and above cloud base, and midaltitude (2–5 km) rotation (Roberts and Wilson [96]).

Evidence from meteorological observations and aircraft data support the idea that microbursts are a very prominent factor in most fatal wind shear accidents. However, it has been proposed (Linden and Simpson [76]) that wind shear and downdraft, at the rear of a vortex ring formed around the microburst and located at its leading edge (i.e., the gust front), are responsible for the danger of flying through thunderstorm downdrafts. Because all fatal thunderstorm-related accidents involving US air carriers from 1970 to date have occurred in or near heavy precipitation, Kessler [66] suggests the most prudent action is to avoid high reflectivity thunderstorms by a sufficient margin.

9.6.2 Single-Doppler Radar

Although the doppler radar measures only the radial wind component, its spatial distribution can signify meteorological events such as tornado cyclones. Moreover, wind can, under certain limitations, be mapped from data obtained with single-doppler radar (Doviak, *et al.* [38]). There are many situations, both in fair as well as stormy weather, where doppler radar can detect coherent echoes over large volumes of space. In rain-free air, the refractive index irregularities (created by turbulent mixing) or insects are responsible for the reflectivity (see Section 9.5.4). Thus, a single-doppler radar offers good promise for both severe-weather warning and *nowcasting* (i.e., forecasting by translating downwind the currently observed weather) as well as wind measurements above the ground, needed for forecasts in numerical weather prediction models.

(a) *Large Scale Weather Systems*

Weather systems of sizes or scales larger than a few hundred kilometers such as those associated with the area of high and low pressures presented on global or continental sized weather maps, do not produce much change in the wind field over the horizontal range surveyed by the radar. Although large changes of wind can occur along the vertical, the wind usually has a weak dependence on horizontal distance. Under such conditions, the atmospheric flow is nearly vertically stratified and then the properties of the *vector* wind associated with large scale weather patterns can be extracted from observations of the *scalar* field of observed radial velocities.

When the antenna beam is scanned in azimuth ϕ while elevation angle θ_e is fixed, the measured radial velocity at constant range r has the ϕ dependence:

$$v_r = (w_z - w_t) \sin \theta_e + w_h \cos \theta_e \cos (\phi - \delta) \tag{9.82}$$

where δ is the wind direction, and w_h is the horizontal wind speed. The w_z and w_h velocities are readily computed from VAD data (Lhermitte and Atlas [75]) under the assumption that air is in rectilinear translation (i.e., w_h, w_z, and δ are independent of ϕ) and w_t can be accurately estimated. (VAD, *velocity azimuth display*, is a presentation of radial velocity versus azimuth at a single range location). Then, $v_r + w_t \sin \theta_e$ has a sinusoidal dependence on ϕ; thus, amplitude and phase of the sine curve are measures of w_h and δ at the height $r \sin \theta_e$ of the sampling circle. Vertical velocity w_z produces an offset of the sine wave. However, when wind is not horizontally homogeneous, v_r is no longer purely sinusoidal.

Caton [22] showed how divergence can be determined from VAD data. Browning and Wexler [18] carried the analyses even further by assuming the wind field was well represented by a linear function over the circle of measurement. With this assumption there are four basic fields of motion that convey air: translational, vortical, divergent, and deformative (Hess [59], p. 198). Fourier analysis of v_r for linear wind reveals that only the vortical motion cannot be measured directly by the VAD method. The first harmonic component gives w_h and δ averaged over the circle, and the second measures deformation (Doviak and Zrnić [33], Sec. 9.3.3). The azimuthal average of v_r, the offset, is

$$\overline{v}_r = \frac{1}{2\pi}\int_0^{2\pi} v_r \, d\phi = (\overline{\text{DIV } w_h})\frac{r}{2}\cos^2\theta_e + (\overline{w}_z - \overline{w}_t)\sin\theta_e, \qquad (9.83)$$

where the bars now denote an average over the sampling circle of radius r. If we have an estimate of $\overline{w}_t(z)$ and a boundary condition on $\overline{w}_z(z)$, we can estimate $\overline{w}_z(z)$ by inserting the mass continuity equation into (9.83), and then solving for $\overline{w}_z$ by integrating the resulting first-order differential equation. If reflectivity is due to refractive index fluctuations (see Section 9.7), then clearly $\overline{w}_t = 0$. Thus, a single-doppler radar can measure the three components of wind averaged over a sampling circle of radius r and produce a vertical profile of wind.

When winds are horizontally uniform while varying with height, unique patterns of doppler velocity fields are formed on PPI displays as the radar beam is scanned at a constant θ_e. These patterns can be easily interpreted by trained personnel to estimate the vertical profiles of wind. PPI displays of doppler velocity contours corresponding to a variety of vertical profiles of wind speed and direction are shown in Figure 9.22, where the center of the PPI display represents the radar location. Photo 9.1 shows a radial velocity field as depicted on a PPI color display for a fair weather situation. Winds that have a component away from the radar (termed "outbound") have positive doppler velocity values indicated by solid contours in Figure 9.22, whereas those toward the radar (termed "inbound") have negative values indicated by dotted contours. Winds directed normal to the beam produce zero doppler velocities. When wind speed is constant with height (left column of Figure 9.22), all contours pass through the center of the display. Also, maximum and minimum doppler velocities occur along the heavier (nonzero) contour lines, rather than at but one point (black dots) as with other wind speed profiles. When the surface speed is zero, only the zero velocity contour (thick, long dashes) passes through the center of the radar display (three rightmost columns of Figure 9.22).

If the wind speed profile has a peak within the height interval on the display, there will be a pair of closed contours 180° from each other; the azimuth of the inbound (or outbound) maximum is the direction from (or toward) which the velocity jet is approaching (or departing) the radar, and the height of the peak value can be computed from the radar antenna's elevation angle and the slant range to that point.

Whereas the wind speed profile controls the overall pattern including the spacing between contours, the vertical profile of wind direction is uniquely specified by the zero velocity contour. Note that the zero contours are identical in each row (a reflection of wind direction profile), although the overall patterns in each row differ significantly (a reflection of wind speed profile).

Since wind direction is perpendicular to a radial line (from display center) at the point where the line intersects the zero velocity contour, wind direction variation with height (range on the PPI display) can be determined by inspection. Air motion is from the negative side of the zero contour toward the positive side. Looking at the middle row of Figure 9.22, we see that there are southerly winds at the ground (practically, just above the ground) because at $r = 0$ the zero velocity line is oriented east-west, and air is approaching from the south and flowing away

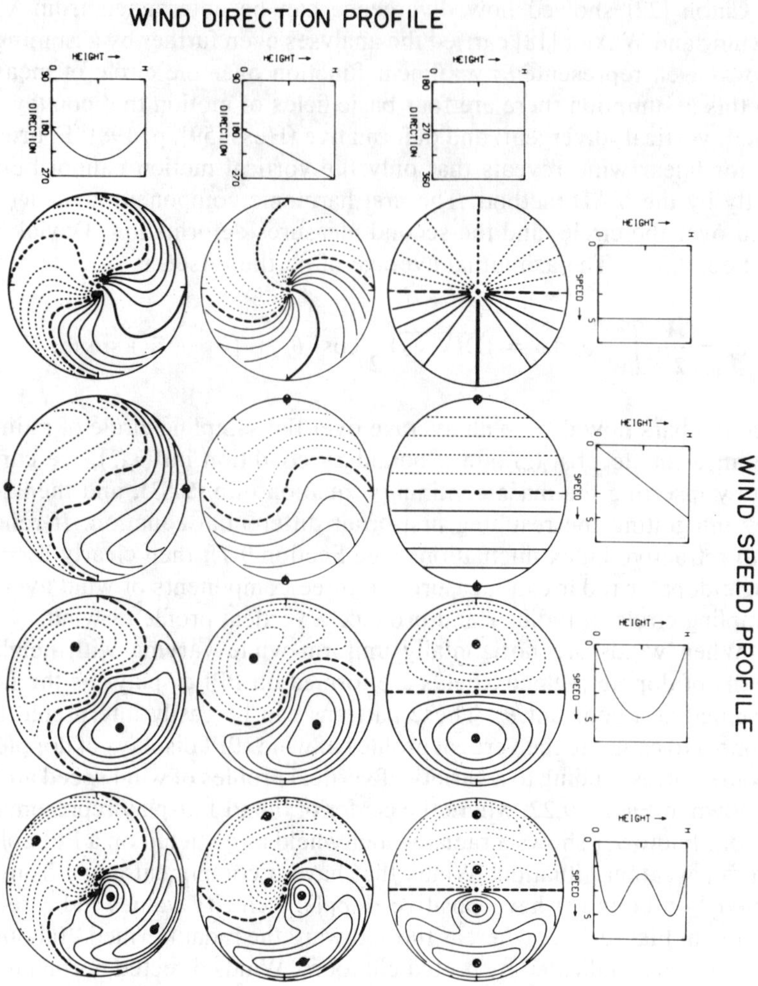

Figure 9.22 Doppler velocity patterns on surfaces at constant elevation angle θ_e for various vertical profiles of horizontally uniform wind. The outer circle of each display depicts doppler velocity at height H and range $r = H/\sin\theta_e$. Contour interval is 0.2S, where S is maximum wind speed (from Wood and Brown [112]).

toward the north. Halfway between the center (zero height) and edge (height = H) of the display, southwesterly winds are perpendicular to the radial line. At the edge of the display, wind is from the west because the radial line intersecting the zero contour is oriented north-south.

Veering winds, representative of warm air advection, produce a striking S-shaped pattern of the zero velocity contours, whereas backing winds, representative of cold air advection, produce a backward S. These features are seen in the middle and bottom rows as well as on Photo 9.1.

Other patterns associated with horizontal variation of wind speed and direction (e.g., fronts, mesocyclone, *et cetera*) are presented by Wood and Brown [112].

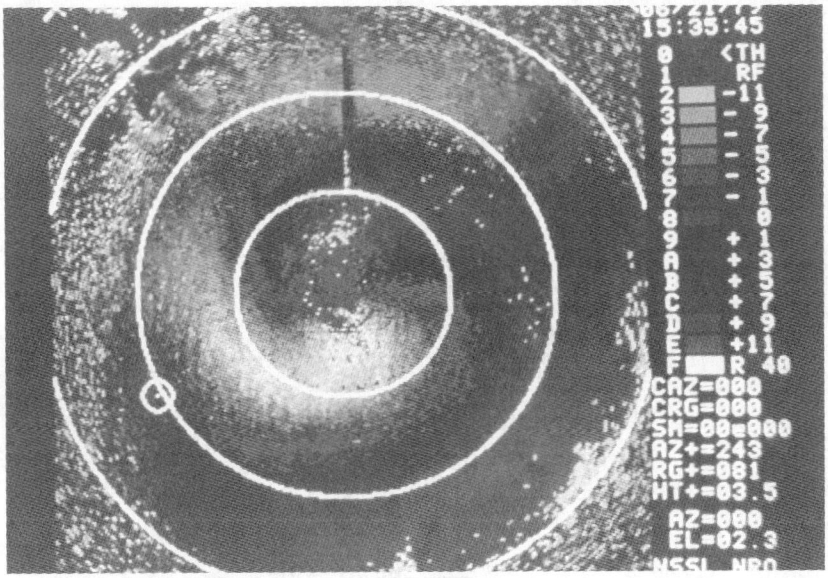

Photo 9.1 Plan position indicator display of doppler velocities in fair weather. Range circles are spaced 40 km apart and the velocity scale is in m s^{-1} with positive values designating target velocity with a component away from the radar. The speckled areas (e.g., outside the 80-km range circle) depicts random velocities estimated in regions where there are no echoes. Elevation angle is 2.3°. (Date: June 21, 1979.)*

(b) *Severe-Storm Cyclone Observations*

Because radar maps spatial distribution of doppler velocities inside storms, significant meteorological events (unseen from outside) such as tornado cyclones produce telltale signatures. It can be shown that nontranslating cyclones have isodops forming a symmetric couplet of closed contours with an equal number of isodops encircling positive- and negative-velocity maxima (Figure 9.23). If the inner portion of the cyclone is a solidly rotating mass of air, its tangential velocity increases linearly with radius to a maximum; outside this maximum, the velocity decreases (roughly) inversely with the radial distance. The isodop contours of such a combined Rankine cyclonic vortex are circular sections connected with straight lines (Figure 9.23). Patterns roughly similar to that in Figure 9.23 have been observed many times, and an example is shown in Figure 9.24a for a tornadic storm that did considerable damage to Stillwater, Oklahoma, in 1975 (Burgess [19]; Brown *et al.* [17]; Zrnić *et al.* [123]). Figure 9.24b shows contours of estimated spectral width σ_v for the same storm, and we immediately see the striking correlation of σ_v with significant radial velocity shear. Regions of large σ_v may also indicate the presence of strong turbulence.

The National Weather Service has, for many years, routinely displayed reflectivity factors on PPI scopes. Although storm reflectivity factor fields cannot be reliably used for tornado detection, they have proven valuable for hydrological studies and severe-weather warnings. Those warnings are primarily based on reflectivity values, storm top heights, and sometimes on circulatory features or "hook" echoes. Tornado cyclone warnings solely based on features seen in reflectivity fields are subject to a high false-alarm rate and missed early detections.

*Original photo in color.

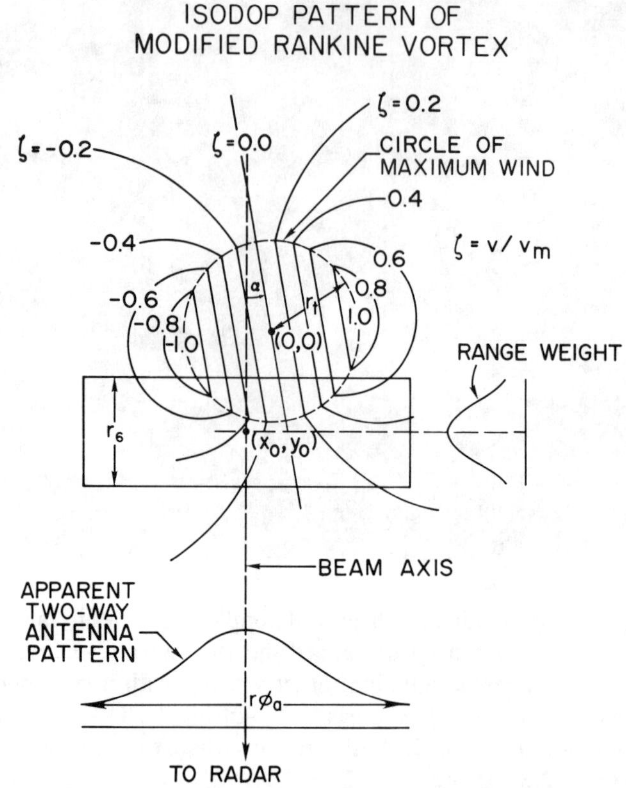

Figure 9.23 Plan view of an idealized isodop pattern for a stationary modified Rankine vortex located at a range large compared with vortex diameter; $\zeta = v/v_m$ is doppler velocity normalized to its peak value. Resolution volume, antenna, and range-weighting functions are depicted. The angular displacement α determines radial inflow ($\alpha < 0$) or outflow ($\alpha > 0$) (from Doviak and Zrnić [33]).

However, PPI displays of doppler velocity reliably show signatures of large circulations called *mesocyclones*. These have diameters of maximum tangential wind as large as 12 km (JDOP [63]) and are often precursors of tornadoes.

For a couplet of closed isodops to be recognized as a signature of a mesocyclone three criteria must be met (JDOP [63]): (1) significant azimuthal shear (i.e., $\geq 5 \times 10^{-3}\ s^{-1}$) must exist between closed isodops of opposite sign, (2) the shear pattern and closed isodops must extend vertically for a height interval comparable with horizontal diameter, and (3) the shear pattern and closed isodops ought to persist for half the period of vortex rotation.

Color displays of reflectivities, radial velocities, and spectral widths allow, in real time, easier quantitative evaluation and better detection of mesocyclones. Photo 9.2 is an example of doppler radar data from a tornadic thunderstorm in central Oklahoma during the evening of April 30, 1978. These data were acquired with NSSL's 10-cm doppler weather radar, and the pictures depict the fields of the three principal moments of the doppler spectra.

At the time of the display (1824:09), there was a tornado on the ground at 335° and 50-km range. Although the reflectivity display showed no evidence of a tornado cyclone (a mesocyclone containing a tornado is also called a tornado cyclone), it was depicted remarkably well by the doppler velocity display (b). The display (c) of spectral width σ_v is an indicator of turbulence and shear within each resolution volume V_6. We note high spectral width values in the vicinity of the

DOVIAK & ZRNIĆ
CHAPTER 9

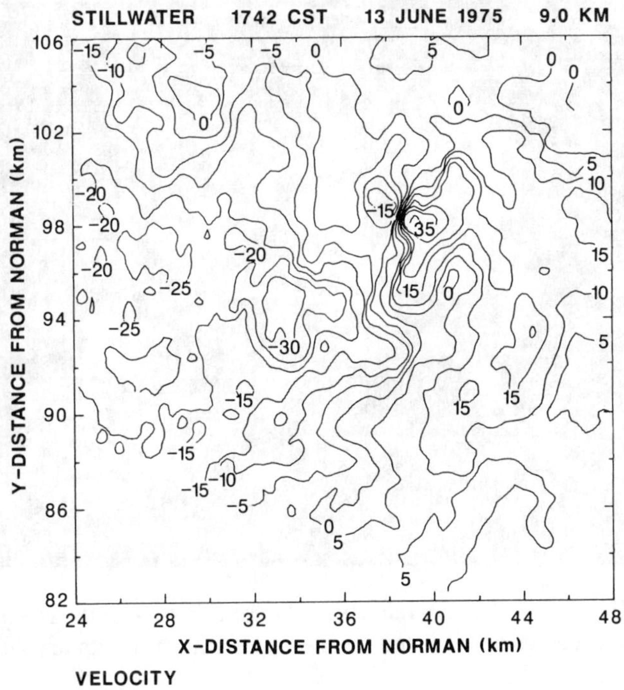

Figure 9.24 (a) Doppler velocity field 9.0 km above ground level (AGL) for the Stillwater, Oklahoma storm (1742 CST on June 13, 1975). Velocity contours (isodops) are in 5 m s^{-1} steps. A mesocyclone is centered at Y = 98 km, X = 38 km.

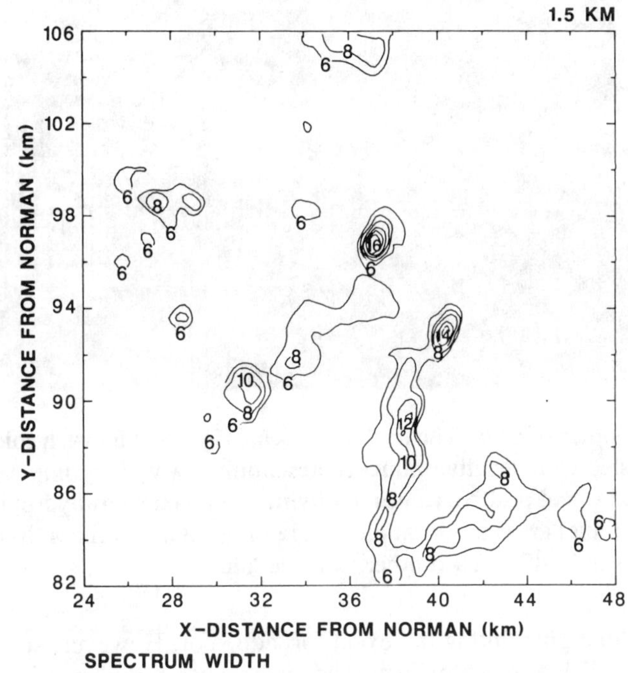

Figure 9.24 (b) Contours of spectral width for the Stillwater, Oklahoma storm at 1.5 km AGL. Values equal to or larger than 6 m s^{-1} are displayed. Large widths at Y = 97 km and X = 37 km are within the tornado mesocyclone; the other region of large widths lies along a gust front (from Doviak and Zrnić [33]).

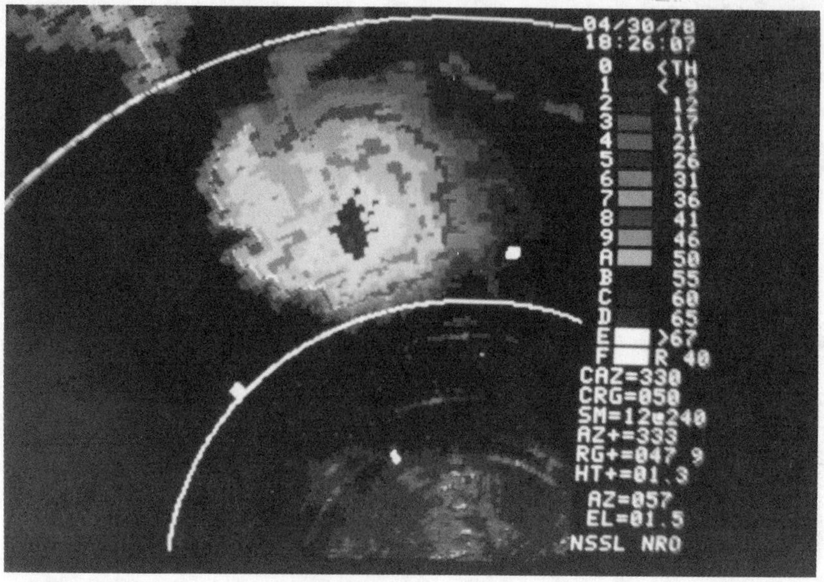

Photo 9.2 **(a) PPI display of storm reflectivity factor *Z* in dBZ units. Median *Z* values for each category are listed on the right. Range marks are 40 km apart**

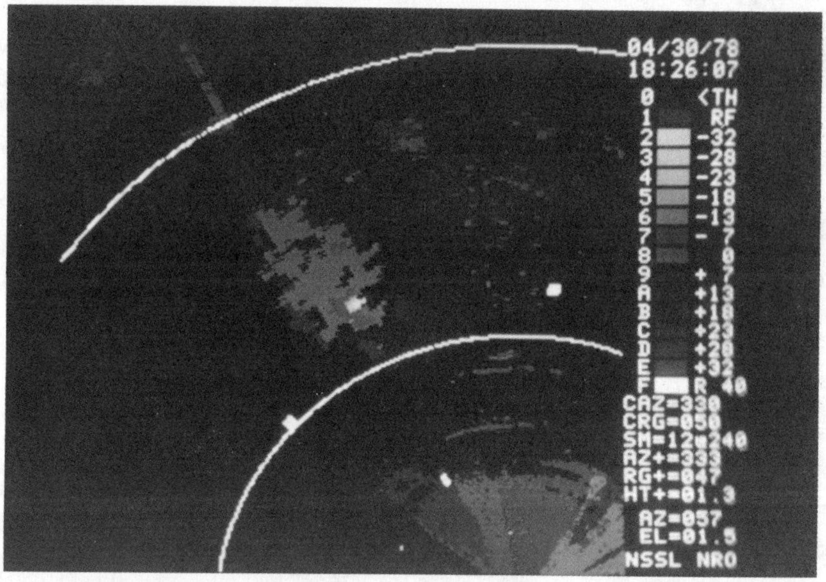

Photo 9.2 **(b) Doppler velocity: The median velocities in m s^{-1} for each color category are listed with negative values representing flow with a component toward the radar and positive values signifying a component away from the radar. Dark (blue) areas are locations where range aliased echoes, having power within ± 10 dB of each other, are overlaid.**

circulation and one pinpointing the exact tornado spot. However, spectral width has not been a reliable indicator of tornadoes because turbulence in storms can produce large values of spectral width and thus may be easily mistaken for tornadoes.

Photo 9.2d is the *multimoment display* (MMD) (Burgess *et al.* [19]); each arrow is a speedometer indicating the doppler speed at its respective location. The tornado circulation is depicted quite nicely by this graphic display, and the presence of wide spectra is evident by the broad arrowhead width, indicating strong shearing winds or turbulence.

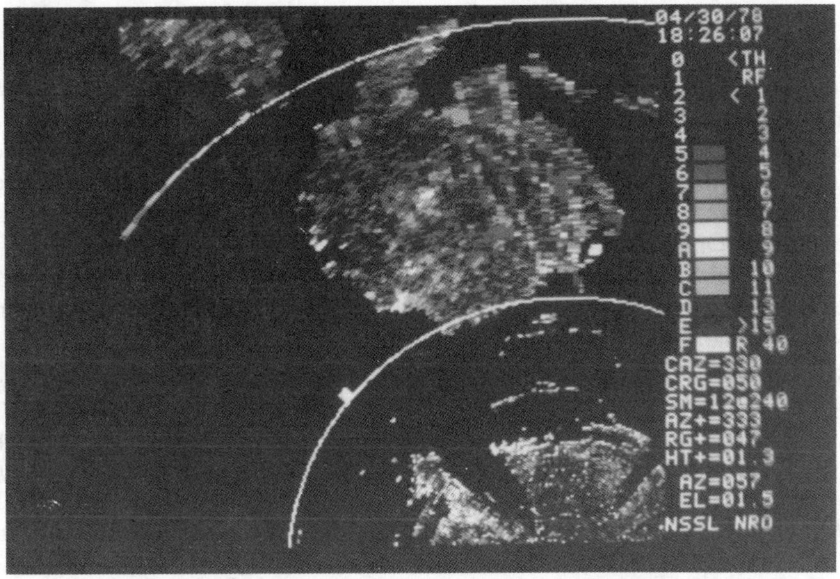

Photo 9.2 **(c) Display of spectral width. Categories are in meters per second.**

Photo 9.2 **(d) Multimoment display that uses arrows to show composites of the three principal doppler spectral moments for each resolution volume. In this B-scan format, azimuthal bearing (degrees) is indicated along the abscissa and range (km) along the ordinate. Zero doppler velocity is an arrow pointing at the 3 o'clock position, and positive radial velocities cause the vector to rotate counterclockwise (*a* + 17 m s^{-1} velocity is at the 12 o'clock position). Doppler velocities that reach the Nyquist limit $\pm\lambda/4T_s$ (34 m s^{-1} in this case) produce arrows at the 9 o'clock position. Reflectivity is proportional to vector length and spectral width is proportional to arrowhead width.**

Note: These four photos were printed originally in color on the cover of *IEEE Trans. Geoscience Electronics*, Vol. GE-17, No. 4, Oct. 1979.

A project involving the National Severe Storms Laboratory, National Weather Service (NWS), Air Weather Service (AWS), Federal Aviation Administration (FAA), and Air Force Geophysical Laboratory was established to conduct experiments to determine the improvement in tornado warnings over that of the present system used by the NWS (JDOP [63]). The present system included *in situ* observations by trained spotters and the NWS's noncoherent WSR-57 radar, whereas warnings with doppler radar were based solely on real-time radar data observations. Operations were conducted during the spring of 1977 and 1978. The MMD used in conjunction with the color display (Photo 9.2) proved most suitable to recognize the doppler velocity signatures of mesocyclones. Although the probability of tornado detection (≈ 0.65) did not improve significantly by using the criteria for mesocyclone recognition, the false-alarm rate decreased from 0.63 to 0.25. Moreover, the average lead time (the time between the issuance of a warning and the time of tornado occurrence) was about 20 minutes when using Doppler data alone, whereas the lead time for the present system was a negative 0.8 minute. However, using criteria for tornado vortex signatures (TVS) discussed in Brown *et al.* [17], the project scientists were able to detect all tornadoes that occurred within a range of less than 115 km (Burgess *et al.* [19]). At ranges beyond 115 km, signatures of small tornadoes are lost because of poor resolution; however, large, destructive tornadoes were detected up to 240 km away by using the 0.8° beam of NSSL's doppler radar. It was established that circulation starts at middle altitudes (6–8 km) and works its way toward the ground.

On the basis of those experiments, the NWS, AWS, and FAA have joined together to develop 10 cm doppler radars for the next generation of weather radars (NEXRAD) to replace the currently outdated nondoppler network. An operational tornado detection system will involve interaction between human operators and an automated scheme in which the velocity pattern of a tornado cyclone is recognized.

(c) *Doppler Spectra of Tornadoes*

In 1961 Smith and Holmes reported a tornado spectrum that was observed with a CW doppler radar [103]; 12 years later, a tornado was first observed by a pulsed doppler radar (Zrnić and Doviak [121]).

The doppler spectrum is principally influenced by the portion of circulation that lies within V_6, so that tracers moving with the same velocities contribute to spectral components according to their reflectivity, isodop density, and illumination function $I(\mathbf{r}_0,r)$ intensity (see Section 9.3.3). Only those targets having spectral power that is above the receiver noise level may be positively identified. However, it has been our experience that tornadoes in V_6 offer enough reflectivity, due to debris and hydrometeors, that a large velocity span can be observed. It is not yet known whether tracers moving at the peak tornado windspeed can be resolved.

When centered on the beam axis, the Rankine vortex model predicts a bimodal spectrum that was observed experimentally several times (Zrnić and Doviak [121]; Zrnić *et al.* [123]). Shown in Figure 9.25 are Rankine model vortex spectra matched to data, of the upper middle two spectra, by a least-squares fit. These examples are from the Stillwater violent tornado (see Figure 9.24 for fields of v_r and σ_v). The two spectra closest to the tornado were simultaneously least-squares fitted. Simulated model vortex spectra and real spectra show very good agreement not only for the two where the fit was made, but also for adjacent gate locations (Figure 9.25). Reflectivity is assumed to be uniform within the resolution volume. Differences in echo power from each resolution volume are accounted for by forcing each simulated spectrum to have power equal to its matching data spectrum. Maximum velocity of 92 m s^{-1} and tornado diameter of 300 m are deduced from the fitting. The deduced maximum velocity is larger than the radar's

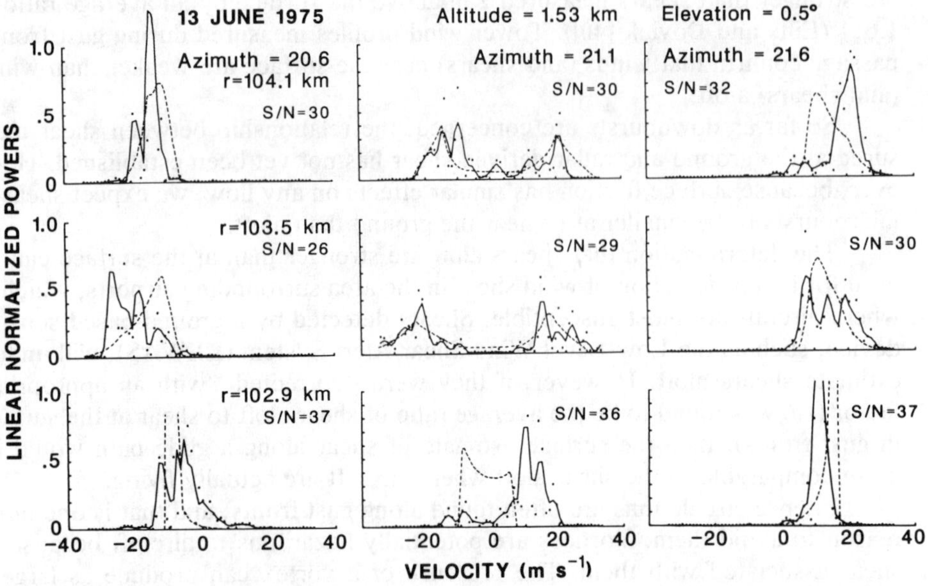

Figure 9.25 Spectra from three consecutive azimuth and range locations of the Stillwater, Oklahoma tornado. Dots show spectrum estimates from recorded time series data weighted with a von Hann window. Solid lines are three-point running averages. Dashed lines are spectra simulated using a tornado model whose parameter values are obtained by minimizing the mean square difference between data and simulated spectra for two spectra (AZ = 21.1° and ranges *r* at 103.5 and 104.1 km). Resolution volume depth is 150 m, range gate spacing 600 m, and antenna beamwidth 0.8°. The tornado is located between the two upper middle resolution volumes. Tornado parameters obtained from these two fitted spectra are used to compute the remaining seven simulated spectra. *S/N* is the signal-to-noise ratio in dB.

unambiguous velocity (± 34.4 m s^{-1}); therefore, aliasing has been introduced in the model spectrum and the estimates are indirect. Asymmetry of spectral peaks (at 21.1°; 104.136-km range) about zero velocity suggests that targets were centrifuged outward at a velocity of 13 m s^{-1}.

(d) *Gust Fronts*

Strong wind shear may also be associated with gust fronts (e.g., Figure 9.24b), which are the leading edge of the mass of precipitation-cooled air, first flowing downward and then outward from the center of a thunderstorm. Gust fronts harbor both vertical and horizontal shear as well as turbulence, and even after propagating several kilometers from the parent thunderstorms and away from the rain-laden air, the gust fronts may retain shear at potentially hazardous levels. This large separation makes it unreliable to discount the gust front as a wind shear hazard merely because of the absence of thunderstorms in close proximity.

Doppler radar offers great promise to measure shears of horizontal wind in gust fronts that are void of precipitation (Zrnić *et al.* [124]). Due to the nature of the instrument, however, measurements are usually representative of heights at beam center so that care must be exercised when comparing these measurements with surface anemometers. Results from a recent study indicate that doppler radar estimated shears measured at heights between 50–600 m in Oklahoma gust fronts

are stronger than shears measured 2 m above the surface by an average ratio of 1.6:1 (Eilts and Doviak [40]). Tower wind profiles measured during gust frontal passage confirm that winds (and shears) near the surface are weaker than winds (and shears) aloft.

So far as downbursts are concerned, the relationship between shear measured at the ground and radar derived shear has not yet been established. However, because surface friction has similar effects on any flow, we expect shear in microbursts to be smaller at or near the ground than aloft.

The determination that shears aloft are stronger than at the surface can be used for better detection of wind shear in the area surrounding airports, which is where aircraft are most susceptible. Shears detected by a ground-based sensing device, such as the Low-Level Wind Shear Alert System (LLWAS), will underestimate shears aloft. However, if they were "corrected" with an appropriate factor (1.6 was found to be the average ratio of shear aloft to shear at the surface in gust fronts), then the surface estimate of shear along a glide path would be more comparable to the shear aloft where aircraft are actually flying.

Intense circulations are often found along gust fronts, and that is one more reason to avoid them. Vortices are potentially hazardous to aircraft because of shear associated with them. The presence of a vortex can produce as large a decrease in headwind as is found in downbursts, if the flight path is tangent to the circle of maximum wind (Doviak and Lee [34]).

Because of a gust front's typical long lifetime (sometimes 1–3 hr) and large horizontal extent (10–90 km), tracking of gust fronts is practical. A gust front appears in a Doppler field as a low-altitude convergence line, which extends horizontally for tens of kilometers. A gust front algorithm has been developed (Uyeda and Zrnić [108]), which locates areas of convergence along a radial and pieces them together to locate lines of convergence which are then identified as gust fronts. The algorithm is able to identify gust fronts and then extrapolate their location up to 20 minutes in advance (Figure 9.26). This 20-minute forecast, when compared with the actual location at the same time, shows good agreement for a number of cases.

(e) *Turbulence*

Both sheared flow in stratified environments and convection can generate turbulence. In either case, the turbulent energy input is at larger sized eddies, which break up into successively smaller ones until turbulent energy is converted into heat by viscosity. If the size of the biggest eddies of isotropic turbulence (i.e., sizes within the inertial subrange of eddy wavelengths, where energy is transferred from larger eddies to smaller ones without net energy loss) are larger than the radar resolution volume, turbulence can be quantified from doppler spectral width measurements (Doviak and Zrnić [33]; Istok and Doviak [62]). So far, there have been few radar measurements to establish how well doppler spectral width in storms characterizes isotropic turbulence. In the next paragraph we present some evidence.

Brewster and Zrnić [16], observing a thunderstorm with a vertically pointed doppler radar, showed excellent agreement in turbulent energy dissipation rates ϵ (which is a parameter with values that are popularly used to indicate turbulence intensity and, consequently, a measure of its hazard to aircraft) computed from spatial spectra of velocity fields with those ϵ computed from doppler spectral widths. Furthermore, ϵ rates determined that longitudinal (S_l) and transverse (S_t) spatial spectra of velocities followed a theoretical relationship (expected for the inertial subrange) up to 2.4-km scales (Figure 9.27).

Thus, it can be argued convincingly that the 2.4-km and smaller scaled eddies are within the inertial range of isotropic turbulence in which kinetic energy is transferred to smaller scaled eddies and to eventual dissipation when viscous

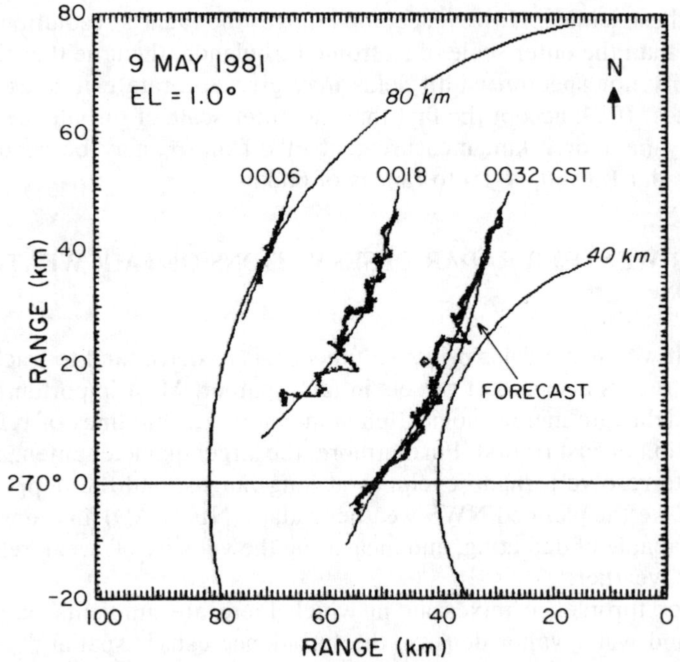

Figure 9.26 **Three consecutive positions of a thunderstorm outflow front on May 9, 1981. The smooth curves are least square fits to data, and the forecast position is indicated.**

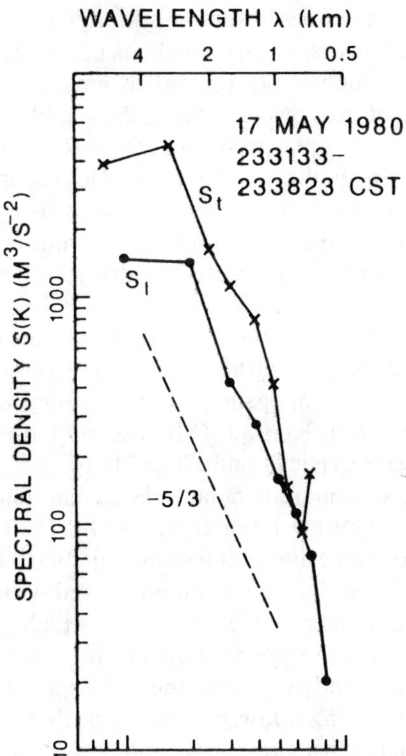

Figure 9.27 **Spatial spectral densities. S_t is the transverse and S_1 is the longitudinal spectrum of vertical velocity variance per unit spatial frequency K. A dashed line represents the $-\frac{5}{3}$ law expected from isotropic turbulence in the inertial subrange.**

forces act on these eddies. However, at long ranges the radar's resolution volume may be larger than the outer scale of isotropic turbulence; then, neither the mean doppler velocity nor spectral width fields may give accurate estimates of eddy dissipation rates. If we accept the fact that the outer scale of turbulence in thunderstorms may be 1 or 2 km, measurement of ϵ from σ_v may be accurate for narrowbeam radar (i.e., $\theta_1 \leq 1°$) to ranges of 60 km.

9.7 DOPPLER WEATHER RADAR OBSERVATIONS OF FAIR WEATHER WIND FIELDS

Although weather radar's principal function is to detect and to track severe storms, these radars also detect targets in fair weather. Most importantly, these targets often delineate meteorological phenomena (e.g., thin lines of reflectivity mark the location of gust fronts). Furthermore, the larger diameter antenna needed by the radar to resolve tornado cyclones at long ranges, and the doppler signal processing, make the planned NWS weather radars (NEXRAD) an order of magnitude more capable of detecting, and measuring the velocity of, weak reflectivity regions in fair weather.

Whenever turbulence mixes air in which there are gradients of potential temperature and water vapor density, the turbulence causes spatial fluctuations in the refractive index n. The fluctuations are small (e.g., one part in one million). Nevertheless, sensitive microwave radars detect the very faint echoes returned from these irregularities in what otherwise (without turbulence) will be a smoothly changing n with negligible backscatter.

Echoes from clear air have been seen almost from the inception of radar observations. These "angel echoes" were at first mystifying, but they were often associated with birds and insects. Clear-air echoes not related to any visible object in the atmosphere were conclusively proved to emanate from refractive index fluctuations through use of multiwavelength radars at Wallops Island (Hardy *et al.* [55]). Simultaneous *in situ* measurements of refractive index fluctuations and reflectivity corroborated this finding (Saxton [98]; Doviak and Berger [32]). These radar studies were preceded by many measurements using tropospheric-scatter communication links that often depend on clear-air refractive index fluctuations to provide reliable wideband circuits between distance points (Rice *et al.* [92]; IEEE [61]).

In the 1960s ultrasensitive incoherent radars were used for remotely detecting and resolving clear-air atmospheric structure, and these studies are well reviewed by Hardy and Katz [54]. The radars showed meteorological phenomena such as convective thermals (Katz [65]; Konrad [68]), sea and land breezes (Meyer [80]), and Kelvin-Helmholtz waves (Hicks and Angell [60]).

Doppler processing of coherent radar echoes can improve target detection by at least an order of magnitude (Hennington *et al.* [57]), and hence medium-resolution weather radars can have a detection capability often associated with the large-aperture (18 to 27-m diameter) antennas used with noncoherent radars. Furthermore, a coherent radar provides a way in which ground clutter can be distinguished from moving atmospheric targets and allows data acquisition at closer ranges, thereby taking advantage of the r^{-2} dependence in echo power.

Chadwick *et al.* [23], using a low-power, 10-cm, FM-CW radar, monitored reflectivity values associated with refractive index fluctuations in the planetary boundary layer, and concluded after one year of observation that winds can always be measured to several hundred meter heights with moderately sensitive radars. Clear-air wind measurement has practical significance because pulsed doppler radars under development by the FAA also can measure wind shear hazards near airports for all weather conditions (Lee and Goff [73]).

Harrold and Browning [56] found that radar can delineate the upper limit of convection before precipitation, showing that it is deeper in some regions than

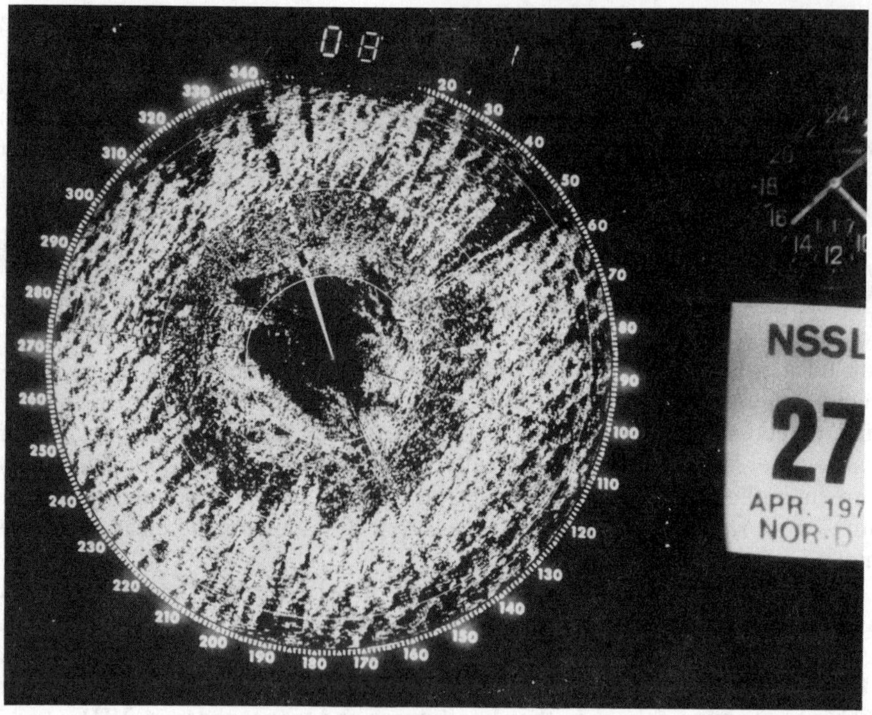

Figure 9.28 PPI display of echo power from clear air as seen by the NSSL doppler radar. The bright areas of higher power are aligned roughly parallel to mean wind. Range marks are 20 km apart. EL = 1.6° (April 27, 1977, 1447 CST).

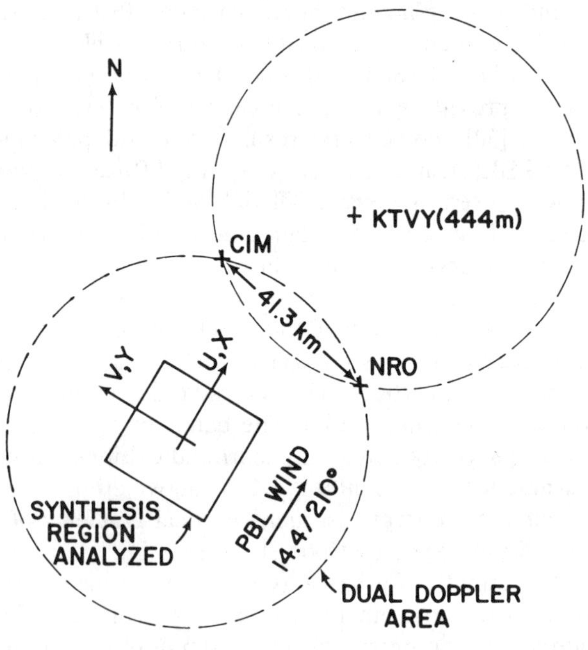

Figure 9.29 The dual-doppler-radar area (dashed lines) within which the angle subtended by the radials from the Cimarron (CIM) and Norman (NRO) radars lies between 30° and 150°. The outlined box is the region from which the doppler velocities were synthesized for detailed wind analyses shown in Figures 9.30 and 9.31. The wind speed and direction are a mean over 1.25-km depth of the planetary boundary layer.

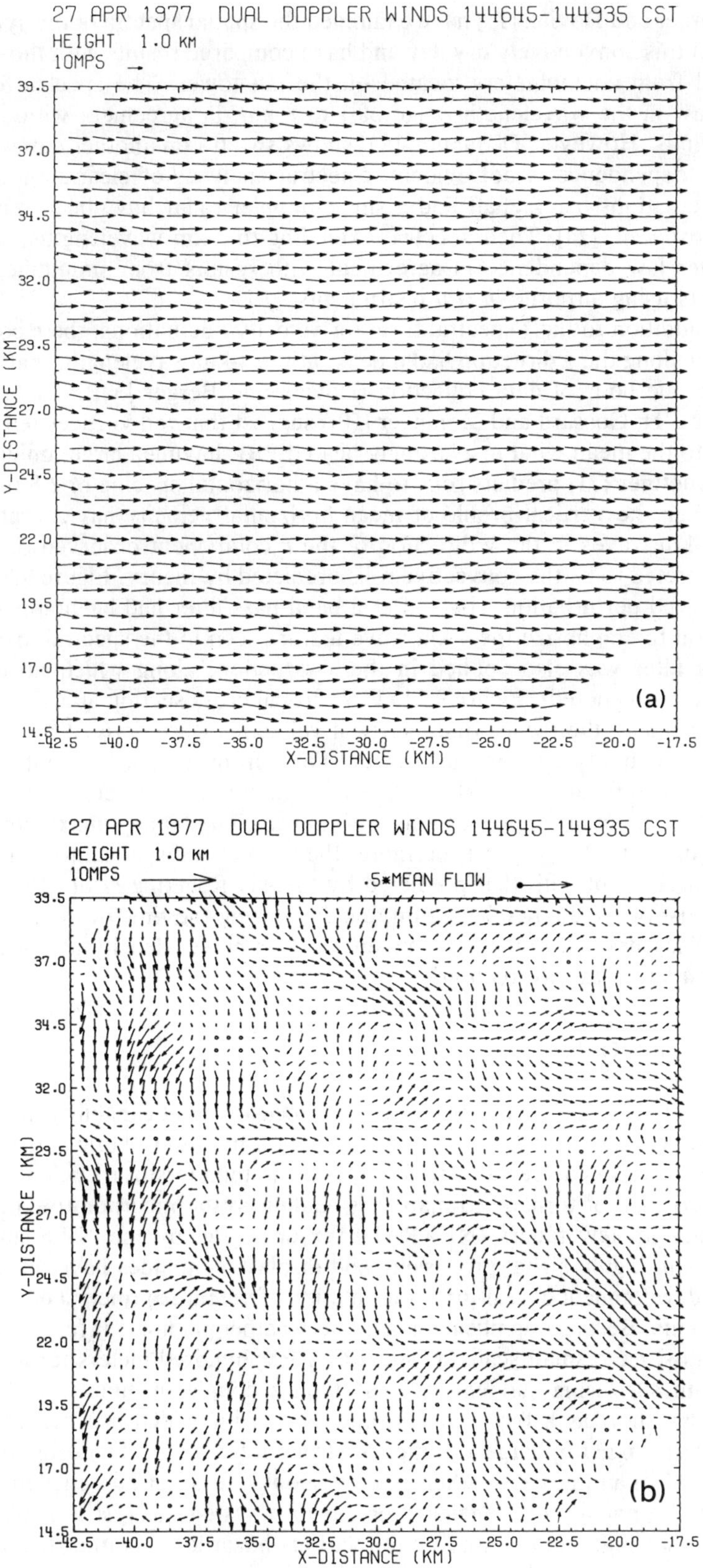

Figure 9.30 **(a) Dual-radar synthesized wind field at 1.0 km above ground for April 1977 at 1448 CST. (b) Wind field with mean removed. Synthesized wind fields were low-pass filtered once in *X* and *Y* direction with a three-point Shuman [100] filter.**

Berger and Doviak [11] have examined the spatial spectra of the synthesized winds on this convectively dry day and have compared results with those spectra obtained from anemometers located on the tall tower. The spectra follow a $\frac{5}{3}$ power law in the wavelength range of 1 to 8 km, in agreement with spectra of tower winds. However, O'Bannon [85] shows spectra on another day when such a power dependence is not evident. Spectral analyses of clear-air longitudinal wind fluctuations using chaff and a single-doppler radar have been reported by Chernikov *et al.* [24]. Their spectra, extending to 3-km wavelengths, also show a $\frac{5}{3}$ power law dependence in agreement with results from simultaneous measurements using airborne *in situ* instruments.

In addition to the $\frac{5}{3}$ spectral form for velocity fluctuations, the v component spectrum along the y direction had a peak, at the 4-km wavelength, that persisted over the one hour of data collection (Doviak and Berger [32]). Theories (e.g., Kuettner [71]; Gossard and Moninger [50]) suggest that roll vortices tend to form parallel to the mean wind in a strongly heated PBL having a large, unidirectional shear. Kuettner [71] predicts rolls to have a horizontal spacing of 2.8 times their depth when the vertical profile of mean horizontal velocity has curvature. The strong 4-km waves in the y direction of the v component observed on April 27 are interpreted to be the convective rolls predicted by theory. Figure 9.31 depicts the observed perturbation winds after a band pass filter had been applied in the y direction to emphasize the 4-km wave feature seen in the spectral displays. (A low-pass filter was also applied in the x direction, along which no dominant wavelength was noted). Figure 9.31a is a vertical cross section at $X = -25.5$ km perpendicular to the mean wind. Vertical velocities were derived by integrating the mass continuity equation using wind fields from the six horizontal surfaces with vertical separation of 250 m. Readily apparent are counter-rotating vortices (roll vortices), having approximately a 4-km wavelength and a maximum vertical velocity of about 1 m s^{-1}. Furthermore, the ratio of roll spacing to height is 2.6, in good agreement with that predicted by theory. Reinking *et al.* [91] analyzed aircraft gust probe data collected on this day during the time of the radar observation. They, too, detected a prominent peak in power density at a wavelength of about 4 km in the y direction.

9.8 CONCLUSIONS

The introduction of doppler frequency measurement capability into weather radars has opened new horizons for exploration by atmospheric scientists. The astounding success achieved with these radars in thunderstorm cyclone detection well in advance of tornado formation is leading to the incorporation of doppler capabilities in new radars (NEXRAD) for operation by the NWS and AWS. Advances in digital signal processing and display techniques have allowed economical development of real-time presentation of the three principal doppler spectral moments. Because significant hazardous weather phenomena (e.g., tornadoes, microbursts) have small scale (as small as a few hundred meters for a tornado), occur with surprising suddenness (less than a couple of minutes), often have lifetimes as short as a few minutes, and can be imbedded in weak reflectivity regions surrounded by large volumes of strong reflectivity, the design of a radar for detailed quantitative weather studies should be a serious challenge to the engineer. An equally challenging problem to a system analyst is automatically detecting, sorting, and quantifying the various hazardous weather features. The techniques are constantly improving, and new scientific disclosures and new operational applications in wind measurement and hazard detection are forthcoming.

Doppler radars at centimeter wavelengths do not have a sufficiently large velocity-range ambiguity product, $r_a v_a$, to match that required to observe, without ambiguity, severe convective storms. No comprehensive data and documentation

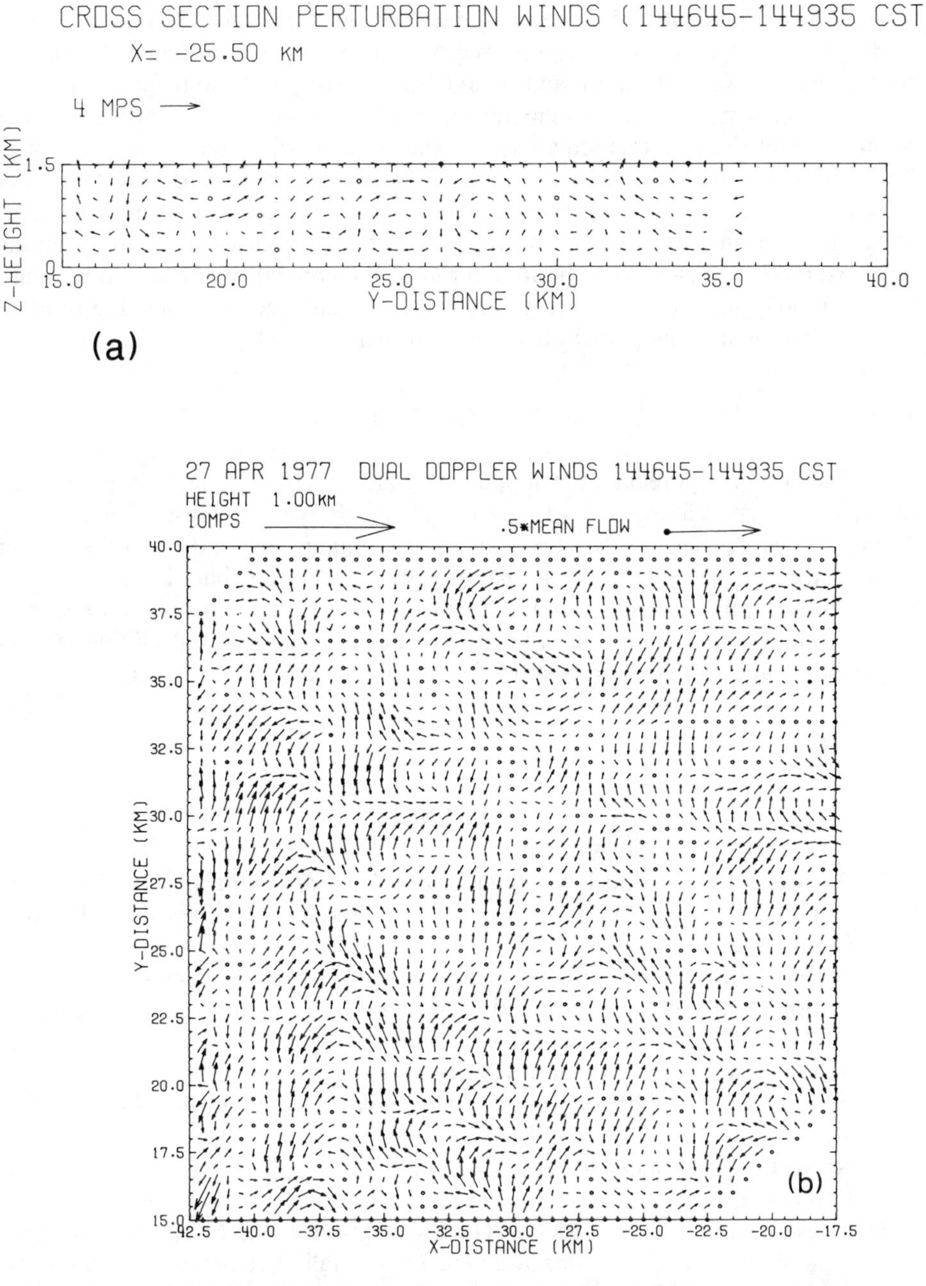

Figure 9.31 Vertical (a) and horizontal (b) cross sections of bandpass filtered wind data that highlight the clear-air roll structure seen in the spectra displays of unfiltered data for April 17, 1977. Vertical cross section on (a) is along the Y direction at $X = -25.50$ km.

show the full extent of the problem, nor is there any foreseeable solution; storm observers apparently will have to accept some limitations in doppler radar weather measurements.

Dual-doppler radar observations of the kinematic structure of severe storms and the planetary boundary layer agree with theoretical models, but much investigation is still required. The doppler weather radar shows promise of greatly increasing our knowledge of thunderstorms and the planetary boundary layer on scales not previously accessible. Furthermore, we can monitor significant mesoscale phenomena of importance to air traffic safety and air pollution control, and we may be able to see the triggering impulses of severe storms that each year cause substantial destruction. Increased power and sensitivity of weather radars and operation at longer wavelengths may soon result in meteorologists being able to observe the wind structure and its evolution throughout the troposphere.

The important advances in meteorological observations brought forth by the application of doppler techniques to weather radars will continue in the future. There is room for further improvement in the radar system to reduce the effect of ambiguities, while lessening data acquisition time for observation of wind, shear, and turbulence in fair or stormy weather. Although weather observations with ground-based radars are limited in range, the capability of radar to measure wind remotely in clear and stormy skies gives a challenge that someday satellite borne radars will allow global observation of winds for all weather conditions.

ACKNOWLEDGMENTS

The authors appreciate the support they have received from their colleagues in the Environmental Research Laboratories, Air Force Geophysics Laboratory, National Weather Service, Nuclear Regulatory Commission, and Federal Aviation Administration, not only in the preparation of this chapter, but also for the continuing development of doppler weather radar technology. We are indebted to Ms. Carole J. Holder for the efficient and accurate typing and editing of this manuscript, and to Ms. Joan Kimpel for the preparation of the figures.

REFERENCES

1. Abramowitz, M., and I.A. Stegun, *Handbook of Mathematical Functions,* National Bureau of Standards, Appl. Math. Ser. 55, US Government Printing Office, Washington, DC, 1964.
2. Ahnert, P.R., W.F. Krajewski, and E.B. Johnson, "Kalman Filter Estimation of Radar—Rainfall Field Bias," Preprints, *Twenty-Third Conference on Radar Meteorology and Conference on Cloud Physics,* American Meteorological Society, Boston, (1986) **3**:JP33–JP37.
3 Armijo, L., "A Theory for the Determination of Wind and Precipitation Velocities with Doppler Radars," *J. Atmos. Sci.* (1969) **26:**570–573.
4 Atlas, D., "Advances in Radar Meteorology," in *Advances in Geophysics,* Landsberg and Mieghem (eds.), Academic Press, New York, 1964, 317–478.
5. Atlas, D., and C.W. Ulbrich, "The Physical Bases for Attenuation-Rainfall Relationships and the Measurement of Rainfall Parameters by Combined Attenuation and Radar Methods," *J. Rech. Atmos.,* (1974) **8**(1–2):275–298; Atlas, D., and C.W. Ulbrich, "Path- and Area-Integrated Rainfall Measurement by Microwave Attenuation in the 1–3 cm Band," *J. App. Meteorol.* (1977) **16**:1322–1331.
6. Atlas, D., R.C., Srivastava, and R.S. Sekhon, "Doppler Radar Characteristics of Precipitation at Vertical Incidence," *Rev. Geophysics Space Physics,* (1973) **2**:1–35.
7. Atlas, D., K.R. Hardy, K.M. Glover, I. Katz, and T.G. Konrad, "Tropopause Detected by Radar," *Science* (1966) **153**:1110–1112.
8. Balsley, B.B., "The Use of Sensitive Coherent Radars to Examine Atmospheric Parameters in the Height Range 1–100 km," Preprints, *Eighteenth Conference on Radar Meteorology,* American Meteorological Society, Boston, (1978) 190–193.
9. Battan, L.J., *Radar Observation of the Atmosphere,* University of Chicago Press, Chicago, 1973.

10. Bean, B.R., and E.J. Dutton, *Radio Meteorology,* National Bureau of Standards Monograph 92, US Government Printing Office, Washington, DC, 1966.
11. Berger, M.I., and R.J. Doviak, "An Analysis of the Clear Air Planetary Boundary Layer Wind Synthesized from NSSL's Dual Doppler-Radar Data," Tech. Memo. ERL NSSL-87, NOAA Environmental Research Laboratories, Boulder, CO, 1979.
12. Blake, L.V., "Prediction of Radar Range," in *Radar Handbook,* M.I. Skolnik (ed.), McGraw-Hill, New York, 1970, 2.1–1.73.
13. Bohne, A., and R.C. Srivastava, "Random Errors in Wind and Precipitation Fall Speed Measurement by a Triple Doppler Radar System," Rep. No. 37, Lab. Atmos. Probing, University of Chicago, Chicago, 1975, 1–10.
14. Born, M., and E. Wolf, *Principles of Optics,* 2nd Ed., Macmillan, New York, 1964, 808 pp.
15. Brandes, E.A., "Optimizing Rainfall Estimates with the Aid of Radar," *J. Appl. Meteorol.* (1975) **14**:1339–1345.
16. Brewster, K.A., and D.S. Zrnić, "Comparison of Eddy Dissipation Rates from Spatial Spectra of Doppler Velocities and Doppler Spectrum Widths," *J. Oceanic and Atmos. Tech.* (1986) **3**:440–452.
17. Brown, R.A., L.R. Lemon, and D.W. Burgess, "Tornado Detection by Pulsed Doppler-Radar," *Mon. Weather Rev.* (1978) **106**:29–38.
18. Browning, K.A., and R. Wexler, "The Determination of Kinematic Properties of a Wind Field Using Doppler Radar," *J. Appl. Meteorol.* (1968) **7**: 105–113.
19. Burgess, D.W., "Single Doppler Radar Vortex Recognition: Part I—Mesocyclone Signatures," Preprints, *Seventeenth Conference on Radar Meteorology,* American Meteorological Society, Boston, (1976) 97–103.
20. Burgess, D.W., L.D. Hennington, R.J. Doviak, and P.S. Ray, "Multimoment Doppler Display for Severe Storm Identification," *J. Appl. Meteorol.* (1976) **15**:1302–1306.
21. Burgess, D.W., J.D. Bonewitz, and D.R. Devore, "Joint Doppler Operational Project: Results, Year 1." Preprints, *Eighteenth Conference on Radar Meteorology,* American Meteorological Society, Boston, (1978) 442–448.
22. Caton, P.A.F., "Wind Measurement by Doppler Radar," *Meteorol. Mag.* (1963) **92**:213–222.
23. Chadwick, R.B., K.P. Moran, and G.E. Morrison, "Measurements Toward a C_n^2 Climatology," Preprints, *Eighteenth Conference on Radar Meteorology,* American Meteorological Society, Boston, (1978) 100–103.
24. Chernikov, A.A., Y.V. Mel'nichuk, N.Z. Pinus, S.M. Shmeter, and N.K. Vinnichenko, "Investigations of the Turbulence in Convective Atmosphere Using Radar and Aircraft," *Radio Sci.* (1969) **4**:1257–1259.
25. Crane, R.K., "Automatic Cell Detection and Tracking," *IEEE Trans. Geosci. Electron.* (1979) **GE-17**(4):250–261.
26. Crawford, K.C., The Design of a Multivariate Mesoscale Field Experiment, Ph.D. dissertation, University of Oklahoma, Norman, OK, 1977.
27. Cressman, G.D., "An Operational Objective Analysis System," *Mon. Weather Rev.* (1959) **87**:367–374.
28. Davies-Jones, R.P., "Dual-Doppler Radar Coverage Area as a Function of Measurement Accuracy and Spatial Resolution," *J. Appl. Meteor.* (1979) **9**: 1229–1233.
29. Donaldson, R.J., Jr., "Vortex Signature Recognition by a Doppler Radar," *J. Appl. Meteorol.* (1970) **9**:661–670.
30. Doviak, R.J., and C.T. Jobson, "Dual Doppler Radar Observations of Clear Air Wind Perturbations in the Planetary Boundary Layer," *J. Geophys. Res.* (1979) **84**:697–702.

31. Doviak, R.J., and D.S. Zrnić, "Receiver Bandwidth Effect on Reflectivity and Doppler Velocity Estimates," *J. Appl. Meteorol.* (1979) **18**:69–76.
32. Doviak, R.J., and M. Berger, "Turbulence and Waves in the Optically Clear Boundary Layer Resolved by Dual-Doppler Radars," *Radio Sci.,* (1980) **15**(2):297–318.
33. Doviak, R.J., and D.S. Zrnić, *Doppler Radar and Weather Observations,* Academic Press, New York, 1984, 458 pp.
34. Doviak, R.J., and J.T. Lee, "Radar for Storm Forecasting and Weather Hazard Warning," *J. Aircraft* (1985) **22**(12):1059–1064.
35. Doviak, R.J., D. Burgess, L. Lemon, and D. Sirmans, "Doppler Velocity and Reflectivity Structure Observed within a Tornadic Storm," *J. Res. Atmos.* (1974) **8**:235–243.
36. Doviak, R.J., P.S. Ray, R.G. Strauch, and L.J. Miller, "Error Estimation in Wind Fields Derived from Dual-Doppler Radar Measurement," *J. Appl. Meteorol.* (1976) **15:**868–878.
37. Doviak, R.J., D. Sirmans, D. Zrnić, and G.B. Walker, "Considerations for Pulse-Doppler Radar Observations of Severe Thunderstorms," *J. Appl. Meteorol.* (1978) **17**:189–205.
38. Doviak, R.J., R.M. Rabin, and A.J. Koscielny, "Doppler Weather Radar for Profiling and Mapping Winds in the Prestorm Environment," *IEEE Trans. Geosci. and Remote Sensing* (1983) **GE-21**(1):25–33.
39. Eilts, M.D., "Low Altitude Wind Shear Detection with Doppler Radar," *J. Climate and Appl. Meteor.,* January 1987.
40. Eilts, M.D. and R.J. Doviak, "Oklahoma Downbursts and Their Asymmetry," *J. Climate and Appl. Meteor.,* January 1987.
41. Evans, J.E., and D. Johnson, "The FAA Transportable Doppler Weather Radar," Preprints, *Twenty-Second Conference on Radar Meteorology,* Zurich, American Meteorological Society, Boston, (1984) 112–113.
42. Foote, G.B., and P.S. duToit, "Terminal Velocity of Raindrops Aloft," *J. Appl. Meteorol.* (1969) **8**:249–253.
43. Frisch, A.S., and S.F. Clifford, "A Study of Convection Capped by a Stable Layer Using Doppler Radar and Acoustic Echo Sounders," *J. Atmos. Sci.* (1974) **31**:1622–1628.
44. Fujita, T.T., "Tornadoes and Downbursts in the Context of Generalized Planetary Scales," *J. Atmos. Sci.* (1981) **38**:1511–1534.
45. Fujita, T.T., *The Downburst, Microburst and Macroburst,* University of Chicago Press, Chicago, 1985, 122 pp.
46. Fukao, S., K. Wakasugi, T. Sato, S. Morimoto, T. Tsuda, I. Hirota, I. Kimura, and S. Kato, "Direct Measurement of Air and Precipitation Particle Motion by Very High Frequency Doppler Radar, *Nature* (1985) **316**: 712–714.
47. Gage, K.S., and J.L. Green, "Evidence of Specular Reflection from Monostatic VHF Radar Observations of the Stratosphere," *Radio Sci.* (1978) **13**: 991–1001.
48. Gage, K.S., W.P. Birkemeyer, and W.H. Jasperson, "Atmospheric Stability Measurement at Tropopause Altitudes Using Forward-Scatter CW Radar," *J. Appl. Meteorol.* (1973) **12**:1205–1212.
49. Gossard, E.E., and A.S. Frisch, "Kinematic Models of a Dry Convective Boundary Layer Compared with Dual Doppler-Radar Observations of Wind Fields," *Boundary Layer Meteorol.* (1976) **10**:311–330.
50. Gossard, E.E., and W.R. Moninger, "The Influence of a Capping Inversion on the Dynamic and Convective Instability of a Boundary Layer Model with Shear," *J. Atmos. Sci.* (1975) **32**:2111–2124.
51. Green, J.L., J.M. Warnock, R.H. Winkler, and T.E. VanZandt, "Studies of Winds in the Upper Troposphere with a Sensitive VHF Radar," *Geophys. Res. Lett.* (1975) **2**:19–21.

others. Some of these areas of deep convection persist for several hours, and, if showers develop, they occur within—and only within—such regions.

Mapping of boundary layer wind over large areas has double significance for severe-storm studies because mapping allows early observation of thunderstorm development—hence storm genesis can be followed from the very beginning of cumulus development—and makes possible the monitoring of thunderstorm outflow, which is usually precipitation free.

Strongest fluctuations in refractive index occur when turbulence mixes large gradients of mean potential temperature and specific humidity (Yaglom [114]). Gage *et al.* [48] measured the height distribution of forward scattered signal strength to show good correlation between it and gradients of potential temperature at high altitudes, where water vapor contributions can be ignored. The *tropopause* is a region where potential temperature always increases with height; this region was first detected by radar in 1966 (Atlas *et al.* [7]). VanZandt *et al.* [109], using backscatter results from a vertically pointed VHF radar, obtained consistent agreement between fluctuations inferred from potential temperature profiles measured with *rawinsonde* and those measured by radar for the clear air above the moist boundary layer.

For many years ionospheric scientists have been employing high-power VHF and UHF radars to observe the ionosphere. Recently, these radars at Jicamarca, Peru (Woodman and Guillen [113]); Arecibo, Puerto Rico; Chatanika, Alaska (Balsley [8]); and Lindau, Germany (Röttger and Liu [97]), have lowered their sights to examine echoes in the stratosphere and troposphere. A VHF (λ = 7 m) pulsed doppler radar specifically designed for tropospheric studies was assembled at Sunset, Colorado, to make continuous observations of winds in the troposphere (Green *et al.* [51]). Some advantages of the VHF radar are that it can sometimes differentiate scatter from hydrometeors and refractive index fluctuations (Green *et al.* [52]; Fukao *et al.* [46]), and may be able to detect coherent scatter from mean refractive index gradients at inversion layers (Gage and Green [47]).

Clear-air winds in the planetary boundary layer (PBL, i.e., surface to heights of about 1.5 km) have been synthesized from dual-doppler-radar measurements (Gossard and Frisch [49]; Kropfli and Kohn [70]), wherein chaff was dispensed over large areas to provide suitable echo levels for detection and processing. Doviak and Jobson [30] showed first results of dual-doppler-radar synthesized wind fields in the PBL clear air, where only the diffuse and intrinsic scatterers of the medium were used as targets. Doviak and Jobson observed mean wind fields at low height to have qualitative agreement with mean wind measured (with conventional anemometers) near the surface.

On April 27, 1977, a day marked by strong nondirectional shear and curvature in the wind profile, NSSL's doppler radar echo power measurements showed evidence of clear-air reflectivity streets (Figure 9.28), an observation that should signify the presence of roll vortices. The first radar detection of clear-air "thermal streets" was reported by Konrad [67]. The bands of enhanced reflectivity seen on Figure 9.28 are spaced roughly 4 km apart and exhibits "streeting" at ranges beyond 40 km, where the beam center is 1.2 km above ground level. This is where roll vortices are mixing the strongest mean vertical gradients of refractive index and where reflectivity is expected to be strongest.

Figure 9.29 locates the Cimarron (CIM) and Norman (NRO) radars and a 444-m tall meteorological instrument-equipped tower in central Oklahoma. Reasonably accurate winds can be synthesized from doppler data in the entire area enclosed by the dashed lines, except for the overlapping region. The winds were fairly uniform from the southwest on this day, but there were small perturbations from the mean wind having a magnitude about one order less than the mean wind itself. The synthesized wind field and its perturbations from the mean are shown in Figure 9.30a, b.

52. Green, J.L., R.H. Winkler, J.M. Warnock, W.L. Clark, K.S. Gage, and T.E. VanZandt, "Observations of Enhanced Clear Air Reflectivity Associated with Convective Clouds," Preprints, *Eighteenth Conference on Radar Meteorology,* American Meteorological Society, Boston, (1978) 88–93.
53. Gunn, K.L.S., and T.W.R. East, "The Microwave Properties of Precipitation Particles," *Q. J. R. Meteorol. Soc.* (1954) **80**:522–545.
54. Hardy, K.R., and I. Katz, "Probing the Clear Atmosphere with High Power, High Resolution Radars," *Proc. IEEE* (1969) **57**:468–480.
55. Hardy, K.R., D. Atlas, and K.M. Glover, "Multiwavelength Backscatter from the Clear Atmosphere," *J. Geophys. Res.* (1971) **71**:1537–1552.
56. Harrold, T.W., and K.A. Browning, "Identification of Preferred Areas of Shower Development by Means of High Power Radar," *Q. J. R. Meteorol. Soc*: (1971) **97**:330–339.
57. Hennington, L., R.J. Doviak, D. Sirmans, D. Zrnić, and R.G. Strauch, "Measurements of Winds in the Optically Clear Air with Microwave Pulse-Doppler Radar," Preprints, *Seventeenth Conference on Radar Meteorology*. American Meteorological Society, Boston (1976) 342–348.
58. Herman, B.M., S.R. Browning, and L.J. Battan, "Tables of the Radar Cross Sections of Water Sphere," Tech. Rep. No. 9, Institute of Atmospheric Physics, University of Arizona, Tucson, 1961.
59. Hess, S.L., *Introduction to Theoretical Meteorology,* Holt, Rinehart and Winston, New York, 1959.
60. Hicks, J.J., and J.K. Angell, "Radar Observations of Breaking Gravitational Waves in the Visually Clear Atmosphere," *J. Appl. Meteorol.* (1968) **7**: 114–121.
61. IEEE (Institute of Electrical and Electronics Engineers), Scatter Propagation Issue, *Proc. IEEE* (1955) **43**:1173–1570.
62. Istok, M.J. and R.J. Doviak, "Analysis of the Relation between Doppler Spectral Width and Thunderstom Turbulence," *J. Atmos. Sci.* (1986) **43**: 2199–2214.
63. JDOP, "Final Report on the Joint Doppler Operational Project (JDOP), 1976–78," NOAA Tech. Memo. ERL NSSL-86, National Severe Storms Laboratory, Norman, OK, 1979, 84 pp.
64. Jordan, E.C., *Electromagnetic Waves and Radiating Systems,* Prentice-Hall, Englewood Cliffs, NJ, 1950.
65. Katz, I., "Sea Breeze Structure as Seen by Radar," Memo BPD 66U-25, Appl. Phys. Lab., Johns Hopkins University, Baltimore, MD, 1966.
66. Kessler, E., "Wind Shear and Aviation Safety," *Nature* (1985) **315**:179–180.
67. Konrad, T.G., "The Alignment of Clear Air Convective Cells," Preprints, *International Conference on Cloud Physics,* Toronto, American Meteorological Society, Boston, (1968) 539–543.
68. Konrad, T.G., "The Dynamics of the Convective Process in the Clear Air as Seen by Radar," Preprints, *Fourteenth Conference on Radar Meteorology,* Tucson, AZ, American Meteorological Society, Boston, (1970) 57–60.
69. Kraus, J.D., *Radio Astronomy,* McGraw-Hill, New York, 1966.
70. Kropfli, R.A., and N.M. Kohn, "Persistent Horizontal Rolls in the Urban Mixed Layer as Revealed by Dual-Doppler Radar," *J. Appl. Meteorol.* (1978) **17**:669–676.
71. Kuettner, J.P., "Cloud Bands in the Earth's Atmosphere," *Tellus* (1971) **23**: 404–425.
72. Laws, J.O., and D.A. Parsons, "The Relation of Raindrop-Size to Intensity," *Trans. Am. Geophys. Union.* (1943) **24**:452–460.
73. Lee, J.T., and C. Goff, "Gust Front Wind Shear and Turbulence—Concurrent Aircraft and Surface Based Observations," Preprints, *Seventh Conference on Aerospace and Aeronautical Meteorology and Symposium on Remote Sensing from Satellites,* Melbourne, FL, American Meteorological Society, Boston, (1976) 48–54.

74. Lhermitte, R.M., "Dual-Doppler Radar Observations of Convective Storm Circulation." Preprints, *Fourteenth Radar Meteorological Conference,* Tucson, AZ, American Meteorological Society, Boston (1970) 153–156; Lhermitte, R.M., "Motions of Scatterers and the Variance of the Mean Intensity of Weather Radar Signals," SRFC-RR-63-57, Sperry-Rand Research Center, Sudbury, MA (1963).
75. Lhermitte, R.M., and D. Atlas, "Precipitation Motion by Pulse Doppler Radar," Preprints, *Ninth Weather Radar Conference,* Kansas City, MO, American Meteorological Society, Boston, (1961) 218–223.
76. Linden, P.F., and J.E. Simpson, "Microbursts: A Hazard for Aircraft," *Nature* (1985) **317**:601–602.
77. McCarthy, J., and R. Serafin, "The Microburst: Hazard to Aviation," *Weatherwise* (1984) **37**(3):120–127.
78. McCarthy, J., and J.W. Wilson, "The Microburst as a Hazard to Aviation: Structure, Mechanisms, Climatology and Nowcasting," Preprints, *Nowcasting II Symposium,* Norrkoping, Sweden (1984) 21–30.
79. McCarthy, J., R. Roberts, and W. Schreiber, "JAWS Data Collection, Analysis Highlights, and Microburst Statistics," Preprints, *Twenty-First Conference on Radar Meteorology,* American Meteorological Society, Boston, (1984) 596–601.
80. Meyer, J.H., "Radar Observations of Land Breeze Fronts at Wallops Island, Virginia," Preprints, *Fourteenth Conference on Radar Meteorology,* Tucson, AZ, American Meteorological Society, Boston, (1970) 61–67.
81. Nathanson, F.E., *Radar Design Principles,* McGraw-Hill, New York, 1969.
82. Nathanson, F.E., and P.L. Smith, "A Modified Coefficient for the Weather Radar Equation," Preprints, *Fifteenth Radar Meteorological Conference,* Champaign-Urbana, IL, American Meteorological Society, Boston, (1972) 228–230.
83. Nelson, S.P., and R.A. Brown, "Multiple Doppler Radar Derived Vertical Velocities in Thunderstorms: Part I—Error Analyses and Solution Techniques," NOAA Tech. Memo ERL NSSL-94, National Severe Storms Laboratory, Norman, OK, 1982, 10 pp.
84. Norbury, J.R., and W.J.K. White, "Microwave Attenuation at 35.8 GHz due to Rainfall," *Electron. Lett.* (1972) **8**:91–92.
85. O'Bannon, T., "A Study of Dual-Doppler Synthesized Clear Air Wind Fields," Preprints, *Eighteenth Conference on Radar Meteorology,* American Meteorological Society, Boston, (1978) 65–69.
86. Papoulis, A., *Probability Random Variables and Stochastic Processes,* McGraw-Hill, 1965, New York.
87. Probert-Jones, J.R., "The Radar Equation in Meteorology," *Q. J. R. Meteorol. Soc.* (1962) **88**:485–495.
88. Rabin, R., and D.S. Zrnić, "Subsynoptic Scale Vertical Wind Revealed by Dual Doppler Radar and VAD Analyses," *J. Atmos. Sci.* (1980) **37**:644–654.
89. Ray, P.S., R.J. Doviak, G.B. Walker, D. Sirmans, J. Carter, and B. Bumgarner, "Dual-Doppler Observation of a Tornadic Storm," *J. Appl. Meteorol.,* (1975) **14**:1521–1530.
90. Ray, P.S., C.L. Ziegler, R.J. Serafin, and W. Bumgarner, "Single- and Multiple-Doppler Observations of Tornadic Storms," *Mon. Weather Rev.* (1980) **108**:1607–1625.
91. Reinking, R.F., R.J. Doviak, and R.O. Gilmer, "Clear-Air Roll Vortices and Turbulent Motions as Detected with an Airborne Gust Probe and Dual-Doppler Radar," *J. Appl. Meteorol.* (1981) **20**:678–685.
92. Rice, P.L., A.G. Langley, K.A. Norton, and A.P. Barsis, "Transmission Loss Predictions for Tropospheric Communication Circuits," National Bureau of Standards Tech. Note 101, Vols. 1–2, US Government Printing Office, Washington, DC, 1967.

93. Rideout, V.C., *Active Networks,* Prentice-Hall, Englewood Cliffs, NJ, 1954, 311 pp.
94. Rinehart, R.E., Internal Storm Motion from a Single Non-Doppler Weather Radar, Ph.D. dissertation, Colorado State University, Ft. Collins, CO, 1979.
95. Rinehart, R.E., and M.A. Isaminger, "Radar Characteristics of Microbursts in the Mid-South," Preprints, *Twenty-Third Conference on Radar Meteorology and Conference on Cloud Physics,* American Meteorological Society, Boston, (1986) **3**:J116–J119.
96. Roberts, R.D., and J.W. Wilson, "Nowcasting Microburst Events Using Single Doppler Radar Data," Preprints, *Twenty-Third Conference on Radar Meteorology,* Snowmass, CO, American Meteorological Society, Boston, (1986) 14–17.
97. Röttger, J., and C.H. Liu, Partial Reflection and Scattering of VHF Radar Signals from the Clear Atmosphere, *Geophys. Res. Lett.* (1978) **5**:357–360.
98. Saxton, A. (ed.), *Advances in Radio Research,* Academic Press, New York, 1964, 226 pp.
99. Sherman, J.W., III, "Aperture-Antenna Analysis," in *Radar Handbook,* M.I. Skolnik (ed.), McGraw-Hill, New York, 1970, 9.1–9.40.
100. Shuman, F.G., "Numerical Methods and Weather Predictions, II," *Mon. Weather Rev.* (1957) **85**:357–387.
101. Silver, S., *Microwave Antenna Theory and Design,* McGraw-Hill, New York, 1949, 623 pp.
102. Skolnik, M.I., *Radar Handbook,* McGraw-Hill, New York, 1970.
103. Smith, R.L., and D.W. Holmes, "Use of Doppler Radar in Meteorological Observations," *Mon. Weather Rev.* (1961) **89**:1–7.
104. Stackpole, J.D., "The Effectiveness of Raindrops as Turbulence Sensors," Preprints, *Ninth Weather Radar Conference,* Kansas City, MO, American Meteorological Society, Boston, (1961) 212–217.
105. Sychra, J., "Relation between Real and Pulse Volume Averaged Fields of Reflectivity and Velocity," Canadian Meteorol. Service Res. Rep. (1972) CMRR 5/72.
106. Taylor, J.W., and J. Mattern, "Receivers," in *Radar Handbook,* M.I. Skolnik (ed.), McGraw-Hill, New York, 1970, 5.1–5.50.
107. Ulbrich, C.W., "Quantitative Assessment of the Effects of Vertical Winds on the Accuracy of Rainfall Rate Measurement by Microwave Attenuation," Preprints, *Twenty-Third Conference on Radar Meteorology,* American Meteorological Society, Boston, (1986) R92–R95.
108. Uyeda, H., and D.S. Zrnić, "Automatic Detection of Gust Fronts," *J. Atmos. and Oceanic Tech.* (1986) **3**:36–50.
109. VanZandt, T.E., J.L. Green, K.S. Gage, and W.L. Clark, "Vertical Profiles of Refractivity Turbulence Structure Constant: Comparison of Observations by the Sunset Radar with New Theoretical Model," *Radio Sci.* (1978) **13**: 819–829.
110. Vasiloff, S.V., E.A. Brandes, R.P. Davies-Jones, and P.S. Ray, "An Investigation of the Transition from Multicell to Supercell Storms," *J. Clim. and Appl. Meteor,* (1986) **25**(7):1022–1036.
111. Wilson, J.W., R.D. Roberts, C. Kessinger, and J. McCarthy, "Microburst Structure and Evaluation of Doppler Radar for Airport Wind Shear Detection," *J. Climate and Appl. Meteor.* (1984) **23**:898–915.
112. Wood, V.T., and R.A. Brown, "Single Doppler Velocity Signatures: An Atlas of Patterns in Clear Air Widespread Precipitation and Convective Storms," Tech. Memo. ERL NSSL-95, NOAA Environmental Research Laboratories, Norman, OK, 1983. See also Wood, V.T., and R.A. Brown, "Single Doppler Velocity Signature Interpretation of Nondivergent Environmental Winds," *J. Atmos. and Oceanic Tech.* (1986) **3**(3):114–128.

113. Woodman, R.F., and A. Guillen, "Radar Observations of Winds and Turbulence in the Stratosphere and Mesosphere," *J. Atmos. Sci.* (1974) **31**: 493–505.
114. Yaglom, A.M., "On the Local Structure of the Temperature Field in a Turbulent Flow," *Dokl. Akad. Nauk. SSSR* (1949) **69**:743.
115. Zawadzki, I., "Factors Affecting the Precision of Radar Measurements of Rain," Preprints, *Twenty-Second Conference on Radar Meteorology,* American Meteorological Society, Boston, (1984) 251–262.
116. Zeoli, G.W., "IF versus Video Limiting for Two-Channel Coherent Signal Processors," *IEEE Trans. Inform. Theory* (1971) **IT-17**:579–587.
117. Ziegler, C.L., "A Dual Doppler Variational Objective Analysis as Applied to Studies of Convective Storms," NOAA Tech. Memo. ERL NSSL-85, National Severe Storms Laboratory, Norman, OK, 1978.
118. Zrnić, D.S., "Estimated Tornado Spectra and Maximum Velocity Statistics," NOAA Final Rep. Grant No. 04-5-022-17, 1975.
119. Zrnić, D.S., "Signal-to-Noise Ratio in the Output of Nonlinear Devices," *IEEE Trans. Inform. Theory* (1975) **IT-21**:662–663.
120. Zrnić, D.S., and B. Bumgarner, "Receiver Chain and Signal Processing Effects on the Doppler Spectrum," Preprints, *Sixteenth Radar Meteorological Conference,* Houston, American Meteorological Society, Boston, (1975) 163–168.
121. Zrnić, D.S., and R.J. Doviak, "Velocity Spectra of Vortices Scanned with a Pulse Doppler Radar," *J. Appl. Meteorol.* (1975) **14**:1531–1539.
122. Zrnić, D.S., and R.J. Doviak, "Matched Filter Criteria and Range Weighting for Weather Radar," *IEEE Trans. Aerosp. Electron. Syst.* (1978) **AES-14**: 925–930.
123. Zrnić, D.S., R.J. Doviak, and D.W. Burgess, "Probing Tornadoes with Pulse Doppler Radar," *Q. J. R. Meteorol. Soc.* (1977) **103**:707–720.
124. Zrnić, D.S., R.J. Doviak, J.T. Lee, and R.S. Ge, "Characteristics of Gust Fronts and Downdrafts from Single Doppler Radar Data," Preprints, *Twenty-First Radar Conference,* American Meteorological Society, Boston, (1983) 650–654.
125. Zrnić, D.S., S.D. Smith, A. Witt, R.M. Rabin, and M. Sachidananda, "Wind Profiling of Stormy and Quiescent Atmospheres with Microwave Radars," NOAA Tech. Memo. ERL NSSL-98, National Severe Storms Laboratory, Norman, OK, (1986) 80 pp.
126. Zrnić, D.S., "Estimation of Spectral Moments for Weather Echoes," *IEEE Trans. Geosci. Electron.* Special Issue Radio Meteorology (1979) **GE-17**(4): 113–128.

List of Symbols

Symbol	Definition
A	Amplitude of the filtered complex video signal (i.e., $\sqrt{I^2 + Q^2}$)
A_c	Effective collection area of the antenna
A_e	Amplitude of the echo voltage
A_l	Amplitude of STALO voltage
A_m	Amplitude of the composite voltage applied to the mixer-detector
B_n	Noise power bandwith of the receiver
B_6	Filter bandwith between half-voltage (one-quarter power) points
c	Speed of light
CW	Continuous wave
D	Diameter of spherical water drop
d	Distance between radars #1 and #2
D_a	Diameter of antenna
dV	Elemental volume
f_d	Doppler frequency (Hz)
f_N	Nyquist frequency
$f^2(\theta, \phi)$	Normalized power density gain function
$G(f)$	Receiver filter's voltage gain *versus* frequency
g_0	Gravitational constant
g_t	Transmitting antenna gain including losses within the antenna
g_t'	Maximum directional gain of the antenna
g_r	Gain of the antenna in reception
I	In-phase video signal
$I(\mathbf{r}_0, \mathbf{r})$	Illumination function
I_0	Prefilter in-phase video signal
i	Current in mixer diode
i_f	Filtered output current of mixer-detector
K	Attenuation rate in decibels per kilometer (dB/km)
k	Boltzmann's constant
k	Specific attenuation, or attenuation coefficient
K_r	Attenuation rate in rain (dB/km)
$\|K_w\|^2$	Parameter related to the refractive index of a hydrometeor
k_g	Specific attenuation of atmospheric gases
k_r, k_ϕ, k_θ	Wind shear in the r, ϕ, and θ directions
l	One-way transmission loss of radiation due to scattering and absorption by atmospheric constituents
l_r	Echo power loss factor due to finite receiver bandwidth
m	An index 1, 2, . . .
N	Noise power
n	An index 1, 2, . . .
$N(D, \mathbf{r})$	Drop size distribution
N_s	Number of scatterers
N_t	Trip number
$P(\tau_s)$	Instantaneous power
P_r	Echo power at the antenna port
P_t	Power into the antenna port
P_t'	Power radiated

DOVIAK & ZRNIĆ
CHAPTER 9

Symbol	Definition
$p(x)$	Probability of x
Q	Quadrature-phase video signal
Q_0	Prefilter quadrature-phase video signal
$R(T_l)$	Autocovariance at lag T_l
r	Range
r_a	Unambiguous range interval
r_6	6 dB range resolution
r_0	Range at which weather targets contribute maximum weight to the I or Q sample at time delay τ_s; it is the range to the resolution volume V_6
$\mathbf{r}_1$	Beam direction of radar #1
$S(r, \theta, \phi)$	Power density radiated
$\overline{S}(f)$	Mean power spectrum density
$S(\mathbf{r}_0, v)$	Doppler power spectrum density (from range location $\mathbf{r}_0$)
$\overline{S}(\mathbf{r}_0, v)$	Mean Doppler power spectrum density (from range location $\mathbf{r}_0$)
SNR	Signal-to-noise power ratio
S_b	Power density along beam axis
S_i	Power density incident on target
S_r	Power density returned to radar antenna
s	Distance parallel to the line connecting two radars
T_l	Time lag
T_s	Pulse repetition period (sampling interval)
T_{sy}	Effective system temperature
t	Time
V	Complex video voltage ($I+jQ$)
V_d	Voltage applied to the mixer-detector
V_e	Echo voltage
V_l	STALO voltage
V_m	Voltage applied to the mixer-detector
V_6	Resolution volume; circumscribed by the surface where the weighting function is 6 dB below its peak
v	Radial velocity
v_a	One-half of the unambiguous Nyquist velocity interval
v_r	Doppler velocity ($= v$ when there is no ambiguity, where v is the wind component along y)
$\overline{\overline{v}}$	Radial velocity estimates $\overline{v}$ interpolated to a grid point
$\overline{v}$	Mean doppler velocity of the power spectrum; an estimate of the mean velocity for V_6
$W(r)$	Range weighting function
w_h	Horizontal wind speed
w_α, w_ρ, w_s	Wind in the α, ρ, and s directions
$w_t\,(D)$	Terminal velocity of a drop of diameter D
w_z	Wind in the z direction
$\overline{x}$	Bar above parameter $x = \eta, v, w_t, w_z$ denotes illumination function (and reflectivity for velocity parameters), weighted value
Z	Reflectivity factor
Z_e	Equivalent reflectivity factor
α	Angle in the cylindrical coordinate system
β	Angle subtended by $\mathbf{r}_1$ and $\mathbf{r}_2$
γ	Air density
$\Delta\alpha$, $\Delta\rho$	Grid spacing in the α, ρ directions
λ	Wavelength

Symbol	Definition
η	Reflectivity, backscatter cross section per unit volume
θ, ϕ	Spherical angles (see Figure 9.1)
θ_0, ϕ_0	Direction of beam axis
θ_e	Elevation angle of the radar beam
θ_1	One-way half-power angular beamwidth (radians)
ψ	$\psi_t + \psi_s$
ψ_e	Phase of echo voltage (radians)
ψ_l	Phase of STALO voltage (radians)
ψ_s	Phase shift produced by the target upon scattering (radians)
ψ_t	Phase of transmitted signal (radians)
ρ	Radial distance in the cylindrical coordinate system
σ	The root-mean-square value of I or Q
σ_α	Variance of w_α
σ_α	Doppler spectral width due to antenna motion
σ_a	Absorption cross section
σ_b	Backscatter cross section of a target
σ_d	Doppler spectral width due to differential terminal velocities
σ_h	Doppler spectral width due to hydrometeor oscillation or rotation
σ_r^2	Second central moment of the weighting function $\lvert W(r)\rvert^2$
σ_s	Doppler spectral width due to shear
σ_t	Doppler spectral width due to turbulence
σ_t	Attenuation or extinction cross section
σ_v	Spectral width of the doppler spectrum
σ_θ^2, σ_ϕ^2	Second central moment of the two-way antenna power pattern along θ, ϕ
σ_s^2, σ_ρ^2	Variance of w_s, w_ρ
σ_{sc}	Total scatter cross section
σ_1^2	Variance of $\overline{v}_1$ (v for radar #1)
τ	Pulse duration
τ_r	Time delay of echo in radar
τ_s	Time delay between transmitted pulse and its echo
ω	Transmitted frequency in radians per second (rad/s)
ω_d	Doppler shift (rad/s)
ω_e	Echo signal frequency (rad/s)
ω_l	STALO frequency (rad/s)

Note: All units are MKS unless otherwise specified.

Index